Lecture Notes in Computer Science 1184

Edited by G. Goos, J. Hartmanis and J. van Leeuwen

Advisory Board: W. Brauer D. Gries J. Stoer

T0171756

Springer

Berlin
Heidelberg
New York
Barcelona
Budapest
Hong Kong
London
Milan
Paris
Santa Clara
Singapore
Tokyo

Jerzy Waśniewski   Jack Dongarra
Kaj Madsen   Dorte Olesen   (Eds.)

Applied
Parallel Computing

Industrial Computation
and Optimization

Third International Workshop, PARA '96
Lyngby, Denmark, August 18-21, 1996
Proceedings

 Springer

Series Editors

Gerhard Goos, Karlsruhe University, Germany

Juris Hartmanis, Cornell University, NY, USA

Jan van Leeuwen, Utrecht University, The Netherlands

Volume Editors

Jerzy Waśniewski
Dorte Olesen
Danish Computing Centre for Research and Education
DTU, UNI•C, Bldg. 304, DK-2800 Lyngby, Denmark
E-mail: Dorte.Olesen@uni-c.dk
        Jerzy.Waśniewski@uni-c.dk

Jack Dongarra
University of Tennessee
107 Ayres Hall, Knoxville, TN 37996-1301, USA
E-mail: dongarra@cs.utk.edu

Kaj Madsen
Technical University of Denmark
DTU, Bldg. 305, DK-2800 Lyngby, Denmark
E-mail: km@imm.dtu.dk

Cataloging-in-Publication data applied for

    Die Deutsche Bibliothek - CIP-Einheitsaufnahme

Applied parallel computing : industrial strength computation
and optimization ; third international workshop ; proceedings /
PARA '96, Lyngby, Denmark, August 18 - 21, 1996. Jerzy
Waśniewski ... (ed.). - Berlin ; Heidelberg ; New York ;
Barcelona ; Budapest ; Hong Kong ; London ; Milan ; Paris ;
Santa Clara ; Singapore ; Tokyo : Springer, 1996
  (Lecture notes in computer science ; Vol. 1184)
  ISBN 3-540-62095-8
NE: Waśniewski, Jerzy [Hrsg.]; PARA <3, 1996, Lyngby-Tårbaek>; GT

CR Subject Classification (1991): G.1-2, G.4, F.1-2, D.1.3, J.1

ISSN 0302-9743
ISBN 3-540-62095-8 Springer-Verlag Berlin Heidelberg New York

© Springer-Verlag Berlin Heidelberg 1996
Printed in Germany

Typesetting: Camera-ready by author
SPIN 10549991   06/3142 – 5 4 3 2 1 0   Printed on acid-free paper

Preface

The Third International Workshop on Applied Parallel Computing in Industrial Problems and Optimization (PARA'96), and two Tutorials, on Wavelets/Signal and Image Processing, and Programming Parallel Computers, were held in Lyngby, Denmark, August 18–21, 1996. The conference was organized and sponsored by the Danish Computing Centre for Research and Education (UNI•C), the Department of Mathematical Modelling (IMM) of the Technical University of Denmark (DTU), and the Danish Natural Science Research Council through a grant for the EPOS project (Efficient Parallel Algorithms for Optimization and Simulation). Support was also received from the IBM, SGI, and DEC computing organizations.

The purpose of the workshop was to bring together scientists working with large computational problems in industrial problems and optimization, and specialists in the field of parallel methods and efficient exploitation of modern high-speed computers. Some classes of methods appear again and again in the numerical treatment of problems from different fields of science and engineering. The aim of this workshop was to select some of these numerical methods and plan further experiments on several types of parallel computers. The key lectures reviewed the most important numerical algorithms and scientific applications on parallel computers. The invited speakers included university and practical industry engineers, as well as numerical analysts and computer experts.

The workshop was preceded on Sunday, August 18th, by two Tutorials, on Wavelets/Signal and Image Processing, and Programming Parallel Computers. During the tutorials, time was allocated for practical exercises on the IBM SP2 and SGI Power Challenge computers. More than 40 people attended the tutorials.

The workshop itself attracted about 100 participants from around the world. Authors from over 20 countries submitted 76 papers, of which 31 were invited and 45 were contributed. The Fourth International Workshop (PARA'98) will be organized in 1998.

Jerzy Waśniewski
Jack Dongarra
Kaj Madsen
Dorte Olesen

September 1996

Table of Contents*

---

* Italic style indicates the speaker.
  Bold style indicates the title of the paper and the invited
  speaker.

Performance Tuning on IBM RS/6000 POWER2 Systems*

R. C. Agarwal, F. G. Gustavson, and M. Zubair

IBM T.J. Watson Research Center,
P.O. Box 218, Yorktown Heights, NY 10598, U.S.A.

Abstract. This tutorial describes the algorithms and architecture approach to produce high-performance codes for numerically intensive computations. In this approach, for a given computation, we design algorithms so that they perform optimally when run on a target machine-in this case, the new POWER2^{TM} machines from the RS/6000 family of RISC processors. The algorithmic features that we emphasize are functional parallelism, cache/register blocking, algorithmic prefetching, and loop unrolling. The architectural features of the POWER2 machine that we describe and that lead to high performance are multiple functional units, high bandwidth between registers, cache, and memory, a large number of fixed- and floating-point registers, and a large cache and TLB (translation lookaside buffer). The paper gives BLAS examples that illustrate how the algorithms and architectural features interplay to produce high performance codes. These routines are included in ESSL (Engineering and Scientific Subroutine Library); an overview of ESSL is also given in the paper.

1 Introduction

The new POWER2^{TM} workstations [2] of the RISC System/6000^{TM}. (RS/6000) family of processors provide multiple fixed-point units (FXUs) and floating-point units (FPUs) which can work in parallel if there are no dependencies. We call this functional parallelism. To achieve functional parallelism requires that the underlying numerical algorithm be parallelized at a very low level (instruction level). The functional parallelism can be achieved if, at the innermost loop level, several computations can be done in parallel. Various algorithmic techniques can be utilized to facilitate functional parallelism. Loop unrolling is an example of such a technique. Multiple FXUs help in prefetching of data into cache. One of the two FXUs can be utilized for cache prefetching, while the second FXU continues to remain available to load/store data (already in cache) into registers. This also requires algorithmic modifications. In many situations, the computation may have to be restructured to get around the serial bottlenecks and arithmetic pipeline delays. POWER2 workstations also provide quad load/store

---

* This paper is a condensation of [3] and is a formal presentation of the concepts presented in the tutorial.

instructions which double the effective bandwidth between floating-point registers (FPRs) and cache. To exploit the quad load/store instructions, the algorithm may have to be modified to access data with stride one. These are all
examples of algorithmic techniques which can help in exploiting the functional
parallelism capability of POWER2 and realize its full performance potential of
four floating point operations (flops) per cycle. The techniques described in this
paper have been heavily utilized in the development of some of the ESSL (Engineering and Scientific Subroutine Library) subroutines for the RS/6000 family
of workstations.

In this paper, we first give an overview of ESSL. We then describe salient
CPU and cache/memory features of POWER2 which can affect optimal algorithm design. Next we describe various algorithmic techniques used to facilitate
utilization of multiple functions units. In each case, we provide appropriate examples and discuss the performance achieved.

2 ESSL overview

ESSL is a high-performance library of mathematical subroutines for IBM RISC
System/6000 workstations (ESSL/6000), and ES/9000^{TM} and ES/3090^{TM} vector
and/or scalar mainframes (ESSL/370) [11]. Currently, ESSL V2.2 consists of 441
subroutines that cover the following computational areas: linear algebra subprograms, matrix operations, linear algebraic equations, eigensystems analysis,
Fourier transforms, convolutions/correlations, and other related signal-processing
routines, sorting and searching, interpolation, numerical quadrature, and random
number generation.

ESSL can be used for both developing and enabling many different types of
scientific and engineering applications, as well as for numerically intensive applications in other areas such as finance. Existing applications can be enabled
by replacing comparable subroutines and in-line code with calls to ESSL subroutines. Because of the availability of a large number of subroutines in ESSL,
the effort required in developing a new application is significantly reduced. This
also makes it easier to move an application to a new platform, because ESSL is
tuned for all platforms where it is available.

In the area of linear algebra, ESSL routines can be used in two different
ways. ESSL provides a set of routines to solve linear equations of various kinds.
Most of these routines are functionally the same as the public domain software
LAPACK [5, 6], though not necessarily in the same call sequence. The user application can be modified to call these ESSL subroutines instead of LAPACK subroutines. This approach provides the highest level of performance. Alternatively,
ESSL also provides a complete set of tuned basic linear algebra subprograms
(BLAS). There are 140 BLAS subroutines in ESSL. BLAS are an industry-wide
standard[12, 10, 9, 8, 7] , providing a uniform functionality and call sequence interface, which makes an application using calls to BLAS highly portable across
high-performance platforms from different vendors. In LAPACK subroutines,
most of the computing is done in BLAS subroutines. LAPACK expects platform

vendors to provide a set of tuned BLAS to achieve high performance. This is the alternative approach, in which the user application calls LAPACK subroutines and links to ESSL for tuned BLAS. This approach provides portability and performance at the same time. The performance achieved in this approach is slightly lower than that obtained by calling ESSL subroutines directly, but it is still very good.

ESSL/6000 provides a set of subroutines tuned for performance on the RISC System/6000 family of workstations, which include the older POWER^{TM} workstations and the newer POWER2^{TM} workstations. These are predominantly compatible with the ESSL/370 product, resulting in easy cross-platform migration between mainframes and workstations. All ESSL/6000 computational subroutines are written in FORTRAN; they are callable from application programs written in FORTRAN and in C. All of the subroutines are fully described in the *ESSL Guide and Reference* [11].

3 POWER2 CPU considerations

POWER2 workstations differ from the original POWER workstations in many respects. POWER workstations have one fixed-point unit (FXU) and one floating-point unit (FPU) and therefore can perform one floating-point multiply-add instruction and a fixed-point instruction every cycle, if there are no dependencies. Additionally, branch instructions can also be overlapped and therefore in effect result in zero-cycle branches. All load/store instructions, including the floating-point load/stores, are handled by the FXU. Additionally, floating-point stores also require the FPU, and therefore these instructions cannot be overlapped with the floating-point arithmetic instructions. See [1] for a full description of a POWER machine.

POWER2 workstations have two FXUs and two FPUs and therefore can perform two fixed-point instructions and two floating-point instructions every cycle, if there are no dependencies. In contrast to POWER, on POWER2 floating-point store instructions can also be overlapped with floating-point arithmetic instructions. Floating-point instructions have a two-to-three-cycle pipeline delay; therefore, to keep both FPUs fully utilized, at least four independent floating-point instructions must be executing at the same time (see [2] for details).

Two additional innovative features of POWER2 are the floating-point quad load/store instructions. The quad load instruction can load two consecutive doublewords of memory into two consecutive FPRs, and the quad store instruction can store two consecutive FPRs into two consecutive doublewords of memory. With both FXUs in use, this results in an effective bandwidth of four doublewords per cycle between the cache and the FPRs. This increased bandwidth is particularly important for those computational kernels where performance would otherwise be limited by the bandwidth between registers and cache. BLAS-1 (level-1 BLAS), for example DDOT and DAXPY, are examples of such kernels. For these kernels, on the same cycle time basis, POWER2 can perform four times faster than POWER. However, it is not always easy for the compiler to generate quad load/store instructions; loops often require additional unrolling. Also, quad load/stores require two consecutive FPRs; this restriction imposes additional constraints on the register assignment logic of the compiler. Please note that quad load/store instructions can be used only if two consecutive data elements from memory are needed in the loop. This may also require a restructuring of the loops.

Individual quad accesses which cross a cache line boundary require additional cycle(s). If the loop performance is limited by the cache bandwidth, quad loads crossing the cache line boundary reduce the available bandwidth. Special coding techniques can be used to handle such a situation. For example, assuming that all double-precision arrays are aligned on doubleword boundaries, there are two possibilities: The array is either aligned on a quadword boundary or on an odd doubleword boundary. Here, we are also assuming that the array is accessed in the loop with stride one using quad load instructions. Then, if the array is aligned on a quadword boundary, a quad load will never cross the cache line boundary. This follows from the fact that the cache line size is a multiple of quadwords. If the array is aligned on an odd doubleword boundary, one load is purposely handled outside the loop, thus making all quad loads inside the loop quadword-aligned. For double-precision two-dimensional arrays, if possible, the user should make their leading dimensions even. This ensures that if the first column is quadword-aligned, all other columns are also quadword-aligned.

4 POWER2 cache considerations

The data cache size on POWER2 workstations is 256 KB; however, only 128 KB may be accessible on machines having fewer than four of memory cards. In this paper, we refer to the data cache as simply the cache. All POWER2 models have 1024 cache lines arranged in four-way associative sets. The cache size of the largest POWER2 models is four times greater than that of the largest POWER models. POWER2 models also have a significantly higher bandwidth to the memory system. The memory system bandwidths on all machines are designed to fetch a complete cache line in eight cycles. Thus, machines with larger caches also have higher bandwidth to the memory system. All POWER2 workstations have a 512-entry, two-way set-associative TLB (table lookaside buffer). This is significantly larger than 128-entry TLB on POWER machines. Furthermore, the number of cycles required for TLB miss processing on POWER2 is considerably smaller than the number of cycles required for TLB miss processing on POWER. This has a very significant impact on considerations of blocking (of large problems) for POWER2. For most problems, the TLB size of POWER2 is not a consideration in blocking. In those computational kernels where arrays are used several times, appropriate cache and TLB blocking is sufficient to give the best possible level of performance. For these kernels, the delay in accessing a cache line is not important because the data are used several times. BLAS-3 routines are examples of such kernels. However, in BLAS-1 and BLAS-2 routines, the arrays are used only once, and cache miss latency becomes the important consideration.

5 Loop unrolling

Loop unrolling is a common technique used to exploit multiple functional units and quad load/store capabilities. In its simplest form, unrolling a loop amounts to a mini-vectorization of the loop. For example, unrolling a loop by eight is equivalent to using a vector length of eight. Generally speaking, vectorization results in independent computations, unless the vector operation is a reduction operation. Inner product (dot product) computation and finding the maximum or minimum of a vector are examples of reduction operations. For these examples also, parallelization is easily achieved if the vectors are long. For the inner product example, outside the loop, we can set up four variables (or registers) for partial sums. The loop is on the length of the inner product and is unrolled by four. Inside the loop, the four partial sum registers are updated with multiply-add operations. This results in four independent computations which can take place in parallel. At the completion of the loop, the four partial sum terms are added together to form the inner product. The final sum computation is only partially parallelized.

Loop unrolling also helps in reusing the data loaded in registers, and thereby reduces the bandwidth (load/store) requirements between registers and cache and cache and main memory. The loop unrolling need not be limited to the innermost loop. A series of nested loops can be unrolled to further facilitate data reuse. Some loop variables can be reused only if outer loops are unrolled. The matrix-matrix multiplication is a good example of nested loop unrolling. By unrolling all three nested loops, all variables in the loop can be reused. The degree of loop unrolling is generally limited by the number of registers available. Heavy loop unrolling may require more registers (floating and/or fixed) and therefore may result in spills. Spills are caused when the number of "logical" registers needed by the compiler exceeds the available number of architected registers. In that case, inside the innermost loop, "logical registers" are saved and restored from temporary memory location, resulting in a large degradation of performance. It is advisable for the user to look at the listings generated by the compiler. If the listings indicate spills in the innermost loop, the degree of unrolling should be reduced.

In many situations, an array is not accessed with stride one in the innermost loop, so quad load/store instructions cannot be used. However, this array may have a stride-one access in one of the outer loops. In that case, to facilitate use of quad/load store instructions, that outer loop should be unrolled. If multiple arrays are accessed in the innermost loop, they may have a stride-one access pattern in different levels of the loop. To facilitate quad access of all arrays, all corresponding loops must be unrolled.

To summarize, loop unrolling serves two goals. The main goal is parallelization of the computation; the secondary goal is reduction in register/cache/memory bandwidth requirements. At some degree of unrolling, we will have achieved a sufficient level of functional parallelism and memory bandwidth reduction to result in peak performance, if the data remain in cache. The question is "Can we get a higher level of performance by further unrolling the loop(s)?" The answer depends on the ratio of the computation cost (number of floating-point operations, called "flops") to data movement cost (number of data items involved in the computation). For the BLAS-3 (matrix-matrix multiplication is an example) kind of computations, where the ratio of flops to data movement cost is large, we do some kind of cache blocking so that data remain in cache. For these computations, occasional cache misses do not seriously affect overall performance. However, for those computations in which flops and data movement cost are about equal (for example, BLAS-1 and BLAS-2), data in cache are not reused, and the cache miss latency becomes very important. In such applications, by reducing the bandwidth requirements (via loop unrolling) between registers and cache, we can free up one of the FXUs to do cache prefetching. This is discussed in detail in the next section.

6 Algorithmic prefetching

On POWER2 machines, the cache miss latency is roughly of the order of 14-20 cycles. The desired data come after 14-20 cycles, followed by the rest of the line in a wraparound fashion in the next seven cycles. There are two FXUs, and each one of them can process only one cache miss at a time. In a typical stride one access code, if a load results in a cache miss, the subsequent load also results in a cache miss to the same line. This ties up both FXUs in fetching the same cache line. Specialized coding gets around this problem and can thus greatly improve memory system performance. We call this algorithmic prefetching [4]; it can significantly improve performance on POWER2 machines. It is fairly easy to implement algorithmic prefetching for the stride-one situation on POWER2. Typically, several cycles before the data from the next cache line are actually needed, a dummy load of an element from the next cache line is done. (A dummy load is a load where the data loaded are not actually used in the computation.) If the next line is already in cache, the load is completed in one cycle, and the FXU becomes available for the next set of instructions. On the other hand, if this load results in a cache miss, it ties up one of the FXUs until the cache miss is processed. However, the other FXU is still available to do the required loads into FPRs to feed both FPUs. When quad loads are extensively used, many computing kernels require only one FXU to feed both FPUs.

Prefetching on POWER is more difficult, because POWER has only one FXU, and a cache miss will stall it. However, in those loops where there are more arithmetic operations then load/store operations, specialized techniques can be used to do prefetching on POWER [4]. The primary concept is to load FPRs with data at the beginning of the loop, so that when the cache miss stalls the FXU, the FPU is kept busy with useful work using data already loaded in the FPRs.

We have used algorithmic prefetching on POWER and POWER2 in many computing kernels. However, prefetching requires extensive loop unrolling. This loop unrolling with dummy loads can actually be done in FORTRAN. In prefetching, there are several different variations, some being quite intricate and complex. If the cache bandwidth is a consideration, the loop is unrolled in such a way that the data loaded during a "dummy load" are actually used in the computation. Depending on the kernel and the number of arrays to be prefetched, this can become complicated, especially if quad loads are to be used and relative quad alignments of different arrays are to be taken into account. On the other hand, if the cache bandwidth is not a consideration (one FXU can handle all of the required loads), a simple prefetching scheme will do a dummy load and not use the data. In this case, if the data are actually in cache, prefetching does not degrade performance. If the data are not in cache, prefetching improves performance significantly.

7 BLAS implementation using loop unrolling and cache prefetching

We now illustrate use of the above techniques in implementing some key BLAS routines.

7.1 BLAS-1 implementation

Because there is very little reuse of data loaded in registers, BLAS-1 performance tends to be limited by the available bandwidth between the cache and the FPRs. Maximum bandwidth is achieved by using quad load/store instructions. An optimal implementation also tries to avoid crossing a cache line boundary on quadword accesses. There is also some limited opportunity for algorithmic prefetching. Here, we describe the implementation and performance of two key BLAS-1 kernels-the DAXPY and DDOT routines. We assume that all arrays have stride one; otherwise, it is not possible to use quad load/store instructions.

DAXPY. In DAXPY, we update a vector y with a scalar multiple of another vector x. The scalar multiplier is loaded outside the loop and remains in a register throughout the computation. For each multiply-add (FMA), DAXPY requires two loads and one store; therefore, its performance is limited by the available bandwidth to the FPRs. Depending on the alignments of the x and y arrays, there are four possibilities. When both arrays are even aligned, the coding to obtain optimal quad load/stores is easy. If both arrays are odd-aligned, one element is computed outside the loop. This makes both the remaining arrays quad-aligned. The difficult case arises when one array is even-aligned and the other is odd-aligned. In this case also, we can restructure the computation so that all quad accesses inside the loop are quad-aligned. This requires accessing one element of the odd-aligned array outside the loop. The loop is unrolled by four and executes in three cycles achieving the peak bandwidth of two quadword accesses between the cache and the FPRs every cycle. On a 50-MHz POWER2, the best performance that can theoretically be expected is 133 MFLOPS. For data in cache, we actually achieve nearly this performance for all four possible alignments of the arrays. By comparison, on a 50-MHz POWER machine, the best possible performance is 33 MFLOPS.

DDOT. The DDOT function computes the dot product of two vectors. Since FMA requires two loads, this seems to match the capabilities of POWER2 ideally, assuming that data are in cache. Thus, on POWER2, DDOT should run at its peak rate of four flops per cycle. However, as a result of the register-renaming implementation, the FXUs cannot perform two load quads every cycle on a continuous basis. The best that can be achieved is eight load quads in five cycles. Thus the best performance that can be expected on a 50-MHz POWER2 is 160 MFLOPS, and we nearly achieved that level of performance. Recall that in the DAXPY case we were doing both quad loads and quad stores; therefore, in that case we did achieve the peak bandwidth of two quad load/stores every cycle.

We developed two versions of the DDOT function, one for data in cache and another for long sequences where data are unlikely to be in cache. For the function which assumes that data are in cache, we unrolled the loop by eight and used four temporary variables to accumulate the partial results. The four subresults were then added together outside the loop. As in the DAXPY case, we had to take into account even-odd quad alignments of both arrays in order to achieve near-peak performance. However, the cost of examining the alignments of both arrays and setting up the unrolling by eight adds extra overhead to the subroutine, which is significant for small n. As an example, on a 50-MHz machine, for a dot product of size $n = 2000$, we measured 148 MFLOPS for all four possible quad alignments of the two arrays.

We also developed a version of DDOT function which does algorithmic prefetching for long sequences. Here we assume that data are not in cache. The subroutine must know the cache line size of the machine in order to do optimal prefetching. When data are not in cache, the performance is limited by the available bandwidth between cache and the memory system, and the prefetching is implemented to maximize it. Recall that during prefetching, one FXU is tied down when a cache miss occurs, but the other FXU remains available to load/store data already in cache. For these routines, we unrolled the loop by 16 and did dummy loads 16 elements apart for both arrays. On a POWER2 machine (50 MHz and 256KB cache size), for data not in cache, this subroutine performed at about 103 MFLOPS, while the subroutine which assumed that the data were in cache performed at about 74 MFLOPS. This represents a 40% performance improvement, due to algorithmic prefetching. However, the prefetched version of this subroutine requires extra loads, and if the data are actually in cache, its performance drops by about 10% compared to the subroutine optimized for data in cache.

If the data loaded in registers (for prefetching) can actually be used in the computation, we can achieve the best performance for data in cache as well as data not in cache. The coding becomes complicated because one must take into account not only relative quadword alignments of the two arrays, but also their relative cache line alignments. This extra logic adds overhead to the routine which is justifiable only for long sequences, say of the order of 2000. We implemented one such version on the above machine. For data in cache, this routine performed in the 150-156-MFLOPS range. For data not in cache, it performed in the 97-103-MFLOPS range. The performance of the subroutine varied in a narrow range depending on the relative alignments of the two arrays.

7.2 BLAS-2 implementation

BLAS-2 computations typically involve a matrix and one or two vectors. In these computations, the matrix elements are generally used only once; for a large matrix, most of it cannot be in cache. When the matrix is not in cache, the best that can be expected is to fully utilize the matrix data brought into cache and simultaneously prefetch the next cache line. During the prefetch, computing is being done on the data just brought into cache. We must also use register, cache, and TLB blocking for the matrix and the vectors, in order to fully use the data before they are swapped out of the cache and the TLB. The work presented in [4] describes algorithmic prefetching as it was implemented for BLAS-2 for POWER. We now consider implementation on POWER2. As pointed out earlier, because of multiple functions units on POWER2, prefetching is easier. It is particularly important to get sufficient reuse of data loaded in registers so as to minimize the load/store requirements. This requires the inner loop to be unrolled by a large factor, and the use of quad load/store capability of POWER2. To illustrate the computational techniques, we describe the example of the matrix-vector multiplication subroutine DGEMV, where the matrix A is stored in the normal form, i.e., column major order.

DGEMV — Normal Case. In DGEMV, a vector x is multiplied by a matrix A and then added to another vector $y(y \leftarrow y + Ax)$. Cache prefetching is the most important consideration, in DGEMV. An optimal implementation of cache prefetching requires knowing the cache line size. Here, we describe the implementation on machines with a 256-byte line size. The outermost blocking was on the number of columns, to minimize the finite cache and TLB effects. Within a vertical block, we unrolled the computation by a large factor (i.e., we implemented a horizontal blocking). The ideal block size (the unrolling factor) corresponds to the cache line size, so that in each subcolumn there is exactly one cache line (32 doublewords). However, because of the floating-point register limitation (there are only 32 FPRs), we restricted the unrolling to 24. The innermost loop operates on the number of columns in a vertical block. Outside this loop, we loaded 24 elements of the y vector into 24 FPRs, T0, T1, ..., T23, corresponding to the horizontal block size or the loop unrolling factor. Within this loop, we processed a subcolumn of the A matrix of size 24. This is like a mini-vectorization with a vector length of 24. We can assume, because of the cache blocking, that a block of the A matrix remains in cache. The code shown in Figure 1 is indicative of the code without prefetching.

```
T0   = Y(I)                    ! load 24  y elements
T1   = Y(I+1)                  ! in 24 FPRs.
 . . .
T23  = Y(I+23)
DO J = J1, J1+JBLK-1
  XJ  = X(J)                   ! load an element of x
  F0  = A(I,  J)               ! one load quad loads
  F1  = A(I+1,J)               ! both F0 and F1
  T0  = T0 + XJ*F0
  T1  = T1 + XJ*F1

   . . .
  F0  = A(I+22,J)
  F1  = A(I+23,J)
  T22 = T22 + XJ*F0
  T23 = T23 + XJ*F1
ENDDO
Y(I)   = T0                    ! store y elements
Y(I+1) = T1                    ! after the loop
 . . .
Y(I+23)  = T23
```

Fig. 1. DGEMV matrix-vector multiplication without prefetching.

Here XJ corresponds to a floating-point register. Note that one quad-load loads two FPRs, feeding both FPUs. Thus, one FXU can feed both FPUs, except for the initial load of x_J into FPR XJ. There are 24 FMAs in the innermost loop (the J loop), requiring 24 loads for the matrix elements which can be performed as 12 quad loads. This gives a 24 times reuse factor for XJ. Thus, for 12 cycles, we can keep one FXU feeding both FPUs at the peak rate of two FMAs per cycle. The other FXU is free to handle cache miss processing by doing a dummy load of an element from the second next column of A, which is not likely to be in cache. The prefetching is accomplished by inserting the following instruction in the inner loop:

```
D = A(I+23,J+2) ! dummy load for prefetch
```

The dummy variable D is not used in the loop. Its sole purpose is to bring the desired section of column $(J + 2)$ into cache if it is not already in cache. The second FXU on POWER2 accomplishes this goal. By prefetching two columns ahead, our measurements show that all of the required data were in cache. If the LDA of the A matrix is even and the initial alignment of the matrix is on an odd doubleword boundary, we process one row outside the main blocking loop, so that each subblock in the main loop is aligned on a quadword boundary. This is to make sure that none of the quad loads inside the inner loop cross a cache line boundary. If LDA of the matrix is odd, for every other column quad loads will cross the cache line boundary, slightly degrading the performance. This is the reason why we recommended earlier that the leading dimensions of multidimensional arrays should be even.

This implementation of DGEMV with algorithmic prefetching is optimal even when the matrix is actually in cache. In that case, the prefetch load does not result in a cache miss and becomes an ordinary load. Since the innermost loop is not limited by the load/store bandwidth, this extra load has no impact on the execution of the loop. For matrices which fit in cache, we achieved 96% of the peak performance. For very large matrices which do not fit in cache, we achieved 81% of the peak (on a 50-MHz machine).

7.3 BLAS-3 routine - DGEMM

For BLAS-3 routines, appropriate cache and TLB blocking is generally sufficient to give the best possible level of performance. For these kernels, the delay in accessing a cache line is not important because the data are used multiple times. DGEMM is typical of BLAS-3 routines. It basically computes the product of two matrices. For DGEMM, fairly good performance can be obtained from the vanilla code, if appropriate preprocessing options are used at the compile time. In most cases, the preprocessor does a reasonable job of cache blocking. The problem arises when the matrix dimensions are powers of two (or related to them). In those cases, because of the cache congruence class conflicts, the effective cache size is reduced. In this case, preprocessor cache blocking is not very effective.

ESSL BLAS-3 routines do cache and TLB blocking customized for the platform on which they are run. They are designed to provide robust performance in almost all situations. If necessary, subarrays are copied into temporary buffers in order to eliminate any problem due to bad leading dimensions. Because the copied data are used many times, the cost of copying becomes insignificant. If the arrays are blocked for cache, we can assume that data remain in cache. In that case, the only consideration is to obtain peak performance at the innermost loop level. For BLAS-3 kernels, the bandwidth between cache and FPRs is not a consideration, because the nested loops can be unrolled in many different ways to get a significant reuse of data loaded into registers. The unrolling of loops also makes it possible to utilize multiple functional units fully, and avoids FPU pipeline delays.

For the previous release of ESSL, which was only for POWER, we implemented a two by two unrolling; i.e., a two-by-two block of the result matrix was computed in the innermost loop. This is equivalent to computing four dot products in the innermost loop. This was sufficient to give the peak performance on POWER. On POWER2, to ensure robust performance in utilizing multiple functional units, we implemented a four-by-two unrolling. This resulted in the peak performance at the innermost loop level for all combinations of the matrix form parameters. Form parameters specify whether the matrices are stored by rows or columns. The four-by-two blocking used for POWER2 is also optimal for POWER. This helps in producing a single source code for POWER and POWER2 machines. The cache and TLB blocking is customized for the platform. This requires different compilations for POWER and POWER2. By using a different compilation for POWER2 machines, we also obtained some additional performance by exploiting quad load/store instructions which are not available on POWER. Blocking for different cache sizes is done at run time. The line size of the machine can be used to determine the cache size at run time. We use the special subroutine IRLINE to determine line size. Once the line size is known, we can set cache size parameters for the particular POWER or POWER2 machine. This determination of cache size parameters is done only once.

DGEMM performance is excellent on POWER2 and POWER. For small values of N, the performance is somewhat uneven, because of our choice of four-by-two blocking. However, even for matrix sizes as small as 20, performance reaches 200 MFLOPS on a 66.6-MHz POWER2 machine. For large-size matrices, including powers of two, the performance is essentially uniform in the range of 90-95% of the peak for the machine.

To summarize, a very efficient DGEMM has been produced on POWER and POWER2 machines by using cache blocking and dot-product-based kernels. DGEMM is the basic computing kernel and building block for almost all of the computing in the area of linear algebra for dense matrices. Therefore, it must demonstrate uniformly good performance for all reasonable choices of parameters and matrix storage formats. ESSL DGEMM has this property.

8 Summary

In this paper, we have described the novel architectural features of the new POWER2 workstations. These features include multiple functional units, quad load/store instructions, and a very high-bandwidth memory system. If one uses the quad load capability of the POWER2 machines, a single FXU can feed both FPUs at the peak rate; therefore, the other FXU can be used to prefetch data into cache. Thus, the multiple functional units of POWER2 allow for the possibility of prefetching data into cache. In other words, POWER2 capabilities can be used to provide functional parallelism, if one develops high-performance numerical algorithms to do so. We have exploited functional parallelism by developing many highly tuned routines for ESSL. The three main techniques we used to exploit functional parallelism were loop unrolling, algorithmic prefetching, and algorithmic restructuring of the computation to serial bottlenecks. We have provided several examples of these techniques. For many BLAS-1 and BLAS-2 routines, on the same cycle time basis, we have demonstrated performance on POWER2 machines that is up to four times higher than that available using POWER. The quadword access facility, along with our use of algorithmic cache prefetching, was primarily responsible for this high level of performance. For BLAS-3 routines, the performance improvement over POWER is slightly more than a factor of two. The two FXUs and two FPUs are responsible for the factor of two. The significantly higher bandwidth of the memory system makes the improvement factor greater than two.

References

1. IBM RISC System/6000 Processor. *IBM Journal of Research and Development*, Volume 34, Number 1, 1–136, January 1990.

2. POWER2 and PowerPC Architecture and Implementation. *IBM Journal of Research and Development*, Volume 38, Number 5, 489–648, September 1994.

3. R. C. Agarwal, F. G. Gustavson, and M. Zubair. Exploiting functional parallelism of POWER2 to design high-performance numerical algorithms. *IBM Journal of Research and Development*, 38(5):563–576, 1994.

4. R. C. Agarwal, F. G. Gustavson, and M. Zubair. Improving performance of linear algebra algorithms for dense matrices using prefetch. *IBM Journal of Research and Development*, 38(3):265–275, 1994.

5. E. Anderson, Z. Bai, C. Bischof, J. Demmel, J. Dongarra, J. Du Croz, A. Greenbaum, S. Hammarling, A. McKenney, S. Ostrouchov, and D. Sorensen. *LAPACK User's Guide*. SIAM, Philadelphia, PA, 2nd edition, 1994. Also available online from http://www.netlib.org.

6. E. Anderson, Z. Bai, C. Bischof, J. Demmel, J. Dongarra, J. Du Croz, A. Greenbaum, S. Hammarling, A. McKenney, and D. Sorensen. LAPACK: A portable linear algebra library for high-performance computers. Technical Report Technical Report CS-90-105 (LAPACK Working Note 20), Computer Science Department, University of Tennessee, Knoxville, Tennessee, 1990. Also available online from http://www.netlib.org/lapack/lawns.

7. Jack J. Dongarra, Jeremy Du Croz, Sven Hammarling, and Ian Duff. Algorithm 679. A set of level 3 basic linear algebra subprograms: Model implementation and test programs. *ACM Transactions on Mathematical Software*, 16(1):18–28, 1990.

8. Jack J. Dongarra, Jeremy Du Croz, Sven Hammarling, and Ian Duff. A set of level 3 basic linear algebra subprograms. *ACM Transactions on Mathematical Software*, 16(1):1–17, 1990.

9. Jack J. Dongarra, Jeremy Du Croz, Sven Hammarling, and Richard J. Hanson. Algorithm 656. An extended set of basic linear algebra subprograms: Model implementation and test programs. *ACM Transactions on Mathematical Software*, 14(1):18–32, 1988.

10. Jack J. Dongarra, Jeremy Du Croz, Sven Hammarling, and Richard J. Hanson. An extended set of FORTRAN basic linear algebra subprograms. *ACM Transactions on Mathematical Software*, 14(1):1–17, 1988.

11. IBM Corporation. *Engineering and Scientific Subroutine Library, Version 2 Release 2: Guide and Reference*, 2nd edition, 1994. Publication number SC23-0526-01.

12. C. L. Lawson, R. J. Hanson, D. R. Kincaid, and F. T. Krogh. Basic linear algebra subprogram for Fortran usage. *ACM Transactions on Mathematical Software*, 5(3):308–323, 1979.

Digital Image Processing: A 1996 Review

Jens Damgaard Andersen

University of Copenhagen, Department of Computer Science, Universitetsparken 1,
DK-2100 Copenhagen, Denmark
E-mail: jda@diku.dk

Abstract. Digital image processing is a fast developing field rapidly gaining importance through spreading of picture-oriented electronic media and the merging of computer and television technology. Digital image processing is used in image analysis, image transmission and image storage and retrieval. Typical application areas are telecommunication, medicine, remote sensing, and the natural sciences and agriculture.
Digital image processing is computationally demanding. Consequently there is a need for fast algorithms which efficiently implement image processing operations. In this 1996 review algorithms for common image transforms, image representations, and image data compression methods will be described. The principles underlying newer image coding techniques such as transform coding (JPEG) and multiresolution wavelet analysis will be outlined. Finally the role of image analysis in computer vision will be considered.

1 Introduction

In 1975 B.R. Hunt published a survey paper entitled "Digital Image Processing" in the Proceedings of the IEEE [8]. At that time this topic was a novelty, the most impressive achievements being the acquisition of digitally processed images of the surface of the moon and the face of Mars. Since then a spectacular development has taken place. Digital image processing is now getting increasingly important in everyday life. Several factors are contributing to this. Most important, technical achievements have made it possible to process images at high speed due to among other things increased computer processing power. Higher network data rates and better compression techniques have made long range transmission of pictorial information (e.g. via *World Wide Web*) ubiquitious. Furthermore, images are increasingly available in digital form, because digital cameras, framegrabbers and scanners are rapidly falling in price due to volume sales to the consumer market, notably of TV equipment and multimedia stations. Accordingly, more and more people will want to manipulate and/or process images.

Although a number of software packages for digital image processing are available it is important that users understand basic image processing techniques in order to know what is feasible and how to apply available software optimally. Furthermore, knowledge of image processing may also make it possible to design targeted software, e.g. for parallel processors, in order to obtain special results or high processsing speeds.

This tutorial covers some recent fundamental techniques in image processing. However, due to the vastness of the subject and space and time limitations some information which is readily available in older survey articles like Hunt [8] or common textbooks such as Jain [9] or Gonzales [7] has not been included. The reader is referred to these sources and the other references mentioned at the end of the article for further information. Thus image acquisition (the optics of the image formation process), the image recording process, and image sampling and quantization and reconstruction and display of digital images will not be dealt with here. It is fairly well covered in the previously mentioned references. On the other hand image transforms are included, since they form a basis for some of the compression techniques discussed in section 3. which treats JPEG encoding of images (now a well established technique). In section 4 we will consider multiresolution analysis and wavelets, and in section 5 image pyramids and linear scale space analysis. Section 6 is dedicated to the coding/segmentation duality and briefly outlines the role of image analysis in computer vision.

2 Image Transforms

Images are characterized by a large degree of redundancy, because there usually is a high correlation between neighbouring image elements. Furthermore, because slow variations in gray scale occur more frequently than rapid oscillations, it is possible to devise coding schemes which perform energy compaction by packing the image energy into a few transform coefficients. Here we will consider transform image coding by the discrete cosine transform (DCT).

We assume that the image is digitized, i.e. converted from the original analog form, usually a time-varying voltage signal, into a sequence of integers. This process of sampling and digitization is performed by electronic circuits (sample-hold and analog-to-digital converter) and need not concern us here. Thus we assume that the signal is available as a matrix of integers, where each matrix element gives the gray scale value of the corresponding image point (pixel). Most often 8 bits are used to encode the gray scale value (255 = white, 0 = black).

2.1 The Discrete Cosine Transform

It is convenient to illustrate the method by first encoding a one-dimensional signal and then subsequently examine the encoding of the two-dimensional images. Thus, instead of a matrix, we first consider an array of integers. This array is divided into fixed-length segments (vectors), which we here take to be of length N, where N is a power of 2.

A discrete image transform is then merely a rotation of the vector x in an N-dimensional vector space [1]. As an example, the discrete cosine transform (DCT) of a vector $\{x_m\}, m = 0, 1 \ldots N - 1$ is defined by

$$\tilde{x} = Ax, \tag{1}$$

where x is the vector containing the signal values, $\tilde{x}$ is a vector of DCT-coefficients, and A is the matrix containing the new system base vectors. The elements of the base vector matrix are given by

$$a_{k+1,m+1} = \sqrt{\frac{2}{N}} \cos \frac{(2m+1)k\pi}{2N}, \tag{2}$$

where $k = 0, 1, 2, \ldots, N-1$ and $m = 0, 1, 2, \ldots, N-1$ and

$$a_{1,m+1} = \frac{1}{\sqrt{N}}, \tag{3}$$

$m = 0, 1, 2, \ldots, N-1$.

For instance, for $N = 8$ we find

$$A = \begin{pmatrix} 0.354 & 0.354 & 0.354 & 0.354 & 0.354 & 0.354 & 0.354 & 0.354 \\ 0.490 & 0.416 & 0.278 & 0.098 & -0.098 & -0.278 & -0.416 & -0.490 \\ 0.462 & 0.191 & -0.191 & -0.462 & -0.462 & -0.191 & 0.191 & 0.462 \\ 0.416 & -0.098 & -0.490 & -0.278 & 0.278 & 0.490 & 0.098 & -0.416 \\ 0.354 & -0.354 & -0.354 & 0.354 & 0.354 & -0.354 & -0.354 & 0.354 \\ 0.278 & -0.490 & 0.098 & 0.416 & -0.416 & -0.098 & 0.490 & -0.278 \\ 0.191 & -0.462 & 0.462 & -0.191 & -0.191 & 0.462 & -0.462 & 0.191 \\ 0.098 & -0.278 & 0.416 & -0.490 & 0.490 & -0.416 & 0.278 & -0.098 \end{pmatrix} \tag{4}$$

It is easy to verify that $A^T A = I_8$, where I_8 is the identity matrix of order 8, i.e. A defines a *unitary* transform.

Thus, in order to perform coding by the DCT the signal must be divided into a sequence of blocks (vectors) and then encoded by means of the set of orthonormal basis vectors.

The one-dimensional DCT can be extended to two-dimensional images. Figure 1 shows a set of 64 two-dimensional cosine basis functions created by multiplying a horizontally oriented set of one-dimensional 8-point basis functions by a vertically oriented set of the same basis functions. The horizontally oriented set of basis functions represents horizontal frequencies and the other set the vertical frequencies. The DC term (constant term) is at top left.

2.2 The inverse transform

The inverse discrete cosine transform (IDCT) is found by premultiplying with A^T in eqn. (1):

$$A^T \tilde{x} = A^T A x \tag{5}$$

or, since A is orthogonal (unitary),

$$x = A^T \tilde{x} \tag{6}$$

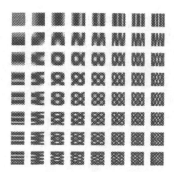

Fig. 1. DCT basis functions. Each 8×8 array of samples shows a different basis function. The horizontal DCT frequency of the basis function increases from left to right and the vertical from top to bottom.

2.3 Two-dimensional DCT

The two-dimensional (forward) discrete cosine transform is defined by

$$F(u, v) = \sum_{i=0}^{N_1-1} \sum_{j=0}^{N_2-1} f(i, j) a(i, j, u, v) \qquad (7)$$

where $N_1 \times N_2$ is the image dimension. If a transform kernel $a(x, y)$ can be expressed as the product of a function in each of the two variables x and y, the transform is *separable*, i.e. $a(x, y) = a_1(x) a_2(y)$.

The twodimensional DCT is separable since the kernel $a(i, j, u, v)$ may be written as

$$a(i, j, u, v) = a_1(i, u) a_2(j, v). \qquad (8)$$

Thus the 2-D DCT may be expressed as

$$F(u, v) = \sum_{i=0}^{N_1-1} \sum_{j=0}^{N_2-1} f(i, j) a_1(i, u) a_2(j, v). \qquad (9)$$

which may be written in matrix form as

$$F = A_1 f A_2 \qquad (10)$$

3 JPEG Coding

JPEG is not a data format but rather a family of algorithms for compression of digitized color still images. This collection of different compression methods has been adopted as an international standard in 1993 under the name ISO 10918. What the standard describe is actually a 'toolbox' from which developers can choose required parts and implement those in hardware or software products. A first step towards reduction of the amount of image data is the use of graphic formats employing internal compression methods such as run length coding, Lempel-Ziv or Huffman-coding in the GIF, PCX or TIFF formats [2]. These methods are lossless; however, the compression ratio achived rarely exceeds three. In the JPEG algorithm it is possible to select a suitable trade-off between compression ratio and reconstruction quality (degree of degradation of the reconstructed image as compared with the original image) for lossy compression modes. In these modes information of lesser significance for reconstruction is discarded. For completeness the standard also defines a lossless coding mode which we will not consider here.

The JPEG algorithms are designed to

- give near state-of-the-art performance regarding compression rate versus image degradation.
- be optimized with respect to photographic pictures with continuous color transitions. It is less suited for other types of image data representing images with sharp contrasts such as cartoons, line graphics or text images, i.e. images containing lage uniform color areas and abrupt changes of color. (For motion pictures a similar standard (MPEG) is defined).
- have a manageable computational complexity
- to define 4 modes of operation: sequential, progressive, lossless and finally hierarchical encoding. In *sequential* encoding each image component is encoded in a single top-bottom, left-right scan. If transmission time is long *progressive* encoding may be used; in this mode the image is reconstructed in multiple coarse-to-fine passes. *Lossless* encoding guarantees exact recovery of the original image. Finally, in *hierarchical* encoding the image is encoded in multiple resolutions so that lower resolution can be viewed without the need for decompressing the image at full resolution.

3.1 The JPEG baseline algorithm

The JPEG standard has defined a "baseline" algorithm which must be implemented in all of the JPEG modes of operation which uses the DCT. Figure 2 shows the basic mode of operation: compression of a single-component (grayscale) image. This mode, which covers a large part of the applications, is called *baseline Codec* (codec = coder + decoder).

The image to be JPEG-encoded (the source image) is divided into 8×8 pixel blocks. Each pixel value is shifted form unsigned integers, e.g. in the range $[0, 255]$ to signed integers in the range $[-128, 127]$ and input to forward DCT (FDCT). At the output from the decoder the inverse DCT (IDCT) outputs 8×8 pixel blocks to form the reconstructed image. The DCT coefficient can be regarded as the relative amount of the 2D spatial frequencies contained in the 64-pixel input signal. The coefficient with zero frequency in both directions is called the "DC coefficient" and the remaining coefficients are called the "AC coefficients". Because image data typically vary slowly from pixel to pixel across the image, most of the spatial frequencies have zero or near-zero amplitude and need not be encoded.

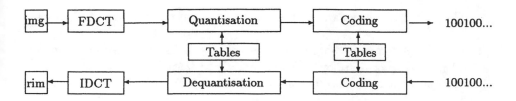

Fig. 2. The baseline codec algorithm. Top: coding of the original image (img.) via forward DCT, quantisation og coding. Bottom: decoding in reverse sequence, resulting in a reconstructed image (rim).

After output from the FDCT each of the 64 DCT coefficients is uniformly quantized according to a 64-element quantization table, which must be specified by the application or the user as input to the encoder. Each table element can be any integer from 1 to 255, which specifies the step size of the quantizer for the corresponding DCT coefficient. By coefficient quantization a further compression is achieved, since no DCT coefficient is then represented with greater precision than is required to obtain the desired image quality, i.e. visually insignificant information is discarded. Quantization is performed by division of each DCT coefficient by its corresponding quantizer step size, followed by rounding to the nearest integer

$$F^Q = \lfloor \frac{F(u,v)}{Q(u,v)} \rfloor \tag{11}$$

where $F(u,v)$ is the DCT coefficient and $Q(u,v)$ the corresponding quatization step.

After quantization the DC coefficients are encoded as the difference from the DC term of the previous block in the encoding order. The final processing step is entropy coding of the quantized DCT coefficients. The JPEG standard specifies two entropy coding methods: Huffmann coding and arithmetic coding [13]. The baseline sequential coding uses Huffman coding (a particular kind of Huffman coding the details of which may be found in the article by Wallace [17]). Decoding is of course the same steps executed in reverse. Accordingly we may state the baseline sequential coding algorithm:

1. DIVIDE THE IMAGE IN 8 × 8 PIXEL BLOCKS
2. PERFORM FORWARD DISCRETE COSINE TRANSFORM
3. DO COEFFICIENT QUANTIZATION, USING A PRESTORED (OR CALCULATED) QUANTIZATION TABLE.
4. ORDER COEFFICIENTS ACCORDING TO THE ZIG-ZAG PATTERN (ZONAL REPRESENTATION)
5. CODE INTERMEDIATE REPRESENTATION
   - FOR AC-COEFFICIENTS: (RUNLENGTH,SIZE)(AMPLITUDE)
   - FOR DC-COEFFICIENTS: (SIZE)(AMPLITUDE)
6. OUTPUT BITSTREAM

The inverse transform are the same steps taken in reverse order, cf. Figure 2.

Availability. A complete C-library with installation guide and Make-files is maintained and may be obtained by ftp from the site `ftp://ftp.uu.net/graphics/jpeg` (the Independent JPEG Group (IJG) distribution).

4 Wavelets and Multiresolution Analysis

Wavelets are used in image analysis (image restoration and compression). We start with the space L^2 of all square integrable functions. Multiresolution analysis is an increasing sequence of closed subspaces $\{V\}_j, j \in \mathbb{Z}$ which approximate $L^2(\mathbb{R})$. The basic theory is found in this volume [10].

The wavelet or multiresolution decomposition divides the image into two parts which we may call "averages" and "details". The details corresponds to the transients at the current scale, while the averages represents transients at larger scales, which appear constant at the current scale. The details are stored and the process is repeated with averages at the next larger scale. The division into details and averages corresponds to high-pass and low-pass filtering. This transformation decorrelates pixel variations by scale, and has been applied succesfully in video compression. It segments the image into layers of increasing detail, so that an image can be described roughly first and with more detail as required.

Wickerhauser has compared different image compression methods, viz. wavelet, wavelet packet, and local transform coding. The results, including reconstructed images for subjective evaluation, is available in proceedings from a recent wavelet conference [18].

5 Image Pyramids and Linear Scale Space Analysis

Interpreting the contents of an image involves selection of a *scale*, i.e. the level at which details are taken into account. A classical example is that of making structure out of an image of branch of a tree, which make sense only at a scale from a few centimeters to at most a few meters. This means that the various image structures could be analyzed at different scales. A physical quantity is measured using measuring devices with certain apertures. At large apertures we perceive the gross structure and with smaller apertures the finer details of the image.

We will call a one-parameter family of derived images for which the parameter has a clear representation a *multiscale representation*. There exists various methods for constructing multiscale representations or one-parameter families of images derived from a given image, for example quad trees, pyramids, and linear scale spaces.

The *pyramid representation*, introduced by Burt [3], is obtained by successively reducing the image size combined with smoothing and subsampling.

The (linear) *scale-space representation* (pioneered by A. Witkin and J.J. Koenderink) corresponding to a given image is obtained by successively convolving the image with 2D Gaussians. By this operation the original image is seen as being "blurred" as if the image were projected by an out-of-focus lens. In general, the linear scale-space representation of a continuous signal is constructed as follows [12]. Let $f : \mathbb{R}^N \to \mathbb{R}$ represent a given signal. Then the scale-space representation $I : \mathbb{R}^N \times \mathbb{R}_+ \to \mathbb{R}$ is defined by setting the scale-space representation at zero scale equal the original signal $I(\cdot; t) = f$ and for $t > 0$

$$I(\cdot; t) = g(\cdot; t) * f, \tag{12}$$

where $t \in \mathbb{R}_+$ is the scale parameter, $g : \mathbb{R}^N \times \mathbb{R}_+ \setminus \{0\} \to \mathbb{R}$ is the Gaussian kernel and $*$ denotes the convolution operator. In arbitrary dimensions ($x \in \mathbb{R}^N, x_i \in \mathbb{R}$) the Gaussian kernel is

$$g(x; t) = \frac{1}{(4\pi t)^{N/2}} e^{-x^T x/(4t)} = \frac{1}{(4\pi t)^{N/2}} e^{-\sum_{i=1}^{N} x_i^2/(4t)} \tag{13}$$

In contrast to other coarse-to-fine structure representing methods the scale space posseses the *semigroup property*: if every smoothing kernel is associated with a parameter value and if two kernels are convolved then the resulting kernel is a member of the same family, viz:

$$h(\cdot; t_1) * h(\cdot; t_2) = h(\cdot; t_1 + t_2) \tag{14}$$

This can also be expressed as the fact that the Gaussian is *self-similar* under autoconvolution.

The continuous semigroup property reveals the *scale invariance* of the representation, since it implies that the "inner structure" of an image, that is, the nesting of structures over scale, is determined only by the data; there is no a priori scale bias or spurious resolution introduced by the filters, as in other multiresolution analysis methods [6].

6 Best Bases and the Coding-Segmentation Duality

The idea underlying the 'best basis' representation is to build out of a set of library functions an orthonormal basis, relative to which a given signal or image has the lowest information cost [5]. The library shall consist of predefined modulated waveforms with some definite orthogonality properties, for example orthogonal wavelet packages or localized trigonometric functions with good time-frequency localization properties. In this respect the representation is related to the principle of universal coding epithomized by Rissanen's MDL [14]. For image representations optimum encoding can be obtained, for example, by spatially segmenting the image and finding the best wavelet package (frequency) tree for each spatial segment. If spatial segments are recursively subdivided and the corresponding information costs are compared with previously calculated

 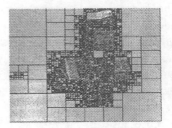

Fig. 3. Best basis example. To the left the original image and to the right a best basis representation (courtesy Klaus Hansen)

costs and at each step the smaller is chosen, we obtain a *best basis representation*. Intuitively each spatial segment of the image is only subdivided if an economic descrption at the current level cannot be found. Thus uniform image areas are not divided whereas high frequency parts of the image ("hot spots") are (see Figure 3).

An image representation which could facilitate extraction of features, or, equivalently, be used for image segmentation, would be advantageous. By representing images by best basis we at the same time obtain a point of departure for the segmentation process. If we have the tree representation it means that image segments which correspond to a deep tree level are the 'hot spots' of the image, places where the interesting things are happening, and which therefore deserve much attention, i.e. allocation of sufficient computing power. On the other hand, image parts corresponding to a low or moderate tree depth are more uniform parts of the image, which can be discarded in the initial analysis.

The first step in analysing an image, about which nothing is known, should be an *uncommitted* analysis. This means that nothing should occur during this analysis which will deter the interpretation process along a wrong track. The problem of uncomitted initial image analysis, also called the *bootstrapping problem* is an important one in image analysis and interpretation.

References

1. N. Ahmed and K.R. Rao: *Orthogonal transforms for digital signal processing.* Berlin: Springer 1975.
2. Born, G.: *The file formats handbook.* London: International Thomson Computer Press 1995.
3. Burt, P., and Adelson, E.H.: The Laplacian pyramid as a compact image code. *IEEE Transactions on Communication,* 31:532-539, April 1983.
4. Cohen, A., and Kovačević: Wavelets: the mathematical background. *Proceedings of the IEEE* 84:514-522, April 1996.
5. Coifman, R.R., Wickerhauser, M.L.: Entropy-based algorithm for best basis selection. *IEEE Transactions on Information Theory,* 32:712-718, March 1992.

6. Florack, L.M.J., Romeny, B.M.H., Koenderink, J.J., and Viergever, M.A.: Linear scale space. *Journal of Mathematical Imaging and Vision* 4:325-351, 1994
7. Gonzales, R.C., Woods, R.E.: Digital Image Processing. Reading, MA: Addison-Wesley 1993.
8. Hunt, B.R.: Digital Image Processing. *Proceedings of the IEEE* 63:693-708, April 1975.
9. Jain, A.K.: *Fundamentals of Digital Image Processing.* Prentice-Hall Int. Editions 1989.
10. Jawerth, B.: Wavelets and filter banks. In: *Proceedings from PARA96*, Lecture Notes in Computer Science. Berlin: Springer 1996 (this book).
11. J.S. Lim: *Two-dimensional signal and image processing.* Prentice-Hall 1990.
12. Lindeberg, T., and Romeny, B.M.t.H.: Linear scale space I-II. In: Geometry-Driven Diffusion in Computer Vision (B.M.H. Romeny, ed). Dordrecht: Kluwer 1994.
13. W.B. Pennebaker og J.L. Mitchell: *JPEG Still Image Data Compression Standard.* Van Nostrand Reinhold, New York 1993.
14. Ramchandran, K., Vetterli, M., and Herley, C.: Wavelets, subband coding, and best bases. *Proceedings of the IEEE* 84:541-560, April 1996.
15. K.R. Rao and P. Yip: *Discrete Cosine Transform: Algorithms, Advantages and Applications.* Academic Press 1990.
16. Strang, G., and Nguyen, T.: Wavelets And Filter Banks. Wellesley, MA: Wellesley-Cambridge Press 1996.
17. G.K. Wallace: The JPEG still picture compression standard. *Communications of the ACM*, vol. **34**, pp. 30-44 (April 1991). Updated postscript version: ftp.uu.net, graphics/jpeg/wallace/ps.gz.
18. Wickerhauser, M.V.: Comparison of picture compression methods: Wavelet, Wavelet Packet, and Local Cosine Transform Coding. In: C. K. Chui et al.: "Wavelets: Theory, Algorithms and Applications". Proc. Int. Conf. on Wavelets, Taormina, Sicily 14-20 october 1993.

Use of Parallel Computations for Account of Options by Monte Carlo Method *

S.S. Artemiev, S.A. Gusev, O.G. Monakhov

Computing Center, Sibirian. Division of Russian Academy of Sciences,
Pr. Lavrentiev, 6, Novosibirsk, 630090, Russia
e-mail: monakhov@comcen.nsk.su

Abstract. Problems of modeling and calculations of dynamics of the price of options by Monte Carlo method on parallel processors are considered. Technique of calculation of some factors, enabling to investigate change of the price of options and to evaluate possible consequences of the made bargains is described. Numerical calculations show the speed up in many cases is close to linear function of number of processors.

Monte Carlo method is a convenient means of numerical modeling of forward financial operations. In particular, all calculations connected with options can be executed by this method. The main difficulty arising by use of Monte Carlo method is it's working time. If calculations with high accuracy are needed, one has to simulate $10^5 - 10^8$ of trajectories of solutions of stochastic differential equations (SDE's), that requires transition from PC to more high-speed computers. It is the most convenient for these purposes to use parallel processors, because statistical algorithms can be parallelized by a natural way. Algorithms for computation of derivatives from option price with respect to parameters for options are indicated in this paper.

An option is the right either to buy or to sell a specified amount or value of a particular underlying interest at a fixed exercise price by exercising the option before its specified expiration date. An option which gives a right to buy is a *call* option, and an option which gives a right to sell is a put option.

Characteristic parameters of an option are:

- Pr — premium of option
- K — exercise price
- T — expiration time
- σ — volatility of the underlying interest
- r — annual riscless interest rate
- S_0 — the price of the underlying interest at the conclusion of the contract

Options are currently available covering four types of underlying interests: equity securities, stock indexes, government debt securities, and foreign currencies. The options are of the european style to a fixed date of execution and of

---

* This work is supported by RFBR projects N96-01-01632, N95-01-00426

the american style, which up to a fixed extreme date of the expiration of the contract can be presented to execution in any moment.

The market cost of an option is determined as a result of auction tenders on optional exchange. The price, on which agree buyer and seller of the option, refers to as the premium. The so the named fair cost of an option is a theoretically justified floor price, at which the subscriber of the option can, using a hedging strategy, to supply by a guaranteed image the optional payments, irrespective of a casual condition of the prices of underlying interest in the market. Hereinafter for simplicity we consider the share without payment of dividends as the unit of the underlying interest.

We set the most general model of dynamics of the price of the share with the help of SDE in a sense Ito of a following kind:

$$DS_t = \mu_t S_t dt + \sigma_t S_t^\gamma dw(t), \quad S(0) = S_0. \tag{1}$$

Here μ_t and σ_t are functions of a time. The kind of them is set from the requirements of the simulated script. In the case of share without dividend payment $\mu_t = r$. γ is a real parameter. At $\gamma = 1$ we have linear SDE with multiplicative noise and variable coefficients.

The exact solution of (??) at $\gamma = 1$ has a view

$$S_t = S_0 exp\left(\int_0^t (\mu_\tau - \frac{\sigma_\tau^2}{2})d\tau + \int_0^t \sigma_\tau dw(\tau)\right). \tag{2}$$

The solution (??) has logarithmically normal one-dimensional distribution for nonrandom functions μ_t and σ_t. The movement of the price of the share at $\gamma = 1$ can be set by the recurrent formula

$$S_{n+1} = S_n exp\left((\mu_n - \frac{\sigma_n^2}{2})h + \sigma_n \sqrt{h}\zeta_n\right). \tag{3}$$

The premium of a standard call option of american style can be calculated as

$$Pr^{call} = \max_{0 \le t \le T} \langle e^{-rt}(S_t - K)^+ \rangle, \tag{4}$$

and the premium of a put option as

$$Pr^{put} = \max_{0 \le t \le T} \langle e^{-rt}(K - S_t)^+ \rangle. \tag{5}$$

The cost of a portfolio of a subscriber changes in a time pursuant to the formula

$$X_0 = Pr^{call}, \quad X_t = \beta_t B_t + \gamma_t S_t,$$

where the factor β_t determines a sum of costs on the bank account on a moment of time t, and the hedging factor γ_t determines a sum in the shares. There are no restrictions on possible values of factors γ_t and β_t in a given way of formation of a portfolio , i.e. employment in the debt is admitted.

The hedging factor is considered as a measure of correlation between the cost of a hedging portfolio and the underlying interest in any moment before expiration date. The minimum hedging is understood a hedging strategy, ensuring guaranteed optional payments at the minimum premium. For a call option of american style the price of the minimum hedging portfolio in any moment of a time can be determined as conditional mathematical expectation

$$X_t^{call} = \max_{t \leq \tau \leq T} \langle e^{-r(\tau-t)}(S_\tau - K)^+ | S_t \rangle, \tag{6}$$

And for a call option it is

$$X_t^{put} = \max_{t \leq \tau \leq T} \langle e^{-r(\tau-t)}(K - S_\tau)^+ | S_t \rangle. \tag{7}$$

For a call option of european style the price of the minimum hedging portfolio in any moment of a time can be determined as conditional mathematical expectation

$$X_t^{call} = \langle e^{-r(T-t)}(S_T - K)^+ | S_t \rangle, \tag{8}$$

And for a put option it is

$$X_t^{put} = \langle e^{-r(T-t)}(K - S_T)^+ | S_t \rangle. \tag{9}$$

Every day changes of the price of underlying interest of option are observed in the market. As a result the price of option changes accordingly. The factor "delta" represents the ratio of the change of the option price to change of the price of the interest:

$$\delta = \frac{\partial Pr(S_0, K, T, r, \mu, \sigma)}{\partial S_0}.$$

the Factor δ shows how the price of option will change at the change of the price of underlying interest on one item. Theoretically, but not in practice, the price of option can not be increased or to decrease in a greater degree than the price of interest of the underlying contract. It means, that the inequalities $0 \leq \delta \leq 1$ for a call option and $-1 \leq \delta \leq 0$ for a put option should be executed. That for a put option the factor δ has negative value means the price of option changes in a opposite direction concerning the price of underlying interest. For a put option a large gain corresponds to $\delta \approx -1$, and a large loss corresponds to $\delta \approx 0$. Comparing factor δ and hedging factor γ_t we see $\delta = \gamma_0$. Except the factor δ there are the other factors connected with the option premium. They are $\Gamma = \frac{\partial^2 Pr}{\partial S_0^2}$, $\omega = \frac{\partial Pr}{\partial \sigma}$ and $\rho = \frac{\partial Pr}{\partial r}$. For SDE (??) with $\gamma = 1$ and with the constant factors μ and σ the maximum likelihood valuation of a historical volatility on a data of discrete observation of the cost or value of underlying interest is well known:

$$\hat{\sigma} = \left\{ \frac{1}{N-1} \sum_{n=0}^{N-1} \frac{(\ln S_{n+1} - \ln S_n - \hat{a}h_n)^2}{h_n} \right\}^{1/2}, \tag{10}$$

Where

$$\widehat{a} = \frac{\ln S_N - \ln S_0}{T_{data}}, \tag{11}$$

t_n - non-uniform grid on a time on a interval $[0, T]$, N - number of discretic supervision S_n on this interval, $h_n = t_{n+1} - t_n$ - interval of a time between the supervision S_{n+1} and S_n. The valuation of parameter μ can be received also simply:

$$\widehat{\mu} = \widehat{a} + \frac{\widehat{\sigma}^2}{2}. \tag{12}$$

When the parameter $\gamma \neq 0$ is given the approximate maximum likelihood valuations of parameters σ and μ are:

$$\widehat{\sigma} = \left\{ \frac{1}{N-1} \sum_{n=0}^{N-1} \frac{(S_{n+1} - S_n - \widehat{\mu} S_n h_n)^2}{h_n S_n^{2\gamma}} \right\}^{1/2}, \tag{13}$$

$$\widehat{\mu} = \left\{ \sum_{n=0}^{N-1} S_n^{1-2\gamma}(S_{n+1} - S_n) \right\} / \left\{ \sum_{n=0}^{N-1} S_n^{2-2\gamma} h_n \right\}. \tag{14}$$

Monte Carlo method (in difference from analytical "martingale" method) permits at pricing of option to use as mathematical model of the price of underlying interest any linear or nonlinear system SDE, but not only scalar linear SDE with multiplicative noise with constant factors of growth and volatility. It also permits to use any non-standard function of payment, any formula of valuation of the premium and any strategy (not just hedging strategy) of formation of a portfolio by the subscriber of option.

In the given paper accounts of a hedging factor and factors δ, Γ, ω, ρ for options of european and american styles are offered. These calculations have been got by using a computing system $MVS - 100$, on the base of eight parallel working processors $Intel860$.

At first we consider calculation of a hedging factor. As the hedging factor γ_t is equal to derivative $\frac{\partial X_t}{\partial S_t}$, for its calculation by the difference method we set on a segment $[0, T]$ a uniform grid $G = \{t_i\}, i = 1, \ldots, n$. And we compute prices $X_{t_i}^{call}, X_{t_i}^{put}$ appropriate to price values of underlying interest S_{t_i}. They are calculated according to the formulas (??), (??) for option of a american style and according to the formulas (??), (??) for option of a european style. Values of the price of a minimum hedging portfolio in the points t_i corresponding to values of underlying interest $S_{t_i} + \Delta S_{t_i}$ are simultaneously calculated. We designate them $\tilde{X}_{t_i}^{call}, \tilde{X}_{t_i}^{put}$. Thus the approximate values of a hedging factor γ_{t_i} are received by difference formula

$$\gamma_{t_i} = \frac{\tilde{X}_{t_i} - X_{t_i}}{\Delta S_{t_i}} \tag{15}$$

In this case the most labour-consuming part of the program is that, which is connected to calculation for each knot of a grid G conditional mathematical expectations according to the formulas (??), (??), (??),(??). Here it is necessary to simulate given number of trajectories of two random processes S_t and $\tilde{S}_t$

determined by the equation (??) at $\gamma = 1$ on each segment $[t_i, T]$ with the initial values S_{t_i} and $S_{t_i} + \Delta S_{t_i}$ accordingly. Then the trajectories S_t and $\tilde{S}_t$ are used for the calculation X_{t_i} and $\tilde{X}_{t_i}$. As far as the simulating of trajectories of the specified random processes can be executed independently, this part of the program can be effectively realized on parallel processors.

As our algorithms contain many calculations of the same type t is convenient to parallelize them as a processor farm.

Thus the common structure of the program for calculation of a hedging factor γ_t consists in following. Let N is a sample volume and let we have in the system P parallel processors with numbers $0, 1, \ldots, P - 1$ connected among themselves by channels . Processor with number 0 we name as "main", and other processors - as "auxiliary". The main processor executes input of the initial information, preparation and transfer on channels of communication of the tasks for auxiliary processors, modeling $N - (P - 1) \times ([\frac{N}{P}] + 1$ trajectories of random processes S_t and $\tilde{S}_t$, receiving from auxiliary processors results of simulating $(P-1) \times ([\frac{N}{P}] + 1)$ trajectories of the random processes and doing processing them, calculation of difference values of derivative for reception of a hedging factor at knots of grid G. Each auxiliary processor accepts from a processor with number 0 initial data for modeling $[\frac{N}{P} + 1]$ trajectories, simulates them and transmits results to a processor 0.

As well as for hedging factor the account of each of the factors δ, Γ, ω, ρ is executed by difference method on a segment (given by user) of change of S_0, S_0, S, r accordingly. Thus on the each segment a uniform grid is entered. As far as the account of factors δ, Γ, ω, ρ for each knot of a grid is executed independently, it is possible to parallelize the calculations over knots of a grid. Thus the partitioning of knots of a grid on processors is conducted similarly that as trajectories on processors are distributed at calculation of a hedging factor.

In the table below the values of speed up, received as a result of accounts of factors γ_t, δ, Γ, ω, ρ are demonstrated. The speed up here is the ratio of calculation time for one processor to calculation time for i $(i = 2, \ldots, 8)$ processors. For each factor in the table the left column contains values of speed up for option of american style, the right column — for european one.

Number	factor									
of proc.	γ_t		δ		Γ		ω		ρ	
2.	2.00	2.00	2.02	2.02	2.02	2.02	1.91	2.06	2.00	2.02
3.	2.98	2.99	3.06	3.06	3.06	3.06	3.02	3.25	3.02	3.06
4.	3.94	3.94	3.85	4.30	4.37	4.37	4.05	4.64	4.23	4.32
5.	4.77	5.00	4.70	5.22	5.85	5.83	5.31	5.64	5.21	5.26
6.	5.38	5.56	5.53	5.98	6.12	6.09	6.06	6.41	6.13	6.03
7.	5.95	6.27	7.23	6.80	8.72	8.57	7.12	7.43	7.14	6.94
8.	6.69	6.76	8.07	7.62	9.43	9.16	8.26	8.24	8.14	7.74

It is easily to see that for considered parallel programs at increase of number of simulated trajectories only expenditures connected with direct simulating of trajectories increase. Therefore at uniform loading of all processors it is possible to expect the speed up will come nearer to linear function of number of processors. The adduced numerical accounts also confirm it.

Wavelets and Differential Equations

Vincent A. Barker

Department of Mathematical Modelling
Technical University of Denmark, Bldg. 305
DK-2800 Lyngby, Denmark
e-mail: vab@imm.dtu.dk

Abstract Wavelet applications to date have been dominated by signal and image processing. While perhaps not immediately appealing as a means of solving differential equations, the growing body of literature in this area indicates that wavelets have a role to play here, too. We give here some of the basic background and an example illustrating how wavelets can be used to solve differential equations.

1. Wavelets

The wavelets most commonly used to solve differential equations are due to I. Daubechies [8]. For every positive even integer N there is a *mother scaling function* $\phi(x)$, $-\infty < x < \infty$, with support on the interval $[0, N-1]$, and a set of real *filter coefficients* $\{a_r\}_{r=0}^{N-1}$ such that ϕ satisfies the *refinement equation*

$$\phi(x) = \sum_{r=0}^{N-1} a_r \phi(2x - r), \quad -\infty < x < \infty \tag{1.1}$$

In addition to ϕ, there is a *mother wavelet* $\psi(x)$, $-\infty < x < \infty$, also supported on the interval $[0, N-1]$, that satisfies the relation

$$\psi(x) = \sum_{r=0}^{N-1} b_r \phi(2x - r), \quad -\infty < x < \infty \tag{1.2}$$

where

$$b_r = (-1)^r a_{N-1-r} \quad r = 0, 1, \ldots, N-1$$

Consider the translations of ϕ defined by

$$\phi(x - l), \quad -\infty < x < \infty, \quad l \in Z$$

where Z is the set of integers. These are orthonormal in the sense that

$$\int_{-\infty}^{\infty} \phi(x - k)\phi(x - l)\, dx = \delta_{kl}$$

A basic property of ϕ is that any monomial of degree less than $N/2$ can be represented exactly by a linear combination of these shifted scaling functions. More precisely,

$$x^k = \sum_{l=-\infty}^{\infty} M_l^k \phi(x-l), \quad k = 0, 1, 2, \ldots, \frac{N}{2} - 1 \qquad (1.3)$$

where the *moment* M_l^k is defined by

$$M_l^k = \int_{-\infty}^{\infty} x^k \phi(x-l)\,dx, \quad l \in Z$$

Taking the mother function ϕ as the point of departure, we can define by the processes of scaling and shifting an infinite set of scaling functions, namely

$$\phi_{pl}(x) = 2^{p/2}\,\phi(2^p x - l), \quad -\infty < x < \infty \quad p, l \in Z \qquad (1.4)$$

The corresponding set of wavelets is given by

$$\psi_{pl}(x) = 2^{p/2}\,\psi(2^p x - l), \quad -\infty < x < \infty \quad p, l \in Z \qquad (1.5)$$

These functions satisfy the following orthonormality relations:

$$\int_{-\infty}^{\infty} \phi_{pk}(x)\phi_{pl}(x)\,dx = \delta_{kl}, \quad p, k, l \in Z \qquad (1.6a)$$

$$\int_{-\infty}^{\infty} \psi_{pk}(x)\psi_{ql}(x)\,dx = \delta_{pq}\delta_{kl}, \quad p, q, k, l \in Z \qquad (1.6b)$$

$$\int_{-\infty}^{\infty} \phi_{pk}(x)\psi_{ql}(x)\,dx = \delta_{kl}, \quad p, q, k, l \in Z, \ q \geq p \qquad (1.6c)$$

In problems involving differential equations, the space domain is often bounded and certain conditions are imposed at the boundary. Now it is an unfortunate fact that scaling functions and wavelets are not easily applicable to some common boundary conditions. (For work on this problem see, for example, [2], [4] and [15]). An important exception is the case of a *periodic boundary condition*. Thus consider a real interval $[0, L]$, where, for simplicity, L is a positive integer. A function f defined on $[0, L]$ is said to be *periodic* if $f(0)=f(L)$. Let p_0 be the algebraically smallest integer such that $L2^{p_0}$ is an integer and $2^{p_0} \geq (N-1)/L$. For every integer $p \geq p_0$ we introduce the grid points $x_l = l/2^p$, $l = 0, 1, 2, \ldots, M_p - 1$, where $M_p = L\,2^p$, and define the *periodized scaling functions*

$$\tilde{\phi}_{pl}, \quad 0 \leq x \leq L, \quad l = 0, 1, 2, \ldots, M_p - 1$$

by requiring that for each $x \in [0, L]$,

$$\tilde{\phi}_{pl}(x) = \begin{cases} \phi_{pl}(x), & x \in \operatorname{supp}(\phi_{pl}) \\ \phi_{pl}(x+L), & \text{otherwise} \end{cases} \qquad (1.7)$$

For $l=0,1,\ldots,M_p-N+1$ the function $\widetilde{\phi}_{pl}$ is simply the restriction of ϕ_{pl} to $[0,L]$. For the remaining values of l, $\widetilde{\phi}_{pl}$ is a 'wrap-around' version of ϕ_{pl}. We note that ϕ_{pl} is supported on the interval $[x_l, x_{l+N-1}]$. Periodized wavelets $\widetilde{\psi}_{pl}$, $l=0,1,\ldots,M_p-1$ are defined analogously.

Let

$$V_p^{per} = SPAN\{\widetilde{\phi}_{pl}, \ l=0,1,\ldots,M_p-1\}, \quad p=p_0, p_0+1, \ldots$$

$$W_p^{per} = SPAN\{\widetilde{\psi}_{pl}, \ l=0,1,\ldots,M_p-1\}, \quad p=p_0, p_0+1, \ldots$$

We have the following basic relations:

$$V_{p_0}^{per} \subset V_{p_0+1}^{per} \subset V_{p_0+2}^{per} \subset \ldots \subset L^2[0,L] \tag{1.8}$$

$$V_p^{per} = V_{p_0}^{per} \oplus W_{p_0}^{per} \oplus W_{p_0+1}^{per} \oplus W_{p_0+2}^{per} \oplus \ldots \oplus W_{p-1}^{per}, \quad p,p_0 \in Z, \ p > p_0 \tag{1.9}$$

and, for any $f \in V_p^{per}$,

$$f(x) = \sum_{l=0}^{M_p-1} c_{pl} \, \widetilde{\phi}_{pl}(x) = \sum_{l=0}^{M_{p_0}-1} c_{p_0,l} \, \widetilde{\phi}_{p_0,l}(x) + \sum_{q=p_0}^{p-1} \sum_{l=0}^{M_q-1} d_{ql} \, \widetilde{\psi}_{ql}(x), \quad 0 \le x \le L \tag{1.10}$$

where

$$c_{ql} = \int_0^L f(x) \widetilde{\phi}_{ql}(x) \, dx, \quad l=0,1,\ldots,Mp-1, \ q=p_0, p$$

$$d_{ql} = \int_0^L f(x) \widetilde{\psi}_{ql}(x) \, dx, \quad l=0,1,\ldots,Mp-1, \ q=p_0, p_0+1,\ldots,p-1$$

With regard to the approximation properties of V_p^{per}, if $f \in C^{N/2}[0,L]$ and

$$\lim_{x \to 0^-} f^m(x) = \lim_{x \to L^+} f^m(x), \quad m=0,1,\ldots,N/2$$

then

$$\| f - P_p(f) \| = O(h^{N/2}), \quad h \to 0 \tag{1.11}$$

where $h=2^{-p}$.

2. The fast wavelet transform

Consider the orthogonal decomposition

$$V_{q+1}^{per} = V_q^{per} \oplus W_q^{per} \tag{2.1}$$

For any $f \in V_{q+1}^{per}$ we may write

$$f(x) = \sum_{l=0}^{M_{q+1}-1} \widetilde{c}_{q+1,l} \, \widetilde{\phi}_{q+1,l}(x) = \sum_{l=0}^{M_q-1} \widetilde{c}_{ql} \, \widetilde{\phi}_{ql}(x) + \sum_{l=0}^{M_q-1} \widetilde{d}_{ql} \, \widetilde{\psi}_{ql}(x)$$

where $M_q = L\,2^q$ and

$$\tilde{c}_{ql} = \int_0^L f(x)\tilde{\phi}_{ql}\,dx\,, \quad \tilde{d}_{ql} = \int_0^L f(x)\tilde{\psi}_{ql}\,dx\,, \quad l = 0,1,\ldots,M_q - 1$$

It is useful to be able to compute the mappings between the two sets of expansion coefficients. The algorithms are the following:

Periodic one-step fast wavelet transform: For $l = 0,1,\ldots,M_q-1$,

$$c_{ql} = 2^{-1/2} \sum_{k=2l}^{N-1+2l} a_{k-2l}\, c_{q+1,\,k\,\mathrm{mod}\,M_{q+1}}$$

$$d_{ql} = 2^{-1/2} \sum_{k=2l}^{N-1+2l} b_{k-2l}\, c_{q+1,\,k\,\mathrm{mod}\,M_{q+1}}$$

Periodic one-step inverse fast wavelet transform: For $l = 0,1,\ldots,M_{q+1}-1$,

$$c_{q+1,l} = 2^{-1/2} \sum_{k=k_1}^{k_2} a_{l-2k}\, c_{q,\,k\,\mathrm{mod}\,M_q} + 2^{-1/2} \sum_{k=k_1}^{k_2} b_{l-2k}\, d_{q,\,k\,\mathrm{mod}\,M_q} \qquad (2.2)$$

where

$$\text{if } l \text{ is even: } \quad k_1 = \frac{2-N+l}{2}\,, \quad k_2 = \frac{l}{2}$$

$$\text{if } l \text{ is odd: } \quad k_1 = \frac{1-N+l}{2}\,, \quad k_2 = \frac{l-1}{2}$$

Consider now the general orthogonal decomposition

$$V_p^{per} = V_{p_0}^{per} \oplus W_{p_0}^{per} \oplus W_{p_0+1}^{per} \oplus W_{p_0+2}^{per} \oplus \ldots \oplus W_{p-1}^{per}$$

Here we want to compute the following mappings, denoted, respectively, the *periodic fast wavelet transform (PFWT)* and the *periodic inverse fast wavelet transform (PIFWT)*.

$$\text{PFWT}: \{c_{p,l}\}_{l=0}^{M_p-1} \rightarrow \{c_{p_0,l}\}_{l=0}^{M_{p_0}-1}\,, \{d_{ql}\}_{l=0}^{M_q-1}\,, \quad q = p_0, p_0+1,\ldots,p-1$$

$$\text{PIFWT}: \{c_{p_0,l}\}_{l=0}^{M_{p_0}-1}\,, \{d_{ql}\}_{l=0}^{M_q-1}\,, \quad q = p_0, p_0+1,\ldots,p-1 \rightarrow \{c_{p,l}\}_{l=0}^{M_p-1}$$

The procedure in each case is a straight-forward recursion based on the corresponding one-step method.

3. Differentiation matrices

3.1. Scaling functions basis

An important reference for this section is [12]. If the mother scaling function is differentiable then, for any $f \in V_p^{per}$,

$$f(x) = \sum_{l=0}^{M_p-1} c_{pl}\,\tilde{\phi}_{pl}(x)\,, \quad 0 \le x \le L \qquad (3.1)$$

$$P_p(f')(x) = \sum_{k=0}^{M_p-1} c_{pk}^{(1)} \tilde{\phi}_{pk}(x), \quad 0 \leq x \leq L \tag{3.2}$$

where $P_p(f')$ is the projection of f' onto V_p^{per}. It is straightforward to show that

$$c^{(1)} = D c \tag{3.3}$$

where the *differentiation matrix* D is of order M_p with entries

$$d_{kl} = \int_0^L \tilde{\phi}_{pk}(x)\tilde{\phi}_{pl}'(x)\,dx\,, \quad k,l = 0,1,\ldots,M_p-1$$

These entries can be expressed in terms of the *connection coefficients*

$$\Lambda_l^{d_1 d_2} = \int_{-\infty}^{\infty} \phi^{(d1)}(x)\,\phi_{0,l}^{(d2)}(x)\,dx\,, \quad l = 0,\pm1,\pm2,\ldots,\pm(N-2) \tag{3.4}$$

where, in this application, $d_1=0, d_2=1$. (The problem of computing connection coefficients is discussed in [3], [5], [13] and [14], for example). The following example serves to illustrate the general appearance of D: For $L=1$, $N=4$ and $p=3$,

$$D = 2^p \begin{bmatrix}
\Lambda_0^{0,1} & \Lambda_1^{0,1} & \Lambda_2^{0,1} & 0 & 0 & 0 & \Lambda_{-2}^{0,1} & \Lambda_{-1}^{0,1} \\
\Lambda_{-1}^{0,1} & \Lambda_0^{0,1} & \Lambda_1^{0,1} & \Lambda_2^{0,1} & 0 & 0 & 0 & \Lambda_{-2}^{0,1} \\
\Lambda_{-2}^{0,1} & \Lambda_{-1}^{0,1} & \Lambda_0^{0,1} & \Lambda_1^{0,1} & \Lambda_2^{0,1} & 0 & 0 & 0 \\
0 & \Lambda_{-2}^{0,1} & \Lambda_{-1}^{0,1} & \Lambda_0^{0,1} & \Lambda_1^{0,1} & \Lambda_2^{0,1} & 0 & 0 \\
0 & 0 & \Lambda_{-2}^{0,1} & \Lambda_{-1}^{0,1} & \Lambda_0^{0,1} & \Lambda_1^{0,1} & \Lambda_2^{0,1} & 0 \\
0 & 0 & 0 & \Lambda_{-2}^{0,1} & \Lambda_{-1}^{0,1} & \Lambda_0^{0,1} & \Lambda_1^{0,1} & \Lambda_2^{0,1} \\
\Lambda_2^{0,1} & 0 & 0 & 0 & \Lambda_{-2}^{0,1} & \Lambda_{-1}^{0,1} & \Lambda_0^{0,1} & \Lambda_1^{0,1} \\
\Lambda_1^{0,1} & \Lambda_2^{0,1} & 0 & 0 & 0 & \Lambda_{-2}^{0,1} & \Lambda_{-1}^{0,1} & \Lambda_0^{0,1}
\end{bmatrix} \tag{3.5}$$

For any $f \in V_p^{per}$ let $f_l = f(x_l)$, $l=0,1,\ldots,M_p-1$. We can identify f by either the set of function values $\{f_l\}_{l=0}^{M_p-1}$ or by the set of expansion coefficients $\{c_{pl}\}_{l=0}^{M_p-1}$ in (3.1). The two sets are related by a system which we denote

$$f = Tc$$

where $c_l = c_{pl}$. When $L=1$, $N=4$ and $p=3$, for example, we have

$$T = \begin{bmatrix}
 & & & & & & \phi(2) & \phi(1) \\
\phi(1) & & & & & & & \phi(2) \\
\phi(2) & \phi(1) & & & & & & \\
 & \phi(2) & \phi(1) & & & & & \\
 & & \phi(2) & \phi(1) & & & & \\
 & & & \phi(2) & \phi(1) & & & \\
 & & & & \phi(2) & \phi(1) & & \\
 & & & & & \phi(2) & \phi(1) &
\end{bmatrix} \tag{3.6}$$

In the general case, the nonzero matrix entries are $\phi(r)$, $r=1, 2, \ldots, N-2$.

Let us define

$$f^{(1)} = T(D(T^{-1}f)) = (TDT^{-1})f$$

The operation $T^{-1}f$ transforms the set of grid values of f to the corresponding set of expansion coefficients. The multiplication by D then computes the coefficients of the projection of f' onto V_p^{per}, as explained above, and the final multiplication by T produces the grid values of this projection. We can call D the *differentiation matrix with respect to coefficient space* and TDT^{-1} the *differentiation matrix with respect to physical space*. Now T, D and T^{-1} are all circulant and products of circulant matrices commute, so

$$TDT^{-1} = DTT^{-1} = D$$

and

$$f^{(1)} = Df \tag{3.7}$$

Thus D is the differentiation matrix with respect to *both* coefficient and physical space.

The matrix D above corresponds to first-order differentiation. The matrix corresponding to m-th order differentiation is analogously derived and we denote it $D^{0,m}$. Its entries are obtained from those of D simply by replacing all superscripts '$0, 1$' in the connection coefficents by '$0, m$'.

3.2. Wavelet basis

It will be recalled that any $f \in V_p^{per}$ can be expanded as

$$f(x) = \sum_{l=-\infty}^{\infty} c_{p_0,l}\, \tilde{\phi}_{p_0,l}(x) + \sum_{q=p_0}^{p-1} \sum_{l=0}^{M_q-1} d_{ql}\, \tilde{\psi}_{ql}(x), \quad 0 \le x \le L \tag{3.8}$$

For simplicity we rewrite this as

$$f(x) = \sum_{l=0}^{M_p-1} d_l\, \tilde{\theta}_l(x), \quad 0 \le x \le L \tag{3.9}$$

Proceeding as in the previous section we find that

$$P_p(f')(x) = \sum_{l=0}^{M_p-1} d_l^{(1)}\, \tilde{\theta}_l(x), \quad 0 \le x \le L \tag{3.10}$$

where

$$d^{(1)} = D_w\, d \tag{3.11}$$

the entries of $\boldsymbol{D}_w$ being given by

$$(\boldsymbol{D}_w)_{kl} = \int_0^L \tilde{\theta}_k(x)\tilde{\theta}'_l(x)\,dx\,, \quad k,l = 0,1,\ldots,M_p - 1$$

Since these connection coefficients contain wavelets as well as scaling functions and moreover involve mixed scales, to compute them directly is a fairly complicated process. However, they can be obtained in an indirect way as follows: Let $\boldsymbol{W}$ be the matrix representation of the PFWT in the preceding section. Then we can write

$$\boldsymbol{d} = \boldsymbol{W}\,\boldsymbol{c}, \quad \boldsymbol{d}^{(1)} = \boldsymbol{W}\,\boldsymbol{c}^{(1)} \tag{3.12}$$

where $\boldsymbol{c}$ and $\boldsymbol{c}^{(1)}$ are the vectors in (3.3). Using (3.3), (3.11) and (3.12) it is straighforward to show that

$$\boldsymbol{D}_w = \boldsymbol{W}\,\boldsymbol{D}\,\boldsymbol{W}^T \tag{3.13}$$

Thus if we have the differentiation matrix $\boldsymbol{D}$ then we can compute $\boldsymbol{D}_w$ via the relation $\boldsymbol{D}_w = \boldsymbol{W}(\boldsymbol{W}\boldsymbol{D}^T)^T$. Each multiplication with $\boldsymbol{W}$ can be implemented columnwise using the PFWT M_p times. In contrast to $\boldsymbol{D}$, matrix $\boldsymbol{D}_w$ is not circulant but rather is characterized by a 'finger-band' pattern.

4. Wavelets for differential equations: An example

We consider the initial-boundary value problem for the 1-D heat equation

$$
\begin{aligned}
u_t &= \nu u_{xx} + f(x), \quad 0 < x < L, \quad t > 0 & \text{(4.1a)} \\
u(x,0) &= h(x), \quad 0 < x < L & \text{(4.1b)} \\
u(0,t) &= u(L,t), \quad t \geq 0 & \text{(4.1c)}
\end{aligned}
$$

where ν is a positive constant and $f(0)=f(L)$. The collocation method with respect to the usual grid on $[0, L]$ yields the system of first-order ordinary differential equations

$$\frac{d}{dt}\hat{u}_l(t) = \nu\,(\hat{u}_{xx})_l(t) + f_l\,, \quad l = 0,1,\ldots,M_p - 1,\ t > 0$$

where $\hat{u}_l(t)=\hat{u}(x_l,t)$ and $f_l=f(x_l)$. The initial conditions are $\hat{u}_l(0)=h_l$, $l=0,1,\ldots,M_p-1$, where $h_l=h(x_l)$. In matrix-vector form this system becomes

$$
\begin{aligned}
\frac{d}{dt}\widehat{\boldsymbol{u}}(t) &= \nu\,\boldsymbol{D}^{0,2}\,\widehat{\boldsymbol{u}}(t) + \boldsymbol{f}, \quad t > 0 \\
\widehat{\boldsymbol{u}}(0) &= \boldsymbol{h}
\end{aligned}
$$

where $\boldsymbol{D}^{0,2}$ is the second-order differentiation matrix (Section 3.1). Using the identity $\boldsymbol{D}^{0,2}=-\boldsymbol{D}^{1,1}$, which is easily derived, we may write this system as

$$
\begin{aligned}
\frac{d}{dt}\widehat{\boldsymbol{u}}(t) &= -\nu\,\boldsymbol{D}^{1,1}\,\widehat{\boldsymbol{u}}(t) + \boldsymbol{f}, \quad t > 0 \\
\widehat{\boldsymbol{u}}(0) &= \boldsymbol{h}
\end{aligned}
$$

In the case $L=3$, $N=6$, $p=2$, one finds that $\boldsymbol{D}^{1,1}/4$ is given by

$$
\begin{bmatrix}
\Lambda^{1,1}_0 & \Lambda^{1,1}_1 & \Lambda^{1,1}_2 & \Lambda^{1,1}_3 & \Lambda^{1,1}_4 & 0 & 0 & 0 & \Lambda^{1,1}_{-4} & \Lambda^{1,1}_{-3} & \Lambda^{1,1}_{-2} & \Lambda^{1,1}_{-1} \\
\Lambda^{1,1}_{-1} & \Lambda^{1,1}_0 & \Lambda^{1,1}_1 & \Lambda^{1,1}_2 & \Lambda^{1,1}_3 & \Lambda^{1,1}_4 & 0 & 0 & 0 & \Lambda^{1,1}_{-4} & \Lambda^{1,1}_{-3} & \Lambda^{1,1}_{-2} \\
\Lambda^{1,1}_{-2} & \Lambda^{1,1}_{-1} & \Lambda^{1,1}_0 & \Lambda^{1,1}_1 & \Lambda^{1,1}_2 & \Lambda^{1,1}_3 & \Lambda^{1,1}_4 & 0 & 0 & 0 & \Lambda^{1,1}_{-4} & \Lambda^{1,1}_{-3} \\
\Lambda^{1,1}_{-3} & \Lambda^{1,1}_{-2} & \Lambda^{1,1}_{-1} & \Lambda^{1,1}_0 & \Lambda^{1,1}_1 & \Lambda^{1,1}_2 & \Lambda^{1,1}_3 & \Lambda^{1,1}_4 & 0 & 0 & 0 & \Lambda^{1,1}_{-4} \\
\Lambda^{1,1}_{-4} & \Lambda^{1,1}_{-3} & \Lambda^{1,1}_{-2} & \Lambda^{1,1}_{-1} & \Lambda^{1,1}_0 & \Lambda^{1,1}_1 & \Lambda^{1,1}_2 & \Lambda^{1,1}_3 & \Lambda^{1,1}_4 & 0 & 0 & 0 \\
0 & \Lambda^{1,1}_{-4} & \Lambda^{1,1}_{-3} & \Lambda^{1,1}_{-2} & \Lambda^{1,1}_{-1} & \Lambda^{1,1}_0 & \Lambda^{1,1}_1 & \Lambda^{1,1}_2 & \Lambda^{1,1}_3 & \Lambda^{1,1}_4 & 0 & 0 \\
0 & 0 & \Lambda^{1,1}_{-4} & \Lambda^{1,1}_{-3} & \Lambda^{1,1}_{-2} & \Lambda^{1,1}_{-1} & \Lambda^{1,1}_0 & \Lambda^{1,1}_1 & \Lambda^{1,1}_2 & \Lambda^{1,1}_3 & \Lambda^{1,1}_4 & 0 \\
0 & 0 & 0 & \Lambda^{1,1}_{-4} & \Lambda^{1,1}_{-3} & \Lambda^{1,1}_{-2} & \Lambda^{1,1}_{-1} & \Lambda^{1,1}_0 & \Lambda^{1,1}_1 & \Lambda^{1,1}_2 & \Lambda^{1,1}_3 & \Lambda^{1,1}_4 \\
\Lambda^{1,1}_4 & 0 & 0 & 0 & \Lambda^{1,1}_{-4} & \Lambda^{1,1}_{-3} & \Lambda^{1,1}_{-2} & \Lambda^{1,1}_{-1} & \Lambda^{1,1}_0 & \Lambda^{1,1}_1 & \Lambda^{1,1}_2 & \Lambda^{1,1}_3 \\
\Lambda^{1,1}_3 & \Lambda^{1,1}_4 & 0 & 0 & 0 & \Lambda^{1,1}_{-4} & \Lambda^{1,1}_{-3} & \Lambda^{1,1}_{-2} & \Lambda^{1,1}_{-1} & \Lambda^{1,1}_0 & \Lambda^{1,1}_1 & \Lambda^{1,1}_2 \\
\Lambda^{1,1}_2 & \Lambda^{1,1}_3 & \Lambda^{1,1}_4 & 0 & 0 & 0 & \Lambda^{1,1}_{-4} & \Lambda^{1,1}_{-3} & \Lambda^{1,1}_{-2} & \Lambda^{1,1}_{-1} & \Lambda^{1,1}_0 & \Lambda^{1,1}_1 \\
\Lambda^{1,1}_1 & \Lambda^{1,1}_2 & \Lambda^{1,1}_3 & \Lambda^{1,1}_4 & 0 & 0 & 0 & \Lambda^{1,1}_{-4} & \Lambda^{1,1}_{-3} & \Lambda^{1,1}_{-2} & \Lambda^{1,1}_{-1} & \Lambda^{1,1}_0
\end{bmatrix}
$$

The entries of $\boldsymbol{D}^{1,1}$ are the connection coefficients in (3.4) for $d_1=d_2=1$.

The backward Euler method may be used for time-stepping. This is defined by the discretization

$$
\frac{\widehat{u}_{n+1} - \widehat{u}_n}{\Delta t} = -\nu \boldsymbol{D}^{1,1}\,\widehat{u}_{n+1} + \boldsymbol{f}
$$

based on time points t_n and t_{n+1} with time step $\Delta t = t_{n+1} - t_n$. Hence we have the recursion

$$
(\boldsymbol{I} + \Delta t\,\nu \boldsymbol{D}^{1,1})\,\widehat{u}_{n+1} = \widehat{u}_n + \Delta t\,\boldsymbol{f}, \quad n = 0, 1, \dots
$$

The matrix on the left-hand side is circulant and symmetric positive definite, and each of these systems can be solved in $O(M_p \log(M_p))$ operations. The recursion is stable for all values of Δt, so this parameter can be chosen on the basis of accuracy alone.

We consider now the Galerkin method for (4.1) with $\hat{u}$ represented by

$$
\hat{u}(x,t) = \sum_{l=0}^{M_{p_0}-1} c_{p_0,l}(t)\,\widetilde{\phi}_{p_0,l}(x) + \sum_{q=p_0}^{p-1}\sum_{l=0}^{M_q-1} d_{ql}(t)\,\widetilde{\psi}_{ql}(x), \quad 0 \le x \le L, \; t \ge 0
$$

For ease of notation we rewrite this as

$$
\hat{u}(x,t) = \sum_{l=0}^{M_p-1} d_l(t)\,\widetilde{\theta}_l(x) \tag{4.2}
$$

The weak formulation of (4.1) is the problem of finding a periodic function $u(x,t)$ that satisfies the relation

$$
\int_0^L u_t v\,dx = -\int_0^L u_x v_x\,dx + \int_0^L fv\,dx, \quad t > 0 \tag{4.3}
$$

for every smooth periodic function $v(x)$. In addition, u must also satisfy the initial condition (4.1b). We discretize (4.3) by substituting (4.2) for u and letting $v = \widetilde{\theta}_k$, $k = 0, 1, \ldots, M_p - 1$. This leads to the initial value problem

$$\frac{d}{dt} d(t) = -\nu D_w^{1,1} d(t) + g$$

with

$$(D_w^{1,1})_{kl} = \int_0^L \widetilde{\theta}_k'(x) \widetilde{\theta}_l'(x) \, dx, \quad g_k = \int_0^L f(x) \widetilde{\theta}_k(x) \, dx, \quad k, l = 0, 1, \ldots, M_p - 1$$

the initial conditions being

$$d_l(0) = \int_0^L h(x) \widetilde{\theta}_l(x) \, dx, \quad l = 0, 1, \ldots, M_p - 1$$

From Section 3.2 we have the relation

$$D_w^{1,1} = W D^{1,1} W^T$$

(see (3.13)), and $D_w^{1,1}$ can be computed from $D^{1,1}$ by $2M_p$ applications of the PFWT.

The backward Euler time-stepping method leads to the recursion

$$(I + \Delta t \, \nu \, D_w^{1,1}) \, d_{n+1} = d_n + \Delta t \, g, \quad n = 0, 1, \ldots$$

Introducing

$$E = I + \Delta t \, \nu \, D_w^{1,1}$$

we have then

$$d_{n+1} = E^{-1} (d_n + \Delta t \, g), \quad n = 0, 1, \ldots$$

This approach to solving (4.1) is discussed in [6]. Matrices $D_w^{1,1}$ and E have finger-band pattern, and it turns out that so does E^{-1} when the numerically small entries of this matrix are removed. Further, working with the multi-scale wavelet basis often yields vectors that have a number of numerically small entries. Thus there is the potential for compressing all of the quantities E^{-1}, g and d_{n+1}.

A computational procedure along these lines is described in [6]. It may be summarized as follows:

Initialization

1. Compute and compress E^{-1} : $E^{-1} \rightarrow (E^{-1})_c$

2. Compute and compress g : $g \rightarrow (g)_c$

3. Compute and compress d^0 : $d^0 \rightarrow (d^0)_c$

Time-stepping: for n = 0,1,...

1. Compute: $d_{n+1} = (E^{-1})_c \left((d_n)_c + \Delta t \, (g)_c \right)$

2. Compress d_{n+1} : $d_{n+1} \to (d_{n+1})_c$

It is essential for the success of this scheme that both the storage of the various vectors and matrices as well as the programming of the operations in which they are involved are based on sparse matrix techniques. See the reference for further details.

REFERENCES

1 K. AMARATUNGA, J. R. WILLIAMS, S. QIAN AND J. WEISS, *Wavelet-Galerkin solutions for one-dimensional partial differential equations*, Int. J. Numer. Methods Eng. **37**, 2703–2716, 1994.

2 P. AUSCHER, *Wavelets with boundary conditions on the interval*, in Wavelets: A Tutorial in Theory and Applications, C. K. Chui (Ed.), Academic Press, San Diego, CA, 217–236, 1992.

3 V. A. BARKER, *Computing connection coefficients*, Technical Report 1996-5, Department of Mathematical; Modelling, Technical University of Denmark, Lyngby, 1996.

4 S. BERTOLUZZA, G. NALDI AND J. -C. RAVEL, *Wavelet methods for the numerical solution of boundary value problems on the interval*, in Wavelets: Theory, Algorithms and Applications, C. K. Chui, L. B. Montefusco and L. Puccio (eds.), Academic Press, San Diego, CA, 425–448, 1994.

5 G. BEYLKIN, *On the representation of operators in bases of compactly supported wavelets*, SIAM J. Numer. Anal. **29**, 1716-1740, 1992.

6 P. CHARTON AND V. PERRIER, *A pseudo-wavelet scheme for the two-dimensional Navier-Stokes equations*, Matemática Aplicada e Computacional (submitted).

7 W. DAHMEN, A. KUNOTH, K. URBAN, *A wavelet Galerkin method for the Stokes equations*, Computing **56**, 259–301, 1996.

8 I. DAUBECHIES, *Ten Lectures on Wavelets*, SIAM, Philadelphia, PA, 1992.

9 M. DOROBANTU, *Wavelet-based Algorithms for Fast PDE Solvers*, Ph.D. Thesis, Technical Report TRITA-NA-9507, Department of Numerical Analysis and Computing Science, Royal Institute of Technology, Stockholm, 1995.

10 B. ENGQUIST, S. OSHER AND S. ZHONG, *Fast wavelet algorithms for linear evolution equations*, SIAM J. Sci. Comput. **15**, 755–775, 1994.

11 R. GLOWINSKI, J. PERIAUX, M. RAVACHOL, T. -W. PAN, R. O. WELLS, JR. AND X. ZHOU, *Wavelet methods in computational fluid dynamics*, in Algorithmic Trends in Computational Fluid Dynamics, M. Y. Hussaini, A. Kumar and M. D. Salas (eds.), Springer, New York, 259-276, 1993.

12 L. JAMESON, *On the Daubechies–based wavelet differentiation matrix*, ICASE Report No. 93-95 NASA Langley Research Center Hampton, VA, 1993.

13 A. KUNOTH, *On the fast evaluation of integrals of refinable functions*, in Wavelets, Images, and Surface Fitting, P.J. Laurent, A. Le Méhauté, L.L. Schumaker (eds.), AKPeters, Boston, 327–334, 1994.

14 A. LATTO, K. L. RESNIKOFF AND E. TENENBAUM, *The evaluation of connection coefficients of compactly supported wavelets*, in Proceedings, French-USA Workshop on Wavelets and Turbulence, Princeton Univ., June 1991, Y. Maday (Ed.), Springer, Berlin 1992.

15 P. MONASSE AND V. PERRIER, *Orthonormal wavelet bases adapted for partial differential equations with boundary conditions*, Preprint 1995.

From First Principles to Industrial Applications

Gert D. Billing

Department of Chemistry, H. C. Ørsted Institute, University of Copenhagen, Denmark

Abstract. The fundamental equations for chemical systems have been known since the twenties. However it is not untill recently that this fact has been used for theoretical predictions of the macroscopic behaviour of chemical non-equilibrium systems. The reason for this developement can be found in the advance in high-speed computing and the construction of new theoretical methods.

1 Introduction

For the motion of electrons and nuclei there is no doubt that the Schrödinger equation gives an accurate and complete description of the dynamics. Eventually relativistic corrections have to be included in some cases. But this does not change the fact that the equations of motion for chemical, physical and bio-chemical systems are in practice known. This has of course been well understood since the equation was first discovered in 1926 by E. Schrödinger. In the years after the equation was discovered it was solved for simple model systems: The harmonic oscillator, the hydrogen atom etc. These solutions are still today the standard text book examples for the use of quantum mechanics. In the thir-ties Dirac claimed that although everything was in principle known not much could be done in terms of actually obtaining numerical accurate numbers for real systems. The reason for this is to be found the delocalized structure of the solution. Thus the solution, the wavefunction is in principle everywhere in space and interacts with everything, i.e. we have to solve for the whole system (all of the many-body problem). This is of course impossible no matter how big a computer one might design. The way out of the problem is to design models. These models should be as realistical as possible, i.e. include as many of the physical effects as possible. At this point it is important to acknowledge the advantage in experimental detection of molecular events. Such detailed experiments are essential for the construction of reliable theoretical methods. Another aspect, which has changed the situation in a favourable way is the enormous increase in computing power. This has made it possible to perform so-called benchmark calculations on systems which are simple enough to allow an accurate solution but also complicated enough to include the physical aspects known to be impor-tant. These benchmark calculations are then the testing material on which more approximate dynamical theories can be evaluated. Theories which in turn can be used for more complicated systems. Thus the benchmark calculations can be viewed as ideal experiments, where all parameters and aspects are completely under control. The more powerful the computers are the more sophisticated such benchmark calculations can be made.

The goal of molecular dynamical calculations is then to solve the fundamental equations in order to obtain information on quantities as: energy transfer, reactivity, transport properties etc. These quantities are as fundamental for the understanding of non-equilibrium chemistry as mass, structure, energy levels etc. for the understanding of chemical equilibrium situations.

Thus the area of interest now is that of chemical change and the understanding of it at the molecular level. Once the fundamental quantities such as: rate constants, diffusion coefficients, heat of conductivity etc. are known these can be used in yet another modeling - namely that of modeling macroscopic systems, i.e. to predict the time-evolution of model systems for e.g. atmospheric and interstellar gas clouds, chemical reactors, fusion-processes, material corrosion and heat resistence just to mention a few of the important and numerous processes of practical interest. The reason that this latter stage of the scheme has become increasingly important has to do with the following: The models have become more realistical, they include more dynamical information and they have become cheaper and faster to perform due to the increase in computational power. In other cases the models are used to predict and test the effect of e.g. pollutants and other changes of the system. Changes one either cannot incorporate in the experiments or changes which are to expensive to test out experimentally. Here one must remember that for such a system the parameter space is large, temperature, pressure, chemical constituents and many other aspects can be varied. It is simply either impossible or too expensive to vary all of these experimentally. Therefore such modeling has become more and more important for many areas of science and technology. But one should of course always have in mind that it is models and not nature itself one is dealing with. Thus constant improvement of the models themselves and the input parameter is needed.

2 Methodology

I have in fig. 1 given a flow-diagram of the road from first principles to the modeling of physical and chemical systems relevant for various technological areas. The electronic structure problem deals with the calculation of the energy of the system for fixed nuclear positions. This separation of the electronic and the nuclear dynamics is based on the so-called Born-Oppenheimer approximation in which the coupling between the electronic and nuclear motion is simplified or ignored. However there are a number of cases where this decoupling cannot be done – or is a bad approximation. In such cases one must account for the socalled non-adiabatic electronic coupling when solving the END (electron-nuclear dynamics). The electronic structure problem cannot be solved without approximations or the introduction of semi-impirical methods for many electron systems. At this point various experimental information - spectroscopic and molecular beam data giving information on e.g. reaction barriers, long range electrostatic interaction, dispersion forces, etc. may be introduced when constructing the potential energy surface to be used when solving the dynamical problem for the nuclear motion. The potential is a function of 3N-6 coordinates, where N is the number of atoms which makes the construction in the general case a non-trivial one. The most widespread approximation is to assume that it can be written as a sum of two-body interaction terms and that potential parameters have a certain degree of universality, i.e. they are the same for systems and molecules which are alike. The potential energy surface is usually represented by an analytical expression involving the bond distances, angles etc. But in general there is not one unique way of representing the interaction potential. The way it is done depends very much one what use one makes of the surface - whether it is used for spectroscopy, energy transfer calculations, reactive scattering etc.

In recent years there has therefore been considerable interest in methods which calculates the potential "on the fly" [1]. However these methods cannot in general be used for molecular dynamics calculations.

Solution of the nuclear dynamics problem poses the next problem. The solution gives the quantities of importance for the modeling of chemical kinetics. Due to the heavy masses of the nuclei it is often sufficiently accurate to use classical mechanics, i.e. to run classical trajectories over the potential energy surface. However some quantum effects remain as non-adiabatic electronic transitions (geometric phase effects [2]) zero-point vibration, tunneling and resonances etc. These effects are quantum mechanical and should be treated as such if reliable rates and cross-sections are needed.

At this point it is often possible to test the dynamical calculations and the potential energy surface by comparing with various type of experiments: Beam-data, relaxation measurements, energy-transfer measurements etc.

3   Modeling of kinetics of complex chemical reactions

The rates for chemical reactions, energy change, diffusion and transport of matter can be used in more or less sophisticated models which simulates the chemistry of the real system as well as possible. In such simulations many hundred reactions may be important. However usually a few of the reactions constitute the rate-determing step - the bottle-neck. In order to test which of the reactions are important one performs a sensitivity analyses. This analyses shows how sensitive the result is to changes in the parameters, e.g. the rate constants. Methods do exist so that this can be performed with just little additional work. For the most important processes one of course need the most accurate data and additional experimental or theoretical work might be necessary in order to obtain better estimates. However even more important is it that the kinetic scheme is complete and that unjustified assumptions as for example equilibrium distributions for degrees of freedom which may not be in equlibrium. Especially for the vibrational degree (and of course the electronic degrees of freedom) it may be necessary to include a labeling of the reacting molecules which incorporates its particular quantum state. In any case this is necessary if the reaction and vibrational relaxation occur on the same time-scale. The reactivity of a vibrational excited

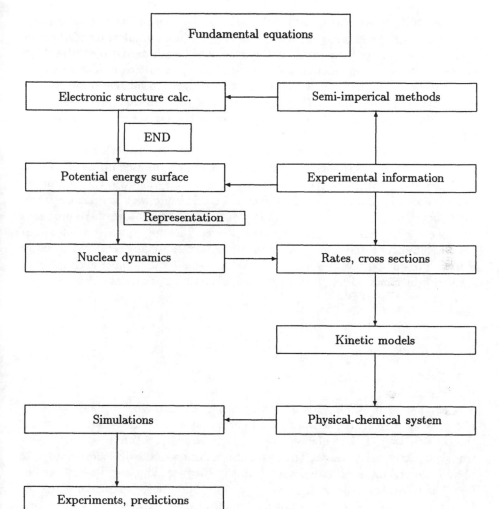

Fig. 1. Flowdiagram for the "ab initio" approach to simulation of non-equilibrium physical-chemical systems

molecule can be very different from that of a Boltzmann molecule. Take for example the photodissociation of ozone

$$O_3 + h\nu \rightarrow O_2(v) + O \tag{1}$$

Here the oxygen molecule is formed in a highly excited vibrational state. This molecule reacts and transfers energy in a very different way than oxygen in the ground state. Hence in order to incorporate this in the modeling it is necessary to work with vibrational labelled concentrations.

For most systems we have also to include the spatial dependence of the concentration $C(x,y,z,t)$ and hence include the transportation of matter from a volume element around x,y,z to one around x',y',z'. That is we have to solve the coupled kinetic reaction scheme simultaneously with the transport or diffusion of matter. If the reaction occur in a container we may have to include the transport of energy to the container walls and the chemical reactions and processes at the surface itself. Sometimes it is even the purpose of the surface to act as a catalyst for the reaction.

Most chemical reactions of interest occur in solution (e.g. water). Here it is important to consider the way the solvent influences the motion of the reactants, the reactivity or whether the solvent merely act as a dielectric continuum [3]. Today molecular dynamics calculations are carried out with various degree of sophistication as far as the interaction potentials and the nature of the processes are concerned. Generally such calculations even the simplest add a bit to our understanding of reactivity. However chemistry and especially reactive chemistry is from a basic point of view a difficult subject, where the number of simplifying features which can off hand be introduced is small if not absent.

4 Discussion

The scheme outlined above is or has been carried through for many processes of technical interest. It has three major computational bottle necks. The first is the construction of the potential energy surfaces by quantum mechanical electronic structure calculations. The next is the solution of the nuclear dynamics to obtain the crucial information needed in the chemical kinetics simulation. The final has to do with the solution of the kinetic equations, the Master-equation, Boltzmann-equation, reaction-diffusion equations etc. There is no doubt that the scheme will be used more and more extensively in the coming years. Today molecular dynamics and kinetics simulations are known in the areas of meteorology (atmospheric chemistry), simulations of the chemistry of interstellar space, combustion processes, catalytic systems, fusion-divertor modeling, space shuttle re-entry problem, biochemical systems etc.

Even decisive political actions are taken as a consequence of some of these results (as for example CO_2 effect on earth-temperature and depletion of ozon). This makes it even more important that the calculations in each of the three steps above are as reliable as possible, that all the important chemical processes are included in the models, that the rates are reliable and relevant for the processes in question etc. The ultimate understanding of the complicated chemistry sometimes has unexpected solutions as e.g. the understanding of the ozon holes at the poles. In any case both chemical and bio-chemical systems evolve as a result of many competing processes and it is necessary with a profound understanding of many details before one can claim that the system is fully understood.

References

1. R. Car and M. Parinello, Phys. Rev. Lett. **55**, 2471(1985).
2. A. Kuppermann, In "Dynamics of Molecules and Chemical Reactions", ed. R. E. Wyatt and J. Z. H. Zhang, Marcel Dekker Inc. New York, 1996, p.411.
3. G. D. Billing anf K. V. Mikkelsen, Chem. Phys. **182**, 249(1994).

Parallel Implementation of a Schwarz Domain Decomposition Algorithm

Petter E. Bjørstad[1], Maksimillian Dryja[2] and Eero Vainikko[1]

[1] Para//ab, Institutt for Informatikk, University of Bergen, N-5020 Bergen, Norway.
Email: petter@ii.uib.no
[2] Department of Mathematics, Warsaw University, Banacha 2, 02-097 Warsaw,
Poland. Email: dryja@mimuw.edu.pl

Abstract. We describe and compare some recent domain decomposition algorithms of Schwarz type with respect to parallel performance. A new, robust domain decomposition algorithm – Additive Average Schwarz is compared with a classical overlapping Schwarz code. Complexity estimates are given in both two and three dimensions and actual implementations are compared on a Paragon machine as well as on a cluster of modern workstations.

1 Introduction

This paper is concerned with parallel algorithms for the numerical solution of equations of the form:
Find $u^* \in V(\Omega)$ such that

$$\sum_{i=1}^{N} \int_{\Omega_i} \rho_i \nabla u^* \cdot \nabla v \, dx = \int_{\Omega} fv \, dx \quad \forall v \in V(\Omega), \tag{1}$$

in a space $V(\Omega) \subset H^1(\Omega)$. Here ρ_i are positive constants and $\bar{\Omega} = \cup_{i=1}^{N} \bar{\Omega}_i$.

This problem is important in many fields of application, an example being the study of flow in porous media [2]. Domain decomposition algorithms have received much attention in the last ten years due to their potential for parallel algorithms as well as overall flexibility within the design of more comprehensive software packages.

A prototype algorithm with good parallel properties is the Additive Schwarz method [12]. Its parallel implementation has been reported on in several papers, see for example [3], [8] and [10]. However, when the parameter ρ_i in (1) has significant jumps across internal boundaries (for example across material layers) the rate of convergence suffers. It turns out that the rate of convergence depends on the distribution of the numerical values of ρ_i (see [5] and [6]) and that the estimates for this behavior are sharp [11], [13]. In Figure 1 we show two different distributions called quasi-monotone and not quasi-monotone respectively. The Additive Schwarz method behaves better in the quasi-monotone case. This motivated an alternative algorithm, the Additive Average method, proposed in [1] which has a rate of convergence independent of any specific distribution of the

ρ_i. Additionally, the new method has an advantage when the discretization is unstructured with a possibly very non-uniform mesh.

Fig. 1. The distribution in A) is not quasi-monotone as there exist no monotone path (through faces) from 1000 to 2000. The distribution in B) is quasi-monotone.

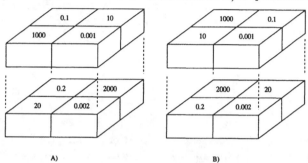

In this paper we compare parallel implementations of the two methods. Originally motivated by robustness, it turns out that the new method also has good parallel characteristics. The rest of this paper is organized as follows. In Section 2 we briefly describe the algorithms that will be considered. The reader is referred to [1] and [12] for a more detailed description. In Section 3 we discuss certain aspects of our parallel implementation of the new method. Section 4 discusses the parallel complexity in 2 and 3 dimensions, while Section 5 reports on experiments on current parallel machines where we can compare actual performance. We end with some conclusions in Section 6.

2   The Methods

Let Ω be a polyhedral region in R^d, $d = 2, 3$. We triangulate Ω and partition this triangulation into a set of possibly overlapping subdomains, Ω_i^h, where $\Omega_i \subseteq \Omega_i^h$. We look for a solution $u \in V^h$ where $V^h \subset V$ is a finite element space on our fine triangulation. To keep the presentation simple we only discuss the case $V = H_0^1$.
  Find $u \in V^h(\Omega)$ such that

$$a(u, v) = f(v) \quad \forall v \in V^h(\Omega), \tag{2}$$

where

$$a(u, v) = \sum_{i=1}^{N} \int_{\Omega_i} \rho_i \nabla u \cdot \nabla v \, dx, \quad f(v) = \int_{\Omega} fv \, dx. \tag{3}$$

We split the space V^h into $N + 1$ subspaces

$$V^h = V_0 + V_1 + \cdots + V_N,$$

where for $i = 1, \ldots N$ $V_i = H_0^1(\Omega_i^h) \bigcap V^h$ and zero outside of Ω_i^h. The space $V_0(\Omega)$ is a coarse space to be defined below.

For each $V_i, i = 0, \ldots, N$ we introduce bilinear forms $b_i(u, v)$ on $V_i \times V_i$ defining subproblems that will be used to form our preconditioner.

Following [12] we can describe the methods in a linear algebra framework where the matrix A denotes our discrete problem to be solved. For $i = 1, \ldots N$ let R_i be the restriction matrix that selects the nodes from subdomain Ω_i^h. Similarly, we define R_0^T to be an interpolation operator from the coarse space V_0 to V^h. An Additive Schwarz preconditioner is then given by

$$B = \sum_{i=0}^{N} B_i \tag{4}$$

where $B_i = R_i^T A_i^{-1} R_i$. Here $A_i = R_i A R_i^T$ for $i = 1, \ldots N$, while A_0 is the matrix representation of a coarse approximation of our problem. We note that the preconditioner (4) can be applied to a vector by applying each term in the sum independently, thus this step can easily use $N + 1$ processors in parallel.

Our methods differ in the choice of subspaces V_i and, in particular, on how the coarse space V_0 and the corresponding bilinear form $b_0(u, v)$ are defined and constructed.

2.1 The Classical Additive Schwarz Method

This method uses overlapping subdomains Ω_i^h. Since previous work [4] has shown that a minimal overlap is preferable, we limit our discussion to the case where Ω_i^h extends precisely one grid block (or one element) outside of Ω_i. The coarse space is defined from a discretization of Ω by using the subdomains Ω_i as discretization elements. We note that this (severely) restricts the shape of Ω_i. A linear interpolation from the coarse triangulation back to our original fine mesh defines R_0^T.

2.2 The Additive Average Method

This method is characterized by subdomains that do not overlap, that is, Ω_i^h is a triangulation of Ω_i. Here we do not have any restrictions on the shape of Ω_i. To simplify the presentation we assume that our fine grid discretization parameter h is constant across all subdomains, our implementation allows a variable h, that is, an unstructured grid across all of Ω. In order to define the coarse space let n_i^b denote the number of nodes on $\partial\Omega_i^h$ and let x be a *nodal point* in our discretization. (Note that we do include external boundary nodes of $\partial\Omega$ with a Dirichlet condition in this count.) We define an interpolation operator I_A as follows:

$$I_A = \begin{cases} u(x) & x \in \partial\Omega_i^h \\ \overline{u}_i & x \in \Omega_i^h \end{cases} \tag{5}$$

where

$$\overline{u}_i = \frac{1}{n_i^b} \sum_{x \in \partial\Omega_i^h} u(x), \tag{6}$$

the average of the nodal values of u on the boundary of Ω_i^h.

Our coarse space V_0 is now simply taken as the range of I_A. The bilinear form on V_0 is defined by

$$b_0(u,v) = c_0 h^{d-2} \sum_i \rho_i \sum_{x \in \partial\Omega_i^h} (u(x) - \overline{u}_i)(v(x) - \overline{v}_i). \tag{7}$$

Note that with this choice we have $A_0 \neq A_C = I_A^T A I_A$. A_C is often called the Galerkin coarse grid approximation. Our A_0 defined by $b_0(u,v)$ approximates A_C and it has more structure making it less expensive to use as part of our preconditioner.

2.3 A Multiplicative Average Method

Just as one can define a multiplicative version of the classical Additive Schwarz method, we may define a new multiplicative variant of the Additive Average method. Since we have no subdomain overlap the individual subdomain solutions can still be performed in parallel, while the coarse solve must proceed sequentially. We begin with the coarse solver followed by the subdomain solvers with an additional residual calculation between the two steps. With conjugate gradient as our preferred Krylov subspace accelerator we must end the iteration step by another residual calculation and a final coarse solve in order to preserve symmetry.

2.4 An Additive Diagonal Scaling Method

This method is similar to the Average Additive method on the fine level, the domain is divided into non-overlapping subdomains and the matrices A_i are defined in the same way as in Section 2.2. The coarse space V_0 is the same as in the Additive Schwarz method from Section 2.1. In addition, we introduce an extra space V_{-1} in order to include the nodes from $\partial\Omega_i^h$ in the preconditioner. Each of these boundary nodes are considered to be the single node of a correspondingly small subdomain surrounding it. In the preconditioning the operation of solving such a single node subdomain problem corresponds to a single division (scaling) by the diagonal element of the stiffness matrix A. Unfortunately, this method is sensitive to the distribution of the coefficients ρ_i [1].

3 Parallel Implementation

Parallel implementations of the Additive Schwarz method are well documented [4], [10], [12]. The Additive Average method lends itself to a parallel implementation and we need only discuss the parallel solution of the coarse problem in more detail.

This problem takes the following form:

$$\begin{pmatrix} A_a & A_{ab} \\ A_{ab}^T & A_b \end{pmatrix} \begin{pmatrix} u_a \\ u_b \end{pmatrix} = \begin{pmatrix} 0 \\ v_b \end{pmatrix} \qquad (8)$$

where we have introduced one extra unknown per subdomain in the vector u_a (of length N) corresponding to the average values $\bar{u}_i$ from (6). We use the subscripts a and b to denote average (values) and boundary (nodes) respectively. The vector u_b has n_b components one for each node on all internal subdomain boundaries.

The $N \times N$ matrix A_a is diagonal with entries

$$(A_a)_{ii} = c_0 h^{d-2} \rho_i n_i^b, \ i = 1, ..., N.$$

We use $c_0 = 1$ in all cases except when having triangular elements in 2 dimensions where $c_0 = 1.7$ was found to give slightly better convergence.

Denote by M_j $(j = 1, \ldots, n_b)$ the number of subdomains that share node number j on a substructure boundary. Let l_k^j, $k = 1, ..., M_j$ be the indices of the corresponding substructures. The $n_b \times n_b$ matrix A_b is also diagonal with the entries

$$(A_b)_{jj} = c_0 h^{d-2} \sum_{k=1}^{M_j} \rho_{l_k^j}.$$

The matrix A_{ab} has the nonzero entries

$$(A_{ab})_{ij} = -c_0 h^{d-2} \rho_i, \ i = 1, ..., N, j = 1, ..., n_b$$

in such a way that $A_a u_a + A_{ab} u_b = 0$ is nothing but the relation (6) between the average and the boundary values in each subdomain.

We note that we can store the matrices A_{ab} and A_b in a compact form. The entries corresponding to the boundary nodes that belong to the same set of substructures have the same value. They can be stored as one floating point number for each combination of boundary node owners and a pointer from each boundary node to the corresponding set.

We next describe how to restrict our residual vector r_{ij}, $(i = 1, ..., N, j = 1, ..., n_i)$. Here n_i is the total number of nodes belonging to substructure Ω_i (excluding exterior boundary value nodes of $\partial\Omega$ with a Dirichlet condition). Let I_i be the index set of nodes interior to substructure Ω_i and let v_b denote the vector of boundary nodes extracted from r_{ij}.

We first compute the components of v_a

$$(v_a)_i = \sum_{j \in I_i} r_{ij}, \ i = 1, ..., N. \qquad (9)$$

Next, we update the vector v_b according to

$$v_b := v_b - A_{ab}^T A_a^{-1} v_a.$$

In order to solve the system (8) we first compute and store its Schur complement

$$S = A_a - A_{ab} A_b^{-1} A_{ab}^T.$$

This computation is easy due to the structure of the blocks involved. The matrix S is $N \times N$, symmetric and sparse with a nonzero diagonal and additional nonzero entries at positions reflecting the connectivity of substructures in Ω. The contribution from each substructure to S can be computed independently and in parallel. The results are communicated to the process(or) assigned to the coarse solver and stored.

For each coarse problem solution step in the conjugate gradient iteration we must solve

$$Su_a = -A_{ab}A_b^{-1}v_b. \tag{10}$$

Note that the right hand side of (10) can be computed in parallel. Each substructure sends only one floating point number to the coarse solver. After the solution of (10) the values of the vector u_a can be assigned to all interior nodes, that is, there is no further interpolation. The substructure boundary nodes get their values from the formula

$$u_b = A_b^{-1}(v_b - A_{ab}^T u_a),$$

which again can be computed in parallel.

4 Complexity

We analyze the complexity of the algorithms for a model problem. Consider the case of a regular domain split into m^d subdomains ($d = 2, 3$), each consisting of n^d blocks (see Figure 2 where $d = 2$, $m = 3$, $n = 4$).

Fig. 2. An example of a regular domain divided into 9 subdomains. The dark bullets denote the boundary nodes while the small circles show the inner nodes of the subdomains that take part in the subdomain solves. The four large circles are the nodes that belong to the coarse space in a classical two-level method.

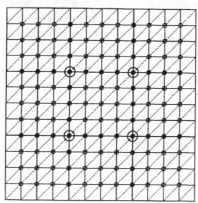

We compare the complexity of one complete conjugate gradient iteration for the Additive and Multiplicative Average method and the classical Additive

Schwarz method with minimal overlap (*i.e.* neighboring subdomains share exactly one node across each common boundary). In Table 1, we give the number

Table 1. Computational complexity for one preconditioning step: classical Additive Schwarz method with minimal overlap compared with the Additive and Multiplicative Average methods.

Task	Classical Additive Schwarz	Additive Average	Multiplicative Average
Subdomain (size)	$(n+1)^d$	$(n-1)^d$	$(n-1)^d$
Coarse prbl. (size)	$(m-1)^d$	m^d	m^d (2 times)
Transfers (flops)	$(18d-3)n^d$	$(2d+15)n^d$	$(6d+19)n^d$
Comm. (words)	$4d\,n^{d-1}$	$2d\,n^{d-1}$	$6d\,n^{d-1}$
Comm. (startups)	$4d+4$	$2d+4$	$6d+5$

of operations for one interior subdomain assuming that all the subdomains are assigned to different processors. We do not include the flops in the coarse problem and the subdomain solution, only the problem sizes are outlined. In this way different solver complexities can be taken into account in any specific implementation. We count the number of words communicated in one send/receive pair and similarly charge such a pair with one communication startup unit. We further only include the highest order terms (in n) in our complexity estimates.

The work in a complete conjugate iteration step can be divided into three parts, the dot products and the vector updates needed in the algorithm itself, the matrix vector product, and the application of the preconditioner. The three vector updates contribute $6n^d$ flops and the three dot products are also $6n^d$ flops and further 3 communication startups. The matrix vector product cost $(2d+1)n^d$ flops. The matrix vector product needs to communicate $2dn^{d-1}$ words to its $2d$ neighbors causing $2d$ startups. These costs are the same for all three methods. The differences in the table are only due to the different cost of the preconditioning step. With no subdomain overlap and no neighbor communication we have only $2n^d$ flops (from equations (4) and (9)) and a single communication startup for the Additive Average method. The standard Additive Schwarz is both more compute intensive in the interpolation/restriction phase and communication intensive with nearest neighbor interaction. The Multiplicative Average method has expensive residual updates that also forces more communication.

The total flop count excluding the solves are listed in the third row of Table 1 called *Transfers*. We have optimized floating point operations over storage, that is, we assume that elementary functions needed in the restriction and the interpolation phase of the classical Additive Schwarz method and some intermediate values in the Average methods are precomputed and stored. Our implementation differs only slightly from this model, and this assumption results in a consistent presentation of the computational complexity of the three methods.

5 Computer Experiments

We have carried out experiments on an Intel Paragon computer and on a cluster of workstations to explore the behavior of the methods from a practical point of view. The standard MPI implementation from Argonne [7] was used for message passing in our algorithms. In practice, it pays to use an inexact solver both for the subdomains as well as for the coarse problem. This can be analyzed by replacing our bilinear forms $b_i(u, v)$ with another form corresponding to the inexact solver. For simplicity we have used two symmetric Gauss-Seidel iterations as our inexact solver throughout this paper. More advanced techniques including a variable number of inner iterations as well as the use of multigrid as an inner solver are possible. Such techniques would reduce the overall computer time since the methods would require fewer iterations to converge. A more accurate subdomain solver tends to be most beneficial for the standard Additive Schwarz algorithm. We note, however, that an accurate solution in this case involves a subdomain problem with a discontinuous coefficient due to the overlap. Such a solution may therefore be more expensive than for the Additive Average method.

Fig. 3. Solution time in seconds for problems with discontinuous coefficients. There are 32×32 gridblocks per subdomain in the 2D and $16 \times 16 \times 16$ gridblocks in the 3D case. The number of subdomains (together with the number of processors) is plotted on the x−axis.

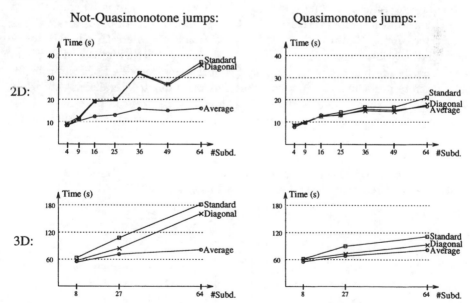

In Figure 3 we plot the time in seconds for the Additive Average, the Additive Diagonal Scaling and the standard Additive Schwarz method with minimal overlap. The figure shows algorithm scalability since we keep the size of a subdomain fixed and increase the number of subdomains along with the number of processors. The first row in the figure shows 2-dimensional results while the second row shows a 3-dimensional computation. Similarly, the first column has non quasi-monotone jumps in the coefficients as in Figure 1A, while the second column has a quasi-monotone distribution of ρ_i as given in Figure 1B. We map only one subdomain to each processor. In the 2-dimensional case a subdomain is a 32×32 grid, while we used a $16 \times 16 \times 16$ grid in 3 dimensions.

One should note that as more and more subdomains are added we increase the complexity of the overall (global) problem since we get a similar growth in the number of discontinuous jumps in the coefficients. The parallel scalability of the algorithms should therefore be interpreted with care. A more proper scalability experiment would keep the continuous problem fixed and then refine the discretization and the number of processors such that each processor solves a constant size subproblem.

We observe that the Additive Average method is considerably faster in the case of non quasi-monotone jumps in the coefficients both in 2 and 3 dimensions. In the quasi-monotone case the differences are smaller, but in 3 dimensions the Additive Average method is still the fastest while in 2 dimensions the time is quite close to that of the Additive Diagonal Scaling method. We conclude from the above that both the Additive Diagonal Scaling method and the standard Additive Schwarz method are sensitive to the quasi-monotonicity property while the Additive Average method, as expected (see [1]), show no such dependency. Due to its lower complexity, the Additive Average method seems to have an advantage over the other methods when implemented on a parallel computer.

The next tests were carried out on four 200 MHz Ultra Sparc workstations connected with a 10Mbit Ethernet. This interconnect has a rather high latency and relatively low communication bandwidth. The Additive and the Multiplicative Average method are compared. Test results are given for a Poisson problem. The discrete problem size is now fixed at 128×128 quadrilateral elements while the number of subdomains is changed. The processors will compute one subdomain at a time when there are more subdomains than processors. Table 2 shows

Table 2. Time, iteration count and condition number for the Additive Average and the Multiplicative Average method. Ω is discretized using 128×128 quadrilateral elements.

#Subd.	Additive			Multiplicative		
	Time (s)	#it.	cnd.#	Time (s)	#it.	cnd.#
2x2	9.7	81	309.2	10.5	58	137.2
4x4	9.0	62	117.4	9.6	40	44.5
8x8	8.7	44	59.7	9.1	26	17.8
16x16	10.7	43	59.7	10.8	22	16.3
32x32	19.1	50	104.8	19.4	25	28.6

the elapsed computing time, the number of iteration steps and the condition number of the preconditioned system for different number of subdomains. The multiplicative variant uses about half the number of iterations, this is in general agreement with theory. Despite this, due to its higher cost per iteration, it is slightly slower than the additive algorithm in all cases. We also note that the lowest condition number and the best overall times are obtained when the size of the coarse problem is roughly the same as each of the subdomain problems.

Fig. 4. Elapsed time when Ω is discretized in a 128×128 mesh partitioned into a variable number of irregular subdomains.

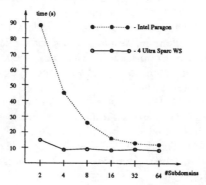

Finally, in Figure 4 we present results from a case where the subdomains are irregular. Again Ω is discretized using a fixed 128×128 mesh of quadrilateral elements, but the subdomains are created using the Chaco version 2.0 software package [9]. The resulting slightly irregular subdomains are processed by our Intel Paragon, a single subdomain to each node and also by a cluster of four Ultra Sparc workstations with multiple subdomains per processor. The total problem elapsed time is plotted, excluding the partitioning time used by Chaco.

A nice speedup can be observed on the Intel Paragon and also on the cluster of workstations when going from 2 to 4 subdomains. We note that the problem solution time on the workstation cluster is fairly robust with respect to subdomain partitioning as long as there is work to be done on all nodes.

6   Conclusions

New domain decomposition algorithms of both additive and multiplicative type have been implemented as parallel algorithms and tested on both a Paragon computer as well as on a cluster of modern workstations. Based on the experiments performed, the so-called Additive Average method seems to have the best performance, in particular, it is faster than the classical Additive Schwarz method combined with inexact solvers. The new method is also more robust with respect to discontinuous coefficients in the partial differential equation and can easily handle both unstructured meshes and irregular subdomains. We also note that by 1996 a cluster of four workstations typically outperforms a 64 node Paragon, a state of art parallel supercomputer only 3 years earlier.

References

1. P. E. BJØRSTAD, M. DRYJA, AND E. VAINIKKO, *Additive Schwarz methods without subdomain overlap and with new coarse spaces*, in Domain Decomposition Methods in Sciences and Engineering, R. Glowinski, J. Périaux, Z. Shi, and O. B. Widlund, eds., John Wiley & Sons, 1996. Proceedings from the Eight International Conference on Domain Dec omposition Metods, May 1995, Beijing.

2. P. E. BJØRSTAD AND T. KÅRSTAD, *Domain decomposition, parallel computing and petroleum engineering*, in Domain-Based Parallelism and Problem Decomposition Methods in Computational Science and Engineering, D. E. Keyes, Y. Saad, and D. G. Truhlar, eds., SIAM, 1995, ch. 3, pp. 39–56.

3. P. E. BJØRSTAD, R. MOE, AND M. SKOGEN, *Parallel domain decomposition and iterative refinement algorithms*, in Parallel Algorithms for Partial Differential Equations, W. Hackbusch, ed., Braunschweig, Germany, 1991, VIEWEG, pp. 28–46. Notes on Numerical Fluid Mechanics, Vol 31, Proceedings of the sixth GAMM-Seminar, Kiel, January 19-21 1990.

4. P. E. BJØRSTAD AND M. D. SKOGEN, *Domain decomposition algorithms of Schwarz type, designed for massively parallel computers*, in Domain Decomposition Methods for Partial Differential Equations, D. E. Keyes, T. F. Chan, G. Meurant, J. S. Scroggs, and R. G. Voigt, eds., SIAM, 1992, pp. 362–375. In the proceedings from the Fifth International Symposium on Domain Decomposition Methods, Norfolk, Virginia 1991.

5. J. H. BRAMBLE AND J. XU, *Some estimates for a weighted L^2 projection*, Math. Comp., 56 (1991), pp. 463–476.

6. M. DRYJA, M. V. SARKIS, AND O. B. WIDLUND, *Multilevel Schwarz methods for elliptic problems with discontinuous coefficients in three dimensions*, Numer. Math., 72 (1996), pp. 313–348.

7. W. D. GROPP, E. LUSK, AND A. SKJELLUM, *Using MPI: Portable Parallel Programming with the Message-Passing Interface*, MIT Press, 1994.

8. W. D. GROPP AND B. F. SMITH, *Experiences with domain decomposition in three dimensions: Overlapping Schwarz methods*, Contemporary Mathematics, (1991).

9. B. HENDRICKSON AND R. LELAND, *The Chaco user's guide, version 2.0*, Tech. Report SAND 94-2692, Sandia National Laboratories, July 1995.

10. D. E. KEYES AND W. D. GROPP, *A comparison of domain decomposition techniques for elliptic partial differential equations and their parallel implementation*, SIAM J. Sci. Stat. Comput., 8 (1987), pp. 166 – 202.

11. P. OSWALD, *On the robustness of the BPX-preconditioner with respect to jumps in the coefficients*, tech. report, Dept. of Math. Texas A&M University, College Station, TX 77843-3368, 1995.

12. B. F. SMITH, P. E. BJØRSTAD, AND W. D. GROPP, *Domain Decomposition: Parallel Multilevel Methods for Elliptic Partial Differential Equations*, Cambridge University Press, 1996.

13. J. XU, *Counter examples concerning a weighted L^2 projection*, Math. Comp., 57 (1991), pp. 563–568.

Practical Experience in the Dangers of Heterogeneous Computing

S. Blackford[*1], A. Cleary[1], J. Demmel[2], I. Dhillon[2], J. Dongarra[1,3], S. Hammarling[1,4], A. Petitet[1], H. Ren[2], K. Stanley[3], R. C. Whaley[2]

[1] University of Tennessee, Knoxville, USA
[2] University of California at Berkeley, USA
[3] Oak Ridge National Laboratory, Oak Ridge, TN, USA
[4] Numerical Algorithms Group Ltd., Oxford, England
[5] Intel SSPD, Beaverton, Oregon, USA

Abstract. Special challenges exist in writing reliable numerical library software for heterogeneous computing environments. Although a lot of software for distributed memory parallel computers has been written, porting this software to a network of workstations requires careful consideration. The symptoms of heterogeneous computing failures can range from erroneous results without warning to deadlock. Some of the problems are straightforward to solve, but for others the solutions are not so obvious, or incur an unacceptable overhead. Making software robust on heterogeneous systems often requires additional communication.

This paper addresses the issue of writing reliable numerical software for networks of heterogeneous computers. We describe and illustrate the problems encountered during the development of ScaLAPACK. Where possible, we suggest solutions to avoid potential pitfalls, or if that is not possible, recommend that the software is not used on heterogeneous networks.

1 Introduction

There are special challenges associated with writing reliable numerical software on networks containing heterogeneous processors, that is processors which may do floating point arithmetic differently. This includes not just machines with completely different floating point formats and semantics (e.g. Cray machines running 'Cray arithmetic' versus workstations running IEEE standard floating point arithmetic), but even supposedly identical machines running with different compilers, or even just different compiler options or runtime environments.

The basic problem occurs when making *data dependent branches* on different processors. The flow of an algorithm is usually data dependent and so slight variations in the data may lead to different processors executing completely different sections of code.

This paper represents the experience of the ScaLAPACK group in developing numerical software for distributed memory message-passing systems, and the

---

[*] formerly S. Ostrouchov

awareness that the software being developed may not be as robust on heterogeneous systems as on homogeneous systems. We briefly describe the work of this group in Section 2.

In Sections 3, 4 and 6 we look at three areas that require attention in developing software for heterogeneous networks: machine parameters, where we discuss what the values of machine parameters, such as machine precision should be; checking global arguments and communicating floating point values; and algorithmic integrity, that is, how can we ensure that algorithms perform correctly in a heterogeneous setting. The particular case of communicating floating point values on IEEE machines is briefly discussed in Section 5.

2 Motivation and Background

The challenges of heterogeneous computing discussed in this paper came to light during the development of ScaLAPACK [2].

ScaLAPACK is a library of high performance linear algebra routines for distributed memory MIMD machines. It is a continuation of the LAPACK project, which has designed and produced an efficient linear algebra library for workstations, vector supercomputers and shared memory parallel computers [1]. Both libraries contain routines for the solution of systems of linear equations, linear least squares problems and eigenvalue problems. The goals of the LAPACK project, which continue into the ScaLAPACK project, are efficiency so that the computationally intensive routines execute as fast as possible; scalability as the problem size and number of processors grow; reliability, including the return of error bounds; portability across machines; flexibility so that users may construct new routines from well designed components; and ease of use. Towards this last goal the ScaLAPACK software has been designed to look as much like the LAPACK software as possible.

Many of these goals have been attained by developing and promoting standards, especially specifications for basic computational and communication routines. Thus LAPACK relies on the BLAS [13, 6, 5], particularly the Level 2 and 3 BLAS for computational efficiency, and ScaLAPACK relies upon the BLACS [7] for efficiency of communication and uses a set of parallel BLAS, the PBLAS [3], which themselves call the BLAS and the BLACS. LAPACK and ScaLAPACK will run on any machines for which the BLAS and the BLACS are available. A PVM ([8]) version of the BLACS has been available for some time and the portability of the BLACS has recently been further increased by the development of a version that uses MPI [14, 15].

ScaLAPACK was developed with heterogeneous environments in mind, as well as standard homogeneous machines. But during development it was realized that we could not guarantee the safe behavior of all our routines in a heterogeneous environment and so, for the time being, both libraries are only fully supported on homogeneous machines, although they are tested on networks of IEEE machines and are believed to work correctly in such environments. It is, though, intended to be able to fully support other heterogeneous environments in the near future. Any known heterogeneous failures are documented in the `errata.scalapack` file on netlib.

3 Machine Parameters

A simple example of where an algorithm might not work correctly is an iteration where the stopping criterion depends on the value of the machine precision. If the precision varies from processor to processor, different processors may have significantly different stopping criteria. In particular, the stopping criterion used by the most accurate processor may never be satisfied if it depends on data computed less accurately by other processors.

Many such problems can be eliminated by using the *largest* machine precision among all participating processors. In LAPACK routine `DLAMCH` returns the (double precision) machine precision (as well as other machine parameters). In ScaLAPACK this is replaced by `PDLAMCH` which returns the largest value over all the processors, replacing the uniprocessor value returned by `DLAMCH`. Similarly, one should use the smallest overflow threshold and largest underflow threshold over the processors being used. In a non-homogeneous environment the ScaLA-PACK routine `PDLAMCH` runs the LAPACK routine `DLAMCH` on each processor and communicates the relevant maximum or minimum value. We refer to these machine parameters as the *multiprocessor machine parameters.*

The present `PDLAMCH` suffers from an obvious weakness, however. As mentioned above, this routine calls `DLAMCH` locally, and then communicates to obtain a machine precision for the entire context. Since this communication is required on every call, `PDLAMCH` may not be called by a subset of processes. There are many examples in ScaLAPACK codes, however, where only a subset of nodes (for instance one column or one row of the process grid) is performing a given computation. This makes `PDLAMCH` uncallable from such computations. Refer to Section 6 for a specific example.

The straightforward solution of performing the communication and storing the values for subsequent calls may not be employed, as the user may be utilizing overlapping process grids (requiring different machine parameters based on which context is calling), or getting the same context handle as grids are freed and allocated.

For this reason, it is expected that the next release of the BLACS will support caching based on the BLACS context. We will then be able to perform the communication once for each context and cache the values on the context. Subsequent `PDLAMCH` calls within the context will then access strictly local data, and thus may be safely called from code performing computations on grid subsets.

It should be noted that if the code contains communication between processors within an iteration, it will not complete if one processor converges before the others. In a heterogeneous environment, the only way to guarantee termination is to have one processor make the convergence decision and broadcast that decision. This is a strategy we shall see again in later sections.

4 Global Arguments and Floating Point Values

The exact definition of the value of "global" variables has not yet been determined for a heterogeneous environment. Global input arguments are not required to be exactly equal in all processes, nor are global output arguments exactly equal on output.

The high level routines in the ScaLAPACK Library check arguments supplied by users for their validity in order to aid users and provide as much reliability as possibility. In particular, global arguments are checked. When these are floating point values they may of course, for the reasons already discussed, have different values on different processors.

This raises the question of how, and even whether, such arguments should be checked, and what action should be taken when a failure occurs. If we compare the values, they may not be the same on different processors, so we need to allow a tolerance based upon the multiprocessor machine precision. Alternatively, we can check a global argument on just one processor and then, if the value is valid, broadcast that value to all the other processors. Of course this alternative approach has extra overhead, but it may be the most reliable solution, since the library routine has algorithmic control, and puts slightly less burden on the user.

Similar issues occur whenever we communicate a floating point value from one processor to another. Unless we have special knowledge, and one such case will be discussed in the next section, we should not assume that the target processor will have exactly the same value as the sending processor and we must write the code accordingly.

5 Communicating Floating Point Values on IEEE Machines

The IEEE standard for binary floating point arithmetic [11] specifies how machines conforming to the standard should represent floating point values. We refer to machines conforming to this standard as *IEEE machines*[6]. Thus, when we communicate floating point numbers between IEEE machines we might hope that each processor has the same value. This is a reasonable hope and will often be realized.

For example, XDR (External Data Representation, [16]) uses the IEEE representation for floating point numbers and so a message passing system that uses XDR will communicate floating point numbers without change[7]. PVM is an example of a system that uses XDR. MPI suggests the use of XDR, but does not mandate its use [15, Section 2.3.3], so presumably we cannot assume that floating point numbers will be communicated without change on IEEE machines when using MPI unless we have additional information about the implementation.

---

[6] It should be noted that there is also a radix independent standard [12].

[7] It should be noted that it is not clear whether or not this can be assumed for denormalized numbers.

As we are expanding our ScaLAPACK Test Suite to encompass more rigorous testing across the bounds of representable numbers (as is present in the LAPACK Test Suite), we are reminded of additional dangers which must be avoided when communicating floating point numbers. Specifically, the implementation of complex arithmetic (unscaled complex division) on certain architectures requires that the range of representable numbers be restricted by a call to the ScaLAPACK routine PxLABAD. PxLABAD is the equivalent of the LAPACK routine xLABAD, which, if desired, takes the square root of the smallest and largest representable numbers for the computation to protect from unexpected overflow or underflow. For obvious reasons, catastrophic results can occur during rigorous testing or merely computations near overflow or underflow if this type of architecture is linked with an architecture that does not require such a restriction.

Another possible danger that we have encountered during testing is in the way that denormalized numbers are handled on certain IEEE architectures. By default, some architectures do not recognize denormalized numbers and set their value to zero. Thus, if the computation involves numbers near underflow and a denormalized number is communicated to such a machine, the computational results will be invalid and the subsequent behavior is unpredictable.

6 Algorithmic Integrity

The suggestions we have made so far certainly do not solve all the problems. We are still left with many of those problems associated with the major concern of varying floating point representations and arithmetic operations between different processors, different compilers and different compiler options. We illustrate the difficulties with just two examples from ScaLAPACK, the second example giving rather more severe difficulties than the first.

Consider the LAPACK routine DLARFG which computes an elementary reflector (Householder transformation matrix) H such that

$$Hx = \beta e_1,$$

where β is a scalar, x is an n element vector and e_1 is the first column of the unit matrix. H is represented in the form

$$H = I - \tau v v^T,$$

where τ is a scalar and v is an n element vector. Since H is orthogonal we see that

$$|\beta| = \|x\|_2.$$

If $|\beta|$ is very small (sub-normal or close to being sub-normal), DLARFG scales x and recomputes $\|x\|_2$. This computation is at the heart of the LAPACK QR, and other, factorizations (see for example [9]).

In the case of the equivalent ScaLAPACK routine PDLARFG, x will usually be distributed over several processors, each of which participates in the computation of $\|x\|_2$ and, if necessary, scales its portion of the vector x and recomputes $\|x\|_2$.

From the previous discussion we can see that we clearly need to take care here, or else, in close cases, some processors may attempt to recompute $\|x\|_2$, while others do not, leading to completely erroneous results, or even deadlock.

There are many other routines in the LAPACK and ScaLAPACK libraries where scaling takes place, either to avoid problems associated with overflow and underflow, or to improve numerical stability, as in the equilibration routines for linear equations.

These problems with scaling can be alleviated as previously described in Section 3.

Another way to ensure correct computation is to put one processor in control of whether or not scaling should take place, and for that processor to request the other processors all either to scale, or not to scale. Having a *controlling processor* is a common way to solve such problems on heterogeneous networks.

As a somewhat harder problem consider the method of bisection for finding the eigenvalues of symmetric matrices performed by the ScaLAPACK routine PDSYEVX. In this algorithm, the real axis is broken into disjoint intervals to be searched by different processors for the eigenvalues contained in each interval. Disjoint intervals are searched in parallel. The algorithm depends on a function, say count(a,b), that counts the number of eigenvalues in the half open interval [a, b). Using count, intervals can be subdivided into smaller intervals containing eigenvalues until the intervals are narrow enough to declare the eigenvalues they contain as being found. The problem here is that two processors may not agree on the boundary between their intervals. This could result in multiple copies of eigenvalues if intervals overlap, or missing eigenvalues if there are gaps between intervals. Furthermore, the count function may count differently on different processors, so an interval [a, b) may be considered to contain 1 eigenvalue by processor A, but 0 eigenvalues by processor B, which has been given the interval by processor A during load balancing. This can happen even if processors A and B are identical in hardware terms, but if the compilers on each one generate slightly different code sequences for count. In this example we have not yet decided what to do about all these problems, so we currently only guarantee correctness of PDSYEVX for networks of processors with identical floating point formats (slightly different floating point operations turn out to be acceptable). See [4] for further discussion. Assigning the work by index rather than by range and sorting all the eigenvalues at the end may give the desired result with modest overhead. Of course, if floating point formats differ across processors, sorting is a problem in itself. This requires further investigation.

The symmetric eigensolvers, PxSYEVX and PxHEEVX, may have trouble on heterogeneous networks when a subset of eigenvalues is chosen by value (i.e. RANGE='V') and one of the limits of that range (VL or VU) is within a couple ulps (PDLAMCH('P')) of an actual eigenvalue. The two processes may then disagree on the number of eigenvalues specified by the range VL and VU and the code breaks with each process returning INFO $\neq$ 0. This situation can happen when running the test code and will be corrected in the next release. In every case that we have seen, the answer is correct despite the spurious error message.

This is not a problem on homogeneous systems.

Redundant work on different processors which is intended to result in identical results, may not do so in a heterogeneous environment. One approach for parallelizing the symmetric eigenproblem is to perform a tridiagonal QR iteration redundantly on all processors, save the plane rotations, and then accumulate the resulting Givens rotations in parallel into the relevant columns of the unit matrix. This results in $O(n^2)$ redundant work, $O(n^3)$ parallel work, and requires no communication. Since QR is not forward stable, slight differences in the underlying arithmetic can lead to completely different rotations and completely incorrect results. This can be solved by having one processor run the QR algorithm and then broadcast the plane rotations to the other processors, but the communication cost is substantial: $O(n^2)$.

7 Closing Remarks

We have tried to illustrate the potential difficulties concerned with floating point computations on heterogeneous networks. Some of these difficulties are straightforward to address, while others require considerably more thought. All of them require some additional level of defensive programming to ensure the usual standards of reliability that users have come to expect from packages such as LAPACK.

We have suggested reasonably straightforward solutions to the problems associated with floating point machine parameters and global values, and have suggested the use of a controlling processor to solve some of the difficulties of algorithmic integrity. This can probably be used to solve most of these problems, but in some cases at the expense of considerable additional overhead.

A topic that we have not discussed is that of the additional testing necessary to give confidence in heterogeneous environments. The testing strategies that are needed are similar to those already employed in reputable software packages such as LAPACK, but it may be very hard to produce actual test examples that would detect incorrect implementations of the algorithms because, as we have seen, the failures are likely to be very sensitive to the computing environment, and in addition may be non-deterministic.

8 Acknowledgments

We wish to thank all of our ScaLAPACK and NAG colleagues for a number of useful discussions on heterogeneous computing and their valuable input to this paper.

References

1. Anderson, E., Bai, Z., Bischof, C., Demmel, J., Dongarra, J., Du Croz, J., Greenbaum, A., Hammarling, S., McKenney, A., Ostrouchov, S., Sorensen, D.: LAPACK Users' Guide, Second Edition. SIAM, Philadelphia, PA, 1995.

2. Choi, J., Demmel, J., Dhillon, I., Dongarra, J., Ostrouchov, S., Petitet, A., Stanley, K., Walker, D., Whaley, R.C.: ScaLAPACK: A Portable Linear Algebra Library for Distributed Memory Computers - Design Issues and Performance. Technical Report UT CS-95-283, LAPACK Working Note #95, University of Tennessee, 1995.

3. Choi, J., Dongarra, J., Ostrouchov, S., Petitet, A., Walker, D., Whaley, R.C.: A Proposal for a Set of Parallel Basic Linear Algebra Subprograms. Technical Report UT CS-95-292, LAPACK Working Note #100, University of Tennessee, 1995.

4. Demmel, J., Dhillon, I., and Ren, H. On the correctness of parallel bisection in floating point, ETNA 3:116–149 (1995).

5. Dongarra, J., Du Croz, J., Duff, I., Hammarling, S.: A Set of Level 3 Basic Linear Algebra Subprograms. ACM Transactions on Mathematical Software, 16(1):1–17, 1990.

6. Dongarra, J., Du Croz, J., Hammarling, S., Hanson, R.: Algorithm 656: An extended Set of Basic Linear Algebra Subprograms: Model Implementation and Test Programs. ACM Transactions on Mathematical Software, 14(1):18–32, 1988.

7. Dongarra, J., Whaley, R.C.: A User's Guide to the BLACS v1.0. Technical Report UT CS-95-281, LAPACK Working Note #94, University of Tennessee, 1995.

8. Geist, A., Beguelin, A., Dongarra, J., Jiang, W., Manchek, R., V. Sunderam, V.: PVM: Parallel Virtual Machine. A User's Guide and Tutorial for Networked Parallel Computing. The MIT Press, Cambridge, Massachusetts, 1994.

9. Golub, G., and Van Loan, C. F.: Matrix Computations, Johns Hopkins University Press, Baltimore, MD, 2nd ed., 1989.

10. Gropp, W., Lusk, E. Skjellum, A.: Using MPI: Portable Programming with the Message-Passing Interface, MIT Press, Cambridge, MA, 1994.

11. IEEE. ANSI/IEEE Standard for Binary Floating Point Arithmetic: Std 754-1985, IEEE Press, New York, NY, 1985.

12. IEEE. ANSI/IEEE Standard for Radix Independent Floating Point Arithmetic: Std 854-1987, IEEE Press, New York, NY, 1987.

13. Lawson, C., Hanson, R., Kincaid, D., Krogh, F.: Basic Linear Algebra Subprograms for Fortran Usage. ACM Transactions on Mathematical Software, 5:308–323, 1979.

14. Message Passing Interface Forum. MPI: A Message Passing Interface Standard. International Journal of Supercomputer Applications and High Performance Computing, 8(3–4), 1994.

15. Snir, M., Otto, S. W., Huss-Lederman, S., Walker, D. W. and Dongarra, J.: MPI: The Complete Reference, MIT Press, Cambridge, MA, 1996.

16. SunSoft. The XDR Protocol Specification. Appendix A of "Network Interfaces Programmer's Guide", SunSoft, 1993.

Coupling the Advection and the Chemical Parts of Large Air Pollution Models

J. Brandt[1], I. Dimov[2], K. Georgiev[2], J. Wasniewski[3] and Z. Zlatev[1]

[1] National Environmental Research Institute,
Frederiksborgvej 399, P. O. Box 358, DK-4000 Roskilde, Denmark
e-mail: lujbr@sun4.dmu.dk, luzz@sun2.dmu.dk
[2] Central Laboratory for Parallel Information Processing, Bulgarian Academy of
Sciences, Acad. G. Bonchev str., Bl. 25-A, 1113 Sofia, Bulgaria;
e-mail: ivdimov@iscbg.acad.bg, georgiev@iscbg.acad.bg
[3] Danish Computing Centre for Research and Education (UNI-C), DTU, Bldg. 304,
DK-2800 Lyngby, Denmark;
e-mail: jerzy.wasniewski@unidhp1.uni-c.dk

Abstract. The discretization of long-range transport models leads to huge computational tasks. The advection (the transport due to the wind) and the chemistry are the most difficult parts of such a model. Normally splitting procedures are used and one tries to develop optimal methods for the advection part and for the chemistry part. Some results obtained in the attempts to design good sets of methods which work well for the coupled advection-chemistry sub-model will be presented. Runs on a Silicon Graphics POWER CHALLENGE computer indicate that the methods perform reasonably well and high speed-ups can be achieved with minimal extra efforts. However, more efforts are needed to get closer to the peak performance of this computer.

1 Mathematical description of an air pollution model

Concentrations and/or the depositions of various harmful air pollutants can be studied, [22], by using a system of partial differential equations (**PDE's**):

$$\frac{\partial c_s}{\partial t} = -\frac{\partial(uc_s)}{\partial x} - \frac{\partial(vc_s)}{\partial y} - \frac{\partial(wc_s)}{\partial z}$$

$$+\frac{\partial}{\partial x}\left(K_x\frac{\partial c_s}{\partial x}\right) + \frac{\partial}{\partial y}\left(K_y\frac{\partial c_s}{\partial y}\right) + \frac{\partial}{\partial z}\left(K_z\frac{\partial c_s}{\partial z}\right)$$

$$+E_s + Q_s(c_1, c_2, \ldots, c_q) - (\kappa_{1s} + \kappa_{2s})c_s, \quad s = 1, 2, \ldots, q. \tag{1}$$

The number of equations q is equal to the number of species that are involved in the model and has been varied from 10 to 168 in our studies. The quantities involved in the mathematical model are defined as follows: (i) the concentrations are denoted by c_s, (ii) u, v and w are wind velocities, (iii) K_x, K_y and K_z are diffusion coefficients, (iv) the emission sources are described by E_s, (v) κ_{1s}

and κ_{2s} are deposition coefficients, (vi) the chemical reactions are denoted by $Q_s(c_1, c_2, \ldots, c_q)$.

1.1 Splitting the large model to simpler sub-models

It is difficult to treat the system of **PDE's** (1) directly. This is the reason for using different kinds of splitting. A simple splitting procedure, based on ideas proposed in [13] and [14], can be defined, for $s = 1, 2, \ldots, q$, by five sub-models, representing respectively the horizontal advection, the horizontal diffusion, the chemistry (together with the emission terms), the deposition and the vertical exchange:

$$\frac{\partial c_s^{(1)}}{\partial t} = -\frac{\partial(uc_s^{(1)})}{\partial x} - \frac{\partial(vc_s^{(1)})}{\partial y} \tag{2}$$

$$\frac{\partial c_s^{(2)}}{\partial t} = \frac{\partial}{\partial x}\left(K_x \frac{\partial c_s^{(2)}}{\partial x}\right) + \frac{\partial}{\partial y}\left(K_y \frac{\partial c_s^{(2)}}{\partial y}\right) \tag{3}$$

$$\frac{dc_s^{(3)}}{dt} = E_s + Q_s(c_1^{(3)}, c_2^{(3)}, \ldots, c_q^{(3)}) \tag{4}$$

$$\frac{dc_s^{(4)}}{dt} = -(\kappa_{1s} + \kappa_{2s})c_s^{(4)} \tag{5}$$

$$\frac{\partial c_s^{(5)}}{\partial t} = -\frac{\partial(wc_s^{(5)})}{\partial z} + \frac{\partial}{\partial z}\left(K_z \frac{\partial c_s^{(5)}}{\partial z}\right) \tag{6}$$

1.2 Space discretization

If the model is split into sub-models as in the previous paragraph, then the discretization methods will lead to five ODE systems ($i = 1, 2, 3, 4, 5$):

$$\frac{dg^{(i)}}{dt} = f^{(i)}(t, g^{(i)}), \qquad g^{(i)} \in R^{N_x \times N_y \times N_z \times N_s}, \qquad f^{(i)} \in R^{N_x \times N_y \times N_z \times N_s}, \tag{7}$$

where N_x, N_y and N_z are the numbers of grid-points along the coordinate axes and $N_s = q$ is the number of chemical species. The functions $f^{(i)}$, $i = 1, 2, 3, 4, 5$, depend on the particular discretization methods used in the numerical treatment of the different sub-models.

2 Testing the performance of the advection-chemistry sub-models

The advection and the chemistry are the most time-consuming processes. Therefore the modellers must be very careful when numerical methods for the treatment of these two processes on high-speed computers are to be selected. It is important to design reliable test-problems and to use them in the selection of numerical methods.

It is easy to test the advection algorithms. One can construct test-examples with known analytical solutions and use them in the verification process, [22]. It is much more difficult to test the chemical algorithms. The commonly used procedure is to calculate a reference solution by running the chemical sub-model with a sufficiently small time-stepsize. The following simple test-example (proposed originally in [12]; see also [22]) can be used to test the accuracy of (a) the advection algorithms, (b) the chemical algorithm and, what is most important, (c) the coupling of advection with chemistry.

$$\frac{\partial c_s}{\partial t} = -(1-y)\frac{\partial c_s}{\partial x} - (x-1)\frac{\partial c_s}{\partial y} + Q_s(c_1, c_2, \ldots, c_q), \qquad (8)$$

$$s = 1, 2, \ldots, q, \qquad 0 \le x \le 2 \qquad and \qquad 0 \le y \le 2. \qquad (9)$$

3   Need for high speed computers

Assume that the full model from Section 1 is to be treated numerically. Then the size of any of the five ODE systems, which arise after the space discretization and the splitting procedure, is equal to the product of the number of the grid-points and the number of chemical species. It grows very quickly when the number of grid-points and/or the number of chemical species is increased; see Table 1.

Number of species	$(32 \times 32 \times 10)$	$(96 \times 96 \times 10)$	$(192 \times 192 \times 10)$
1	10240	92160	368640
2	20480	184320	737280
10	102400	921600	3686400
35	358400	3225600	12902400
56	573440	5160960	21381120
168	1720320	15482880	61931520

Table 1
The number of equations per system of ODE's that are to be treated at every time-step. The typical number of time-steps is 3456 (when meteorological data covering a period of one month + five days to start up the model is to be handled). The number of time-steps for the chemical sub-model is even larger, because smaller step-sizes have to be used in this sub-model.

In the special test-problem introduced in Section 2 by (8)-(9), the number of systems of ODE's is reduced to two, but the number of equations per system remains the same. Such large problems can be solved **only** if new and modern high-speed computers are used. Moreover, it is necessary to select the right numerical algorithms (which are most suitable for the high speed computers available) and to perform the programming work very carefully in order to exploit fully the great potential power of the vector and/or parallel computers; see [22]. Some results obtained on the Silicon Graphics Power Challenge will be presented in this paper.

4 Numerical methods

The mathematical terms in the five sub-models have different properties. Therefore it is natural to apply different numerical methods in the treatment of the different sub-models in an attempt to optimize globally the computational process. The problem of choosing numerical methods for the two most difficult sub-models will be discussed in this section.

4.1 Numerical treatment of the horizontal advection

The horizontal advection causes a lot of difficulties during the numerical treatment. Different methods have been proposed: pseudospectral algorithms, finite differences, finite elements, special algorithms producing non-negative solutions, semi-Lagrangian algorithms. It may be appropriate to try wavelets.

The pseudospectral algorithm is often based on the use of trigonometric interpolation to calculate approximations to the spatial derivatives, but other formulations of this method can also be applied ([8]. The algorithm is accurate and a good implementation is not very expensive. However, it requires periodic boundary conditions and extra efforts are needed in order to deal with this problems. A pseudospectral algorithm, based on the use of trigonometric interpolants which are truncated Fourier series, is used in the Danish Eulerian Model [22].

The **finite differences** are still very popular. It is easy to implement such methods. However, the horizontal advection deals very often with sharp gradients. This causes problems for the finite difference methods. One must apply high-order differences to resolve the gradients. The simpler low-order differences (used in the up-wind methods) produce smooth solutions, but also introduce a lot of artificial diffusion.

Finite elements algorithms are becoming more and more popular. Different finite elements can be used in the different parts of the space domain, and this is a great advantage of the finite elements methods. Smaller and/or more accurate finite elements could be used in difficult sub-domain. However, it is rather difficult to implement such a method in conjunction with the other physical processes. The fact is that the finite element algorithms in air pollution models are still used with the same finite elements in the whole region (an exception being the use of finite elements to handle the vertical exchange; see below). Linear one-dimensional finite elements are popular. Such methods have been used in some of our experiments as well as in the model described in [2].

Algorithms producing non-negative solutions are also very popular. The previous three groups of methods could produce negative concentrations, which leads to a disaster when the chemical reactions are handled. There are different algorithms which produce non-negative concentrations. Some experiments with the methods developed by Bott [1] and Holm [11] have been carried out. These methods produce smooth concentration fields with no negative concentrations, but they are in general more expensive than many of the methods from the previous three groups.

The semi-Lagrangian methods are accurate and have good stability properties. They have been used in several air pollution models. We have experimented with a semi-Lagrangian scheme, which has been more expensive than the other methods we used. However, the implementation of this scheme can probably be improved considerably.

We have not carried out experiments with methods based on the use of **wavelets**. It would be interesting to see how such methods will perform.

Many of the algorithms listed above may produce **negative** concentrations If this is the case, then some kind of smoothing has to be applied before the treatment of the chemical reactions. It is difficult to justify theoretically the application of a smoothing procedure. The selective filter proposed by Forester [7] seems to work rather well in practice. This filter performs smoothing only in regions around detected oscillations. Therefore, the high concentrations are normally not affected by the use of this filter.

4.2 Numerical treatment of the chemical transformations

The quasi-steady-state-approximation, QSSA, is commonly used in air pollution modelling since 1978, [10]. Recently updated versions of this method have been used, [21], [3]. This method may have difficulties in the periods around sun-rises and sun-sets when the photochemical reactions are activated or disactivated, but it is very cheap computationally.

Classical methods (mainly the Backward Euler Formula, the Trapczoidal Rule and some Runge-Kutta methods) have also been used; see [4], [16], [17], [19]. These methods produce more accurate results than the QSSA, but they are more expensive.

Recently, the classical methods have been used with different kinds of partitioning ([9], [20]). The use of partitioning improves the performance of the methods without affecting too much the accuracy. This is a promising approach, but more experiments are needed in order to justify its usefulness in the case where both advection and chemistry take place.

Extrapolation techniques are very popular in other fields of science and engineering. It may be useful to try such techniques also in the treatment of the chemical parts of large air pollution problems; these methods have been used for solving some other chemical problems in [6].

The chemical sub-models are sometimes described by a system of differential-algebraic equations. Techniques similar to those used in [18] can be applied in this situation.

The QSSA is still used in the Danish Eulerian Model ([22]. This method is also used in the experiments discussed in the next section.

5 Testing the accuracy

5.1 Testing the accuracy of some advection schemes

If the chemical terms are removed and if it is assumed that $q = 1$, then the test-example defined by (8)-(9) is reduced to the classical Molenkampf-Crowley rotation test ([15], [5]). The wind velocities are defined in a special way, so that the wind trajectories are concentric circles, whose common centre is the point with coordinates $x = 1$, $y = 1$. Moreover the motion is with a constant angular velocity. If we have some given distribution of the concentrations in the starting time t_{start}, then at some time t_{end} a full rotation of the concentration field around the point $x = 1$, $y = 1$ will be accomplished and, thus, at t_{end} the concentrations must be distributed in the same way as at the beginning (at t_{start}). Results, obtained after one rotation with (i) the up-wind method, (ii) the finite elements algorithm, (iii) the Bott's scheme and (iv) the scheme proposed by Holm, are given in Fig. 1. These results should be compared with the upper left plot in Fig. 2 (in the ideal case, where there are no errors, the results obtained by the numerical methods should be identical with the initial distribution). It is seen that the Holm's scheme is the best one among these methods However, it is also the most time-consuming. E. Holm has improved the performance of his scheme and now it is much more efficient (private communication). The Bott's scheme gives also good overall accuracy, however, this scheme seems to have some problems on the base of the cone. The finite elements algorithm produces some oscillations, but it is considerably cheaper than the other two algorithms. Finally, the simple and cheap up-wind method has problems with preserving the high concentrations at the top of the cone. The pseudospectral algorithm gives much more accurate results for this test (see the upper right plot in Fig. 2), However, it requires periodic boundary conditions. The conclusion is that all methods have both positive and negative properties and no method is perfect. The problem of finding an optimal advection method is still open, The solution of this problem depends both on the computer that is to be used and on the particular implementation of the numerical method chosen in the air pollution model under consideration.

5.2 Testing the accuracy of the QSSA

Consider the case where only the chemical terms are retained in (8). In this case the chemistry is carried out independently for each grid-point. Moreover, at many grid-points the calculations are identical. Therefore, only a few points are important, and the chemical transformations in these points can be performed (with a small time-step) in a very accurate manner. After that the whole solution field at the end-point of the time interval can be reproduced. This allows us to check the accuracy of the results obtained by the chemical algorithm chosen (at least at the end-point of the time-interval). This procedure has been used to determine a time-stepsize by which good results can be produced by using QSSA. The time-stepsize so found was 30 seconds (the same time-stepsize has been recommended in [10]).

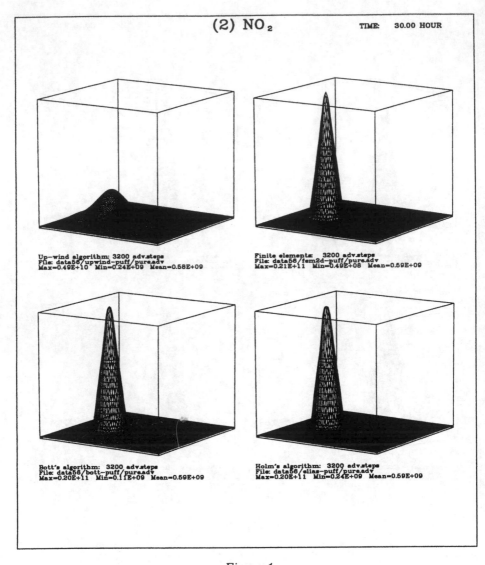

Figure 1
Pure advection test: (a) the up-wind method (upper, left), (b) finite elements (upper, right), (c) Bott's algorithm (lower, left) and (d) Holm's algorithm (lower, right).

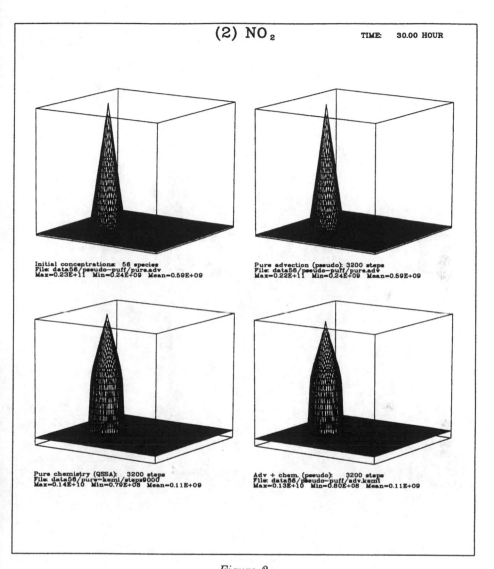

Figure 2
Nitrogen di-oxide concentrations at: (a) the beginning (upper, left), (b) the end
of the pure advection test (upper, right), (c) the end of the pure chemistry test
(lower, left) and (d) the end of the advection-chemistry test (lower, right).

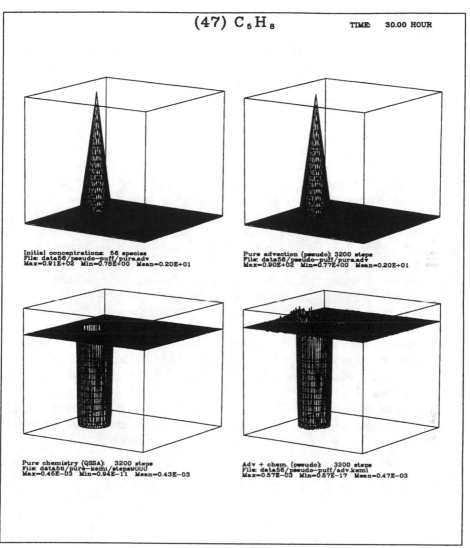

Figure 3
Isoprene concentrations at: (a) the beginning (upper, left), (b) the end of the
pure advection test (upper, right), (c) the end of the pure chemistry test (lower,
left) and (d) the end of the advection-chemistry test (lower, right).

5.3 Coupling advection with chemistry

Consider now the coupling of the chemical sub-model with the other sub-models in a large air pollution model. To simplify the discussion only the coupling of chemistry and advection will be considered. Similar discussion can be carried out when the other processes (the diffusion, deposition and vertical exchange processes) are added to the advection-chemistry combination.

Assume that both the advection algorithm and the chemistry algorithm have been carefully tested. Assume also that the tests indicate that both algorithms perform satisfactorily well. Then the big question is: **Will also the combination of these two algorithms perform well?** It will be shown that, unfortunately, the answer to this question is in some cases negative.

Consider Fig. 2 and Fig. 3. The upper plot on the left hand side in each of these two figures represents the initial distribution of the concentrations. The upper plot of the right hand side represents the distribution of the concentrations after a full rotation in the case when a pure advection test is carried out. **In the ideal case, when there are no errors, the upper two plots must be identical.** The lower plot on the left hand side represents the case where only a pure chemical test is run over the time-interval needed to perform a full rotation. The lower plot on the right hand side represents the distribution of the concentrations after one rotation for the most general case that can be treated by this test; the case where both the transport and the chemical reactions are activated. **In the ideal case, when there are no errors, the lower two plots must be identical.**

The results shown on Fig. 2 are very good. However, the results shown on Fig. 3 indicate that problems could appear. In this particular case the difficulties are due to the sharp gradients caused by the chemical reactions.

6 Performance on Silicon Graphics POWER CHALLENGE

The Silicon Graphics POWER CHALLENGE is a parallel machine with shared memory. The number of processors of the particular machine used by us (at UNI-C, Lyngby, Denmark) is 16, but a special permission is needed when more than 8 processors are to be used. The test-problem (8) has been used in these experiments. It has been discretized on a (96×96) grid. A chemical scheme containing 56 species has been used. Results obtained by running the test-problem on 1 and 8 processors are shown in Table 2.

In **the advection part** a parallel task is an advection time-step for a given chemical species. In the case where 8 processors are used, each processor receives 7 tasks per time-step. There is no communication and the loading balance is perfect. Therefore the speed-up (the ratio of the computing time needed on 1 processor and the computing time needed on 8 processors) is rather good.

In the **chemical part** a parallel task is the performance of the chemical reactions (at a given time-step) at a given grid-point. In the case where 8 processors are used, each processor receives 1152 tasks per time-step. There is no communication and the loading balance is perfect. Therefore also here the speed-up is rather good.

It is clear that if both the advection part and the chemical part parallelize well, then good results should be expected for the **coupled advection-chemistry sub-model**. The results in Table 2 show that the speed-up for the coupled sub-model is also rather good.

Characteristic measured	Advection	Chemistry	Adv. + Chem.
Comp. time on 1 processor	1999.2	4501.0	6027.6
Comp. time on 8 processors	387.4	696.3	952.4
Speed-up	5.2	6.5	6.3
Efficiency (speed-up)	65%	81%	79%
MFLOPS on 8 processors	197.9	172.5	215.1
Efficiency (peak-performance)	10%	8%	10%

Table 2

Results obtained in the runs of the test-problems with 56 chemical species on a (96×96) grid when a CHALLENGE POWER computer is used.

The results in Table 2 indicate that it is necessary to work on the code in order to improve the efficiency with regard to the peak-performance. It should be mentioned here that the efficiency achieved in the runs of the same test-problem on CRAY C92A is about 50%. If the cache memory available on POWER CHALLENGE is exploited as much as possible, then it may be possible to achieve the same efficiency with regard to the peak-performance as that achieved on CRAY C92A.

Acknowledgements

This research has been supported by grants from the Bulgarian Ministry of Education (I-505/95), the Danish Natural Sciences Research Councill (SNF), the Danish Academy of Research and NATO (Grant OUTR.CRG.960312).

References

1. Bott, A., "A positive definite advection scheme obtained by nonlinear renormalization of the advective fluxes", Mon. Weather Rev., 117(1989), 1006-1015.
2. Brandt, J., Mikkelsen, T., Thykier-Nielsen, S. and Zlatev, Z., "Using a combination of two models in a tracer model", Mathematical and Computer Modelling, Vol 23 No. 10 (1996), 99-115.
3. Brandt, J., Wasniewski, J. and Zlatev, Z., "Handling the chemical part in large air pollution models", Appl. Math. and Comp. Sci., 6 (1996), 101-121.
4. Chock, D. P., Winkler, S. L. and Sun, P, "Comparison of stiff chemistry solvers for air quality models", Environ. Sci. Technol., 28 (1994), 1882-1892.

5. Crowley, W. P., "Numerical advection experiments", Mon. Weath. Rev., 96 (1968), 1-11.
6. Deuflhard, P., Nowak, U. and Wulkow, M., "Recent development in chemical computing", Computers Chem. Engng., 14 (1990), 1249-1258.
7. Forester, C. K., "Higher order monotonic convective difference schemes", J. Comput. Phys., 23(1977), 1-22.
8. Fornberg, B., "A practical guide to pseudospectral methods", Cambridge Monographs on Applied and Computational Mathematics, Vol. 1, Cambridge University Press, Cambridge-New York-Melburne, 1996.
9. Hertel, O., Berkowicz, R., Christensen, J. and Hov, Ø., "Test of two numerical schemes for use in atmospheric transport-chemistry models", Atmos. Environ., 27A (1993), 2591-2611.
10. Hesstvedt, E., Hov, Ø. and Isaksen, I. A., "Quasi-steady-state approximations in air pollution modelling: comparison of two numerical schemes for oxidant prediction", Internat. J. Chem. Kinetics, 10 (1978), 971-994.
11. Holm, E., "High-order numerical methods for advection in atmospheric models". PhD Thesis, Department of Meteorology, Stockholm University, 1993.
12. Hov, Ø., Zlatev, Z., Berkowicz, R., Eliassen, A. and Prahm, L. P., "Comparison of numerical techniques for use in air pollution models with non-linear chemical reactions", Atmos. Environ., 23 (1988), 967-983.
13. Marchuk, G. I., "Mathematical modeling for the problem of the environment", Studies in Mathematics and Applications, No. 16, North-Holland, Amsterdam, 1985.
14. McRae G. J., Goodin, W. R. and Seinfeld, J. H., "Numerical solution of the atmospheric diffusion equations for chemically reacting flows", J. Comp. Phys., 45 (1984), 1-42.
15. Molenkampf, C. R., "Accuracy of finite-difference methods applied to the advection equation", J. Appl. Meteor., 7 (1968), 160-167.
16. Odman, M. T., Kumar, N. and Russell, A. G., "A comparison of fast chemical kinetic solvers for air quality modeling", Atmos. Environ., 26A (1992), 1783-1789.
17. Peters, L. K., Berkowitz, C. M., Carmichael, G. R., Easter, R. C., Fairweather, G., Ghan, S. J., Hales, J. M., Leung, L. R., Pennell, W. R., Potra, F. A., Saylor, R. D. and Tsang, T. T., "The current state and future direction of Eulerian models in simulating the tropospherical chemistry and transport of trace species: A review", Atmos. Environ., 29 (1995), 189-221
18. Petzold, L. R., "Order results for implicit Runge-Kutta methods applied to differential-algebraic systems", SIAM J. Numer. Anal., 23 (1986), 837-852.
19. Shieh, D. S., Chang, Y. and G. R. Carmichael, G. R., "The evaluation of numerical techniques for solution of stiff ordinary differential equations arising from chemical kinetic problems", Environ. Software, 3 (1988), 28-38.
20. Skelboe, S. and Zlatev, Z., "Using partitioning in the treatment of the chemical part of air pollution models", Springer, Berlin, to appear.
21. Verwer, J. G. and Simpson, D., "Explicit methods for stiff ODE's from atmospheric chemistry", Appl. Numer. Math., to appear.
22. Zlatev, Z., "Computer treatment of large air pollution models", Kluwer Academic Publishers, Dordrecht-Boston-London, 1995.

Advanced Optimizations for Parallel Irregular Out-of-Core Programs*

Peter Brezany and Minh Dang

Institute for Software Technology and Parallel Systems
University of Vienna, Liechtensteinstrasse 22, A-1090 Vienna, Austria
E-mail: {brezany,dang}@par.univie.ac.at

Abstract. Large scale irregular applications involve data arrays and other data structures that are too large to fit in main memory and hense reside on disks. This paper presents a method for implementing this kind of applications on distributed-memory systems. The method is based on a runtime system that has been built on top of the CHAOS library.

1 Introduction

Parallelizing irregular codes for distributed-memory systems is a challenging problem and is of growing importance. In such codes, access patterns to major data arrays are only known at runtime, which requires preprocessing in order to determine the data access patterns and consequently, to find what data must be communicated and where it is located. One strategy used for parallelization irregular codes transforms each irregular data-parallel loop into three phases, called he *work distributor*, the *inspector*, and the *executor* [6]. The work distributor determines how to spread the work (iterations) among the available processors. The inspector analyzes the communication patterns of the loop, computes the description of the communication, and derives translation functions between global and local accesses, while the executor performs the actual communication and executes the loop iterations. All the phases are supported by an appropriate runtime library. For example, the CHAOS [5] library is a well-known system which supports the handling of irregular computations on massively parallel systems.

Large scale irregular applications involve large arrays and other data structures. Runtime preprocessing provided for these applications results in construction of additional large data structures which increase the memory usage of the program substantially. Consequently, a parallel program may quickly run out of memory. Therefore, some data structures must be stored on disks; such applications are called *out-of-core (OOC)* applications. The performance of an OOC program strongly depends on how fast the processors, the program runs on, can access data from disks. However, it is difficult for the programmer to manage the memory-disk interface explicitly, and moreover, programmer-inserted

---

* The work described in this paper is being carried out as part of the research project "Language, Compiler and Advanced Data Structure Support for Parallel I/O Operations" supported by the Austrian Research Foundation (FWF Grant P11006-MAT).

I/O significantly reduces portability. Therefore, the development of appropriate parallelization methods for OOC irregular programs that are based on portable libraries is an important research issue.

This paper presents a method for parallelization of OOC irregular problems that is based on a new runtime system that is built on top of the CHAOS library. The runtime system provides support for buffering, caching, and prefetching of irregular array sections from disks. Its primitives can either be used together with a compiler to translate irregular data parallel OOC programs written in a language like HPF, or used directly by application programmers. The parallelization method and performance are illustrated on a template taken from a real irregular application.

```
INTEGER, PARAMETER :: NNINTC = ..., NNCELL = ...
INTEGER, PARAMETER :: icoef = ..., imesh = ...
DOUBLE PRECISION, DIMENSION (NNINTC) :: D2, BP, BS, BW,
                                          BL, BN, BE, BH
INTEGER LC(NNINTC,6)
DOUBLE PRECISION D1(NNCELL)
OPEN (icoef, FILE = 'coefficients.dat', STATUS = 'OLD')
OPEN (imesh, FILE = 'mesh.dat', STATUS = 'OLD')
READ (icoef) BP, BS, BW, BL, BN, BE, BH
READ (imesh) LC
      ...
L:  DO  nc = 1, NNINTC
        D2(nc) = BP(nc) * D1(nc) − BS(nc)  * D1(LC(nc,1))
               − BW(nc) * D1(LC(nc,4)) − BL(nc) * D1(LC(nc,5))
               − BN(nc)  * D1(LC(nc,3)) − BE(nc) = D1(LC(nc,2))   &
               − BH(nc)  * D1(LC(nc,6))
    END DO
      ...
```

Fig. 1. Kernel loop of the GCCG solver

The code template in Fig. 1 will be used as a running example throughout the paper. This example is derived from the FIRE benchmark solver GCCG [1, 3] whereby loop L represents the kernel loop of this solver.

2 Execution Model for Irregular OOC Programs

2.1 Basic Approach

The simplest way to view the OOC execution model is to assume that all OOC arrays are stored in an *array database (AD)*, and the application programmer or the parallelizing compiler only see the interface to the procedures of a runtime system that operates on AD. All computations are performed on the data in processors' local memories. During the course of the computation on a processor, sections of the operand arrays are fetched from AD into the local memory of the processor, the new values are computed and the sections updated are stored back into AD if necessary. On each processor, the computation is performed in phases where each phase operates on only so large parts of the arrays which can fit into the local memory. The portion of the array which is in processor's memory is called the *In-Core Local Array Portion (ICLAP)* [2]. The computation performed in a phase corresponds to the execution of a *tile* of loop iterations. To determine the phases, work of the loop can be distributed among the processors, for example, in the following way: the global iteration space is first partitioned into a set of tiles and a number of tiles is then assigned to each processor.

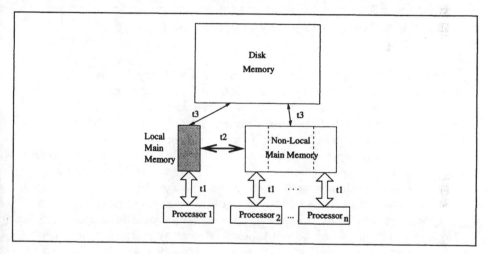

Fig. 2. Memory Hierarchy.

2.2 Optimizations

From the point of view of data access costs incurred during the execution of an OOC program, a *three level hierarchy memory model* can be considered which is depicted in Fig.2. It is assumed that processors reference their *local primary memories* in unit time, $t_1 = 1$. The model further stipulates that a processor's reference to data that is not in its own internal memory requires a latency of t_2 time units fulfilled if the data is currently in memory of a remote processor and t_3 time units if the data is in disk memory. It is assumed that $t_3 \gg t_2 \gg t_1$. These memory hierarchy features force programmers and compiler developers to attend to the matter of: (1) locality and data reusing, i.e. how to arrange computations so references to data in local internal memory are maximized, and the time-consuming references to data on disks are minimized, and (2) hiding communication and I/O latency by computations.

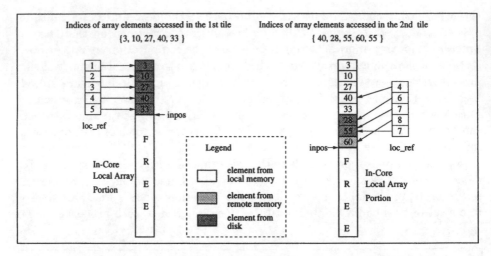

Fig. 3. Caching and Addressing Array Elements in the ICLAP.

Optimizing the transfer between memory hierarchy levels. The execution model discussed in Subsection 2.1 does not consider reusing of the data already fetched to the local or remote memory instead of reading them again from disk or communicate between processors. Irregular accesses to arrays may exhibit some sort of *temporal locality*; if a data element is referenced in one tile, it will tend to be referenced again soon. It is possible to utillize this feature, when optimizing the OOC execution model. To achieve this goal, the storage allocated on individual processors to *ICLAP* is managed like a *cache*. In Fig. 3, we show the case study of managing this cache for a simple example.

Let the length of tiles of a one-dimensional iteration space be 5, and {3,10,27, 40,33} and {40,28,55,60,55} be the index sets of array sections of an array A referred during the course of the first two tiles on a processor p. The first set of references results in a *total miss* - the elements with indices {3,10,27,40,33} are neither in the local nor in the remote memory; they are fetched from disk. Their local indices on processor p are put in the index conversion table *loc_ref*. Accessing the elements with indices {40,28,55,60,55} in the second tile results in a *hit* for the element with index 40, a *partial hit* for the element with index 60 which will be fetched from memory of a remote processor and a *total miss* for the elements with indices 28 and 55 which will be fetched from disk. Note that the element with index 55 is accessed twice but only one copy is fetched from disk. The table *loc_ref* is updated. The variable *inpos* points at the header of a list that registers the free positions in the cache.

Hiding I/O latency. Overlapping memory accesses with computations is a standard technique for improving performance on modern architectures which have deep memory hierarchies. To overlap time expensive I/O accesses, we distinguish between two types of processes: *application processes* and *I/O processes*. According to the SPMD programming paradigm each computing processor is executing the same application process but on different parts of the data space. The application processes implement the computation specified by the original program. The I/O processes run independently on all or a number of dedicated nodes and perform the data requests of the application processes. The application and I/O processes are loosely coupled; all synchronization necessary between them at runtime is achieved through message passing. The I/O process simulates the control structure of the application process to be able to prefetch in advance the data sections required. The I/O process can either be placed on the same computing node as the appropriate application process that is served by it, or the I/O process may be placed on an I/O node; it depends on the underlying hardware architecture.

Data sieving. A simple way of reading an irregular array section is to read each its element individually (it is called the *direct read method*) which issues into large number of I/O calls and low granularity of data transfer. It is possible to use a much more efficient method called *data sieving* [4] to read a regular section of the array that includes the original irregular section. The required data elements are then filtered out by the I/O system and passed to the application process.

3 Implementation

```
INTEGER, PARAMETER :: NUMBER_OF_TILES = ...
INTEGER, PARAMETER :: BUFFSIZE = ...,block = 1
INTEGER, DIMENSION(NUMBER_OF_TILES) : tilelwbs, tileupbs
INTEGER    tt_D1, tile, max_size, tile_size, elmondisk, n_disk,
             inpos,inposold, sched
TYPE (DIST_DESC) dd_D1
DOUBLE PRECISION, DIMENSION(:), ALLOCATABLE :: D1, D2, BP,
                                    BS, BE, BN, BW, BL, BH
INTEGER, DIMENSION(:,:), ALLOCATABLE :: LC
INTEGER, DIMENSION(:), ALLOCATABLE :: glob_ref, loc_ref, loc, proc,
                    disk, foffsets, permutation, new_indices
WD:CALL wd_block_tiles(1,NNINTC,NUMBER_OF_TILES,tilelbs,tileupbs,max_size)
O1: CALL io_wd_block (NUMBER_OF_TILES, tilelbs, tileupbs, max_size)
    ALLOCATE (D1(max_size), BP(max_size), BS(max_size), BE(max_size),&
            BW(max_size), BL(max_size), BH(max_size), LC(max_size,6),&
            BN(max_size),glob_ref(max_size*7), loc_ref(max_size*7),&
            loc(max_size*7), proc(max_size*7), disk(max_size*7),&
            foffsets(max_size*7), new_indices(max_size*7),&
            permutation(max_size*7),D2(max_size+BUFFSIZE))
T1: tt_D1 = build_iotranslation_table(block, dd_D1)
```

Fig. 4. Out-of-Core Form of the GCCG code (Part 1).

```
         INTEGER, PARAMETER :: NUMBER_OF_TILES = ...; inpos = 1
LT:  DO tile = 1, NUMBER_OF_TILES
I1:      CALL io_get_regular1 (LC); k = 1; tile_size = tileupbs(tile) - tilelwbs(tile) + 1
G1:      DO i = 1, tile_size
G2:          glob_ref(k) = local_to_global(i); glob_ref(k+1) = LC(i,1);
G3:          glob_ref(k+2) = LC(i,2); glob_ref(k+3) = LC(i,3);
G4:          glob_ref(k+4) = LC(i,4); glob_ref(k+5) = LC(i,5);
G5:          glob_ref(k+6) = LC(i,6); k = k + 7
G6:      END DO
D1:      CALL io_dereference (tt_D1, glob_ref, loc, proc, disk, k-1, perm, elmondisk)
DS:      CALL disk_schedule(glob_ref,disk,elmondisk,foffsets,new_indices,n_disk)
O2:      CALL io_wantlist(foffsets, n_disk)
I2:      CALL io_get_regular7 (BP, BS, BE, BN, BW, BL, BH)
M1:      CALL mem_schedule (sched,glob_ref(elmondisk+1),proc,loc,k-elmondisk,&
                             inpos+n_disk, new_indices(n_disk+1),nnonlocal)
         inposold = inpos
         IF (iscachefull(inpos, max_size+BUFFSIZE)) THEN
DE1:         inpos = 1; CALL free_iotranslation_table(tt_D1)
         ELSE
R1:          CALL remap_iotran_table(tt_D1,new_indices,nnonlocal+n_disk,inpos)
R2:          inpos = inpos + nnonlocal + n_disk
         END IF
MG:      CALL gather(sched,D1(inposold+n_disk+1),D1); CALL free_sched(sched)
PE:      loc_ref(permutation(1:7*tile_size)) = glob_ref(1:7*tile_size)
I3:      IF (elmondisk > 0) THEN CALL io_get_irregular (D1(inposold)) END IF
         k = 1
E1:      DO i = 1, tile_size
E2:          D2(i) = BP(i)*D1(loc_ref(k))-BS(i)*D1(LC(loc_ref(k+1),1)) &
E3:              - BW(i)*D1(LC(loc_ref(k+2),4))-BL(i)*D1(LC(loc_ref(k+3),5))&
E4:              - BN(i)*D1(LC(loc_ref(k+4),3))-BE(i)*D1(LC(loc_ref(k+5),2))&
E5:              - BH(i)*D1(LC(loc_ref(k+5),6)); k = k + 7
E6:      END DO
O3:      CALL io_write(D2,tile_size)
     END DO
```

Fig. 5. Out-of-Core Form of the GCCG code (Part 2).

The most of the complexity of the implementation of an irregular OOC program is in the runtime system. Our runtime system is built on top of the CHAOS library. The Fortran code which specifies the application process is introduced in Fig. 4 and 5.

The *work distributor* is realized by the procedure *wd_block_tiles* (line *WD* of Fig.4). It accepts the lower and upper bounds of the loop iteration space and the number of required tiles as input parameters. It returns the description of the tiles (*tilelwbs* and *tileupbs*) to be executed on the calling processor and their maximum size. The tile description is then sent to the I/O process (line *O1*) which can start the pre-fetching of the sections of *BP, BS, BE, BN, BW, BL,* and *BH* that will be needed in the first tile.

The procedure *build_iotranslation_table* (line *T1*) builds a lookup table called the *translation table* which lists for each array element its address on the disk and the processor number and memory offset, if the element is in main memory.

The body of the *DO* loop *LT* from Fig. 5 represents the *inspector* and *executor* for an individual tile. After receiving the appropriate section of indirection array *LC* for a tile of iterations (line *I1*) from the I/O process, the global reference list *glob_ref* is built (lines *G1-G6*). This list is passed to the procedure *io_dereference* (line *D1*) which determines the memory addresses (owner processors and memory offsets) for the elements that are already in memory, and disk offsets for the elements that need to be fetched from disk. This information is returned by means of arrays *proc*, *local*, and *disk*. The number of disk element references (*elmondisk*) is computed by this procedure as well. The procedure sorts *glob_ref*, *proc*, *local*, and *disk* in such a way that the items corresponding to disk references precede the items that correspond to memory references. The original order can be established by the permutation array *perm*. Two data structures are then computed that describe the required I/O and communication: the I/O schedule (procedure *disk_schedule* at line *DS*) and communication schedule (procedure *mem_schedule* at line *M1*). Both procedures use a hash table to remove any duplicated references to non-local elements so that only a single copy for each non-local element is fetched from disk or transmitted from another processor. These procedures also perform the global to local index conversion. The incoming data elements are stored in the buffer; the first part of buffer is filled by the elements from disk and the second one by the elements from other processors. This buffer immediately follows the portion of memory allocated for local data (a similar approach was outlined in [4]). Moreover, the global indices of incoming elements are stored in *new_index* that is later used for updating the translation table. The procedure *disk_schedule* returns the array (*foffsets*) of offsets of elements that need to be fetched from disk. This array is sent to the I/O process (line *O2*) which can now start the appropriate I/O operations. The procedure *mem_schedule* returns the communication schedule (*sched*) that is used by the procedure *gather* (line *MG*) to gather the requested non-local elements from other processors.

The variable *inpos* points at the first free position in the cache. In the first implementation, the cache is reinitiated if it is full (line *DE1*). The procedure *remap_iotran_table* (line *R1*) updates the entries of translation table for the elements that will be communicated or transferred from disk.

After the requested disk-elements have been received from the I/O process (line *I3*), the tile iterations are carried out by the executor (lines *E1-E6*).

4 Preliminary Performance Results

The performance of the OOC GCCG solver on the Intel Paragon System for different number of processors and tiles is given in Tab.1. We compare the performance of the implementation which includes overlapping I/O with communication and computation with the solution in which the application process is blocked while waiting on I/O. Further, we show the significant impact of data sieving on the performance.

Direct Read Method							
Number of Procs	2	2	4	4	4	8	8	16
Number of Tiles	4	8	2	4	8	2	4	2
non-Overlapping	343.2	354.6	374.1	479.1	351.3	354.3	340.5	425.1
Overlapping	362.4	388.1	311.8	389.5	333.2	334.5	328.3	414.5
Using Data Sieving							
Number of Procs	2	2	4	4	4	8	8	16
Number of Tiles	4	8	2	4	8	2	4	2
non-Overlapping	74.1	41.1	42.5	28.2	30.9	21.8	27.1	32.7
Overlapping	79.4	42.8	44.9	27.3	27.6	22.9	25.6	31.2

Table 1. Performance of OOC GCCG Solver (time in seconds).

5 Conclusions

The approach presented in this paper is a promising technique for the parallelization of irregular programs that operate on huge data structures. The performance results show that overlapping I/O accesses with communication and computation and applying data sieving increase the performance. Applying cache mechanisms enables reusing of data and consequently decreases the number of transfers between memory hierarchies. In our future research, we are going to investigate the impact of various caching mechanisms on the overall performance.

References

1. G. Bachler, R. Greimel. *Parallel CFD in the Industrial Environment.* Unicom Seminars, London, 1994.
2. R.R. Bordawekar, A.N. Choudhary, K. Kennedy, C. Koelbel, M. Paleczny. *A Model and Compilation Strategy for Out-of-Core Data Parallel Programs.*
3. P. Brezany, V. Sipkova, B. Chapman, R. Greimel, *Automatic Parallelization of the AVL FIRE Benchmark for a Distributed-Memory System*, PARA95.
4. A. Choudhary et al. *PASSION: Parallel And Scalable Software for Input-Output.* Technical Report CRPC–TR94483-S, Syracuse University, Syracuse, NY, 1994.
5. R. Ponnusamy et al. *A manual for the CHAOS runtime library.* Technical report, University of Maryland, May 1994.
6. J. Saltz, K. Crowley, R. Mirchandaney, , and H. Berryman. *Run-time scheduling and execution of loops on message passing machines.* Journal of Parallel and Distributed Computing, 8(2):303–312, 1990.

A Software Architecture for Massively Parallel Input-Output*

Peter Brezany[a], Thomas A. Mueck[b] and Erich Schikuta[b]

University of Vienna
[a]Institute for Software Technology and Parallel Systems
Liechtensteinstr. 22, A-1092 Vienna, E-mail: brezany@par.univie.ac.at
[b]Department of Data Engineering, Rathausstrasse 19/4
A-1010 Vienna, Austria, E-mail: {mueck,schiki}@ifs.univie.ac.at

Abstract. For an increasing number of data intensive scientific applications, parallel I/O concepts are a major performance issue. Tackling this issue, we provide an outline of an input/output system designed for highly efficient, scalable and conveniently usable parallel I/O on distributed memory systems. The main focus of this paper is the parallel I/O runtime system support provided for software-generated programs produced by parallelizing compilers in the context of High Performance FORTRAN efforts. Specifically, our design is presented in the context of the Vienna Fortran Compilation System.

1 Introduction

The main issue for I/O subsystems in supercomputing environments is to feed arrays of computing processors with huge amounts of raw data (nowadays typically in the Tbyte range) in such a way that neither the processor execution time nor the program turnaround time increase significantly.

On the hardware side, disk arrays (RAID), HIPPI interfaces, data vaults, etc. are the enabling technology approaches towards high-performance Input/Output systems. However, on the software side, current I/O subsystems suffer from a lack of a high-level programming support. This issue (including language, compiler and runtime system aspects) is currently a hot research topic at a variety of scientific and industrial institutions.

In this paper, we investigate the I/O problem from a runtime system support perspective. We focus on the design of an advanced parallel I/O support, called VIPIOS (VIenna Parallel I/O System), to be targeted by language compilers supporting the same programming model like High Performance Fortran (HPF) [12]. The VIPIOS design is partly influenced by the concepts of parallel database technology.

Section 2 introduces a formal model, which forms a theoretical framework on which the VIPIOS design is based. Here, we assume that the reader is familiar

* The work described in this paper is being carried out as part of the research project "Language, Compiler and Advanced Data Structure Support for Parallel I/O Operations" supported by the Austrian Research Foundation (FWF Grant P11006-MAT).

with the basic concepts of HPF. The VIPIOS architecture and the interface
to a particular HPF compilation system, i.e., the Vienna Fortran Compilation
System [2], are addressed in Section 3 which is the core of the paper. Section 4
discusses related work and Section 5 concludes with a brief summary.

2 Data Mapping Model

The model introduced in this section describes the mapping of the problem
specific data space starting from the application program data structures down
to the physical layout on disk across several intermediate representation levels.

Parallelization of a program Q is guided by a mapping of the *data space* of
Q to the processors and disks. The data space $\mathcal{A}$ is the set of declared arrays of
Q.

Definition 1 Index Domain

Let $A \in \mathcal{A}$ denote an arbitrary array. The index domain *of A is denoted by
I^A. The* extent *of A, i.e. the number of array elements is denoted by ext^A.*

The index domain of an array is usually determined through its declaration.

In a lot of languages, only array elements or whole arrays may be referenced.
On the other hand, some languages like Fortran 90 provide features for computing subarrays. A subarray can either be characterized as an array element or
an array section.

Definition 2 Index Domain of an Array Section[2]

*Let us consider an array A of rank n with index domain $I^A = [l_1 : u_1, \ldots, l_n : u_n]$
where the form $[l_i : u_i]$ denotes the sequence of numbers $(l_i, l_i + 1, \ldots, u_i)$. The
index domain of an array section $A' = A(ss_1, \ldots, ss_n)$ is of rank n', $1 \leq n' \leq n$,
iff n' subscripts ss_i are section subscripts (subscript triplet or vector subscript)
and the remaining subscripts are scalars. For example the index domain of the
Fortran 90 array section $A(11 : 100 : 2, 3)$ is given by $[1 : 45]$.*

The extended index domain *of the array section A', denoted by $\bar{I}^{A'}$, is given
by $\bar{I}^{A'} = [ss'_1, \ldots, ss'_n]$ where $ss'_i = ss_i$, if ss_i is a section subscript, and $ss'_i :=
[c_i : c_i]$ if ss_i is a scalar subscript that is denoted by c_i. For example the extended
index domain of the array section $A(11:100:2,3)$ is given by $[11:100:2,3:3]$.*

2.1 Mapping Data Arrays to Computing Processors

Data distribution is modelled as a mapping of array elements to (non-empty)
subsets of processors. The notational conventions introduced for arrays above
can also be used for *processor arrays*, i.e., I^H denotes the index domain of a
processor array H.

A *distribution* of an array maps each array element to one or more processors which become the *owners* of the element, and, in this capacity, store the
element in their local memory. Distributions are modelled by functions between
the associated index domains.

---

[2] We borrowed this definition from [1].

Definition 3 Distributions

Let $A \in \mathcal{A}$, and assume that H is a processor array. A distribution of the array A with respect to H is defined by the the mapping:

$$\mu_H^A : \mathsf{I}^A \to \mathcal{P}(\mathsf{I}^H) - \{\phi\}, \text{ where } \mathcal{P}(\mathsf{I}^H) \text{ denotes the power set of } \mathsf{I}^H$$

2.2 Logical storage model

Definition 4 File
A file F_L consists of a sequence of logically connected information units, which we call records r.

Definition 5 Record
A record r consists of a fixed number m of typed data values (attributes).

$$r = (A_1, A_2, ...A_m)$$

One record can store, for example, one float number or a complete data structure.

Definition 6 File View
The sequence of data elements which results from a write operation of an array A to a file F_L and is assumed by subsequent read operations is described by the function:

$$\alpha^A \; : \; \mathsf{I}^A \to [1 : ext^A]; \; (the \; standard \; Fortran \; record \; ordering \; is \; assumed)$$

2.3 Physical storage model

Definition 7 Disk
A disk d is the physical storage medium of interest. It is organized into storage units of fixed size (buckets). Let

$$D = \{d_1, ..., d_N\}$$

denote the set of all available disks.

Definition 8 Bucket
A bucket is a physical, contiguous sequence of bytes of a fixed size. It is the smallest unit, which the operating system of the underlying machine can transfer at once between the external storage medium and the main memory of a processor. Let B_j denote the set of all buckets of a disk j. The set B of all buckets available for storing users' data is defined by the union of all buckets B_j of all disks. Therefore,

$$B = \bigcup B_j, \; where \; 1 \le j \le |D| \; and \; |D| \; denotes \; the \; number \; of \; disks$$

Definition 9 Physical file
A physical file F_P is defined by a fixed number of physical blocks or buckets, which are stored on the external storage medium.

Definition 10 Declustering function
A declustering function δ defines the record distribution scheme for D, according to one or more attributes and/or any other distribution criteria,

$$\delta : R \rightarrow D, \text{ where } R \text{ is the set of records and } D \text{ the set of disks.}$$

In other words, the declustering function δ partitions the set of records R into disjoint subsets R_1, R_2, ... $R_{|D|}$, which are assigned to disks.

Definition 11 Organization function
The organization function o_d gives the buckets B_d of the disk d, where a specific record is stored.

$$o_d : R \rightarrow B_d, \text{ where } B_d \text{ are the buckets of disk } d$$

Each disk has its own organization function. Therefore a family of organization functions exists, i.e. $o_1, o_2, ...o_{|D|}$.

Definition 12 Array area
An array area E^A denotes the set of file records that store the elements of an array A or array section A'.

It is now possible to redefine the declustering and organization functions for array areas.

Definition 13 Areal declustering function
An areal declustering function Δ defines the set of disk drives D, which is spanned by a given area E^A. In other words, with $\mathcal{E}$ denoting the set of all possible areas of an array A

$$\Delta : \mathcal{E} \rightarrow \mathcal{P}(D) - \{\phi\}$$

Definition 14 Areal organization function
The areal organization function O_d defines the set of buckets of disk d, where the records covered by the given area E^A are stored, i.e.

$$O_d : \mathcal{E} \rightarrow \mathcal{P}(B_d) - \{\phi\}, \text{ where } B_d \text{ is the set of buckets of disk } d$$

Every disk has its own area organization function. Therefore a family of area organization functions exists, i.e. O_1, O_2, ... $O_{|D|}$.

Definition 15 Query mapping function
The query mapping function q gives the set of buckets B over all disks defined by a given area, i.e.

$$q : \mathcal{E} \rightarrow \mathcal{P}(B), \text{ where } B \text{ is the union of buckets over all disks}$$

The query mapping function can be defined by a combination of the areal declustering Δ and the areal organization function O, i.e.

$$q(E^A) = \bigcup O_i(E^A), \text{ where } i \text{ in } \Delta(E^A)$$

3   Design of the Parallel I/O Runtime Support

The I/O runtime system VIPIOS introduced in this section provides support for parallel access to files for read/write operations, optimization of data-layout on disks, redistribution of data stored on disks, communication of out-of-core[3] (OOC) data and many optimizations including data prefetching from disks based on the access pattern knowledge extracted from the program by the compiler or provided by a user specification. In the first implementation, VIPIOS will provide a mass storage support for VFCS [2].

3.1   Design Objectives

The goals of VIPIOS can be summarize as follows.

1. Efficiency. The objective of runtime optimizations is to minimize the disk access costs for file I/O and OOC processing. This is achieved by a suitable data layout on disks and data caching policies. Specifically, VIPIOS aims at optimizing data locality, storing data as close as possible to the processing node, and efficient software-controlled prefetching to partially hide memory latency by explicit (i.e., runtime system driven) execution of disk and/or network prefetch operations. Prefetching moves data (e.g., file blocks, OOC array sections) as close as possible to the processor which is going to need the data.
2. Parallelism. All file data and meta-data (description of files) are stored in a distributed and parallel form across multiple I/O devices. The user and the compilation system have the ability, in the form of hints, to influence the distribution of the file data.
3. Scalability. The architecture of VIPIOS is inherently portable. Control of the whole system is decentralized.
4. Portability. VIPIOS is able to provide support for any HPF compilation system that supplies all the interface information in the required format. In order to interface a large class of computer systems, VIPIOS strongly relies on standards, like MPI [13] and MPI-IO [9].

3.2   Coupling VIPIOS to VFCS

The VFCS overall architecture including VIPIOS is depicted in Fig. 1. VIPIOS is coupled to the modules of VFCS that deal with I/O. The source program is processed by the *frontend* and transformed into an internal representation (IR) which is kept in the *program database*. The frontend is able to process HPF programs (as depicted in the figure) and Vienna Fortran [15] programs as well. All transformations are performed on IR. The parallelization process can conceptually be divided into four steps: (1) data distribution, (2) work distribution,

---

[3] Mass storage I/O is also necessary to handle accesses to huge arrays whose parts must be kept on disks due to main memory constraints.

(3) communication generation, and (4) insertion of I/O operations. Based on extensive analyses powerful optimizations are performed to reduce communication and I/O costs. VFCS produces explicitly parallel Fortran SPMD program with calls to the selected message-passing library (e.g., MPI), parallel I/O library (e.g., VIPIOS) and other components of the runtime system. This program is finally translated by the *Fortran compiler* into the object program. Due to the interactive component, the user may receive feedback from the system, supply information to the system, and select transformation strategies.

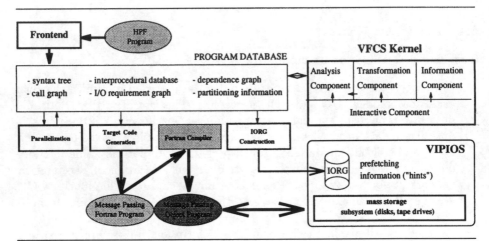

Fig. 1. VIPIOS Operational in a VFCS Environment

The input program is annotated with respect to I/O requirements. In [4, 5], we proposed language constructs to specify OOC structures and I/O operations for distributed data structures. These constructs can be used by the programmer to provide information helping VFCS and VIPIOS to operate the underlying I/O subsystem in an efficient way. Information selected from the annotation and static information extracted from the input program by VFCS at compile time is passed to VIPIOS through VIPIOS procedure calls from the target SPMD program. Placement of these calls is the result of program optimization. Program analysis and performance prediction phases extend the IR with information used for optimizing transformations and construction of an I/O requirement graph (IORG) which is used for guiding the prefetching process at runtime. Section 3.6 describes the structure of the IORG. This graph is incrementally constructed in the program database of VFCS during the compilation process and written to a file at its end. The IORG is passed to VIPIOS as an information base for prefetching heuristics. The generated SPMD program passes to VIPIOS the actual requirements on array section transfer as well as additional runtime data replacing symbolic information in the IORG (e.g., actual values for variable names).

From the VFCS point of view, VIPIOS can be considered as an abstract mass storage I/O machine. Its interaction with the compilation and runtime environment is depicted in Fig. 2 where the term *application process* denotes a running instance of the SPMD program. Some of the heuristics which VIPIOS applies for runtime optimizations are discussed later.

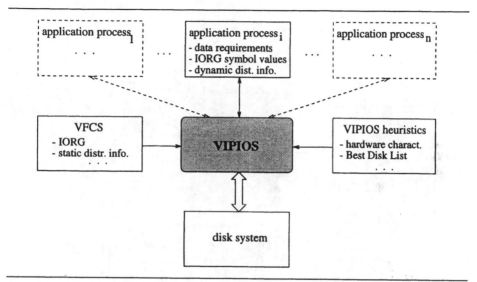

Fig. 2. VIPIOS Viewed as Abstract Mass Storage I/O Machine

3.3 Process Model

The framework VIPIOS distinguishes between two types of processes: application processes and VIPIOS servers. The application processes are created by the VFCS according to the SPMD paradigm each processor executing the same program on different parts of the data space. The VIPIOS servers run independently on all, a number or dedicated nodes and perform the data and other requests of the application processes. The number and the location of the VIPIOS servers are dependent on the underlying hardware architecture (disk arrays, local disks, specialized I/O nodes, etc.), the system configuration (number of available nodes, types of available nodes, etc.), the VIPIOS system administration (number of serviced nodes, disks, application processes, etc.) or (simply) the user needs (I/O characteristics, regular, irregular problems, etc.).

These VIPIOS servers are similar to data server processes in database systems. For each application process one unique VIPIOS server is assigned and accomplishes the data requests. It is also possible that one VIPIOS server services a number of application processes. In other words one-to-one or one-to-many relationships are supported.

The process model is depicted by Fig. 3. The VIPIOS call interface VI, which is linked with the application process AP, handels the communication with the assigned VIPIOS server VS.

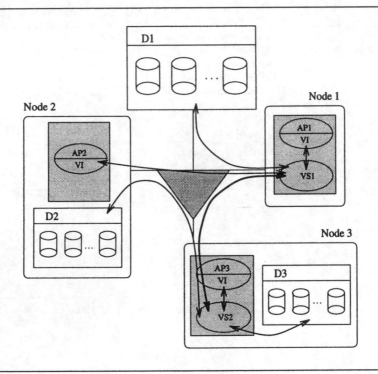

Fig. 3. Process Model: Application Processes and VIPIOS Servers on top of a Disk Architecture with Local and Global Disk Systems

In this figure, different architectural models are summarized. Both local disk architectures, as disk D2 and D3 (nodes 2 and 3), and global disk systems as disk D1 are supported (the surrounding frame depicts a physical processing node). Further, it can be seen that varying process models are feasible. On the one hand application process AP3 on node 3 is serviced by VIPIOS server VS2 running on the same node, which in turn administrates the local disk D3 and the remote disk D2 (local to node 2). On the other hand the server process V1 handles the requests of the local application process AP2 and accesses the global disk only. The connection between the 2 servers illustrates that both can request each other in accessing remote data.

For each application process, all data requests are transparently caught by the assigned VIPIOS processes. Locally or remotely retrieved data are accessed by these processes and ensure that each application process has access to its requested data items.

The VFCS provides information about the problem specific data distribution, the stride size of the slabs of the out-of-core data structures and the presumed data access profile. Based on this information, the VIPIOS organizes the data and tries to ensure high performance access to stored data. Additional data distribution and usage information can be provided by the Vienna Fortran programmer using new language constructs. This type of information allows the VFCS/VIPIOS system to parallelize read and write operations for out-of-core arrays by selecting a well-suited data organization in the files.

An important advantage of the proposed framework is the transparence of the architecture to the application programmer as well as to the VFCS compiler developer.

Summing up, the VIPIOS, as a component of the VFCS, is responsible for the organization and maintenance of all data held on the mass storage devices.

3.4 Data Locality

The main design principle of the VIPIOS to achieve high data access performance is *data locality*. This means that the data requested by an application process should be read/written from/to the best-suited disk.

We distinguish between logical and physical data locality.

Logical data locality denotes to choose the best suited VIPIOS server for an application process. This server is defined by the topological distance and/or the process characteristics. In most cases the access time is proportional to the topological distance of the application process to the VIPIOS server in the system network. It is also possible that special process characteristics can influence the VIPIOS server performance, like available memory, disk priority list of the underlying node (see the following), etc. Therefore it is also possible that a remote VIPIOS server could provide better performance than a closer one. However only one specific VIPIOS server is chosen for each application process, which handles the respective requests. This process is called the *buddy server*, while all other servers are called *foe servers*. The buddy and foe relation is illustrated by Fig. 4.

The *physical data locality* principle aims to determine the disk set providing the best (mostly the fastest) data access. For each node an ordered sequence of the accessible disks of the system is defined (the *best disk list, BDL*), which orders the disk according to their access behavior. Disks with good access characteristics precede disks with bad access characteristics in this list (Fig. 4).

3.5 Two-Phase Data Administration Process

The data administration process of a VIPIOS servers can be divided into two phases, the preparation and the administration phase. The *preparation phase* prepares the administrated data according to the data layout of the data structure, the presumed access profile and the physical restrictions of the system (available main memory, disk space, etc.). The input data for this phase are partly prepared in the program compilation process. The phase is performed

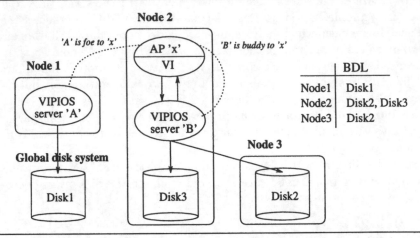

Fig. 4. "Buddy" and "Foe" Relation Between Application Processes and VIPIOS Servers

during the system startup time and precedes the execution of the application processes. In this phase, the physical data layout schemas are defined, the actual VIPIOS server process for each application process and the disks for the stored data according to the locality principles are chosen. Further, the data storage areas are prepared, the necessary main memory buffers allocated, etc.

The *administration phase* accomplishes the I/O requests of the application processes. It is obvious that the preparation phase is the basis for good I/O performance.

3.6 Support for Efficient Software-Controlled Prefetching

Software-controlled prefetching ([10]) in a distributed processing environment is a technique to reduce memory latency by explicit (i.e., runtime system driven) execution of disk and/or network prefetch operations in order to move data (e.g., file blocks, OOC array slabs, startup/restart volumes) as close as possible to the processor which is going to need the data.

In this subsection, we outline the application of software-controlled prefetching to optimization of OOC programs.

In our approach, the compile-time knowledge about loops and other out-of-core code parts is represented by a graph structure called I/O Requirement Graph. This graph is constructed incrementally during the compilation process and written to a file at its end. The IORG is used by VIPIOS to direct the prefetching and data reorganization process.

We denote the IORG by the triple $G=(N,E,s)$ where (N,E) is a directed graph, $s \in N$ is the start node with indegree 0, and each $n \in N$ is in the transitive cover s^*. The set of nodes N consists of nodes represented in-core program parts and out-of-core ones. In the IORG, code segments are condensed into single nodes encompassing the following components:

- *Branch label.* In general, IORG structures contain patterns with multiple branches including symbolic information to be resolved at runtime. A unique label is assigned to each branch which is included in all nodes belonging to the branch. The VIPIOS is informed at runtime about the branch selection by a compiler-inserted call from the SPMD program. In the IORG these call points are annotated. When the executing program goes through the corresponding instruction sequence, the call is issued. Runtime data for the substitution of symbolic information is transferred in a similar way. Synchronisation points are determined in the context of dataflow analysis thus finding process states in which values can be bound to these symbols in the IORG.
- *Time cost estimation.*
It estimates how much processor time is used by a program fragment represented by this node.
- *Descriptions of array sections handled by a node.*
I/O array sections are read from or written to the so called OOC array database contained in the VIPIOS repository.

The construction of an IORG is described in the technical report [6].

4 Related Work

There has been a lot of effort to improve parallel I/O performance by software means. However, many systems, for example the Intel's Concurrent File System, provide a quite unsatisfactory level of programming abstraction. Consequently, application programmers and compiler developers have to make themselves familiar with each developer's parallel I/O interface in order to deal with technical details such as buffering or physical disk layout which clearly should be hidden from the application programming level. A higher level abstraction is provided by several parallel I/O libraries like Vesta [8] PIOUS [11], PASSION [7] and Jovian [3]. Very little work has examined database techniques for storing multidimensional arrays on disks. Panda [14] is a server-directed I/O architecture that provides a high-level interface for array I/O.

5 Conclusions

VIPIOS provides a new abstraction of parallel files, which is suitable for highly parallel mass storage systems. It is in the first place designed to provide support for parallelizing compilers. Four basic requirements influenced the design of VIPIOS: efficiency, parallelism, scalability and portability. Efficient data prefetching is achieved through passing the global information about the data access profile from compile time to runtime. The VIPIOS/compilation system interface presented here has been proposed in the context of VFCS, however, it can be easily integrated into any HPF compilation system.

References

1. S. Benkner. *Vienna Fortran 90 and its Compilation.* Ph.D. Thesis, University of Vienna, September, 1994.
2. S. Benkner et al. *Vienna Fortran Compilation System - Version 1.2 - User's Guide*, University of Vienna, October 1995.
3. R. Bennett, K. Bryant, A. Sussman, R. Das, and J. Saltz. *Jovian: A Framework for Optimizing Parallel I/O.* In Proceedings of the 1994 Scalable Parallel Libraries Conference, IEEE Computer Society Press, Oct. 1994
4. P. Brezany, T. Mueck, E. Schikuta. *Language, Compiler and Parallel Database Support for I/O Intensive Applications.* Proc. HPCN Europe 1996, Milan, Italy, May 1995, Springer-Verlag, pp. 14–20.
5. P. Brezany, T. Mück, and E. Schikuta: *Mass storage support for a parallelizing compilation system.* In Proceedings of the Conference EUROSIM – HPCN Challenges 1996, North Holland, Elsevier, June 1996.
6. P. Brezany, T. Mück, and E. Schikuta: *A Software Architecture for Massively Parallel Input-Output.* Technical Report, Department of Data Engineering, University of Vienna, September 1996.
7. A. Choudhary et al. *PASSION: Parallel And Scalable Software for Input-Output.* Technical Report CRPC–TR94483-S, Syracuse University, Syracuse, NY, 1994.
8. P. F. Corbett, D. G. Feitelson. *Design and Implementation of the Vesta Parallel File System.* In Proc. Scalable High Performance Computing Conference, Knoxville, May 1994, pp. 63–70.
9. P. Corbett et al. *MPI-IO: A parallel file I/O Interface for MPI. Version 0.3.* Technical Report NAS-95-002, NAS, January 1995.
10. T.C. Mowry. *Tolerating Latency Through Software-Controlled Data Prefetching.* Ph.D.Thesis, Standford University, March 1994.
11. S. A. Moyer, V. S. Sunderam. *PIOUS: A Scalable Parallel I/O System for Distributed Computing Environments.* In Proc. Scalable High Performance Computing Conference, Knoxville, May 1994, pp. 71–78.
12. High Performance Fortran Forum. *High Performance Fortran Language Specification.* Scientific Programming, Vol. 2, No. 1, 2, 1993.
13. Message Passing Interface Forum. *MPI: A Message-Passing Interface Standard.* April 1994.
14. K. E. Seamons. *Panda: Fast Access to Persistent Arrays Using High Level Interfaces and Server Directed Input/Output.* Ph.D.Thesis, University of Illinois at Urbana-Champaign, 1996.
15. H. Zima, P. Brezany, B. Chapman, P. Mehrotra, and A. Schwald. *Vienna Fortran – a Language Specification.* ACPC Technical Report Series, University of Vienna, Vienna, Austria, 1992.

Spatial Genetic Algorithm and Its Parallel Implementation[1]

Andrzej Broda and Witold Dzwinel

Institute of Computer Science AGH, Al. Mickiewicza 30, 30-059 Cracow, Poland.
dzwinel@uci.agh.edu.pl

Abstract. The spatial genetic algorithm (SGA) is presented. Locality is realized by mapping GA population on a cellular automata. The role of neighborhood in genetic search is shown by comparing SGA with the parallel recombinative simulated annealing (PRSA) approach proposed by Mahfoud and Goldberg in [1]. It appears, that not optimized SGA outdoes PRSA in *loose* and is only slightly worse in *tight* optimization problems defined in [1]. However, because of high potential parallel speedup for SGA and the possibility of domain decomposition due to SGA locality, the efficient parallel realization of SGA seems to be more perspective than PRSA. The SGA opens the way for investigations of more sophisticated neighborhood definitions based on the lattice gas and molecular dynamics paradigms, which are well scalable (in parallel computing sense) and can simulate the natural space-and-time environment lacking in pure GAs.

1 Introduction

Disruption of good subsolutions by mutation and crossover operators is the main disadvantage of genetic algorithms (GA). A tradeoff between diversity and performance consists in the control of frequency of disruption events. Maintaining diversity is a principal demand. Although, crossover and mutation generate new solutions, the certain limitations come from the facts that:

- crossing nearly identical strings yields offspring similar to the parents,
- mutation examines the full solution space but may take an extensively long time to gain an acceptable solution.

Parallel recombinative simulated annealing (PRSA) approach proposed by Mahfoud and Goldberg in [1], can be interpreted as a method, which goal is to raise diversity, simultaneously limiting the number of worthless solutions (population members). It is done by maintaining the Boltzmann distribution in the population and using SA (simulated annealing) paradigm of cooling. Introducing a

[1]This work is sponsored by the Polish Committee of Scientific Research (KBN), Grant No. KBN 8 S503 006 07.

minimization of cost function (energy) in addition to fitness, reinforces the process of search, making the recombination and mutation operators "energy directed". As proved in [1], this makes the process of search convergent to the global solution (for problem dependent suitable cooling schedule). What is interesting, for relatively large populations the solution is obtained before the cooling threshold is achieved. Otherwise, for smaller populations, "energy directed" mutation is able to find the proper solution. The tradeoff between "GA factor" and "SA factor" let us to find the conditions of the best performance of PRSA algorithm. The priceless advantages of PRSA over both pure GA and SA are as follows:

1. Unlike GAs, it provides the possibility to obtain a good solution for problems characterized by long strings. GAs require huge populations to obtain a reasonable result. As it is shown in [1], PRSA is able to find the solution of the 24-bit, eight-subfunction, *loose* version of Goldberg's order-3 deceptive problem for small population (n is order of 10) in a reasonable period of time, while pure GA failed even for $n=5000$ [1].
2. It introduces the population and *implicit parallelism* to SA paradigm, which via polynomial increase of the best partial solutions (schemata) speedups the process of search. Unlike SA, PRSA is inherently parallel, which enables it to gain a high performance for parallel implementation of the algorithm [1].

However, diversity introduced by SA paradigm implementation is computationally expensive and requires by hand optimization not only GA (probabilities of recombination, mutation, number of population members etc.), but SA parameters like: cooling period, cooling constants, starting, switching and final temperatures as well.

Another approach, complementary to PRSA, is to introduce locality by using a mapping to a cellular automata. This idea, was presented in [2,3] as a way of facilitating parallelization. In fact, despite the intrinsic parallelism of GAs, they are still difficult to parallelize to a level of high scalability. This is a consequence of the fact, that GAs (PRSA included) use global knowledge for their selection process. Therefore, their parallel versions (synchronous) use *master-slave* approach, where *master* sends-and-receives the current population to-from *slaves* and randomly shuffle pointers to population elements. Taking into account that for large scale computations in multiprocessor environment all members of populations have to be sent from-to *slaves* to-from *master*, both the limited memory of *master* processor and transmission delay will be the bottlenecks of efficient massive parallel processing. On the other hand, asynchronous parallel processing (very often use in GAs), in which the population members migrate to-from other processor, suffers from slow convergence due to migration [1] and scales wrong with increasing number of processors. Mapping population onto spatial structure resolves the problem of scalability, what will be shown in course of this paper. Moreover, mapping provides a new technique for assuring diversity in GAs population.

In the paper SGA (spatial genetic algorithm) is proposed and compared with PRSA for the 24-bit, eight-subfunction, *loose* and *tight* versions of Goldberg's order-3 deceptive problem. Both serial and potential parallel speedups are compared. Two parallel versions of SGA are discussed. Finally, possible directions for future research to obtain more adaptive and problem independent algorithm are suggested.

2 The algorithm

Let us assume that N is the number of population members, G - the number of generations, $\Re$ - the square array of $n=Sz \times Sz$ size. It is assumed that the array is closed by the periodic boundary conditions, i.e., the closest neighbors of edge elements are located on the opposite sides of the array. In Fig.1 the neighborhood of an element C is shown. It consists of eight S - the closest neighbors of C.

X	X	X	X	X
X	S	S	S	X
X	S	C	S	X
X	S	S	S	X
X	X	X	X	X

Fig.1. The $\Re(5 \times 5)$ array. 25 population members are mapped into all the elements of the array. The closest neighbors of C element are marked by S.

Spatial genetic algorithm

(A) Generate N members of an initial population and map each element of the population on the randomly selected, void element of $\Re$ array.

(B) **Repeat** $i=1, G$

inspect all the elements j (population members) of $\Re$ array and supplement current genotype C in j by the genotype found as follows:

(1)   Out of the closest neighbors of j (see Fig.1) and j itself, select two individuals A and B. The selection probability is an increasing function of fitness factors of neighboring individuals and j.

(2) Substitute C by a better individual chosen from A and B.

(3) Substitute C by one out of two possible results of A and B crossover (optional one) with pc probability ($pc=0.5$ will be considered).

(4) Mutate C with pm probability ($pm=0.5$ will be considered).

Owing to disruptive effects of genetic operators, GAs are incline to a loss of solutions and their substructures or bits [4]. Upon disruption, the pure GA will not maintain old but better solutions, especially those, for which below-average but perspective schemata must be discarded. This effect can be overcome only by using larger population sizes. However, the perspective, low order, wide span schemata have no chance of surviving.

On one hand, after introducing neighborhood, the exponential increase in the number of above-average schemata - typical of simple GA - may no longer be valid (the population member is compared only to its neighbors but not to the whole population). On the other hand, higher diversity of population can be expected. In general, better schemata will not die out, but dominate their surroundings. Their propagation will be stable but inevitable and not dependent on their advantage in fitness. In order to prevent fast domination of a better solution above the others, the fitness function for pure GA is usually rescaled to balance the individuals. It may happen, that above-average individual will be lost after such "equalization". Unlike in GAs, here an individual is safe until a better one can be found in its vicinity. This gives time for self-improvement and - if it is possible - to elaborate better solution by the moment of confrontation. Similarly, the best individuals do not die out, because they are better than their neighbors. As a result, there is little danger of the population being dominated by the superior individual too fast, and little risk of a loss of an individual which is only slightly better than its neighbors.

Because long schemata are easier for disruption, only short schemata play a considerable role in classical GAs. Since the SGA individual is surrounded by its "relatives", a process of reconstruction of disrupted schemata is more likely. This fact was confirmed in SGA tests and experiences with loose problems presented in course of this paper. As it is stated in [1] about problems with good solutions represented by schemata of wide span, GAs usually miss them. It will be shown in the next section, that SGA not only use wide span schemata to find the solution but also makes a process of search faster than PRSA proposed by Mohfoud and Goldberg.

3 SGA against PRSA

The tests were carried out for global minimum search of the 24-bit, eight-subfunctions, of Goldberg's order-3 deceptive problem defined in [1]. The 24-bit function is a sum of eight 3-bit subfunctions. Each of the subfunctions represents order-3 deceptive problem. This choice is justified by the fact, that deceptive problems present the hardest challenge for GAs [4]. Two representations were tested:

tight: **aaa**bbbcccdddeeefffggghhh
loose: abcdefghabcdefghabcdefgh

The *loose* version extremely separates the bits of subfunctions, forcing the algorithm to work on long schemata. Because the classical GAs are not able to solve this problem [1] only PRSA will be considered in confrontation with SGA.

In Fig.2 and Tabs.1,2 the comparison between PRSA and SGA serial implementations is shown. "FE" stands for the number of function evaluations and "SS" denotes the "serial speedup" in relation to pure SA algorithm. The results for PRSA (see Fig.2 and Tab.1) presented in [1] were obtained after tuning the algorithm and testing different cooling speeds. Therefore, they represent the best PRSA realization. However, this fact is a potential source of troubles with self-adaptability of the algorithm to other optimization tasks and may result in considerable speed deceleration. In contrast, SGA does not involve any tuning.

Moreover, apart from mutation (p_m) and crossover (p_c) likelihood, it does not posses any parameters to be set. Concerning the two parameters mentioned, their values are arbitrary ($p_c=0.5$ and $p_m=0.5$), which means they may not to be optimal. This fact and certain convergence for populations $n \geq 36$ (this limit is intuitively obvious due to the algorithm specificity) make SGA convenient and versatile tool for constructing adaptive genetic algorithm in the future (alike adaptive simulated annealing proposed by Ingber [5]).

Tab.1. FE and SS for PRSA for different populations   **Tab.2.** FE and SS for SGA for different populations

n	Tight problem		Loose problem	
	FE	SS	FE	SS
2	195791	2.69		
4	127492	4.13		
8	112341	4.69		
16	94634	5.56	205456	2.56
32	53280	9.88	92192	5.71
64	16672	31.5	158784	3.31
128			169600	3.10
256			193690	2.72
512			394752	1.33

n	Tight problem		Loose problem	
	FE	SS	FE	SS
36	46933	11.21	75107	7.01
64	32154	16.37	60659	8.68
100	21740	24.21	64150	8.20
144	17323	30.38	71280	7.38
196	22070	23.85	74500	7.06
225	17055	30.86	67073	7.05
256	18355	28.68	89446	5.88
324	18468	28.50	88614	5.94
400	19000	27.70	107360	4.90
484	26088	20.18	125501	4.19
625	32813	16.04	118500	4.44

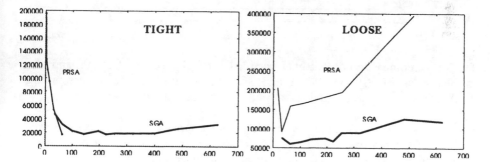

Fig.2. PRSA versus SGA for thigh and loose problems (axis y - number of function evaluations needed to obtain the global minimum, axis x - population).

As shown in Fig.2 and Tabs.1,2, the best realization of PRSA (for n=64) is slightly better than SGA one (for n=144) for tight problem. As it was told in [1], for PRSA and n>64 the number of FE increases considerably (unfortunately Mahfoud and Goldberg do not give any FE evaluation for n>64), while for SGA FE is relatively stable for a wide range of population capacities. However, the striped version of PRSA with fast quenching cooling schedule [1] gives FE=6000 for n=128, which shows the strength of PRSA tuning.

Nevertheless, the strength of SGA consists in efficient processing of wide span schemata and explicit highly scalable geometric (domain) parallelism. The former approves the results obtained by SGA for loose problem. As we can see, the best SGA result outdoes that obtained for PRSA on about 50% (according to [1], pure GA is not able to solve the loose problem even for n=5000!!). Moreover, the performance of SGA is stable for a wide range of n, whereas the PRSA leader is almost twice as good as the second best (see Fig.2).

Taking into account parallel realization, SGA proves even better. The maximum potential parallel speedup PS is defined in [1] as the serial speedup multiplied by the maximal number of processors which can be theoretically engaged in computations for given n. Assuming that the number of processors is equal to 256, the PS values for tight and loose problems for SGA and PRSA are shown in Fig.3.

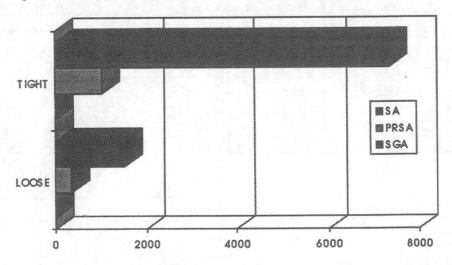

Fig.3. The comparison of maximal potential parallel speedups for tight and loose problems for SA (PS=1), PRSA and SGA algorithms on 256-processors ideal (without memory access problems) shared memory computer.

However, such the theoretical shared memory approach is currently impossible because of not resolved simultaneous memory access. Therefore, message-passing realization on massively parallel computers has to be considered.

4 Parallel realization of SGA

As was told in the introduction, GAs are difficult to parallelize to a level of high scalability. It comes from the fact, that GAs (PRSA included) use global knowledge in their selection process. Usually, the parallel message-passing versions (synchronous) use *master-slave* approach, where *master* sends-and-receives the current population to-from *slaves* and randomly shuffle pointers to population elements. For large scale computations all population members have to be sent from-to *slaves* to-from *master*. Both the limited memory of *master* processor and transmission delay will be bottlenecks of efficient massive parallel processing. On the other hand, asynchronous parallel processing (very often use in GAs), where the population members migrate to-from other processor, suffers from slow convergence due to migration [1] and scales wrong with increasing number of processors. In contrast, parallelization of SGA is straightforward because of spatial component enabling domain decomposition. Two approaches were examined (see Fig.4).

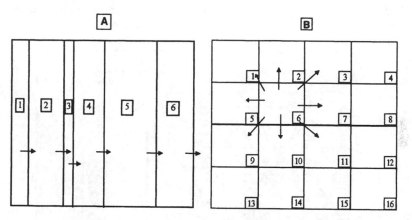

Fig.4. Two approaches to SGA domain parallelism. The width of strips can be regulated to assure load balancing. Similarly the number of cells is assumed to be greater than the number of processors. The number of cells sent to the processor depends on its current load. The arrows show neighboring strips and cells, edge elements of which have to be added to each of strip and cell send to the separate processor.

The first one consists in division of computational box ($\Re$ array, see Fig.1) onto strips. They are mapped, one to one, onto multiprocessor system. In result, each of strip consists of population members located on a single processing element (processor + memory). Additionally, boundary elements of the neighboring strip (shown by the arrow in Fig.4A) are allocated in the processor memory. As soon as a new generation is computed, the boundary elements of strips are transmitted to the processors where the neighboring strips are allocated. Additionally, the widths of strips can be regulated, what is reflected by transmission of boundary population elements between processor to enable proper load balancing. As shown in [6,7], such approach assures the high level of scalability and straightforward increase of n with increasing number of processors. Due to implicit parallelism of GAs, the exponential increase of probability of global minimum finding can be achieved with superlinear speedup. It was observed on 5 networked RISC 6000 workstations using SGA program running under PVM 3.6 for n=3600.

The second version of parallel realization of SGA uses cell decomposition of computational box (see Fig.4B) and *master-slave* paradigm. However, in comparison with PRSA parallel realization (synchronous algorithm), SGA *master* processor does not use scheduling procedure. This enables us to construct real (unlike PRSA asynchronous algorithm) asynchronous program with simple procedure of load balancing. Namely, the number of cells assumed is greater than the number of processors. Cells (each of equal size, with their population members and boundary elements of neighboring cells) are queued and sent to the processors in accordance with their current load (overloaded processors obtain less cells for computations than others). Because the queue is never empty (the number of cells is distinctly greater than the number of processors) the processors do not wait for the slowest one, having every time new cells for processing (see Fig.5). A special procedure takes care about the population's integrity. Interesting enough, as a result of locality, parts of a few subsequent generations can be processed at the same time.

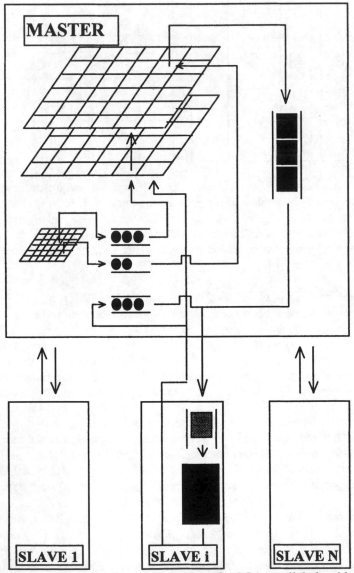

Fig.5. The scheme of *master-slave* computations for SGA parallel algorithm.

For n=3600 on the network of 5 IBM RISC 6000 workstations in PVM 3.6 environment, an average idle time for a single processor never exceeded 10% of CPU time for the simple test functions mentioned above, and fell below 1% for more complex optimization problems (i.e. graph partitioning for 30 nodes). Currently, the SGA parallel algorithms are being tested on the multiprocessor system CONVEX SPP 1200/XA (32 processors).

5 Conclusions

PRSA and SGA represent two different strategies of diversity and performance balance for GAs. The first method promises diversity by SA sampling. This gives priceless possibility to operate on small populations, simultaneously assuring the convergence to the global solution. It is especially valuable for long strings. Pure GAs involve extremely large populations for such the problems without any warranty to obtain a good solution. Disadvantage of PRSA consists in drop of performance caused by SA sampling and cooling paradigm involving careful choice of parameters.

SGA method assures diversity by mapping population on cellular automata and introducing locality. This prevents destruction of long schemata, what is reflected by the good results obtained for the loose problem. The algorithm efficiency is very stable for broad range of n because of small number of free parameters involved. Additionally, SGA gives the possibility for domain decomposition, which yields in highly scalable parallel algorithm. However, SGA operates on larger populations than PRSA, which may be unprofitable for long string problems. Therefore, it would be important to examine, both for PRSA and SGA, how the population size should grow (to obtain the solution) for increasing length of strings. Both methods should be tested for real long string problems to assess the profits of parallelization and to compare their real possibilities.

The combination of two methods seems to be an interesting idea. It can be done, for example, using molecular dynamics paradigm. Unlike for cellular automata, the movable particles would represent population members. The particles will move in computational box in accordance to the Newton laws and crossover in case of collisions. Introducing initial temperature of particle system and dumping force (dissipating kinetic energy from the system), the temperature will drop decreasing collisions frequency and realizing, in fact, SA model with built-in process of cooling. Such approach is very easy to parallelize in the way similar to that for SGA. The preliminary tests made on feature selection problem in pattern recognition (the problem was presented in [8]), show the main profit of this method: faster convergence to the global minimum.

Acknowledgments

Thanks are due to Dr M.Bubak and Professor Dr J. Kitowski for comments and discussions.

References

1. Mahfoud, S.,W., i Goldberg, D.,E., 1995, "Parallel Recombinative Simulated Annealing: Agenetic Algorithm", *Parallel Computing*, **21**, 1, 1.
2. Sloot, P., M., A., Kaandrop, J., A., Schoneveld, A., 1995, "Dynamic Complex Systems (DCS): a new approach to parallel computing in physics, *Technical Report of Department of Computer Science, University of Amsterdam*, **CS-95-08**.
3. Tomassini, M., 1993, "The parallel genetic cellular automata: application to global function optymization". In R.F. Albrecht, C.R., Reeves, and N.C. Steele, editors, *Artificial neural nets and genetic algorithms*, 385-391, Wien, 1993. Springer-Verlag.
4. Goldberg, D., E., 1989, *Genetic Algorithms in Search, Optimization and Machine Learning*, Addison-Wesley Pub.
5. Ingber, L., 1995, "Adaptive Simulated Annealing (ASA): Lessons Lerned", *Control and Cybernetics*, in print.
6. Boryczko, K., Bubak, M., Gajęcki, M., Kitowski, J., Mościński, J. i Pogoda, M., 1994, "Transmission rates and performance of a network of computers", *Lecture Notes in Computer Science*, High Performance Computing and Networking, W.Gentzsch i U.Harms, (Eds.), **797**, 142, Springer-Verlag, Berlin.
7. Bubak, M., Mościński, J., Pogoda, M., i Słota, R., 1995, "Load Balancing for Lattice Gas and Molecular Dynamics Simulations on Networked Workstations", *Lecture Notes in Computer Science*, High Performance Computing and Networking, Hertzberger, B., Serazzi, G., (Eds.). **919**, 118-123, 329, Springer-Verlag, Berlin 1995.
8. Dzwinel, W., 1995, "In Search for the Global Minimum in Problems of Features Extraction and Selection", *Proceedings of the Third Congress on Intelligent Techniques and Soft Computing, EUFIT'95*, 28-31 Sierpień 1995, Aachen, Niemcy, **3**, 1326.

Addressing Algebra as a Tool for Parallel Program Development

John Brown

Silicon Graphics, Inc., 2011 N. Shoreline Blvd., Mountain View, CA 94043-1389

Abstract: Address algebras are a powerful set of basic tools for the design and implementation of parallel programs. Address algebras give insightful shorthands for mapping abstract algorithms with their explicit problem dimensions and data communications to real machines with all the implications of communications topologies, fixed processor dimensions and non-uniform data access. These techniques are quite general and should be a fundamental tool for any parallel programmer today. The basic ideas of address algebra will be discussed along with many examples of their practical application.

1. Introduction

The design of effective and efficient parallel programs for distributed memory parallel computers is a complex interplay of parallel algorithm design, data placement strategies and efficient communications implementations. Efficient communications strategies require regular communication patterns that can exploit the particular topologies of the target architectures. With many communication possibilities and the inherent complexity that communications introduce, often, designing and debugging data placement and communications for parallel machines are a formidable challenge. The use of addressing algebras is a fundamental tool for both designing efficient communications as well as providing a powerful aid for understanding the effect of any particular communication.

What is addressing algebra? Addressing algebra is a collective term for a variety of related techniques that examine and explain algorithms, communication operations and data layout, all, involving large data arrays, by simple manipulations of the address fields of the indices of the arrays. These simple manipulations are equivalent to an entire operation on all elements of the array. This reduces the problem of dealing with large numbers of data elements and processors to simple manipulations of a few address field components.

We will look as several of the basic issues associated with using addressing algebras for algorithm design. The first issue we examine is how basic algorithms can be understood from an addressing perspective. We will then look at data layout and data communications on distributed memory computers examining how data structures can be efficiently mapped to memory using addressing notions and how address transformations correspond to data communication patterns. Finally, we look at the interesting and surprising issue of unusual address interpretations and how these can lead to efficient algorithm designs by examining classic Gray coding.

2. Algorithms and Address Representation

Abstract Algorithm and Address Representation
Consider an algorithm operating on abstract data structures. Some of the data structures are large linear or multi-dimensional arrays referenced by index numbers. Different operands in the algorithm operate with one or more of the elements of the arrays. Generally, all of the elements of some particular array will participate in some operation which is quite similar to the participation of all the other elements of that same array in similar operations. There is a pattern to the operations that is generally expressed by coding loops in the source. These patterns operate systematically on the indices of the arrays. Quite often, we can map these patterns of access to particular specific active digits in some basis of expressing the value of the index. By far the most common and useful basis is the common binary representation, but other bases are quite appropriate for different algorithms.. A simple example will illustrate this idea clearly. Let's consider the classic Fast Fourier Transform (FFT). The FFT algorithm is an acceleration of the naive Fourier Transform of an array of data. The classic definition of the Discrete Fourier Transform of an array a of N elements is given in Equation (1) where A_k is the kth element of the Fourier transform of a .

$$A_k = \sum_{j=0}^{N} e^{i\frac{2\pi}{N}jk} a_j \tag{1}$$

Splitting the input elements a_k into odd and even parts and starting from Equation (2) it is easy to show that A_k can be formed from the individual Fourier transforms of the odd and even members of the input array a as shown in Equation (3).

$$A_k = \sum_{j=0}^{\frac{N}{2}} e^{i\frac{2\pi}{N}2jk} a_{2j} + e^{i\frac{2\pi}{N}(2j+1)k} a_{2j+1} \tag{2}$$

$$A_k = \sum_{j=0}^{\frac{N}{2}} e^{i\frac{2\pi}{\frac{N}{2}}jk'} a_{2j} + e^{i\frac{2\pi}{\frac{N}{2}}k'} e^{i\frac{2\pi}{\frac{N}{2}}jk'} a_{2j+1} \qquad \text{: if } k<\tfrac{N}{2}, \text{ then } k'=k \tag{3}$$

$$= \sum_{j=0}^{\frac{N}{2}} e^{i\frac{2\pi}{\frac{N}{2}}jk'} a_{2j} - e^{i\frac{2\pi}{\frac{N}{2}}k'} e^{i\frac{2\pi}{\frac{N}{2}}jk'} a_{2j+1} \qquad \text{: if } k\geq\tfrac{N}{2}, \text{ then } k'=k-\tfrac{N}{2}$$

The term $e^{i\frac{2\pi}{\frac{N}{2}}k'}$ is the so-called "twiddle factor" of an FFT butterfly. Of course, the smaller individual transforms can use the same trick recursively and over $Log(N)$ stages the whole FFT is completed with $O(N)$ operations at each stage. The communication pattern of data elements based on Equation (3) for an 8-point Fourier transform is shown in Figure 1. and the communication pattern of a recursively implemented FFT is shown in Figure 2.

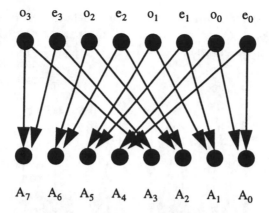

Figure 1. *Fourier transform constructed from odd and even Fourier transformed subsets.*

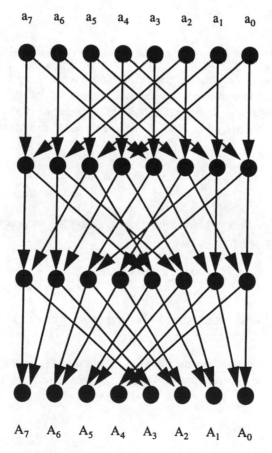

Figure 2. *Fast Fourier Transform communication pattern.*

Since, at any point in the algorithm, only a pair of data elements is exchanging data, it is possible to save storage by overwriting the pair of data elements with the resulting output. This is shown in Figure 3.

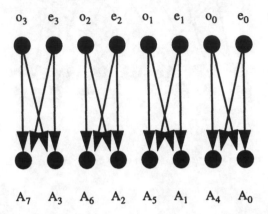

Figure 3. In-place single pass Fourier transform communication pattern.

However, we have now scrambled the output and have to keep that in mind. How have we scrambled the result? We have simply interleaved the upper half of the data output with the lower half. If we look at how the address bits of the output are stored in memory the interleaving looks like taking the standard representation of memory location as:

$$\alpha_2 2^2 + \alpha_1 2^1 + \alpha_0 2^0 \rightarrow \alpha_1 2^2 + \alpha_0 2^1 + \alpha_2 2^0 \tag{4}$$

Alternatively, in the standard binary bit representation the address bit α_i is simply some positional binary digit and we write:.

$$\alpha_2 \alpha_1 \alpha_0 \rightarrow \alpha_1 \alpha_0 \alpha_2 \tag{5}$$

The simple address manipulation is to perform a left circular shift of 1 bit on the address bits.

If we systematically replace output using the positions of the interacting pair of input data elements as we linearly recurse, we get the communication patterns shown in Figure 4.

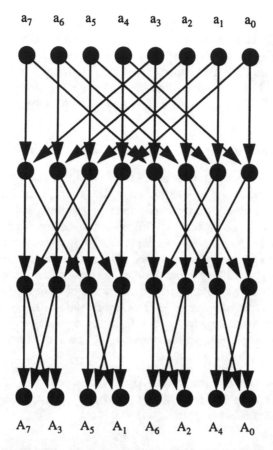

Figure 4. The Cooley-Tukey FFT communication pattern.

However, now, the recursion has caused the output bits to be presented in reversed order. This is the famous Cooley-Tukey FFT algorithm [1]1 with the bit-reversed output address representation shown in Equation (6).

$$\alpha_2\alpha_1\alpha_0 \rightarrow \alpha_0\alpha_1\alpha_2 \qquad (6)$$

Since there is a great deal of parallelism in the FFT, many different mapping structures are possible between the different butterflies of the FFT. Different mapping structures have performance implications since they imply different access patterns that affect vector strides, bus and bank conflicts, cache line access patterns and cache blocking. Any high performance implementation of an FFT libraries must exercise great care to choose appropriate FFT variants that map well to the target platform. The representation of these different mappings is greatly simplified by tracking the particular representation of address bits at any butterfly operation. Of course, many of these same considerations apply to other divide-and-conquer algorithms.

The extension of address representation to more complex structures such as multi-dimensional arrays is quite straightforward since the implicit mapping of these structures to memory is well known. A classic example is the matrix in FORTRAN where the contiguous direction of memory storage corresponds to a column for most classic algorithms. For the most general case, Chinese arithmetic representations of the addresses are often appropriate for arrays with non-power of 2 leading dimensions.

3. Mapping Abstract Data Structures to Distributed Memory

With distributed memory systems, memory itself is fragmented into individual pieces with important implications for performance. In general, distributing data arrays across the different memory locations promotes greater total system memory access bandwidth. However, poorly laid out memory can promote excessive expensive memory accesses or may lead to resource contentions that reduce performance considerably. There are also a wide variety of options to lay out data that can cause confusion for the applications programmer as he tracks perhaps tens of thousands of individual memory heaps holding many thousands of data elements. In addition, with many different configurations of processors and memory for any particular platform, developing good portable code for even a single platform can be quite challenging. Here, addressing representations become very effective tools for even the most elementary parallel programs.

The two most common distributed data layouts are cyclic and block, examples of which are shown in Figure 5(a) and Figure 5(b), respectively. Here, we assume 4 processors with 32 data elements in an array.

Cyclic Layout			
P_0	P_1	P_2	P_3
0	1	2	3
4	5	6	7
8	9	10	11
12	13	14	15
16	17	18	19
20	21	22	23
24	25	26	27
28	29	30	31

(a)

Block Layout			
P_0	P_1	P_2	P_3
0	8	16	24
1	9	17	25
2	10	18	26
3	11	19	27
4	12	20	28
5	13	21	29
6	14	22	30
7	15	23	31

(b)

Figure 5. Cyclic and block data layouts for 32 elements on 4 processors.

The layout of this 32 element array for the cyclic case consists of the address data mapping shown in Equation (7) where m_i corresponds to a memory address bit and p_i corresponds to a processor address bit.

$$a_4 a_3 a_2 a_1 a_0 \rightarrow m_2 m_1 m_0 p_1 p_0 \tag{7}$$

The corresponding mapping for the block case is shown in Equation (8).

$$a_4 a_3 a_2 a_1 a_0 \rightarrow p_1 p_0 m_2 m_1 m_0 \tag{8}$$

We can ask which is the right mapping and the right algorithm for an FFT on this data array. Using one of the above with our original formulation of the FFT will inevitably require interprocessor communication at many levels since input blocks are accessed with a fundamentally different stride from output. The Cooley-Tukey variant offers hope for efficient computations since the strides for input and output are the same at any stage. Since we proceed by combining input array elements which are separated by half the array length, it makes sense to use the cyclic data layout to initiate the FFT. We observe that, for the first 3 butterflies, the cyclic data layout causes the butterfly operations to only access local data. In the language of the address representation for the first 3 butterflies we convert input bit field to output bit fields in bit-reversed order as in Equation (9).

$$
\begin{array}{ccccc}
a_4 & a_3 & a_2 & a_1 & a_0 \\
m_2 & m_1 & m_0 & p_1 & p_0
\end{array}
\rightarrow
\begin{array}{ccccc}
A_0 & A_1 & A_2 & a_1 & a_0 \\
m_2 & m_1 & m_0 & p_1 & p_0
\end{array}
\tag{9}
$$

The last two butterflies will require some interprocessor communication. We will complete the algorithm in the following section.

4. Address Transformations and Associated Data Communications

Changing the explicit layout of some abstract data structure implicitly requires data movement and very often interprocessor data movement. Issues affecting data communications are blocking factors and network topologies and hierarchies. Blocking factors arise from cache line size, cache block size, the number of processors, the number of memory banks or the length of vector operations, as well as the fundamental lengths of the problem being solved. Network topologies will create classes of efficient communications that can utilize much of the available peak interprocessor bandwidth, and other classes of communications with modest or poor communication bandwidths. In address representation language, blocking factors will typically correspond to runs of address digits and network communications will roughly correspond to sweet spots for address digit positions in an appropriate representation. We have already seen one of the most common manipulations of addresses: the bit circular shift. Bit circular shifts are easy to understand since they correspond to some sort of interleaving, or un-interleaving of data. Bit swaps are another common address manipulation where two runs of bits are swapped. Bit swaps are easy to understand since they correspond to axis transposition for some hypercube grid of data points. Since bit circular shifts preserve more of the total order of the original digits, they tend to be more useful and lead to more aesthetic algorithms.

Let's return to the unfinished FFT. A naive completion of the algorithm would be to complete the last two butterflies by interprocessor communications for each of the last two butterfly operations. A minimum of one complete data movement is required for each butterfly using this approach. Since data communication is already being done, why not rearrange the data so that all the remaining data pairs will be local to the processor that requires them. This single rearrangement will require only one data communication of the whole array.

5. Address Representations: The Gray Code

One of the most surprising concepts in addressing algebra is that unusual interpretation of the meaning of a digit can lead to effective communication strategies. This is the case with Gray code representations where a binary digit value of 0 may, or may not, correspond to the first position in an ordering of a pair of numbers! Consider the Ising problem of a ring of atoms that interact with each other only through their nearest neighbor and that we have a ring of 8 atoms. Each neighbor interacts with only its nearest neighbor. We want to map this problem to a multiprocessor built of 8 processors arranged in a cubical network topology and numbered as in Figure 6.

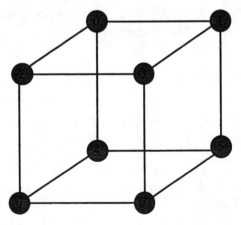

Figure 6. Cubic network of processors with processor labels.

The problem is to map the processes for any atom to an appropriate processor so that communications with neighboring atoms are fast and load-balanced. We observe immediately that the simple mapping of atoms to processors according to processor number does not have good communication properties since some neighboring pairs have long communication paths. An example is the pair of processors 3 and 4 that require 3 hops for a communication between each other. We also observe a property of the cube that any pair of processors separated by only a single link will differ in the binary representation of their processor addresses by a single bit. This is immediately clear from a binary address analysis, since communication in any of the three directions of the cube maps to a simple flipping of a single bit of the address. Communications would be perfect if all neighbors differed from each other by only a single bit. Such a representation is known as a binary Gray code. An example of such a code is the sequence: 0, 1, 3, 2, 6, 7, 5, 4. This results in the mapping shown in Figure 7. A simple nearest neighbor communication is indicated by the arrows.

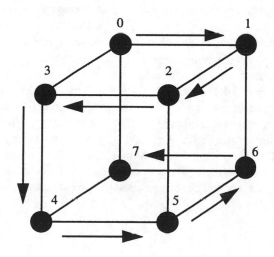

Figure 7. Data layout of ring of atoms on corresponding processors.

There are many possible Gray codes, but the most common is constructed by the following procedure. Start with the binary sequence 0, 1. This is the current sequence. Form the next sequence members by first prepending a binary 0 to the current sequence and then prepend a 1 to the reversed current sequence and replace the current sequence with this new pair of sequences: 00, 01, 10, 11. Then repeat the procedure as necessary: 000, 001, 010, 011, 111, 110, 101, 100, etc. Since we are butting together the same sequences that differ, by construction, at the join by only a single bit, we guarantee that any number in the sequence differs from its prior value by 1 and only 1 bit. In this common Gray coding representation, the ordinal meaning of 0 and 1 at any bit position depends on the whether the value of the next highest bit is 0 or 1. This astonishing property leads to simple algorithms to convert a Gray code representation to a standard binary presentation and vice-versa. From this simple property it is easy to construct a one-line expression to convert a standard address into a corresponding Gray code which is shown in the following C code fragment:

$$gray = (index^\wedge(index<<1))>>1;$$

The Gray code representation can be used to embed any low dimensional data structure onto a higher dimensioned hypercube of processors and, so, finds many uses for hypercube-like multi-processor topologies. Further details for using Gray codes are available in [2]2

6. Summary

We have examined a variety of techniques that involve mapping either algorithmic operations, data layout, or data communications to manipulations of the addresses of the fundamental data array elements. This tutorial is intended only as an introduction to these techniques and there are many texts available to the reader that will explore many of these techniques in great detail.[3][4]34 Today's high performance computers are both complex and powerful. The problems they solve often involve millions of data elements while the hardware has many characteristic lengths that affect performance. Tracking large data arrays through a program's execution poses an extreme challenge to HPC programmers. Addressing algebras can be quite effective for both simplifying the programming and debugging challenge as well as suggesting a variety of alternate formulations that may have attractive performance benefits. These basic techniques comprise a powerful core technology who's mastery will benefit any parallel programmer .

References

1.   1 J.W. Cooley and J.W. Tukey, *Math. Comp.* **19**, 1965, pp. 297-301.

2.   2 G.C. Fox, et al.,*Solving Problems on Concurrent Processors*, Vol. 1,Prentice Hall,1988.

3.   3 Knuth uses these techniques extensively in his classic: D.E. Knuth, *The Art of Computer Programming*,Addison-Wesley, 1981.

4.   4 An excellent text is P. Fletcher, *Regular Mapping of Multi-Dimensional Data on Parallel Processors*, Technical Report TR-HJ-93-05, CSIRO Division of Information Technology, Canberra, Australia, May, 1993.

Optimizing the NAS Parallel BT Application for the POWER CHALLENGEarray

John Brown and Marco Zagha

Silicon Graphics, Inc., 2011 N. Shoreline Blvd., Mountain View, CA 94043-1389

Abstract: The POWER CHALLENGEarray is a coarse-grained collection of large processor SMP nodes. This creates interesting parallelization opportunities for scalable applications. The NAS BT benchmark is a classical ADI-like application with non-trivial communication requirements. The coarse-grained distributed feature of the POWER CHALLENGEarray provides unique parallelization strategies. We explore the implementation of this benchmark on this machine and discuss the general implications for scalable application development

1 Introduction

Clusters of coarse-grained SMP nodes offer interesting opportunities for developing effective parallel applications. Just as powerful microprocessors are attractive central building blocks for Massively Parallel Processors (MPPs), for similar reasons, large SMP nodes can be attractive building blocks for assembling large multiprocessor systems. Examples of these machines are the Silicon Graphics POWER CHALLENGEarray and the DEC TruCluster. We have implemented one of the NAS parallel benchmark applications (BT) on the Silicon Graphics POWER CHALLENGEarray. The algorithmic design and implementation of the BT benchmark show many of the unique algorithmic opportunities of these coarse-grained systems, as well as the inherent challenges for practical programming of these unique architectures.

Arrays of SMP nodes can be very effective architectures for large compute centers. They are both powerful and economically attractive throughput engines for large multi-user computer centers. Typically, they support both shared memory and message passing programming models. They offer many parallelized commercial applications, albeit, that these applications generally exploit only modest levels of parallelism. However, the vast majority of today's commercially available parallel applications are just of this type, and, by far, there are more parallel applications that have been developed for SMP platforms than for any other multiprocessor architecture. Large SMP nodes are very effective parallel development platforms with sophisticated support of debuggers, profilers and parallelizing compilers. Effectively using shared resources, SMP clusters provide powerful and robust throughput price/performance. In addition, the coarse-grained parallelism of these nodes can be exploited for many applications.

There are two chief parallelization benefits to coarse-grained nodes. Many parallel applications exhibit data locality of one sort or another. Large coarse-grained nodes offer the advantage of decreased communication bandwidth requirements through surface to volume effects. Large coarse-grained nodes also allow parallelism with small degrees of freedom to be utilized effectively. This kind of modest parallelism is often ignored with the current concern with scalability, but is very practical for real problems and real machines. This is the kind of parallelism that we exploit for the work reported here.

SMP clusters also pose significant programming challenges. The network technology tying these clusters together creates a distinct memory hierarchy with very non-uniform memory access. The typical programming model will be fundamentally a message passing model, and the highest levels of performance will be achieved with hybrid shared memory and message passing schemes. Since the underlying memory architecture is complex, building general portability with good performance portability is also complex. However, this is also true for many other MPP designs and quite often the programming challenge is justified by the productivity results.

2 The POWER CHALLENGEarray

The Silicon Graphics POWER CHALLENGEarray used in this study consists of eight SMP POWERnodes with up to 18 MIPS R8000 90 MHz processors per POWERnode [1][2]. Each processor has a peak performance of 360 Mflops. The nodes are connected by 2 bi-directional HiPPI channels per POWERnode through a fully connected HiPPI crossbar. A variety of programming styles are possible with the POWER CHALLENGEarray: HPF, PVM and MPI using standard FORTRAN or C. In particular, MPI has been highly optimized for both internode and intranode communication performance. Intranode bandwidths of > 60 MBytes/s and latencies of ~18 μs have been observed. The corresponding internode bandwidth and latency are > 95 MBytes/s and ~110 μs, respectively

3 The Benchmark

The BT application is one of 3 Alternating Direction Implicit (ADI)-like applications which emulate typical CFD applications used by NASA. The application benchmarks consist of a 3-dimensional array of points that are updated successively in the x, y and z directions by solving a system of equations per planar gridpoint. They find a numerical solution to a system of five nonlinear three-dimensional Partial Differential Equations (PDEs) using an ADI method. The NAS Block Tridiagonal (BT) benchmark solves systems with a 5x5 block structure, where the non-zeros of each system are formed along pencils (rows, columns, layers) of the 3-dimensional grid. For an NxNxN grid, NxN independent block tridiagonal systems are solved for each grid dimension. With 2 different grid sizes, we report results on the larger of the two, the Class B problems with 200 iterations on a 102x102x102 3-dimensional mesh. This benchmark is an interesting test of supercomputer performance since it is both compute- and communication intensive, and the 3-dimensional structure offers a lot of flexibility in choosing data-partitioning strategies. Further details about the benchmark are presented in [3].

4 Parallel ADI Algorithms

ADI Algorithms (including the BT benchmark) can be implemented using a wide variety of algorithms, including the following:

Transpose

> The simplest algorithm conceptually is one based on transposing two of the three grid dimensions after each sweep. The main disadvantage of this algorithm is that the communication costs are proportional to the problem size.

Pipelined Gaussian Elimination

> This algorithm uses a one-dimensional decomposition of the 3D grid. Sweeps along two   dimensions use no communication, and sweeps along the third dimension are pipelined. In order to reduce the load imbalance, work is divided into small blocks, where the blocksize is chosen as a compromise between the costs of communication and load imbalance. This technique has been used on the IBM SP [4].

3D Coloring

> An algorithm due to Bruno and Capello [5] uses overpartitioning of the grid into $p^{1/2}*p^{1/2}*p^{1/2}$ blocks. Each processor is assigned $p^{1/2}$ blocks in such a way that for every X, Y, and Z plane of the cube, there is one block per processor. This algorithm eliminates the load-balancing problems of Pipelined Guassian Elimination, but requires extra communication due to the overpartitioning.

5 A New Algorithm

The SGI POWER CHALLENGEarray implementation uses a new algorithm that takes advantages of the unique capabilities of the array. One tremendous advantage that the array has over a classic distributed memory system is that each node of the array is itself a powerful shared-memory system. This greatly simplifies the data distribution problem since a system of up to 144 processors is constructed from only eight array nodes. In fact, with a 3-dimensional problem, the eight nodes can be arranged logically into a cube divided in half along the three dimensions.

Having only two nodes on any slice of the cube enables the use of a new algorithm which takes advantage of symmetry in the underlying tridiagonal linear systems. Though normally LU decomposition proceeds from the upper left corner of the matrix, any corner will do equally well. This observation leads to a simple 2-way parallelism to balance the work. In each grid dimension, the nodes in the array work on half of their linear systems in one direction, and half of them in the other direction. When half of the tridiagonal matrix is factored, the processors exchange a small amount of data and the factoring continues on the opposing processors in the active direction. This algorithm communicates only $O(n^2)$ data rather than the $O(n^3)$ data communicated by a transpose-based algorithm. And in contrast to the Pipelined Guassian Elimination algorithm, this algorithm is perfectly load-balanced. An additional modification of the algorithm to maximize performance is two divide planes into sub-blocks to minimize cache data movements.

6 Performance Model

Performance can be accurately predicted based on a simple performance model. We assume linear speedup in computation and a sustained transfer rate of 30 Mbytes/sec (a conservative estimate including significant application data-movement overheads). Message latencies are negligible. Table 1estimates the communication (I/O) and computation costs for various machine sizes. The border elements column indicates the number of grid points per node on the communication surface. The Mbytes column indicates the number of Mbytes per array node transferred per iteration. It is calculated as the product of several quantities: number of border elements, number of bytes per word (8), number of elements transferred for the u vector (5 * 2 * 2), and number of elements transferred for the tridiagonal solves (5*5 + 5*2). With a baseline time of 32.5 seconds per iteration for a single processor, we obtain performance predictions shown in Table 1.

Table 1. Communication and Computation Complexity

Array Nodes	CPUs	Border Elements	Mbytes Per CPU Per Iter.	Pred. I/O Time	Pred. Compute Time	Pred. Total Time	Meas. Time
1	13	0	0	0	500	500	495
2	26	n*n*1	4.4	29	250	279	280
4	52	n*(n/2)*2	4.4	29	125	154	157
8	104	(n/2)*(n/2)*3	3.3	22	62	84	84

7 Class B Results on Various Machines

Times for Class B results for various configurations of the array are presented in Table 2. Figure 1. shows both performance and relative scaling for the POWER CHALLENGEarray with other various MPP platforms from the December 1995 NAS benchmark performance report [6]. As can be readily seen, the absolute performance per processor and scalability are quite good.

Table 2. POWER CHALLENGEarray NAS BT Parallel Class B Performance Results

Processors	Seconds
8	673
16	392
26 (2 x 13)	280
52 (4 x 13)	157
104 (8 x 13)	84

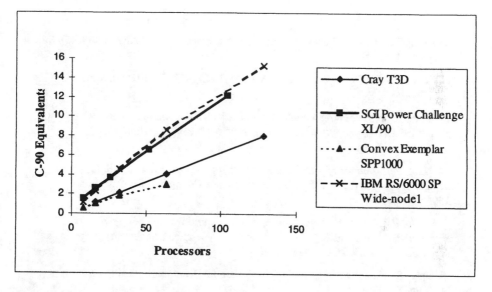

Figure 1. NAS BT Class B Performance Results

8 Conclusions

We have presented an implementation of the Class B NAS BT benchmark targeted to the SGI POWER CHALLNGEarray. This approach shows several fundamental advantages of using arrays of coarse-grained SMP systems to build powerful compute platforms. The small number of nodes and the power of each individual nodes help both to reduce communication to computation requirements as well as enable unique algorithmic approaches that exploit parallelism with inherently small degrees of freedom. While such algorithms will not have scalability, per se, they are quite practical and effective and often used in practice. Coupled with the excellent base support that SMP platforms offer for development of parallel applications as well as the simplifications of programming that coarse nodes provide, SMP arrays built with high bandwidth networks with low latency provide powerful, practical and effective computational platforms for Grand Challenge class problems.

Acknowledgments

Special thanks are given to Jeff McDonald for his original implementation of the NAS BT benchmark for the POWER CHALLENGE. Eric Salo and Jim Pinkerton gave us early access to MPI implementations and helped with performance tuning, and Ira Pramanick and Brian Totty helped with MPI programming questions. Leo Dagum pointed us to several research papers on parallel ADI.

References

1. POWER CHALLENGEarray Technical Report, PWR-CHALL-ARA-TR (1/96), Silicon Graphics, Inc.,1996.

2. POWER CHALLENGE Technical Report, PWR-CHALL-TR (07/94), Silicon Graphics, Inc.,1994.

3. D. Bailey, et al., THE NAS Parallel Benchmarks,RNR Technical Report RNR-94-007, March 1994, or see http://www.nas.nasa.gov/NAS/NPB/Specs/RNR-94-007/npbspec.html.

4. V. Naik, *Performance of NAS Parallel Application-Benchmarks on IBM SP1*,Proceeding of the Scalable High Performance Computing Conference, May 1994,pp 121-128.

5. J. Bruno, P.R. Cappello, *Implementing the Beam and Warming Method on the Hypercube*, Proceedings of 3rd Conference on Hypercube Concurrent Computers and Applications, Pasadena, CA, Jan. 19-20, 1988.

6. Subhash Saini and David H. Bailey, NAS Parallel Benchmark Results 12-95,Report NAS-95-021, Dec. 1995.

Providing Access to High Performance Computing Technologies

Shirley Browne[1], Henri Casanova[1], and Jack Dongarra[2]

[1] University of Tennessee, Knoxville TN 37996 USA
[2] University of Tennessee and Oak Ridge National Laboratory

Abstract. This paper describes two projects underway to provide users with access to high performance computing technologies. One effort, the National HPCC Software Exchange, is providing a single point of entry to a distributed collection of domain-specific repositories. These repositories collect, catalog, evaluate, and provide access to software in their specialized domains. The NHSE infrastructure allows these repositories to interoperate with each other and with the top-level NHSE interface. Another effort is the NetSolve project which is a client-server application designed to solve computational science problems over a network. Users may access NetSolve computational servers through C, Fortran, MATLAB, or World Wide Web interfaces. An interesting intersection between the two projects would be the use of the NetSolve system by a domain-specific repository to provide access to software without the need for users to download and install the software on their own systems.

1 The National HPCC Software Exchange

1.1 Overview of the NHSE

The National HPCC Software Exchange (NHSE) is an Internet-accessible resource that facilitates the exchange of software and of information among research and computational scientists involved with High Performance Computing and Communications (HPCC) [1] [3]. The NHSE facilitates the development of discipline-oriented software repositories and promotes contributions to and use of such repositories by Grand Challenge teams, as well as other members of the high performance computing community.

The expected benefits from successful deployment of the NHSE include the following:

- Faster development of better-quality software so that scientists can spend less time writing and debugging programs and more time on research problems.
- Reduction of duplication of software development effort by sharing of software.
- Reduction of time and effort spent in locating relevant software and information through the use of appropriate indexing and search mechanisms and domain-specific expert help systems.

---

[3] http://www.netlib.org/nhse/

- Reduction of duplication of effort in evaluating software by sharing software review and evaluation information.

The scope of the NHSE is software and software-related artifacts produced by and for the HPCC Program. Software-related artifacts include algorithms, specifications, designs, and software documentation. The following three types of software being made available:

- Systems software and software tools. This category includes parallel processing tools such as parallel compilers, message-passing communication subsystems, and parallel monitors and debuggers.
- Basic building blocks for accomplishing common computational and communication tasks. These building blocks will be of high quality and transportable across platforms. Building blocks are meant to be used by Grand Challenge teams and other researchers in implementing programs to solve computational problems. Use of high-quality transportable components will speed implementation, as well as increase the reliability of computed results.
- Research codes that have been developed to solve difficult computational problems. Many of these codes will have been developed to solve specific problems and thus will not be reusable as is. Rather, they will serve as proofs of concept and as models for developing general-purpose reusable software for solving broader classes of problems.

1.2 Domain-specific Repositories

The effectiveness of the NHSE will depend on discipline-oriented groups and Grand Challenge teams having ownership of domain-specific software repositories. The information and software residing in these repositories will be best maintained and kept up-to-date by the individual disciplines, rather than by centralized administration. Domain experts are the best qualified to evaluate, catalog, and organize software resources within their domain.

Netlib – Mathematical Software An example of a domain-specific repository is the Netlib mathematical software repository, which has been in existence since 1985 [2]. Netlib differs from other publicly available software distribution systems, such as Archie, in that the collection is moderated by an editorial board and the software contained in it is widely recognized to be of high quality. Netlib distributes freely-available numerical libraries such as EISPACK, LINPACK, FFTPACK, and LAPACK that have long been used as important tools in scientific computation. The Netlib collection also includes a large number of newer, less well-established codes. Software is available in all the major numerical analysis areas, including linear algebra, nonlinear equations, optimization, approximation, and differential equations. Most of the software is written in Fortran, but programs in other languages, such as C and C++, are also available. Netlib uses the Guide to Available Mathematical Software (GAMS) classification system [3] to help users quickly locate software that meets their needs.

A branch of Netlib specialized to high performance computing, called HPC-netlib, is currently under development. HPC-netlib will provide access to algorithms and software for both shared memory and distributed memory machines, as well as to information about performance of parallel numerical software on different architectures.

PTLIB – Parallel Tools Another domain-specific repository that is under development is the PTLIB parallel tools repository. PTLIB will provide access to high-quality tools in the following areas: communication libraries, data parallel language compilers, automatic parallelization tools, debuggers and performance analyzers, parallel I/O, job scheduling and resource management.

1.3 Repository Interoperation

In addition to providing access to its own software, a repository may wish to import software descriptions from other repositories and make this software available from its own interface. For example, a computational chemistry repository may wish to provide access to mathematical software and to parallel processing tools in a manner tuned to the computational chemistry discipline. A repository interoperation architecture is shown in Figure 1.

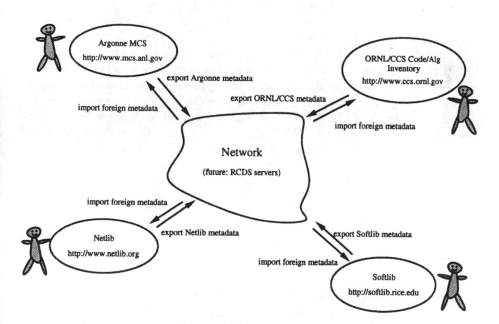

Fig. 1. Repository Interoperation Architecture

The NHSE is using the Reuse Library Interoperability Group's Basic Interoperability Data Model (BIDM) as its interoperability mechanism [4]. Participating HPCC repositories and some individual contributors have placed META and LINK tags in the headers of HTML files that describe their software resources. This information may then be picked up by other repositories and incorporated into their own software catalogs. The NHSE is developing a toolkit called Repository in a Box (RIB) that will assist repository maintainers in creating and maintaining software catalog records, in exchanging these records with other repositories (including the top-level virtual NHSE repository), and in providing a user interface to their software catalog. The Resource Cataloging and Distribution System under development at the University of Tennessee will provide a scalable substrate for repository interoperability by providing catalog and location servers that map resource names to catalog and location information.

1.4   Software Review Framework

The NHSE has designed a software review policy that enables easy access by users to information about software quality, but which is flexible enough to be used across and specialized to different disciplines. The three review levels recognized by the NHSE are the following: Unreviewed, Partially reviewed, and Reviewed. The *Unreviewed* designation means only that the software has been accepted into the owning repository and is thus within the scope of HPCC and of the discipline of that repository. The *Partially reviewed* designation means that the software has been checked by a librarian for conformance with the scope, completeness, adequate documentation, and construction guidelines. The *Reviewed* designation means that the software has been reviewed by an expert in the appropriate field, for example by an author of a review article in the electronic journal *NHSE Review* [4], and found to be of high quality. Domain-specific repositories and expert reviewers are expected to refine the NHSE software review policy by adding additional review criteria, evaluation properties, and evaluation methods and tools. The NHSE also provides for soliciting and publishing author claims and user comments about software quality. All software exported to the NHSE by its owning repository or by an individual contributor is to be tagged with its current review level and with a pointer to a review abstract which describes the software's current review status and includes pointers to supporting material.

2   The NetSolve project

An ongoing thread of research in scientific computing is the efficient solution of large problems. Various mechanisms have been developed to perform computations across diverse platforms. The most common mechanism involves software libraries. Unfortunately, the use of such libraries presents several difficulties. Some software libraries are highly optimized for only certain platforms and do not provide a convenient interface to other computer systems. Other libraries

---

[4] http://nhse.cs.rice.edu/NHSEreview/

demand considerable programming effort from the user, who may not have the time to learn the required programming techniques. While a limited number of tools have been developed to alleviate these difficulties, such tools themselves are usually available only on a limited number of computer systems. MATLAB [5] is an example of such a tool.

These considerations motivated the establishment of the NetSolve project. NetSolve is a client-server application designed to solve computational science problems over a network. A number of different interfaces have been developed to the NetSolve software so that users of C, Fortran, MATLAB, or the World Wide Web can easily use the NetSolve system. The underlying computational software can be any scientific package, thus helping to ensure good performance through choice of an appropriate package.. Moreover, NetSolve uses a load-balancing strategy to improve the use of the computational resources available. Some other systems are currently being developed to achieve somewhat similar goals. Among them are the Network based Information Library for high performance computing (Ninf) [6] project which is very comparable to NetSolve in its way of operation, and the Remote Computation System (RCS) [7] which is a remote procedure call facility for providing uniform access to a variety of supercomputers.

We introduce the NetSolve system, its architecture and the concepts on which it is based. We then describe how NetSolve can be used to solve complex scientific problems.

2.1 The NetSolve System

The NetSolve system is a set of loosely connected machines. By *loosely* connected, we mean that these machines can be on the same local network or on an international network. Moreover, the NetSolve system can be running in a *heterogeneous* environment, which means that machines with different data formats can be in the system at the same time.

The current implementation sees the system as a completely connected graph without any hierarchical structure. This initial implementation was adopted for simplicity and is viable for now. Our current idea of the *NetSolve world* is of a set of independent NetSolve systems in different locations, possibly providing different services. A user can then contact the system he wishes, depending on the task he wants to have performed and on his own location. In order to manage efficiently a pool of hosts scattered on a large-scale network, future implementations might provide greater structure (e.g., a tree structure), which will limit and group large-range communications.

Figure 2 shows the global conceptual picture of the NetSolve system. In this figure, a NetSolve client send a request to the NetSolve agent. The agent chooses the "best" NetSolve resource according to the size and nature of the problem to be solved.

Several instances of the NetSolve agent can exist on the network. A good strategy is to have an instance of the agent on each local network where there are NetSolve clients. Of course, this is not mandatory; indeed, one may have only a single instance of the agent per NetSolve system.

Every host in the NetSolve system runs a NetSolve *computational* server (also called a *resource*, as shown in Figure 2). The NetSolve resources have access to scientific packages (libraries or stand-alone systems).

An important aspect of this server-based system is that each instance of the agent has its own *view* of the system. Therefore, some instances may be aware of more details than others, depending on their locations. But eventually, the system reaches a stable state in which every instance possesses all the available information on the system.

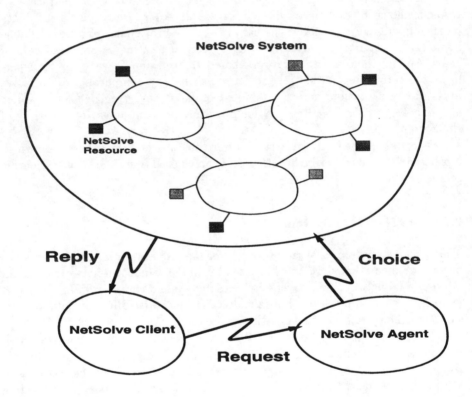

Fig. 2. The NetSolve System

Communication within the NetSolve system is achieved through the TCP/IP socket layer and heterogeneous environments are supported thanks to the XDR protocol [8].

2.2 Problem Specification

To keep NetSolve as general as possible, we needed to find a formal way of describing a problem. Such a description must be carefully chosen, since it will affect the ability to interface NetSolve with arbitrary software.

A problem is defined as a 3-tuple: $< name, inputs, outputs >$, where

- *name* is a character string containing the name of the problem
- *inputs* is a list of input objects
- *outputs* is a list of output objects

An object is itself described as follows: $< object, data >$, where *object* can be 'MATRIX', 'VECTOR' or 'SCALAR' and *data* can be any of the standard FORTRAN data types. This description has proved to be sufficient to interface NetSolve with numerous software packages. The NetSolve administrator can then not only choose the best platform on which to install NetSolve, but also select the best packages available on the chosen platform.

The current installation of NetSolve at the University of Tennessee uses the BLAS [9], [10], [11], LAPACK [12], ItPack [13], LINPACK [14] and FitPack [15]. These packages are available on a large number of platforms and are freely distributed. The use of ScaLAPACK [16] on massively parallel processors would be a way to use the power of high-performance parallel machines via NetSolve.

2.3 Client Interfaces

One of the main goals of NetSolve is to provide the user with a large number of interfaces and to keep them as simple as possible.

The MATLAB Interface :

We developed a MATLAB interface which provides interactive access to the NetSolve system. Interactive interfaces offer several advantages. First, they are easy to use because they completely free the user from any code writing. Second, the user still can exploit the power of software libraries. Third, they provide good performance by capitalizing on standard tools such as MATLAB. Let us assume, for instance, that MATLAB is installed on one machine on the local network. It is possible to use NetSolve via the MATLAB interface on this machine and in fact use the computational power of another more powerful machine where MATLAB is not available.

Within MATLAB, NetSolve may be used in two ways. It is possible to call NetSolve in a *blocking* or *nonblocking* fashion. Here is an example of the MATLAB interface to solve an linear system computation using the blocking call:

```
>> a = rand(100); b = rand(100,1);
>> x = netsolve('ax=b',a,b)
```

This MATLAB script first creates a random 100×100 matrix, a, and a vector b of length 100. The call to the **netsolve** function returns with the solution. This call manages all the NetSolve protocol, and the computation may be executed on a remote host.

Here is the same computation performed in a nonblocking fashion:

```
>> a = rand(100); b = rand(100,1);
>> request = netsolve_nb('send','ax=b',a,b)
>> x = netsolve_nb('probe',request)
   Not Ready Yet
>> x = netsolve_nb('wait',request)
```

Here, the first call to netsolve_nb() sends a request to the NetSolve agent and returns immediately with a request identifier. One can then either *probe* for the request or *wait* for it. Probing always returns immediately, either signaling that the result is not available yet or, if available, stores the result in the user data space. Waiting blocks until the result is available and then store it in the user data space. This approach allows user-level parallelism and communication/computation overlapping (see Section 2.4).

Other functions are provided, for example, to obtain information on the problems available or on the status of pending requests.

C and FORTRAN interfaces :

In addition to the MATLAB interface, we have developed two programming interfaces, one for Fortran and one for C. Unlike the interactive interfaces, programming interfaces require some programming effort from the user. But again, with a view to simplicity, the NetSolve libraries contain only a few routines, and their use has been made as straightforward as possible. As in MATLAB, the user can call NetSolve *asynchronously*.

A very attractive feature of these interfaces is that NetSolve preserves the original calling sequence of the underlying numerical software. It is then almost immediate to convert a code to NetSolve, as shown in the short FORTRAN example below :

```
C     Linear system solve : Call to LAPACK

      call DGESV(N,1,A,MAX,IPIV,B,MAX,INFO)


C     Linear system solve : Call to NetSolve

      call FNSOLVE('DGESV',NSINFO,
                   N,1,A,MAX,IPIV,B,MAX,INFO)
```

2.4 Performance

One of the challenges in designing NetSolve was to combine ease of use and excellence of performance. Several factors ensure good performance without increasing the amount of work required of the user. In addition to the availability of diverse scientific packages (as discussed in a preceding section), these factors include load balancing and the use of simultaneous resources.

Load Balancing :

Load balancing is one of the most attractive features of the NetSolve project. NetSolve performs computations over a network containing a large number of machines with different characteristics, and one of these machines is the most suitable for a given problem, meaning the one yielding the shortest response time. NetSolve provides the user with a "best effort" to find this *best* resource.

As seen on figure 2, a NetSolve client sends a request to an instance of the NetSolve agent. This instance of the agent has some knowledge about the computational resources in the system. Hopefully this knowledge is not too out of date (which is ensured by a set of protocols we do not have space to describe here) and, for each resource M, allows a fairly accurate computation of :

- T_n : the time to send the data to M and receive the result over the network, and
- T_c : the time to perform the computation on M.

All the details about the protocols involved in this computations are given in [17]. The whole idea behind this scheme is that it would be too inefficient to have the agent compute exact values for T_n and T_c for each incoming request. Instead, we prefer to have a quick estimate, which might not be as accurate.

Simultaneous resources :

Using the nonblocking interfaces to NetSolve, the user can design a Net-Solve application that has some parallelism. Indeed, it is possible to send asynchronously several requests to NetSolve. The load balancing strategy described above insures that these problems will be solved on different machines, in parallel. The client has then just to wait for the results to come back.

Here is another strength of NetSolve : as soon as a new resource is started, it takes part in the system, and can be used. Therefore, without modifying his code or knowing in fact anything about the servers, a user can see the performance of his application greatly improved.

2.5 Fault Tolerance

Fault tolerance is an important issue in any loosely connected distributed system like NetSolve. The failure of one or more components of the system should not cause any catastrophic failure. Moreover, the number of side effects generated by such a failure should be as low as possible, and the system should minimize the drop in performance. We tried to make NetSolve as fault tolerant as possible.

A first aspect of fault-tolerance in NetSolve takes place at the server level. It is possible to stop a NetSolve server (resource or instance of the agent) at any time, and restart it safely at any time. In fact, every NetSolve server is an independent entity. This insures that the NetSolve system will remain coherent after any kind of network/machine problem. In the installation of NetSolve at the University of Tennessee, the whole system is managed by a 'cron' job, and servers are restarted automatically after machines go down for back-ups for instance.

Another aspect of fault tolerance is that it should minimize the side effects of failures. To this end, we designed the client-server protocol as follows. When the NetSolve agent receives a request for a problem to be solved, it sends back a list of computational servers sorted from the most to the least suitable one. The client tries all the servers in sequence until one accepts the problem. This strategy allows the client to avoid sending multiple requests to the agent for the same problem if some of the computational servers are stopped. If at the end of the list no server has been able to answer, the client asks another list from the agent. Since it has reported all the encountered failures, it will receive a different list.

Once the connection has been established with a computational server, there still is no guarantee that the problem will be solved. The computational process on the remote host may die for some reason. In that case, the failure is detected by the client, and the problem is sent to another available computational server. This process is transparent to the user but, of course, lengthens the execution time. The problem is migrated between the possible computational servers until it is solved or no server remains.

3   Conclusions and Future Work

The NHSE is providing a means for the HPCC community to share software and information and thus broaden and accelerate the use of high performance computing technologies in scientific and engineering applications. By supplying the tools and mechanisms for HPCC repositories to interoperate, the NHSE is enabling different HPCC agencies and research groups to leverage each others efforts. During the next year, the NHSE will be bringing online several new domain-specific repositories as well as promoting the review and evaluation of software in these domains.

The NetSolve project is still at an early development stage and there is room for improvement at the interface level as well as at the conceptual level. At the interface level, we are thinking of providing a Java interface to NetSolve. At the conceptual level, the load-balancing strategy must be improved in order to change the "best guess" into a "best choice" as much as possible. The challenge is to come close to a best choice without flooding the network. The danger is to waste more time computing this best choice than the computation would have taken in the case of a best guess only. All these improvements are intended to combine ease of use, generality and performance, the main purposes of the NetSolve project.

We plan to investigate extending the NHSE Repository in a Box toolkit with a remote execution facility based on NetSolve. This facility would allow repository maintainers to provide remote access to software, instead of having users download and install the software on their own systems. We will also investigate how to provide server safe execution environments for user code so that users may upload functions for execution on a remote server. This capability is important for software packages that require user-defined functions to be provided as input.

References

1. Shirley Browne, Jack Dongarra, Stan Green, Keith Moore, Tom Rowan, Reed Wade, Geoffrey Fox, Ken Hawick, Ken Kennedy, Jim Pool, Rick Stevens, Robert Olsen, and Terry Disz. The National HPCC Software Exchange. *IEEE Computational Science and Engineering*, 2(2):62–69, 1995.
2. Jack J. Dongarra and Eric Grosse. Distribution of mathematical software via electronic mail. *Communications of the ACM*, 30(5):403–407, May 1987.
3. Ronald F. Boisvert, Sally E. Howe, and David K. Kahaner. GAMS: A framework for the management of scientific software. *ACM Transactions on Mathematical Software*, 11(4):313–355, December 1985.
4. Shirley Browne, Jack Dongarra, Kay Hohn, and Tim Niesen. Software repository interoperability. Technical Report CS-96-329, University of Tennessee, 1996.
5. Inc The Math Works. *MATLAB Reference Guide*. 1992.
6. *Ninf : Network based Information Library for Globally High Performance Computing*. Proc. of Parallel Object-Oriented Methods and Applications (POOMA), Santa Fe, 1996.
7. W. Gander P. Arbenz and M. Oettli. The remote computational system. *Lecture Note in Computer Science, High-Performance Computation and Network*, 1067:662–667, 1996.
8. Sun Microsystems, Inc. XDR: External Data Representation Standard. RFC 1014, Sun Microsystems, Inc., June 1987.
9. D. Kincaid C. Lawson, R. Hanson and F. Krogh. Basic linear algebra subprograms for fortran usage. *ACM Transactions on Mathematical Software*, 5:308–325, 1979.
10. S. Hammarling J. Dongarra, J. Du Croz and R. Hanson. An extended set of fortran basic linear algebra subprograms. *ACM Transactions on Mathematical Software*, 14(1):1–32, 1988.
11. I. Duff J. Dongarra, J. Du Croz and S. Hammarling. A set of level 3 basic linear algebra subprograms. *ACM Transactions on Mathematical Software*, 16(1):1–17, 1990.
12. E. Anderson, Z. Bai, C. Bischof, J. Demmel, J. Dongarra, J. Du Croz, A. Greenbaum, S. Hammarling, A. McKenney, S. Ostrouchov, and D. Sorensen. *LAPACK Users' Guide*. SIAM Philadelphia, Pennsylvania, 2 edition, 1995.
13. David M. Young David R. Kincaid, John R. Respess and Roger G. Grimes. Itpack 2c: A fortran package for solving large sparse linear systems by adaptive accelerated iterative methods. Technical report, University of Texas at Austin, Boeing Computer Services Company, 1996.
14. C. B. Moler J. J. Dongarra, J. R. Bunch and G. W. Stewart. *LINPACK Users' Guide*. SIAM Press, 1979.
15. A. Cline. Scalar- and planar-valued curve fitting using splines under tension. *Communications of the ACM*, 17:218–220, 1974.
16. J. Dongarra and D. Walker. Software libraries for linear algebra computations on high performance computers. *SIAM Review*, 37(2):151–180, 1995.
17. *NetSolve: A Network Server for Solving Computational Science Problems*. To appear in Proc. of Supercomputing '96, Pittsburgh, 1996.

Parallel Object-Oriented Library of Genetic Algorithms

Marian Bubak[1,2], Waldemar Cieśla[1], and Krzysztof Sowa[1]

[1] Institute of Computer Science, AGH, al. Mickiewicza 30, 30-059 Kraków, Poland
[2] Academic Computer Centre – CYFRONET, Nawojki 11, 30-950 Kraków, Poland
email: bubak@uci.agh.edu.pl

Abstract. This paper describes a parallel genetic algorithms library which enables easy development of parallel programs exploiting genetic algorithm approach. The library is based on TOLKIEN C++ sequential library. SPMD paradigm is applied and PARA++ library is used for message passing. Island and global population models of parallelism have been implemented. The library may be used on virtual network computer and on parallel machines.

1 Introduction

Genetic algorithms (GA) are particularly useful when methods based on domain specific knowledge are not developed yet or are not available because of the nature of a problem [1, 2, 3]. GAs have proven to be useful in hard optimisation problems, classification, control of very large scale and varied data, for financial and economic modelling and decision making, as well as in so called bio-informatics [4], engineering and industry [5]. As parallelism is an intrinsic feature of GAs, many implementation of parallel genetic algorithms are reported [6, 7].

After analysis of publicly available libraries and systems of GAs [8, 9] we have decided to develop parallel genetic algorithms library which should enable easy creation of parallel programs exploiting GA approach. The library should be flexible, easy maintainable, with maximal attainable performance of its routines and portable to wide range parallel computing environments including virtual network computers and massively parallel machines.

2 Major design features

We have decided to reuse existing code which implements sequential GAs and to apply SPMD paradigm as a base of the library architecture to ensure ease of programming. As the result of tradeoff between promise to improve the programming process offered by object-oriented techniques and computation efficiency requirements we have decided to use C++ language [10].

Parallel GA library is based on TOLKIEN [11, 12] – C++ sequential GAs library with clear and flexible architecture and extensive documentation, and on

PARA++ [13, 14] – C++ library for streams-based message passing which is very convenient layer of abstraction between parallel GA code and underlying lower-level environments PVM and MPI.

Two models of GA parallelism have been chosen for implementation: the island model and the global population one [7, 15]. In the island model individuals are divided into several subpopulations assigned to concurrent processes which perform local genetic operations and solutions are exchanged by migration. In the global population model a temporary global population is created once for every several generations by gathering all individuals in one process and redistributing them randomly.

In order to reduce execution time of parallel GA programs on networked workstations the library should be equipped with a load balancing facility.

3   Classes representing individuals

A part of a genetic algorithm program closest to the problem being solved includes the way of representing individuals specific to the application domain and appropriate genetic operators. C++ classes responsible only for this range of functionality are derived from TGAObj class of TOLKIEN library [11]. The best example of such a class is BinHaploid which implements the most frequently used binary string representation. Requirements concerning classes in the case of parallel genetic algorithms must be significantly extended. Namely, such classes should have the ability of being sent between processes working in parallel (on networked workstations or on parallel computers). As an additional feature they should offer the possibility of saving and restoring their complete internal state on files. To fulfil these requirements retaining clear and uniform design, an abstract base class called MobileTGA has been derived from TGAObj. Inheritance hierarchy of all classes discussed here is presented in Fig. 1. Such approach, when there is one class from which all the individual classes used must be derived, has an advantage of possibility to apply compile-time static type checking to ensure that initial population state is valid. Thus, every new created class representing program should have defined two fundamental groups of member functions (which are pure virtual in MobileTGAObj class):

- genetic algorithm specific functions:
  void randomise();
  used for random initialisation of individual representation;
  TGAObj *oddPtCrossover(const TGAObj &, int) const;
  TGAObj *evenPtCrossover(const TGAObj &, int) const;
  TGAObj *uniformCrossover(const TGAObj &, float) const;
  which implement basic techniques of crossover operators appropriate to the representation used;
  void mutate(float);
  performing mutation operator;

– functions supporting space and time persistence:
`ParaStreamOut &sendCopy(ParaStreamOut &) const;`
`ParaStreamIn &receiveCopy(ParaStreamIn &) const;`
used to send internal state of objects using member functions of `ParaStream-Out` and `ParaStreamIn` classes from Para++ library, which enable only sending primitive data types;
`void printAllOn(ostream &) const;`
`void readAllFrom(istream &);`
which implement writing and reading internal object state to and from streams.

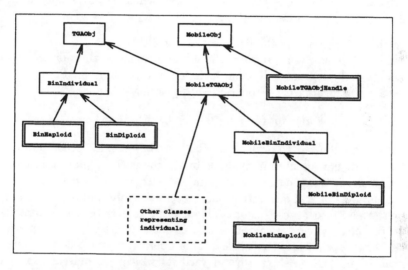

Fig. 1. Inheritance hierarchy of classes representing individuals.

Two basic classes: `MobileBinHaploid` and `MobileBinDiploid` which represent binary individuals and obey the rules described above are included into the library and may be used in typical GA applications. All classes derived from `MobileTGAObj` should also support run-time type information mechanisms which are used mainly by the class `MobileTGAObjHandle`. This solution enables to use heterogeneous population of individuals, what in turn means that objects of different types may coevolve in the same run of program provided that crossover operators between them are implemented. To make it possible, objects of class `MobileTGAObjHandle` act as wrappers to objects of arbitrary concrete subclass of `MobileTGAObj` sending their internal state through a network or writing it to files together with unique type identifiers. When reverse operations are executed (receiving from the network and reading from the file), default objects of appropriate class are created by the `MobileTGAObjHandle` member functions using received identifier value and then their internal state is initialised according to flowing in data.

4 Local operations

In any model of parallel genetic algorithm evolving individuals are distributed among available processors. In result, in any run of a parallel GA there are two parts which are realized alternately:

- local operations performed independently in every parallel process; they include evaluations of values of objective function for elements of subpopulation, transformation of these values into fitness factors, selection of individuals, crossover, mutation and replacement of old population elements with new-created individuals;
- exchange of individuals between processes which requires communication and synchronisation between them.

This section describes only the first part of parallel GA.

User developing PGA program should implement evaluation of objective function for every type of individual included into population. Routine written should have the form of C++ function with prototype like

```
double ObjectiveFunction(const TGAObj &individual);
```

This routine is used to evaluate initial population and to compute objective function for new-created individuals or for the entire population, e.g. when function characteristics change with time or with generation number. In the next step of local operations during one generation, values of objective function are transformed into fitness factors. This is done to ensure that fitness factors used in the next phases are positive and to modify them slightly in order to achieve better optimisation efficiency. This operation and following ones, are implemented in TOLKIEN in several variants which are reused also in PGA library. Natural way of expressing such relationship in object-oriented languages is inheritance. Abstract base class ScalingScheme was created with several derived classes: NonNegScaling, LinearScaling and SigmaScaling; their names are self-explanatory [1].

When fitness factors are calculated, one may choose a pool of individuals that will undergo further operations [1, 15]. This may be done with several new classes which have been derived from common abstract base called SelectionScheme:

- RW_Select (roulette wheel selection with replacement);
- RWwoR_Select (roulette wheel selection without replacement);
- SUS_Select (stochastic universal sampling);
- TournSelect;
- LinearRanking.

All variants of crossover and mutation operations are implemented as member functions of classes describing individuals because they must match the specific representation used. Class Crossover is therefore created only to choose appropriate operation during run time and execute it with given parameters (crossover probability). It has two subclasses: MultiPtCrossover and UniformCrossover.

As a result of genetic operations a temporary pool of new individuals is created and it must replace a part of the old population. In TOLKIEN library subclasses of Population hold collection of individuals of types derived from TGAObj together with objective function, and they implement several strategies of replacement of old elements. As in PGA program every individual may be sent to subpopulation residing in other process, elements in local population must be restricted to some subclass of MobileTGAObj. That's why abstract base class Subpopulation was introduced with the following subclasses:

- SimpleSubpopulation – replaced individuals are chosen randomly;
- EliteSubpopulation – elements with the worst fitness are deleted first;
- CrowdingSubpopulation – De Jong's crowding model.

5 Island model

This and the next sections which describe library of parallel genetic algorithms address all aspects of communication between processes. First, two models of interchange of individuals between parts of population belonging to different processes are described: an island model and a global population one. The abstract base class SPMD_PGA is the root of inheritance hierarchy of all models of PGA implemented in the library 2.

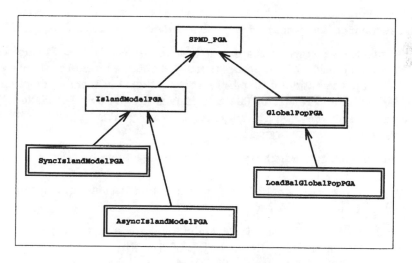

Fig. 2. Inheritance hierarchy of PGA models.

Prototype of protected constructor of class SPMD_PGA is

```
SPMD_PGA(Subpopulation *subpopP, SelectionScheme *locSelP,
    Crossover *crossP, float mutRate, float genGap,
    ScalingScheme *scalP = NULL, BOOL reEval = FALSE);
```

Arguments subpopP, locSelP, crossP and scalP are pointers to variants of classes described in previous section. mutRate is mutation probability, genGap (generation gap) is a parameter which determines number of individuals that undergo genetic operations. When reEveal is TRUE then the whole population is evaluated in every generation. Parameters described here are also put on argument lists of constructors of all concrete classes derived from SPMD_PGA. The most important member function of SPMD_PGA is

```
void generation();
```

It executes single complete step of an genetic algorithm. This step includes local genetic operations and exchange of individuals between subpopulations. Local operations are organised in the same way for all subclasses of SPMD_PGA.

In the island model data which represent individuals are sent according to predefined migration topology [7, 15]. The topology may be defined as a graph with subpopulations as nodes and two parameters for each edge: migration interval and sent fraction of subpopulation. Its representation has the form of class MigrationTopology which may be initialised with static parameters using one of predefined functions, e.g.:

```
void    bidirectionalRing(int interval,  float  subpopFrac);
```

Object of class MigrationTopology describes only edges adjacent with local subpopulation and their parameters may be set with functions:

```
void  setDestination(int  destNum, int interval, floatsubpopFrac);
void  setSource(int  sourceNum, int  interval, float subpopFrac);
```

Member functions of this class are based on the object of class ParaProcess from Para++ library. TIDs of processes may be obtained from the same class.

Actual implementation of island model is contained in member functions of three classes: base IslandModelPGA and derived from it SyncIslandModelPGA and AsyncIslandModelPGA. Their constructors are similar to that of SPMD_PGA and have two additional parameters:

- SelectionScheme *migrSelP – pointing to object used to select individuals for migration;
- MigrationTopology *migrTopP – pointing to object describing migration topology to use in PGA run.

In both variants of island model individuals are selected and sent in the same way according to description in object of class MigrationTopology. In the synchronous version of the island model, local process has access to numbers of individuals that are sent from other processes and local program waits for all of them. In this algorithm processes are partially synchronised during migration and there is guarantee that every individual enters destination subpopulation in the same generation in which it has left its origin. In asynchronous variant, individuals are received in non-blocking manner. This means that local program doesn't wait even for a moment and inserts in the subpopulation all individuals that has arrived to the process so far.

6 Global population model

Global population model implemented in this library has similar distribution of individuals as the island one. This may seem strange at first, however, actual work of this algorithm is very close to those traditional, widely described in literature as centralised ones or algorithms with global population [6, 7]. The advantage of this approach is that uniform design of both models is retained in the library. Interchange of individuals between processes has form of creation of a temporary global population. In first phase all individuals are sent to one process which establishes global population. Then it is randomly permutated, and elements are redistributed and sent back to processes. This action may take place in every generation or rarely to avoid too large communication overheads. The model described here is implemented with the class GlobalPopPGA derived from SPMD_PGA so its constructor has the same parameters as in the base class besides two additional parameters:

- ParaProcess *paraProcP – pointing to object describing TIDs of processes taking part in PGA run;
- int interval – parameter determining how often global population is created.

7 Load balancing

Functionality of the library described so far enables to assign to processes a given number of individuals which remain unchanged during the whole run of a PGA program. This is a serious drawback when GA program is running on most frequently accessible computational systems, namely networked workstation and parallel machines which are shared with other users. Nonuniform and changing with time load across available processors leads to situations when processes temporarily faster have to wait for their slower partners. This problem may be solved by adding load balancing facilities [16] to the library. As the first possibility one may consider a very simple and natural extension of the global population model. During redistribution phase pool of individuals is partitioned accordingly to relative temporary speeds of processors. New number od individuals n_k^{s+1} assigned to process k is calculated as

$$n_k^{s+1} = n_k^s + \lambda \cdot \left(\frac{\frac{n_k^s}{t_k^s}}{\sum_{i=1}^N \frac{n_i^s}{t_i^s}} \cdot \sum_{i=1}^N n_i^s - n_k^s \right)$$

where n_i^s denote the distribution of individuals in the previous stage of PGA program, t_i^s are recent computation times of processes and $\lambda \in [0,1]$ is a parameter (*transfer*) which allows to reduce too violent changes of number of individuals on processors. In order to apply this load balancing algorithm a programmer should use the class LoadBalGlobalPopPGA instead of the base class GlobalPopPGA. The constructor of LoadBalGlobalPopPGA class has one additional parameter – transfer.

The second approach is more universal and flexible, and it may be applied with each model implemented in the library. Algorithm of finding out the new distribution is the same as the previous one. The implementation is contained in the class LoadBalancing with its member function balance() which calculates new distribution of individuals and sends to processors only required part of individuals. Element of subpopulation are accessed with methods of class SubpopInterface which provide operations of adding and removal of individuals.

8 File and I/O operations

Large part of PGA program consists of saving and restoring population state, reading in parameters and storing obtained results. These operations are especially hard to program because processes are executed on different machines. Therefore some facilities have been developed to arrange access to a common file and input/output streams through a single process.

They include two classes: CommonOutputStream and CommonInputStream which provide overloaded write() and read() member functions for several data types e.g. double, MobileTGAObj, MobileTGAObjHandle and Subpopulation. These operations are realized with star topology communication.

9 Efficiency of the library

Two test problems have been chosen: minimisation of fourth De Jong function [1] (F4) and solving of the travelling salesman problem for 48 points (*att48* [17]) (TSP). Most important parameters are given in Tab. 1.

First, test computations were done on homogeneous cluster of IBM RS/6000 workstations with g++ v2.7.2 compiler, PVM v3.3.8 and Para++ v1.03. Tests comprise mainly comparison of runs of two PGA models for different number of processors. A few experiments were done to find out the influence of parameters governing the intensity of individuals exchange on results of optimisation. Evaluation of obtained results is done using several criteria which allow to focus on different aspects of PGA runs. They include:

- execution speed expressed as time of computation or one of the quantities commonly used in parallel computing, e.g. speed-up, efficiency;
- search quality which show how the best value of objective function obtained by optimising PGA changes with generation number;
- search efficiency being the most important criterion because it expresses the dependence of the best solution found on the computation time.

Fig. 3 and Fig. 4 show exemplary execution times for PGA solving both test problems as a function of number of processors for island and global population models, respectively. Obtained speed-ups are rather far from linear, efficiency for 4 processors is 0.6 – 0.7. This is due to large communication between processes which is required to the keep rate of individuals exchange high enough.

Parameter	minimisation of F4	T S Problem
number od subpopulations	4	4
subpopulation class	elite	elite
size of population	200	1000
number of generations	200	500
scaling scheme	non-negative values	non-negative values
generation gap	0.7	0.7
selection scheme	roulette-wheel	roulette-wheel
crossover scheme	one point	PMX
crossover rate	0.5	0.5
mutation rate	0.03	0.03
selection for migration	roulette-wheel	roulette-wheel
migration topology	bidirectional ring	bidirectional ring
migration interval	4	16
migration fraction	0.2	0.2
global population interval	4	16

Table 1. Values of parameters for F4 and TSP test problems.

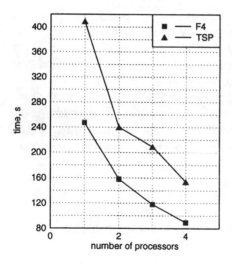

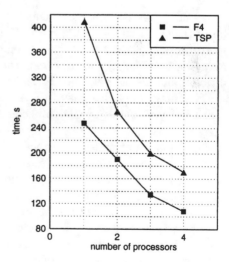

Fig. 3. Computation time of island model algorithm: F4 - minimisation of F4 De Jong function, TSP - att48 from TSPLIB.

Fig. 4. Computation time of global population model algorithm: F4 - minimisation of F4 De Jong function, TSP - att48 from TSPLIB.

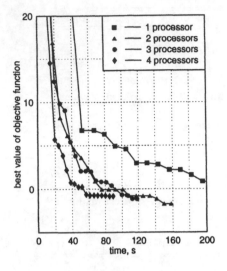

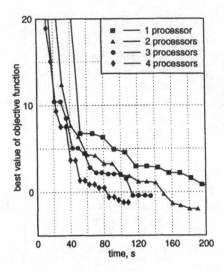

Fig. 5. Search efficiency of island model algorithm for F4 problem.

Fig. 6. Search efficiency of global population model algorithm for F4 problem.

Results presented in Fig. 5 and Fig. 6 confirm that acceptable solutions can be obtained in shorter time with PGA on larger number of processors. It is worth noticing that for stochastic algorithms like GA, when there is no exact mapping between sequential and parallel program courses of computation, traditional metrics of computation quality are less useful.

In practical applications of PGA optimal choice of parameters governing intensity of individuals exchange becomes an important issue. Results like these presented in Fig. 7 and Fig. 8 show that there is a certain value which is the best for a given optimisation process. The reason of this is that for larger rates of exchange communication overhead increases while diversity decreases due to tighter coupling between subpopulations. On the other hand too strong separation of subpopulations leads to worse search efficiency.

We have implemented the library also on Exemplar SPP1200/XA. Para++, TOLKIEN and GA library as well as the test program were compiled and linked on HP 712/80 workstation with g++ v2.7.2 and PVM v3.3.10, and next the program was moved on Exemplar.

10 Concluding remarks

With the parallel GA library presented in this paper a user may easily start experimentation with a genetic program without going into details of communication, synchronisation, parallel debugging and testing. The library is portable; it was implemented on networked IBM RS/6000 and SUN SPARCstation workstations as well as on the parallel computer HP Exemplar SPP1200/XA.

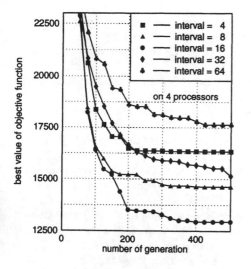

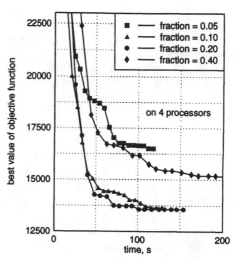

Fig. 7. Search quality of island model algorithm solving TSP for different values of migration interval.

Fig. 8. Search efficiency of island model algorithm solving TSP for growing migrating fraction of subpopulation.

The results of investigation of the efficiency for travelling salesman problem and for minimisation of De Jong function have demonstrated usefulness of the library. Nethertheless, further studies are necessary in order to improve its performance.

We are going to develop a new class which should be a skeleton of a typical GA program based on this library. This skeleton should also include input/output operations. We plan to extend the set of classes representing individuals of different types what should shorten elaboration of GA programs for broader area of applications. On Exemplar SPP1200 the native C++ will be applied when it becomes available. It seems reasonably to test MPI as an internal communication layer of the library.

Acknowledgements
We thank Prof. Jacek Mościński for comments, Mr Marek Pogoda for discussions, and ACC CYFRONET staff for help.
The work was partially supported by KBN under grant 8 S503 006 07.

References

1. D. E. Goldberg. Genetic Algorithms in Search, Optimization and Machine Learning. Addison-Wesley, 1989.
2. Z. Michalewicz, "Genetic algorithms + data structures = evolution program", Springer-Verlag, 1992.

3. D. Beasley, R.D. Bull, R.R. Martin, "An overview of genetic algorithms", *University Computing* **15** (1993) 58-69 (Part 1) and 170-181 (Part 2).

4. J. Stender, E. Hillebrand, and J. Kingdom, (Eds.), "Genetic algorithms in optimisation, simulation and modelling", IOS Press, 1994.

5. G. Winter, J. Périaux, M. Galan, P. Cuesta (eds). "Genetic Algorithms in Engineering and Computer Science", Wiley, 1995.

6. R. Bianchini, C. M. Brown, "Parallel Genetic Algorithms on Distributed-Memory Architectures", Technical Report 436, University of Rochester. Computer Science Department, 1993. (`ftp://ftp.cs.rochester.edu/pub/papers/systems/93.tr436.parallel_genetic_algorithms.ps.Z`)

7. E. Cantú-Paz, "A Summary of Research on Parallel Genetic Algorithms", Report No. 95007, Illinois Genetic Algorithms Laboratory, 1995. (`ftp://gal4.ge.uiuc.edu/pub/papers/IlliGALs/95007.ps`)

8. J. Heitkoetter, D. Beasley (Eds.), "The Hitch-Hiker's Guide to Evolutionary Computation: A list of Frequently Asked Questions (FAQ)", USENET: comp.ai.genetic, 1996. (`ftp://rtfm.mit.edu/pub/usenet/news.answers/ai-faq/genetic/`)

9. J.R. Filho, C. Alippi, P. Treleaven, "Genetic Algorithm Programming Environments", Department of Computer Science, University College, London, 1994. (ENCORE: `.../EC/GA/papers/ieee94.ps.gz`)

10. B. Stroustrup, "The C++ Programming Language", Second Edition. Addison-Wesley, 1993.

11. A. Yiu-Cheung Tang, "TOLKIEN: Toolkit for Genetics-Based Applications" Department of Computer Science, The Chinese University of Hong Kong, 1993-94. (ENCORE: `.../EC/GA/src/tolkien-1.5.tar.gz`)

12. A. Yiu-Cheung Tang, "Constructing GA Applications Using TOLKIEN" Department of Computer Science, The Chinese University of Hong Kong, 1994. (ENCORE: `.../EC/GA/src/tolkien-1.5.tar.gz`)

13. O. Coulaud, E. Dillon, "Para++: C++ Bindings for Message Passing Libraries", Institut National de Recherche en Informatique et en Automatique, 1995. (`ftp://ftp.loria.fr/pub/loria/numath/para++-v1.0.tar.Z`)

14. O. Coulaud, E. Dillon, "PARA++: C++ bindings for message passing libraries" in: Dongarra, J., Gengler, M., Touraucheau, B., and Vigouroux, X. (Eds), EuroPVM'95, Hermès, Paris 1995.

15. D. Whitley, "A Genetic Algorithm Tutorial", Technical Report CS-93-103. Colorado State University. Department of Computer Science, 1993. (ENCORE: `.../EC/GA/papers/tutor93.ps.gz`)

16. M. Bubak, J. Mościński, M. Pogoda, R. Słota, "Load balancing for lattice gas and molecular dynamics simulations on networked workstations", in: *Hertzberger, B., Serazzi, G., (eds.), Proc. Int. Conf. HPCN, Milan, Italy, May 1995*, Lecture Notes in Computer Science **796**, pp. 329-334, Springer-Verlag, 1995.

17. G. Reinelt, "TSPLIB95", (`http://www.iwr.uni-heidelberg.de/iwr/comopt/soft/TSPLIB95/DOC.PS.gz`)

Monitoring of Performance of PVM Applications on Virtual Network Computer

Marian Bubak[1,2], Włodzimierz Funika[1], Jacek Mościński[1,2]

[1] Institute of Computer Science, AGH, al. Mickiewicza 30, 30-059, Kraków, Poland
[2] Academic Computer Centre – CYFRONET, Nawojki 11, 30-950 Kraków, Poland
email: {bubak,funika,jmosc}@uci.agh.edu.pl

Abstract. We present a toolkit of monitoring and performance analysis facilities together with sample monitoring sessions and results of
performance analysis of three parallel programs running under PVM on
networked workstations. A trace file collected by Tape/PVM is converted into SDDF metaformat and then it is visualized with Pablo based
tool. Performance is analyzed also with a set of metrics.

1   Introduction

Development of parallel distributed computing has resulted in demand for powerful performance analysis and visualization systems which should provide information indispensable to get insight into behavior of a program. The first, very
rough estimation of a distributed program performance may be done with speed-
up and efficiency indices. The more refined insight is needed the harder can it
be obtained [1].

Nowadays, there is a number of monitoring and performance analysis tools
which may be classified as *PVM-oriented* (XPVM, Xab, EASYPVM, HeNCE),
environment independent (Pablo, JEWEL, AIMS, VAMPIR, PARvis, SIMPLE,
TATOO) and *environment dependent* (PICL, Express-tools, Linda-tools) [2], [3].

There are two important issues in monitoring and performance analysis. The
first one is the problem of representative quality of trace data. For example, till
now XPVM [4] traces represent potentially perturbed applications due to on-line
event message routing. The second issue addresses the problem of choosing and
processing the data that could give information about the locations in a program
and the causes of poor performance. The problem is whether there should be
a standard set of visualization graphs and analysis modules or they should be
ajustable by the user. Two extreme examples of these viewpoints are ParaGraph
[5] and Pablo [7].

Our goal is to use existing tracing and performance analysis tools to build up
a toolkit that can supply realistic time-stamped trace data to synthetic and easy
to interpret representations of behavior of programs running on virtual network
computer under PVM.

Previously we used XPVM monitoring tool to collect trace data for off-line
performance analysis. The data after error pruning, transforming into the pure

SDDF format, were analyzed and visualized with a tool built of Pablo modules [3]. At present, in order to have reliable time stamp data, we use Tape/PVM [6].

Section 2 describes the structure of the monitoring toolkit, Section 3 presents the way the trace data are processed, and Section 4 contains examples of performance analysis of three typical parallel applications running under PVM on networked workstations. Our ongoing and future works are briefly outlined at the end of the paper.

2 Monitoring using Tape/PVM trace facility

Our monitoring tool consists of an instrumentor comprised of a graphical editor and a parser for inserting monitoring points into a source code, and of a monitoring facility comprised of the Tape/PVM monitoring library and daemon task (Fig. 1).

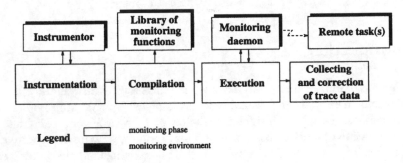

Fig. 1. Monitoring tool and flow of data

Following elements are instrumented:

- PVM function calls,
- procedure calls,
- loops,
- user selected variables.

Features of Tape/PVM that motivated our choice in its favour were as follows:

- capability to correct time data in order to compensate clock drifts,
- compact format of trace data that enables to spare buffer room,
- library of functions for handling trace data,
- tunable parameters of monitoring.

The first step of monitoring process of a parallel program is instrumentation of a source code with the instrumentor. Next, the program is compiled and linked with monitoring library and then it is executed. The monitoring daemon collect trace data, and after time-stamps correction the data are written onto a file in the Tape/PVM format which can be handled with library functions.

3 Off-line performance analysis

After a raw event trace data have been written down onto a file they are converted to the SDDF format, so they can be used for analysis and visualization with Pablo based modules and other tools.

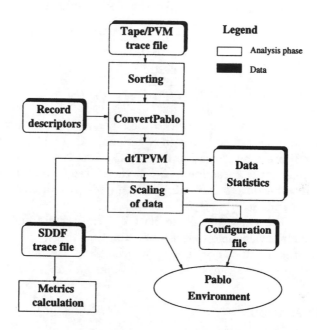

Fig. 2. Trace data processing.

The Pablo features that inspired us to adapt it were the following:

- portability for a wide range of platforms,
- extensibility (it can be extended by user-written modules),
- flexibility (it can be tuned to parameters of trace data),
- convenient format of trace data (human readable SDDF),
- powerful library of methods and classes for handling trace data.

Processing of trace file consists of the following steps (see Fig.2):

- trace data are globally sorted according to time-stamps,
- *ConvertPablo* transforms data into SDDF format,
- *dtTPVM* supplements data by the program state transitions information and computes ranges of trace data,
- *scaling program* sets up the corresponding data ranges in the Pablo graph configuration file.

In the next phase of trace data processing following metrics are calculated:

- cumulative utilization U,
- utilization (u),
- weighted utilization $(\bar{u})$.

Utilization and weighted utilization are calculated as follows:

$$u(i,j) = \frac{t_e(i,j) - t_c(i,j) - t_s(i,j)}{t_e(i,j)}, \tag{1}$$

$$\bar{u}(j) = \frac{\sum_{i=1}^{i=n} \frac{u(i,j)}{\tau(i,j)}}{\sum_{i=1}^{i=n} \frac{1}{\tau(i,j)}}, \tag{2}$$

where n is the number of nodes (workstations), $t_e(i,j)$ is the execution time over a given set of program actions j (e.g. actions within a step of simulation), $t_c(i,j)$ and $t_s(i,j)$ are communication time and system time due to usage of the parallel environment over the action, $\tau(i,j)$ is the average time per characteristic program unit (e.g. particle, lattice site) on i-th node in j-th step, $u(i,j)$ is the utilization of i-th node in j-th step.

Cumulative utilization is based on the PVM calls traces and is a ratio of accumulated computation time to execution time within the lifetime of a program. We have also developed profiles of programs and subroutines: time usage by nodes, time usage by routines, time used by routines on individual nodes.

4 Sample sessions

For performance analysis we have chosen three typical applications running on networked workstations under PVM. They are: 2-D molecular dynamics simulation (MD2), lattice gas automata (LGA) simulation, and finite-elements method based calculation of plastical deformations (FEM). The experiments were carried out on a heterogenous network of workstations comprised of HP9000/712 (HP712), IBM RS/6000-520 (RS520), IBM RS/6000-320 (RS320) and SUN IPX (IPX).

4.1 2-D molecular dynamics simulation

MD2 is a molecular dynamics program for simulation of large 2-D Lennard-Jones systems. Computational complexity is reduced with distributed neighbor and link (Hockney) lists. Parallel algorithm is based on geometric decomposition. The computational box is divided along y axis into domains with several layers of Hockney cells. Each process keeps complete data set of its domain, local neighbor list and the positions of particles in the boundary layer of the previous domain. Load balancing is based on repartitioning the computational box [8].

In Fig.3 we present a sample of monitoring session with Pablo based displays for a run on 4 workstations (HP712, RS520, RS320 and IPX). The following performance data are presented: synchronized time of every event in a

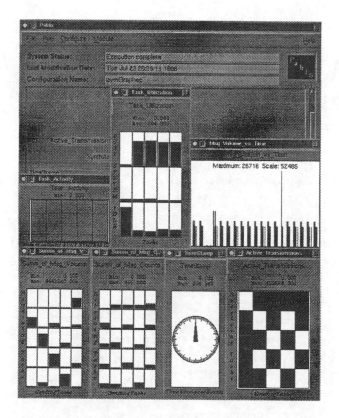

Fig. 3. Sample of monitoring session of MD2 on RS320, RS520, HP712 and IPX.

parallel program (Time_Stamp), matrix of current connections with the trace of recent message volume (Active_Transmissions), current computation time spent by every task between two consecutive barriers (Barrier_Mean_Time), cumulative time spent in each state by every task (Task_Utilization), cumulative number of messages sent between tasks (Summ_of_Msg_Counts), cumulative volume of messages sent between tasks (Summ_of_Msg_Volume), current total volume of messages in a program as a function of time (Msg_Volume_vs_Time), state (busy, system, idle) of every task (Task_Activity).

In Fig. 4 volume of messages received by nodes as a function of time is shown. At the beginning processes are intensively communicating – this is data distribution phase. Next, three message volumes dominate: 10 bytes messages correspond to exchange of data between neighboring domains in each simulation step, messages close to 10^4 bytes are received when neighbor lists are rebuilt, and those over 10^4 correspond to load balancing.

In the Fig.5 and Fig.6 utilization for two configurations - 2 nodes (RS320 and RS520) and 4 nodes (HP712, RS520, RS320 and IPX) is shown. In the case of 4 nodes load balancing is more distinguishable on the slower workstations.

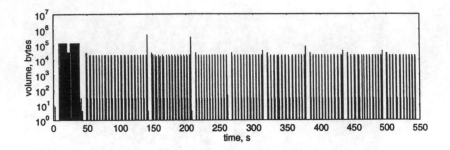

Fig. 4. Receive data volume of MD2 on 2 RS/6000-320 workstations.

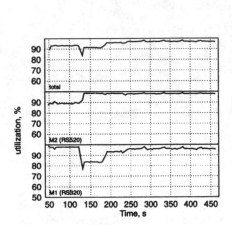

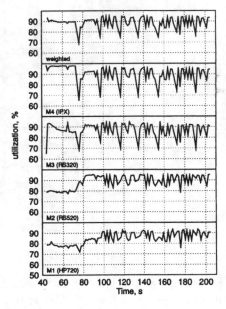

Fig. 5. Utilization for MD2 on 2 processors; 131072 particles.

Fig. 6. Utilization for MD2 on 4 processors; 131072 particles.

4.2 Lattice gas automata

For parallel lattice gas automata (LGA) simulation lattice is divided into domains along the y axis [9]. Master sends to workers the geometrical description
of a system simulated and receives averaged values from workers. Workers generate initial states of domains and after the computation phase send the averaged
values to master. Communication among neighbors can be optionally asynchronous that results in reduced communication overhead. The communication topology is ring. The dominant communication patterns are short messages during
evolution while the averaging and load-balancing phases are dominated by long
messages.

In Fig.7 and Fig.8 utilization for LGA program running on two virtual net-

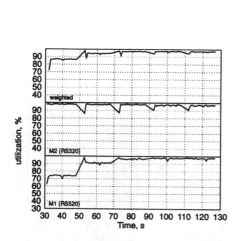

Fig. 7. Utilization for LGA on 2 processors; 480 × 2880 sites.

Fig. 8. Utilization for LGA on 4 workstations; 480 × 2880 sites

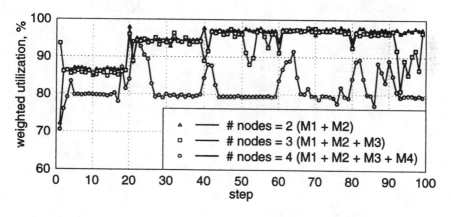

Fig. 9. Utilization for LGA for different number of workstations; lattice: 480 × 2880 sites; M1 = RS320, M2 = RS520, M3 = IPX, M4 = HP712.

work computers is shown. One can observe that on heterogeneous network utilization is lower because load balancing actions are realized more often. More illustrative is Fig. 9 in which weighted utilization for 3 configurations of networked workstations is presented. Utilization decreases considerably when network becomes strongly heterogeneous.

4.3 Plastic material deformation

This program is designed for simulation of plastic metal deformation using the finite element method. The mathematical model consists of two coupled components: mechanical and thermal one. Parallelization is realized by division of a mesh of elements into strips assigned to different processes (workstations) [10].

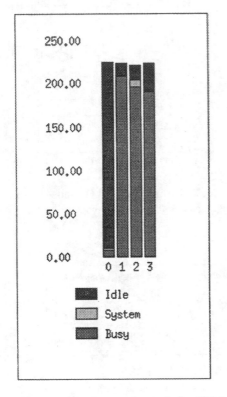

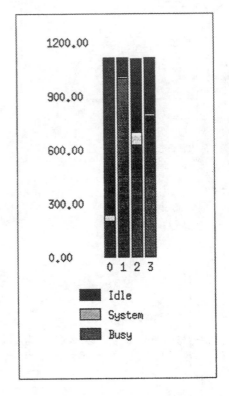

Fig. 10. Time usage by node for FEM on 3 RS320s; mesh 256 × 9: strips with 95, 68, 95 columns

Fig. 11. Time usage by node for FEM on 3 RS320s; mesh 48 × 48: strips with 10, 20 and 10 columns

As an alternative for time-lined metrics we have developed profiles of programs and routines. The *time usage by node* profile presents the lifetime of a program, idle time due to blocking receive operations, system time due to parallel execution, e.g. send operations. The *time usage by routines* presents the same data related to routines a program is comprised of.

In Fig.10 we show the "time usage by node" profile for strips with 95, 68, 95 columns on 3 RS320s. The bar 0 corresponds to master program while other bars are allocated to node (worker) programs. The master program spends most time in waiting. The blocking receive and system times of worker programs are shown as two upper layers of the bars. The bar heights as well as these of their com-

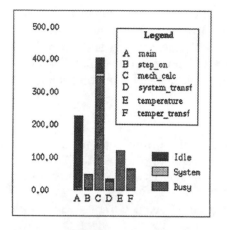

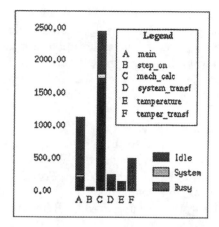

Fig. 12. Time usage by routines for FEM on 3 RS320s; mesh 256 × 9: strips with 95, 68, 95 columns

Fig. 13. Time usage by routines for FEM on 3 RS320s; mesh 48 × 48: strips with 10, 20 and 10 columns

ponents are almost equal. It suggests that the node programs are relatively well balanced. By contrast, for the problem with the strips with 10, 20, 10 columns (Fig.11), i.e. a square mesh, the profile reveals that two nodes have increased communication/execution ratio. The "time usage by routines" profile (see Fig. 12 and 13) indicates which routine is responsable for most communication. In the latter case "mech_calc" routine spends more time in communication than in the former case.

5 Concluding remarks

We have presented the tool for monitoring and performance analysis of parallel applications running on networked workstations. This tool supplies realistic time-based trace data that may be interpreted by evaluation and analysis modules. It enables to choose points of interest in a program, get profiles of program and routines, visualize performance values in their evolution. The utilization and weighted utilization metrics give valuable insight into behavior of a program. The portability of the monitoring tool is provided by source code instrumentation.

Monitoring and performance analysis process was illustrated with three typical applications. From the performance analysis it follows that MD2 and LGA parallel programs are well tuned for distributed execution on virtual network computer while the FEM program exhibits worse performance which may be improved by proper load balancing. The utilization while executing LGA program on a strongly heterogenous network is worth of special interest.

We are working on a profiling facility that could make it possible to spare disk space. The monitoring tool needs to be equipped with control panel. An important issue we work on is a possibility to explore application behavior on various levels of detail: that of whole program, procedure, loop, code block or individual statement. The elaboration of new metrics will enhance our possibilities to analyze the application behavior from different points of view. This will necessitate the elaboration of a language for extracting and clustering the relevant data from trace information. Location, presentation of parallel program structures which are responsible for poor performance by relating them to the source code is of great importance.

Acknowledgements

We would like to thank Eric Maillet for his help in relieving our acquaintance with Tape/PVM features. We are very grateful to Ms Renata Chrobak, Ms Renata Słota and Mr Marek Pogoda for providing us with their programs and for valuable comments.
The work was supported by KBN under grant 8 S503 006 07.

References

1. Hollingsworth, J., Lumpp, J.E., Jr. and Miller, B.P., "Performance measurement of parallel programs", in: Tvrdik, P., *Proc. of ISIPALA'93*, Prague, July 1993, pp. 239-254.
2. Cheng, D.Y., "A Survey of Parallel Programming Languages and Tools", Computer Sciences Corporation, NASA Ames Research Center, *Report RND-93-005* March 1993.
3. Bubak M., Funika W., Mościński J., Tasak D. "Pablo based monitoring tool for PVM applications" in: Jack Dongarra, Kaj Madsen, Jerzy Waśniewski (Eds), *PARA95 - Applied Parallel Computing*, Springer-Verlag Lecture Notes in Computer Science **1041** (1996) pp. 69-78.
4. Geist, A., Beguelin, A., Dongarra, J., Jiang, W., Manchek, R., Sunderam, V., "PVM: Parallel Virtual Machine", MIT Press, 1994.
5. Heath, M.T. Etheridge, J.A., "Visualizing the performance of parallel programs", *IEEE Software* September 1991, pp. 29-39.
6. Maillet, E., "TAPE/PVM an efficient performance monitor for PVM applications. User's guide", *IMAG-Équipe APACHE, Grenoble*, 1995.
7. Reed, D.A. Aydt, R., Madhyastha, T.M., Noe, R.J., Shields, K.A., and Schwartz, B.W., "The Pablo Performance Analysis Environment", *Technical Report, Dept. of Comp. Sci., University of Illinois*, 1992.
8. Bubak, M., Mościński, J., Pogoda, M., "Distributed 2-D molecular dynamics simulation on networked workstations and multiprocessors", in: Liddell, H., Colbrook, A., Hertzberger, B., Sloot, P., (eds.), *Proc. Int. Conf. High Performance Computing and Networking, Brussels, Belgium, April 1996*, Lecture Notes in Computer Science **1067**, pp. 243-250, Springer-Verlag, 1996.
9. Bubak, M., Mościński, J., Słota, R., "FHP lattice gas on networked workstations", in: Gruber, R., Tomassini, M., (eds) *The 6th Joint EPS-APS International Conference on Physics Computing*, Lugano, Switzerland, August 22-26, 1994, pp. 427-430.
10. Bubak, M., Chrobak, R., Kitowski, J., Mościński, J., Pietrzyk, M., "Parallel finite element calculations of plastic deformations on Exemplar SPP1000 and on networked workstations", *Journal of Materials Processing Technology* **60** (1996) 409-413.

A Parallel Version of the Quasi-Minimal Residual Method Based on Coupled Two-Term Recurrences

H. Martin Bücker and Manfred Sauren

Zentralinstitut für Angewandte Mathematik
Forschungszentrum Jülich GmbH, 52425 Jülich, Germany
E-mail: m.buecker@kfa-juelich.de and m.sauren@kfa-juelich.de

Abstract. For the solution of linear systems of equations with unsymmetric coefficient matrix, Freund and Nachtigal (SIAM J. Sci. Comput. 15 (1994), 313–337) proposed a Krylov subspace method called Quasi–Minimal Residual method (QMR). The two main ingredients of QMR are the unsymmetric Lanczos algorithm and the quasi-minimal residual approach that minimizes a factor of the residual vector rather than the residual itself. The Lanczos algorithm spans a Krylov subspace by generating two sequences of biorthogonal vectors called Lanczos vectors. Due to the orthogonalization and scaling of the Lanczos vectors, algorithms that make use of the Lanczos process contain inner products leading to global communication and synchronization on parallel processors. For massively parallel computers, these effects cause delays preventing scalability of the implementation. Consequently, parallel algorithms should avoid global synchronization as far as possible. We propose a new version of QMR with the following properties: Firstly, the Lanczos process is based on coupled two-term recurrences; secondly, both sequences of Lanczos vectors are scalable; and finally, there is only a single global synchronization point per iteration. The efficiency of this algorithm is demonstrated by numerical experiments on a PARAGON system using up to 121 processors.

1 Introduction

Since engineering and scientific problems arising from "real" applications are typically becoming larger and larger, there is a strong demand for ever increasing computing power. Besides the tremendous progress in computer architecture, the main feature of delivering highest performance is the design of sophisticated algorithms. While using parallel computing platforms the need for adequate algorithms is particularly important because, in addition to computation, communication among the processing elements has to be controlled. The implementation of sequential algorithms on parallel computers generally tends to achieve only a low degree of efficiency such that new algorithms specifically-designed for parallel processing are required. Unfortunately, a trade-off between parallelism and stability has been observed in numerical computing [2].

In this paper, a parallel version of the quasi-minimal residual method (QMR) is presented. This iterative method for the solution of systems of linear equations with unsymmetric coefficient matrix originally involves the unsymmetric Lanczos algorithm with look-ahead [4], a technique developed to prevent the process from breaking down in case of numerical instabilities. We here assume that such breakdowns do not occur and propose a new version of the Lanczos algorithm without look-ahead as the underlying process of QMR. The parallelization of QMR as well as for the whole class of Krylov subspace methods on distributed memory processors is straightforward [11] and consists in parallelizing the three kinds of operations: vector updates, matrix-vector products, and inner products. Vector updates are perfectly parallelizable and, for large sparse matrices, matrix-vector products can be implemented with communication between only nearby processors. The bottleneck is usually due to inner products enforcing global communication, i.e., communication of all processors. There are two strategies to remedy the performance degradation which, of course, can be combined. The first is to restructure the code such that most communication is overlapped with useful computation. The second is to eliminate data dependencies such that several inner products are computed simultaneously. This paper is concerned with the latter strategy in order to reduce the number of global synchronization points. A *global synchronization point* is defined as the locus of an algorithm at which all local information have to be globally available in order to continue the computation.

The organization of the paper is as follows. Section 2 is devoted to the unsymmetric Lanczos algorithm, especially to a sketch of a new variant used as the underlying process of QMR that is described in Sect. 3. Before making some concluding remarks Sect. 4 presents numerical experiments giving some comparisons of the new QMR version with the standard approach.

2 Unsymmetric Lanczos Algorithm

In the early 1950's, Cornelius Lanczos published two closely related papers [8, 9] in which he proposed an algorithm called *method of minimized iterations*. Nowadays, this algorithm is referred to as the *unsymmetric Lanczos algorithm* or *Lanczos biorthogonalization algorithm*. Given a general matrix $\mathbf{A} \in \mathbb{C}^{N \times N}$ and two starting vectors $\mathbf{v}_1, \mathbf{w}_1 \in \mathbb{C}^N$ satisfying $\mathbf{w}_1^T \mathbf{v}_1 = 1$, the algorithm generates two finite sequences of vectors $\mathbf{v}_n$ and $\mathbf{w}_n$ with the following three properties:

$$\mathbf{v}_n \in \mathcal{K}_n(\mathbf{v}_1, \mathbf{A}) , \tag{1}$$
$$\mathbf{w}_n \in \mathcal{K}_n(\mathbf{w}_1, \mathbf{A}^T) ,$$

where $\mathcal{K}_n(\mathbf{y}, \mathbf{A}) = \mathrm{span}\{\mathbf{y}, \mathbf{A}\mathbf{y}, \dots, \mathbf{A}^{n-1}\mathbf{y}\}$ denotes the nth Krylov subspace generated by the matrix $\mathbf{A}$ and the vector $\mathbf{y}$, and finally

$$\mathbf{w}_m^T \mathbf{v}_n = \begin{cases} 0 & \text{if } n \neq m , \\ 1 & \text{if } n = m . \end{cases}$$

If the vectors $\mathbf{v}_n$ and $\mathbf{w}_n$ called *Lanczos vectors* are put as columns into matrices

$$\mathbf{V} = [\mathbf{v}_1\ \mathbf{v}_2\ \cdots\ \mathbf{v}_N] \quad \text{and} \quad \mathbf{W} = [\mathbf{w}_1\ \mathbf{w}_2\ \cdots\ \mathbf{w}_N]$$

the Lanczos algorithm is characterized—up to scaling—by the three equations

$$\mathbf{W}^T\mathbf{V} = \mathbf{I} \tag{2a}$$

$$\mathbf{A}\mathbf{V} = \mathbf{V}\mathbf{T} \tag{2b}$$

$$\mathbf{A}^T\mathbf{W} = \mathbf{W}\mathbf{T}^T , \tag{2c}$$

where $\mathbf{I} \in \mathbb{C}^{N\times N}$ is the identity matrix and $\mathbf{T} \in \mathbb{C}^{N\times N}$ is a tridiagonal matrix. We stress that these formulae have to be interpreted as an iterative process; i.e., in each iteration the Lanczos process generates a further column of each of the matrices $\mathbf{V}, \mathbf{W}$ and $\mathbf{T}$. The process typically terminates with far less iterations than the order of $\mathbf{A}$ suggests. The tridiagonal structure of $\mathbf{T}$ leads to *three-term recurrences* for the generation of the Lanczos vectors which is immediately apparent if the algorithm is formulated in vector notation rather than matrix notation; see [6] for details. We further state that (2) involves two global synchronization points per iteration which readily can be reduced to a single one by a simple reorganization of the statements [7]. Furthermore, this version opens the alternative either to scale the $\mathbf{v}_n$'s or $\mathbf{w}_n$'s but not both.

It has been pointed out by several authors [12, 10, 3, 5] that in practice one should scale both sequences of Lanczos vectors appropriately, e.g.,

$$\|\mathbf{v}_{n+1}\| = \|\mathbf{w}_{n+1}\| = 1, \qquad n = 0, 1, 2, \ldots$$

in order to avoid over- or underflow. This can only be achieved by giving up the strict biorthogonality $\mathbf{W}^T\mathbf{V} = \mathbf{I}$ and setting $\mathbf{W}^T\mathbf{V} = \mathbf{D}$ instead, where $\mathbf{D} = \mathrm{diag}(\delta_1, \delta_2, \ldots, \delta_N)$ is a diagonal matrix with $\delta_i \neq 0$ for $i = 1, 2, \ldots, N$. The corresponding matrix formulation is

$$\mathbf{W}^T\mathbf{V} = \mathbf{D} \tag{3a}$$

$$\mathbf{A}\mathbf{V} = \mathbf{V}\mathbf{T} \tag{3b}$$

$$\mathbf{A}^T\mathbf{W} = \mathbf{W}\mathbf{D}^{-1}\mathbf{T}^T\mathbf{D} . \tag{3c}$$

The resulting algorithm is still based on three-term recurrences and one easily finds an implementation with a single global synchronization point per iteration.

Since its introduction most of the work around the unsymmetric Lanczos algorithm is concerned with the above three-term recurrences although, in addition to that version, Lanczos himself mentioned a different variant involving *coupled two-term recurrences* [9]. Freund and Nachtigal [5] claimed that, at least when the Lanczos algorithm is used as the underlying process of QMR, the latter variant of the Lanczos algorithm may be numerically more stable. This is the reason why we here propose a new version of the unsymmetric Lanczos algorithm based on coupled two-term recurrences with the following additional properties: Both sequences of Lanczos vectors are scalable and there is only a single global synchronization point per iteration. We remark that each iteration of the coupled two-term procedure with unit length scaling of both Lanczos vectors given in [5] consists of three global synchronization points which can be reduced to two without considerable difficulties—but not to a single one.

Throughout the rest of the paper, we assume that the tridiagonal matrix $\mathbf{T}$ has an LU decomposition

$$\mathbf{T} = \mathbf{LU} \ , \tag{4}$$

where the factors $\mathbf{L}$ and $\mathbf{U}$ are of lower and upper bidiagonal form, respectively. It is this bidiagonal structure of $\mathbf{L}$ and $\mathbf{U}$ that results in coupled two-term recurrences. The principal idea of the new approach is to start from (3) by using (4) as well as introducing $\mathbf{P} = \mathbf{VU}^{-1}$ and $\tilde{\mathbf{Q}} = \mathbf{WD}^{-1}\mathbf{U}^T$ which leads to

$$\mathbf{W}^T\mathbf{V} = \mathbf{D} \tag{5a}$$

$$\mathbf{V} = \mathbf{PU} \tag{5b}$$

$$\tilde{\mathbf{Q}} = \mathbf{WD}^{-1}\mathbf{U}^T \tag{5c}$$

$$\mathbf{AP} = \mathbf{VL} \tag{5d}$$

$$\mathbf{A}^T\mathbf{W} = \tilde{\mathbf{Q}}\mathbf{L}^T\mathbf{D} \ . \tag{5e}$$

The derivation of the vector notation of (5) is similar to the one given in [1] leading to Algorithm 1 having only a single global synchronization point per iteration while offering the possibility to scale both sequences of Lanczos vectors.

Input $\mathbf{A}, \tilde{\mathbf{v}}_1$ and $\tilde{\mathbf{w}}_1$ satisfying $\tilde{\mathbf{w}}_1^T \tilde{\mathbf{v}}_1 \neq 0$

$\mathbf{p}_0 = \mathbf{q}_0 = 0$

Choose γ_1 and ξ_1; $\varrho_1 = \tilde{\mathbf{w}}_1^T \tilde{\mathbf{v}}_1$, $\varepsilon_1 = \left(\mathbf{A}^T\tilde{\mathbf{w}}_1\right)^T \tilde{\mathbf{v}}_1$, $\mu_1 = 0$, $\tau_1 = \dfrac{\varepsilon_1}{\varrho_1}$

$\mathbf{v}_1 = \frac{1}{\gamma_1}\tilde{\mathbf{v}}_1$ and $\mathbf{w}_1 = \frac{1}{\xi_1}\tilde{\mathbf{w}}_1$

for $n = 1, 2, 3, \ldots$ do

    $\mathbf{p}_n = \mathbf{v}_n - \mu_n\mathbf{p}_{n-1}$

    $\mathbf{q}_n = \mathbf{A}^T\mathbf{w}_n - \dfrac{\gamma_n\mu_n}{\xi_n}\mathbf{q}_{n-1}$

    $\tilde{\mathbf{v}}_{n+1} = \mathbf{A}\mathbf{p}_n - \tau_n\mathbf{v}_n$

    $\tilde{\mathbf{w}}_{n+1} = \mathbf{q}_n - \tau_n\mathbf{w}_n$

    Choose γ_{n+1} and ξ_{n+1}

    $\varrho_{n+1} = \tilde{\mathbf{w}}_{n+1}^T\tilde{\mathbf{v}}_{n+1}$

    $\varepsilon_{n+1} = \left(\mathbf{A}^T\tilde{\mathbf{w}}_{n+1}\right)^T \tilde{\mathbf{v}}_{n+1}$

    $\mu_{n+1} = \dfrac{\gamma_n\xi_n\varrho_{n+1}}{\gamma_{n+1}\tau_n\varrho_n}$

    $\tau_{n+1} = \dfrac{\varepsilon_{n+1}}{\varrho_{n+1}} - \gamma_{n+1}\mu_{n+1}$

    $\mathbf{v}_{n+1} = \frac{1}{\gamma_{n+1}}\tilde{\mathbf{v}}_{n+1}$ and $\mathbf{w}_{n+1} = \frac{1}{\xi_{n+1}}\tilde{\mathbf{w}}_{n+1}$

endfor

Algorithm 1: Lanczos algorithm based on coupled two-term recurrences

3 Quasi-Minimal Residual Method

The unsymmetric Lanczos algorithm in its form (5) is now used as an ingredient to a Krylov subspace method for solving a system of linear equations

$$\mathbf{A}\mathbf{x} = \mathbf{b} \ , \quad \text{where} \quad \mathbf{A} \in \mathbb{C}^{N \times N} \quad \text{and} \quad \mathbf{x}, \mathbf{b} \in \mathbb{C}^N \ . \tag{6}$$

The nth iterate of any Krylov subspace method is given by

$$\mathbf{x}_n \in \mathbf{x}_0 + \mathcal{K}_n(\mathbf{r}_0, \mathbf{A}) \tag{7}$$

where $\mathbf{x}_0$ is an initial guess for the solution of (6) and $\mathbf{r}_0 = \mathbf{b} - \mathbf{A}\mathbf{x}_0$ is the corresponding residual vector. According to (1), the Lanczos vectors $\mathbf{v}_n$ can be used to span the Krylov subspace in (7) if the Lanczos algorithm is started appropriately, e.g.,

$$\mathbf{v}_1 = \frac{1}{\gamma_1}\mathbf{r}_0 \quad \text{with} \quad \gamma_1 = \|\mathbf{r}_0\|_2 \ . \tag{8}$$

Hence, there exist a vector $\mathbf{z}_n \in \mathbb{C}^n$ such that

$$\mathbf{x}_n = \mathbf{x}_0 + \mathbf{V}_n\mathbf{z}_n \ , \tag{9}$$

where $\mathbf{V}_n = [\mathbf{v}_1 \ \mathbf{v}_2 \ \cdots \ \mathbf{v}_n]$ is the collection of the first n Lanczos vectors. Besides the generation of the Krylov subspace, the actual iterates $\mathbf{x}_n$ have to be defined, i.e., the task is to fix the vector $\mathbf{z}_n$ in (9). While spanning the Krylov subspace by the Lanczos algorithm in its three term or coupled two-term form, the quasi-minimal residual method [4, 5] determines $\mathbf{z}_n$ by minimizing a factor of the residual vector $\mathbf{r}_n$ as follows.

At the very beginning, let $\mathbf{L}_n \in \mathbb{C}^{(n+1) \times n}$ and $\mathbf{U}_n \in \mathbb{C}^{n \times n}$ be the leading principal $(n+1) \times n$ and $n \times n$ submatrices of the bidiagonal matrices $\mathbf{L}$ and $\mathbf{U}$ generated by the Lanczos algorithm. The Lanczos relations (5b) and (5d) are then expressed as

$$\mathbf{V}_n = \mathbf{P}_n\mathbf{U}_n$$
$$\mathbf{A}\mathbf{P}_n = \mathbf{V}_{n+1}\mathbf{L}_n \ ,$$

where $\mathbf{P}_n$ consists of the first n column vectors of $\mathbf{P}$. By introducing $\mathbf{y}_n = \mathbf{U}_n\mathbf{z}_n$ the iterates are given by

$$\mathbf{x}_n = \mathbf{x}_0 + \mathbf{P}_n\mathbf{y}_n \ . \tag{10}$$

Thus, the definition of the iterates proceeds by fixing $\mathbf{y}_n$ rather than $\mathbf{z}_n$. The residual vector in terms of $\mathbf{y}_n$ is obtained by the above scenario, namely

$$\mathbf{r}_n = \mathbf{b} - \mathbf{A}\mathbf{x}_n = \mathbf{r}_0 - \mathbf{A}\mathbf{P}_n\mathbf{y}_n = \mathbf{r}_0 - \mathbf{V}_{n+1}\mathbf{L}_n\mathbf{y}_n \ .$$

On account of (8), the residual vector can be factorized leading to

$$\mathbf{r}_n = \mathbf{V}_{n+1}\left(\gamma_1 \mathbf{e}_1^{(n+1)} - \mathbf{L}_n\mathbf{y}_n\right) \ ,$$

where $e_1^{(n+1)} = (1, 0, \ldots, 0)^T \in \mathbb{C}^{n+1}$. The quasi-minimal residual method determines y_n as the solution of the least-squares problem

$$\left\| \gamma_1 e_1^{(n+1)} - L_n y_n \right\|_2 = \min_{y \in \mathbb{C}^n} \left\| \gamma_1 e_1^{(n+1)} - L_n y \right\|_2 .$$

In [1] we avoid the standard approach of computing a QR factorization of L_n by means of Givens rotations and show instead that y_n is given by the coupled iteration

$$y_n = \binom{y_{n-1}}{0} + g_n , \tag{11}$$

$$g_n = \theta_n \binom{g_{n-1}}{0} + \kappa_n e_n^{(n)} , \tag{12}$$

where the scalars θ_n and κ_n are computed from simple recurrences too; see [1] for details.

Inserting the first recurrence relation (11) into (10) yields

$$\begin{aligned} x_n &= x_0 + P_{n-1} y_{n-1} + P_n g_n \\ &= x_{n-1} + P_n g_n \\ &= x_{n-1} + d_n , \end{aligned} \tag{13}$$

where $d_n = P_n g_n$ is introduced. Using the second recurrence (12), the vector d_n is updated by

$$\begin{aligned} d_n &= \theta_n P_{n-1} g_{n-1} + \kappa_n P_n e_n^{(n)} \\ &= \theta_n d_{n-1} + \kappa_n p_n . \end{aligned} \tag{14}$$

Note that the vector p_n is produced by the unsymmetric Lanczos process. By defining $s_n = A d_n$, the residual vector is then obtained by inserting (13) into $r_n = b - A x_n$ giving

$$r_n = r_{n-1} - s_n . \tag{15}$$

Multiplying (14) through by A, gives a recursion for the vector s_n as follows

$$s_n = \theta_n s_{n-1} + \kappa_n A p_n . \tag{16}$$

The result is a QMR version based on coupled two-term recurrences with scaling of both sequences of Lanczos vectors which involves only a single global synchronization point per iteration. Note that QMR delivers an upper bound for $\|r_n\|$ which should be considered in implementations of the stopping criterion rather than (15) and (16); see [4] for details.

4 Numerical Experiments

In this section, the parallel variant of QMR using Algorithm 1 as the underlying Lanczos process is compared to the original QMR based on coupled two-term recurrences, i.e., Algorithm 7.1 of [5] with additional assignments for the recursive computation of the residual vector analogous to (15) and (16). So, the number of operations per iteration of both algorithms is the same. In the example below we scale the Lanczos vectors by choosing $\gamma_{n+1} = \|\tilde{\mathbf{v}}_{n+1}\|$ and $\xi_{n+1} = \|\mathbf{A}^T \tilde{\mathbf{w}}_{n+1}\|$ in Algorithm 1. The experiments were carried out on Intel's PARAGON XP/S 10 at Forschungszentrum Jülich with the OSF/1 based operating system, Rel. 1.4. The double precision FORTRAN code is translated using Portland-Group compiler, Rel. 5.0.3, with optimization level 3.

As an example consider the partial differential equation

$$Lu = f \qquad \text{on} \quad \Omega = (0, 1) \times (0, 1)$$

with Dirichlet boundary condition $u = 0$ where

$$Lu = -\Delta u - 20 \left(x \frac{\partial u}{\partial x} + y \frac{\partial u}{\partial y} \right)$$

and the right-hand side f is chosen so that the solution is

$$u(x, y) = \frac{1}{2} \sin(4\pi x) \sin(6\pi y) .$$

We discretize the above differential equation using second order centered differences on a 440×440 grid with mesh size $1/441$, leading to a system of linear equations with unsymmetric coefficient matrix of order $193\,600$ with $966\,240$ nonzero

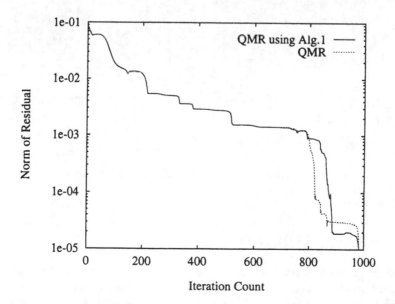

Figure 1. Convergence history

entries. Diagonal preconditioning is used. For our test runs we choose $\mathbf{x}_0 = 0$ as initial guess and stop as soon as $\|\mathbf{r}_n\| < 10^{-5}$. Smaller tolerances do not yield better approximations to the exact solution u; the absolute difference between the approximations and the exact solution stagnates at $9 \cdot 10^{-5}$.

To partition the data among the processors the parallel implementation subdivides Ω into square subdomains of equal size. Thus, each of the processors holds the data of the corresponding subdomain. Due to the local structure of the discretizing scheme, a processor has to communicate with at most 4 processors to perform a matrix-vector product.

Figure 1 shows the convergence history of both QMR versions where $\|\mathbf{r}_n\|$ is computed recursively. A similar behavior of both variants is recognized. We state that there is hardly any difference to the true residual norm $\|\mathbf{b} - \mathbf{A}\mathbf{x}_n\|$ in both versions. The parallel performance results are given in Fig. 2. These results are based on time measurements of a fixed number of iterations. The speedup given on the left-hand side of this figure is computed by taking the ratio of the parallel run time and the run time of a serial implementation. While the serial run times of both variants are almost identical there is a considerable difference concerning parallel run times. For all numbers of processors, the new parallel variant is faster than the original QMR. The saving of run time grows with increasing number of processors in comparison to the original QMR; see right-hand side of Fig. 2. More precisely, the quantity depicted as a percentage is $1 - T_A(p)/T_B(p)$, where $T_A(p)$ and $T_B(p)$ are the run times on p processors of QMR using Alg. 1 and Alg. 7.1 of [5], respectively.

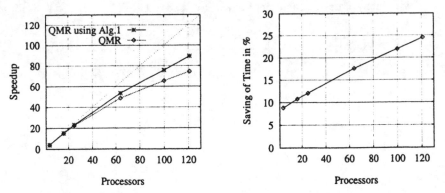

Figure 2. Results of the parallel implementations

5 Concluding Remarks

For the quasi-minimal residual method that usually uses the unsymmetric Lanczos algorithm in its three-term form, there is an alternative formulation based on coupled two-term recurrences. The task concerning parallelism in Krylov subspace methods is to minimize the number of global synchronization points per iteration by making inner products independent. This paper presents a parallel version of QMR based on coupled two-term recurrences having a single global synchronization point per iteration. Moreover, the possibility to scale both sequences of Lanczos vectors improves numerical stability. The parallel version demonstrates high scalability on a parallel computer with distributed memory.

There is room for further improvement in numerical stability by designing techniques to change the scaling factors of the Lanczos vectors at run time or to integrate look-ahead techniques into the version presented.

References

1. H. M. Bücker and M. Sauren. A Parallel Version of the Unsymmetric Lanczos Algorithm and its Application to QMR. Internal Report KFA–ZAM–IB–9605, Research Centre Jülich, Jülich, Germany, March 1996.
2. J. W. Demmel. Trading Off Parallelism and Numerical Stability. In M. S. Moonen, G. H. Golub, and B. L. R. De Moor, editors, *Linear Algebra for Large Scale and Real-Time Applications*, volume 232 of *NATO ASI Series E: Applied Sciences*, pages 49–68. Kluwer Academic Publishers, Dordrecht, The Netherlands, 1993. Proceedings of the NATO Advanced Study Institute on Linear Algebra for Large Scale and Real-Time Applications, Leuven, Belgium, August 1992.
3. R. W. Freund, G. H. Golub, and N. M. Nachtigal. Iterative Solution of Linear Systems. In *Acta Numerica 1992*, pages 1–44. Cambridge University Press, Cambridge, 1992.
4. R. W. Freund and N. M. Nachtigal. QMR: A Quasi–Minimal Residual Method for Non–Hermitian Linear Systems. *Numerische Mathematik*, 60(3):315–339, 1991.
5. R. W. Freund and N. M. Nachtigal. An Implementation of the QMR Method Based on Coupled Two–Term Recurrences. *SIAM Journal on Scientific Computing*, 15(2):313–337, 1994.
6. G. H. Golub and C. F. Van Loan. *Matrix Computations*. The Johns Hopkins University Press, Baltimore, second edition, 1989.
7. S. K. Kim and A. T. Chronopoulos. An Efficient Nonsymmetric Lanczos Method on Parallel Vector Computers. *Journal of Computational and Applied Mathematics*, 42:357–374, 1992.
8. C. Lanczos. An Iteration Method for the Solution of the Eigenvalue Problem of Linear Differential and Integral Operators. *Journal of Research of the National Bureau of Standards*, 45(4):255–282, 1950.
9. C. Lanczos. Solutions of Systems of Linear Equations by Minimized Iterations. *Journal of Research of the National Bureau of Standards*, 49(1):33–53, 1952.
10. B. N. Parlett, D. R. Taylor, and Z. A. Liu. A Look–Ahead Lanczos Algorithm for Unsymmetric Matrices. *Mathematics of Computation*, 44(169):105–124, 1985.
11. Y. Saad. Krylov Subspace Methods on Supercomputers. *SIAM Journal on Scientific and Statistical Computing*, 10(6):1200–1232, 1989.
12. D. R. Taylor. *Analysis of the Look Ahead Lanczos Algorithm for Unsymmetric Matrices*. Ph. D. dissertation, Department of Mathematics, University of California, Berkeley, CA, November 1982.

A Proposal for Parallel Sparse BLAS

Fabio Cerioni[2] and Michele Colajanni[2] * and Salvatore Filippone[1] ** and
Stefano Maiolatesi[2]

[1] IBM Semea
P.le G. Pastore 6
Roma, Italy
[2] Dipartimento di Informatica,
Sistemi e Produzione
Università degli Studi di
Roma – Tor Vergata, Roma, Italy

Abstract. In this paper we propose a set of parallel interfaces that
extends the sparse BLAS presented in [8] to distributed memory parallel
machines with message passing programming interfaces. Our main target
is the implementation of iterative methods with domain-decomposition
based preconditioners in an effective and structured way, while keeping
the benefits of the serial sparse BLAS. The structure of our framework
has also been influenced by the dense PBLAS proposal [5].

1 Introduction

Our aim in this work is to provide an interface for the convenient implementation
of iterative methods for sparse linear systems on distributed memory computers;
therefore the kind of applications we have in mind closely follows those forming
the framework of the serial Sparse BLAS (SpBLAS) proposal [8].

The current proposal should be viewed as complementary to the "templates"
concept provided in [2], addressing convenient implementation of those templates.

One of the key points of this proposal is to maintain the benefits of the Sp-
BLAS proposal by focussing on those issues arising from the distributed memory
nature of our target computational engines, while keeping the local parts of the
computations strictly adherent to the SpBLAS interface.

In handling data distribution and communications this proposal has also been
influenced by the PBLAS proposal developed in the context of ScaLAPACK [4];
however the specific nature of sparse linear systems has forced a number of
changes in the semantics of the operations.

Our main contribution is the definition of an adequate set of parallel operations that are able to capture the computational requirements of common sparse
computations, while guaranteeing convenient packaging of data structures, isolation from low-level message passing details and preservation of the advantages of

---

* colajanni@utovrm.it
** filippon@vnet.ibm.com

the serial SpBLAS software. This is achieved by providing a dual Fortran 90 and FORTRAN 77 interface; to allow maximal reuse of existing serial software [8] the internal implementation is based on the FORTRAN 77 interface.

Our approach assumes that the application is parallelized through an index space decomposition; this approach is convenient and consistent with common usage for most applications we are aware of, with the user keeping control over the choice of the data decomposition.

We have been influenced by ideas expressed at the BLAS Technical Workshop held at the University of Tennessee in November 1995 [6]; a previous version of this paper has appeared in [3].

2 General Overview

The main operations included in our library, called "Parallel Sparse BLAS" (PSBLAS), are:

- Sparse matrix by dense matrix product;
- Sparse triangular systems solution;
- Vector and matrix norm;
- Dense matrix sums;
- Dot products;
- Data exchange and update;
- Data structure preprocessing and initialization.

A more detailed description of these operations will be given in Sec. 4. Some of the operations do not appear in the serial SpBLAS because they are trivially implemented either through open code or calls to the dense BLAS; they do appear here, however, and are different from the similar PBLAS because of the peculiarity of the adopted data distribution. The domain decomposition features of the underlying data distribution also prompts the need for specialized data exchange routines; we have chosen the BLACS [7] to perform the underlying communication operations in a portable and efficient way.

2.1 Data Decomposition

In any distributed memory application the data structure used to partition the problem is essential to the viability of the entire approach; the criteria guiding the decomposition choice are:

1. Load balancing;
2. Communication costs reduction;
3. Efficency of the serial parts of the computation;

In the case of dense linear algebra algorithms [4] a block-cyclic data distribution of the index space is sufficiently general and powerful to achieve a good compromise between the different factors affecting performance; this distribution also has the advantage of being supported by HPF, thus enabling a very high-level interface with the application codes. On the other hand, since the target algorithms for the PSBLAS library are parallel sparse iterative solvers (typically arising in the numerical solution of partial differential equations), to achieve acceptable performance we have to pay special attention to the structure of the problem they come from.

The matrix nonzero pattern is influenced by the shape of the computational domain, by the discretization strategy and by the equation/unknown ordering; the matrix itself can be interpreted as the adjacency matrix of the graph identifying the computational grid. If we assign each variable of the problem to a processor, we are inducing a partition of the sparse matrix which, in turn, is logically equivalent to a partition of the discretization graph; in this framework the parallelization of the linear system solver gives rise to an auxiliary graph partitioning problem. The general graph partition problem is $\mathcal{NP}$-complete; methods for obtaining reasonably good decompositions in an effective way form the subject of a vast body of literature, and are outside the scope of the present work (see [11, 10, 13] and references therein). The quality of the partition will influence the performance of the computational kernels; however for most problems of interest a surface to volume effect takes place, and moreover it is often the case that the partition of the computational grid can be guided by physical considerations. For these reasons we designed our library in such a way as to leave to the user total control over the specification of data allocation.

Within the operations of PSBLAS we classify elements of the index space of the computational domain in a more detailed way by making use of the following sets of points:

Internal: These are points belonging to a given domain and only interacting with points belonging to the same domain, i.e. requiring no data communication;

Boundary: These are points belonging to a given domain and having some interaction with points belonging to other domains, thus requiring communication;

Halo: The halo of a given domain is the set of points belonging to other domains whose value is needed to proceed with the computation; it defines the communication requirements;

Overlap: We have an overlap when a (boundary) point belongs to multiple domains; this is a peculiar situation that may give some advantages in preconditioning for certain problems [9].

We do not require that the number of domains be identical to the number of processes: more than one domain may be assigned to a given process, whith redundant information being cleaned up in the preprocessing stage.

2.2   Parallel Environment

To support our parallel environmnent we choose to define a uni-dimensional BLACS process grid, shaped as a $P \times 1$ process array; the BLACS context is subsequently used in all internal communications.

3 Data structures

The description of the data allocation details is contained in a set of integer arrays; all matrices, both sparse and dense, are assumed to be distributed conformal to the index space partition specified in the descriptor vectors. Parameters in the descriptors are labeled as global when their value must be equal on all processes, local otherwise. The definition of the descriptors is as follows:

MATRIX_DATA The main descriptor vector, containing:
   DEC_TYPE_ Decomposition type (global);
   M_ Total number of equations (global);
   N_ Total number of variables (global);
   N_ROW_ Grid variables owned by the current process (local);
   N_COL_ Grid variables used by the current process (local);
   CTXT_ Communication context as returned by the BLACS (global).
HALO_INDEX_ A list of the halo elements to be exchanged with other processes; for each process it contains the process identifier, the number of points to be received, the indices of points to be received, the number of points to be sent and the indices of points to be sent; The list may contain an arbitrary number of groups; its end is marked by a -1;
OVRLAP_INDEX_ A list of the overlap elements, organized in groups exactly like the halo index descriptor;
OVRLAP_ELEM For all local overlap points contains index of the point and the number of domains sharing it; the end is marked by a -1.

The use of the Fortran 90 language allows a convenient packaging of the data structures needed to describe the data allocation; the interface for the descriptors is defined as:

```
MODULE TYPE_PSP
   INTEGER, PARAMETER :: DEC_TYPE_=1, M_=2, N_=3&
        & N_ROW_=4, N_COL_=5, CTXT_=6,&
        & PSBLAS_DECOMP_TYPE=0
   TYPE DECOMP_DATA_TYPE
      INTEGER, POINTER :: MATRIX_DATA(:), HALO_INDEX(:),&
           & OVRLAP_ELEM(:,:), OVRLAP_INDEX(:)
   END TYPE DECOMP_DATA_TYPE
END MODULE TYPE_PSP
```

For the local storage of sparse matrices we use the interface proposed in [8]:

```
MODULE TYPESP
   TYPE D_SPMAT
      INTEGER M, K
      CHARACTER*5 FIDA
      CHARACTER*1 DESCRA(10)
```

```
    INTEGER     INFOA(10)
    REAL(KIND(1.D0)), POINTER :: ASPK(:)
    INTEGER, POINTER :: IA1(:), IA2(:), PL(:), PR(:)
  END TYPE D_SPMAT
END MODULE TYPESP
```

4 Computational kernels

In this paper we will describe only the Fortran 90 interface to our subroutines; however the computational kernels are built as wrappers around a FORTRAN 77 layer, described in more details in [3]. We choose to describe the Fortran 90 interface as the use of advanced features provides greater readability and usability of the resulting codes.

All subroutines are prefixed with the characters F90_PS and are generic interfaces for real and complex data types; there is also support for dense arrays of rank one and two.

In all the subroutines, the parameters are ordered in such a way that those positioned after DECOMP_DATA are always optional.

F90_PSDOT Compute the dot product (x, y) of two vectors X(IX:IX+N-1,JX), Y(IY:IY+N-1,JY).

$$DOT = F90\_PSDOT \ (X, \ Y, \ DECOMP\_DATA, \ IX, \ JX, \ IY, \ JY)$$

F90_PSAXPBY Compute

$$Y \leftarrow \alpha X + \beta Y$$

with matrices X(IX:IX+M-1,JX:JX+N-1),Y(IY:IY+M-1,JY:JY+N-1).

CALL F90_PSAXPBY
        (ALPHA, X, BETA, Y, DECOMP_DATA, M, N, IX, JX, IY, JY)

F90_PSAMAX Compute the maximum absolute value, or infinity norm:

$$amax \leftarrow \max_i\{|X(i)|\} = \|X\|_\infty$$

with vector X(IX:IX+N-1,JX).

$$AMAX = F90\_PSAMAX \ (X, \ DECOMP\_DATA, \ N, \ IX, \ JX)$$

F90_PSASUM Compute the 1-norm of a vector:

$$asum \leftarrow \sum_i |X(i)| = \|X\|_1$$

of vector X(IX:IX+N-1,JX).

$$ASUM = F90\_PSASUM \ (X, \ DECOMP\_DATA, \ N, \ IX, \ JX)$$

F90_PSNRM2 Compute the 2-norm

$$nrm2 \leftarrow \|X\|_2$$

of vector X(IX:IX+N-1,JX).

$$NRM2 = F90\_PSNRM2 \ (X, \ DECOMP\_DATA, \ N, \ IX, \ JX)$$

F90_PSNRMI Compute the infinity-norm of a distributed sparse matrix

$$nrmi \leftarrow \|A\|_\infty$$

NRMI = F90_PSNRMI $(A, DECOMP\_DATA, M, N, IA, JA)$

F90_PSSPMM Compute the sparse matrix by dense matrix product

$$Y \leftarrow \alpha P_R A P_C X + \beta Y$$
$$Y \leftarrow \alpha P_R A^T P_C X + \beta Y$$

with matrices X(IX:IX+N-1,JX:JX+K-1), Y(IY:IY+M-1,JY:JY+K-1). The permutation matrices P_R and P_C are purely local, since they are needed to handle details of the local storage format, and are specified inside the local sparse matrix data structure.

CALL F90_PSSPMM

(ALPHA, A, X, BETA, Y, DECOMP_DATA, TRANS, M, N, K, IA,
JA, IX, JX, IY, JY)

F90_PSSPSM Compute the triangular system solution:

$$Y \leftarrow \alpha P_R T^{-1} P_C X + \beta Y$$
$$Y \leftarrow \alpha D P_R T^{-1} P_C X + \beta Y$$
$$Y \leftarrow \alpha P_R T^{-1} P_C D X + \beta Y$$
$$Y \leftarrow \alpha P_R T^{-T} P_C X + \beta Y$$
$$Y \leftarrow \alpha D P_R T^{-T} P_C X + \beta Y$$
$$Y \leftarrow \alpha P_R T^{-T} P_C D X + \beta Y$$

The triangular sparse matrix T is constrained to be block-diagonal; this is equivalent to the application of local ILU or IC preconditioning; the permutation matrices are local, as in the matrix-matrix product. The *IOPT* parameter specifies whether a cleanup of the overlapped elements is required on exit. Uses dense matrices X(IX:IX+M-1,JX:JX+N-1),
Y(IY:IY+M-1,JY:JY+N-1).

CALL F90_PSSPSM

(ALPHA, T, X, BETA, Y, DECOMP_DATA, TRANS, UNITD,
IOPT, D, M, N, IT, JT, IX, JX, WORK)

5 Iterative Methods

To illustrate the use of the library routines we report in Table 1 the template for the CG method from [2], with local ILU preconditioning and normwise backward error stopping criterion [1]. The example shows the high readability and usability features of the PSBLAS with Fortran 90 interface, as can be readily seen comparing the mathematical formulation of the algorithm with its PSBLAS implementation. Efficiency improvements can also be obtained from this reference implementation simply by making use of the optional parameters available in the subroutine interfaces.

Template	PSBLAS Implementation

Compute $r^{(0)} = b - Ax^{(0)}$	`call f90_psaxpby(one,b,zero,r,decomp_data)`
	`call f90_psspmm(-one,A,x,one,r,decomp_data)`
	`rho = zero`

for $i = 1, 2, \ldots$
    solve $Mz^{(i-1)} = r^{(i-1)}$
	`do  it = 1, itmax`
	`    call f90_psspsm(one,L,r,zero,w,decomp_data)`
	`    call f90_psspsm(one,U,w,zero,z,decomp_data)`

$\rho_{i-1} = r^{(i-1)^T} z^{(i-1)}$	`    rho_old = rho` `    rho = f90_psdot(r,z,decomp_data)`

if $i = 1$
    $p^{(1)} = z^{(0)}$
else
    $\beta_{i-1} = \rho_{i-1}/\rho_{i-2}$
    $p^{(i)} = z^{(i-1)} + \beta_{i-1}p^{(i-1)}$
endif
	`    if (it == 1) then`
	`        call f90_psaxpby(one,z,zero,p,decomp_data)`
	`    else`
	`        beta = rho/rho_old`
	`        call f90_psaxpby(one,z,beta,p,decomp_data)`
	`    endif`

$q^{(i)} = Ap^{(i)}$
$\alpha_i = \rho_{i-1}/p^{(i)^T} q^{(i)}$
	`    call f90_psspmm(one,A,p,zero,q,decomp_data)`
	`    sigma = f90_psdot(p,q,decomp_data)`
	`    alpha = rho/sigma`

$x^{(i)} = x^{(i-1)} + \alpha_i p^{(i)}$ $r^{(i)} = r^{(i-1)} - \alpha_i q^{(i)}$ Check convergence:	`    call f90_psaxpby(alpha,p,one,x,decomp_data)` `    call f90_psaxpby(-alpha,q,one,r,decomp_data)`

$\|r^{(i)}\|_\infty \leq \epsilon(\|A\|_\infty \cdot \|x^{(i)}\|_\infty +$
$+ \|b\|_\infty)$
	`    rni = f90_psamax(r,decomp_data)`
	`    xni = f90_psamax(x,decomp_data)`
	`    bni = f90_psamax(b,decomp_data)`
	`    ani = f90_psnrmi(A,decomp_data)`
	`    err = rni/(ani*xni+bni)`
	`    if (err.le.eps) return`

end	`enddo`

Table 1. Sample CG implementation

6 Auxiliary routines

To help the user setting up the computational environment we have defined a set of tools to define the parallel data structure, allocate and assemble the various matrices involved.

The user is required to provide a subroutine defining the allocation of the index space to the various processes; the subroutine returns, for a given global index, the process(es) owning that index. The tools library contains:

F90_PSSPALLOC Allocate Global Sparse Matrix; builds the descriptor vectors based on the information provided by the user defined subroutine PARTS, and allocates the data structures for the sparse matrix.

CALL F90_PSSPALLOC
    (M, N, PARTS, DECOMP_DATA, A, ICONTXT, IERRV, NNZV, NNZ)

F90_PSSPINS Insert Sparse Matrix; a local sparse matrix $BLCK$ in coordinate format is inserted in the global sparse matrix A; may be called repeatedly on different processes.

CALL F90_PSSPINS
    (M, N, PARTS, DECOMP_DATA, A, IA, JA, BLCK, WORK, IERRV, IBLCK, JBLCK)

F90_PSSPASS Assembly Sparse Matrix; finalize the global sparse matrix A; the user has the ability to specify how to treat replicated matrix entries. Inconsistencies introduced in the insertion phase are detected here.

CALL F90_PSSPASS
    (PARTS, A, DECOMP_DATA, WORK, IERRV, OVERLAP)
F90_PSSPFREE Deallocate a sparse matrix.
    CALL F90_PSSPFREE $(A,\ DECOMP\_DATA)$

Similar routines are provided for building dense matrices distributed conformally to the sparse matrix.

7 Performance data

In the current implementation of PSBLAS we have mostly strived to achieve correctness and usability; we believe that many performance improvements are still possible, especially in the local computations that are handled by the serial Sparse BLAS code.

Nonetheless the efficiency of the library is already quite satisfactory; we have tested two methods, Conjugate Gradient and Bi-CGSTAB on an IBM SP2 with the SPSwitch interconnection.

For the Conjugate Gradient test we have generated a standard Poisson matrix on a cubic domain, and we have partitioned it in a very simple fashion, just assigning blocks of rows of equal sizes to the various processes; we show in Table 2 the time per iteration for various cube edge dimensions, as the number of processors changes. As we can see the speedup is acceptable, provided that the number of grid points is large enough to offset the system overhead and the communication costs. The number of iterations to convergence changes over the various test cases, but typically by no more than 10 %.

For the Bi-CGSTAB tests we have used a 3-D problem from the seven-point difference discretization of an operator similar to the one mentioned in [12]:

$$(Lu) = -(u_{xx} + u_{yy} + u_{zz}) +$$
$$a_1 u_x + a_2 u_y + a_3 u_z + bu = f$$

Time per iteration			
NP $L = 30$	$L = 40$	$L = 50$	$L = 80$
1 0.0549	0.1165	0.2261	0.9304
2 0.0369	0.0674	0.1246	0.4841
4 0.0334	0.0463	0.0769	0.2826
8 0.0388	0.0364	0.0528	0.1659
16 0.0373	0.0333	0.0441	0.0990

Table 2. CG: Constant problem size, cube of size L

Time per iteration			
NP $N = 8K$	$N = 27K$	$N = 64K$	$N = 125K$
1 0.0307	0.0870	0.2104	0.4124
2 0.0743	0.1051	0.2321	0.4455
4 0.0608	0.1170	0.2515	0.4663
8 0.0617	0.1312	0.2707	0.4884

Table 3. Bi-CGSTAB: Scaled problem size, with N grid points per processor

The computational grid in this case is scaled so that as we add processors we keep the number of grid points per processor constant, i.e. we refine the computational grid as we have more computational power available. The grid is partitioned into subdomains, then all grid points belonging to a subdomain are assigned to a processor. As we can see from Table 3 the time per iteration depends almost exclusively on the number of grid points per processor, provided that this number is sufficiently large.

8 Conclusions

We have presented a proposal for parallel sparse BLAS tailored for the implementation of iterative methods on distributed memory computers; the prototype library is built on top of other existing standard or proposed standard library, and makes extensive use of advanged language features provided by Fortran 90. Future development directions include the implementation of a wider range of iterative methods and of support for different preconditioning techniques.

Acknowledgements

We thank Prof. Salvatore Tucci of the Università di Roma Tor Vergata for his help in setting up this cooperative work.

References

1. Mario Arioli, Iain Duff, and Daniel Ruiz. Stopping criteria for iterative solvers. *SIAM J. Matrix Anal. Appl.*, 13:138–144, 1992.
2. Richard Barrett, Michael Berry, Tony Chan, James Demmel, June Donat, Jack Dongarra, Victor Eijkhout, Roldan Pozo, Charles Romine, and Henk van der Vorst. *Templates for the solution of linear systems*. SIAM, 1993.
3. F. Cerioni, M. Colajanni, S. Filippone, and S. Maiolatesi. A proposal for parallel sparse BLAS. Technical Report RI.96.05, University of Roma Tor Vergata, Department of Computer Science, Systems and Industrial Engineering, March 1996.
4. J. Choi, J. Demmel, J. Dhillon, J. Dongarra, S. Ostrouchov, A. Petitet, K. Stanley, D. Walker, and R. C. Whaley. ScaLAPACK: A portable linear algebra library for distributed memory computers. LAPACK Working Note #95, University of Tennessee, 1995.
5. J. Choi, S. Ostrouchov, A. Petitet, D. Walker, and R. Clint Whaley. A proposal for a set of parallel basic linear algebra subprograms. LAPACK Working Note #100, University of Tennessee, May 1995. http://www.netlib.org/lapack/lawns.
6. Jack Dongarra, Sven Hammarling, and Susan Ostrouchov. BLAS technical workshop. LAPACK Working Note #109, University of Tennessee, May 1995. http://www.netlib.org/lapack/lawns.
7. Jack J. Dongarra and R. Clint Whaley. A user's guide to the BLACS v1.0. LAPACK working note #94, University of Tennessee, June 1995. http://www.netlib.org/lapack/lawns.
8. Iain S. Duff, Michele Marrone, Giuseppe Radicati, and Carlo Vittoli. A set of level 3 basic linear algebra subprograms for sparse matrices. Technical Report RAL-TR-95-049, Computing and Information System Department, Atlas Centre, Rutherford Appleton Laboratory, Oxon OX11 0QX, September 1995. SPARKER Working Note # 1,
   Ftp address seamus.cc.rl.ac.uk/pub/reports.
9. Salvatore Filippone, Michele Marrone, and Giuseppe Radicati di Brozolo. Parallel preconditioned conjugate-gradient type algorithms for general sparsity structures. *Intern. J. Computer Math.*, 40:159–167, 1992.
10. Bruce Hendrickson and Robert Leland. An improved spectral graph partitioning algorithm for mapping parallel computations. *SIAM J. Sci. Comput.*, 16(2):452–469, 1995.
11. George Karypis and Vipin Kumar. *METIS: Unstructured Graph Partitioning and Sparse Matrix Ordering System*. University of Minnesota, Department of Computer Science, Minneapolis, MN 55455, August 1995. Internet Address: http://www.cs.umn.edu/~karypis.
12. C. T. Kelley. *Iterative Methods for Linear and Nonlinear Equations*. SIAM, 1995.
13. Alex Pothen, Horst D. Simon, and Kang-Pu Liou. Partitioning sparse matrices with eigenvectors of graphs. *SIAM J. Matrix Anal. Appl.*, 11:430–452, 1990.

Parallel Search-Based Methods in Optimization

Jens Clausen

DIKU - Department of Computer Science
University of Copenhagen
Universitetsparken 1
DK 2100 Copenhagen Ø
Denmark

Abstract. Search-based methods like Branch and Bound and Branch and Cut are essential tools in solving difficult problems to optimality in the field of combinatorial optimization, and much experience has been gathered regarding the design and implementation of parallel methods in this field.

Search-based methods appear, however, also in connection with certain continuous optimization problems and problems in Artificial Intelligence, and parallel versions hereof have also been studied.

Based on parallel Branch and Bound algorithms, the advantages as well as the difficulties and pitfalls in connection with parallel search-based methods are outlined. Experiences with parallel search-based methods from the three fields mentioned are then described and compared in order to reveal similarities as well as differences across the fields.

1 Introduction

The key characteristic of combinatorial optimization problems is the finiteness of the solution space. A problem Q can be defined as the minimization of a function $f(x)$ of variables $(x_1 \cdots x_n)$ over a region of *feasible solutions*, S: $\min_{x \in S} f(x)$, where S contains a finite but usually very large number of elements. The function f is called the *objective function* and may be of any type. The set of feasible solutions is usually a subset of a finite set of potential solutions P determined partly by general conditions on the variables, e.g. that these must be non-negative and integer or binary, and partly by problem specific special constraints defining the structure of the feasible set.

The finiteness of S implies that a trivial algorithm for the solution of a given problem is immediately at hand: finding the optimal solution by explicitly enumerating the set of potential solutions. The number of potential solutions is, however, in almost all cases exponential in the structural parameters of the problem (e.g. the number of binary variables), and hence explicit enumeration is only a theoretical possibility. For some combinatorial optimization problems the exponentiality of the size of P presents no problem - efficient (polynomially bounded) methods taking advantage of the structure of the problem are know. For others - the $\mathcal{NP}$-complete problems - no polynomial method exploiting the problem structure is known, and for these search based methods as Branch-and-Bound (B&B) are used.

A B&B algorithm searches the complete space of solutions for a given problem for the best solution. The use of *lower bounds* for f over the subspaces of S combined with the value of the *current best solution* enables the algorithm to search parts of the solution space only implicitly. If for a given subspace the value of the lower bound function $lb(\cdot)$ exceeds the value of the best solution found so far, the subspace cannot contain the optimal solution and hence the solutions in the subspace need not be explicitly enumerated - the subspace is *discarded* or *cut off*. Otherwise *branching* takes place: The subspace is subdivided hoping that for each of the generated subspaces either the lower bound function increases enabling a cut-off, or the optimal value in the subspace with respect to f is easy to determine. Of course the worst-case behavior of B&B-algorithms is exponential, however practice shows that in many cases such algorithms are feasible though computationally demanding.

A B&B algorithm is often described in terms of the traversal of a *search tree*, in which the nodes represent subspaces of the solution space. Each iteration then processes a node in the search tree representing a not yet explored subspace of the solution space (called a *live node* or *live subspace*) as described above, and the search ends when no live nodes are left.

At first glance global optimization of continuously differentiable functions seems to have very little in common with the B&B method just described. If, however, finding the minimum value and the set of minimum points of a given function f is done using *interval methods*, the resemblance becomes obvious. Here the goal is to identify the global minimum value of a given function f over an n-dimensional box X in $\mathcal{R}^n$ as well as the set of all minimum points in X. Classical methods for global optimization are sometimes not feasible. An alternative is the interval based method, i.e. to find an enclosure (in terms of a box) of the minimum points and a set of boxes of sufficiently small size enclosing all minimum points. Here an *inclusion function* F is used, which for each subbox Y in X returns an interval containing all values $f(y)$ for $y \in Y$. If a function value corresponding to a point in X is known, F can be used to rule out the possibility that an optimal point is present in a given subbox: If inf $F(Y)$ is larger than the given value, Y does not contain a minimum point. If such a decision cannot be made, a subdivision of Y into at least two subboxes takes place hoping that inf $F(\cdot)$ increases for each subbox or that the minimum for f is easy to find by classical methods (e.g. using monotonicity of f). This process is also called Branch-and-Bound, even if the search space is not by nature finite.

Tree search and graph search are also classical solution methods in Artificial Intelligence and game theory. Here the problem is to identify a feasible solution - termed a *goal node* - in a tree or graph given implicitly by an initial node and a function, which generates the descendants of a given node in the tree/graph - the process is called *node expansion*. The depth of the tree/graph may be very large, e.g. in case the nodes represent states of a game. Often a cost function c assigns costs to the edges of the graph and one searches for the goal node minimizing the cost of the path from the initial node to the goal node. The dependency between the path followed to a node and the cost of the node implies that it is generally not possible to give a good estimate of the minimum cost of a goal node. Hence cut-off is possible only after a first goal node has been found, and the potential large depth of the tree/graph makes the choice of the next node to expand essential for the efficiency of the search. A *heuristic* function h estimating the minimal cost of a path from a given non-goal node to a goal node is then often given, and the values of the *evaluation* function $f(\cdot) = c(\cdot) + h(\cdot)$ for unexpanded nodes are applied in the selection of the next node to expand to guide the search. The algorithms usually referred to in this context are DFS (Depth First Search), A and A*. *Iterative deepening* versions (IDFS and IDA*) are used to cope with the problem of finding the first goal node without getting stuck in a part of the search tree containing no goal node. Here the search tree/graph is in each iteration developed until all unexpanded nodes have depth or value of f above some limit, which increases from iteration to iteration.

In all three cases described, the sequential algorithm proceeds in iterations. Each iteration corresponds to the processing of a tree or graph node or a box, and the processing of different nodes/boxes is independent and uses only very limited amounts of global information (the best function value found so far). Parallelization hence seems easy, and one would expect good results wrt. scalability properties. Experiences show, however, that a number of different issues have to be taken into account when parallelizing the algorithms.

Rather than treating one of the fields described in depth, I will in the following compare concepts and experiences across the subjects in order to reveal similarities and differences, and to identify issues, for which experiences from one of the subjects may lead to increased efficiency and new insight in one or both of the others. My basis will be parallel B&B for combinatorial optimization, since my personal experiences stem from this field. Hence Section 2 briefly reviews the concepts and central issues hereof, and Section 3 and 4 relate results from the two other fields to the material of Section 2. Finally Section 5 contains conclusions and suggestions for further research.

2 Parallel Branch-and-Bound for Combinatorial Optimization problems

2.1 Components of B&B algorithms

Both sequential and parallel B&B algorithms for minimization problems have three main components:

1. a *bounding function* providing for a given subspace of the solution space a lower bound for the best solution value obtainable in the subspace,
2. a *strategy for selecting* the live solution subspace to be processed in the current iteration, and
3. a *branching* rule to be applied if a subspace after investigation cannot be discarded, hereby subdividing the subspace considered into two or more subspaces to be investigated in subsequent iterations.

Additionally a method for finding the first (good) feasible solution is usually provided.

Most B&B algorithms also have a problem specific part, which in each iteration attempts to exploit the available information as much as possible. Different tests ruling out subspaces as candidates to contain the optimal solution may be performed in addition to the bound calculation, and methods for finding the optimal solution of the subspace at hand directly may be tried. Such methods usually lead to increased efficiency of the algorithm, but they also heavily influence the relative computational efficiency (in terms of the total number of nodes bounded) of the different strategies for selecting the next live subspace to process, cf. [7].

The B&B algorithm develops the search tree dynamically during the search. The tree initially consists of only the root node. In each iteration of a the algorithm, a live node is selected for exploration using some strategy to select between the current live subspaces. Then either an *eager* or a *lazy* processing strategy is used. For the eager strategy a branching is first performed: The subspace is subdivided into smaller subspaces and for each of these the bound for the subspace is calculated. In case the node corresponds to a feasible solution or the bound is the value of an optimal solution, the value hereof is compared to the current best solution keeping the best. If the bound is larger than or equal to the current best solution, or no feasible solution exists in the subspace, this is discarded. Otherwise, the possibility of a better solution in the subspace cannot be ruled out, and the node (with the bound as part of the information stored) is then joined to the pool of currently live subspaces. In case a *lazy* strategy is used, the order of bound calculation and branching is reversed, and the live nodes are stored with the bound of their father as part of the information.

The main virtue of the eager processing strategy is that it facilitates Best First Search (BeFS) - the nodes of the search tree is processed in the order of increasing lower bound. Using BeFS B&B all nodes with a bound less than the optimum (called *critical nodes*) and some of the nodes with bound equal to the optimum are branched, thereby in some sense minimizing the number of bound calculations performed. In general BeFS B&B is infeasible due to memory problems, since the number of subspaces to be stored in practice grows exponentially with the problem size. This also holds for the Breadth First Strategy (BFS), in which all nodes on one level of the search tree are processed before any node on succeeding levels.

The eager strategy supports also DFS B&B, in which the nodes are processed in order of depth in the search tree. DFS B&B may develop nodes of the search tree with a lower bound larger than the optimum and hence perform a number of bound calculations larger than necessary, however, note that if the initial solution value equals the optimum, then any selection strategy will lead to the processing of the same number of nodes. DFS B&B does not lead to memory problems and is in general used when solving large scale problems.

The lazy strategy postpones bound calculation as long as possible. Sequentially the advantage hereof is not obvious, however, in a parallel setting with different processors working on different nodes of the search tree, a processor may find a solution, which enables cut-off of nodes already bounded in other parts of the search tree, and hence also of the descendants of these nodes. The lazy strategy supports only DFS B&B, whereas true BeFS B&B is not possible since each node is stored with the bound of its generating node.

2.2 Parallel B&B algorithms

The goals of parallelizing a B&B algorithm are to increase the solution speed of already solvable problems and to make feasible the solution of hitherto unsolved problems, either through increased computational power or through increased memory size of the system available. A key issue in the parallelization is to ensure that no processor is idle if work is available at other processors, i.e. *load balancing*. Also, the amount of work done by the parallel version of the algorithm should not exceed (or exceed only marginally) the work performed by the corresponding sequential algorithm. A large body of literature on various aspect of parallel B&B is available, and [11, 12] are good surveys with extensive reference lists.

Parallelism in connection with B&B introduce the possibility of *anomalies*, i.e. that increasing the number of processors in the parallel system does not lead to a corresponding increase in speed of computation. If the running time increases with an increase in number of processors, a *detrimental* anomaly has been observed, whereas if the time decreases by a factor larger than the ratio between the number of processors in the new and the old parallel system, we have an *acceleration* anomaly. The reason for such behavior is that the number of nodes developed in the search tree usually varies with the number of processors - the discovery of good or optimal solutions may take place both earlier and later than presently. One usually want to increase the probability of observing acceleration anomalies and decrease the probability of detrimental anomalies.

Parallelizing B&B algorithms usually amounts to parallelizing the processing of the search tree nodes. Since bound calculations for nodes in different levels of the search tree normally differ in number of operations, SIMD architectures are in general not used for B&B. However, both shared memory and distributed memory MIMD architectures have been used leading essentially to two different parallelization paradigms: The *master-slave* paradigm and the *distributed* paradigm.

The idea of the master-slave paradigm is to make the pool of live subspaces a shared data structure. In a distributed MIMD setting this is achieved by dedicating one processor, the *master*, to administration of the subproblem pool and current best solution. Communication takes place by message passing. In the shared memory setting the same effect is accomplished through variable sharing.

The node processing is performed by the slave processors, each of which iterates a loop consisting of getting a live subspace from the master, performing one or more sequential B&B iterations storing the results locally (new live subspaces and (possibly) a new best solution), and communicating results to the master.

Termination problems become trivial to solve, and any problem selection strategy can be used by the master so that the sequential B&B can be followed as close as possible. This seems desirable both from a practical and a theoretical point, since it can be shown that speed-up anomalies can be prevented if a set of fairly restrictive general conditions are met by the problem and bounding function in question. The master data structure or processor however easily becomes a bottleneck in the system since all slaves need to communicate with the master frequently.

The distributed paradigm solves the bottleneck problem by decentralizing the pool of live subspaces to the individual processors. This on the other hand implies that no central control of selection strategy is possible, and that load balancing becomes a difficult issue: Not only do processors need to exchange subspaces in order to avoid processor idling, but much care has to be taken that all processors in general work on parts of the search space, which are equally likely to contain the optimal solution. Essentially, there are two extreme methods for distribution termed *on request* resp. *without request* in [11]. In the first, idle processors request work from other processors. Essential design questions are which processors to ask, what to do if a request for work cannot be met, and which problems to send if the request can be met. If a without request scheme is used, a processor sends work to other processors based on its own local information about the workload situation, and here questions of frequency of messages, recipients, and which problems to send are key design issues.

Finally the possibility of static workload distribution should be mentioned: Initially to generate a number of subspaces large enough that each processor receives a substantial number of these, and then letting the processors work independently on their own subspaces of the entire solution space. The loss of efficiency due to idle processors should in this case be balanced against the implementationwise very simple communication protocol. If subspaces with almost equal values of $lb(\cdot)$ are generated, experience shows that for some problems the efficiency loss is small, cf. [3].

2.3 Experiences with Parallel B&B

My personal experiences with parallel B&B stem from work with the Graph Partitioning Problem (GPP), the Quadratic Assignment Problem and the Job Shop Scheduling Problem (JSS), cf. [4, 5, 6, 15] for details. The following is a short summary of the experiences gained, and contains points on general parallel processing, B&B in general, and parallel B&B.

a: Do not use parallel processing if the problem is too easy.
b: Choose the right hardware for your problem (or problem for your hardware, if you are a basic researcher).
c: Centralized control is only feasible in systems with a rather limited number of processors.
d: If the problem in question has a bound calculation function providing strong bounds, then the number of live subproblems at any time might be small.

Then only a few of the processors of a parallel system can be kept busy with useful work at the same time.

e: If on the other hand the bound calculation gives rise to large search trees in the sequential case, parallel B&B will most likely be a very good solution method. Here static workload distribution may lead to an easily programmable and efficient algorithm if the system used is homogeneous.

f: The problem of anomalies is overestimated. In most cases detrimental anomalies occur only if the problem at hand is very easy. Acceleration anomalies occur frequently for some problems and seldomly for others, and are closely related to the identification of good feasible solutions.

g: When using dynamic workload distribution, the time spent on programming, testing, and tuning sophisticated methods may not pay off well. Often good results are possible with relatively simple schemes.

h: When exchanging subspaces, try to facilitate that the quality of the subspaces processed (e.g. in terms of lower bound value) is uniform over the processors.

i: Do not send subspaces, on which no processing have been done, to other processors requesting work.

j: The importance of finding a good initial solution cannot be overestimated, and the time used for finding such one is often only few percentages of the total running time of the parallel algorithm.

k: In case an initial solution very close to the optimum is expected to be known, the choice of node selection strategy and processing strategy makes little difference.

l: With a difference of more than few percent between the value of the initial solution and the optimum the theoretically superior BeFS B&B shows inferior performance compared to both lazy and eager DFS B&B. This is in particular true if the pure B&B scheme is supplemented with problem specific efficiency enhancing test for e.g. supplementary exclusion of subspaces, and if the branching performed depends on the value of the current best solution.

3 Parallel Branch-and-Bound for Global Optimization

Search based combinatorial optimization and global optimization using interval methods possess a lot of similarities and some differences. In both cases one seeks to identify a minimum value of a function over a feasible set using bounding information to avoid an untractable number of function evaluations, and problem specific/tailored methods are used in addition to bounding to increase efficiency. However, in global optimization the solution space in terms of number of potential subboxes is in theory infinite, and one seeks the set of *all* minimum points

The elements of a B&B method for global optimization are very similar to those encountered for combinatorial optimization - bounding function, selection strategy and branching rule:

1. an *inclusion function* F providing for a given subbox of the initial box an interval containing the function values obtainable in the subbox,

2. a *strategy for selecting* the live subbox to be processed in the current iteration, and

3. a *subdivision* rule to be applied if a subbox after investigation cannot be discarded, hereby subdividing the subbox into two or more subboxes.

Due to the similarity between the concepts and methods, parallelization of B&B for global optimization closely resembles the parallelization described in Section 2, and also theoretical results are similar. The body of literature is, however, much more limited - examples of parallelizations are [1, 2, 8, 9, 10, 13].

The strategies for selecting the next subbox to be investigated are similar in the two fields: oldest-first (BFS), depth-first (DFS) and best-first (BeFS). However, it appears as if the question of lazy vs. eager processing strategy has not been addressed. The effect of the value of the initial solution on the number of subboxes investigated for different strategies is not discussed. A subbox is divided in one of it dimensions, and subdivision rules are usually composed of a dimension selection rule and a rule specifying how many subboxes to create. Subdivision rules making use of the current bounds for the minimum value are not mentioned.

The results from parallelizations match those described in Section 2.3. Similar load balancing schemes have been investigated in the two fields, and the main trends are the same. The possibility of static workload distribution being almost as efficient as dynamic workload distribution for some types of functions is discarded without experiments. Anomalies are also observed, however, in global optimization these occasionally stem from sequential algorithmic inefficiency, and the field seems not as mature as is the case for combinatorial optimization. Rules of thumb regarding the characteristics of problems/functions for which parallelization works/fails are similar in the fields.

Regarding theoretical results cf. [1], it is possible to show that a best-first-strategy for selecting the next subbox to be investigated leads to a minimal set of investigated subboxes in the same sense as it leads to a minimal number of investigated nodes. The fact that all minimum points are sought for enables a theoretical limitation on the superlinear speed-up possible - a result which is not true if only a single minimum point is to be found.

4 Parallel Search Methods in AI.

Search problems in AI resembles search problems in optimization with respect to problem setting - even if a cost function for the nodes is not given, a cost function of equal value for all goal nodes turns the given problem into an optimization problem. Hence parallel search methods for AI problems are often described together with methods for optimization problems, cf. [12, 14]. However, bound functions are in general rudimentary consisting of the hitherto incurred cost of the node in question, and the search space is represented implicitly by the expansion function. This may generate a graph (possibly with cycles) or a tree, and the depth of the generated graph/tree may be very large. Since the cost function of a node depends on the path followed to the node, a solution consist not only of the node but includes also the path. Hence care is necessary when the search structure is a graph.

The elements of a search method for an AI problem can thus be described as:

1. a *strategy for selecting* the unexpanded node to be processed in the current iteration, and
2. a *expansion* rule to be applied if a node after investigation cannot be discarded, hereby generating new unexpanded nodes.

Strategies for selecting the next node to expand correspond to those described previously with the heuristic function f playing the role of the lower bound lb in BeFS search. The lack of any initial solution and the iterative deepening approach, however, adds a new dimension to the search algorithms. Also, the possibility of search overhead in terms of expansion of nodes with subspaces containing no minimal goal nodes is much larger than encountered in combinatorial optimization implying an increased instability with respect to anomalies.

With respect to load balancing in parallel search methods, again similar ideas have been pursued. The idea of static workload distribution is in [14] modified so that processors initially get each an even share of generated nodes with same evaluation function value v (corresponding to same lower lower bound in B&B), but if a node gets idle it requests unexpanded nodes with value v from other processors. Especially in combination with IDA* good results are reported. Predicting which components and parameters are optimal to use in a search method for a given problem seems more difficult than in the two other fields.

Regarding theoretical results, it is also here possible to show superiority of BeFS provided that the heuristic function c does not overestimate the remaining cost of a path to a goal node.

5   Conclusions

Parallel search methods and experiences with these in combinatorial optimization, AI, and global optimization by interval methods are to a large extent overlapping, the differences in general stemming from differences in the problem fields - search for one vs. all minimal solutions, lack of good initial solutions, cost dependence of path in the search tree, etc.

In general, combinatorial optimization and AI seem with respect to parallelization to have evolved with a reasonable interplay between the disciplines. The interplay between these two and global optimization on the other hand has been limited, and several ideas seem worthwhile further investigation in the setting of global optimization, e.g. eager/lazy node processing, static workload distribution, subbox selection strategy in combination with subbox subdivison methods, and iterative deepening methods.

References

1. S. Berner, "Ein paralleles Verfahren zur verifizierten globalen Optimierung", Fachbereich Mathematik, Univ. Wuppertal, 1995.
2. O. Caprani, B. Godthaab and K. Madsen, "Use of a real-values local minimum in parallel interval global optimization", Interval Computations 2, p. 71 - 82, 1993.
3. J. Clausen, "Parallel Branch-and-Bound - Principles and Personal experiences", 20 p., presented at Nordic Summer Course on Parallel Algorithms in Mathematical Programming, Linköping, August 1995. DIKU report 95/29, to be published by Kluwer.
4. J. Clausen and J.L. Träff, "Implementation of Parallel Branch-and-Bound Algorithms - Experiences with the Graph Partitioning Problem", Annals of Operations Research 33 p. 331 - 349, 1991.
5. J. Clausen and J.L. Träff, "Do Inherently Sequential Branch-and-Bound Algorithms Exist ?", Parallel Processing Letters 4,1-2, p. 3 - 13, 1994.
6. J. Clausen and M. Perregaard, "Solving Large Quadratic Assignment Problems in Parallel", DIKU report 1994/22, 14 p., to appear in Computational Optimization and Applications.
7. J. Clausen and Michael Perregaard, "On the Best Search Strategy in Parallel Branch-and-Bound - Best-First-Search vs. Lazy Depth-First-Search", DIKU report 96/14, 11 pages.
8. L.C. Dixon and M. Jha, "Parallel algorithms for global optimization", Journal of Optimization Theory and Applications 79, p. 385 - 395, 1993.
9. J. Eriksson, "Parallel Global Optimization Using Interval Analysis", Dissertation, University of Umeå, Sweden, 1991.
10. T. Henriksen and K. Madsen, "Use of a depth-first strategy in parallel global optimization", Report no. 92-10, Inst. for Numerical Analysis, Tech. Univ. of Denmark, 1992.
11. B. Gendron and T. G. Cranic, "Parallel Branch-and-Bound Algorithms: Survey and Synthesis", Operations Research 42 (6), p. 1042 - 1066, 1994.
12. A. Grama and V. Kumar, "Parallel Search algorithms for Discrete Optimization Problems", ORSA Journal of Computing 7 (4), p. 365 - 385, 1995.
13. R.E. Hansen, E. Hansen and A. Leclerc, "Rigorous methods for global optimization", in C.A. Floudas and P.M. Pardalos (eds.) "Recent Advances in Global Optimization", Princeton University Press, 1992.
14. R. Lüling, B. Monien, A. Reinefeld and S. Tshöke, "Mapping Tree-Structured Combinatorial Optimization Problems onto Parallel Computers" in A. Ferreira and P.M. Pardalos (eds.) "Solving Combinatorial Optimization Problems in Parallel - Methods and Techniques", Springer LNCS 1054, 1996.
15. M. Perregaard and J. Clausen , "Solving Large Job Shop Scheduling Problems in Parallel", DIKU report 94/35, under revision for Annals of OR.

An Hierarchical Approach for Performance Analysis of ScaLAPACK-Based Routines Using the Distributed Linear Algebra Machine

Krister Dackland and Bo Kågström

Department of Computing Science, University of Umeå, S–901 87 Umeå, Sweden.
Email addresses: dacke@cs.umu.se and bokg@cs.umu.se.

Abstract. Performance models are important in the design and analysis of linear algebra software for scalable high performance computer systems. They can be used for estimation of the overhead in a parallel algorithm and measuring the impact of machine characteristics and block sizes on the execution time. We present an hierarchical approach for design of performance models for parallel algorithms in linear algebra based on a parallel machine model and the hierarchical structure of the ScaLAPACK library. This suggests three levels of performance models corresponding to existing ScaLAPACK routines. As a proof of the concept a performance model of the high level QR factorization routine PDGEQRF is presented. We also derive performance models of lower level ScaLAPACK building blocks such as PDGEQR2, PDLARFT, PDLARFB, PDLARFG, PDLARF, PDNRM2, and PDSCAL, which are used in the high level model for PDGEQRF. Predicted performance results are compared to measurements on an Intel Paragon XP/S system. The accuracy of the top level model is over 90% for measured matrix and block sizes and different process grid configurations.

1 Introduction

At PARA95 we presented an algorithm for reduction of a regular matrix pair (A, B) to block Hessenberg-triangular form (H, T) [3]. It was shown how to reorganize the elementwise algorithm to perform blocked factorizations and higher level BLAS operations. The objective was to develop an algorithm that showed good performance on computers with hierarchical memory and to make use of this effort to design an efficient parallel implementation. Presently, we are extending the blocked algorithm to a scalable and portable program by expressing most computations in terms of existing factorization and update routines from the ScaLAPACK library [1]. To facilitate the design and analysis of the algorithm we decided to develop a performance model.

Since the ScaLAPACK routines are intended to be black boxes for the user it would be appropriate to build a model on a library of performance models corresponding to existing and future ScaLAPACK routines. Indeed, we suggest that any library of routines for scalable high performance computer systems should also include a corresponding library of performance models. In this contribution we present an approach for design of performance models based on the

Distributed Linear Algebra Machine (DLAM) [1], and the hierarchical structure of the ScaLAPACK library. By use of lower level ScaLAPACK models it is straightforward to design new higher level models for new routines and applications based on ScaLAPACK.

The rest of the paper is organized as follows. Section 2 introduces the parallel machine model and the data layout for matrices and vectors. In Section 3 different level 1, 2 and 3 performance models are presented. Finally, in Section 4 we show predicted and measured timing results that verify the hierarchical approach for performance modeling.

2 Parallel Machine Model and Data Layout

Parallel Machine Model. DLAM is a theoretical model of a parallel computer dedicated to dense linear algebra. The machine characteristics of a target architecture are specified with a few parameters that define the costs for computations and communications.

A P-process DLAM consists of P BLAS processes that communicate through a $P_r \times P_c$ BLACS network, which is a logical 2D grid with $P_r \cdot P_c \leq P$. Data are exchanged between BLAS processes by calling BLACS primitives. The processes can only perform BLAS [9, 7, 6] and BLACS [4, 5] operations.

In the DLAM model BLAS operations of the same level (1, 2 or 3) are assumed to have approximately the same performance measured in millions of floating point instructions per second (Mflop/s). So from a performance analysis perspective DLAM only distinguishes three different BLAS instructions and the execution times for level 1, 2, or 3 BLAS instructions are denoted γ_1, γ_2, and γ_3, respectively.

BLACS support point-to-point communication, broadcast and combine operations along a row or column of the process grid. The time to transfer n items between two processes is modeled as

$$T_s(n, \alpha, \beta) = \alpha + n\beta,$$

where α is the start up cost and β is the per item cost (1/bandwidth). The corresponding models for broadcast and combine operations are

$$T_b(Top, p, n, \alpha, \beta) = K('bcast', Top, p, n)T_s(n, \alpha, \beta),$$
$$T_c(Top, p, n, \alpha, \beta) = K('combine', Top, p, n)T_s(n, \alpha, \beta),$$

where p is the number of processes involved, Top is the network topology to be emulated during communication and $K(\cdot)$ is a hardware dependent function. If the physical network supports a 2D mesh and collisions are ignored then $K('bcast', '1 - tree', p, n)$ is approximated as $\log_2(p)$. The parameter $Top = '1 - tree'$ means that there is only one branch at each step of the broadcast tree. The corresponding number for $K('combine', '1 - tree', p, n)$ is $2\log_2(p)$.

Data Layout. Matrices of size $m \times n$ are block cyclic distributed [1] over a $P_r \times P_c$ BLACS network. In general the size of a distributed block is $r \times c$ with $r = 1$ or $c = 1$ in the vector case. In the following all blocks are square $(r = c)$, resulting in $M_r = \lceil m/r \rceil$ and $N_c = \lceil n/r \rceil$ blocks in the row and column directions, respectively.

3 Level 1, 2 and 3 Performance Models

We distinguish three levels of performance models. Models for algorithms that operate on vectors (a single column or row) and perform computations using at most level 1 BLAS are classified as level 1 models. Models for algorithms that are designed mainly to operate on a single block column (or block row) and to perform computations using level 2 BLAS are classified as level 2 models. Finally, models for algorithms that are designed to operate on a matrix consisting of several block columns (or block rows) and to perform computations using level 3 BLAS and possibly lower level routines are classified as level 3 models.

3.1 The Modeled High Level Algorithm

Here we give a brief description of the ScaLAPACK routine PDGEQRF [2], which implements a blocked right looking QR factorization algorithm.

The first iteration of the algorithm factorizes the first block column of the matrix using elementary Householder transformations. The trailing block columns are updated with respect to the factorized block using level 3 BLAS operations. To factorize the remaining block columns the same steps are applied recursively to the remaining submatrix.

PDGEQRF is built on top of BLAS, BLACS and lower level ScaLAPACK routines. In the algorithm description below the hierarchical structure of PDGEQRF is displayed with indented pseudo code.

for $j = 1 : m : r$
    PDGEQR2, QR-factorize $A_{j:m,j:j+r-1}$:
        for $i = 1 : r$
        PDLARFG, generate elementary reflector v_i:
            PDNRM2, compute 2-norm of $A_{j+i-1:m,j+i-1}$.
            PDSCAL, scale $A_{j+i-1:m,j+i-1}$.
        PDLARF, update remaining columns in current panel $A_{j+i:m,j+i:j+r-1}$.
        end
    PDLARFT, compute triangular factor T of block reflector.
    PDLARFB, apply block reflector to $A_{j:m,j+r:n}$.
end

3.2 Level 1 Performance Models

We present level 1 performance models for the ScaLAPACK routines PDSCAL, PDNRM2, and PDLARFG, called by PDGEQRF.

PDSCAL multiplies an m-element vector x by a real scalar. The time to perform this operation with x distributed across a process column of the grid is modeled as

$$T_{\text{pdscal}}(m, P_r, r) = r\lceil M_r/P_r \rceil \gamma_1,$$

where $\lceil M_r/P_r \rceil$ is the maximum number of $r \times 1$ blocks held by any process, $M_r = \lceil m/r \rceil$, and γ_1 is the level 1 BLAS speed. A similar expression holds for a vector distributed over a process row.

PDNRM2 returns the 2-norm of a vector x. The time to perform this operation with x distributed across a process column of the grid is modeled as

$$T_{\text{pdnrm2}}(m, P_r, r) = T_c(Top, P_r, 1, \alpha, \beta) + 2r\lceil M_r/P_r \rceil \gamma_1.$$

Also here $\lceil M_r/P_r \rceil$ is the maximum number of $r \times 1$ blocks held by any process and each block involves r multiply and add operations ($2r$ flops). Since partial results are distributed across a process column a combine operation of one single element is required to get the final result.

PDLARFG generates an $m \times 1$ vector v in the elementary reflector H, such that $Hx = (y_1, 0, \cdots, 0)^T$, where x is a distributed vector. H is represented in the form $H = I - \tau v v^T$, where τ is a real scalar ($1 \leq \tau \leq 2$). The computation of the Householder vector v with $v_1 = 1$ can be expressed as

$$\beta = x_1 + sign(x_1)\|x_{1:m}\|_2, \quad v_{2:m} = x_{2:m}/\beta.$$

The time for generating v is approximated by

$$T_{\text{pdlarfg}}(m, P_r, r) = $$
$$T_b(Top, P_r, 1, \alpha, \beta) + T_{\text{pdnrm2}}(m, P_r, r) + T_{\text{pdscal}}(m, P_r, r),$$

where the first two components originate from the computation of β and the last component from the computation of $v_{2:m}$. $T_b(Top, P_r, 1, \alpha, \beta)$ is the time to broadcast x_1 across the active process column. Moreover, PDNRM2 and PDSCAL are used for the computation of the 2-norm of x and the generation of v. Since computing τ is only a few (local) scalar operations we ignore the cost in the performance model for PDLARFG.

3.3 Level 2 Performance Models

We present level 2 performance models for the ScaLAPACK routines PDLARF, PDGEQR2 and PDLARFT, called by PDGEQRF.

PDLARF applies an elementary reflector $H = I - \tau v v^T$ to an $m \times n$ distributed matrix A from either the left or the right hand side. The main operations involved are

$$w = A^T v, \quad A \leftarrow A + \tau v w^T.$$

We assume that the matrix A is distributed over a $P_r \times P_c$ process grid and that the vector v is distributed over the first process column. Each process holds at most $M = r\lceil M_r/P_r \rceil$ rows and $N = ((n \div r) \div P_c)r + \min(n \bmod P_c r, r)$ columns of A. The time to perform PDLARF is modeled as

$$T_{\text{pdlarf}}(m, n, P_r, P_c, r) =$$
$$T_b(Top, P_c, M, \alpha, \beta) + T_c(Top, P_r, M, \alpha, \beta) + 4MN\gamma_2.$$

Components originating from computing $w = A^T v$ include broadcasting v across the rows of processes, the DGEMV operation to a cost of $2MN$ level 2 flops, and combining the locally computed parts of w. The final component corresponds to the local rank-1 update $A \leftarrow A + \tau vw^T$ using DGER to a cost of $2MN$ level 2 flops.

PDGEQR2 computes a QR factorization of a distributed $m \times n$ matrix A by applying a series of Householder transformations $H_i = I - \tau_i v_i v_i^T, i = 1, \ldots, n$ from the left hand side. PDLARFG is used to generate the vector v_i that annihilates the elements below entry i of $A_{:,i}$. PDLARF is used to update the trailing $n - i$ columns of A with respect to v_i. Using the hierarchical approach, the cost for performing these operations are modeled as

$$T_{\text{pdgeqr2}}(m, n, P_r, P_c, r) =$$
$$\sum_{i=0}^{n-1} T_{\text{pdlarfg}}(m - i, P_r, r) + T_{\text{pdlarf}}(m - i, n - (i + 1), P_r, P_c, r).$$

Notice, when $i = n - 1$, PDLARF is not called and there is no contribution to the cost from the second term in the summation.

PDLARFT is used to form the triangular factor T in the block reflector $H = I - VTV^T$, where the columns of V are the Householder vectors $v_1, \ldots, v_r$. In our case, these originate from a QR factorization of a column panel using PDGEQR2. The algorithm used to extract T can be expressed as [10]:

```
for i = 1 : r − 1
    W₁:ᵢ,ᵢ ← V^T_:,1:ᵢ V:,i+1        % Use level 2 DGEMV
end
T ← [−2]
for j = 1 : r − 1
    z ← −2TW₁:ᵢ,ᵢ                   % Use level 2 DTRMV
    T ← [ T  z ]
        [ 0 −2 ]
end
```

The modeled cost for performing these computations is

$$T_{\text{pdlarft}}(m, P_r, r) =$$
$$T_c(Top, P_r, r^2/2, \alpha, \beta) + \{r^2(r - 1)\lceil M_r/P_r \rceil + (2r^2 - r)(r - 1)/6\}\gamma_2.$$

The first γ_2-component corresponds to the local DGEMV steps. The number of rows of V^T in each step is $1, 2, \ldots, r-1$, giving $r(r-1)/2$ rows in total and the number of columns to be operated on for any process is at most $r\lceil M_r/P_r\rceil$. We approximate the total number of flops to $2(\#\text{rows}\cdot\#\text{columns})$. Then the contributions to the triangular matrix W of size $r \times r$ are combined to the top of the current column of processes. The cost for the combine operation is $T_c(Top, P_r, r^2/2, \alpha, \beta)$. The second γ_2-component corresponds to the computations performed in the local DTRMV steps. Each of these involve T of size $i \times i$, corresponding to i^2 flops. A summation over $i = 1 : r - 1$ results in $(2r^2 - r)(r - 1)/6$ level 2 flops.

3.4 Level 3 Performance Models

PDGEQRF is a level 3 routine that calls the ScaLAPACK level 3 auxiliary routine PDLARFB. Performance models for these two algorithms are presented below.

PDLARFB applies a block reflector $H = I - VTV^T$ or its transpose to an $m \times n$ matrix A from the left or right hand side. A is block cyclic distributed across a $P_r \times P_c$ grid with square blocks of size $r \times r$. V is block cyclic distributed across the first column of the active part of the grid (also denoted the current column) and the top-left process of this grid holds the $r \times r$ T.

Here we focus on the left hand transposed case, i.e., $H^T A = A - VT^TV^T A$. This update of A is realized by the following three level 3 operations:

$$W \leftarrow A^T V, \quad W \leftarrow WT, \quad A \leftarrow A - VW^T.$$

First V and T are sent to all processes that take part in the computation. Parts of V of size at most $r\lceil M_r/P_r\rceil \times r$ are broadcasted rowwise to the cost $T_b(Top, P_c, r^2\lceil M_r/P_r\rceil, \alpha, \beta)$. The triangular matrix T of size $r \times r$ is broadcasted to all processes in current process row to the cost $T_b(Top, P_c, r^2/2, \alpha, \beta)$.

The cost for computing $W \leftarrow A^T V$ using the level 3 BLAS DGEMM is a function of the number of rows of V and columns of A $(r\lceil M_r/P_r\rceil \times r\lceil N_c/P_c\rceil)$ that each process holds, resulting in $2r^3\lceil M_r/P_r\rceil\lceil N_c/P_c\rceil$ level 3 flops. The result of this operation is distributed across the active grid and must be combined columnwise to the current process row to the cost $T_c(Top, P_r, r^2\lceil N_c/P_c\rceil, \alpha, \beta)$.

The $W \leftarrow WT$ step of the update is performed by processes in the current row using the level 3 BLAS DTRMM. No communication is needed in this step since required parts of T $(r \times r)$ and W $(r\lceil N_c/P_c\rceil \times r)$ are already local to these processes. The number of level 3 flops performed in DTRMM is modeled as $r^3\lceil N_c/P_c\rceil$.

To compute $A \leftarrow A - VW^T$ using DGEMM the local parts of W are broadcasted columnwise to the cost $T_b(Top, P_r, r^2\lceil N_c/P_c\rceil, \alpha, \beta)$. The cost for the final DGEMM part of the computation is modeled as $2r^3\lceil M_r/P_r\rceil\lceil N_c/P_c\rceil\gamma_3$.

After summing all these contributions and collecting similar terms, we get the total modeled cost for PDLARFB:

$$
\begin{aligned}
T_{\text{pdlarfb}}(m, n, P_r, P_c, r) = \\
T_b(Top, P_c, r^2(\lceil M_r/P_r\rceil + 1/2), 2\alpha, \beta) + T_b(Top, P_r, r^2\lceil N_c/P_c\rceil, \alpha, \beta) + \\
T_c(Top, P_r, r^2\lceil N_c/P_c\rceil, \alpha, \beta) + r^3(4\lceil M_r/P_r\rceil + 1)\lceil N_c/P_c\rceil\gamma_3.
\end{aligned}
$$

PDGEQRF computes a QR factorization of a distributed $m \times n$ matrix A using a level 3 algorithm. We illustrate the first step with a block-partitioned

$$A = \begin{bmatrix} A_{11} & A_{12} \\ A_{21} & A_{22} \end{bmatrix}.$$

In this right-looking implementation the panel $[A_{11}, A_{21}]^T$ is first factorized using a level 2 algorithm followed by an update of the remaining matrix $[A_{12}, A_{22}]^T$:

$$\begin{bmatrix} A_{12} \\ A_{22} \end{bmatrix} \leftarrow Q^T \begin{bmatrix} A_{12} \\ A_{22} \end{bmatrix} = (I - VT^TV^T) \begin{bmatrix} A_{12} \\ A_{22} \end{bmatrix}.$$

Before the block reflector Q is applied, the triangular T is extracted. To complete the QR factorization the same steps (factor, extraction of T and block reflector update) are applied recursively to the trailing submatrix.

The ScaLAPACK routines used in this factorization and the block sizes in different operations are listed below.

- PDGEQR2: Compute $N_c = \lceil n/r \rceil$ level 2 factorizations on panels of A of decreasing size, $(m - i \cdot r) \times r$ for $i = 0, \ldots, N_c - 1$.
- PDLARFT: Extract triangular T in the block reflector corresponding to each of the N_c level 2 factorizations. The length of the Householder vectors in step i is $m - i \cdot r$ for $i = 0, \ldots, N_c - 1$.
- PDLARFB: Update the trailing submatrix with the block reflector from each step in the level 3 algorithm. The size of the submatrix of A to be updated in step i is $(m - i \cdot r) \times (n - (i+1)r)$ for $i = 0, \ldots, N_c - 1$.

Using the hierarchical approach, the aggregate cost for PDGEQRF is modeled as:

$$T_{\text{pdgeqrf}}(m, n, P_r, P_c, r) = \sum_{i=0}^{N_c - 1} T_{\text{pdgeqr2}}(m - i \cdot r, r, P_r, 1, r) +$$

$$T_{\text{pdlarft}}(m - i \cdot r, P_r, r) + T_{\text{pdlarfb}}(m - i \cdot r, n - (i+1)r, P_r, P_c, r).$$

3.5 Expansion of Hierarchical Models

The hierarchical models are primarily used for predicting the performance of parallel algorithms. Typically, a model is used for estimating the overhead in a parallel algorithm and measuring the impact of machine characteristics, grid configurations and block sizes on the execution time. By expanding individual components in a hierarchical model we can derive compact models that are appropriate for a theoretical complexity analysis as well. We illustrate with the model for the level 2 QR factorization.

By approximating $r\lceil M_r/P_r \rceil$ and $r\lceil N_c/P_c \rceil$ by m/P_r and n/P_c, respectively, and assuming $Top = $ '1 − tree' and expanding all components in $T_{\text{pdgeqr2}}(\cdot)$, the model for PDGEQR2 can be written

$$T_{\text{qr2}}(m, n, P_r, P_c, r) =$$
$$\{5n \log_2(P_r) + n \log_2(P_c)\}\alpha + \{(2M + 3n) \log_2(P_r) + M \log_2(P_c)\}\beta +$$
$$3M\gamma_1 + \frac{2}{P_r P_c}\{mn(n - 1) - (n^3 - n)/3\}\gamma_2,$$

where $M = (mn - n(n+1)/2)/P_r$ is the number of annihilated elements divided by the number of process rows.

Since PDGEQR2 is a level 2 routine it is mainly used by level 3 routines for performing a factorization of a block column or block row. In our case PDGEQR2 is called with $P_c = 1$ and $n = r$, which means that the $\log_2(P_c)$ terms in the communication overhead disappear.

			PDLARFB			PDGEQRF		
$m = n$	r	$P_r \times P_c$	Model	XP	M/XP	Model	XP	M/XP
512	32	1×1	0.76	0.75	1.01	4.62	4.81	0.96
512	64	1×1	1.54	1.51	1.02	5.26	5.25	1.00
1024	32	1×1	3.01	2.93	1.03	34.3	34.4	1.00
1024	64	1×1	6.06	5.88	1.03	36.8	36.1	1.02
1024	32	2×2	0.78	0.76	1.03	10.3	10.9	0.94
1024	64	2×2	1.57	1.53	1.03	12.1	12.3	0.98
2048	32	2×2	3.04	2.98	1.02	72.4	72.5	1.00
2048	64	2×2	6.13	5.96	1.03	79.1	77.8	1.02
1024	32	4×2	0.41	0.41	1.00	6.16	7.13	0.86
1024	64	4×2	0.85	0.83	1.02	7.44	8.24	0.90
2048	32	4×2	1.57	1.53	1.03	39.5	40.7	0.97
2048	64	4×2	3.19	3.09	1.03	44.2	44.7	0.99
1024	32	2×4	0.39	0.39	1.00	6.21	6.85	0.91
1024	64	2×4	0.80	0.78	1.03	8.00	8.32	0.96
2048	32	2×4	1.53	1.51	1.01	40.1	40.7	0.99
2048	64	2×4	3.09	3.02	1.02	46.6	45.9	1.02
2048	16	4×4	0.39	0.39	1.00	20.4	22.3	0.91
2048	32	4×4	0.79	0.77	1.03	22.3	23.9	0.93
2048	64	4×4	1.61	1.56	1.03	26.4	27.8	0.95
2048	128	4×4	3.31	3.21	1.03	35.8	36.3	0.99
4096	16	4×4	1.53	1.51	1.01	143.2	145.1	0.99
4096	32	4×4	3.08	3.01	1.02	150.6	151.1	1.00
4096	64	4×4	6.20	6.02	1.03	165.9	164.2	1.01
4096	128	4×4	12.6	12.3	1.02	198.6	194.6	1.02

Table 1. Predicted vs. measured results for PDLARFB and PDGEQRF

4 Verification of Performance Models

We have compared the level 1, 2, and 3 models of the ScaLAPACK routines to measured performance results on an Intel Paragon XP/S system. Only a few of these are presented here, including results for the level 3 models and the model for the level 2 QR factorization.

$m \times n$	P_r	r	PDGEQR2 ($1/\beta = 30$ Mbytes/s)			PDGEQR2 ($1/\beta = 18$ Mbytes/s)		
			Model	*XP*	*M/XP*	*Model*	*XP*	*M/XP*
1024×32	1	32	0.09	0.10	0.90	0.09	0.10	0.90
1024×64	1	64	0.35	0.34	1.03	0.35	0.34	1.03
2048×32	1	32	0.18	0.20	0.90	0.18	0.20	0.90
2048×64	1	64	0.70	0.67	1.04	0.70	0.67	1.04
1024×32	2	32	0.06	0.08	0.75	0.07	0.08	0.88
1024×64	2	64	0.20	0.23	0.87	0.22	0.23	0.96
2048×32	2	32	0.11	0.13	0.85	0.13	0.13	1.00
2048×64	2	64	0.40	0.39	1.03	0.42	0.39	1.08
1024×32	4	32	0.04	0.07	0.57	0.05	0.07	0.71
1024×64	4	64	0.13	0.17	0.76	0.14	0.17	0.82
2048×32	4	32	0.07	0.09	0.78	0.09	0.09	1.00
2048×64	4	64	0.23	0.26	0.88	0.26	0.26	1.00

Table 2. Predicted vs. measured results for PDGEQR2

When nothing else is explicitly stated we model the performance in Mflop/s of the level 1, 2 and 3 BLAS instructions as $1/\gamma_1 = 10$, $1/\gamma_2 = 25$ and $1/\gamma_3 = 45$, respectively. The level 3 speed is chosen as the peak performance of double precision general matrix-matrix multiply and add (DGEMM) given in [8], while γ_1, γ_2 are estimated from experiments. Moreover, the latency and bandwidth of the system's network are modeled as $\alpha = 3.7e - 5$ seconds and $1/\beta = 30.0$ Mbytes/s for large messages. These numbers have been empirically determined through experiments using the NX communication library.

In the result tables we display the problem size ($m \times n$), processor grid configuration ($P_r \times P_c$) and the square block size (r). The columns denoted *Model* and *XP* show the time in seconds for the performance model and the Paragon system, and finally the *M/XP* column displays the ratio of the two results.

Predicted and measured results for the level 3 routines are displayed in Table 1. The accuracy of the PDLARFB model is very high. The error is less than 4% for all measured problem sizes, block sizes, and grid configurations. Since the major fraction of the total execution time for PDGEQRF is consumed by PDLARFB, the accuracy of PDGEQRF is high too.

Predicted and measured results for PDGEQR2 are displayed in Table 2. The column with $1/\beta = 30$ Mbytes/s shows that the model mainly underestimates the measured time when $P_r > 1$. Since PDGEQR2 communicates small messages of data it is reasonable to assume that the obtained communication rate is below 30 Mbytes/s. By decreasing $1/\beta$ to 18 Mbytes/s we obtain improved results. Since most computations in PDGEQRF are performed using level 3 operations, the impact of the level 2 models are less significant, explaining the good results in Table 1.

Acknowledgements

We are grateful to Ken Stanley for providing information regarding performance modeling of ScaLAPACK.

References

1. J. Choi, J. Demmel, I. Dhillon, J. Dongarra, S. Ostrouchov, A. Petit, K. Stanley, D. Walker, and R.C. Whaley. ScaLAPACK: A Portable Linear Algebra Library for Distributed Memory Computers - Design Issues and Performance. *Technical Report UT CS-95-283, LAPACK Working Note 95*, 1995.
2. J. Choi, J. Dongarra, S. Ostrouchov, A. Petit, D. Walker, and R.C. Whaley. The Design and Implementation of the ScaLAPACK LU, QR, and Cholesky Factorization Routines. To appear in *Scientific Programming*, 1996.
3. K. Dackland and B. Kågström. Reduction of a Regular Matrix Pair (A, B) to Block Hessenberg-Triangular Form. In Dongarra et. al., editor, *Applied Parallel Computing: Computations in Physics, Chemistry and Engineering Science*, pages 125–133, Berlin, 1995. Springer-Verlag. Lecture Notes in Computer Science, Vol. 1041, Proceedings, Lyngby, Denmark.
4. J. Dongarra and R. van de Geijn. Two dimensional Basic Linear Algebra Communication Subprograms. *Technical Report UT CS-91-138, LAPACK Working Note 37, University of Tennessee*, 1991.
5. J. Dongarra and R. C. Whaley. A Users Guide to BLACS v1.0. *Technical Report UT CS-95-281, LAPACK Working Note 94, University of Tennessee*, 1995.
6. I. Duff, S. Hammarling, J. Dongarra, and J. Du Croz. A Set of Level 3 Basic Linear Algebra Subprograms. *ACM Transactions on Mathematical Software*, 16(1):1–17, 1990.
7. S. Hammarling, R. Hanson, J. Dongarra, and J. Du Croz. Algorithm 656: An extended Set of Basic Linear Algebra Subprograms: Model Implementation and Test Programs. *ACM Transactions on Mathematical Software*, 14(1):18–18, 1988.
8. Intel Corporation. Paragon System Basic Math Library Performance Report. *Order Number 312936-003*, 1995.
9. D. Kincaid, F. Krogh C. Lawson, and R. Hanson. Basic Linear Algebra Subprograms for Fortran Usage. *ACM Transactions on Mathematical Software*, 5(3):308–323, 1979.
10. R. Schreiber and C. Van Loan. A Storage Efficient WY Representation for Products of Householder Transformations. *SIAM J. Sci. and Stat. Comp.*, 10:53-57, 1989.

MeDLey: An Abstract Approach to Message Passing

E. Dillon[1], J. Guyard[1], G. Wantz[2]

[1] CRIN/INRIA, 615, rue du jardin botanique, 54602 VILLERS-les-NANCY,
FRANCE
[2] CRP-CU, Luxembourg

Abstract. Message Passing has become one of the most popular parallel
programming paradigms because of its ease of use. Conversely, using it
efficiently often requires lots of efforts, when designing a distributed application. Finally, such an application often requires optimization, since
performances are not always as good as expected.

With MeDLey, we propose an abstract approach to message passing.
The basic idea is to provide a language that might be used to specify all
communications within a distributed application. With this specification,
the user does not have to care about the implementation any more, and
leaves it to the compiler's charge. Finally, the aim of this approach is to
achieve more efficiency by generating specific communication primitives
rather than using generic message passing libraries.

MeDLey has followed two main steps: MeDLey-0 was the first version
of the compiler. It was able to generate C++ and MPI communications
primitives, mainly to help the user in the design process. MeDLey-1
now tries to gain more efficiency by implementing communications with
dedicated devices like shared-memory or AAL5 for ATM-based clusters
of workstations.

1 Introduction

In the scope of task parallelism, two models of data exchanges have actually
emerged and are widely used on different architectures:

- A model based on *distributed memory*, where co-operating tasks have to call
  explicitly send and receive functions to exchange data through a medium of
  communication. The common programming paradigm in this case is "Message Passing".
- A model based on *shared memory*, where co-operating tasks simply make
  memory accesses, without using explicit communications.

These two models can be seen from different points of view. On one hand,
when the *shared memory* case is applied to a dedicated architecture, it can lead to
high performance during data exchanges between tasks, since the only limitation
is introduced by the memory bandwidth. However, the price to pay for this high
performance is that the user has to cope with data accesses which often implies
contention.

On the other hand, the *distributed memory* model fits more kinds of hardware architectures, as well as providing an easier programming paradigm, since the user does not have to cope with memory contention any more. As a consequence, however, performances may definitely drop due to all buffering operations that may appear during a data exchange [Dillon 95].

However, from the user point of view, a parallel application is still represented by a set of tasks, that are exchanging data, whatever underlying model of data exchange is used between tasks. That is the reason why we propose a new "abstract" approach to define the data exchange between tasks. This approach is based on a specification language called "MeDLey", which stands for *Message Definition Language*.

In this paper, we present the first two steps of this project:

- MeDLey-0: the definition of a language to specify the data exchanges between co-operating tasks. This language was designed to provide a wide set of data exchange possibilities in spite of being too complex. In this first step, the aim was to provide a way to enhance the developing efficiency of the user, by introducing a language that would be compiled into a portable MPI implementation of the specified data exchanges.
- MeDLey-1: based on the former specification of a parallel application, the MeDLey compiler should not only be able to generate a portable MPI implementation, but also a specific implementation by taking into account a hardware architecture, to change portability to best efficiency on a particular architecture (currently shared-memory machines and ATM-based CoWs[3].)

The rest of the paper is organized as follows: first, we will briefly recall the most common way to exchange data between tasks according to the "message passing" programming paradigm in section 2. Section 3 will then focus on the first step of the project, by presenting the syntax and semantics of MeDLey-0, through a little example. The first extensions to improve the efficiency of MeDLey-0 are presented in section 4. We will finally conclude in section 5.

2 Data exchanges: the message passing paradigm

In the scope of control parallelism, a parallel application is defined by a set of tasks that may exchange data. A lot of message passing environments allow to implement different ways to exchange data. They are however all based on the same operations. These operations are represented by figure 1.

Since all data are exchanged through messages, the first operation to do, is to build the message content out of the local data structures of the sending task. The aim of this "packing" operation is to gather non-contiguous data of the local memory into a packed buffer thus representing the content of the message.

---

[3] Cluster of Workstations

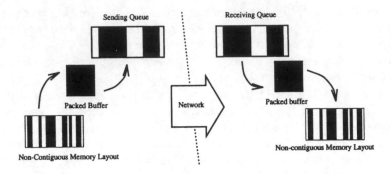

Fig. 1. The different steps occuring during a message exchange

The second operation on the sending side is to actually send the message. This is generally done through a so-called "send" operation. This operation may be declined with various semantics depending on the level of asynchronism used.

On the receiving side, two explicit operations are required to complete the data exchange. First, an actual receive of the message and then the unpacking of it into the local memory.

All those operations remain general, in the sense that they may apply to any kind of data (scalars, structs, etc.), with any kind of semantics (synchronously, asynchronously, etc.), and are explicit, since each of them corresponds to a specific function call in the message passing library.

In order to provide a simple view of communications between tasks, we introduce the MeDLey language, so that the user only has to cope with the data exchanges between tasks, letting the compiler cope with the actual communications, including all error prone operations such as packing/unpacking and sending/receiving function calls.

3 MeDLey-0: an abstract definition of data exchanges

The MeDLey language was designed to specify communications within distributed applications based on control parallelism. As a consequence, a MeDLey specification of such an application is split into modules, each of them corresponding to a task of the application.

3.1 The MeDLey language

This section gives an overview of the MeDLey [Dillon 96] language through a little example. This example consists of the implementation of the π computation with the Montecarlo method. Basically, the application is organized as follows:

– A "father" task is responsible for the gathering of partial results, computed by a set of "son" tasks.

– A set of "son" tasks is responsible for the actual computation of the partial results.

Example 1 shows the MeDLey specification of the "father" task[4]. This example shows us that the *connected with* statement must be used to specify the inter-connection between tasks, i.e. which task will communicate with which other task. Moreover, this example defines that only one "father" task will communicate with several "sons" tasks, thanks to the *setof* keyword.

Example 1. Example of a task definition.

```
// The father task                                              1
//
task father connected with setof(son)                          3
uses                                                            4
  {                                                             5
     double box,  coeff;                                        6
     int number;   // The number of shoot sent to a son         7

     int in;      // a received shoot from a son                9
     int who;     // the rank of the son who sent the shoot    10
  }                                                             11
sends                                                           12
  {                                                             13
     shootMessage to son = sequence {number};                  14
  }                                                             15
receives                                                       16
  {                                                             17
     startMessage from son = sequence{box, coeff};             18
     inMessage from son = sequence {in, who};                  19
  }                                                             20
```

A MeDLey module is split into 3 main parts:

– the **uses** part: this part contains all the data that will be involved in data exchanges, i.e. all message contents will be defined out of these data declarations.
– the **sends** part: this part contains the definition of all outgoing messages. These definitions are made using the variable declared in the uses part. Finally, for each message, a name should be given (*shootMessage* e.g.) and a target task (*to son*, e.g.).

---

[4] the "son" task specification is not given in this paper

– the **receives** part: this part contains the definition of all incoming messages. More clearly, each definition maps an incoming message into local variables. Here again, all variables must have been defined in the **uses** part.

So, the MeDLey syntax remains very simple. All data declarations follow the C rules, and allow all scalar datatypes (int, float, double, etc.), as well as built-in *Matrix* and *Vector* datatypes.

Finally, it is possible to define whether a sending or receiving operation must be synchronous or asynchronous, by using one of the *sync* or *async* keywords before each definition. By default, all sending operations are *asynchronous*, whereas all receiving operations are *synchronous*.

3.2   The generated code

We have developed a compiler for the MeDLey language. This compiler has been called MeDLey-0.

Basically, it takes a set of MeDLey modules, corresponding to a MeDLey specification of an application, and generates all sending and receiving operations for each task.

Within MeDLey-0, the target language is currently C++ and C, whereas all message passing operations are implemented with MPI [MPI Forum 95]. Example 2 shows the result of the MeDLey compiler on the previous task definition.

For each MeDLey module, a C++ class is generated. All outgoing/incoming messages are mapped into methods, whereas all data declarations of the *uses* part are mapped into public instance variables.

Finally, some other functions are generated by MeDLey-0. These functions include an *init()* and *end()* method, that must be called at runtime in order to set things right to work with MeDLey-0, plus other useful functions for the user (giving the local instance of task, the total number of running tasks, etc.)

Example 2. Generated code for "father" module (fatherMDL.hh).

```
#ifndef __fatherMDL_H__
#define __fatherMDL_H__
#include "MeDLeyClass.hh"

class father: public MeDLeyClass
{
public:
  // Some useful methods
  int MDL_MyRank();       // My rank
  int MDL_FamilySize();   // Number of father running
  int MDL_End();
  int MDL_Init(int& argc, char**& argv);
```

```
// Data declarations
int who,in, number;
double coeff, box;

// Sending functions from 'sends'
int MDL_SendTo_son_shotMessage(int to);
// Receiving functions from 'receives'
int MDL_RecvFrom_son_startMessage(int from);
int MDL_RecvFrom_son_inMessage(int from);
[...]}
```

3.3   How to use it ?

Finally, the inclusion of the MeDLey generated code remains very simple.

Example 3 shows how to use the "father" class generated by the MeDLey compiler.

The first thing to do is to instantiate the object corresponding to the task ("father" in the example above), and to call the "init" method on it.

After this, all message definitions should have been mapped to a method on the "father" object. Moreover, since the specification mentioned that a "father" task was connected to a *setof* "son" tasks, all sending or receiving functions linked to sons are broadcasting functions or functions that accept a message from any "son".

Finally, all declared variable (in the *uses* part) are now instance variables of the "father" object. They are used for all communications, and so must be set by the user during his computations.

Example 3. Final code that calls MeDLey-generated code.

```
#include "fatherMDL.hh"
#include "mpi.h"

main(int argc, char **argv)
{ [...]

  // the MeDLey object instantiation
  father t;
  t.MDL_Init(argc, argv);
  [...]

  // receives box and coeff
  t.MDL_RecvFrom_son_startMessage();

  do  // the computation
    { [...]
```

```
        // let's receive a shoot
        t.MDL_RecvFrom_son_inMessage();
        IN += t.in; // we use the declared variable
        TOTAL += BATCH;
        ++taskSharing[t.who]; // we use the declared variable
     } while (TOTAL<MINIMUM);
  double I = t.box * double(IN)/TOTAL;
  [...]
  // The end !!
  t.MDL_End();
}
```

4 MeDLey-1: portability vs efficiency

The first step of the MeDLey project was to provide a simple but complete language to specify the data exchanges within a distributed application. This was achieved by the MeDLey-0 implementation: the user does not have to care anymore about the way things will be implemented, he just has to care about which data exchanges must be defined.

The second step of the MeDLey project is now not only to provide ease of use for message passing, but also efficiency, without introducing more complexity in the language. This leads to the second version of the MeDLey compiler, called MeDLey-1.

When talking about Message Passing, people always think about the overhead due to both synchronization and bufferization steps of exchanges [Stricker 95b] [Stricker 95a]. MPI-based code generated by MeDLey-0 has the advantage to be portable thanks to MPI's portability, but in some cases it should be possible to get more efficiency, by taking into account some specificities of the hardware environment and generating dedicated code for a particular hardware.

Indeed, in MeDLey-1, the main extension to MeDLey-0 was to add the ability to generate C++ or C code not only based on MPI but also on other devices. In particular, MeDLey-1 will generate the same communication primitives as MeDLey-0, but data exchanges will be implemented with:

− **shared memory** operations
− **aal5 network interface** calls

More clearly, with such an implementation, the MeDLey compiler should be able to by-pass the MPI implementation, and to reach directly the "communication" device, in order to get more efficiency during data exchanges.

Finally, to avoid most data copies, the data declarations of the *uses* part will be revisited, so that data are contiguously stored in memory for most outgoing/incoming messages.

5 Conclusion

We have presented a new language to specify the data exchanges within a distributed application. Based on this language, we have emphasized efficiency at two levels.

With MeDLey-0, we gave a new abstraction level for the design of a parallel application. Thanks to a MeDLey specification, MeDLey-0 is able to automatically generate a portable MPI-based implementation for all data exchanges.

References

[Dillon 95]   Eric Dillon, Carlos Gamboa Dos Santos and Jacques Guyard. Homogeneous and Heterogeneous Networks of Workstations: Message Passing Overhead. In *MPI Developers Conference '95*, June 1995.

[Dillon 96]   E. Dillon and G. Wantz. Medley: Efficient communications for distributed applications. Technical report, INRIA-Lorraine, 1996. To appear.

[MPI Forum 95] MPI Forum. MPI: A Message Passing Interface Standard. University of Tennessee, June 1995.

[Stricker 95a]   T. Stricker and T. Gross. Optimizing memory system performance for communication in parallel computers. In *Proceedings of the 22nd International Symposium on Computer Architecture*, Santa Marguerita di Ligure, Italy, June 1995.

[Stricker 95b]   T. Stricker, J. Stichnoth, D. O'Hallaron, S. Hinrichs and T. Gross. Decoupling synchronization and data transfer in message passing systems of parallel computers. In *Proceedings of the 9th International Conference on Supercomputing*, Barcelona, Spain, July 1995.

ScaLAPACK Tutorial *

Jack Dongarra[1,2] and L. Susan Blackford[**1]

[1] Department of Computer Science, University of Tennessee, Knoxville,
TN 37996-1301, USA
[2] Mathematical Sciences Section, Oak Ridge National Laboratory, Oak Ridge,
TN 37831, USA

Abstract. ScaLAPACK is a library of high performance linear algebra routines for distributed memory MIMD computers. It is a continuation of the LAPACK project, which designed and produced analogous
software for workstations, vector supercomputers, and shared memory
parallel computers. The goals of the project are efficiency (to run as
fast as possible), scalability (as the problem size and number of processors grow), reliability (including error bounds), portability (across all
important parallel machines), flexibility (so users can construct new routines from well-designed parts), and ease-of-use (by making LAPACK
and ScaLAPACK look as similar as possible). Many of these goals, particularly portability, are aided by developing and promoting *standards*,
especially for low-level communication and computation routines. We
have been successful in attaining these goals, limiting most machine dependencies to two standard libraries called the BLAS, or Basic Linear
Algebra Subroutines, and BLACS, or Basic Linear Algebra Communication Subroutines. ScaLAPACK will run on any machine where both the
BLAS and the BLACS are available.

This tutorial will begin by reviewing the fundamental design principles of the BLAS and LAPACK and their influence on the development
of ScaLAPACK. The two dimensional block cyclic data decomposition
will be presented, followed by a discussion of the underlying building
blocks of ScaLAPACK, the BLACS and the PBLAS. The contents of
the ScaLAPACK library will then be enumerated, followed by example programs and performance results. And finally, future directions and
related projects will be described.

1 Introduction

Much of the work in developing linear algebra software for advanced architecture computers is motivated by the need to solve large problems on the fastest

** formerly L. Susan Ostrouchov

* This work was supported in part by the National Science Foundation Grant No.
ASC-9005933; by the Defense Advanced Research Projects Agency under contract DAAL03-91-C-0047, administered by the Army Research Office; by the Office of Scientific Computing, U.S. Department of Energy, under Contract DE-AC05-
84OR21400; and by the National Science Foundation Science and Technology Center
Cooperative Agreement No. CCR-8809615.

computers available. In this tutorial, we focus on the development of standards for use in linear algebra and the building blocks for a library and the aspects of algorithmic design and parallel implementation.

The linear algebra community has long recognized the need for help in developing algorithms into software libraries, and several years ago, as a community effort, put together a *de facto* standard for identifying basic operations required in linear algebra algorithms and software. The hope was that the routines making up this standard, the Basic Linear Algebra Subprograms (BLAS), would be implemented on advanced-architecture computers by many manufacturers, making it possible to reap the portability benefits of having them efficiently implemented on a wide range of machines. This goal has been largely realized.

The key insight of our approach to designing linear algebra algorithms for advanced architecture computers is that the frequency with which data are moved between different levels of the memory hierarchy must be minimized in order to attain high performance. Thus, our main algorithmic approach for exploiting both vectorization and parallelism is the use of block-partitioned algorithms, particularly in conjunction with highly-tuned kernels for performing matrix-vector and matrix-matrix operations (BLAS). In general, block-partitioned algorithms require the movement of blocks, rather than vectors or scalars, resulting in a greatly reduced startup cost because fewer messages are exchanged.

A second key idea is that the performance of an algorithm can be tuned by a user by varying the parameters that specify the data layout. On shared memory machines, this is controlled by the block size, while on distributed memory machines it is controlled by the block size and the configuration of the logical process grid.

Sections 2 and 3 review the fundamental design principles of the BLAS and LAPACK. The two dimensional block cyclic data decomposition is presented in Section 4 as the basis for matrix distribution in ScaLAPACK. The building blocks of the ScaLAPACK library, the BLACS and the PBLAS, are then presented in Sections 5.1 and 5.2. The contents and performance of ScaLAPACK and described in Section 5. And finally, Section 6 summarizes and discusses future directions and related projects.

2   The Basic Linear Algebra Subprograms (BLAS)

The Basic Linear Algebra Subprograms are key to portability and efficiency across sequential and parallel environments. There are three levels of BLAS:

Level 1 BLAS [2]: for vector operations, such as $y \leftarrow \alpha x + y$
Level 2 BLAS [3]: for matrix-vector operations, such as $y \leftarrow \alpha A x + \beta y$
Level 3 BLAS [4]: for matrix-matrix operations, such as $C \leftarrow \alpha A B + \beta C$.

Here, A, B and C are matrices, x and y are vectors, and α and β are scalars.

The Level 1 BLAS are used in LAPACK, but for convenience rather than for performance: they perform an insignificant fraction of the computation, and they cannot achieve high efficiency on most modern supercomputers.

The Level 2 BLAS can achieve near-peak performance on many vector processors, such as a CRAY Y-MP, or Convex C-2 machine. However, on other vector processors such as a CRAY-2, the performance of the Level 2 BLAS is limited by the rate of data movement between different levels of memory [11].

The Level 3 BLAS overcome this limitation. They perform $O(n^3)$ floating-point operations on $O(n^2)$ data, whereas the Level 2 BLAS perform only $O(n^2)$ operations on $O(n^2)$ data. The Level 3 BLAS also allow us to exploit parallelism in a way that is transparent to the software that calls them. While the Level 2 BLAS offer some scope for exploiting parallelism, greater scope is provided by the Level 3 BLAS, as Table 1 illustrates.

Table 1. Speed (Megaflops) of BLAS Operations on a CRAY Y-MP. All matrices are of order 500.

Number of processors:	1	2	4	8
Level 2: $y \leftarrow \alpha A x + \beta y$	311	611	1197	2285
Level 3: $C \leftarrow \alpha A B + \beta C$	312	623	1247	2425
Peak	333	666	1332	2664

3 The Linear Algebra Package (LAPACK)

LAPACK [1] provides routines for solving systems of simultaneous linear equations, least-squares solutions of linear systems of equations, eigenvalue problems, and singular value problems. The associated matrix factorizations (LU, Cholesky, QR, SVD, Schur, generalized Schur) are also provided, as are related computations such as reordering of the Schur factorizations and estimating condition numbers. Dense and banded matrices are handled, but not general sparse matrices. In all areas, similar functionality is provided for real and complex matrices, in both single and double precision.

The original goal of the LAPACK project was to make the widely used EISPACK [17] and LINPACK libraries run efficiently on shared-memory vector and parallel processors. On these machines, LINPACK and EISPACK are inefficient because their memory access patterns disregard the multi-layered memory hierarchies of the machines, thereby spending too much time moving data instead of doing useful floating-point operations. LAPACK addresses this problem by reorganizing the algorithms to use block matrix operations, such as matrix multiplication, in the innermost loops [6, 1]. These block operations can be optimized for each architecture to account for the memory hierarchy [5], and so provide a transportable way to achieve high efficiency on diverse modern machines.

LAPACK can be regarded as a successor to LINPACK and EISPACK. It has virtually all the capabilities of these two packages and much more besides. It improves on them in four main respects: speed, accuracy, robustness and functionality. While LINPACK and EISPACK are based on the vector operation kernels of the Level 1 BLAS, LAPACK was designed at the outset to exploit the matrix-matrix operation kernels of the Level 3 BLAS. Because of the coarse granularity of these operations, their use tends to promote high efficiency on many high-performance computers, particularly if specially coded implementations are provided by the manufacturer.

Extensive performance results for LAPACK can be found in the LAPACK Users' Guide [1].

3.1 A Block Partitioned Algorithm Example

We consider the Cholesky factorization algorithm, which factorizes a symmetric positive definite matrix as $A = U^T U$. To derive a block form of Cholesky factorization, we partition the matrices into blocks, in which the diagonal blocks of A and U are square, but of differing sizes. We assume that the first block has already been factored as $A_{00} = U_{00}^T U_{00}$, and that we now want to determine the second block column of U consisting of the blocks U_{01} and U_{11}.

$$\begin{pmatrix} A_{00} & A_{01} & A_{02} \\ \cdot & A_{11} & A_{12} \\ \cdot & \cdot & A_{22} \end{pmatrix} = \begin{pmatrix} U_{00}^T & 0 & 0 \\ U_{01}^T & U_{11}^T & 0 \\ U_{02}^T & U_{12}^T & U_{22}^T \end{pmatrix} \begin{pmatrix} U_{00} & U_{01} & U_{02} \\ 0 & U_{11} & U_{12} \\ 0 & 0 & U_{22} \end{pmatrix}.$$

Equating submatrices in the second block of columns, we obtain

$$A_{01} = U_{00}^T U_{01}$$
$$A_{11} = U_{01}^T U_{01} + U_{11}^T U_{11}.$$

Hence, since U_{00} has already been computed, we can compute U_{01} as the solution to the equation

$$U_{00}^T U_{01} = A_{01}$$

by a call to the Level 3 BLAS routine STRSM; and then we can compute U_{11} from

$$U_{11}^T U_{11} = A_{11} - U_{01}^T U_{01}.$$

This involves first updating the symmetric submatrix A_{11} by a call to the Level 3 BLAS routine SSYRK, and then computing its Cholesky factorization. Since Fortran 77 does not allow recursion, a separate routine must be called (using Level 2 BLAS rather than Level 3), named SPOTF2 in Figure 1. In this way, successive blocks of columns of U are computed. The LAPACK-style code for the block algorithm is shown in Figure 1.

```
do j = 0, n-1, nb
  jb = min( nb, n-j )
  call strsm( 'left', 'upper', 'transpose', 'non-unit', j, jb, one,
              a, lda, a(0,j), lda )
  call ssyrk( 'upper', 'transpose', jb, j, -one, a(0,j), lda, one,
              a(j,j), lda )
  call spotf2( 'upper', jb, a(j,j), lda, info )
  if( info .ne. 0 ) go to 20
end do
```

Fig. 1. The body of the "LAPACK-style" routine for block Cholesky factorization. In this code fragment, nb denotes the width of the blocks.

4 Block Cyclic Data Distribution

The way in which a matrix is distributed over the processes has a major impact on the load balance and communication characteristics of the concurrent algorithm, and hence largely determines its performance and scalability. The block cyclic distribution provides a simple, yet general-purpose way of distributing a block-partitioned matrix on distributed memory concurrent computers. It has been incorporated in the High Performance Fortran standard [14].

The block cyclic data distribution is parameterized by the four numbers P_r, P_c, r, and c, where $P_r \times P_c$ is the process template and $r \times c$ is the block size.

Suppose first that we have M objects, indexed by an integer $0 \le m < M$, to map onto P processes, using block size r. The m-th item will be stored in the i-th location of block b on process p, where

$$\langle p, b, i \rangle = \left\langle \left\lfloor \frac{m}{r} \right\rfloor \bmod P, \left\lfloor \frac{\lfloor \frac{m}{r} \rfloor}{P} \right\rfloor, m \bmod r \right\rangle .$$

In the special case where $r = 2^{\hat{r}}$ and $P = 2^{\hat{P}}$ are powers of two, this mapping is really just bit extraction, with i equal to the rightmost $\hat{r}$ bits of m, p equal to the next $\hat{P}$ bits of m, and b equal to the remaining leftmost bits of m. The distribution of a block-partitioned matrix can be regarded as the tensor product of two such mappings: one that distributes the rows of the matrix over P_r processes, and another that distributes the columns over P_c processes. That is, the matrix element indexed globally by (m, n) is stored in location

$$\langle (p, q), (b, d), (i, j) \rangle =$$

$$\left\langle (\lfloor \frac{m}{r} \rfloor \bmod P_r, \lfloor \frac{n}{c} \rfloor \bmod P_c), \left(\left\lfloor \frac{\lfloor \frac{m}{r} \rfloor}{P_r} \right\rfloor, \left\lfloor \frac{\lfloor \frac{n}{c} \rfloor}{P_c} \right\rfloor \right), (m \bmod r, n \bmod c) \right\rangle .$$

The nonscattered decomposition (or pure block distribution) is just the special case $r = \lceil M/P_r \rceil$ and $c = \lceil N/P_c \rceil$. Similarly a purely scattered decomposition (or two dimensional wrapped distribution) is the special case $r = c = 1$.

5 ScaLAPACK

The ScaLAPACK software library is extending the LAPACK library to run scalably on MIMD, distributed memory, concurrent computers [9]. For such machines the memory hierarchy includes the off-processor memory of other processors, in addition to the hierarchy of registers, cache, and local memory on each processor. Like LAPACK, the ScaLAPACK routines are based on block-partitioned algorithms in order to minimize the frequency of data movement between different levels of the memory hierarchy. The fundamental building blocks of the ScaLAPACK library are a set of Basic Linear Algebra Communication Subprograms (BLACS) [8] for communication tasks that arise frequently in parallel linear algebra computations, and the Parallel Basic Linear Algebra Subprograms (PBLAS), which are a distributed memory version of the sequential BLAS. In the ScaLAPACK routines, all interprocessor communication occurs within the PBLAS and the BLACS, so the source code of the top software layer of ScaLAPACK looks very similar to that of LAPACK.

Figure 2 describes the ScaLAPACK software hierarchy. The components below the dashed line, labeled Local, are called on a single processor, with arguments stored on single processors only. The components above the line, labeled Global, are synchronous parallel routines, whose arguments include matrices and vectors distributed in a 2D block cyclic layout across multiple processors.

ScaLAPACK Software Hierarchy

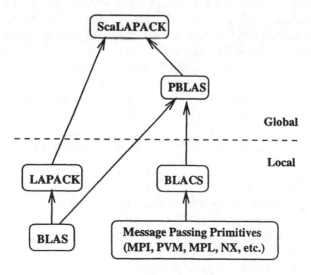

Fig. 2. ScaLAPACK Software Hierarchy

5.1 The Basic Linear Algebra Communication Subprograms (BLACS)

The **BLACS** (Basic Linear Algebra Communication Subprograms) [8] are a message passing library designed for linear algebra. The computational model consists of a one or two dimensional grid of processes, where each process stores matrices and vectors. The BLACS include synchronous send/receive routines to send a matrix or submatrix from one process to another, to broadcast submatrices to many processes, or to compute global reductions (sums, maxima and minima). There are also routines to construct, change, or query the process grid. Since several ScaLAPACK algorithms require broadcasts or reductions among different subsets of processes, the BLACS permit a process to be a member of several overlapping or disjoint process grids, each one labeled by a *context*. Some message passing systems, such as MPI [15], also include this context concept. The BLACS provide facilities for safe inter-operation of system contexts and BLACS contexts.

5.2 PBLAS

In order to simplify the design of ScaLAPACK, and because the BLAS have proven to be very useful tools outside LAPACK, we chose to build a Parallel BLAS, or PBLAS [10], whose interface is as similar to the BLAS as possible. This decision has permitted the ScaLAPACK code to be quite similar, and sometimes nearly identical, to the analogous LAPACK code. Only one substantially new routine was added to the PBLAS, matrix transposition, since this is a complicated operation in a distributed memory environment [7].

We hope that the PBLAS will provide a distributed memory standard, just as the BLAS have provided a shared memory standard. This would simplify and encourage the development of high performance and portable parallel numerical software, as well as providing manufacturers with a small set of routines to be optimized. The acceptance of the PBLAS requires reasonable compromises among competing goals of functionality and simplicity.

The PBLAS operate on matrices distributed in a 2D block cyclic layout. Since such a data layout requires many parameters to fully describe the distributed matrix, we have chosen a more object-oriented approach, and encapsulated these parameters in an integer array called an *array descriptor*. An array descriptor includes

(1) the descriptor type,
(2) the BLACS context (see Section 5.1),
(3) the number of rows in the distributed matrix,
(4) the number of columns in the distributed matrix,
(5) the row block size (r in Section 4),
(6) the column block size (c in Section 4),
(7) the process row over which the first row of the matrix is distributed,
(8) the process column over which the first column of the matrix is distributed,
(9) the leading dimension of the local array storing the local blocks.

By using this descriptor, a call to a PBLAS routine is very similar to a call to the corresponding BLAS routine.

```
CALL DGEMM ( TRANSA, TRANSB, M, N, K, ALPHA,
             A( IA, JA ), LDA,
             B( IB, JB ), LDB, BETA,
             C( IC, JC ), LDC )

CALL PDGEMM( TRANSA, TRANSB, M, N, K, ALPHA,
             A, IA, JA, DESC_A,
             B, JB, DESC_B, BETA,
             C, IC, JC, DESC_C )
```

DGEMM computes C = BETA * C + ALPHA * op(A) * op(B), where op(A) is either A or its transpose depending on TRANSA, op(B) is similar, op(A) is M-by-K, and op(B) is K-by-N. PDGEMM is the same, with the exception of the way in which submatrices are specified. To pass the submatrix starting at A(IA,JA) to DGEMM, for example, the actual argument corresponding to the formal argument A would simply be A(IA,JA). PDGEMM, on the other hand, needs to understand the global storage scheme of A to extract the correct submatrix, so IA and JA must be passed in separately. DESC_A is the array descriptor for A. The parameters describing the matrix operands B and C are analogous to those describing A. In a truly object-oriented environment matrices and DESC_A would be synonymous. However, this would require language support, and detract from portability.

The presence of a context associated with every distributed matrix provides the ability to have separate "universes" of message passing. The use of separate communication contexts by distinct libraries (or distinct library invocations) such as the PBLAS insulates communication internal to the library from external communication. When more than one descriptor array is present in the argument list of a routine in the PBLAS, it is required that the individual BLACS context entries must be equal. In other words, the PBLAS do not perform "inter-context" operations.

We have not included specialized routines to take advantage of packed storage schemes for symmetric, Hermitian, or triangular matrices, nor of compact storage schemes for banded matrices [10].

5.3 ScaLAPACK sample code

Given the infrastructure described above, the ScaLAPACK version (PDGETRF) of the LU decomposition is nearly identical to its LAPACK version (DGETRF).

SEQUENTIAL LU FACTORIZATION CODE	PARALLEL LU FACTORIZATION CODE

```
DO 20 J = 1, MIN( M, N ), NB
   JB = MIN( MIN( M, N )-J+1, NB )


   Factor diagonal and subdiagonal blocks and test for exact
   singularity.

   CALL DGETF2( M-J+1, JB, A( J, J ), LDA, IPIV( J ),
$               IINFO )

   Adjust INFO and the pivot indices.

   IF( INFO.EQ.0 .AND. IINFO.GT.0 ) INFO = IINFO + J - 1
   DO 10 I = J, MIN( M, J+JB-1 )
      IPIV( I ) = J - 1 + IPIV( I )
10 CONTINUE

   Apply interchanges to columns 1:J-1.

   CALL DLASWP( J-1, A, LDA, J, J+JB-1, IPIV, 1 )


   IF( J+JB.LE.N ) THEN

      Apply interchanges to columns J+JB:N.

      CALL DLASWP( N-J-JB+1, A( 1, J+JB ), LDA, J, J+JB-1,
$                  IPIV, 1 )

      Compute block row of U.

      CALL DTRSM( 'Left', 'Lower', 'No transpose', 'Unit',
$                 JB, N-J-JB+1, ONE, A( J, J ), LDA,
$                 A( J, J+JB ), LDA )
      IF( J+JB.LE.M ) THEN

         Update trailing submatrix.

         CALL DGEMM( 'No transpose', 'No transpose',
$                    N-J-JB+1, N-J-JB+1, JB, -ONE,
$                    A( J+JB, J ), LDA, A( J, J+JB ), LDA,
$                    ONE, A( J+JB, J+JB ), LDA )
      END IF
   END IF
20 CONTINUE
```

```
DO 10 J = JA, JA+MIN(M,N)-1, DESCA( 6 )
   JB = MIN( MIN(M,N)-J+JA, DESCA( 6 ) )
   I = IA + J - JA

   Factor diagonal and subdiagonal blocks and test for exact
   singularity.

   CALL PDGETF2( M-J+JA, JB, A, I, J, DESCA, IPIV, IINFO )


   Adjust INFO and the pivot indices.

   IF( INFO.EQ.0 .AND. IINFO.GT.0 )
$     INFO = IINFO + J - JA




   Apply interchanges to columns JA:J-JA.

   CALL PDLASWP( 'Forward', 'Rows', J-JA, A, IA, JA, DESCA,
$                J, J+JB-1, IPIV )


   IF( J-JA+JB+1.LE.N ) THEN

      Apply interchanges to columns J+JB:JA+N-1.

      CALL PDLASWP( 'Forward', 'Rows', N-J-JB+JA, A, IA,
$                   J+JB, DESCA, J, J+JB-1, IPIV )

      Compute block row of U.

      CALL PDTRSM( 'Left', 'Lower', 'No transpose', 'Unit',
$                  JB, N-J-JB+JA, ONE, A, I, J, DESCA, A, I,
$                  J+JB, DESCA )
      IF( J-JA+JB+1.LE.M ) THEN

         Update trailing submatrix.

         CALL PDGEMM( 'No transpose', 'No transpose',
$                     N-J-JB+JA, N-J-JB+JA, JB, -ONE, A,
$                     I+JB, J, DESCA, A, I, J+JB, DESCA,
$                     ONE, A, I+JB, J+JB, DESCA )
      END IF
   END IF
10 CONTINUE
```

5.4   ScaLAPACK – Contents and Documentation

The ScaLAPACK library provides routines for the solution of linear systems of equations, symmetric positive definite banded linear systems of equations, condition estimation and iterative refinement, for LU and Cholesky factorization, matrix inversion, full-rank linear least squares problems, orthogonal and generalized orthogonal factorizations, orthogonal transformation routines, reductions to upper Hessenberg, bidiagonal and tridiagonal form, reduction of a symmetric-definite generalized eigenproblem to standard form, the symmetric, generalized symmetric and the nonsymmetric eigenproblem. Similar functionality is provided for real and complex matrices, in both single and double precision.

A comprehensive Installation Guide is provided, as well as test suites for all ScaLAPACK, PBLAS, and BLACS routines.

5.5   ScaLAPACK – Performance

The main factors that affect the performance of linear algebra software on distributed-memory machines are the block size and the configuration of the process grid.

The ScaLAPACK codes run efficiently on a wide range of distributed memory MIMD computers, such as the IBM SP series, the Cray T3 series, and the Intel series. Figure 3 presents a variety of performance results on the IBM SP-2 and the Intel Paragon. Extensive performance results can be found in [9]. On 8 wide nodes of an IBM SP-2 for example, a 13000 × 13000 LU factorization runs at 1.6 Gflop/s. A 2000 × 2000 LU factorization on the same machine reaches already 1.0 Gflop/s. These performance results correspond to a very efficient use of this machine. The ScaLAPACK library can also be used on clusters of workstations, and any system for which PVM [12] or MPI [13, 15, 16] is available. Performance results on the TMC CM-5, however, have been disappointing because of the difficulty of using the vector units in message passing programs.

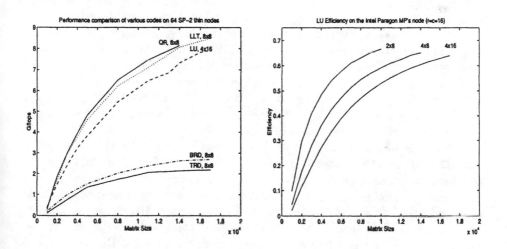

Fig. 3. IBM SP-2 and Intel Paragon Performance

6 Conclusions and Related Projects

Both LAPACK and ScaLAPACK are slated for updated software and documentation releases toward the end of 1996. Alternative interfaces for the libraries are under development. A BLAS Technical Forum is being established to consider expanding the BLAS for serial and parallel computation. And finally, a number of other aspects of the ScaLAPACK project, including sparse and out-of-core activities, are available or underway.

The upcoming LAPACK release (version 3.0) will introduce routines for the singular value decomposition computed by the divide-and-conquer method, new simple and expert drivers for the generalized nonsymmetric eigenproblem, a faster QR decomposition with column pivoting, a faster solver for the rank-deficient least squares, and a blocked version of the reduction of a upper trapezoidal matrix to upper triangular form. The third edition of the LAPACK Users' Guide will coincide with this release.

Future releases of the ScaLAPACK library will extend the flexibility of the PBLAS and increase the functionality of the library to include routines for the solution of general banded linear systems, general and symmetric positive definite tridiagonal systems, rank-deficient linear least squares problems, generalized linear least squares problems, and the singular value decomposition. A draft of the ScaLAPACK Users' Guide is currently available on *netlib* and we plan to have the final draft ready for publication at the end of 1996.

A Fortran 90 interface for the LAPACK library is currently under development. This interface will provide an improved user-interface to the package by taking advantage of the considerable simplifications of the Fortran 90 language such as assumed-shape arrays, optional arguments, and generic interfaces.

As HPF compilers have recently become available, work is currently in progress to produce an HPF interface for the ScaLAPACK library. HPF provides a much more convenient distributed memory interface than can be provided in processor-level languages such as Fortran77 or C. This interface to a subset of the ScaLA-PACK routines will be available at the time of the next ScaLAPACK release. With the increased generality of the PBLAS to operate on partial first blocks, ScaLAPACK will be fully compatible with HPF [14].

Also underway is an effort to establish a BLAS Technical Forum to consider expanding the BLAS in a number of ways in light of modern software, language, and hardware developments. The goals of the forum are to stimulate thought and discussion, and define functionality and future development of a set of standards for basic matrix and data structures. Dense and sparse BLAS are considered, as well as calling sequences for a set of low-level computational kernels for the parallel and sequential settings. For more information on the BLAS Technical Forum refer to the URL:

http://www.netlib.org/utk/papers/blast-forum.html

The EISPACK, LINPACK, LAPACK, BLACS and SCALAPACK linear algebra libraries are in the public domain. The software and documentation can be retrieved from *netlib* (http://www.netlib.org).

7 Acknowledgments

We wish to acknowledge Antoine Petitet for assembling the previous version of this tutorial, and thank him for his advice and contributions to this effort.

References

1. Anderson, E., Bai, Z., Bischof, C., Demmel, J., Dongarra, J., Du Croz, J., Greenbaum, A., Hammarling, S., McKenney, A., Ostrouchov, S., Sorensen, D.: LAPACK Users' Guide, Second Edition. SIAM, Philadelphia, PA, 1995.

2. Lawson, C., Hanson, R., Kincaid, D., Krogh, F.: Basic Linear Algebra Subprograms for Fortran Usage. ACM Transactions on Mathematical Software, 5:308–323, 1979.
3. Dongarra, J., Du Croz, J., Hammarling, S., Hanson, R.: Algorithm 656: An extended Set of Basic Linear Algebra Subprograms: Model Implementation and Test Programs. ACM Transactions on Mathematical Software, 14(1):18–32, 1988.
4. Dongarra, J., Du Croz, J., Duff, I., Hammarling, S.: A Set of Level 3 Basic Linear Algebra Subprograms. ACM Transactions on Mathematical Software, 16(1):1–17, 1990.
5. Anderson, E., Dongarra, J.: Results from the initial release of LAPACK. LAPACK working note 16, Computer Science Department, University of Tennessee, Knoxville, TN, 1989.
6. Anderson, E., Dongarra, J.: Evaluating block algorithm variants in LAPACK. LAPACK working note 19, Computer Science Department, University of Tennessee, Knoxville, TN, 1990.
7. Choi, J., Dongarra, J., Walker, D.: Parallel matrix transpose algorithms on distributed memory concurrent computers. In Proceedings of Fourth Symposium on the Frontiers of Massively Parallel Computation (McLean, Virginia), pages 245–252. IEEE Computer Society Press, Los Alamitos, California, 1993. (also LAPACK Working Note #65).
8. Dongarra, J., Whaley, R.C.: A User's Guide to the BLACS v1.0. Technical Report UT CS-95-281, LAPACK Working Note #94, University of Tennessee, 1995.
9. Choi, J., Demmel, J., Dhillon, I., Dongarra, J., Ostrouchov, S., Petitet, A., Stanley, K., Walker, D., Whaley, R.C.: ScaLAPACK: A Portable Linear Algebra Library for Distributed Memory Computers - Design Issues and Performance. Technical Report UT CS-95-283, LAPACK Working Note #95, University of Tennessee, 1995.
10. Choi, J., Dongarra, J., Ostrouchov, S., Petitet, A., Walker, D., Whaley, R.C.: A Proposal for a Set of Parallel Basic Linear Algebra Subprograms. Technical Report UT CS-95-292, LAPACK Working Note #100, University of Tennessee, 1995.
11. Dongarra, J., Duff, I., Sorensen, D., Van der Vorst, H.: Solving Linear Systems on Vector and Shared Memory Computers. SIAM Publications, Philadelphia, PA, 1991.
12. Geist, A., Beguelin, A., Dongarra, J., Jiang, W., Manchek, R., V. Sunderam, V.: PVM: Parallel Virtual Machine. A User's Guide and Tutorial for Networked Parallel Computing. The MIT Press, Cambridge, Massachusetts, 1994.
13. Gropp, W., Lusk, E. Skjellum, A.: Using MPI: Portable Programming with the Message-Passing Interface, MIT Press, Cambridge, MA, 1994.
14. Koebel, C., Loveman, D., Schreiber, R., Steele, G., Zosel, M.: The High Performance Fortran Handbook. The MIT Press, Cambridge, Massachusetts, 1994.
15. Message Passing Interface Forum. MPI: A Message Passing Interface Standard. International Journal of Supercomputer Applications and High Performance Computing, 8(3–4), 1994.
16. Snir, M., Otto, S. W., Huss-Lederman, S., Walker, D. W. and Dongarra, J.: MPI: The Complete Reference, MIT Press, Cambridge, MA, 1996.
17. Wilkinson, J., Reinsch, C.: Handbook for Automatic Computation: Volume II - Linear Algebra. Springer-Verlag, New York, 1971.

Bulk Synchronous Parallelisation of Genetic Programming

Dimitris C. Dracopoulos and Simon Kent

Brunel University
Department of Computer Science
and Information Systems
London, UK

Abstract. A parallel implementation of Genetic Programming (GP) is described, using the Bulk Synchronous Parallel Programming (BSP) model, as implemented by the Oxford BSP library. Two approaches to the parallel implementation of GP are examined. The first is based on global parallelisation while the second implements the island model for evolutionary algorithms. It is shown that considerable speedup of the GP execution can be achieved and that the BSP model is very suitable for parallelisation of similar algorithms.

1 Introduction

Genetic Programming (GP) is a relatively new discipline which offers a method for the automatic discovery of computer programs to solve problems which are difficult, or impossible to solve through conventional methods. Genetic programming [3] is an extension of the genetic algorithm (GA) [1] function optimisation method. Both methods are based on the evolutionary process of Darwinian natural selection.

Despite its advantages, one drawback of Genetic Programming is the considerable execution time which can be required to produce a solution. The process is, however, very amenable to parallelisation. Koza[4] has carried out some work on parallel GP using transputers, however there has been relatively little work in this area (unlike genetic algorithms for which several parallel implementations exist).

This paper investigates two implementations of parallel Genetic Programming, using the Bulk Synchronous Parallel (BSP) model of parallel computation which is relatively simple to use, and can be run on machines varying from advanced supercomputers to a simple network of workstations as were used for this paper.

The first implementation uses a global approach where the GP process is explicitly parallelised by dividing the task amongst a number of processors. The second method adopts a coarse grained approach in which the ratio between communication and computation is reduced, in an attempt to further reduce execution time.

2 Genetic Programming

Genetic Programming applies the Darwinian principle of natural selection to produce solutions to problems. Evolution is captured in a computer by a repetitive computational

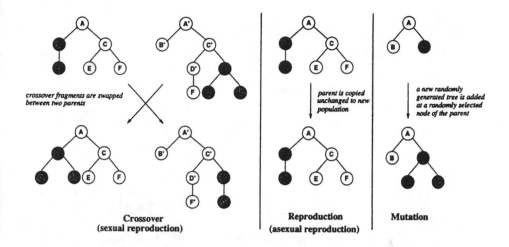

Fig. 1. Basic Genetic Operators used in Genetic Programming

process. An initial population of computer programs is generated randomly as a starting point for the process. Each individual is executed to measure its fitness or ability to solve the problem. A new population is then generated from the previous generation, using the genetic operators which are shown in Figure 1. Natural selection is present because the programs which are best at solving the problem are most likely to be chosen to participate in genetic operations.

This process of fitness evaluation and evolution is repeatedly applied until a solution is found or a timeout occurs. The whole process is illustrated in Figure 2.

3 BSP

The Bulk Synchronous Parallel Programming model was proposed by Leslie Valiant[6] for general-purpose parallel computing. The model provides a very basic set of operations which allow synchronisation of processors, and communication between them. This may be implemented using a dedicated language, or by means of a library of routines.

Each processor executes the same program code, therefore conforming to a single program, multiple data (SPMD) model. Processor allocation is static, meaning that processors are allocated at the beginning of the program with no subsequent change to the allocation during execution. Communication may be achieved using one of a variety of methods such as TCP/IP or shared memory.

Processors within a BSP computer do not proceed in lock-step. Synchronisation is achieved by using *supersteps* which are units of parallel execution during which a processor may perform computations on its own personal data, or initiate stores or fetches between itself and other processors. At the end of a given *superstep*, a BSP process must wait until all other processors have finished their computation and communication. To

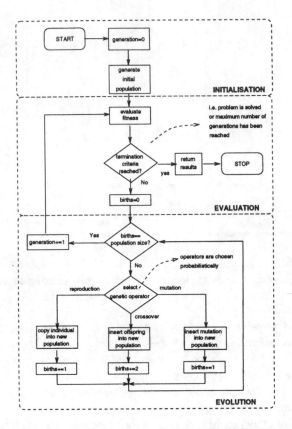

Fig. 2. Flowchart of the process used in Genetic Programming

ensure deterministic behaviour, it should never be assumed that communication which is started in a particular *superstep* will be completed in the same *superstep*. However, non-local stores and fetches are guaranteed to have finished after the end of a *superstep*.

A BSP computer can be defined in terms of the following parameters:

p = number of processors
s = CPU speed of processors
L = minimum elapsed time between successive synchronisations
g = global computation / communication balance

The units of s, L and g are arbitrary, although they must obviously be the same to allow meaningful comparisons to be made. These parameters can be useful when making predictions as to what speedup can be expected from a particular BSP configuration. L and g are indicators of the overhead which may be experienced when using BSP, where L represents the overhead involved in completing a *superstep* (barrier synchronisation) and g represents the amount of computation which could have been carried out locally while inter-processor communication takes place.

The implementation used for this paper was the Oxford BSP Library which consists of only six operations as shown in Table 1.

BSP_START	start of the BSP program
BSP_FINISH	end of the BSP program
BSP_SSTEP_START(n)	start of superstep n
BSP_SSTEP_END(n)	end of superstep n
BSP_STORE(to, from_data, to_data, length)	store from local to remote processor
BSP_FETCH(from, from_data, to_data, length)	fetch from remote to local processor

Table 1. The Oxford BSP Library basic operations.

4 The Problem

The problem chosen to test the BSP GP implementation was the Artificial Ant problem as used by Koza[3]. The problem involves moving a robot ant along a trail of food which lies on a grid. This trail contains 157 pieces of food as shown in Figure 3. In this figure, a black square represents a piece of food, and a grey square represents a gap in the trail. The figure shows the top left corner of the full grid which is 100×100 squares.

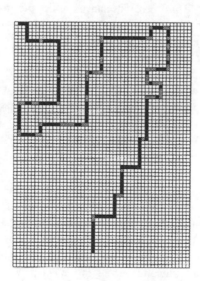

Fig. 3. The Artificial Ant trail.

The ant is capable of a simple set of actions, these being:

- Move forward one grid square
- Turn Left 90°
- Turn Right 90°

The ant also has a limited decision making ability through a function:

- If food in square immediately ahead then action a else action b

For this problem, a population of 2,000 individual programs are initially generated and subsequently evaluated and evolved by each process. When the programs are evaluated against the trail, a constraint is placed on them, whereby they are only allowed 3,000 'moves' in which to traverse the trail. This is to prevent the ant finding all the food on the trail simply by randomly wandering over all 10,000 grid squares.

5 Implementation

5.1 Global Parallelisation

The first approach taken to produce a parallel GP implementation involved explicitly parallelising part of what is otherwise the same process as for a sequential run of the GP process. A master-slave paradigm was adopted whereby the master process performs the standard sequential GP process. The slaves assist the master only during fitness evaluation which is the most computationally expensive part of the process.

During fitness evaluation the slaves collect equal portions of the population from the master, evaluate them, and return the fitness of each individual to the master. Unfortunately, there is a limitation in version 1.2 of the Oxford BSP Library which only allows non-local access to statically declared data[1]. As the program data is stored in dynamic data structures, some additional overhead is incurred when data is transferred between processes. Before communication occurs, dynamic data must be packed into a statically declared buffer, thus enabling data transfer to occur. The process shown in Figure 4.

5.2 Coarse Grained Parallelisation

The alternative method used to produce a parallel GP implementation was to use a coarse grained approach. In this context, the granularity refers to the ratio between communication and computation. In the global parallelisation method, the processes were tightly coupled, with communication occurring every generation. The coarse grained parallelisation method attempts to reduce the communication between processes, thus reducing overhead, and improving speedup. The approach adopted is known as the Island Model and has been used, for example, by Gordon[2] to produce implementations of a parallel genetic algorithm and also by Koza[4].

---

[1] A future release of the BSP Oxford library will allow communication between dynamic data objects. Therefore in this paper, we present some additional results, assuming dynamic communication is possible.

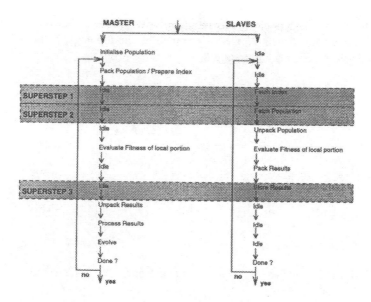

Fig. 4. Architecture of global parallel GP system

A number of processes are started, each with their own private population, which they are responsible for initialising, evaluating and evolving. Unlike global parallelisation, all processes have equal status. The standard GP process is modified by the addition of a migration operator as shown in Figure 5. Each process is considered to be an island, detached from the other processes. Every few generations, after the fitness evaluation phase, migration occurs, whereby certain individuals are moved between processes, thus distributing the genetic material throughout all the process `islands'. For this paper, the top 10% of individuals from each island were migrated every 10 generations.

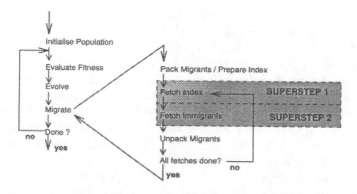

Fig. 5. Architecture of coarse parallel GP system

An important consideration when using the island model is the nature of the connections between the islands. This is determined by the topology. The two examples used for this paper are shown in Figure 6.

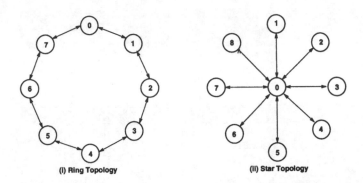

(i) Ring Topology   (ii) Star Topology

Fig. 6. Topologies used with the island model GP implementation.

6 Speedup Prediction

Previous reference was made in section 3 to the BSP parameters, and the fact that they could be used to make approximate predictions of the speedup which could be achieved by parallelising the GP process.

In the Oxford BSP user guide[5], it is suggested that a conservative upper bound for the overhead involved in performing one of the *supersteps* is:

$$C \leq L + g \times h . \tag{1}$$

The variables L and g are as defined previously in section 3, and h is the number of data-words communicated (sent + fetched) in the *superstep* by a particular process.

Values for the BSP parameters were determined by running a short program which produced estimated values of s, L, g_∞ and $N_{\frac{1}{2}}$. The variable g_∞ represents the asymptotic value of g which would be approached by very large data objects, and $N_{\frac{1}{2}}$ is the number of words communicated to achieve half the asymptotic bandwidth. The value of g can be approximated by $g_\infty \times (1 + N_{\frac{1}{2}}/n)$. It is less efficient to communicate small data objects than it is to send large ones. Table 2 shows the values for this paper.

Predictions for global parallelisation were compared against a standard sequential run. For the island model predictions, the comparison was made with a single sequential run with population which is $n \times p$, where n is the number of processes, and p is the population size. The two systems therefore have the some overall population size. The cost of a sequential run is taken to be the cost of a sequential run with population size p is multiplied by the number of processors used. Collecting true times for such runs was impractical because the large populations required a large amount of memory, resulting in the workstations swapping data between physical memory and disc, thus slowing down the execution.

223

p	s	L	g_∞	$N_{\frac{1}{2}}$
2	1.75	8000	35	2.3
4		13000	80	2.0
8		22000	180	1.6
16		55000	350	1.5

Table 2. Estimated values of BSP parameters for a network of SPARCstation 5 workstations communicating using TCP/IP.

Predictions were made in two ways, one accounting for the additional packing overhead, and the other ignoring it.

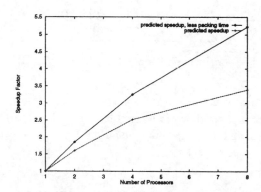

Fig. 7. Graph of predicted speedup for global parallelisation

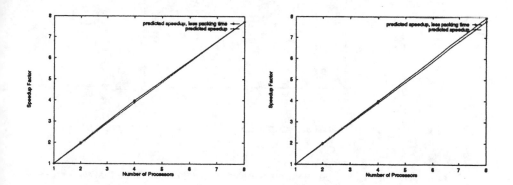

Fig. 8. Graph of predicted speedup for island model - ring topology

Fig. 9. Graph of predicted speedup for island model - star topology

7 Results

The workstations used to run the software were a number of SUN SPARCstation 5 machines with 70MHz microSPARC-II processors, and running SOLARIS 2.4, each with 32Mb of RAM. These machines were connected on a common subnet with ethernet cabling. Communication between processors was achieved using TCP/IP.

The three implementations were run for 50 generations, and the average elapsed time was recorded. The results are shown in Table 3.

processors	elapsed time (sec)		
	global	ring	star
1	4786	4980	5100
2	3380	5280	5400
4	2268	6180	6120
8	1878	7020	6430

Table 3. Elapsed time for runs of 50 generations for parallel GP implementations

The recorded speedups are shown in three graphs. The global parallelisation speedup is shown in Figure 10, and the ring topology and star topology speedups are shown in Figures 11 and 12 respectively.

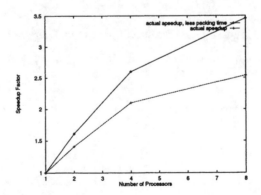

Fig. 10. Graph of actual speedup achieved with global implementation

Although the size of the problem was not very large, one can see that the achieved speedup of the GP run is significant. As the size of the problem increases, one can expect to see that for the global parallelisation version the achievable speedup of the BSP GP system will be improved, as the communication overhead will be small compared with the actual parallel computational time. A prediction of such a speedup can be easily calculated by applying the cost modelling procedure of the previous section.

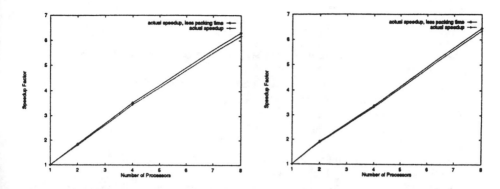

Fig. 11. Graph of actual speedup achieved with ring topology

Fig. 12. Graph of actual speedup achieved with star topology

Apparently, given a problem of fixed size, there is an optimum number of processors with which a maximum speedup can be achieved. Therefore, to optimise the performance of the BSP GP system, the choice of the number of processors must follow the suggested number p of the cost modelling procedure.

8 Conclusions

A parallel implementation of a Genetic Programming system was described, using the Bulk Synchronous Programming (BSP) model of parallel computation. Two approaches to parallelisation were examined. The first was based on global parallelisation while the second implemented a coarse grained parallelisation. The island model parallel version achieved better speedups as it required less communication. Although the size of the test problem was quite small, the speedups achieved were significant. The implementation is portable to a number of different platforms, and the performance of the BSP GP system can be predicted for different parallel architectures.

Using the implemented BSP GP system, higher speedups can be achieved in large problems. In addition, the time required to parallelise evolutionary algorithms with the BSP model is insignificant, compared with the achieved speedups, when the problem is a complex and time consuming. The Oxford BSP library used for this paper proved to be very suitable and efficient for the parallelisation of the GP process, so as to suggest its use for algorithms of similar nature.

9 Acknowledgements

This work was partly supported by EPSRC award no 95700741.

References

1. David E. Goldberg. *Genetic Algorithms in Search, Optimization and Machine Learning*. Addison Wesley, 1989.
2. V. Scott Gordon and Darrell Whitley. Serial and parallel genetic algorithms as function optimizers. In Stephanie Forrest, editor, *Proceedings of the 5th International Conference on Genetic Algorithms*. Morgan Kaufman, 1993.
3. John R. Koza. *Genetic Programming: on the Programming of Computers by means of Natural Selection*. MIT Press, 1993.
4. John R. Koza and David Andre. Parallel genetic programming on a network of transputers. Technical Report CS-TR-95-1542, Stanford University, 1995.
5. Richard Miller and Joy Reed. The Oxford BSP Library users' guide. Technical report, University of Oxford, 1993.
6. Leslie G. Valiant. A bridging model for parallel computation. *Communications of the Association for Computing Machinery*, 33(8):103–111, 1990.

Frontal Software for the Solution of Sparse Linear Equations

Iain S. Duff and Jennifer A. Scott

Rutherford Appleton Laboratory, Didcot, Oxon OX11 0QX, England

Abstract. We discuss the power and limitations of frontal solvers for the solution of large sparse systems of linear equations. We describe their design and user interface. We compare sparse frontal codes from the Harwell Subroutine Library (HSL) against other HSL sparse solvers and consider the effect of ordering on the frontal solver. We consider both the case of assembled and unassembled systems for both symmetric positive-definite and unsymmetric matrices. We use problems arising in real engineering or industrial applications in our tests.

1 Introduction

The frontal method [4, 14, 15, 16, 18] is a technique for the direct solution of

$$\mathbf{A}\mathbf{x} = \mathbf{b}, \tag{1.1}$$

where $\mathbf{A}$ is a large sparse matrix. This approach has the merit of rather simple logic and relatively little data movement and integer overhead. The floating-point arithmetic is performed using dense linear algebra kernels so that the computational rate, measured in Mflop/s, is high. However, frontal methods require many more floating-point operations for factorization than other approaches unless the matrix can be ordered into a band form with few zeros in the band. It is thus interesting to see how this trade-off works in practical applications, and this is the main theme of this paper.

Although it is possible to input the matrix $\mathbf{A}$ in either assembled or unassembled form (as, for example, in the code `MA42` in the Harwell Subroutine Library [12]), the power of the frontal method, as we illustrate in Section 5, is more apparent when the matrix $\mathbf{A}$ comprises contributions from the elements of a finite-element discretization. That is,

$$\mathbf{A} = \sum_{k=1}^{m} \mathbf{A}^{(k)}, \tag{1.2}$$

where $\mathbf{A}^{(k)}$ is nonzero only in those rows and columns that correspond to variables in the kth element.

In Section 2, we discuss salient features of the frontal method and show how the computational kernel consists of the Level 3 Basic Linear Algebra Subprogram (BLAS) _GEMM, that implements **dense** matrix-matrix multiplication. We briefly describe, in Section 3, the interface of our frontal solvers and discuss the codes available in the Harwell Subroutine Library for frontal solution and associated computation. We use test matrices from real problems occurring in engineering or industry in the subsequent experiments in our paper and discuss their origins and characteristics in Section 4. We consider the effect of reordering the system and the preassembly of element problems in Section 5. We compare our frontal codes with other Harwell Subroutine Library [17] codes in Sections 6 and 7, considering symmetric positive-definite and unsymmetric systems respectively. We present some concluding remarks in Section 8.

2   Frontal methods

A key feature of frontal methods is that the matrix $\mathbf{A}$ is never assembled explicitly but the assembly and Gaussian elimination processes are interleaved, with each variable being eliminated as soon as its row and column are *fully summed*, that is, after the last occurrence in a matrix $\mathbf{A}^{(k)}$. This allows all intermediate working to be performed in a dense matrix, termed the *frontal matrix*, whose rows and columns correspond to variables that have not yet been eliminated but occur in at least one of the elements or equations that have been assembled.

We now describe the method for element entry in more detail. A full discussion of the equation entry can be found in [4]. If all the fully summed variables are permuted to the first rows and columns of the frontal matrix, we can partition the frontal matrix $\mathbf{FA}$ as

$$\mathbf{FA} = \begin{pmatrix} \mathbf{F}_{11} & \mathbf{F}_{12} \\ \mathbf{F}_{21} & \mathbf{F}_{22} \end{pmatrix}, \tag{2.1}$$

where $\mathbf{F}_{11}$ is a square matrix of order k and $\mathbf{F}_{22}$ is of order $k_1 \times k_1$. Note that $k + k_1$ is equal to the current order of the frontal matrix, and $k \ll k_1$, in general. The rows and columns of $\mathbf{F}_{11}$, the rows of $\mathbf{F}_{12}$, and the columns of $\mathbf{F}_{21}$ are fully summed; the variables in $\mathbf{F}_{22}$ are not yet fully summed. Pivots may be chosen from anywhere in $\mathbf{F}_{11}$. $\mathbf{F}_{11}$ is then factorized, multipliers are stored over $\mathbf{F}_{12}$ and the Schur complement $\mathbf{F}_{22} - \mathbf{F}_{12}^T \mathbf{F}_{11}^{-1} \mathbf{F}_{12}$ is formed. At the next stage, further entries from the original matrix are assembled with this Schur complement to form another frontal matrix. In the symmetric positive-definite case, the frontal matrix is symmetric (so $\mathbf{F}_{21} = \mathbf{F}_{12}^T$) and one can always choose the k diagonal entries of $\mathbf{F}_{11}$ as pivots. However, in the unsymmetric case, numerical considerations can prevent us choosing k pivots and so the Schur update that is passed to the next stage may be correspondingly larger.

If only the frontal matrices are held in-core, the frontal method can solve quite large problems with modest amounts of high-speed memory and can use dense linear algebra kernels, in particular the Level 3 Basic Linear Algebra Subprograms (BLAS) [3] for the numerical factorization. If the matrix $\mathbf{F}_{11}^{-1}$ is held in factored form, then it is clear from the discussion in the previous paragraph that the Schur complement can be formed using the Level 3 BLAS subroutines _TRSM and _GEMM. Level 2 and 3 BLAS can also be used in the factorization of $\mathbf{F}_{11}$. We remark that, because the size of the frontal matrix increases when a variable appears for the first time and decreases whenever it is eliminated, the order in which the elements are assembled is critical. Elements should be preordered to reduce the order of the frontal matrices. Various algorithms have been developed for achieving this and we discuss the effect of reordering in Section 5.

3 Harwell Subroutine Library frontal solvers

The two frontal codes that we use in the following comparisons are the code MA62 for symmetric positive-definite systems and the code MA42 for unsymmetric problems. In this section, we describe characteristics of their user interface and briefly discuss associated software in the Harwell Subroutine Library (HSL).

The symmetric positive-definite code, MA62, only permits entry by elements, whereas the unsymmetric code, MA42, allows entry by either elements or equations.

Both codes use reverse communication to obtain information from the user. The structure of problem is first input by calling a subroutine for each element or equation, the primary reason for these calls being to establish when variables can be fully summed and hence are candidates for use as pivots. Thereafter, for the positive-definite solver MA62, a further set of calls enables an accurate forecast to be made for the length of user supplied arrays and the maximum order of the frontal matrix so that numerical factorization can be run efficiently and reliably. For the unsymmetric MA42 code, the use of such a prediction routine is optional and will only give lower bounds on the relevant quantities because of the possibility of numerical pivoting in the factorization. In these *symbolic phases* only the integer indexing information need be passed. Both codes can use direct access files for the matrix factors, and the user must define these by a simple subroutine call if this option is required.

The numerical factorization is then performed with the user required to call a subroutine for each element or equation. The information from the earlier symbolic phases is used to control the pivot selection and elimination within the current frontal matrix. Optionally, forward elimination can be performed on a set of elemental right-hand side vectors, in which case a final back substitution phase yields appropriate solutions. Subsequent right-hand sides can be solved using the matrix factors, in which case the right-hand sides are supplied as dense vectors and a single subroutine call is all that is required. In the unsymmetric case, the same factors can be used by MA42 to solve the equation $\mathbf{A}^T\mathbf{x} = \mathbf{b}$.

The calling sequence for MA42 can be rather daunting and so a simpler interface has been designed for matrices that are already assembled. The resulting routine, MA43, does not use direct access files and moreover just calls MA42 internally. We have therefore not used this code in the comparisons in this paper. Since our symmetric code, MA62, only accepts entry by elements, we use the HSL code MC37 to generate an element problem from assembled matrices.

Ordering routines have been developed for the frontal solvers and use similar logic to bandwidth minimization. The HSL code MC43 gives the user the choice of basing the ordering on the element structure or the usual sparse matrix pattern [11]. There is little difference in the quality of the ordering between these two approaches but, for the case where we generate elements from the assembled matrix, we used the sparse matrix pattern because many small elements were generated so that the former option was much slower. The HSL code MC40 can be used to preorder symmetrically structured equations. Since MC40 provides an ordering on the variables, for MA42, we needed to scan the equations to determine an ordering for the rows, first ordering all rows in which the first variable in the ordering appears, then those unordered rows in which the second variable appears, and so on.

The other main auxiliary routines for the HSL frontal codes concern their use in a parallel computing environment. In this case, MA42 must be used in conjunction with MA52 and the variables should be classified as either interior or interface variables [13]. Reordering schemes have been developed to exploit this different structure [19].

4 Test problems

In this section, we describe the test problems that we use for the comparisons in this paper. In all cases, they arise in real engineering and industrial applications. A brief description of each of the unassembled finite-element test problems is given in Table 4.1. The first six problems are from the Harwell-Boeing Collection [5, 6], the RAMAGE01 problem is from Alison Ramage of the University of Strathclyde, the problem AEAT is from Andrew Cliffe of AEA Technology, and the remaining problems (TRDHEIM, CRPLAT2, OPT1, and TSYL201) were supplied by Christian Damhaug of Det Norske Veritas, Norway. For all the problems, values for the entries of the matrix were generated using the HSL random number generator FA01. For the symmetric positive-definite test cases, each element was made symmetric and diagonally dominant. The elements were resequenced using MC43 before the frontal solvers were called.

The other (assembled) matrices are shown in Table 4.2 and are all from the Harwell-Boeing Collection. For element entry to the frontal solvers, the symmetric problems were first converted to element form by MC37.

All the HSL codes that we use in our numerical experiments have control parameters which are set to default values. We use these defaults in every case, even if different codes sometimes choose a different value for essentially the same parameter.

The experimental results in this paper were obtained on a CRAY J932 using 64-bit floating-point arithmetic, and the vendor-supplied BLAS. In separate runs on the Level 3 BLAS subroutine SGEMM, we found that its peak performance (r_∞) was 90 Mflop/s attained on matrices of order greater than 500 and that the vector length for half peak $(n_{1/2})$ was 50, although we have observed that SGEMM avoids operations with explicit zeros and so can apparently perform faster than peak if account is not taken of this. All codes were compiled using the CRAY Fortran compiler cf77-7, with compiler option -Zv. All times are CPU times in seconds and include the i/o overhead for the codes which use direct access files.

Identifier	Order	Number of elements	Description/discipline
CEGB3306	3222	791	2.5D Framework problem
CEGB2919	2859	128	3D cylinder with flange
CEGB3024	2996	551	2D reactor core section
LOCK1074	1038	323	Lockheed gyro problem
LOCK2232	2208	944	Lockheed tower problem
LOCK3491	3416	684	Lockheed cross-cone problem
RAMAGE01	1476	128	3D Navier-Stokes
AEAT	5081	800	Double glazing problem
TRDHEIM	22098	813	Mesh of the Trondheim Fjord
CRPLAT2	18010	3152	Corrugated plate field
OPT1	15449	977	Part of condeep cylinder
TSYL201	20685	960	Part of condeep cylinder

Table 4.1. The unassembled finite-element test problems

Identifier	Order	Number of entries	Description/discipline
BCSPWR10	5300	13571	Eastern US Power Network
BCSSTK15	3948	60882	Model of an offshore platform
BCSSTK18	3948	80519	Nuclear power station
BP 1600	1600	4841	Basis matrix from LP problem
GRE 1107	1107	5664	Simulation studies in computer systems
JPWH 991	991	6027	Circuit physics modelling
LNS 3937	3937	25407	Fluid flow modelling
LNSP3937	3937	25407	Fluid flow modelling
NNC1374	1374	8606	Nuclear reactor core modelling
ORANI678	2529	90158	Economic model of Australia
OSRREG 1	2205	14133	Oil reservoir simulation
PORES 3	532	3474	Oil reservoir simulation
PSMIGR 3	3140	543162	Population migration
SHERMAN 3	5005	20033	Oil reservoir simulation
WEST2021	2021	7353	Chemical engineering

Table 4.2. The assembled test problems. First three problems are symmetric.

5 The effect of ordering and elemental form

¿From the earlier discussion, it should be clear that the ordering of the elements or equations will have a significant effect on the performance of the frontal solver. We have already reported on the effect of this in [11] and here confine ourselves to examining the effect of ordering on our symmetric positive-definite solver, MA62.

We show the results of using the MC43 ordering in Table 5.1. In some cases, a significant reduction in the maximum front size is obtained and this is reflected in the much lower times for factorization. We note that, in all cases, the original order is that provided by the application and, in most instances, this was believed to be a "good" ordering. In this table and the later ones in which the number of floating-point operations are quoted, we count all operations (+,-,*,/) equally and assume that there are no zeros in the frontal matrices. We note that substantial savings in time and "ops" can be obtained by profile reduction on some of our problems.

Identifier	Max front size		Number of ops $(*10^8)$		Time seconds	
	Before	After	Before	After	Before	After
CEGB3306	354	78	2.06	0.15	2.1	1.0
CEGB2919	348	291	1.22	1.00	2.8	2.3
CEGB3024	152	132	0.37	0.23	1.4	1.1
LOCK1074	810	126	2.90	0.08	2.0	0.4
LOCK2232	1266	72	12.56	0.07	3.9	0.7
LOCK3491	834	217	11.60	0.67	11.9	1.8
RAMAGE01	457	372	1.78	1.16	3.3	2.1
AEAT	150	161	1.13	0.70	3.5	2.3
TRDHEIM	276	348	5.37	4.92	14.4	12.8
TSYL201	1200	540	154.3	55.33	208.0	82.6
OPT1	2681	983	633.8	56.57	327.3	78.2
CRPLAT2	1564	538	252.2	25.97	279.2	41.6

Table 5.1. The results of using a profile reduction ordering with MA62.

Several of the codes that we are testing accept entry by elements (MA42, MA46, and MA62) and one, MA42, has both an equation and an element entry. We can thus compare the efficiency of a frontal code on the original element problem and its assembled form by runs using MA42, preordering with MC43 and MC40, respectively. We show the results from these runs in Table 5.2. It is quite obvious from these results that we should avoid preassembling an element problem.

Identifier	Factorization time Seconds		Storage Kwords	
	elements	assembled	elements	assembled
CEGB3306	1.58	3.69	380	794
CEGB2919	4.33	12.65	1014	1949
CEGB3024	1.89	3.81	482	805
LOCK1074	0.65	1.14	160	232
LOCK2232	0.87	1.70	220	335
LOCK3491	3.37	9.00	888	1598

Table 5.2. The effect of preassembly on **MA42**.

6 A comparison of the frontal code MA62 with other HSL codes

In this section, we examine the performance of the frontal code **MA62** and compare it with the HSL code **MA27** which uses a multifrontal algorithm [7, 8]. We note that, although **MA27** will handle symmetric indefinite problems, by setting the threshold pivoting parameter to zero the matrix is assumed to be positive definite and a logically simpler path in the code is invoked.

We compare the two codes, first on our set of element problems (see Table 4.1), then on the symmetric assembled matrices (first three problems in Table 4.2). In the former case, the matrices are assembled before calling **MA27**; in the latter case, an element problem is first created from the assembled problems using **MC37**. In neither case is the cost of this preprocessing included. The results of both sets of comparisons are given in Table 6.1. Note that the "Total" storage figures for **MA62** are for the amount of memory required to hold the frontal matrices.

It is clear from these results that the frontal code is usually better on unassembled finite-element problems, particularly in the analyse and numerical factorization phases. The benefit of using a frontal method to reduce the maximum amount of storage needed to perform the factorization is evident both in the case of assembled and unassembled matrices. However, storage for the factors is usually far greater for the frontal method and this is reflected in the much poorer times for subsequent solution, although it should be mentioned that **MA62** is more efficient if multiple right-hand sides are being solved at the same time. It is not generally advisable to use a frontal method on an assembled matrix although the in-core storage required is low.

Identifier	Code	Time (seconds)			Factor ops $(*10^6)$	Storage (Kwords)	
		Analyse	Factorize	Solve		Total	Factors
CEGB3306	MA27	0.40	0.56	0.028	2.10	114	81
	MA62	0.20	0.79	0.125	14.56	6	222
CEGB2919	MA27	1.82	3.60	0.037	57.27	590	384
	MA62	0.11	2.27	0.107	99.48	85	542
CEGB3024	MA27	0.60	0.93	0.033	7.10	175	146
	MA62	0.20	0.94	0.154	23.12	18	295
LOCK1074	MA27	0.27	0.56	0.012	4.90	109	71
	MA62	0.14	0.36	0.044	7.79	16	94
LOCK2232	MA27	0.43	0.67	0.021	2.72	133	83
	MA62	0.28	0.51	0.088	7.30	5	130
LOCK3491	MA27	0.89	2.06	0.040	20.21	336	254
	MA62	0.25	1.70	0.165	67.33	47	518
RAMAGE01	MA27	1.25	4.18	0.028	94.00	549	345
	MA62	0.11	2.17	0.069	116.20	139	414
AEAC5801	MA27	1.31	3.20	0.067	44.37	526	430
	MA62	0.31	2.21	0.229	69.54	24	626
TRDHEIM	MA27	10.55	18.34	0.250	211.03	2893	2002
	MA62	0.56	11.86	0.815	491.55	121	3601
TSYL201	MA27	13.01	101.08	0.423	4285.00	8922	7069
	MA62	0.71	80.28	1.306	5532.58	292	10964
OPT1	MA27	12.30	90.12	0.386	3648.86	7741	5975
	MA62	0.77	73.60	1.139	5657.32	966	8936
CRPLAT2	MA27	5.34	44.31	0.291	1623.76	4554	3815
	MA62	1.23	42.20	1.304	2596.97	290	7521
BCSPWR10	MA27	0.55	0.39	0.064	0.27	56	56
	MA62	1.19	1.27	0.764	14.80	4	411
BCSSTK15	MA27	2.67	7.59	0.068	219.38	951	788
	MA62	2.89	8.27	0.678	441.28	144	1780
BCSSTK18	MA27	4.97	7.30	0.157	142.63	890	797
	MA62	4.86	25.37	1.939	1714.71	236	5709

Table 6.1. A comparison of MA62 and MA27 on symmetric positive-definite systems

7 A comparison of the frontal code MA42 with other HSL codes

In this section, we examine the performance of the frontal code MA42 for unsymmetric problems. In this case, we have several alternatives in the Harwell Subroutine Library.

For unassembled problems, we compare directly with an in-core multifrontal code for unsymmetric finite-element input MA46 [2] and first assemble the matrices to perform runs with the multifrontal code MA41 [1] and the general sparse code **MA48** [9, 10]. For the unsymmetric assembled problems, the Harwell Subroutine Library does not contain the equivalent of a profile minimizer, so the performance of **MA42** will depend on the original ordering. We note that the *Solve* times given in Tables 7.1 and 7.2 do not include the time required to perform iterative refinement, although iterative refinement is sometimes needed for accurate solutions. It is again clear that it is better to use a code that accepts entry by elements for the unassembled problems and that, for assembled problems, the relative performance depends on the particular problem.

Identifier	Code	Time (seconds)			Factor ops $(*10^6)$	Storage (Kwords)	
		Analyse	Factorize	Solve		Total	Factors
BP1600	MA42	0.06	0.44	0.077	18.70	44	341
	MA41	0.27	0.13	0.012	0.72	71	35
	MA48	0.12	0.04	0.006	0.02	29	12
JPWH 991	MA42	0.07	0.62	0.116	21.24	164	372
	MA41	0.23	0.32	0.015	4.68	146	77
	MA48	0.69	0.34	0.006	7.87	142	129
PORES 3	MA42	0.04	0.26	0.046	5.05	11	106
	MA41	0.05	0.07	0.007	0.16	35	16
	MA48	0.19	0.09	0.003	0.98	26	18
GRE 1107	MA42	0.08	0.97	0.158	84.58	75	813
	MA41	0.30	0.74	0.025	20.47	271	192
	MA48	0.76	0.41	0.007	6.50	144	131
NNC1374	MA42	0.10	0.59	0.069	8.18	5	201
	MA41	0.20	0.66	0.027	9.19	96	154
	MA48	0.96	0.49	0.010	5.01	155	135
ORSREG 1	MA42	0.15	8.76	0.469	1265.66	390	4419
	MA41	0.70	0.88	0.039	14.44	339	205
	MA48	3.45	2.07	0.018	60.02	597	564
LNS 3937	MA42	0.27	29.52	1.275	9511.09	2284	16523
	MA41	0.94	1.79	0.078	32.65	486	415
	MA48	12.65	6.03	0.041	194.75	1287	1231
LNSP3937	MA42	0.27	2.29	0.398	75.76	11	1288
	MA41	0.99	1.81	0.078	32.64	486	414
	MA48	10.06	4.53	0.037	97.46	981	926
PSMIGR 3	MA42	1.49	166.18	0.597	18385.63	7783	11346
	MA41	19.89	148.27	0.164	8947.92	21415	6389
	MA48	38.89	134.81	0.175	10512.12	13924	12831
SHERMAN3	MA42	0.33	5.73	0.945	601.34	298	5596
	MA41	1.62	1.23	0.067	18.03	479	276
	MA48	5.25	2.34	0.029	52.29	630	590
WEST2021	MA42	0.13	1.21	0.164	70.21	54	774
	MA41	0.34	0.23	0.025	0.35	108	47
	MA48	0.31	0.11	0.011	0.05	44	25

Table 7.1. A comparison of codes on unsymmetric assembled problems.

Identifier	Code	Time (seconds)			Factor ops ($*10^6$)	Storage (Kwords)	
		Analyse	Factorize	Solve		Total	Factors
CEGB3306	MA42	0.19	1.60	0.123	23.29	6	450
	MA46	0.56	0.58	0.127	2.29	235	172
	MA41	0.57	0.77	0.075	4.96	416	176
	MA48	2.70	1.22	0.020	6.66	450	294
CEGB2919	MA42	0.10	4.39	0.100	183.28	85	1087
	MA46	2.04	2.81	0.132	56.36	813	775
	MA41	2.28	4.46	0.087	96.38	1582	720
	MA48	16.10	10.39	0.043	237.81	2463	1814
CEGB3024	MA42	0.19	1.95	0.145	39.31	18	593
	MA46	0.68	0.82	0.165	5.23	316	269
	MA41	0.69	1.18	0.068	11.96	532	268
	MA48	5.00	2.47	0.023	19.41	754	539
LOCK1074	MA42	0.14	0.68	0.040	12.83	16	188
	MA46	0.48	0.44	0.045	4.29	167	138
	MA41	0.37	0.67	0.030	0.88	291	136
	MA48	2.02	1.22	0.010	15.28	397	292
LOCK2232	MA42	0.27	0.89	0.081	10.97	5	261
	MA46	0.67	0.55	0.089	2.72	234	173
	MA41	0.58	0.87	0.054	7.59	418	189
	MA48	2.68	1.48	0.016	7.79	487	322
LOCK3491	MA42	0.24	3.40	0.155	120.57	47	1039
	MA46	1.25	1.58	0.158	19.03	572	503
	MA41	1.18	2.41	0.107	38.70	991	482
	MA48	9.88	5.24	0.035	60.23	1359	1031
RAMAGE01	MA42	0.10	4.35	0.066	220.78	140	829
	MA46	1.67	3.59	0.093	93.45	716	690
	MA41	1.53	5.05	0.056	186.41	1515	684
	MA48	9.41	10.01	0.036	438.52	2227	1819
AEAC5081	MA42	0.31	4.61	0.215	122.26	25	1257
	MA46	1.60	3.15	0.297	50.05	1002	919
	MA41	1.58	3.83	0.143	71.35	1430	498
	MA48	19.69	10.63	0.060	187.39	2327	1922
TRDHEIM	MA42	0.55	22.19	0.750	883.72	121	7221
	MA46	10.47	11.86	0.884	178.72	4119	3847
	MA41	13.77	21.74	0.689	344.26	8497	3734
	MA48	119.69	49.73	0.241	492.39	11613	7700
TSYL201	MA42	0.71	171.71	1.230	10752.72	295	21955
	MA46	15.71	113.74	1.380	3630.16	13587	13317
	MA41	17.91	124.03	0.848	7160.12	22297	13119
	MA48	672.04	424.48	0.651	15115.17	37623	32673
OPT1	MA42	0.78	154.47	0.996	11098.17	966	17891
	MA46	14.49	95.15	1.194	3098.87	11272	11059
	MA41	14.72	111.86	0.663	6549.02	18961	11079
	MA48	454.16	322.14	0.474	12508.19	29556	25665
CRPLAT2	MA42	1.19	82.72	1.180	4995.81	292	15056
	MA46	7.15	30.50	0.983	845.03	6355	6018
	MA41	7.32	41.15	0.682	1806.52	9821	6130
	MA48	256.24	153.21	0.344	5123.96	18131	16175

Table 7.2. A comparison of codes on unsymmetric element problems.

8 Conclusions

We have shown that frontal codes can be a very powerful approach for the solution of large sparse systems and are particularly efficient for unassembled finite-element problems when a good ordering can be found. We notice that, in this case, although other approaches may result in much less fill-in, the frontal code is often better in terms of the analyse and factorize times and, if the factors are held in direct access files, is far superior in terms of main memory. For most assembled problems, the use of other approaches might be better and frontal methods can perform badly.

We would expect some differences if runs were made on other computing platforms. However, our experience in this respect leads us to believe that the relative performance would be similar for any machine supporting an efficient version of the Level 3 BLAS.

References

1. Patrick R. Amestoy and Iain S. Duff. Vectorization of a multiprocessor multifrontal code. *Int. J. of Supercomputer Applics.*, 3:41–59, 1989.
2. A. C. Damhaug and J. K. Reid. MA46, a Fortran code for direct solution of sparse unsymmetric linear systems of equations from finite-element applications. Technical Report RAL-TR-96-10, Rutherford Appleton Laboratory, 1996.
3. Jack J. Dongarra, Jeremy Du Croz, Iain S. Duff, and Sven Hammarling. A set of Level 3 Basic Linear Algebra Subprograms. *ACM Trans. Math. Softw.*, 16:1–17, 1990.
4. Iain S. Duff. Design features of a frontal code for solving sparse unsymmetric linear systems out-of-core. *SIAM J. Scientific and Statistical Computing*, 5:270–280, 1984.
5. Iain S. Duff, Roger G. Grimes, and John G. Lewis. Sparse matrix test problems. *ACM Trans. Math. Softw.*, 15(1):1–14, March 1989.
6. Iain S. Duff, Roger G. Grimes, and John G. Lewis. Users' guide for the Harwell-Boeing sparse matrix collection (Release I). Technical Report RAL 92-086, Rutherford Appleton Laboratory, 1992.
7. Iain S. Duff and John K. Reid. MA27 – A set of Fortran subroutines for solving sparse symmetric sets of linear equations. Technical Report AERE R10533, Her Majesty's Stationery Office, London, 1982.
8. Iain S. Duff and John K. Reid. The multifrontal solution of indefinite sparse symmetric linear systems. *ACM Trans. Math. Softw.*, 9:302–325, 1983.
9. Iain S. Duff and John K. Reid. MA48, a Fortran code for direct solution of sparse unsymmetric linear systems of equations. Technical Report RAL 93-072, Rutherford Appleton Laboratory, 1993.
10. Iain S. Duff and John K. Reid. The design of MA48, a code for the direct solution of sparse unsymmetric linear systems of equations. *ACM Trans. Math. Softw.*, 22(2):187–226, 1996.
11. Iain S. Duff, John K. Reid, and Jennifer A. Scott. The use of profile reduction algorithms with a frontal code. *Int J. Numerical Methods in Engineering*, 28:2555–2568, 1989.

12. Iain S. Duff and Jennifer A. Scott. MA42 – a new frontal code for solving sparse unsymmetric systems. Technical Report RAL 93-064, Rutherford Appleton Laboratory, 1993.

13. Iain S. Duff and Jennifer A. Scott. The use of multiple fronts in Gaussian elimination. In J. G. Lewis, editor, *Proceedings of the Fifth SIAM Conference on Applied Linear Algebra*, pages 567–571, Philadelphia, 1994. SIAM Press.

14. Iain S. Duff and Jennifer A. Scott. The design of a new frontal code for solving sparse unsymmetric systems. *ACM Trans. Math. Softw.*, 22(1):30–45, 1996.

15. Iain S. Duff and Jennifer A. Scott. MA62 – a frontal code for sparse positive-definite symmetric systems from finite-element applications. Technical Report To appear, Rutherford Appleton Laboratory, 1996.

16. P. Hood. Frontal solution program for unsymmetric matrices. *Int J. Numerical Methods in Engineering*, 10:379–400, 1976.

17. HSL. *Harwell Subroutine Library. A Catalogue of Subroutines (Release 12)*. AEA Technology, Harwell Laboratory, Oxfordshire, England, 1996. For information concerning HSL contact: Dr Scott Roberts, AEA Technology, 552 Harwell, Didcot, Oxon OX11 0RA, England (tel: +44-1235-434714, fax: +44-1235-434136, email: Scott.Roberts@aeat.co.uk).

18. B. M. Irons. A frontal solution program for finite-element analysis. *Int J. Numerical Methods in Engineering*, 2:5–32, 1970.

19. J. A. Scott. Element resequencing for use with a multiple front algorithm. Technical Report RAL-TR-95-029, Rutherford Appleton Laboratory, 1995. To appear in *Int J. Numerical Methods in Engineering*.

Parallel Heuristics for Bandwidth Reduction of Sparse Matrices with IBM SP2 and Cray T3D

A. Esposito, L. Tarricone

Ist.di Elettronica, Via G. Duranti, 1/A-1, 06131, Perugia, Italy

Abstract

The solution of a sparse linear system of equations is the core of the problem in many mathematical models. In this paper a strategy is proposed to reduce the bandwidth of the system matrix, so that direct banded solvers can be used with high efficiency. A combinatorial optimization method is implemented on an IBM SP2 (8 nodes) and on a Cray T3D (64 nodes) to perform a global bandwidth minimization, superior to previous constructive approaches. Examples of applications of the strategy in microwave circuit design demonstrate the efficiency of the method and its parallel implementation, as well as its versatility.

1 Introduction

The use of numerical techniques is perhaps the most common way to approach complex scientific problems with a high degree of accuracy. The solution of a linear system of equations

$$Ax=B \qquad (1)$$

which is solved many times, with different right-hand-sides **B,** is often the computational core of numerical methods. In many applications, the system matrix **A** is sparse and its pattern generally depends on the problem domain decomposition, or on the numbering chosen for the problem unknowns. In some cases, as already demonstrated [1], the most robust and efficient strategy to solve the system couples an appropriate numbering of unknowns, reducing the bandwidth of **A,** with a banded direct factorize-and-solve algorithm. In fact, even though very efficient direct sparse solvers are available on the market, they are often prone to the fill-in problem, depending on the matrix pattern [2]. This problem is severe, expecially when the "bandwidth reducer" should be a part of an automatic package, with strong requirements of flexibility.

In this paper, the problem of bandwidth reduction is addressed, and solved with a new approach, based on combinatorial optimization. The problem being NP-hard [3], parallel platforms are used to achieve good computing times. Comparisons with previous approaches based on graph theory are given, demonstrating the effectiveness and versatility of the method, and results demonstrate the practical advantages in real industrial applications.

2 Sparse Matrix Bandwidth Reduction

In the literature, several constructive heuristics have been proposed for sparse matrices with symmetric zero-non-zero structure. One of the most effective methods is the modified reverse Cuthill-MacKee (MRCM) [4], based on graph theory, with some recent enhanced versions [3]. This method is not completely suitable for every problem. First of all, it is suited to matrices with a symmetric zero-non-zero structure. Some tricks could be used to deal with non-symmetric matrices using MRCM, but they severely affect the matrix condition number. Moreover, on some matrix patterns, MRCM is not effective enough. In the following we demonstrate this. Finally, the bandwidth reduction performed by MRCM is dependent on the initial row/column numbering. In fact, MRCM performs a suboptimal numbering optimization, and the existence of local optima causes this sensitivity to the starting point in the search of the best row/column permutation.

Because of the above mentioned drawbacks, a need arises, in some applications, for a global optimization of matrix permutation, achieving the minimum bandwidth and working also on matrices with a non-symmetric zero-non-zero pattern.

3 TS Bandwidth Reduction

TS is a combinatorial optimization method performing global optimization through a fast search in the whole search space. Using heuristics and intelligent problem solving, it dynamically inhibits some partitions of the search space (called "*tabu*"), thus enhancing the search speed. The search is iterative, and at each iteration a set of *neighbours* for the current solution is defined. With a chosen cost function, a score is computed for each neighbour, and the best one is selected as the solution for the next iteration. As cost function, a weighted sum of bandwidth and profile is chosen. A complete description of TS can be found in [5].

3.1 Parallel TS Implementation : Platforms

As TS is computationally costful, a parallel implementation is mandatory. It was implemented on the IBM SP2 at Perugia University, Italy, and on the Cineca CRAY T3D in Bologna, Italy. The SP2 in Perugia is an 8 thin-node configuration, each node comprising a POWER2 processor with 66.7 MHz clock speed, 266 MFLOPS peak performance and 64 kB Data Cache. The interprocessor communication is supported by an apposite hardware, the so called high performance switch which has a 40 MB/sec peak bi-directional bandwidth and a latency of about 40 microsecs.

The Cineca T3D contains 64 processing elements (PEs) grouped into 32 nodes communicating via a 3-D torus interconnection network. Each PE is a DEC Alpha chip 21064 with 150 MFLOPS peak performance and 150 MHz clock speed and 8KB data cache. The peak bandwidth is 300 MB/sec and latency is about 1 microsec. The memory is physically distributed among the processors but it is globally addressable.

3.2 Parallel TS Implementation : Programming Paradigms

On the SP2 we experienced both an host-node programming paradigm, and a Single Program Multiple Data (SPMD) approach (using standard PVM [6] and PVMe, PVM specifically tuned for IBM platforms).

On the T3D the programming paradigm was an SPMD shared memory one. The communication among processes was implemented with the CRAY Shared Memory Access library for CRAY T3D version [7]. This implementation, using the logical shared memory mapping feature, completely outperformed on the T3D the standard PVM version.

On both platforms, there are several approaches to the parallelization of the TS algorithm. One possible pathway consists in running several searches in parallel and periodically compare the results. The search is then continued by all the processes from the best solution found so far. Each search can differ from another for some tuning parameters. Another common technique consists in distributing demanding partitions of the problem among the processes and merging the results of their computation.

We experienced both strategies: in the first strategy (we call PS since now on) different search paths are followed in parallel by different processes. In the second strategy (we call PN since now on) the neighbours computation and their cost evaluation is distributed among the available processors. Both implementations follow an SPMD philosophy. From several benchmarks, it has been observed that PN is more effective and scalable, and ensures a good load-balancing. Fig. 1 shows computational times (elapsed) to perform a 500-iteration search on 500x500 and 1000x1000 matrices. In Fig. 2, a comparison is proposed between two different implementations on the T3D: the shared-memory one and the PVM message-passing one. It is easily observed that the former outperforms the latter, this justifying the choice for the shared-memory programming paradigm. In this discussion, we just refer to results attained on the T3D, as the number of processors available on the SP2 is not suited to a satisfactory scalability analysis. As for the IBM SP2, a comparison has been made between PVM and PVMe implementations. A 15% improvement on communication times has been observed.

Fig. 1. A graph describing the scalability of the implemented TS bandwidth minimizer on the Cray T3D. Curves refer to different sizes of the matrix.

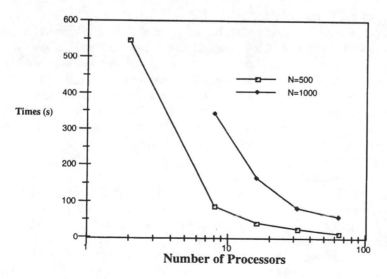

Fig. 2. A comparison on the T3D between the PVM message-passing TS implementation and the shared memory one.

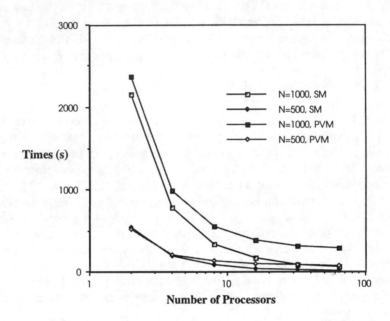

4 Results on real Applications

Up to now, the proposed TS solution has been applied to different numerical methods for the analysis and design of electromagnetic circuits, as well as to other problems (computational chemistry, VLSI implementation of circuits and neural networks). In the Computer Aided Design of electromagnetic microwave circuits, the use of a parallel TS, in cooperation with entry level workstations, has allowed a decrease in the designing time of up to one order of magnitude, with CPU costs still quite affordable. This has been experienced using two different numerical methods, which are both enhanced by using a TS strategy on a linear system of equations.

4.1 Mode-Matching Analysis of EM Circuits

The former numerical technique, called Mode-Matching (MM), can experience great speed-ups, using a renumbering strategy in appropriate manners, as described in [1]. In Tab. I, timings for the solution of the linear system in a MM simulation of a complex circuit, without and with TS renumbering of unknowns, as well as timing for renumbering, are shown. Consider that in MM, the best permutation to achieve bandwidth minimization is computed just once, and the linear system is solved several times, with different rigth-hand sides. Therefore, a cooperative environment is used, where the parallel platform computes the optimum permutation, while the whole analysis of the circuits (with many linear system solutions) is performed on an entry level workstation, with low CPU costs.

Numbering	Solution Time (s, IBM 250 T)	TS time (s) on SP2 (8 procs)	TS time (s) on T3D (64 proc)
Heuristical Choice	950	-	-
Optimized with TS	46	19.5	4.7

Table I : Timings for a MM analysis of a complex circuit without and with TS renumbering strategy on matrix rows/columns

It can be observed that the analysis takes (46+4.7)s using a cooperative environment composed of a serial RISC and a parallel computer Cray T3D, and (46+19.5)s using the SP2 to perform renumbering. The row/column numbering optimization is performed on the parallel computer, while the whole analysis is performed on the serial workstation. It can be observed that the renumbering can reduce simulation times of about 20 times. For industrial applications of tuning and trimming of MW devices, where a circuit can easily be simulated thousands of times, this means saving several days.

Moreover, it should be pointed out that on the same matrix of Tab. I MRCM was less satisfactory than TS, as its capability of reducing the bandwidth was strongly influenced by the initial choice for the numbering, this being quite dangerous for an automatic tool. Moreover, on some cases MM matrices have a non-symmetric structure, and this severely limits MRCM effectiveness on this application.

4.2 EM Circuit Analysis with Finite Element Methods (FEM)

The users of FEM are traditionally familiar with bandwidth reduction methods [8]. MRCM and similar approaches have been used since their appearance. Many FEM packages already offer some "bandwidth reducers" tuned for the typical pattern of matrices emerging from FEM analyses, and a benchmark of the TS performance on these matrices is therefore quite interesting.

We discuss here the FEM analysis of a class of electromagnetic circuits, composed of rectangular waveguides inhomogeneously filled with dielectric. A public-domain FEM code, called EMAP1 [9], based on a variational scalar formulation, has been used. It has been interfaced with our TS module performing a renumbering of the unknowns of the linear system (1) representing the numerical core of the FEM simulation. Even in this case, the permutation matrix $\mathbf{P}$, so that $\mathbf{PAP}^T$ has minimum bandwidth, is computed only once, while the linear system is generally solved many times with different right-hand sides .

We present in Tab. II results referred to a specific circuit. The standard solution in EMAP1 generates a system matrix $\mathbf{A}$ whose size is 639 and bandwidth is 151. On this matrix, MRCM is uneffective, as its application results in a final bandwidth wider than the initial one (this behaviour has been observed on several different cases, and with different implementations, both developed by the authors and included in commercial or freeware packages). Two TS optimization levels are discussed, a so-called "intermediate" optimization (IO) and a "fine" optimization (FO), differing basically in the number of iterations performed and on the stop criteria. IO is performed looking for a good balancing between bandwidth reduction and computational effort, with loose convergence criteria and appropriate choices for tuning parameters. FO aims at finding out the global optimum whatever the computational cost be.

Numbering	Matrix Bandwidth	Solution Time (s) (100 Freq. Points)	TS Time (s) T3D (64 proc)	TS Time (s) SP2 (8 proc)
EMAP1	151	1370	-	-
Optimized (FO)	102	234	220	-
Optimized (IO)	128	715	28	105

Table II : Application of TS strategy to a FEM analysis of a simple circuit.

Results show that TS is able to optimize the unknown numbering so that bandwidth is reduced up to 128 with IO, and 102 with FO. Only times attained with parallel TS implementation are shown, as serial times are too long to be advantageous (this holds for the previous MM case as well). Solution times for banded EMAP1 solver on an IBM 250 T are shown. The possibility of selecting different optimization levels looks quite important for FEM problems. It can be observed that computational costs are rather high, and the tunability of optimization requirements makes the environment really cost-effective. Moreover, it is interesting to look at Fig. 3, showing the value of cost function (proportional to **A** matrix bandwidth) during the FO run. It can be observed the importance of a global optimization strategy, able to come out from local minima, as demonstrated by the non-monothonic shape of the cost function in Fig. 3. The starting point was a local minimum, and in a few iterations this has been understood by TS, which has diversified its search in a correct direction to reach the global optimum.

In conclusion, as for MM, a cooperative environment is used, with a parallel platform computing only once the optimum permutation of unknowns, and a serial RISC workstation receiving this information and accomplishing the FEM analysis. This, as before, keeps CPU costs affordable, and guarantees a reduction in computing times of more than 45% using an IO, and about 66% with FO. Attention should be paid when using FO, as its computational costs are high, and in some cases the time enhancements are not worth. Anyway, the presented results demonstrate the usefulness of the TS approach for FEM problems, and its superior performance to previous methods.

Fig. 3. Values for the TS cost function during the search whose timings are given in Tab. II. It is easily observed that the starting numbering is a local minimum. TS is able to understand it in a few iterations and diversifies its search looking for a global optimum.

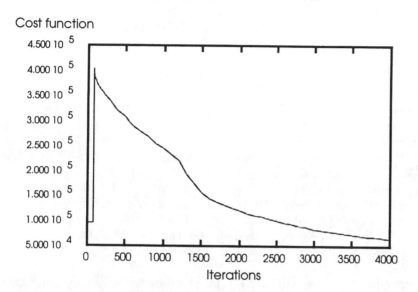

5 Conclusions and Future Developments

In this work a parallel solution to the bandwidth reduction of sparse matrices is suggested. An implementation on Cray T3D and IBM SP2 of a metaheuristic strategy based on Tabu Search is experienced. Results on real industrial applications in the CAD of electromagnetic circuits demonstrate the efficiency of the method, which is currently extended to many scientific problems, such as VLSI implementation of neural networks for image processing and design of high power electric networks.

New strategies are also investigated, based on stochastic iterative constructive approaches [10] as well as on efficient domain-decomposition [11] or graph partitioning [12].

Acknowledgment

This work has been partially sponsored by CINECA, Bologna, Italy, which granted its parallel platform Cray T3D. Authors are also grateful to CINECA professionals for their kind cooperation.

References

[1] L. Tarricone, M. Dionigi, R. Sorrentino, A Strategy for the Efficient Mode-Matching Analysis of Complex Networks, Int. Journ. On MW and MM Wave C.A.E., 6, 3, (1996), pp. 183-196.

[2] I. S. Duff, A. M. Erisman, J. K. Reid, *Direct Methods for Sparse Matrices*, Oxford Univ. Press, 1986.

[3] D. Kuo, G. J. Chang, 'The Profile Minimization Problem in Trees', SIAM J. Comput., **23**, 1, (1994) 71-81.

[4] N. E. Gibbs, W. G. Poole, P. K. Stockmeier, "An algorithm for reducing the bandwidth and profile of a sparse matrix", SIAM J. of Numer. Anal., Vol. 13, 2, (1976) 236-250.

[5] F. Glover, M. Laguna, 'Tabu Search', in Modern Heuristic Techniques for Combinatorial Problems, Ed. C. Reeves, 70-150, Blackwell Sc. Publ., Oxford, 1993.

[6] A. Beguelin, J. Dongarra, A. Geist, R. Manchek, V. Sunderam, *PVM User's Guide*, Oak Ridge Nat. Lab., May 1994.

[7] R. Barriuso, A. Knies, *Cray SHMEM User Library*, Cray Research, 1994.

[8] P. P. Silvester, R. L. Ferrari, *Finite Elements for Electrical Engineers*, Cambridge: Cambridge Univ. Press, 1990.

[9] T. H. Hubing, M. W. Ali, G. K. Bhat, "EMAP: a 3-D finite element Modeling Code", *J. Appl. Comp. Electromagnetics Soc*, vol. 8, 1, 1993.

[10] A. Esposito, S. Fiorenzo Catalano, F. Malucelli, L. Tarricone , "Parallel heuristics for matrix bandwidth reduction problems emerging in waveguide network simulations", AIRO 96, 17-20 Sept., Perugia, 1996

[11] A. I. Zecevic, D.D. Sijiak, "Balanced Decomposition of Sparse Systems for Multilevel Parallel Processing", IEEE Trans on Circ. and Syst., 41,3, (1994), 220-233.

[12] G. Karypis, V. Kumar, METIS : Unstructured Graph Partitioning and Sparse Matrix Ordering System, Univ. of Minnesota, 1995

Parallel Libraries on Distributed Memory Architectures: The IBM Parallel ESSL

Salvatore Filippone *

IBM SEMEA
P.le Giulio Pastore 6,
I-00144, Roma, Italy

1 Introduction

The use of libraries in software development activities is related to structured design of applications, to obtain software with better reliability, better performance, improved maintainability and lower costs. All these factors are most relevant when considering parallel computing machines with distributed memory: the use of libraries can effectively and dramatically improve the efficiency of usage of such advanced computational engines.

In this paper we will review some design principles and usability features, with specific reference to the IBM Parallel Engineering and Scientific Subroutine Library (PESSL) [6], a library specifically designed for IBM RISC System/6000 Scalable POWERParallel Systems (SP).

We begin by reviewing the salient features of the the underlying hardware and system software; then we discuss some design principles that guided the development of the most recent version of this library; we will then describe the interfaces currently available, and we conclude with a brief discussion performance issues and usage hints.

The IBM Parallel ESSL has been developed by the IBM RISC System/6000 Divsion, Poughkeepsie, N.Y., the T.J. Watson Research Center, Yorktown Heights, N.Y., and the European Center for Scientific and Engineering Computing, Rome, Italy. The linear algebra part of Parallel ESSL is based on the ScaLAPACK [2] public domain software, jointly developed by the University of Tennessee at Knoxville, the Oak Ridge National Laboratory and the University of California, Berkeley.

2 IBM SP Hardware and Software

The SP series is a line of parallel machines based on the processor technology of the IBM RISC System/6000 [7]. The current product is the SP2 machine; it consist of a collection of up to 128 computing nodes based on the POWER2 processor architecture.

Processing nodes for the SP2 are available in three models: the "Wide" nodes, the "Thin" nodes and the "Thin-2" nodes; salient machine characteristics are summarized in table 1.

---

* filippon@vnet.ibm.com

	Wide	Thin	Thin-2
Clock	66.7 or 77 Mhz	66.7 Mhz	66.7 Mhz
Peak MFLOPS	266	266	266
Memory bus	256	64	128
Data cache	256 KB	64 KB	128 KB Level 1
			0-2 MB Level 2

Table 1. SP2 Node characteristics

Each POWER2 CPU has dual fixed-point units, dual floating point units and an instruction/branch unit, all capable of operating simultaneously; since each floating-point unit supports multiply-and-add operations the processor is capable of delivering 4 floating-point operations per clock cycle.

Nodes can communicate by means of different network interconnections; the main ones are:

- An internal Ethernet network, used for system purposes;
- A fast data communication adapter for the Scalable POWERparallel Switch, intended for parallel application support;
- A network connection to the external world; supported adapters include Ethernet, Token Ring and FDDI;

The Scalable POWERparallel Switch (SP Switch) is the latest result in the development of interconnection network, with a scalable structure supporting a point-to-point peak bidirectional bandwidth of 150 MB/sec, for a peak aggregate bandwidth of 19.2 GB/sec on 128 nodes; at the application level it is possible to obtain more than 100 MB/sec. The architecture of the switch is based on a multi-stage omega-network, and was originally developed at the T.J. Watson Research Center.

The system is managed by a single RISC System/6000 control workstation, with special connections to perform hardware monitoring; the control workstation is in charge of managing the users, accessing and mounting external file systems, and handling software and hardware configuration files. Each machine node runs a full AIX operating system.

The parallel communication interfaces include the AIX implementation of the Message Passing Interface (MPI [8]), upon which Parallel ESSL 1.2 is based; it provides basic point to point and collective communication functions. The communication interface can run in IP mode over both the SP Switch and the Ethernet, and in dedicated User Space mode on the SP Switch for best performance. The parallel environment includes a parallel debugger and a trace driven Program Visualization tool.

The programmability of the SP2 machine has been greatly enhanced by the introduction of the IBM XL High Performance Fortran (HPF) compiler [5], allowing users to write programs in a high-level language with directives, shielding them from the details of explicit message passing programming.

3 Parallel ESSL design features

The Parallel ESSL is a mathematical library specifically tuned for the SP machines; it allows programmers to access the computational power of the underlying hardware, with little development effort on the user side, scaling in performance from a small number of processors up to the full machine configuration.

The Parallel ESSL library has been designed to fully exploit the computational capabilities of the hardware, while maintaining usability and compatibility with existing standards. It is also designed to support existing standards in parallel computations, such as the BLACS and ScaLAPACK, while adding new features of usability and performance tuning.

One of the most important design features of a parallel library is to provide optimal usage of the individual computational nodes of the parallel machines, in the serial parts of the computation. To this end Parallel ESSL library is based on the serial ESSL/6000 library for local node computations; this ensures full exploitation of the computational capabilities of the POWER2 processing nodes, by using such techniques as cache managements, efficient memory usage through data blocking and contiguous storage accesses, and minimized paging. Detailed information on tuning codes for the POWER2 processor may be found in [7].

The latest relase of Parallel ESSL also includes a major step in usability by providing a full interface for use with HPF.

3.1 Environment and communication interface – Message passing

The user-level communication provided by PESSL is conformant to the BLACS standard [4], implemented on top of the AIX Parallel Environment Message Passing Interface.

The BLACS interface provides primitives for setting up 2-dimensional logical process grids; therefore they provide a convenient way for the user to organize the parallel application structure. The user has the freedom of defining different logical grids and using them at the same time over the same physical processor set. Communication is also handled via BLACS calls, including both point-to-point and collective operations; the BLACS used in Parallel ESSL are implemented on top of the MPI interface.

3.2 Usability features

The introduction of an HPF interface in PESSL 1.2 marks a major step forward in usability; PESSL computational kernels are accessed as EXTRINSIC subroutines. The interface is designed so as to combine usage of such Fortran 90 features as overloading with respect to the data types, assumed shape arrays and optional arguments to achieve maximal readability; it is also compatible with the current proposal for the BLAS and LAPACK Fortran 90 interface [3].

All PESSL routines, whether HPF or message passing, provide extensive parameter checking and error reporting; parameters are checked both for legality and for consistency across different processes. As an extension over public domain ScaLAPACK, whenever a subroutine requires a user-provided work area, the corresponding PESSL subroutine is also able to supply its own memory allocation. Finally the extensive documentation provided, including performance tips and sample programs, enhances the usability of the library.

4 Computational areas

Computational areas supported in the Parallel ESSL include linear algebra, Fourier and analysis random number generation. The Linear Algebra part of Parallel ESSL is compatible with the most established public domain library, ScaLAPACK [2]; therefore application programs using ScaLAPACK routines available in PESSL need only be linked with the new library. The Linear Algebra HPF interface has been developed to conform to the proposed standard for the Fortran 90 interface to LAPACK [3].

Parallel BLAS Parallel BLAS of level 2 and 3 from ScaLAPACK are supported; in the HPF interface multiple level 2 and 3 BLAS have been coalesced into single, simplified calling sequences. As an example, the HPF calling sequence for GEMM provides all the functionalities of the message-passing subroutines PDGEMM, PDGEMV and PDGER.

Linear Algebraic Equations The Linear Algebraic Equations section contains solvers for general, symmetric or hermitian positive definite, banded symmetric positive definite and tridiagonal matrices.

In the HPF interface the calling sequences for real and complex matrices are unified in a single high-level calling sequence, e.g. the HPF GETRF corresponds to both PDGETRF and PZGETRF message passing subroutines, according to the data type of the actual arguments.

Eigensystems and Singular Value Analysis Currently supported eigensystems routines include a complete solver for the symmetric eigenvalue problem, and reduction routines for general matrices; the HPF interface is the major addition to this section in the current product release.

Fourier Transforms Fourier transforms include 2 and 3 dimensional transforms for complex, real-to-complex and complex-to-real data; all subroutines are available in short and long precision.

The new HPF interface maps all the subroutines to the same calling sequence FFT; the interface will choose the appropriate specific version according to the data types and array shapes of the actual arguments.

Random Number Generation Parallel ESSL provides a uniform pseudo-random number generator based on the multiplicative congruential method:

$$s_i = (a(s_{i-1}) \bmod (m) = (a^i s_0) \bmod (m)$$

with s_0 the user-specified initial seed, and with parameters

$$a = 44485709377909.0$$
$$m = 2^{48}$$

5 Performance tuning considerations

Tuning a parallel application for best performance requires some action be taken by the user on a number of parameters.

The first and foremost tuning issue facing the user is the choice of the appropriate parallel machine size: given a problem size there will be a definite number of processors guaranteeing the best trade-off between raw performance and utilization. Adding more processors may even cause a slow-down, especially for small problems; as an extreme example consider solving a 100×100 linear system on 8 processors.

Taking as a reference linear algebra computations there are two main choices to be made when running the application:

1. Choice of the processor grid configuration;
2. Choice of the matrix distribution parameters.

As a general rule processor grids as close to square as possible are better suited to linear algebra computations. A notable exception to this rule is the *LU* factorization code in PDGETRF and PZGETRF; in this case partial pivoting imposes additional requirements onto the communication network, and therefore a grid $P \times Q$ with $P < Q$ is to be preferred.

The blocking size used to distribute the matrices also influences performance; many factors are involved in the optimal choice, including:

- Cache utilization;
- Communication and synchronization costs;
- Load balancing.

The Parallel ESSL Guide and Reference provides some advice on these factors, with recommended values providing consistent performance across a wide range of cases; however no general rule can substitue experimental data if the objective is to achieve the *absolute best* choice of parameters for the specific problem instance at hand.

6 HPF Performance considerations

The HPF language provides a great advance in usability for the library functions by freeing the user from such concerns as explicit details of data placement. The HPF version of the library is provided as a set of interfaces that ultimately call the message passing subroutines; therefore all the considerations about the message passing interface also apply to the HPF interface.

The HPF language adds some overhead because of its bookkeeping activities, and thus it has some impact on the overall performance; as a general rule the HPF version of a given subroutine will be slower than the corresponding message passing version. The extent of the overhead is dependent on the specific subroutine, and is also changing with different releases of the HPF compiler.

7 An Example: Linear Algebraic Equations in HPF

An an example of usage we show a skeleton HPF code to solve a linear system given two functions to generate "on the fly" the coefficient matrix and the right hand side matrix; the example shows how to align arrays with each other, and makes use of user provided "pure" functions to generate matrix coefficients. More examples can be found in the PESSL manual [6].

```
PROGRAM GEHPF
  USE PESSL_HPF
  REAL(KIND(1.D0)), PARAMETER :: ZERO=0.D0, ONE=1.D0
  INTEGER, PARAMETER  :: NPROWS=2, NPCOLS=2, NB=60, OUTUNIT=8
  REAL(KIND(1.D0)), ALLOCATABLE :: A(:,:), B(:,:)
  INTEGER, ALLOCATABLE ::        IPIV(:)
!HPF$ PROCESSORS PROCS(NPROWS,NPCOLS)
!HPF$ ALIGN B(:,:) WITH A(:,:)
!HPF$ ALIGN IPIV(:) WITH A(:,*)
!HPF$ DISTRIBUTE (CYCLIC(NB),CYCLIC(NB)) ONTO PROCS :: A
INTERFACE
    FUNCTION MATEL(I,J,N)
      !HPF$ PURE MATEL
      REALD(KIND(1.D0)) MATEL
      INTEGER I,J,N
    END FUNCTION MATEL
    FUNCTION RHSEL(I,J,N,NRHS)
      !HPF$ PURE RHSEL
      REALD(KIND(1.D0)) RHSEL
      INTEGER I,J,N,NRHS
    END FUNCTION RHSEL
END INTERFACE
PRINT *,'Linear system sizes ?'
READ (*,*) N, NRHS
ALLOCATE(A(N,N),STAT=IRCODE)
IF (IRCODE.NE.0) STOP'Memory allocation failure '
ALLOCATE(B(N,NRHS),STAT=IRCODE)
```

```
IF (IRCODE.NE.0) STOP'Memory allocation failure '
ALLOCATE(IPIV(N),STAT=IRCODE)
IF (IRCODE.NE.0) STOP'Memory allocation failure '
!       Generate two matrices A and B
FORALL (I=1:N,J=1:N)     A(I,J) = MATEL(I,J,N)
FORALL (I=1:N,J=1:NRHS)  B(I,J) = RHSEL(I,J,N,NRHS)
!       Factor the coefficient matrix
CALL GETRF(A,IPIV,IRET)
!       Solve for the RHS.
IF (IRET.EQ.0) THEN
   CALL GETRS(A,IPIV,B,TRANS)
ELSE
   PRINT *, 'Error during factorization: zero pivot at column',IRET
ENDIF
DEALLOCATE(IPIV)
DEALLOCATE(B)
DEALLOCATE(A)
STOP
END PROGRAM GEHPF
```

8   Performance Data

Obtaining best performance has been one of the main objectives of PESSL development; we show some performance data for selected routines. In figure 1 we show performance in MFLOPS for the solution of a real linear system with the message passing routines PDGETRF and PDGETRS.

Table 2 lists performance data for the PDCFT2 complex two-dimensional Fourier transform; the data were obtained on 66 MHz Wide nodes with the High Performance Switch, the previous version of the communication interconnet, capable of a peak point-po-point bandwidth of 40 MB/sec.

	Matrix order				
Nodes	1008	1024	2048	4096	5040
1	105	128			
2	133	154	202		
4	271	314	335		
8	503	586	650	695	
16	845	911	1003	1085	961

Table 2. Performance in MFLOPS for PDCFT2 with Wide node 66.7 MHz and HPS switch

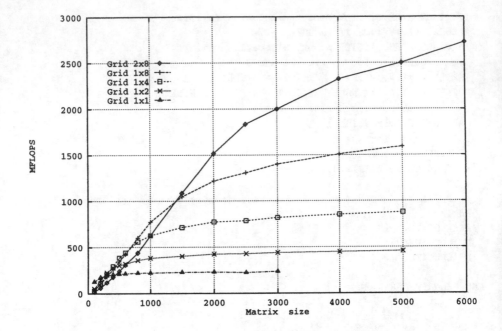

Fig. 1. PDGETRF/PDGETRS performance with Thin-2 nodes and SP Switch

9 Conclusions

We have discussed some issues arising in the design and use of parallel libraries on distributed memory computer architectures; as a reference case we have examined in detail various features of the IBM Parallel Engineering and Scientific Subroutine Library.

References

1. E. Anderson, Z. Bai, C. Bischof, J. Demmel, J. Dongarra, J. Du Croz, A. Greenbaum, S. Hammarling, A. McKenney, S. Ostrouchov, and D. Sorensen. *LAPACK Users' Guide Second Edition.* SIAM Pub., 1995.
2. Choi, J., Demmel, J., Dhillon, J., Dongarra, J., Ostrouchov, S., Petitet, A., Stanley, K., Walker, D. and Whalley, R. C. ScaLAPACK: A Portable Linear Algebra Library for Distributed Memory Computers Tech. Rep. CS-95-283, University of Tennessee, 1995, and LAPACK working note 95, available from http://www.netlib.org/lapack/lawns.
3. J. Dongarra, J. Du Croz, S. Hammarling, J. Wasniewski and A. Zemla A Proposal for a Fortran 90 Interface for LAPACK Tech. Rep. CS-95-303, University of Tennessee, 1995, and LAPACK working note 101, available from http://www.netlib.org/lapack/lawns.

4. Dongarra, J., and Whaley, R. C. A User's Guide to the BLACS Tech. Rep. CS-95-281, University of Tennessee, 1995, and LAPACK working note 94, available from http://www.netlib.org/lapack/lawns.
5. IBM XL HPF Language Reference. IBM Corporation, SC09-2226, 1996.
6. IBM Parallel Engineering and Scientific Subroutine Library Release 2. Guide and Reference. IBM Corporation, GC23-3836, 1996.
7. POWERPC and POWER2: Technical Aspects of the New IBM RISC System/6000 IBM Corporation, SA23-2737-00, 1994.
8. Message Passing Inteface Forum, *MPI: A Message Passing Interface Standard, Version 1.1* , University of Tennessee, June, 1995. Available from http://www/mcs.anl.gov/Projects/mpi/index.html

Parallel Least Squares Estimates
of 2-D SPECT Image Reconstructions
on the SGI Power Challenge

Stefano Foresti[1] and Larry Zeng[2] and Grant Gullberg[2] and Ron Huesman[3]

[1] Utah Supercomputing Institute, University of Utah, Salt Lake City, Utah 84112
[2] Department of Radiology, University of Utah, Salt Lake City, Utah 84112
[3] Lawrence Berkeley Laboratory, Berkeley, CA

Abstract. Single photon emission computed tomography (SPECT) has become an important diagnostic tool in nuclear medicine, because of the important clinical role that it can offer. Reconstruction methods, necessary to compensate several effects that severely degrade the quality of the image, are computationally very intensive.

The estimate of reconstruction errors in SPECT reconstructions with weighted least squares includes the singular value decomposition of 180 projection matrices of size 4096×4096. We have developed a very efficient parallel implementation on an 8 processor SGI PowerChallenge.

1 Introduction

Single photon emission computed tomography (SPECT) has become an important diagnostic tool in nuclear medicine. Over the last decade cardiac SPECT has seen a significant growth in the number of procedures [5]. The reason for the rapid growth is the increasing awareness of the important clinical role that SPECT can offer in the detection of myocardial infarcts and the diagnosis of ischemic heart disease [6, 7]. In addition to the heart, SPECT will continue to be a valuable diagnostic tool for the brain, liver, bone, and other organs, because it is inexpensive and noninvasive, in addition to the fact that nuclear medicine procedures with SPECT can answer physiological questions that cannot be answered with any other diagnostic modality such as X-ray CT or magnetic resonance imaging (MRI).

SPECT is not without problems: the accuracy of SPECT images is degraded by several physical factors, including image noise, photon attenuation, scatter of photons, and a spatially varying point response, which is the direct result of the geometric response of the collimator and photon scatter. Reconstruction methods are necessary to compensate these effects, and improve the quality of the image. Significant improvements have been made over the last few years [8].

All reconstruction methods are computationally very intensive [4], and involve the solution of very large systems of linear equations. While the solution of these systems has been traditionally carried out with iterative methods, together with the recomputation of matrix components, the power and capacity of symmetric multiprocessor workstations (SMP) now enables a timely solution of 2-D problems via direct solvers.

This allows one to incorporate the reconstruction errors into the estimation of kinetic parameters derived from time-activity-curves generated from recon-structed dynamic SPECT data. This may be important in application to dy-namic SPECT where reconstructions are formed from very low count projection measurements. Considering the low counts in dynamic SPECT, it is hypothe-sized that more efficient estimates would be obtained by weighting the estimates of kinetic parameters using reconstruction values and errors estimated from time sequences of weighted least squares reconstructions than from unweighted least squares reconstructions.

A detailed description of the background, methodology and results for this re-search is in [3]. In this paper we focus on the numerical methods and the strategy to efficiently estimate reconstruction errors in SPECT weighted reconstructions, which involve the singular value decomposition of 180 projection matrices of size 4096 x 4096.

2 Theory

Let the imaging model be $P = FX$ where X is the source distribution vector, P is the projection vector, and F is the $m \times n$ projection operator. The weighted least squares solution for X is the solution of the matrix equation: $F^T \Phi^{-1} P = MX$, with $M = F^T \Phi^{-1} F$, where F^T is the transposed matrix of F, and Φ is a diagonal matrix of projection measurement errors: $\Phi = diag\{\sigma_1^2, \sigma_2^2, \ldots, \sigma_N^2\}$. The estimate of the source distribution X is then $X = M^+(F^T \Phi^{-1} P)$, where M^+ is a generalized inverse of M. In order to find M^+, we first find the singular value decomposition of M as $M = U \Sigma U^T$, where U is an orthogonal matrix and σ is a diagonal matrix of singular values: $\Sigma = diag\{\lambda_1, \lambda_2, \ldots, \lambda_N\}$, with $\lambda_1 \geq \lambda_2 \geq \ldots \geq \lambda_N$, A diagonal matrix $D = diag\{1/\lambda_1, 1/\lambda_2, \ldots, 1/\lambda_r, 0, \ldots, 0\}$, is defined where r is determined by regularization. Then the generalized inverse of M is obtained as $M^+ = UDU^T$.

For the least squares reconstruction, the estimated source distribution is obtained by $X = UDU^T(F^T P)$ i.e. $X = \sum_{i=1}^{r}(u_i(F^T P)/\lambda_i)u_i$, where u_i is the i-th column vector of U. The k-th pixel of the estimated source distribution can be expressed as

$$x_k = \sum_{l=1}^{L} \sum_{m=1}^{r} \sum_{n=1}^{N} u_{km} u_{lm} f_{nl} p_n / \lambda_m \ . \tag{1}$$

In our study there are two regions of interest: a heart tissue region and a blood pool region in the ventricular cavity, each containing 10 voxels. We define:

$$H = \sum_{k \in Heart} x_k \tag{2}$$

and

$$B = \sum_{k \in Blood} x_k \ , \tag{3}$$

to be the sum total of counts in each region. The covariance between the two random variables H and B are

$$cov(H, B) = \sum_{i \in Heart} \sum_{j \in Blood} \sum_{n=1}^{N} \left(\sum_{l=1}^{L} \sum_{m=1}^{r} \frac{u_{im} u_{lm} f_{nl}}{\lambda_m} \right) \left(\sum_{l'=1}^{L} \sum_{m'=1}^{r} \frac{u_{jm'} u_{l'm'} f_{nl'}}{\lambda_{m'}} \right) p_n \tag{4}$$

Similarly, the variances of H and B are given by

$$\sigma^2(H) = \sum_{i \in Heart} \sum_{j \in Heart} \sum_{n=1}^{N} \left(\sum_{l=1}^{L} \sum_{m=1}^{r} \frac{u_{im} u_{lm} f_{nl}}{\lambda_m} \right) \left(\sum_{l'=1}^{L} \sum_{m'=1}^{r} \frac{u_{jm'} u_{l'm'} f_{nl'}}{\lambda_{m'}} \right) p_n \tag{5}$$

$$\sigma^2(B) = \sum_{i \in Blood} \sum_{j \in Blood} \sum_{n=1}^{N} \left(\sum_{l=1}^{L} \sum_{m=1}^{r} \frac{u_{im} u_{lm} f_{nl}}{\lambda_m} \right) \left(\sum_{l'=1}^{L} \sum_{m'=1}^{r} \frac{u_{jm'} u_{l'm'} f_{nl'}}{\lambda_{m'}} \right) p_n \tag{6}$$

For the weighted least squares reconstruction, the source distribution is obtained by $X = UDU^T(F^T F^{-1} P)$ i.e. $X = \sum_{i=1}^{r} (u_i (F^T \Phi^{-1} P)/\lambda_i) u_i$, where u_i is the i-th column vector of U. The k-th pixel of the estimated source distribution is

$$x_k = \sum_{l=1}^{L} \sum_{m=1}^{r} \sum_{n=1}^{N} \left(\frac{u_{km} u_{lm} f_{nl} p_n}{\lambda_m \sigma_n^2} \right) \tag{7}$$

or

$$x_k = \sum_{l=1}^{L} \sum_{m=1}^{r} \sum_{n=1}^{N} \left(\frac{u_{km} u_{lm} f_{nl}}{\lambda_m} \right) \tag{8}$$

The covariance between the two random variables H and B is

$$cov(H, B) = \sum_{i \in Heart} \sum_{j \in Blood} \sum_{n=1}^{N} \left(\frac{u_{in} u_{jn}}{\lambda_n} \right) \tag{9}$$

Similarly, the variances of H and B are given by

$$\sigma^2(H) = \sum_{i \in Heart} \sum_{j \in Heart} \sum_{n=1}^{N} \left(\frac{u_{in} u_{jn}}{\lambda_n} \right) \tag{10}$$

$$\sigma^2(B) = \sum_{i \in Blood} \sum_{j \in Blood} \sum_{n=1}^{N} \left(\frac{u_{in} u_{jn}}{\lambda_n} \right) \tag{11}$$

Please refer to [3] for more information on the compartment model and the theory for estimation of the kinetic parameters.

call the routine DGEMM in the BLAS library. We used the optimized BLAS library for the SGI PowerChallenge, running at an average speed of 273 Mflops per processor, which is close to peak (300 Mflops).

We chose to use the LAPACK [1] library for the Schur decomposition, motivated by its design with block methods calling BLAS, and the optimal performance that BLAS may achieve on the PowerChallenge. We experimented two different LAPACK routines for the Schur decompoosition: DSYEV (based on the QR factorization) and DSYEVD (based on the Cuppen divide and conquer algorithm).

DSYEV requires $n^3 + o(n)$ operations and may only use BLAS 2 routines, which do not have the best data locality, and may not approach peak performance. On the other hand, DSYEVD requires $4n^3 + o(n)$ operations, but it uses the most efficient BLAS 3 for a major portion of the computations. Given the size of our matrices, DSYEVD proved to be the more effective choice: after studying optimal settings for the block sizes, we could speed-up the decomposition time by an average factor of 2.5 with respect to DSYEV.

The decomposition of a matrix is independent of other matrices: it is desirable to perform 180 decompositions concurrently. However, the total memory required to decompose a 4096×4096 matrix is abour 800 MBytes, which exceeds the memory currently available on desktop workstations. We decided to exploit the parallelism contained in each matrix-multiply and Schur decomposition. In both cases the easiest and most efficient method is to call the parallel BLAS library optimized for the SGI PowerChallenge. Indeed, the block design of the LAPACK defers number crunching and parallelism to the BLAS.

We studied optimal parameter settings: while DSYEVD calling serial BLAS best performs with a block size in the range of 48-64, DSYEVD calling parallel BLAS best performs with a block size in the range of 80-96.

While the matrix-multiply with BLAS 3 has a nearly ideal speed-up with respect to the number of processors used, this is not true for DSYEVD. This is because the portion of DSYEVD spent in BLAS 3 is not higher than 50 %, and it is dependent on the eigenvalues of the matrix in consideration. We found that 4 processors speeded up DSYEVD at most 3 times with respect to 1 processor, but 8 processors speeded up no more than 3.5 times.

5 Results and conclusions

The average computation time of the weighted least squares reconstruction, relative to one frame, was about 50 minutes on an 8 processors SGI PowerChallenge.

In order to minimize the elapsed time to compute 180 reconstructions, we scheduled 2 concurrent reconstructions, running on 4 processors each. Considering that the SGI PowerChallenge at the Utah Supercomputing Institute has 8 processors with 2 GB of memory, this proved to be the optimal combination of concurrent jobs versus processors dedicated to each of them. In fact 2 concurrent reconstructions fit in core memory (1.6 GB), and the extra processors speed up each reconstruction by a factor of up to 3.

3   Methododology

Dynamic cardiac SPECT data were acquired using a PRISM 3000 (Picker International) to image a canine after injection of $99mTc$–teboroxime. Sequential $5s$ tomographic acquisitions of 60 projections were acquired for $17m$. The data were reconstructed into a sequence of $5s$ 64×64 images for one transaxial slices using for each reconstruction a separate singular value decomposition to calculate the weighted reconstructed estimate, the variance of the estimate, and the covariance between tissue and blood ROIs. A set of 180 singular value decompositions of size 4096×4096 needed to be carried out.

The following summarizes the data steps that were performed:

1. Form the projector matrix F.
2. Determine weighted least squares reconstructions using SVD matrix inversion (180 64×64 transaxial slices every $5s$). The weighted least squares reconstruction is obtained using Equation (8) and the Unweighted least squares reconstruction was obtained using Equation (1).
3. Generate blood and heart regions of interests (10 pixel regions) using Equations (2) and (3).
4. Calculate covariance $cov(H, B)$ between heart and blood regions of interest using Equation (4) and variances $\sigma^2(H)$ and $\sigma^2(B)$ using Equations (5) and (6).
5. Determine weighted least squares estimates of the kinetic parameters.

The computations required to form the matrix F, the regions, covariance and least squares were negligible: we want to focus the discussion on the methods to compute the 180 SVDs, that have large memory and computational requirements.

4   Computational methods

We consider the normal equations of the unweighted least squares problem

$$\mathbf{F}^T \, \mathbf{F} \, \mathbf{x} \; = \; \mathbf{F}^T \, \mathbf{p} \tag{12}$$

and the corresponding weighted least squares problem:

$$\mathbf{F}^T \, \Lambda(\mathbf{p}) \, \mathbf{F} \, \mathbf{x} \; = \; \mathbf{F}^T \, \Lambda(\mathbf{p}) \, \mathbf{p}$$

The matrices arising from the normal equations are symmetric positive semi-definite: their SVD can be computed with the numerically more efficient Schur decomposition, which is used for the symmetric eigenvalue problem.

We are interested in the SVD of the matrix $\mathbf{M}(0) = \mathbf{F}^T\mathbf{F}$ and the 180 matrices $\mathbf{M}(p) = \mathbf{F}^T\Lambda(\mathbf{p})\mathbf{F}$, for $\mathbf{p} = 1, \ldots, 180$.

The formation of a matrix $\mathbf{M}(p)$ involves a matrix multiply, which requires $2n^2m$ operations, where the number of rows and columns are respectively $m = 4096$ and $n = 3360$. The most efficient way to implement this operation is to

We computed the study of 180 SPECT 2-D reconstructions with weighted least squares and estimation of dymanic cardiac SPECT kinetic parameters in less than 3 days. This is a considerable result: on one hand, this problem was never attempted because it was estimated to take several months to compute on workstations; on the other hand, it opens new light on the feasibility of 3-D direct reconstructions.

6 Acknowledgements

We thank Prof. Jim Demmel at University of California, Berkeley, for providing insight on the most recent LAPACK software for symmetric eigenvalue problems.

This research work was partially supported by NIH Grant RO1 HL 39792, The Whitaker Foundation, Picker International and the Utah Supercomputing Institute. We also thank the Utah Supercomputing Institute for providing access to the SGI PowerChallenge.

References

1. E. Anderson et al. *LAPACK User's Guide.* SIAM, Philadelphia, 1992.
2. S. Foresti and L. Zeng. Parallel methods for the reconstruction of SPECT images. In D.H. Bailey, P.E. Bjorstad, J.R. Gilbert, M.V. Mascagni, R.S. Schreiber, H.D. Simon, V.J. Torczon, and L.T. Watson, editors, *Proceedings of the Seventh SIAM Conference on Parallel Processing for Scientific Computing, February 15-17, 1995, San Francisco*, pages 56–61, Philadelphia, PA, USA, 1995. SIAM.
3. G. Gullberg, R. Huesman, L. Zeng, and S. Foresti. Efficient estimation of dynamic cardiac SPECT kinetic parameters using weighted least squares estimates of dynamic reconstruction. In *Conference Record of the 1995 IEEE Nuclear Science Symposium and Medical Imaging Conference, Oct. 21-28, 1995, San Francisco Airport*, (to appear).
4. K. Lang and R. Carson. EM reconstruction algorithms for emission and transmission tomography. *J. Comput. Assist. Tomogr.*, 8:306–316, 1984.
5. C. D. Maynard. Medical imaging in the nineties: new direction for nuclear medicine. *J. Nucl. Med.*, 34:157–164, 1993.
6. S. J. McPhee and D. W. Garnick. Imaging the heart: Cardiac scintigraphy and echocardiography in U.S. hospitals (1983). *J. Nucl. Med.*, 27:1653–1641, 1986.
7. H. W. Strauss and E. L. Palmer. Editorial: Cardiovascular nuclear medicine – training for the future. *J. Nucl. Med.*, 27:1642–1643, 1986.
8. B. M. W. Tsui, G. T. Gullberg, E. R. Edgerton, J. G. Ballard, J. R. Perry, W. H. McCartney, and J. Berg. Correction of nonuniform attenuation in cardiac SPECT imaging. *J. Nucl. Med.*, 30:497–507, 1989.

The Prospect for Parallel Computing in the Oil Industry

Bengt Fornberg

Department of Applied Mathematics, University of Colorado
Boulder, CO 80309-0526, USA

Abstract. About 10% of all 'supercomputers' have been delivered to the oil industry. Its two big computational tasks are seismic data processing (to deduce underground geological structures from surface-based probing with elastic waves), and reservoir modeling (to simulate the flows within a producing field, in order to optimize the amount of hydrocarbons that can be recovered). Both tasks are extremely CPU- and, for seismic modeling, also I/O intensive.

Trying to forecast the future is always chancy, and especially so for parallel computing. To stay on a somewhat safe ground, we will focus on the two main tasks just mentioned, and on the computational demands they entail. A change of direction towards large-scale parallelism is well on the way in the former area, but held back a few years in the latter (mainly by coding complexities).

1 Introduction

The last decade has seen a long sequence of brilliantly innovative commercial parallel (or vector) supercomputer ventures e.g. (in no special order) Denelcor (HEP), FPS, Saxpy Supercomputer, SCS, Control Data (with STAR-100, Cyber 205), ETA, Alliant, Evans and Sutherland Supercomputer, BBN (butterfly), Multiflow Trace Company, Thinking Machine (with Connection Machine), Cray Computer Company, Cray Research, Intel (with Paragon) and Kendall Square. What do all the entries in this list have in common? They either

- abandoned the area of supercomputing,
- got sold out to competitors, or
- went outright 'bust'

All were based on very promising new concepts, and it would have seemed that the market has been ready to absorb as much computer power as could be provided (at steadily falling costs). This teaches us to be very careful with predictions. Since it hardly would be helpful to anybody if I added my personal guesses to the list of failed predictions, we will keep on safer grounds by looking at the main computational tasks are in the oil industry. This will leave no doubt that parallel computing is the trend for the future.

When we here talk about parallelism, we refer to the use of separate processors working on separate flows of data - not the hidden parallelism that has provided several orders of magnitude improvements without users hardly needing to be concerned by any changes in architecture (e.g. the EDSAC, with a clock period of 2 μs produced 100 flops in 1949; in 1986, one processor Cray X-MP achieved 172 Mflops with a clock period of 9.5 ns. With a 200 times faster clock, nearly 2,000,000 more floating point operations were performed).

About 10% of all 'supercomputers' have been delivered to the oil industry. If the comparison is limited to other industrial areas (such as aerospace, automotive, electronics, pharmaceuptics, chemical, construction, etc.), the oil industry's proportion stands at about 25 %. Although parallel computing is on the rise, it remains still well below its potential. Factors that have held back a more widespread use include

- severe decline of basic research within the oil industry, and
- cost involved in making changes (an order of magnitude improvement is often needed to justify replacing large working codes).

However, strong factors favor of an increased use of parallelism

- trend towards 3-D modeling (resulting in enormous both data sets and operation counts),
- increased need to obtain results quickly, and
- increased pressure towards highest cost-effectiveness.

Besides seismic modeling and reservoir simulation, visualization is also critically important. It will not be discussed here, since it typically does not require massively parallel computing resources.

In the next sections, we will briefly discuss oil supplies, and then focus on seismic modeling and reservoir simulation, in order to provide a better understanding of the character of their computational demands. In conclusion, we will make some comments on the current status, and note some trends.

2 Origin and Availability of Hydrocarbons

The generally accepted view regarding the origin of the worlds hydrocarbon (HC) resources is that they come from decayed marine biomass, accumulated in calm ocean basins, and trapped by silt and sediments. This process has been ongoing ever since the origin of life, but has been of significance only after the 'cambrian explosion' in marine life some 465 M years ago. Most major reservoirs are more recent - often between 100 M and 1 M years old (the latter corresponding to the typical time for oil to 'mature').

In some cases, oil fields appear to be slowly refilling themselves - and oil has also been spotted in places where there hardly can be any organic sediments present. This has caused some speculations whether HC in rare cases also can be of volcanic (non-organic) origin, and migrate up from the earth's mantle;

For example, traces of oil have been spotted by Lake Siljan in Sweden, where a meteorite a very long time ago fractured the granite bedrock (which has no sedimentary history).

Most of the HC that the earth once possessed has, over the millennia, slowly risen to the surface and bio-degraded. What we explore for, and now extract, are the few remains which, by chance, have become trapped on their way up, due to

- impenetrable, dome-shaped rock layers (known as anticlines) and, to a lesser extent,
- 'fault traps' - sealed corners between inclined layers and geological fault lines.

No present exploration method (short of expensive drilling) can effectively detect deep HC deposits directly, either down in the ground, on the rise, or when reaching the surface (nowadays at minute rates). The dominant technique - seismic surveying - aims instead at detecting these geological trap structures within areas that have a promising sedimentary character. Very heavy HC (tar sands, asphalts etc.) rise only extremely slowly, and can still be found in vast quantities.

Many individual oil fields have 'run dry' (or, to be more precise, most of the HC are still left in them, but further extraction has becomes too costly). However, for the last several years, more oil has been discovered than has been extracted. Proven, recoverable reserves are now increasing - not decreasing. They stand now at about 700 bbl (i.e. 700×10^9 barrels) - about half of this in the Middle East. The annual world consumption is about 18 bbl. This trend of rising reserves will likely come to an end fairly soon. When the recoverable liquid HC supplies start to run low, prices will rise. At the moment, these are around 20 - 25 $/bl. There is a long-term upper ceiling at about 40 - 45 $/bl. This is the break-even level beyond which it becomes cheaper to make synthetic oil from coal. Synthetic oil was widely used in Nazi Germany, and has also been produced on a large scale in South Africa. Easily accessible coal supplies (e.g. surface deposits) will suffice for maybe 500 - 1000 years at present consumption.

Another massive fossil energy resource is oil shale (near-solid deposits). Although estimates are uncertain, it is believed that US, Canada and Venezuela each have over 3,000 bbl in oil-equivalent reserves (compared to 30 bbl of proven, untapped oil in the US). However, both extraction methods and conversions to lighter fuels are costly or difficult. Ways to extract include

- surface mining,
- in situ combustion (pump down air - get out low-grade HC with the smoke), and
- make slurry (pump down water under high pressure to release grain pressures and mobilize it; this technique is at present barely experimental).

The most serious problem for both coal and shale useage may well prove to be the CO_2 greenhouse effect. For this, no satisfactory solutions appear to be in sight. No presently used methods for coal or shale exploration or extraction pose any major computational challenges.

This discussion of energy supplies is not meant to imply that the next big energy sources will be coal and shale - the conclusion is only that our society collapsing because oil very suddenly runs out, is not a likely future scenario. (The problems with a renewable energy resource such as sunlight are of technical/ engineering type - not of supply. All fossil fuels accumulated throughout the history of the earth -coal, oil shale, oil, natural gas etc. - contain the energy equivalent of about 20 days of sunlight falling on the earth).

We can quite safely assume that the very biggest oil fields have now been found. Searching for smaller deposits in increasingly difficult areas and at increasing depths (e.g. deep under water or in remote areas with hostile climate) requires advances in exploration. Reservoir modeling will also become more essential as the questions regarding the economical feasibility of exploiting marginal reservoires will arise more often. These increasing needs, combined with the falling costs of computing, are the key factors that drive the trend towards very large-scale parallel computing.

3 Seismic Modeling

In contrast to many other mineral deposits, HC reservoirs do not produce any distinctive electric, magnetic, or gravity anomalies that can be detected from the surface. Apart from drilling, the only widely useable exploration option is to use seismic waves. Such waves are

- easy to generate (explosions, truck-mounted vibrators, air guns towed after ships etc.),

- able to penetrate many kilometers of earth, water, and rock with only moderate energy loss,

- strongly influenced by properties of the medium - in particular interfaces cause reflections which bring relatively strong return signals back to the surface, and

- easy to detect on their return to the surface.

Some approaches involving electromagnetic waves have been tried, but none has the range and reliability of regular seismic techniques.

Ground penetrating radar in the 100 MHz range can reach down about 20-40 meters, and is therefore better suited for finding ground pollutants or buried objects than mineral deposits. Attempts have also been made to exploit the fact that seismic signals can cause electromagnetic responses (via weak ionization and pore flow) or reversely, that strong electromagnetic signals can cause a weak acoustic responses.

Ideas from geometrical optics can be used to follow seismic ray paths if the wavelengths are short compared to the size of the layers that are studied (a questionable assumption - both are often in the 10-100 meter range). In the case of interfaces, energy is transferred between different types of waves (primarily P- and S-waves; pressure- and shear waves respectively). Part of the energy is transmitted and part is reflected. Already with only a few layers, a vast number of waves will arise. The goal is to somehow invert the surface response data into a picture of the underground structures. The crudest possible method to find the depth of the first interface would be to only measure the travel time for the first reflected signal. To get a more complete picture, a large number of corrections / refinements must be made

- *Normal Moveout* - correction for the varying thickness of the weathered layer nearest the surface,

- *Stacking* - combining data from many source and receiver positions to reduce noise and cancel effects from multiple reflections,

- *Migration* - a correction procedure for inclined interfaces,

The seismic images resulting from these three steps provide skilled exploration geophysicists with a rough picture of the major interfaces. Sometimes a fourth step is added

- *Elastic wave solution*: the elastic (or acoustic) wave equation is repeatedly solved in an iterative fashion in which the assumptions about underground structures are altered in order to reconcile measured and computed seismic returns. The governing equations lend themselves extremely well to parallel (domain decomposition-type) implementations.

  Curiously, an analog technique for elastic wave modeling emerged briefly in the 1980's: small (about 10 cm side) plastic models with many distinct layers could be molded automatically. When subjected to ultrasound, these models displayed the full range of 3-D elastic wave phenomena.

Normal moveout requires little computing. Stacking reduces the massive raw data sets (all the original geophone return signals) in preparation for the migration step. In modern practice, the stacking and migration steps are often combined into *pre-stack migration* codes. These computations (requiring vast amounts of both I/O and CPU power) tend to parallelize very well. At present, maybe 90% of all production parallel computing in the industry is applied towards pre-stack migration, with the remaining 10% split between reservoir modeling and some elastic wave calculations.

The rest of this section intends to convey a very brief impression of the concepts involved in stacking and migration (in the simple case when these two stages are carried out consecutively, and in 2-D).

Figure 1 indicates how pressure waves would travel if the source and the receiver were at the same location (not a good idea with explosive sources), the layers were perfectly horizontal, and each medium constant - needless to say an unrealistically simplified case. The timing of the first few returned signals would correspond to interface locations, but soon after, the return would get very complicated due to multiple reflections.

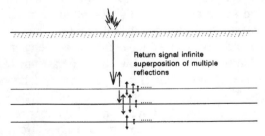

Figure 1.   Illustration of multiple reflections arising when signals bounce between interfaces.

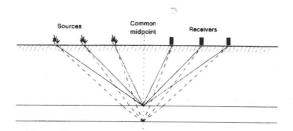

Figure 2.   Illustration of common midpoint principle.

Figure 2 suggests a more realistic idea; have sources and receivers well separated. Simple trigonometry permits the elapse times to be compensated in order to account for the extra horizontal travel. Averaging the results from many cases with a *common midpoint* will provide a much better signal-to-noise ratio; both random noise and multiple reflections are likely to partly cancel out. In practical usage (on land), many geophones (receivers) are placed out, and a source is moved (in the sea, an air gun source is towed near a ship, with long streamers of receivers trailing). The collected data is then re-arranged - *stacked* - to achieve a lot of common midpoint averages. Data volumes are reduced, and the (corrected, averaged) time traces at different locations (plotted with physical space horizontally and time in the negative vertical direction) can resemble a vertical physical slice down through the ground.

If the layers are inclined (in particular if they feature dips (synclines)) rays can cross, focus, diffract etc., and the common midpoint process has almost no justification. When stacking is carried out, the result will bear little resemblance to the layer structure. Synclines and domes (a formation that might have trapped oil) often look nearly identical. *Migration* intends to compensate for these errors caused by inclined layers. As a thought experiment, one can assume that the sources were set off, not one at the time and at the surface, but simultaneously, and along the top interface. We also assume the strengths of these imagined sources to be proportional to the reflection coefficients there, and that the resulting waves towards the surface travel with double speed. The surface response would resemble the received signal. Assuming that the constant coefficient acoustic 2-D wave equation $\frac{\partial^2 u}{\partial t^2} = \frac{\partial^2 u}{\partial x^2} + \frac{\partial^2 u}{\partial y^2}$ describes how these waves travel through the top layer to reach the surface, we can reverse the time, and solve this same equation backwards (with the measured surface responses as initial conditions). These reversed waves will, after the appropriate time, give an approximation of the interface, i.e. the surface data has been migrated back to the interface. In reality, a vast number of refinements are called for, e.g. to handle multiple layers, slowly variable media, 3-D effects etc.

Traditionally, migration was performed first after the massive geophone data volumes had been reduced through stacking. However, since the justification for stacking is basically absent in case of inclined layers, the data reduction of stacking destroys genuine data. *Pre-stack migration* is possible on modern large computers.

Different versions go under names such as finite difference-, frequency-wave number-, and Kirchhoff migration.

4 Reservoir Modeling

Let us first remove a couple of common misconceptions - that oil is floating around in underground lakes, and when a well is drilled, it is likely to spurt out (in the style of Spindletop, Texas 1901, or as we saw from Kuwait during the Gulf War). Figure 3 depicts schematically gas, oil and water being trapped under an impervious rock layer. The main constituent everywhere is rock. Oil, gas and water occupy only microscopic pores within heavily compressed sandstone (or sand and clay, fused under enormous pressures; the pore volume may range between 1-20%). Pore structures are of the order 1 - 100 μm (difficult to study even if rock samples are polished/sliced). Significant questions remain on how such pores connect in order to make channels (cracks in the rock form another HC migration path).

Most oil fields range in area between 1/10th to 10 km², and in thickness of oil-bearing rock between 1 and 100 m. Deposits are usually small; of about 10,000 tapped in the US, the top 60 produce half of the total oil.

Figure 3 illustrates the phenomenon of *coning*, a consequence of the fact that both gas and water are more mobile than oil. As soon as either of these reach a production well, very little further oil can then be obtained. *Reservoir modeling* was first developed to better understand this phenomenon of coning.

Historically, the use of oil goes back to early civilizations (rare surface deposits were used for waterproofing boats, cloth, and sometimes as fuel for lights). The concept of trapped underground reservoirs wasn't appreciated until large-scale production began in the 1860s. The need to do modeling to better understand coning became obvious around 1930. The goal of reservoir modeling has remained to compare the economical returns of changes in recovery procedures, and to forecast recovery potentials.

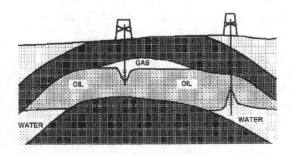

Figure 3.   Schematic illustration of an HC reservoir with gas, oil, and water saturating a region of porous rock.

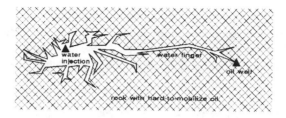

Figure 4.   Fingering - instability when trying to force out oil by water pressure.

Recovery is carried out in three main stages:

Primary: Just let the oil flow, or pump it out; this 'black oil' extraction leaves 70 - 85 % behind,

Secondary: Mainly 'water sweep'; inject water to force oil out. This idea was discovered by accident as a sometimes effective method, but must be used with great care. Water flows easier than oil, and *fingering* (Figure 4) can create water channels that make residual oil near impossible to extract. About 65 - 70% of the original oil still remains left behind.

Tertiary: (Also known as EOR: Enhanced oil recovery); the task is now to mobilize as much of the trapped left-overs as possible. The key trappers are surface tension and viscosity. The former can be reduced by surfactants (detergents). It can be possible to extract another 5-10%, but cost and technical complexity are both very high. The many variations include

-    Polymers are micron-length very thin strands which, already in concentrations as low as 1 part in 400, can make water thick like saturated sugar. Because of various non-Newtonian effects (e.g. polymers tending to curl up in the presence of brine), large Monte-Carlo simulations are used together with physical experiments in current efforts to establish governing equations for polymer flow. Polymers can be used both to push oil (with reduced risk of fingering), and to block water pathways,

-    Pump in turn surfactant, polymer, and water (as we want to recover some of the more expensive ingredients),

-    Inject foam (made up of about 75% air and 25% liquid). Foams can be stable for weeks, and are surprisingly rigid when under pressure. They can be generated in-situ, or be pumped down,

-    Inject CO_2: We often want to get rid of CO_2 (at up to 40% of the volume, it makes some very rich natural gas fields impossible to use), but it is expensive both to transport and to compress (or to cool to liquid). However, it can join oil to form an easier-flowing one-phase fluid.

-    Increase the temperature: inject hot water or steam or pump down air or oxygen to maintain a controlled in-situ combustion.

Modeling of these EOR techniques is a formidable task; key issues include the formulation of governing equations, finding effective numerical algorithms, dealing with complicated and unknown geometries, the presence of dual flow mechanisms (pores and fractures) for at least 3 fluid media (oil, gas, water), phase transitions oil and gas etc.

The historical evolution of reservoir simulation is sketched below:

- *Sand-packed boxes* (with oil on top, water below) began to be used around 1930 to simulate coning, but were unwieldy to work with.

- *Electrolytic models* (from about 1940) exploit that one can approximate Laplace's equation in complicated geometries by measuring electrical voltages at various places of a badly conducting object. For modeling Darcy's law (slow flow through a porous medium), one built physical models in the shape of reservoirs. The surface was molded such that its depressions matched the thickness times the permeability at various field locations. This surface then served as the bottom of a tank with electrolyte. One can then move around electrodes in the fluid, and measure potentials.

- Around 1950, *analog computers* were introduced, which could simulate unsteady, compressible flows by what amounted to a finite difference solution advanced continuously in time (not needing to be concerned with time step stability conditions; these enter only when time is also discretized). Networks of resistors and capacitors were connected in such a way that electrical quantities matched quantities to be modeled: voltages - pressure; current - fluid flow rate, resistance - permeability times thickness, and capacitance - porosity times thickness times compressibility. But in the early 1950s, a new era was arriving:

- *Digital computers*: Ever since their origin, the steady increases in computer power (and also in numerical methods) has greatly increased the range of phenomena that can be included, and the complexity of geometries that can be handled.

Although reservoir modeling is an enormous consumer of computational power, the complexity of the codes has so far kept large-scale parallel computing on the sidelines. However, many parallel test codes have been developed, and in the next year or two, their big breakthrough in this filed can be expected.

Two examples can illustrate the importance of reservoir modeling:

- Alaska: Prudhoe Bay field - largest in North America:

Estimated total reserves:	24 bbl
Recoverable (typically 1/3rd)	8 bbl
Gain by modeling (about 2 %)	160 Mbl
Gain in market value (at 20$/bl)	3200 M$

- Siberia: Short term incentives sometimes encouraged hasty production to meet 5-year plan deadlines. Too much water flooding caused fingering, and may have reduced recoverable volumes in some fields by as much as 30%.

5 Conclusions

The tasks of seismic exploration and reservoir modeling require vast computer resources. At present, the major part of the MPP resources in the industry (about 90%) are devoted to seismic, and then primarily for migration. The trend here is to shift from 2-D post-stack to 3-D and pre-stack. A big driving force behind this change to MPP (apart from cost-effectiveness) is the importance in keeping data turn-around time as short as possible. For instance, several IBM PS2-systems are now operating on board exploration vessels - permitting the ships to navigate and search in response to information that has just been collected. In the past, data was written to tapes, which had to be brought to shore, shipped to processing centers, worked on in batch mode etc. When special sections of interest were discovered, another cycle in this process had to commence, i.e. scheduling a new marine expedition etc.

Of the four primary machine structures (SISD, SIMD, MISD and MIMD), only the latter, in conjunction with distributed memory, would seem to have the potential to meet the needs in terms of scalability to the sizes needed. With Intel pulling out of the MPP field as a systems manufacturer (focusing instead on supplying chips to other ventures, such as the Sandia T-flop machine based on 9,000 Intel P-6 chips), the major commercial systems are the Cray T^3D and IBM SP2. Of these two, IBM is clearly the commercially most successful one. Of the 1,300 IBM SP2-systems delivered, about 100 can be found in the oil industry (in 30 companies). In contrast, only about 40 Cray T^3D have been delivered, 4 of them to oil companies (CTG, Exxon, Mobil, and Phillips).

Looking forward, there is nothing foreseeable that could halt the MPP's increasing dominance in the area of seismic processing. Regarding reservoir simulations, the transition to MPP is likely to start in earnest next year, when production-class massively parallel simulators will be coming on-line.

Performance Evaluation and Modeling of Reduction Operations on the IBM RS/6000 SP Parallel Computer

Paraskevi Fragopoulou[1], and Ole H. Nielsen[1,2]

[1] UNI•C, Technical University of Denmark, Bldg. 304, DK-2800 Lyngby, Denmark.
[2] Center for Atomic-scale Materials Physics (CAMP), Physics Dept., Technical University of Denmark, Bldg. 307, DK-2800 Lyngby, Denmark.

Abstract. We discuss algorithms for global reduction (or combine) operations (e.g., global sums) for numbers of processors that need not be a power of 2, and implement these using standard message-passing techniques on distributed-memory parallel computers. We present performance results measured on an IBM RS/6000 SP parallel computer at UNI•C. Significant performance improvements are obtained by using a recursive doubling method with a vector splice/gather approach.

1 Introduction

Consider a parallel code where, initially, each of n processors holds a vector of data items; finally the result of the *reduction* of the n vectors over a commutative and associative operator is available on all of the participating processors. Such combine operations are also synonymously denoted as *combine* operations. This type of operation has special importance because it constitutes a building block in several types of production codes. The performance issue is often important because global communication operations tend to impose a heavy load upon the inter-processor communication network.

We have implemented and compared the performance of three different global reduction algorithms on the IBM RS/6000 SP parallel computer. The implemented algorithms have been used to improve significantly the performance and scalability of a parallel Car-Parinello *ab-initio* molecular dynamics code [1].

2 Reduction algorithms

Assume that there are n processors, labeled p_i, $0 \leq i < n$, each one holding a vector v_i, $0 \leq i < n$, respectively, of size m. Given an associative and commutative operator, denoted by $\oplus$, the goal of the global reduction is to reduce the n vectors element-wise, i.e., the operation $v_0 \oplus v_1 \oplus \ldots \oplus v_{n-1}$. The result, which is also a vector of size m, should be available on all the participating processors. An example of such operation would be a global sum of the vectors v_i, $0 \leq i < n$.

In this paper, we describe three different global reduction algorithms, and model the theoretical communication complexity of the algorithms as follows:

We assume that each processor can communicate directly with any other processor as in a fully-connected network, which is the case for the IBM RS/6000 SP. The communication is composed of distinct stages or rounds and each processor can send and simultaneously receive one message during each round of the algorithm, i.e., the *one-port* communication model. In most message-passing systems, the time to transmit a message of length m between two processors is modeled as $t = t_s + mt_c$, where t_s is the start-up time (latency) and t_c is the per message unit (byte, integer, floating point etc.) transmission time. The communication performance of the algorithms is modeled with respect to the number of communication rounds they require, denoted by C_1, and the sum of the longest message sent or received by any processor during each round, denoted by C_2. Thus, an algorithm has an estimated communication complexity of $T = C_1 t_s + C_2 t_c$ [2].

2.1   Algorithm 1

In what follows, all subscript and index operations are modulo n and all logarithms are base 2.

The first global reduction algorithm is based on a simple master/slave approach. One of the n processors, designated to be the master, takes the responsibility to receive one-by-one the vectors located in each one of the other processors (slaves) and reduces them with the vector it holds. As soon as the master processor completes the reduction of the vectors, it broadcasts the result to the slaves. In our implementation the final vector is transmitted by the master to each one of the $n - 1$ slaves in sequence, since the PVM library's broadcast is believed to be implemented in this way.

The number of communication rounds required for this algorithm is $n - 1$ for both the collection phase and the broadcasting phase. During each round, a message of length m is transmitted by each processor, and thus the theoretical communication complexity of this algorithm is $2(n - 1)(t_s + mt_c)$. The maximum number of operations performed by a processor during the course of this algorithm is $(n - 1)m$, since all the vector reductions are performed by the master.

2.2   Algorithm 2

The second algorithm was first presented by Bruck and Ho [2] and its theoretical communication complexity was analyzed under the k-*port* communication model, i.e., assuming that each processor can send and simultaneously receive distinct messages from k other processors during a communication round. Here, we give a description of this algorithm under the more restricted *one-port* model, which is the appropriate one for the IBM RS/6000 SP.

This algorithm is a variation of the recursive doubling approach [2], extended to handle the situation where the number of processors n is not necessarily a power of 2. Assume that the binary representation of $n - 1$ is $n - 1 = n_{k-1} \ldots n_1 n_0$, where $k = \lceil \log n \rceil$. Here, $\lceil x \rceil$ denotes the smallest integer greater than or equal to x (the "ceiling" function). The basic idea of this algorithm is to treat $n - 1$ according to Horner's rule as follows:

$$n - 1 = 2 * (2 * (\ldots (2 * (2 * n_{k-1} + n_{k-2}) + n_{k-3}) \ldots) + n_1) + n_0.$$

This algorithm completes in $k = \lceil \log n \rceil$ rounds. At the end of round $r, 0 \leq r < k$, processor p_i holds the following three variables:

1. $c = 2 * (\ldots (2 * (2 * n_{k-1} + n_{k-2}) + n_{k-3}) \ldots) + n_{k-r-1}$ denotes the number of vectors that have already been reduced (initially $c = 0$).
2. v_i^0 holds the reduction result of the c vectors located in the c processors cyclically preceding p_i, i.e., $v_i^0 = v_{i-1} \oplus v_{i-2} \oplus \ldots \oplus v_{i-c}$
3. v_i^1 holds the reduction result of $c + 1$ vectors, namely, vector v_i and the c vectors located in the c processors cyclically preceding p_i, i.e., $v_i^1 = v_i \oplus v_{i-1} \oplus v_{i-2} \oplus \ldots \oplus v_{i-c} = v_i \oplus v_i^0$

In the standard recursive doubling approach, which is valid only for n equal to a power of 2, the number of vectors which have been reduced doubles with each round. In this variation of the algorithm, the number of vectors c which have been reduced changes with each round r as $c = 2c + n_{k-r-1}$, which means that it either doubles (if $n_{k-r-1} = 0$), or it becomes double plus one (if $n_{k-r-1} = 1$).

The following pseudo-code describes the steps executed by processor p_i during round $r, 0 \leq r < k$:

```
if (n_{k-r-1} .eq. 0) then
        Send v_i^0 to processor p_{i+c}
        Receive v_{i-c}^0 from processor p_{i-c} and store in v_{tmp}
else if (n_{k-r-1} .eq. 1) then
        Send v_i^1 to processor p_{i+c+1}
        Receive v_{i-c-1}^1 from processor p_{i-c-1} and store in v_{tmp}
endif
v_i^0 = v_i^0 + v_{tmp}
v_i^1 = v_i^1 + v_{tmp}
c = 2 * c + n_{k-r}
```

In the beginning of round $r, 0 \leq r < k$, each processor holds the reduction result of c vectors in v^0. During round r, each processor receives either vector v^1 (if $n_{k-r-1} = 1$), which is the reduction result of $c + 1$ other vectors, or vector v^0 (if $n_{k-r-1} = 0$), which is the reduction result of c other vectors, from another appropriate processor. Thus, at the end of the round, each processor holds in v^0 the reduction result of $2 * c + n_{k-r-1}$ vectors. At the end of the algorithm, the final reduction result is stored in variable v_i^1 of each processor p_i. An inductive proof for the correctness of this algorithm can be found in the paper by Bruck and Ho [2].

During each round, m data elements are received by each processor. Therefore, the communication performance of this algorithm is $\lceil \log n \rceil (t_s + m t_c)$. The total number of operations performed on each processor is $(2 \lceil \log n \rceil - 1)m$. This factor of 2 in the number of operations per processor is the penalty paid for extending the algorithm to work when the number of processors is not necessarily a power of two, since two vectors v^0 and v^1 need to be updated at each round.

An example of this algorithm for $n = 6$ can be seen in Table 1. Since all index operations are modulo n, the final v^1 vector on all processors are identical, given that $\oplus$ is an associative operator.

	Before Round 1 $(c = 0)$		Round 1 $(c=1, n_2=1)$ $p_i \xrightarrow{v_i^1} p_{i+1}$		Round 2 $(c=2, n_1=0)$ $p_i \xrightarrow{v_i^0} p_{i+1}$		Round 3 $(c=5, n_0=1)$ $p_i \xrightarrow{v_i^1} p_{i+3}$
	v^0	v^1	v^0	v^1	v^0	v^1	v^1
p_0	-	0	5	5:0	4:5	4:0	1:0
p_1	-	1	0	0:1	5:0	5:1	2:1
p_2	-	2	1	1:2	0:1	0:2	3:2
p_3	-	3	2	2:3	1:2	1:3	4:3
p_4	-	4	3	3:4	2:3	2:4	5:4
p_5	-	5	4	4:5	3:4	3:5	0:5

Table 1. An example of algorithm 2 for $n = 6$. The entry $i : j$ for v^0 or v^1 at a specific round means that the corresponding processor holds the reduction result of vectors $v_i, v_{i+1}, \ldots v_j$ in v^0 or v^1, respectively, at the end of this round.

2.3 Algorithm 3

The third algorithm was also first presented by Bruck and Ho in [2], and as for algorithm 2 we describe it under the *one-port* model. This algorithm is based on a two-phase approach. Each processor splices its vector into n blocks, each of size $\lceil \frac{m}{n} \rceil$ or $\lfloor \frac{m}{n} \rfloor$. We denote by $b_{i,j}$, $0 \le i, j < n$, the jth block of vector v_i placed in processor p_i. The first phase of the algorithm is a partial reduction that results in each processor p_i holding the final reduction result only for block b_i. During the second phase of the algorithm, each processor has to inform all other processors of the partial result that it holds. In other words, the second phase is nothing more than a concatenation algorithm.

The first phase requires $k = \lceil \log n \rceil$ communication rounds. For simplicity, we describe this phase of the algorithm for the case where n is a power of 2. If this is not the case, a single extra initial round is required [2]. We first show how processor p_i computes the final reduction result for block b_i at the end of this phase. The reduction of each individual block is performed using a simple recursive doubling technique. For example, for processor p_{n-1} to receive the result for block b_{n-1} the reduction proceeds as follows: in the first round processors $p_0, p_1, \ldots p_{\frac{n}{2}-1}$ send block b_{n-1} to processors $p_{\frac{n}{2}}, p_{\frac{n}{2}+1}, \ldots, p_{n-1}$, respectively. The receiving processors reduce the block they receive with their b_{n-1} block. In the second round only the last $\frac{n}{2}$ processors, which received a block in the previous round, participate in the reduction of block b_{n-1}. In general, in round r,

$r = \log n, (\log n) - 1, \ldots, 1$, the last 2^r processors participate in the reduction of block b_{n-1} and from these, the first half processors send b_{n-1} to the second half processors, respectively. The same process is simultaneously performed for each other block. The recursive doubling for block b_i with destination p_i, $0 \leq i < n$, is described with the following pseudo-code:

for r=(log n)-1 down to 0 do
    1. Processors $p_{i-2^r}, \ldots p_{i-2^{r+1}-1}$ send block b_i to processors
       $p_i, \ldots p_{i+2^r-1}$, respectively.
    2. Each processor that receives a b_i block reduces it with the
       b_i block it currently holds.
end for

In Table 2 we can see the partial reduction phase for each block separately on $n = 4$ processors.

	b_0 reduction	b_1 reduction
Round 1	$p_2 \xrightarrow{b_{2,0}} p_0 \; (b_{0,0} \oplus b_{2,0})$ $p_1 \xrightarrow{b_{1,0}} p_3 \; (b_{3,0} \oplus b_{1,0})$	$p_3 \xrightarrow{b_{3,1}} p_1 \; (b_{1,1} \oplus b_{3,1})$ $p_2 \xrightarrow{b_{2,1}} p_0 \; (b_{0,1} \oplus b_{2,1})$
Round 2	$p_3 \xdashrightarrow{b_{3,0} \oplus b_{1,0}} p_0 \; (b_{0,0} \oplus b_{2,0} \oplus b_{3,0} \oplus b_{1,0})$	$p_0 \xdashrightarrow{b_{0,1} \oplus b_{2,1}} p_1 \; (b_{1,1} \oplus b_{3,1} \oplus b_{0,1} \oplus b_{2,1})$

	b_2 reduction	b_3 reduction
Round 1	$p_0 \xrightarrow{b_{0,2}} p_2 \; (b_{2,2} \oplus b_{0,2})$ $p_3 \xrightarrow{b_{3,2}} p_1 \; (b_{1,2} \oplus b_{3,2})$	$p_1 \xrightarrow{b_{1,3}} p_3 \; (b_{3,3} \oplus b_{1,3})$ $p_0 \xrightarrow{b_{0,3}} p_2 \; (b_{2,3} \oplus b_{0,3})$
Round 2	$p_1 \xdashrightarrow{b_{1,2} \oplus b_{3,2}} p_2 \; (b_{2,2} \oplus b_{0,2} \oplus b_{1,2} \oplus b_{3,2})$	$p_2 \xdashrightarrow{b_{2,3} \oplus b_{0,3}} p_3 \; (b_{3,3} \oplus b_{1,3} \oplus b_{2,3} \oplus b_{0,3})$

Table 2. The first phase of algorithm 3 for $n = 4$ and each block separately.

During round $r = (\log n)-1, \ldots, 0$, a total of 2^r processors participate in the reduction of each block. The partial reduction phase is performed simultaneously for all blocks. Thus, during round $r = (\log n)-1, \ldots, 0$, each processor participates in the reduction of 2^r blocks. More specifically, during round $r = (\log n)-1, \ldots, 0$, processor p_i sends to processor p_{i+2^r} the following 2^r blocks: $b_{i+2^r}, \ldots, b_{i+2^{r+1}-1}$. In Table 3 we can see the complete (for all blocks) partial reduction phase on $n = 4$ processors.

The concatenation phase also requires $k = \lceil \log n \rceil$ rounds. In the beginning of this phase, processor p_i holds a single final block b_i. During each one of the first $\lfloor \log n \rfloor$ rounds, each processor doubles the number of the final blocks it holds. In the beginning of round r, $0 \leq r < \lfloor \log n \rfloor$, processor p_i holds the final blocks $b_i, \ldots, b_{i+2^r-1}$. During round r, processor p_i receives from processor p_{i+2^r} all the final blocks that this processor currently holds, namely, blocks

Round 1 (all blocks)	Round 2 (all blocks)
$p_0 \xrightarrow{b_{0,2},b_{0,3}} p_2$ $(b_{2,2} \oplus b_{0,2},\ b_{2,3} \oplus b_{0,3})$	$p_0 \xrightarrow{b_{0,1}\oplus b_{2,1}} p_1$ $(b_{1,1} \oplus b_{3,1} \oplus b_{0,1} \oplus b_{2,1})$
$p_1 \xrightarrow{b_{1,3},b_{1,0}} p_3$ $(b_{3,3} \oplus b_{1,3},\ b_{3,0} \oplus b_{1,0})$	$p_1 \xrightarrow{b_{1,2}\oplus b_{3,2}} p_2$ $(b_{2,2} \oplus b_{0,2} \oplus b_{1,2} \oplus b_{3,2})$
$p_2 \xrightarrow{b_{2,0},b_{2,1}} p_0$ $(b_{0,0} \oplus b_{2,0},\ b_{0,1} \oplus b_{2,1})$	$p_2 \xrightarrow{b_{2,3}\oplus b_{0,3}} p_3$ $(b_{3,3} \oplus b_{1,3} \oplus b_{2,3} \oplus b_{0,3})$
$p_3 \xrightarrow{b_{3,1},b_{3,2}} p_1$ $(b_{1,1} \oplus b_{3,1},\ b_{1,2} \oplus b_{3,2})$	$p_3 \xrightarrow{b_{3,0}\oplus b_{1,0}} p_0$ $(b_{0,0} \oplus b_{2,0} \oplus b_{3,0} \oplus b_{1,0})$

Table 3. The complete partial reduction phase of algorithm 3 for $n = 4$.

$b_{i+2^r}, \ldots, b_{i+2^{r+1}-1}$. Thus, at the end of round r processor p_i holds the 2^{r+1} final blocks $b_i, \ldots, b_{i+2^{r+1}-1}$. After $\lfloor \log n \rfloor$ rounds, processor p_i holds the $2^{\lfloor \log n \rfloor}$ final blocks $b_i, \ldots, b_{i+2^{\lfloor \log n \rfloor}-1}$. At this point the algorithm completes if n is a power of 2. In any other case, an extra round is required to complete the concatenation of the final result. During this last round, processor p_i receives from processor $p_{i+2^{\lfloor \log n \rfloor}}$, the $n - 2^{\lfloor \log n \rfloor}$ blocks $b_{i+2^{\lfloor \log n \rfloor}}, \ldots, b_{i+n}$. An example of the second phase of this algorithm on $n = 6$ processors can be seen in Table 4.

	Round 1 $p_i \xrightarrow{b_i} p_{i-1}$						Round 2 $p_i \xrightarrow{b_i,b_{i+1}} p_{i-2}$						Round 3 $p_i \xrightarrow{b_{i+2},b_{i+3}} p_{i+2}$					
	b_0	b_1	b_2	b_3	b_4	b_5	b_0	b_1	b_2	b_3	b_4	b_5	b_0	b_1	b_2	b_3	b_4	b_5
p_0	0:5	0:5					0:5	0:5	0:5	0:5			0:5	0:5	0:5	0:5	0:5	0:5
p_1		0:5	0:5					0:5	0:5	0:5	0:5		0:5	0:5	0:5	0:5	0:5	0:5
p_2			0:5	0:5					0:5	0:5	0:5	0:5	0:5	0:5	0:5	0:5	0:5	0:5
p_3				0:5	0:5		0:5			0:5	0:5	0:5	0:5	0:5	0:5	0:5	0:5	0:5
p_4					0:5	0:5	0:5	0:5			0:5	0:5	0:5	0:5	0:5	0:5	0:5	0:5
p_5	0:5					0:5	0:5	0:5	0:5			0:5	0:5	0:5	0:5	0:5	0:5	0:5

Table 4. An example of the concatenation phase of algorithm 3 on $n = 6$ processors.

This algorithm completes in $2\lceil \log n \rceil$ communication rounds. The number of message transmissions performed is $C_2 = 2\lceil \frac{m}{n} \rceil (n - 1)$. Thus, the theoretical communication complexity of the algorithm is $2\lceil \log n \rceil t_s + 2\lceil \frac{m}{n} \rceil (n - 1)t_c$. The number of operations performed by each processor is $\lceil \frac{m}{n} \rceil (n - 1)$. Although this algorithm requires double the rounds of algorithm 2, it performs better with respect to the number of message transmissions and the number of operations performed.

A further advantage of algorithm 3 is that all processors will hold identical results upon its completion, since each block b_i is uniquely determined by processor p_i prior to the concatenation phase. Algorithm 2 does not possess this property, in case numerical roundoff errors violate the assumption of associativity (e.g., addition on binary computers is not necessarily associative).

3 IBM RS/6000 SP message-passing performance

The IBM RS/6000 SP parallel computer system [4] (formerly named SP2), contains from 2 up to 512 standard IBM RS/6000 workstation CPUs mounted in *frames* containing from 2 to 16 CPUs. All of the CPU nodes may be interconnected by an *SP Switch* communications network, which allows the user to view all nodes in the system as being equidistant, owing to very low hardware latencies in the switch.

We have measured the node-to-node communication time using IBM's PVMe implementation of the PVM 3.3 library. In addition, we tested IBM's recent MPI message-passing library [5] subroutines: the basic MPI_Send, the buffered MPI_Bsend, the synchronous MPI_Ssend and the combined send-and-receive MPI_-Sendrecv. For both PVMe and MPI we noticed distinct steps in the timing for messages of about 200, 400, 640, and 880 Bytes. The zero-length timing defines the *latency*, as given in Table 5. For MPI we found that the latency for mpi_send, mpi_bsend and mpi_sendrecv is roughly half of that found for PVMe. The MPI_Ssend is distinguished by a higher latency than that of the PVMe.

Send operation	Latency (μsec)	Bandwidth (MB/s)
pvmfsend/PvmDataDefault	92	26.4
pvmfsend/PvmDataInPlace	92	29.4
pvmfpsend	104	29.4
MPI_Sendrecv	72	31.1
MPI_Bsend	64	29.4
MPI_Send	53	33.2
MPI_Ssend	134	33.2

Table 5. Latency (in microseconds) and asymptotic bandwidth for very long messages (in MBytes/sec) for different implementations of *send* operations (with the reciprocal receive operations).

The measured bandwidths are displayed in Fig. 1. A bandwidth decrease from 60 to 200 kBytes is evident in mpi_bsend and mpi_sendrecv, as for PVMe. The asymptotic bandwidths are given in Table 5. We performed measurements of single messages of length up to 20 MBytes, but saw no deviation whatsoever from the asymptotic values in Fig. 1.

4 Reduction performance on IBM RS/6000 SP

We implemented the three algorithms described in section 2 on an IBM RS/6000 SP parallel computer. We chose the IBM PVMe message-passing library (PVM version 3.3) which is highly tuned for the IBM RS/6000 SP Switch. Furthermore, we used the combined pack-and-send pvmfpsend and the reciprocal pvmfprecv subroutines for transmitting and receiving messages.

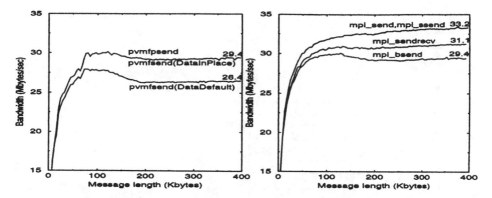

Fig. 1. Bandwidth of node-to-node communication in MBytes/sec as a function of the message length. The PVMe subroutines pvmfsend and pvmfpsend (and corresponding receive subroutines) are shown. The bandwidth of the pvmfsend subroutine using the PvmDataInPlace data encoding is identical to this of the pvmfpsend subroutine. The MPI subroutines are: MPI_Send, MPI_Bsend, MPI_Ssend and MPI_Sendrecv. Notice that the measurements for the mpi_send and mpi_ssend subroutines are identical. The timings are averages over 500 iterations.

The relative performance of algorithms 1, 2 and 3 was experimentally established. We compared only the message passing time of the algorithms, since this is the dominating factor in their performance, and since we mainly want to establish the communication performance. The top three curves in Fig. 2(a) display the ratio of the message passing time required for algorithm 3 over the message passing time required for algorithm 2. The tests were performed on 4, 6, 8, 10, 12, and 16 processors and for vector sizes of up to 400 kBytes. The ratios on 6 and 8 processors were similar (with small deviations), and thus we plotted a single curve. The same was true for the tests on 10,12, and 16 processors. From Fig. 2(a), we notice that for small vector sizes of up to 10 kBytes, algorithm 2 performs slightly better than algorithm 3. However, as the vector size increases, algorithm 3 performs better than algorithm 2. On 4 processors, algorithm 3 is asymptotically 25% more efficient than algorithm 2, on 6 and 8 processors, it is 44% more efficient, and on 10, 12, and 16 processors algorithm 3 is up to 54% more efficient than algorithm 2. These performance measurements can be verified through the theoretical communication model established in section 2. The theoretical communication performance of these algorithms, if we neglect the start-up time, which is insignificant at the level provided on the IBM RS/6000 SP for long messages, is $\lceil \log n \rceil m t_c$ for algorithm 2, and $2\lceil \frac{m}{n} \rceil (n-1)t_c$ for algorithm 3. Thus, their ratio is:

$$\text{Ratio} = \frac{2\lceil \frac{m}{n} \rceil (n-1) t_c}{\lceil \log n \rceil m t_c} = \frac{2(n-1)}{n\lceil \log n \rceil}$$

This ratio for different numbers of processors can be seen in the following table:

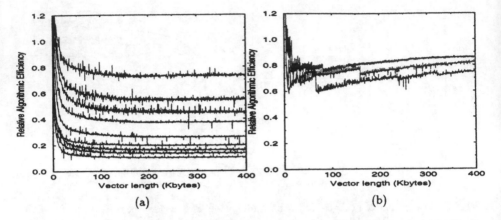

Fig. 2. (a) Ratio of the message passing time required for algorithm 3 over the message passing time required for algorithms 1 and 2 for vector size of up to 400 kBytes. The top three curves are the ratio of algorithm 3 over algorithm 2 on $n = 4$, $n = 6, 8$ and $n = 10, 12, 16$ processors (from top to bottom). The lower six curves are the ratio of algorithm 3 over algorithm 1 on 4, 6, 8, 10, 12, and 16 processors (from top to bottom). (b) Ratio of the total time required for algorithm 3 over the total time required for the MPI_Allreduce subroutine for vector sizes of up to 400 kBytes. From top to bottom, the three lines correspond to timings on 4, 8, and 16 processor. The timings of all algorithms are averages over 500 iterations.

n	4	6	8	10	12	16
Ratio	0.750	0.555	0.583	0.450	0.478	0.469

The ratios on different numbers of processors, derived from the theoretical model, agree asymptotically with the relative efficiency of algorithm 3 over algorithm 2, which we observed experimentally in Fig. 2(a). This relative efficiency would be even more dramatic if the total time required by the algorithm were compared, including the time spent for the reduction operations. Algorithm 3 performs only $\lceil \frac{m}{n} \rceil (n - 1)$ reduction operations, compared to the $2(\lceil \log n \rceil - 1)m$ reduction operations performed by algorithm 2, which has to update two buffers with intermediate results at each round.

The bottom 6 curves in Fig 2(a) are the ratio of the message passing time required for algorithm 3 over the message passing time required for algorithm 1 on 4, 6, 8, 10, 12, and 16 processors, respectively from top to bottom, and for vector sizes of up to 400 kBytes. We notice as expected that algorithm 3 performs much better than the naive algorithm 1, and for 16 processors algorithm 3 is asymptotically a factor of 10 more efficient.

The performance of the most efficient of the algorithms described in section 2, i.e. algorithm 3, was further compared with the performance of the MPI_Allreduce subroutine from the IBM MPI library. This subroutine implements the reduction over a commutative and associative operator of a number of vectors, each one located on a different processor, and in the end the re-

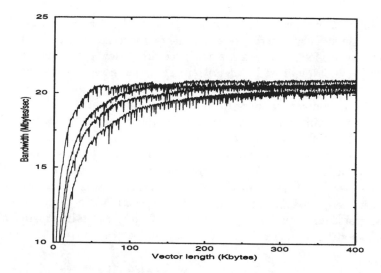

Fig. 3. The top curve is the bandwidth (in MBytes/sec) achieved by algorithm 2 on any number of processors for vector sizes of up to 400 kBytes. The three lower curves are the bandwidth achieved by algorithm 3 on 4, 6, and 12 processors, respectively, for the same vector sizes. The timings of all algorithms are averages over 500 iterations.

duction result is available on all participating processors. The implementation details of this subroutine, as well as the algorithm used, are not documented by IBM, and as a consequence we can only compare the total execution time of our implementation of algorithm 3, with the total execution time of this subroutine, rather than their message passing time. In Fig. 2(b) is displayed the ratio of the total execution time of algorithm 3, over the total execution time of the MPI_Allreduce subroutine. We notice that for vector sizes of more than 5 kBytes, algorithm 3 is more efficient than the MPI_Allreduce subroutine. On 16 processors and for vector sizes of 100-200 kBytes, algorithm 3 is about 40% more efficient than the MPI subroutine. Finally, on 16 processors algorithm 3 is asymptotically 25% more efficient than the MPI_Allreduce subroutine.

The same tests were performed for all three algorithms and the MPI_Allreduce subroutine, for vector sizes of up to 20 MBytes, but no deviations from their asymptotic behavior shown in Fig. 2 was observed.

In Fig. 3 we display the bandwidth in MBytes/sec achieved by algorithms 2 and 3 for vector sizes of up to 400 kBytes. The bandwidth is the ratio of the total size of the messages transmitted over the total time spent for these transmissions. The bandwidth for algorithm 2 was plotted only as one curve, because the measurements on different numbers of processors produced the same curve with almost no deviations. For algorithm 2 we notice that the asymptotic bandwidth, which is about 21 MBytes/sec was achieved for vector sizes of more than 50 kBytes.

For algorithm 3, we notice that a slightly different bandwidth curve was obtained for different numbers of processors: For $n = 4$, $n = 6, 8$, and $n = 10, 12, 16$ three different curves were produced. The reason is that the effect of the latency is more obvious for algorithm 3, since it requires $2\lceil \log n \rceil$ rounds, rather than the $\lceil \log n \rceil$ rounds required by algorithm 2. Furthermore, because of the message splicing technique employed in algorithm 3, messages are transmitted in smaller chunks, in contrast to algorithm 2 where only messages of the initial vector size m are transmitted. For this reason, for algorithm 3 the asymptotic bandwidth is achieved slightly slower as the number of processors increases. Soon the effect of the latency becomes negligible and eventually almost the same asymptotic bandwidth of 21 MBytes/sec is achieved on any number of processors. We tested both algorithms for vector sizes of up to 20 MBytes and we did not observe any deviation from the asymptotic bandwidth behavior.

The behavior of algorithms 2 and 3 can be easily verified from the theoretical communication model established in section 2 as follows:

$$(\text{algorithm 2}) \qquad \text{Bandwidth} = \frac{\lceil \log n \rceil \, m}{\lceil \log n \rceil \, t_s + \lceil \log n \rceil \, m \, t_c} \tag{1}$$

$$(\text{algorithm 3}) \qquad \text{Bandwidth} = \frac{2 \lceil \frac{m}{n} \rceil \, (n-1)}{2 \lceil \log n \rceil \, t_s + 2 \lceil \frac{m}{n} \rceil \, (n-1) \, t_c} \tag{2}$$

Equation (1) verifies that the number of processors does not affect the bandwidth performance of algorithm 2, since the $\lceil \log n \rceil$ factor cancels out, which is exactly what we observe experimentally. This is not the case for algorithm 3, where we see that the number of processors affects the performance, so that the asymptotic bandwidth is reached slightly slower as the number of processors increases. If we take $t_s = 140 \mu sec$ and $1/t_c = 21$ MBytes/sec, equations (1) and (2) accurately fit the corresponding curves in Fig. 3. This means that our implementations achieve a satisfactory bandwidth of 21 MBytes/sec, compared to the 29 MBytes/sec achieved through the round trip tests (see Fig. 1).

5   Conclusions

In this paper we have investigated in detail three algorithms for the global combine operations on distributed-memory parallel computers. Two of the algorithms scale as the logarithm of the number of processors. Such efficient algorithms are important for important classes of production codes on parallel computers. An important property of algorithm 3 is that it would lead to identical result on all processors even if the reduction operator were not associative, in contrast to algorithm 2. This can be of practical importance, for example, when the *summation* operator is used.

On an IBM RS/6000 SP, the implemented combine algorithms achieved a satisfactory bandwidth of 21 MBytes/sec. The performance measurements on the IBM RS/6000 SP show that algorithm 3 is 25-44% faster than algorithm 2, and that the naive algorithm 1 is far inferior to the other algorithms when the vector length exceeds a few kBytes. Algorithm 3 implemented using IBM's

PVMe implementation of the PVM library is about 30% more efficient than the IBM MPI library's **MPI_Allreduce** subroutine on the IBM RS/6000 SP.

References

1. B. Hammer and Ole H. Nielsen, *Parallel Ab-Initio Molecular Dynamics* in proceedings of the *Workshop on Applied Parallel Computing in Physics, Chemistry and Engineering Science* (PARA'95), August 21-24, 1995, ed. J. Wasniewski, Springer Lecture Notes in Computer Science, vol. 1041, pp. 295.
2. J. Bruck and C.-T. Ho, *Efficient Global Combine Operations in Multi-Port Message-Passing Systems*, Parallel Processing Letters vol. 3(4), pp. 335, 1993.
3. J. Bruck, C.-T. Ho, S. Kipnis, and D. Weathersby, *Efficient Algorithms for All-to-All Communications in Multi-Port Message Passing Systems*, manuscript.
4. IBM's RS/6000 SP documentation is available on WWW:
   URL:http://www.rs6000.ibm.com/software/sp_products/sp3.html.
5. IBM's Parallel Environment (PE) product, which includes the MPI library, is described on WWW:
   URL:http://www.rs6000.ibm.com/software/sp_products/pe.html.

A Vectorization Technique for a Family of Finite Difference Formulae and Its Performance Evaluation

Seiji Fujino[1] and Ryutaro Himeno[2]

[1] Hiroshima City University, Faculty of Information Sciences, 151-5 Ozuka, Numata-cho, Asaminami-ku, Hiroshima, 731-31 Japan
[2] Nissan Motor Co., Ltd., Nissan Research Center, 1 Natsushima-cho, Yokosuka, Kanagawa, 237 Japan

Abstract. In this contribution the vectorization technique on the current vector supercomputer are derived from the relation between the number of gridpoints in the x, y directions in three dimension. This technique is applied to vectorization of the SOR method. Moreover the actual efficiency on the vector supercomputer is examined, and it is shown that the SOR method vectorized by this technique has a high efficiency as more than 90% of the maximum speed of the vector supercomputer.

1   Introduction

Finite Difference Formulae are commonly derived from the relation between the updated gridpoint and several related gridpoints in the neighborhood. The number of neighboring gridpoints increases as higher accuracy of finite difference formulae[5]. This fact means that the number of gridpoints which can be treated independently on the computers decreases. That is, this is major disadvantage in view of utilizing effectively the current vector supercomputer.

On the other hand, it is generally known that data in main memory of the computer should be accessed continuously, and length of vector is preferable to be as long as possible on the vector supercomputer. Moreover, when many difference formulae are used in the various applications, the analytic domain is commonly a rectangle or rectangular parallelpiped. Therefore, in the contribution we impose the number of gridpoints in the x, y directions so as to be kept with a certain relation for gaining a higher performance of the vector supercomputer.

For example, in case of the stencil with 9 points in two dimension, the number of gridpoints in the x direction: N_x is determined as $N_x \equiv 2 \pmod 4$. All the gridpoints are supposed to be ordered lexicographically. Owing to this remedy, longer vector length is gained as order of $O(\frac{N}{4})$, where N is the total number of the gridpoints, and the access strides to main memory become always one[3]. In this article, we apply the proposed method to vectorization of the SOR(Successive Over-Relaxation) method on the vector supercomputer, and investigate its high efficiency [1] [4] [9].

2 Repeated Lexicographical Coloring method

Let N_x, N_y and N_z be the numbers of gridpoints in the x, y and z direction, respectively. When all the gridpoints in the analytic region are ordered lexicographically, the ordering number of updated gridpoints (i, j, k) can be written as follows.

$$S(i, j, k) = i + N_x(j - 1) + N_x N_y(k - 1). \tag{1}$$

Let c be the number of coloring of gridpoints. When all the gridpoints in the analytic region are colored, it is clear that the gridpoints with same color cannot be updated simultaneously on the vector supercomputers. Therefore we can see that the color of the updated gridpoint (i, j, k) with m points stencil differs from that of $(m - 1)$ gridpoints in the neighborhood only if the following condition is satisfied.

$$S(i_l, j_l, k_l) - S(i, j, k) \not\equiv 0 \pmod{c}, \quad (l = 1, 2, \cdots, m - 1). \tag{2}$$

In this case, the update of gridpoint (i, j, k) can be made independently from the neighboring $(m - 1)$ gridpoints. That is, the vectorization of gridpoint (i, j, k) can be realized. It is well known that the so-called multicolor ordering method for iterative methods are very efficient on the vector supercomputers [2] [3] [6] [7] [8]. Therefore in this article we call it the R.L.C. (Repeated Lexicographical Coloring) method with including the decision of the number of the gridpoints in the x, y directions (see Fig.2).

Moreover, we investigate an efficiency of the R.L.C. method on the vector supercomputer. It is desirable that the number of coloring: c is as small as possible for gaining longer vector length, and it is the most efficient when an array is asigned per every color. In the next subsection the relation on the number of gridpoints in the x, y directions on two kinds of stencils with 9 points in 3D and 7 points in 2D are introduced.

2.1 Application of R.L.C. to stencil with 9 points in 2D

In Fig.1 it is shown that the addresses of stencil with 9 points : $(i-1, j-1), (i, j-1), (i+1, j-1), (i-1, j), (i, j), (i+1, j), (i-1, j+1), (i, j+1)$ and $(i+1, j+1)$ in 2D as shown in Fig. 4(b). The address of updated gridpoint (i, j) is written as $S(i, j) = i + N_x(j - 1)$.

$M + N_x - 1$	$M + N_x$	$M + N_x + 1$
$M - 1$	$M^{(*)}$	$M + 1$
$M - N_x - 1$	$M - N_x$	$M - N_x + 1$

$$(^*)M = i + N_x(j - 1)(= S(i, j))$$

Fig. 1. The addresses of stencil with 9 points in 2D.

Utilizing the characteristic that the stencil is symmetric, we consider about only half of the eight gridpoints in the neighborhood. The differences of the address between the four gridpoints: $(i+1,j)$, $(i-1,j+1)$, $(i,j+1)$ and $(i+1,j+1)$ and the updated gridpoint (i,j) are written using the number of coloring c as follows.

$$S(i+1,j) - S(i,j) = 1 \not\equiv 0, \tag{3}$$
$$S(i-1,j+1) - S(i,j) = -1 + N_x \not\equiv 0 \ (\text{mod } c), \tag{4}$$
$$S(i,j+1) - S(i,j) = N_x \not\equiv 0 \quad (\text{mod } c), \tag{5}$$
$$S(i+1,j+1) - S(i,j) = 1 + N_x \not\equiv 0 \ (\text{mod } c). \tag{6}$$

From eqns. (4), (5) and (6), it can be derived that $N_x \equiv 2 \ (\text{mod } 4)$ for the necessary condition of the R.L.C. method with the least number of coloring. In Fig.2 we show an example of the 4-colored gridpoints by the R.L.C. method on two dimensional analytic region when N_x is six. In this case it is trivial that $N_x \equiv 2 \ (\text{mod } 4)$. As you can see from Fig.2, four colos ($\bullet$, $\circ$, $\clubsuit$ and $\diamondsuit$) which are ordered lexicographically from the left-under corner are distributed repeatedly.

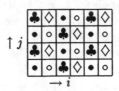

Fig. 2. An example of the 4-colored gridpoints by the R.L.C. method on two dimensional analytic region when N_x is six.

2.2 Application of R.L.C. to stencil with 7 points in 3D

In the same way, it is shown that the addresses of stencil with 7 points : $(i,j,k-1)$, $(i,j-1,k)$, $(i-1,j,k)$, (i,j,k), $(i+1,j,k)$, $(i,j+1,k)$ and $(i,j,k+1)$ in 3D as shown in Fig.5. The address of updated gridpoint (i,j,k) is written as $S(i,j,k) = i + N_x(j-1) + N_x N_y(k-1)$.

		$\boxed{M + N_x}$			
$\boxed{M - N_x N_y}$	$\boxed{M-1}$	$M^{(*)}$	$\boxed{M+1}$		$\boxed{M + N_x N_y}$
		$\boxed{M - N_x}$			
$k-1$		k			$k+1$

Fig. 3. The addresses of stencil with 7 points in 3D as shown in Fig. 5.
$(*)M(= S(i,j,k)) = i + N_x(j-1) + N_x N_y(k-1)$.

As the stencil is symmetric, we consider also about only half of the six gridpoints in the neighborhood. The differences of the addresses between three gridpoints: $(i+1,j,k)$, $(i,j+1,k)$ and $(i,j,k+1)$ and the updated gridpoint (i,j,k)

are shown also using the number of coloring c as follows.

$$S(i+1, j, k) - S(i, j, k) = 1 \not\equiv 0, \tag{7}$$
$$S(i, j+1, k) - S(i, j, k) = N_x \not\equiv 0 \pmod{c}, \tag{8}$$
$$S(i, j, k+1) - S(i, j, k) = N_x N_y \not\equiv 0 \pmod{c}. \tag{9}$$

From eqns. (8) and (9), we can gain the condition: $N_x \equiv 1$ and $N_y \equiv 1$ (mod 2) for 7 points in 3D as the necessary condition for the R.L.C. method with the least number of coloring c .

3 Application of R.L.C. method to various stencils

In this section the conditions on N_x and N_y to various stencils in 2D and 3D in the R.L.C. method are introduced.

3.1 Stencils in 2D

In Fig.4 we show four kinds of stencils with 5, 9, 9 and 13 points in two dimension. In Table 1 the conditions on N_x for these stencils in the R.L.C. method are presented.

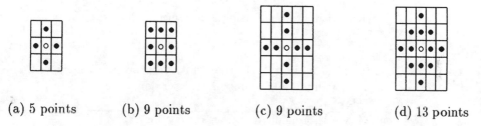

(a) 5 points   (b) 9 points   (c) 9 points   (d) 13 points

Fig. 4. Stencils with 5, 9, 9 and 13 points in two dimension.

Table 1. Conditions on N_x in the R.L.C. method for four kinds of stencils with 5, 9, 9 and 13 points.

no. of points	Stencil	Conditions on N_x
5	Fig. 4(a)	$N_x \equiv 1 \pmod 2$
9	Fig. 4(b)	$N_x \equiv 2 \pmod 4$
9	Fig. 4(c)	$N_x \equiv 1$ or $N_x \equiv 2 \pmod 3$
13	Fig. 4(d)	$N_x \equiv 2$ or $N_x \equiv 3 \pmod 5$

3.2 Stencils in 3D

In Fig. 5, 6 and 7 we show stencils with 7, 19 and 27 points using three z-planes indexed with $k-1$, k and $k+1$ in three dimension. In Table 2 the conditions on N_x and N_y for these stencils in the R.L.C. method are presented, respectively.

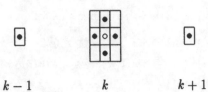

$$k-1 \qquad\qquad k \qquad\qquad k+1$$

Fig. 5. Stencil with usual 7 points using 3 z-planes in three dimension.

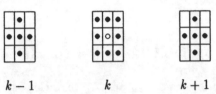

$$k-1 \qquad\qquad k \qquad\qquad k+1$$

Fig. 6. Stencil with 19 points using 3 z-planes in three dimension.

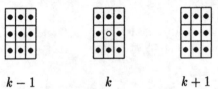

$$k-1 \qquad\qquad k \qquad\qquad k+1$$

Fig. 7. Stencil with 27 points using 3 z-planes in three dimension.

Table 2. Conditions on N_x and N_y in the R.L.C. method for stencils with 7, 19 and 27 points in three dimension.

no. of points	Stencil	Conditions on N_x and N_y
7	Fig. 5	$N_x \equiv 1$ and $N_y \equiv 1 \pmod 2$
19	Fig. 6	$N_x \equiv P$ and $N_y \equiv \pm P \pmod 7$ $(P = 2, 3, 4, 5)$
27	Fig. 7	$N_x \equiv Q$ and $N_y \equiv Q \pmod 8$ $(Q = 2, 6)$

In Fig. 8 an example of seven coloring for 19 points on the three z-planes: $k-1$, k and $k+1$ in the neighborhood for updating the gridpoint (i, j, k) marked with o is shown.

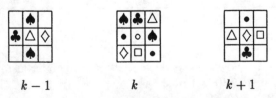

$$k-1 \qquad\qquad k \qquad\qquad k+1$$

Fig. 8. An example of seven coloring for 19 points in 3D.

4   Numerical Experiments

In this section the performance of the multicolored SOR method vectorized by the proposed R.L.C. method is shown. All the computations were carried out on NEC SX-4/2C with single processor, capable of 2 Gflops, in double-precision floating-point arithmetic. We have chosen two stencils as shown in Fig.4(b) and Fig.6. Here we consider the model elliptic partial differential equations defined in the boundary-fitting coordinate

$$-\nabla^2 u(x, y) = b \quad \text{on} \quad [0, 1]^2, \tag{10}$$

$$-\nabla^2 u(x, y, z) = b' \quad \text{on} \quad [0, 1]^3. \tag{11}$$

with Dirichlet boundary conditions.

We solved the discretized equation $Au = b$ of eqn. (10) using the 4-colored SOR method, and the discretized equation $A'u = b'$ of eqn. (11) using the 7-colored SOR method, where A and A' is nonsymmetric matrices, and b and b' are the constant vectors, respectively. The stopping criterions for successful convergence were $\|r_k\|_2/\|b\|_2 \leq 10^{-5}$, $\|r'_k\|_2/\|b'\|_2 \leq 10^{-5}$, where r_k, r'_k are the residual for the k-th iteration, and the Euclidian norm is used, and the components of the initial vector u_0 are chosen all zeros.

In Table 3(a),(b) we show the performance of the 4-colored SOR method and that of the 7-colored SOR method. In these Tables, column 1 lists the vector length (V.L.), column 2 lists the number of gridpoints N_x, N_y and N_z which represent in the x, y, and z directions, column 3 lists the computational rates in Mflops and column 4 lists the ratios to the maximum speed of single processor of SX-4/2C. Some observations can be made from these Tables as follows. In addition N_x, N_y were determined as being satisfied with the conditions as shown in Table 1 and 2.

1. When the vector length is over 90, more than half performance of the maximum speed can be attained.
2. The performance of the 4-colored SOR method vectorized by the R.L.C. method for the stencil with 9 points in two dimension is more than 90% of the maximum speed at the vector length of about 12000.
3. The performance of the 7-colored SOR method vectorized by the R.L.C. method for the stencil with 19 points in three dimension is 94% of the maximum speed at the vector length of about 15000.

Table 3. Performance of (a)the 4-colored SOR method for stencil with 9 points and that of (b)the 7-colored SOR method for stencil with 19 points.

(a)4-colored SOR method

V.L.[*]	(N_x, N_y)	Mflops	Ratio
90	(18, 18)	1.080	0.540
143	(23, 23)	1.271	0.636
210	(28, 28)	1.331	0.666
380	(38, 38)	1.435	0.718
870	(58, 58)	1.606	0.803
1023	(63, 63)	1.686	0.843
1368	(73, 73)	1.703	0.852
1980	(88, 88)	1.769	0.885
4160	(128, 128)	1.783	0.892
8010	(178, 178)	1.790	0.895
11990	(218, 218)	1.799	0.900
13110	(228, 228)	1.794	0.897
14883	(243, 243)	1.802	0.901

(*)V.L.: Vector Length

(b)7-colored SOR method

V.L.[*]	(N_x, N_y, N_z)	Mflops	Ratio
831	(23,23,11)	1.748	0.874
1414	(30,30,11)	1.792	0.896
1587	(23,23,21)	1.752	0.876
2151	(37,37,11)	1.834	0.917
2342	(23,23,31)	1.817	0.909
4087	(51,51,11)	1.862	0.931
6062	(37,37,31)	1.871	0.934
8018	(37,37,41)	1.866	0.933
11339	(44,44,41)	1.877	0.939
15234	(51,51,41)	1.880	0.940

5 Conclusions

We examined efficiencies of the proposed R.L.C. method for various stencils with many gridpoints. Numerical experiments were done on the vector supercompter SX-4/2C. As a result, it turned out that the R.L.C. method for vectorization is extremely efficient as more than 90% of the maximum speed of the vector supercomputer.

References

1. Adams, L.M., Jordan, H.F.: Is SOR color-blind ?. SIAM J. Sci. Stat. Comput. **7**(1986) 490–506
2. Brand, C.W.: An incomplete-factorization preconditioning using repeated red-black ordering. Numer. Math. **61**(1992) 433–454
3. Fujino, S., Tamura, T., and Kuwahara, K.: Multicolored Poisson Solver for Fluid Flow Problems. Proc. of the 8-th GAMM Conference on Numerical Methods in Fluid Mechanics The Netherlands (1989) 148–158
4. Hageman, L.A., Young, D.M.: Applied Iterative Methods Academic Press, New York 1981
5. Iwatsu, R., Hyun, J.M., and Kuwahara, K.: Driven Cavity Flow with Stabilizing Temperature Stratification. AIAA Paper 92-0713 Reno NV 1992
6. Jones, M.T., Plassmann, P.E.: The effect of many color orderings on the convergence of iterative methods. Proc. of Copper Mountain Conference on Iterative Methods **2** Colorado 1992
7. Poole, E., Ortega, J.M.: Multicolor ICCG Methods for Vector Computers. SIAM J. Numer. Anal. **24**(1987) 1394–1418

8. Ramdas, M., Kincaid, D.R.: Parallelizing ITPACKV 2D for the CRAY Y-MP. Proc. of IMACS Int. Conference on Iterative Methods in Linear Algebra 1992 323–347
9. Young, D.M.: Iterative Solution of Large Linear Systems. Academic Press New York 1971

Parallel Simulation of Finishing Hot Strip Mills

Daniel García*[1], Francisco J. Suárez[1], Javier García[1], Jose M. López[1],
Faustino Obeso[2], Jose A. González[2]

[1] University of Oviedo, Campus de Viesques, 33204 Gijón Spain
[2] Integral Siderurgic Corporation SA, Apartado 93, 33080 Avilés Spain

Abstract. This paper[3] deals with the development of a simulator that generates the appropriate signals to test and validate high performance monitoring systems for hot strip mills under a controlled environment. The paper begins with an introduction to the model and its objectives. Next a presentation of the physical installation is done. The third section is devoted to explain the mathematical model of the hot strip mill, splitting the global model in several subparts. The following section explains briefly the design of the simulator and its implementation on a distributed memory multiprocessor system. Finally, several simulation results of the current implementation of the simulator are presented.

1 Introduction

A strategy for developing real-time observers for complex industrial processes consists in the developing of a simulator of the process. The observer and simulator models are based normally in the same mathematical principles but they are fed with different sets of data. The simplest mathematical models [1, 2, 3] of the industrial process which proportionate the desired precision should be used in the simulator, because the same mathematical models will be run in real-time in the observer.

On the other hand, in the development of high performance specialized real-time observers and monitors for large industrial installations, such as hot strip mills, the proper operation of observers before placing the field-instrumentation in the industrial installation and conditioning the rooms for computing equipment should be ensured.

To debug and validate the proper operation of observers and monitors, sets of signals coming from the industrial installation should be used. But there are some problems in using signals retrieved directly from the industrial installations:

- Exact knowledge of all operational parameters of a large installation during signal capturing is an almost impossible task.
- Execution of controlled experiments to analyze the influence of the parameters of the installation over the final quality of manufactured products is an unaffordable task due to the economic losses originated by the production of poor quality strips during the experimentation.

---

* Corresponding author. Email: daniel@etsiig.uniovi.es
[3] This work has totally been funded by the ESPRIT HPC project 8169 ESCORT

For all these reasons, the necessity of a detailed simulator of the signals generated by hot strip mills arose in the development of high performance monitoring systems for this type of large industrial installation.

2 The physical process: the hot-strip mill installation

The finishing hot strip mill area is composed of seven stand with loopers in between (see figure 1) which produce a progressive thickness reduction on the strip.

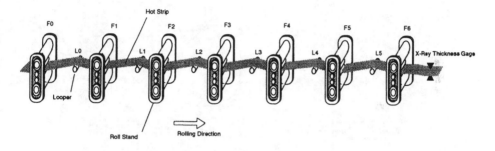

Fig. 1. General view of the finishing hot strip mill

Each stand is formed by two pairs of rolls. One pair are the work rolls of small diameter, mechanically coupled, moved by a motor-reduce unit and working in direct touch with the strip. The another pair are the backup rolls of greater diameter, no motorized and with the mission of pressing on the work rolls to obtain the desired strip thickness reduction. The bottom backup roll is fixed in the vertical axis while the top one is operated by an electromechanical screwdown system for performing strip thickness adjustments.

The loopers are six swinging rolls which keep the strip tension between the rolling stands constant and absorb the differences of material flow, being its main aim to avoid jams because of speed differences between the output of one stand and the input of the following one. Moreover, each looper acts as a speed regulator for the upstream stand, that is, generates speed references signals for all the stands upstream to the looper which try to keep equal the entry strip speed and the exit strip speed of all the upstream stands.

The entry material and output product of the finishing hot strip mill is a long and thin strip of steel at very high temperature. As the strip get through the roll stands its thickness is reduced while it becomes longer.

Along the hot strip mill there also is complex instrumentation, which allows the measurement of working variables: load cells to measure the roll separating force at each stand, X-ray thickness gauge at the end of the mill, two temperature sensors at the beginning and at the end of the mill, etc. With these measured variables the control system of the mill tries to keep the strip thickness as constant as possible moving the screwdown system on each stand.

3 The mathematical model

In order to simulate the mill, a set of equations or tables for each of the basic elements that conform the mill is used. Three different submodels can be easily identified: the stand, strip and looper submodels. In figure 2 the relations between the basic submodels are shown.

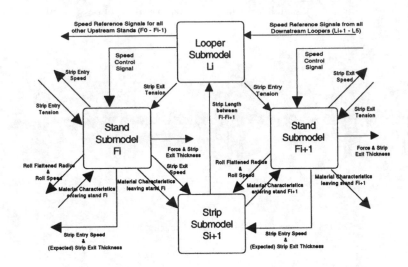

Fig. 2. Submodel interconnections

3.1 The stand submodel

This is the component of the model that determines the state of the physical stand. The equation that rules the relation between the thickness of the strip at the exit of a stand h_2, the roll separating force F and the mechanical state of the stand S [4, pages 615–618] is:

$$h_2(x) = \int_0^F \frac{1}{K_s}\, dF + S(x) \tag{1}$$

where x is the length along the roll central axis, which indicates that the strip can have different thicknesses along its width. Although the stand frame is a very rigid structure, the roll separating force F stretches it by the amount $\frac{F}{K_s}$ (the stand behaves like a spring). The stand structural stiffness K_s (force divided by stand stretch) is not constant, but depends on the load of the stand. The relation $\frac{F}{K_s}$ for different stands can be seen in figure 3. These graphs are obtained from calibration tests carried out on the CSI[4] hot strip mill. The term S (see figure 4) is also known as the stand open gap and can be expressed as:

[4] CSI: Corporación Siderúrgica Integral (Integral Siderurgic Corporation)

$$S = dbb(t) - [\, r_{tw}(\pi/2 - \omega_w t, x) + r_{tw}(3\pi/2 - \omega_w t, x) + 2\Delta R_{twt}(x) +$$
$$r_{bw}(\pi/2 + \omega_w t, x) + r_{bw}(3\pi/2 + \omega_w t, x) + 2\Delta R_{bwt}(x) +$$
$$r_{tc}(3\pi/2 + \omega_{tb} t, x) + r_{ts}(\pi/2 + \omega_{bb} t, x) - d_{ty} +$$
$$r_{bc}(\pi/2 - \omega_{tb} t, x) + r_{bs}(3\pi/2 - \omega_{bb} t, x) + d_{by} +$$
$$h_{toil} + h_{boil} - y_{tw}(x) - y_{bw}(x) - \delta_w(x)] \tag{2}$$

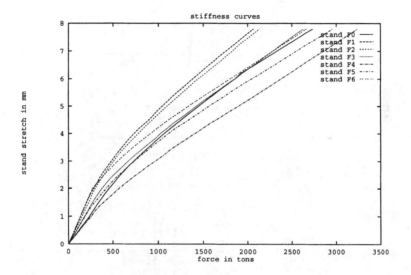

Fig. 3. Stiffness K_s curves for different stands

The roll radii of equation 2 are functions of the angular position and the location along the roll axis x, that is, the radii may change around the periphery and the peripheries may also change along the roll length. The radii r_{tw}, r_{bw}, r_{tc}, r_{bc}, r_{ts} and r_{bs} take into account the wear and the constructive crown which may be different along the x dimension of the roll.

The terms d_{ty} and d_{by} represent the eccentricities (different center) between the backup shafts and the backup contours. In fact, d_{ty} and d_{by} are only the projection of the eccentricities in the vertical axis, as can be seen in figure 4. The equations for these parameters are:

$$d_{ty} = -ecc_{tb}\sin(\theta_{t0} - \omega_{tb}t) \tag{3}$$
$$d_{by} = ecc_{bb}\sin(\theta_{b0} + \omega_{bb}t) \tag{4}$$

where ecc_{tb} and ecc_{bb} are the distance between the centers of the shaft and the contour of the top and bottom backup rolls and θ_{t0} and θ_{b0} are the initial angle (with the horizontal) of the line that joins those points.

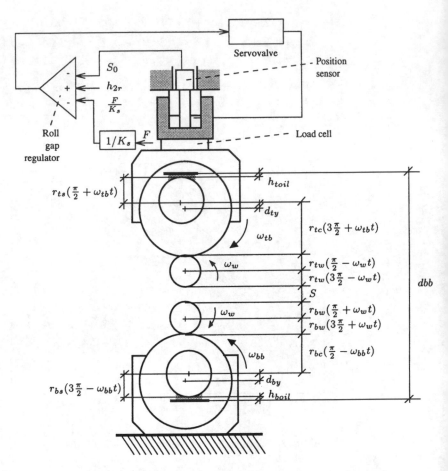

Fig. 4. Stand mechanical elements and screwdown controller

The top and bottom work rolls are mechanically coupled, having the same speed ω_w. The speed of top ω_{tb} and bottom ω_{bb} backup rolls are given by equation 5, where R_{tw}, R_{tb}, R_{bw} and R_{bb} are the mean values of the radii of each roll.

$$\omega_{tb} = \omega_w \frac{R_{tw}}{R_{tb}} \qquad \omega_{bb} = \omega_w \frac{R_{bw}}{R_{bb}} \tag{5}$$

S depends also on the distance between bearings dbb which is changed by the position of the actuators, the height of the oil-film layer in the backup rolls, h_{toil} and h_{boil}, and on those effects that change the geometry of the rolls , including the change due to thermal dilatation of the work rolls (ΔR_{twt} and ΔR_{bwt}) and their deflection and contraction due to load (y_{tw}, y_{bw} and δ_w).

The terms y_{tw} and y_{bw} determine the vertical displacement of the center of the work rolls due to load but they do not consider the contraction of the surface of the work roll in touch with the strip. The term δ_w is the contraction of the surface of the work rolls due to the load produced by the roll separating force.

3.2   The looper submodel

Changes in screw positions and entering strip thickness may cause changes in the speeds of the strip in and out of the stand, so control of the speed of adjacent stands is necessary. This is done using the position of the loopers. If any slack is formed between adjacent stands the looper will rise. If the strip becomes tighter, the looper will be pulled down trying to keep the strip tension constant. With the position of the loopers downstream of given stand, the speed control system of the mill generates a speed control signal for the stand. The parts that compose the looper model are: looper position, strip tension and speed regulation signal.

Looper position The angle θ of figure 5 is calculated solving the transcendental equation 6, where K_1 to K_4 are constants based on the looper geometry.

$$K_1 + K_2 \cos^2 \theta + K_3 \cos \theta + K_4 \sin \theta = 0 \tag{6}$$

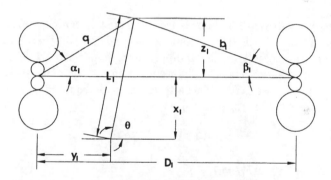

Fig. 5. Looper geometry

Strip tension This mathematical model describes the strip tension between two consecutive stands. This tension will be given by the actuation of the looper between the stands, according with the equation:

$$St_{1_{i+1}} = St_{2_i} = \frac{k_l I_l}{L_l h_{2_i} w C_l} \tag{7}$$

where I_l is the looper motor current, $k_l I_l$ is the looper motor torque and C_l is a constant based on the looper geometry.

Speed regulation signal The position of the looper θ is read and compared with the equilibrium position θ_{equi}. The resultant error is sent to the speed control system which will modify the speed reference in order to maintain the loopers in their equilibrium position.

3.3 The strip submodel

The strip submodel has three basic components: calculation of material resistance, determination of the change of material temperature and tracking of material characteristics from one stand to the next.

Material resistance This simulator uses the method of Alexander-Ford [4, pages 269–290] to calculate the material resistance according with the equation:

$$K_w = 0.5 \left(\pi + \frac{2\sqrt{R_w(h_1 - h_2) - \frac{(h_1 - h_2)^2}{4}}}{h_1 + h_2} \right) k \tag{8}$$

where $k = f(T, \lambda, \alpha_{ch})$ is the shear yield stress of the strip, T its temperature, α_{ch} is the alloy chemical composition and λ is the strain rate defined as:

$$\lambda = \omega_w \sqrt{\frac{R_w}{h_1}} \left[1 + \frac{h_1 - h_2}{4h_1} \right] \sqrt{\frac{h_1 - h_2}{h_1}} \tag{9}$$

Temperature rundown The knowledge of local strip temperature is necessary for the determination of the shear yield stress and the roll thermal crown. The thermal model chosen relates the temperature change from one stand to the next in terms of radiation loss ΔT_w, an energy input ΔT_H due to rolling and a conduction or water loss ΔT_W. These terms may be equated as follows (model developed by Lee et al [5, chapter 18]):

$$\Delta T_R = -19.304 \frac{t}{h_2} T^4 \times 10^{-12} \,^\circ C \tag{10}$$

$$\Delta T_H = 1.653 \frac{P_D}{Q} \,^\circ C \tag{11}$$

$$\Delta T_W = -(0.3 \Delta T_H + 17.5) \,^\circ C \tag{12}$$

where t is the time for any part of the strip to move from one stand to the next, T is the temperature immediately ahead of any stand, h is the strip thickness leaving the stand, P_D is the power applied to the drive motor and Q is the throughput of the mill.

Strip speed and material tracking During the simulation, the strip characteristics (temperature and thickness) are forwarded from one stand to the next according with a calculated movements based on a strip speed model. The equations for entry(i=1)/leaving(i=2) strip speeds and the length of strip accumulated between stands are:

$$V_i = \omega_w R_w \cos\left(\frac{1}{2} \sqrt{\frac{h_1 - h_2}{R_w}} \right) \frac{3h_2 + h_1}{4h_i} \quad i = 1, 2 \tag{13}$$

$$d_{i,i+1} = \int_0^t (V_{1_i} - V_{2_{i+1}}) \, dt \tag{14}$$

4 Simulator Design and Implementation

This hot strip mill model simulates two different working modes of a real hot strip mill installation:

- Calibration mode.
- Production mode.

In the first case no material is rolled. The work rolls of each stand are tightly pressed together while they are turned. Different combinations of tightening and speed produce three calibration tests: elasticity calibration test, oil-film calibration test and initial eccentricity calibration test.

In production mode the simulator must generate the signals which a hot strip mill would generate during the lamination process of a strip.

The simulator is designed to take advantage of the possibilities of code parallelization. The nature of the problem leads to an implementation based on replicated processes running in parallel, conforming to a typical data pipeline where data generated by a process upstream feeds a process immediately downstream. Three basic processes have been implemented which simulate different parts of the hot strip mill: stand process, strip process and looper process.

The global model will replicate these submodels, seven times for the stand submodel (from F0 to F6) and six times for the looper submodel (from L0 to L5). The strip model is also replicated seven times (from the entrance to stand F0, S0 to the entrance to stand F6, S6).

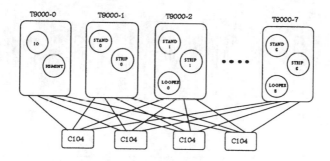

Fig. 6. Allocation of processes to processors

Each of the basic processes implements a mathematical model, which involves a set of equations which must be solved simultaneously. There are physical aspects of the hot strip mill that have not yet been modeled, but currently seven systems of 44 non linear equations must be solved for each sampling time.

The program is run on a PARSYS SN9500 distributed memory machine with 8 T9000 CPUs. The nature of the problem leads naturally to the mapping of one stand, one strip and one looper process per CPU, with the free CPU devoted to communication with the host, as is shown in figure 6.

5 Simulation results

The implementation of the model allows a significant degree of freedom to characterize the behavior of each of the components of the mill. The simulation program takes the following configuration parameters from files: the profile of each roll of the mill, the curves force-speed-oil_height and force-stretch of each stand, the initial speed and position of the actuators, the characteristics of the alloy with the curves temperature-strain_rate-shear and finally length, thickness, width and temperature of the strip that enters the first stand. Using data supplied by the CSI technicians, the parameters of the model have been tuned to simulate the behavior of the CSI mill.

Figure 7 presents the force simulated during an initial eccentricity test. It can be seen the typical beat phenomenon [6] which is produced under "kiss rolling" conditions. During each beat cycle, the roll force varies from a minimum to maximum value and then back to a minimum value. The roll force is maximum when a mutual angular position of the backup rolls is such that the eccentricities of the top and bottom backup rolls are added, and is minimum when the eccentricities are subtracted.

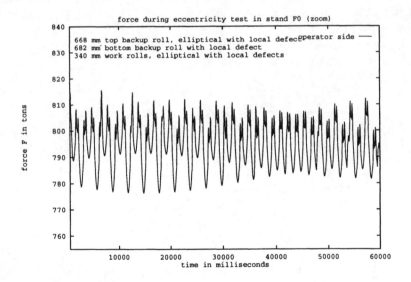

Fig. 7. force during eccentricity test

The curves in figure 8 present the simulated roll separating forces of the seven stands of the mill during the rolling of a single coil. In this production simulation, the entry strip to the first stand is 100 meters long, 20±0.1 mm thick, 1.5 m wide, has a head temperature of 1060°C with a rundown from head to tail of 20°C, and a variation around the mean value of ±2°C. The alloy characteristics are of a steel type number one. The rolls are chosen from a set of ten typical roll profiles.

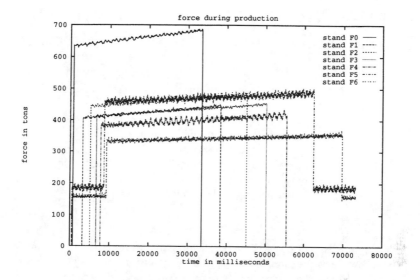

Fig. 8. forces during production

In this case, the stand F6 has work rolls of 340 mm of radius, one perfect and the other with a local defect of 0.04 mm. The top backup roll has a 708–708.04 mm elliptical profile. The bottom backup roll has a diameter of 682 mm and a local defect of 0.04 mm. All the rolls have a constructive crown of 0.25 mm.

References

1. Vladimir B. Ginzburg: "Basic principles of customized computer models for cold and hot strip mills". Iron and Steel Engineer, September 1985, pp. 21–35
2. A. Lage, J Pellicer et al: "Modeling and Simulation of a Hot Strip Rolling Mill". Proceeding of the IECON 94: 20th International Conference on Industrial Electronics Control and Instrumentation. September 1994, pp. 2017–2022
3. Vladimir B. Ginzburg: "High-quality steel rolling: theory and practice". Marcel Dekker Inc, 1993.
4. Vladimir B. Ginzburg: "Steel Rolling Technology: theory and practice". Marcel Dekker Inc, 1989.
5. William L. Roberts: "Hot Rolling of Steel". Marcel Dekker Inc, 1983.
6. Vladimir B. Ginzburg: "Rolling Mill Technology Series, Volume 1: Roll Eccentricity". United Engineering, Inc. International Rolling Mill Consultants, Inc. Pittsburg, Pennsylvania, 1992.

PEPE: A Trace-Driven Simulator to Evaluate Reconfigurable Multicomputer Architectures*

José M. García[1], José L. Sánchez[2], Pascual González[2]

[1] Universidad de Murcia, Facultad de Informática
Campus de Espinardo, 30071 Murcia, Spain
jmgarcia@dif.um.es
[2] Universidad de Castilla-La Mancha, Escuela Politécnica
Campus Universitario, 02071 Albacete, Spain
{jsanchez,pgonzalez}@info-ab.uclm.es

Abstract. Recent research on parallel systems with distributed memory has shown that the most difficult problem for system designers and users is related with the interconnection network. In this paper, we describe a programming and evaluating tool for multicomputers, named PEPE. It allows the execution of parallel programs and the evaluation of the network architecture. PEPE takes a parallel program as input and generates a communication trace obtained from this program. Next, PEPE simulates and evaluates the behaviour of a multicomputer architecture for this trace. The most important parameters of the multicomputer can be adjusted by the user. PEPE generates performance estimates and quality measures for the interconnection network. Another important feature of this tool is that it allows us to evaluate networks whose topology is reconfigurable. Reconfigurable networks are good alternatives to the classical approach. However, only recently this idea became the focus of much interest, due to technological developments (optical interconnection) that made it more viable. A reconfigurable network yields a variety of possible topologies for the network and enables the program to exploit this topological variety to speed up the computation.

1 Introduction

The growing demand for high processing power in various scientific and engineering applications has made multiprocessor architectures increasingly popular. This is exemplified by the proliferation of a variety of parallel machines with some diverse design philosophies. This diversity in architectural design has created a need for developing performance models and simulators for multiprocessors, not only to analyze the effectiveness of a design, but also to reduce the design time.

Distributed memory multiprocessors (often called multicomputers) are increasingly being used for providing high levels of performance for scientific applications. The distributed memory machines offer significant advantages over their shared memory counterparts in terms of cost and scalability, but it is a

---

* This work was supported in part by CICYT under Grant TIC94-0510-C02-02

widely accepted fact that they are much more difficult to program than shared memory machines. As a result, the programmer has to distribute code and data on processors himself, and manage communication among processes (or tasks) explicitly.

Simulators provide many advantages over running directly on a multiprocessor, including the cost effectiveness of workstations, the ability to exploit powerful sequential debuggers, the support for non-intrusive data collection and invariant checking, and the versatility of simulation. Several simulators as Pie [12] or Paret [11] have been developed. Usually these projects are aimed at developing programming environments and tools for the programmer, rather than tools designed to evaluate the architecture of the system. Newer high-performance simulators such as TangoLite [4] or Proteus [2] consider the architectural support of the system showing several results about the performance of parallel machines.

A very interesting way of simulation is to evaluate the behaviour of an architecture using a trace taken from adequated algorithms. Trace-driven simulations, which evaluate network performance on actual communication streams taken from characteristic programs, are the most reliable way for network design evaluation. These simulations require a great computation power, because of the many different design possibilities that must be simulated, and because of the length of the communication traces that drive the simulation. We have developed a simulator to evaluate the main features of the interconnection network, because in this way it can more faithfully represent the hardware implementation, taking into account details like channel multiplexing, partial buffering and delays in blocked messages. Furthermore, we are very interested in evaluating some new features in parallel machines, mainly reconfigurable architectures, that is, multicomputers whose networks can change their topology dynamically.

In this paper we describe our environment called PEPE (this acronym stands for **P**rogramming **E**nvironment for **P**arallel **E**xecution). PEPE provides a user-friendly visual interface for all phases of parallel program development, i.e. parallel algorithm design, coding, debugging, task mapping, execution control and evaluation of some architecture parameters.

Our environment is different from previous work because of two major features. First, PEPE allows us to evaluate the network for a communication trace taken from proper scientific problems we have previously coded in the environment. That is, we can evaluate and modify the behaviour of the interconnection network for real problems and not only for predetermined workloads. Second, it allows us to evaluate the performance of a multicomputer with a reconfigurable interconnection network. A completely connected interconnection network can match the communication requirements of any application, but it is too expensive to build even for a moderate number of nodes. Reconfigurable interconnection networks are alternatives to complete connections. This paradigm is suitable with either electronic or optical interconnections, which are applicable to a large class of networks. Some researchers believe that the immediate goal in the development of computer networks should be hybrid optical-electronic systems which combine the advantages of both electronic and optical technologies while avoiding their disadvantages. Reconfigurable networks are especially suitable when these technologies are combined.

Up to now, we only know another related environment for parallel programming on reconfigurable multicomputers. It is described in [1]. This environment is devoted to phase-reconfigurable programs, that is, programs must be implemented as series of phases, and each phase is assumed to be separated from another one by synchronization - reconfiguration points. These points select the adequated topology for each phase. Additionally, this tool is language-oriented. Our environment focuses on testing the different parameters of dynamically reconfigurable networks. A dynamically reconfigurable network means that a network can vary its topology arbitrarily at runtime. In this approach, any arbitrary topology is allowed, so that the interconnection network can easily match the communication requirements of a given algorithm. In our environment, we can study in depth the main concepts and options about dynamically reconfigurable networks, their limitations and tradeoffs.

The rest of the paper is arranged as follows. In section 2 the overall environment is described and its different parts are shown. The programming style and several issues related to it are detailed in section 3. The structure of the traces is discussed in section 4. In section 5, we present the network simulator and explain the main results we can obtain with it. Finally, we outline some conclusions.

2 An Overview of PEPE

In this section we are going to present PEPE. The key features of this environment are simplicity and completeness. Our main goal has been to obtain a flexible system which allows us an easy and efficient way to evaluate the multicomputer reconfigurable architecture.

Our environment has been developed on a workstation using C-language. PEPE provides a user-friendly visual interface for all phases of parallel program development and tunning. This graphical interface (figure 1) has been designed with the aim of keeping it really comfortable to the user, following the styles adopted nowadays by most of the human-oriented interfaces [9].

In our environment, the user gets tools for easy experimentation with both, different parallelization possibilities and different network parameters. With this methodology, the programmer can analyze very quickly several parallelization strategies and evaluate these strategies with tools for performance analysis.

PEPE simulates the execution of a parallel algorithm at two levels. At the first level, we are interested in verifying the behaviour of the parallel algorithm and studying the different strategies of parallelization, as well as the problems that arise when the parallel algorithms are coded and executed, such as deadlock, livelock, etc. For this, the execution of the parallel algorithm is simulated over a virtual architecture. At the second level, the behaviour of the reconfigurable network is studied. For this, we start with the communication pattern produced by the parallel algorithm and the performing of the network for this communication pattern is simulated. At this level, PEPE allows us to use different parameters

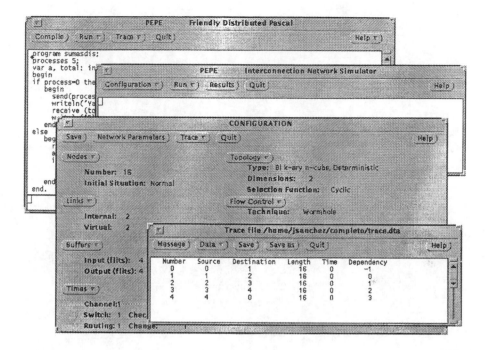

Fig. 1. PEPE's graphical interface

for the reconfigurable network. These levels give rise to the two phases of the simulator.

PEPE has two main phases and several modules within it. The first phase is more language-oriented, and it allows us to code, simulate and optimize a parallel program. In this phase, interactive tools for specification, coding, compiling, debugging and testing were developed. This phase is architecture independent. The second phase has several tools for mapping and evaluating the reconfigurable architecture. We can vary several network parameters such as interconnection topology and routing algorithm. In figure 2 we show the modules of PEPE. The link between the phases is an intermediate code that is generated as an optional result of the first phase. This intermediate code is a trace of the communication pattern of the source program. This allows the user to use the environment as a whole or each phase singly. For example, we can execute only the first phase for testing the parallel behaviour of an algorithm on an ideal multicomputer. We can also obtain an intermediate code from a key parallel algorithm. Then, we can execute several times the second phase from this trace with different network parameters to evaluate and tune the network for this key algorithm.

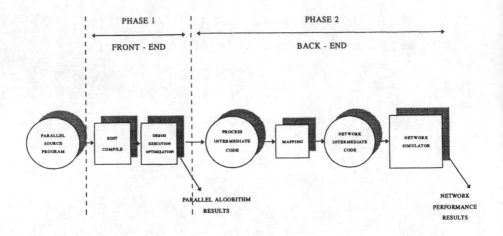

Fig. 2. Modules of PEPE

3 The Programming Tool

In this section we outline some important features of the first phase of PEPE. In this module we try to overcome the difficulty of programming multicomputers with an integrated approach and virtual concepts. The programming tool aims at increasing programming productivity. It takes a user parallel program as its input and tries to simulate its parallel execution on a virtual multicomputer. In this way, the user can test his parallel algorithm and, in case (s)he is not satisfied, come back to code a new parallel version.

In this phase it is important to abstract from specific architectural details such as topology, number of processors, etc. This module supports the concepts of virtual processor and virtual network. The user must take this into account, since the correct execution of a program must not depend upon the topology of the interconnection network.

Usually, the parallel programming style for most of these systems corresponds to the SPMD model [10], in which each processor asynchronously executes the same program but operates on distinct data items. PEPE uses the SPMD model and an extension of Pascal for coding parallel algorithms. This parallel language [7] which we have developed, is based on standard Pascal with some extensions to allow an easy and elegant programming of parallel algorithms, consisting of processes which communicate by means of message-passing. Finally, PEPE uses static scheduling. All processes must be created before starting execution.

For debugging parallel programs, PEPE's environment offers a parallel debugger. With this debugger the programmer gets a global view of the parallel system. Some features are breakpoints, the inspection of program states, displaying the contents of data structures and the state of each process or the modification of the contents of data structures.

4 The Communication Trace

At the end of the first phase, we can optionally generate an intermediate code (or trace) to evaluate the network performance for the parallel algorithm. That is, the network performance can be evaluated from the communication pattern obtained from a parallel algorithm. The trace records the set of messages that must be sent through the network. As we will see, this trace is independent of the timing parameters of the network.

The trace contain a complete information for each message. It consists of five fields whose meaning is detailed below. Additionally, each message has a message identifier not included in the trace. It is equal to the row number where it appears in the trace.

a) Source Process. Indicates the process that sent this message.
b) Destination Process. Indicates the process for which the message is destined.
c) Message Length. This field indicates the number of data bytes. It does not include control information.
d) Injection Time. Indicates the instant at which the message was injected into the network. This value is given in clock cycles (clock frecuency is a simulator parameter). This value can be absolute or relative as detailed below.
e) Dependency. This field indicates if a message is dependent on the reception of another message or independent. If it is independent, the value of this field is -1; otherwise, its value indicates the identifier of the message on which it depends.

Next, we are going to explain the last two fields in detail. When a parallel program starts execution, some processes send messages. Upon reception of those messages, some processes perform some computations, eventually sending more messages. Thus, there are two types of messages in a parallel algorithm: dependent and independent. A message is independent of any other message if a process can send this message without having to wait for the arrival of another message. On the opposite side, a message m is dependent on another message m' when a process p receives message m', performs some computations and sends message m. This dependency arises either because message m makes use of the information contained in message m', or because process p has no way to reach the instruction that sends message m without receiving message m' before, even if m does not use the information in m'.

Please, note that in many cases this dependency cannot be statically resolved by the compiler, and it is necessary to wait until execution time to know which ones are the dependencies between messages. Thus, communication traces must be generated during the simulation of the execution of the parallel algorithm.

Figure 3 shows the code for a process in three different cases. In the first case, the message is sent independently of any other message, because the process does

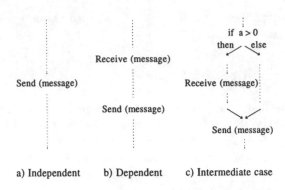

Fig. 3. Types of messages: dependent and independent messages

not need to receive any other message to execute the send instruction. Case b) shows a dependent message; the process has to receive a message. After processing it, the send instruction is executed. In case c), the dependency between messages is determined at run time. In this case, depending on the value of the variable a, the message to send will be dependent or independent. As the communication trace of the algorithm is generated during the simulated execution, this trace will always be correctly generated for the different input values.

The value of the field that contains the time at which a message is sent can be absolute or relative. An absolute value indicates the moment at which a message is sent. At this time, the network simulator will inject the message into the network. Obviously, for independent messages, this field will always contain an absolute value. On the other hand, a relative value indicates the time since the arrival of a certain message until the departure of the message on which it depends. Therefore, the network simulator will spend a time equal to this value between the arrival of a message at a node and the injection of the dependent message. For dependent messages, the injection time is always taken as a relative value.

The injection time is obtain by computing the time that the simulated processor needs to execute the instructions before the instruction send (absolute value), or between a pair of dependent instructions send and receive (relative value). Our environment allows us to choose among the execution times of some commercial processors like transputers and others. The use of dependent messages allows us to use the same traces to simulate different network parameters.

5 The Network Simulator

Next, we are going to describe the second phase of our environment, the interconnection network simulator. The unique feature included in our simulator is that it permits the network to be dynamically reconfigured. A reconfigurable network presents some advantages, the most interesting one being that it can easily match the network topology to the communication requirements of a given program, properly exploiting the locality in communications; moreover, programming a parallel application becomes more independent of the target architecture because the interconnection network adapts to the application dynamically.

In our simulator we can evaluate the performance of the interconnection network for parallel applications and not only for synthetic workloads. This allows us to vary the parameters of the reconfigurable network and study how to improve its behaviour in real cases. This phase of the simulator consists of two modules, the mapping module and the network simulator module. We are going to detail the features of each one of them.

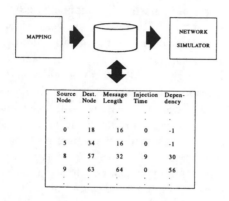

Source Node	Dest. Node	Message Length	Injection Time	Dependency
.	.	.	.	.
.	.	.	.	.
.	.	.	.	.
0	18	16	0	-1
5	34	16	0	-1
8	57	32	9	30
9	63	64	0	56
.	.	.	.	.

Fig. 4. Intermediate code after the mapping

The mapping module is the first one of the second phase and is responsible for translating the process-oriented communication trace to a processor-oriented intermediate code according to some pre-defined mapping functions.

The intermediate code output by this module is slightly different from the communication trace. Now, the first two fields are related to processors (source and destination processor) instaed of processes. The remaining code is unchanged. Figure 4 shows an example of this intermediate code. With this module, we can evaluate several mapping algorithms for a given parallel algorithm.

The last module is properly the network simulator. It is an improved version of a previous simulator [6] that supports network reconfiguration. It can simulate at the flit level different topologies and network sizes up to 16K nodes. The topology of the network is definable by the user. Each node consists of a processor, its local memory, a router, a crossbar and several channels. Message reception is buffered, allowing the storage of messages independently of the processes that must receive them. The simulator takes into account memory contention, limiting the number of messages that can be sent or received simultaneously. Also, messages crossing a node do not consume any memory bandwidth.

Wormhole routing is used. In wormhole routing [5] a message is descomposed into small data units (called flits). The header flit governs the route. As the header advances along the specified route, the remaining flits follow in a pipeline fashion. The pipelined nature of wormhole routing makes the message latency largely insensitive to the distance in the message-passing network.

The crossbar allows multiple messages to traverse a node simultaneously without interference. It is configured by the router each time a successful routing is made. It takes one clock cycle to transfer a flit from an input queue to an output queue. Physical channels have a bandwidth equal to one flit per clock cycle and can be split into up to four virtual channels. Each virtual channel has queues of equal size at both ends. The total queue size associated with each physical channel is held constant. Virtual channels are assigned the physical channel cyclically, only if they can transfer a flit. So, channel bandwidth is shared among the virtual channels requesting it. It should be noted that blocked messages and messages waiting for the router do not consume any channel bandwidth.

The most important performance measures obtained with our environment are delay, latency and throughput. Delay is the additional latency required to transfer a message with respect to an idle network. It is measured in clock cycles. The message latency lasts since the message is injected into the network until the last flit is received at the destination node. An idle network means a network without message traffic and, thus, without channel multiplexing. Throughput is usually defined as the maximum amount of information delivered per time unit.

The network reconfiguration is transparent to the user, being handled by several reconfiguration algorithms that are executed as part of the run-time kernel of each node. This class of reconfiguration is not restricted to a particular application, being very well suited for parallel applications whose communication pattern varies over time. The goal of the network reconfiguration is to reduce the congestion of the network. For this, when the traffic between a pair of nodes is intense, the reconfiguration algorithm will try to put the source node close to the destination node to reduce the traffic through the network and, therefore, to reduce the congestion that may have been produced. The reconfiguration algorithm decides when a change must be carried out by means of a cost function. The network reconfiguration is carried out in a decentralized way, that is, each node is responsible for trying to find its best position in the network depending on the model of communication. Also, reconfiguration is limited, preserving the original topology. There are several different parameters [8] that can be varied to adjust how the reconfiguration is performed.

With reconfigurable networks, we want to reduce the latency and delay and to increase the throughput. Also, we want to have a small number of changes to keep the reconfiguration cost low. The quality of each reconfiguration is measured by the simulator.

6 Conclusions

Application development via high-performance simulation offers many advantages. Simulators can provide a flexible, cost-effective, interactive debugging environment that combines traditional debugging features with completely nonintrusive data collection.

We have presented an environment for evaluating the performance of multicomputers. Our environment, unlike most of the earlier work, captures both the communication pattern of a given algorithm and the most important features of the interconnection network, allowing even the dynamic reconfiguration of the network. This feature is a valid alternative to solve the communication bottleneck problem. By means of using a reconfigurable topology, the principle of locality in communications is exploited, leading to an improvement in network latency and throughput.

Until now, the study of these features was difficult because there were no tools that permited varying the different parameters of reconfigurable networks. In this paper, a system that solves this problem and opens a way for studying reconfigurable networks has been presented. A reconfigurable network yields a variety of possible topologies for the network and enables the program to exploit this topological variety in order to speed up the computation [3].

References

1. Adamo, J.M., Trejo, L.: Programming environment for phase-reconfigurable parallel programming on supernode. Journal of Parallel and Distributed Computing. **23** (1994) 278–292
2. Brewer, E.A., Dellarocas, C.N., Colbrook, A., Weihl, W.E.: Proteus: A high-performance parallel-architecture simulator. In Proc. 1992 ACM Sigmetrics and Performance '92 Conference, (1992) 247–248
3. Ben-Asher, Y., Peleg, D., Ramaswami, R., Schuster, A.: The power of reconfiguration. Journal of Parallel and Distributed Computing, **13** (1991) 139–153
4. Davis, H., Goldschmidt, S.R., Hennesy, J.: Multiprocessor simulation and tracing using Tango. In Proc. of the Int. Conf. on Parallel Processing, (1991) II99–II107
5. Dally, W.J., Seitz, C.L.: Deadlock-free message-routing in multiprocessor interconnection networks. IEEE Transactions on Computers, **C-36**, No. 5 (1987) 547–553
6. Duato, J.: A new theory of deadlock-free adaptive routing in wormhole networks. IEEE Transactions on Parallel and Distributed Systems, 4, No. 11 (1993) 1–12
7. García, J.M.: A new language for multicomputer programming. SIGPLAN Notices, **6** (1992) 47–53
8. García, J.M., Duato, J.: Dynamic reconfiguration of multicomputer networks: Limitations and tradeoffs. Euromicro Workshop on Parallel and Distributed Processing, IEEE Computer Society Press, (1993) 317-323
9. Hartson, H.R., Hix, D.: Human-computer interface development: Concepts and systems. ACM Computing Surveys, **21**, No. 1 (1989)
10. Karp, A.: Programming for parallelism. IEEE Computer, (1987) 43–57
11. Nichols, K.M., Edmark, J.T.: Modeling multicomputer systems with PARET. IEEE Computer, (1988) 39–48
12. Segall, Z., Rudolph, L.: PIE: A programming and instrumentation environment for parallel processing. IEEE Software, (1985) 22–37

Data Acquisition and Management in BEPC

X. Geng, J. Zhao, Y. Yan, Y. Yu, W. Liu, J. Xu

Institute of High Energy Physics, Chinese Academy of Sciences
P.O.Box 918-10, Beijing 100039, China
Email: gengxs@bepc3.ihep.ac.cn

Abstract. This paper describes a method in upgrading of the BEPC control system, that the dedicated adapter VAXCAMACChannel (*VCC*) was replaced by a commercial adapter KSC2922/3922, Qbus CAMAC interface. All low level I/O driver routines have been changed without changing whole CAMAC hardware system. The upgraded control system adopts a distributed architecture and several hierarchical databases are installed in FEC computers, so the data flow should be controlled. Once raw data in any node have been refreshed, the changing will be transferred to other nodes to keep uniformity of the data in those databases.

1 Introduction

The original BEPC control system takes a centralized structure which mainly is composed of console, VAX750 computer, the intelligent channel VCC, the CAMAC system and hardware devices. The structure of the system is shown in figure 1. VCC is a dedicated product from SLAC, which is no longer produced and short of spare parts. Now VCC has been replaced by the commercial product KSC2922/3922, a Qbus−CAMAC adapter produced by Kinetic Systems Corporation, which serves as data communication interface to the CAMAC hardware.

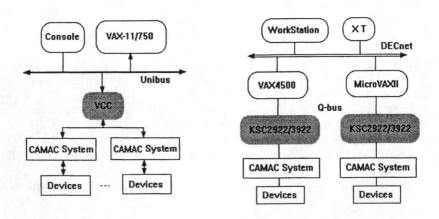

Fig.1 Original control system structure   Fig.2 Upgraded BEPC control system

The upgraded control system adopts a distributed architecture based on Ethernet and it has been put into use since Oct. 1994 (Figure 2). Because of the shortage of the upgrading period, the low level CAMAC system were preserved.

2 Hardware Structure

VAX−11/750 takes VCC as its Unibus-CAMAC interface, but VAX4500 and MicroVAXII computers in the improved system use Q-bus and their interface KSC2922/3922. KSC 2922 Computer Bus Adapter provides an interface between the DEC Q-Bus and up to eight 3922 dedicates crate controllers through a byte wide parallel bus. The 2922/3922 combination provides four DMA modes and a programmed transfer mode. All modes of operation are capable of transferring 16, or 24 bit CAMAC data words. DMA data rates up to 0.77 Mbytes per second can be achieved.

3 System Software

Data acquisition software consists of packet creation program PBZ, data I/O program XCAMAC, device on/off program DCOUT, the digital voltage acquisition program of the main B and Q magnet power supplies SPRDVM and beam position monitor program BPM.

In the control system, there are nearly 7,000 signals which can be classed into 7 kinds of type:

Input signals	DM (digital monitor)	1 bit digital input
	AM (analog monitor)	analog input
	DI (digital input)	16 bits digital input
	DV (digital volte)	R*4 analog input
Output signals	DC (digital control)	1 bit digital output
	AC (analog control)	analog output
	DO (digital output)	16 bits digital output

For acquisition of the signals mentioned above, the following programs were rewritten: The VCC packets are made by subroutine PBZ. PBZ takes the CAMAC I/O address of each signal from database and assembles it to the control word and VCC packet. Then the control words and VCC packets are sent to the database to be used by XCAMAC and other processes. The sequence of the VCC packets in the database is: DMAMDI packets, ACOUT pockets, ACIN packets, DCDO packets, IPSC packets and SAM packets. The system of BEPC have about 1350 VCC packets.

Data acquisition process XCAMAC refreshes the database at a rate of 2 times per second, and acquires about 4000 signals every time. What is more, the process also carries out the ramp operation of magnet power supplies during particle acceleration of BEPC.

The subroutine SPRDVM acquires the digital voltage signals (DV), so that Operators can monitor the present current status of magnet power supplies. DVM3456A is connected to VAX computer via 3388 GPIB interface.

4 Improvement

There is a hierarchical database in original control system which has a static area and a dynamic area. In the static area, there are information of CAMAC interface and machine parameters. The original data from the accelerator equipment are stored in the dynamic area which are refreshed in 2 HZ. For the distributed architecture of the new system, the original database has to be modified. First of all, we installed the database in each FEC computer with same data structure and created the 3922 packets area in the dynamic area of the databases. To keep the uniformity of the data records in these databases, a network communication program was developed to exchange the data between those databases and several new sections were inserted the dynamic area of each database to hold the raw data from other node through network which are refreshed once per second. As figure 3 shown, when the database on node 1 received the raw data from its input/output port, the network communication manager is notified by a event flag to fetch the data and send them to the database in node 2. In order to prevent the alteration of the high level application program, 3922 data area in dynamic area are mapped onto the original VCC data area and the local index of the database is replaced by a new global index in each database when the control system startup, so that the raw data from all of the accelerator devices can be read in each node.

The format of data and command packet differs from that for VCC, therefore the main work is changing the packet chains from VCC format to 3922 format. Another difference is data bit format. VCC requires 16 high bits to be valid, but 2922 need 16 low bits.

New packet organization program QPBZ acquires the CAMAC I/O address of every signal from the database, assembles them to CAMAC control words by calling 3922 driver subroutine such as Cainit, Caopen, Caclos, Canaf, Cainaf, Cablk, Cahalt, Caexew and Camsg, etc. Since block transfer operation is need for acquiring analog signals by SAM module, so we wrote a new program for the organization of SAM packets.

When testing of the 2922/3922, we found that I/O speed of KSC2922/3922 is lower than that of VCC. To transfer same number of data when 30 main power supplies are ramping, the minimum interval between two QIO is 30 ms for VCC and 44 ms for 3922. Because VCC is a intelligent module which can assemble F17 command sending to CAMAC modules, but 3922 is a dumb module, the F17 Command should be sent by VAX computer, so the length of the package chain of 3922 is 1.5 times as long as that of VCC. It may take more time to transfer the data, therefore, we don't acquire device status information during

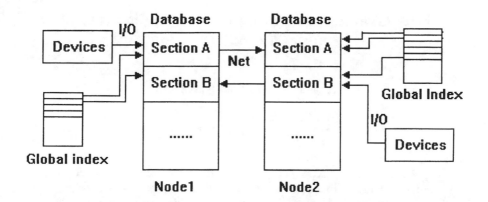

Fig. 3 The upgraded database

the ramping of main PS to enhance the speed of it. In accordance with 3922 packet rules, the output data are placed in the packet chains and readback is mapped onto old VCC area.

5 Conclusion

The upgraded BEPC control system has been completed in Oct. 1994. The dedicated adapter VCC has been replaced by the commercial product successfully and the real-time response speed and the reliability of the system are improved. In near future, we are upgrading the DVM acquisition with a analog scanning module.

6 Acknowledgment

The author would like to thank Prof. Huiying Luo and Chunhong Wang, who gave us a lot of supports.

References

1. . Zhao, X. Geng et al., Proc. of Conf. on High Energy Acc. (1995)
2. . Zhao, Nation Conf.
3. . Geng, Y. Yan, Nation Conf.

First Graph Partitioning and Its Application in Sparse Matrix Ordering

Anshul Gupta, Fred Gustavson

IBM T.J. Watson Research Center,
P.O. Box 218, Yorktown Heights, NY 10598, U.S.A.

Abstract. Graph partitioning is a fundamental problem in several scientific and engineering applications, including task partitioning for parallel processing. In this paper, we describe heuristics that improve the state-of-the-art practical algorithms used in graph partitioning software in terms of both partitioning speed and quality. An important use of graph partitioning is in ordering sparse matrices for obtaining direct solutions to sparse systems of linear equations arising in engineering and optimization applications. The experiments reported in this paper show that the use of these heuristics results in a considerable improvement in the quality of sparse matrix orderings over conventional ordering methods. In addition, our graph-partitioning based ordering algorithm is more parallelizable than minimum-degree based orderings algorithms and it renders the ordered matrix more amenable to parallel factorization.

1 Introduction

Graph partitioning is an important problem with extensive application in scientific computing, optimization, VLSI design, and task partitioning for parallel processing. The graph partitioning problem, in its most general form, requires dividing the set of nodes of a weighted graph into k disjoint subsets or partitions such that the sum of weights of nodes in each subset is nearly the same (within a user supplied tolerance) and the total weight of all the edges connecting nodes in different partitions is minimized. In this paper, we describe heuristics that improve the state-of-the-art practical graph partitioning algorithms in partitioning speed and, for small number of parts, also in partitioning quality. We have developed a graph partitioning and sparse matrix ordering package (WGPP) [10][1] based on the heuristics described in this paper.

Graph partitioning plays an important role in computing fill-reducing orderings of sparse matrices for solving large sparse systems of linear equations. Finding an optimal ordering is an NP-complete problem and heuristics must be used to obtain an acceptable non-optimal solution. Improving the run time and quality of ordering heuristics has been a subject of research for almost three decades. Two main classes of successful heuristics have evolved over the years: (1) Minimum Degree (MD) based heuristics, and (2) Graph Partitioning (GP) based heuristics. MD based heuristics are local greedy heuristics that reorder

---

[1] The package is available to users in the form of a linkable module.

the columns of a symmetric sparse matrix such that the column with the fewest nonzeros at a given stage of factorization is the next one to be eliminated at that stage. GP based heuristics regard the symmetric sparse matrix as the adjacency matrix of a graph and follow a divide-and-conquer strategy to label the nodes of the graph by partitioning it into smaller subgraphs.

The initial success of MD based heuristics prompted intense research [7] to improve its run time and quality and they have been the methods of choice among practitioners. The Multiple Minimum Degree (MMD) algorithm by George and Liu [6, 7] and the Approximate Minimum Degree (AMD) algorithm by Davis, Amestoy, and Duff [3] represent the state of the art in MD based heuristics. Recent work by the author [10, 11], Hendrickson and Rothberg [15], Ashcraft and Liu [1], and Karypis and Kumar [17] suggests that GP based heuristics are capable of producing better quality orderings than MD based heuristics for finite-element problems while staying within a small constant factor of the run time of MD based heuristics.

A detailed version of this paper is available as [9].

2 Application in Parallel Solution of Linear Equations

Both, graph partitioning and GP based sparse matrix ordering, have applications in parallel solution of large sparse systems of linear equations. Partitioning the graph of a sparse matrix to minimize the edge-cut and distributing different partitions to different processors minimizes the communication overhead in parallel sparse matrix-vector multiplication [21]. Sparse matrix-vector multiplication is an integral part of all iterative schemes for solving sparse linear systems.

GP based ordering methods are more suitable for solving sparse systems using direct methods on distributed-memory parallel computers than MD based methods in two respects. There is strong theoretical and experimental evidence that the process of graph partitioning and sparse matrix ordering based on it can be parallelized effectively [19]. On the other hand, the only attempt to perform a minimum degree ordering in parallel that we are aware of [8] was not successful in reducing the ordering time over a serial implementation. In addition to being parallelizable itself, a GP based ordering also aids the parallelization of the factorization and triangular solution phases of a direct solver. Gupta, Karypis, and Kumar [12, 16] have proposed a highly scalable parallel formulation of sparse Cholesky factorization. This algorithm derives a significant part of its parallelism from the underlying partitioning of the graph of the sparse matrix. In [13], Gupta and Kumar present efficient parallel algorithms for solving lower- and upper-triangular systems resulting from sparse factorization. In both parallel factorization and triangular solutions, a part of the parallelism would be lost if an MD based heuristic is used to preorder the sparse matrix.

3 Multilevel Graph Partitioning

Recent research [14, 18] has shown *multilevel algorithms* to be fast and effective in computing graph partitions. A typical multilevel graph partitioning algorithm has four components, namely *coarsening, initial partitioning, uncoarsening*, and *refining*. In the following subsections, we briefly discuss these components of graph partitioning and the new heuristics and improvements over the current techniques.

3.1 Coarsening

The goal of the coarsening phase is to reduce the size of a graph while preserving its properties that are essential to finding good a partition. The original graph is regarded as a weighted graph with a unit weight assigned to each edge and each node. In a coarsening step, a maximal set of edges of the graph is identified such that no two edges have a vertex in common. This set of edges is known as a *matching*. The edges in this set are now removed. The two nodes that an edge in the matching connects are collapsed into a single node whose weight is the sum of the weights of the component nodes. Note that coarsening also results in some edges being collapsed into one, in which case the collapsed edge is assigned a weight equal to the sum of weights of the component edges.

Given a weighted graph after any stage of coarsening, there are several choices of matchings for the next coarsening step. A simple matching scheme [14] known as *random matching (RM)* randomly chooses pairs of connected unmatched nodes to include in the matching. In [18], Karypis and Kumar describe a heuristic known as *heavy-edge matching* (HEM) to aid in the selection of a matching that not only reduces the run time of the refinement component of graph partitioning, but also tends to generate partitions with small separators. The strategy is to randomly pick an unmatched node, select the edge with the highest weight among the edges incident on this vertex that connect it to other unmatched vertices, and mark both vertices connected by this edge as matched.

HEM can miss some heavy edges in the graph because the nodes are visited randomly for matching. For example, consider a node i, the heaviest edge incident on which connects it to a node j. If i is visited before j and both i and j are unmatched, the edge (i, j) will be included in the matching. If there exists an edge (j, k), such that i and k are not connected, it will be excluded from the matching even if it is much heavier than (i, j) because j is no longer available for matching. To overcome this problem, after the first few coarsening steps, we switch to what we call the *heaviest-edge matching*. We sort the edges by their weights and visit them in the decreasing order of their weights to inspect them for their possible inclusion in the matching. Ties are broken randomly.

HEM and its variants reduce the number of nodes in a graph by roughly a factor of two at each stage of coarsening. Therefore, the number of coarsening, uncoarsening, and refining steps required to partition an n-node graph into k parts is $\log_2(n/k)$. If r (instead of 2) nodes of the graph are coalesced into one at each coarsening step, the total number of steps can be reduced to about

$\log_r(n/k)$. Fewer steps are likely to reduce to overall run time. However, as r is increased, the task of refining after each uncoarsening step (see Section 3.3 for the description of uncoarsening and refinement) becomes harder. This affects both the run time and the quality of refinement. In our experiments, we observed that increasing r from 2 to 3 results in more than 20% time savings with only a minor compromise in partitioning quality. In WGPP, we use a combination of heavy-edge matching, heaviest-edge matching, and *heavy-triangle matching (HTM)*; the latter matching coalesces three nodes at a time. HTM picks an unmatched node at random and matches it with two of its neighbors such that the sum of the weights of the three edges connecting the three nodes is maximized over all pairs of neighbors of the selected node.

3.2 Initial Partitioning

All multilevel graph partitioning schemes described in the literature stop the coarsening phase when the graph has been reduced to a few hundred to a few thousand nodes and use some heuristic to compute an initial partition of the coarse graph at a substantial run time cost. In WGPP, we have completely eliminated the initial partitioning phase, thereby simplifying and speeding up the overall partitioning process.

One of the effective heuristics for initial partitioning is *graph growing* [18]. The basic function of graph growing is to form a cluster of highly connected nodes. Note that coarsening with HEM or HTM also strives to achieve the same goal. In fact, graph growing is the bottom-up equivalent of coarsening. Therefore, for a k-way partition of an n-node graph, WGPP continues to coarsen the graph until it contains exactly k nodes. This coarse k-node graph serves as a good initial partitioning, provided that the coarsening algorithm does not allow a node to participate in matching if its weight substantially exceeds n/k.

3.3 Uncoarsening and Refinement

The uncoarsening and refining components of graph partitioning work together. Initially, the k nodes of the coarsest graph are assigned different tags indicating that they belong to different initial partitions. They are then split into the nodes that were collapsed to form them during the coarsening phase. This reversal of coarsening is carried out one step at a time. The nodes of the uncoarsened graph inherit their tags from their parent nodes in the coarser graph. At any stage of uncoarsening, the edges connecting a pair of nodes belonging to different partitions constitute an edge-separator of the graph. The removal of the edge-separator from the graph breaks it into k disconnected subgraphs. After each step of uncoarsening, the separator is refined. While refining an edge-separator, an attempt is made to minimize its total weight by switching the partitions of some nodes if this switching reduces the separator size and does not make the weights of the parts too imbalanced.

Figure 1 gives an overview of our uncoarsening and refinement scheme for partitioning into k parts. After coarsening the original n-node graph to k nodes,

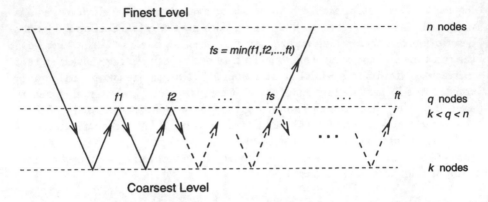

Finest Level

Fig. 1. An overview of our uncoarsening and refining scheme for finding a k-way partition. f_i is a weighted average of the edge-separator size and the inverse of weight imbalance between the parts in the i-th partition.

we uncoarsen it to q nodes, where $k < q < n$, while refining the edge-separator after each step of uncoarsening. At this stage we save the partition, coarsen the q-node graph back to k nodes, and repeat the uncoarsening and refining process. This process is repeated a few times and of all the partitions of the q-node graph generated, we chose the best for further uncoarsening to obtain a partition of the original n-node graph. The best partition is selected on the basis of a weighted average of the size of edge-separator and the inverse of the weight imbalance between the parts.

Depending on the number of partitions k, we either use a variation of the popular Kernighan-Lin heuristic [20] or a greedy refinement scheme [18] for refining the edge-separators. The linear time Fiduccia-Mattheyses variation [5] of the Kernighan-Lin heuristic is used for small number of partitions and the greedy algorithm is used if the number of partitions required is large. The Kernighan-Lin and Fiduccia-Mattheyses heuristics were originally developed to refine two partitions, but can be adapted for refining multiple partitions [14] by using multiple priority queues.

4 Ordering Sparse Matrices using Multiple Multilevel Recursive Bisections

In this section, we describe how graph bisection (partition into two parts) by the process described in Section 3 is used to compute a fill-reducing ordering of a symmetric sparse matrix. The overall ordering algorithm involves several stages, a preprocessing step, and a postprocessing step. In Section 4.1, we describe the core algorithm and its various stages. The preprocessing and postprocessing steps are described in detail in [9].

Any GP based ordering algorithm regards the matrix as the adjacency matrix of a graph and assigns labels to the nodes of the graph. These labels specify the sequence in which the matrix columns corresponding to the nodes are eliminated during numerical factorization. The overall approach of our ordering algorithm follows the fundamental technique of generalized nested dissection [6]. The graph is bisected by finding and removing a node-separator, labeling the nodes of the two resulting subgraphs by applying the same technique recursively, and labeling the nodes of the separator after the nodes of the subgraphs have been labeled (i.e. the separator nodes get a higher label than any of the nodes of the subgraph). The recursion terminates when the subgraphs become too small, at which stage they are labeled using a minimum degree heuristic.

4.1 Graph Bisection

A key step in our ordering algorithm is finding a small node bisector of a graph. This can be accomplished by the heuristics described in Section 3 with some modifications to the coarsening and refinement strategies.

The heavy- and heaviest-edge coarsening strategies discussed in Section 3.1 may occasionally fail to satisfactorily coarsen the graph of an LP matrix. An example of such a graph is one that has a star-like structure. Since RM and HEM disregard the possibility of matching disconnected nodes, they fail to preserve the properties of such a graph upon coarsening and may result in a very unbalanced coarse graph. Moreover, RM and HEM reduce the size of a star shaped graph by only one node in one coarsening step. In order to overcome these problems, conceptually, WGPP regards every graph as a completely connected graph with a weight zero assigned to every edge (i, j) that does not really exist in the physical graph. It then attempts to maximize a modified weight function of any two unmatched nodes i and j when considering them to include in the matching as a pair. The modified edge weight $ew'(i, j)$ between two nodes i and j with node weights $nw(i)$ and $nw(j)$, respectively, and connected by an edge of (possibly zero) weight $ew(i, j)$ is given by

$$ew'(i, j) \; = \; \alpha \times ew(i, j) + \frac{\beta}{nw(i) + nw(j)}.$$

Here α and β are constants defined at the beginning of the given level of coarsening. WGPP uses $ew'(i, j)$ instead of the real edge weight $ew(i, j)$ within the HEM framework. The modified edge weight criterion yields a balanced coarsening in fewer steps while preserving the benefits of HEM. To save coarsening time in the actual implementation, WGPP switches to using the modified edge weights only if heavy- and heaviest-edge matchings fail to reduce the size of the graph sufficiently.

The second modification required to adapt the graph partitioning algorithm to bisection for ordering is in the refinement phase. A typical graph partitioning applications requires the total weight of the edge-separator to be minimized. However, while computing a fill-reducing ordering of a sparse matrix, it is the size of the node-separator that must be minimized. Current graph partitioning based ordering algorithms follow two different approaches to finding a small node-separator. Karypis and Kumar [17] refine the edge-separator between the two subgraphs after each step of uncoarsening so that few edges connect nodes of different subgraphs in the final partitioning of the original graph. Then they use an algorithm for finding a minimum cover [4, 22] to compute a node-separator from the edge-separator. This approach relies heavily on the assumption that the size of a node-separator is proportional to the size of the edge-separator containing it. This assumption is often incorrect, especially for the highly unstructured LP matrices.

Ashcraft and Liu [1] find a node-separator within the coarse graph and refine it into a node-separator of the original graph in order to overcome the drawback of the edge-separator approach. However, a heavier node-separator in the coarse graph can result in the smaller node-separator in the original graph after uncoarsening (and vice versa). The weight of a node in the coarse graph is the number of nodes of the original graph that are collapsed during the coarsening steps to form the coarse node. Not all the component nodes of a coarse separator node may have edges crossing partition boundaries. As the graph is uncoarsened, such nodes are eliminated from the node-separator by the refining process. How many such nodes are eliminated from a node-separator depends on the connectivity of these nodes to the nodes outside the separator. Therefore, unlike in the original graph, the size of the edge-separator is not completely irrelevant in coarse versions of the graph. In fact, the coarser the graph, the more relevant the edge-separator size is in the predicting size of the node-separator in the original graph.

With these motivations, we slightly modify the graph partitioning strategy illustrated in Figure 1 to compute bisections for sparse matrix ordering. We coarsen the original n-node graph to two nodes. We then repeatedly uncoarsen it to q nodes, where $2 < q < n$, and re-coarsen it to two nodes to select an initial partition. The uncoarsening from two to q is accompanied by refining the edge-separator. We select the best bisection of the q-node graph based on a weighted average of the size of edge-separator, the size of the node-separator, and the inverse of the imbalance between the two parts. The selected partition of the q-node graph is uncoarsened to the original n-node graph one step at a time. After each uncoarsening step, we refine the node-separator; i.e., in the final steps of uncoarsening, the subject of refinement is changed from the edge-separator to a node-separator. For refining the node-separator, we use Ashcraft and Liu's modification [2] of the Fiduccia-Mattheyses algorithm [5].

5   Experimental Results

In this section, we present experimental results on the run time and quality of graph partitions and sparse matrix orderings generated by WGPP. To the best of our knowledge, the Metis [17] software package represents the state of the art in graph partitioning—both in terms of partitioning time and quality. In Section 5.1, we compare WGPP with Metis. Currently the best minimum-degree based code available for computing fill-reducing ordering for sparse matrices is that of approximate minimum degree (AMD) [3]. Metis is one of the well known graph-partitioning based sparse matrix ordering software. Recent work by Ashcraft and Liu [1] and by Hendrickson and Rothberg [15] report graph-partitioning based ordering heuristics that are better than Metis, but the corresponding software is not available for direct comparison with WGPP. Therefore, in Section 5.2, we compare WGPP's sparse matrix orderings with those of AMD and Metis. Results for the traditional multiple minimum degree (MMD) algorithm are also included in for reference. The details of the origin and the number of nodes and edges in the graphs and sparse matrices on which the experimental data is presented in Sections 5.1 and 5.2 can be found in [9]. All the codes being compared were compiled with -O3 option using XLF 3.2 and run on an IBM RS6000/590 work-station.

5.1 Graph Partitioning

Graph	PMETIS(HEM+BKL)		WGPP(HEM+BKL)		$\frac{T_{METIS}}{T_{WGPP}}$	$\frac{|EC|_{METIS}}{|EC|_{WGPP}}$
	T (sec.)	$|EC|$	T (sec.)	$|EC|$		
1. BCSSTK-15	0.452	1558	0.139	1528	3.25	1.02
2. COPTER-2	2.394	2252	1.012	2068	2.37	1.09
3. CRONE	0.487	183	0.304	164	1.60	1.12
4. CUBE-35	1.267	1387	0.766	1295	1.65	1.07
5. GRID-127	0.467	395	0.285	379	1.64	1.04
6. HSCT16K-B	2.763	6508	0.817	6488	3.38	1.00
7. HSCT88K	9.621	10933	2.659	8327	3.62	1.31
8. PILOT	0.464	3408	0.130	3379	3.57	1.01
Total	17.92	26624	6.20	23628	2.93	1.13
Average					2.64	1.08

Table 1. Run time and edge-cut comparison between WGPP and Metis for graph bisection.

The quality of a partition of an unweighted graph is measured in terms of the edge-cut, or the total number edges between nodes belonging to different partitions, and the balance in the number of nodes assigned to each part. In this section, we compare the edge-cuts of the partitions produced by WGPP and Metis. In Metis, the upper bound on the imbalance between the weights of the partions is 3%; i.e., the number of nodes in any partition does not exceed $1.03 \times n/k$, where n is the total number of nodes in the graph and k is the number of parts. In WGPP, the user has the option of specifying the maximum tolerable imbalance. The experiments in this section were conducted with this option set at 3%. Although well below 3%, the actual imbalance was observed to be somewhat higher for WGPP than for Metis. The comparisons in this section are made for 2, 24, and 160 parts and the respective comparisons are

Graph	KMETIS(HEM+BGR)		WGPP(HTM+BKL)		$\frac{T_{METIS}}{T_{WGPP}}$	$\frac{|EC|_{METIS}}{|EC|_{WGPP}}$
	T (sec.)	$|EC|$	T (sec.)	$|EC|$		
1. BCSSTK-15	0.977	12886	0.650	12563	1.50	1.03
2. COPTER-2	3.434	25801	3.047	25566	1.13	1.01
3. CRONE	0.747	1838	0.894	1811	0.84	1.01
4. CUBE-35	1.806	9527	2.960	8917	0.61	1.07
5. GRID-127	0.758	3167	1.050	2922	0.72	1.08
6. HSCT16K-B	3.762	53541	1.299	68356	2.90	0.78
7. HSCT88K	10.95	90182	5.558	88482	1.97	1.02
8. PILOT	1.222	39534	0.384	42027	3.18	0.94
Total	23.65	236476	15.46	250644	1.53	0.94
Average					1.61	0.99

Table 2. Run times and edge-cut comparison between Metis and WGPP for partitioning graphs into 24 parts.

representative of partitioning graphs into small, medium, and large number of parts. Both Metis and WGPP offer a few different choices of coarsening and refinement algorithms. In each case, the best choices of Metis are compared with the default choices of WGPP. Although the choices are different for WGPP in each case, WGPP makes these choices itself depending on the size of the graph and the number of partitions.

Graph	KMETIS(HEM+BGR)		WGPP(HEM+BGR)		$\frac{T_{METIS}}{T_{WGPP}}$	$\frac{|EC|_{METIS}}{|EC|_{WGPP}}$
	T (sec.)	$|EC|$	T (sec.)	$|EC|$		
1. BCSSTK-15	1.624	32197	0.348	31797	4.67	1.01
2. COPTER-2	5.271	61802	2.118	62302	2.49	0.99
3. CRONE	1.459	6120	0.543	6894	2.69	0.89
4. CUBE-35	3.241	20213	1.590	20962	2.04	0.96
5. GRID-127	1.421	8885	0.487	9126	2.92	0.97
6. HSCT16K-B	6.231	178985	1.688	180410	3.69	0.99
7. HSCT88K	13.560	250598	3.993	271210	3.40	0.92
8. PILOT	2.075	54872	0.321	54777	6.39	1.00
Total	34.88	613672	11.09	637478	3.15	0.96
Average					3.54	0.97

Table 3. Run time and edge-cut comparison between WGPP and Metis for partitioning graphs into 160 parts.

Tables 1 through 3 give the various results. In these tables, HEM refers to heavy-edge matching in the context of Metis and heaviest-edge matching in the context of WGPP, HTM refers to heavy-triangle matching, BKL refers to boundary Kernighan-Lin refinement, and BGR refers to boundary greedy refinement. PMETIS is a variation of Metis that uses recursive bisection for partitioning and KMETIS is a variation that uses a single coarsening and refining cycle to generate the entire partition.

Table 1 compares 2-way partitions using Metis and WGPP. WGPP produced a smaller edge-cut for each graph and was much faster than Metis—about 2.64 times on an average. Table 2 compares the run times and edge-cuts for 24-way partitions. WGPP compares quite favorably with Metis; however, it is significantly worse on one graph, HSCT16K-B, as a result of which, the total edge-cut of all 8 graphs (250644) is worse than that of Metis (234308). Table 3 compares Metis and WGPP for 160-way partitions. WGPP turns out to be more than thrice as fast as Metis on an average for 160-way partitions while producing edge-cuts that are a few percent higher.

Matrix	MMD		AMD		METIS		WGPP	
	T	OPC	T	OPC	T	OPC	T	OPC
1. ALLGRADE	20.7	1438	(8.1)	1334	24.9	2042	16.3	(1226)
2. BCSSTK-13	0.2	57	(0.1)	54	0.6	64	0.6	(53)
3. BCSSTK-15	0.3	173	(0.1)	169	1.1	101	1.1	(89)
4. BCSSTK-31	2.1	2651	(0.9)	2579	13.8	1615	5.5	(1410)
5. BCSSTK-32	2.1	1262	(0.8)	(953)	22.2	1995	4.5	1630
6. COMP-1	662.9	2676	337.5	2753	–	–	(16.6)	(2181)
7. COPTER-2	3.6	11741	(2.7)	12225	13.5	6812	12.5	(5584)
8. CUBE-35	2.5	14251	(1.7)	14198	7.4	9692	6.8	(8788)
9. FLEET-12	90.0	6294	36.8	4926	–	–	(3.0)	(3307)
10. GISMONDI	11.5	303770	(2.7)	304937	4.3	133882	5.6	(128431)
11. GRID-127	(0.2)	48	0.3	(46)	2.5	49	2.1	52
12. HSCT16K-A	1.0	772	(0.3)	(720)	7.3	830	4.0	792
13. HSCT16K-B	1.5	1093	(0.4)	1131	9.2	952	7.1	(609)
24. HSCT44K	5.5	(2257)	(1.1)	2276	39.8	5046	4.6	3576
25. HSCT88K	8.8	12975	(1.4)	10872	41.8	12637	6.7	(9871)
26. K-8	14.7	2153	15.4	2460	8.8	1335	(3.1)	(492)
27. KEN-18	38.6	(186)	14.1	204	–	–	(6.2)	204
28. KK-6	6601.2	(1872)	2371.5	2054	–	–	(6.4)	1881
29. PDS-20	393.4	7069	(5.9)	7907	6.9	5683	7.1	(2130)
30. PILOT	0.9	47	(0.2)	50	0.8	50	0.7	(39)

Table 4. Run time (in seconds) and factorization operation count (in millions) comparison between MMD, AMD, Metis, and WGPP. The best time and operation count appears in parentheses for each matrix. A "–" implies that the ordering failed to complete.

5.2 Sparse Matrix Ordering

Table 4 compares MMD, AMD, Metis, and WGPP for the ordering time and the number floating point operations (in millions) on 20 sparse matrices from various sources. The best time and ordering for each matrix is enclosed in parentheses. AMD is the fastest of all orderings for a majority of the problems, but its run time is very inconsistent for the linear programming matrices. In the worst case, it is 370 times slower than WGPP. In this suite of test problems, WGPP produces the best ordering for two-thirds of the matrices.

6 Concluding Remarks

This paper presents heuristics that improve the run time and the quality of the state-of-the-art methods for graph partitioning and sparse matrix ordering. We have developed a graph partitioning and sparse matrix ordering package (WGPP) [10] based on the heuristics described in this paper. For graph partitioning, WGPP is considerably faster than the well known package Metis, while generating partitions of almost comparable quality. A comparison of WGPP with three other widely used sparse matrix ordering codes shows it to be the best in quality and consistent in run time on a suite of 20 randomly selected sparse matrices.

References

1. Cleve Ashcraft and Joseph W.-H. Liu. Robust ordering of sparse matrices using multisection. Technical Report CS 96-01, Department of Computer Science, York University, Ontario, Canada, 1996.
2. Cleve Ashcraft and Joseph W.-H. Liu. Generalized nested dissection: Some recent progress. In *Proceedings of Fifth SIAM Conference on Applied Linear Algebra*, Snowbird, Utah, June 1994.
3. Timothy A. Davis, Patrick Amestoy, and Iain S. Duff. An approximate minimum degree ordering algorithm. Technical Report TR-94-039, Computer and Information Sciences Department, University of Florida, Gainesville, FL, 1994.
4. Iain S. Duff and Torbjorn Wiberg. Remarks on implementations of $O(n^{1/2}\tau)$ assignment algorithms. *ACM Transactions on Mathematical Software*, 14:267–287, 1988.
5. C. M. Fiduccia and R. M. Mattheyses. A linear time heuristic for improving network partitions. In *Proceedings of the 19th IEEE Design Automation Conference*, pages 175–181, 1982.
6. Alan George and Joseph W.-H. Liu. *Computer Solution of Large Sparse Positive Definite Systems*. Prentice-Hall, Englewood Cliffs, NJ, 1981.
7. Alan George and Joseph W.-H. Liu. The evolution of the minimum degree ordering algorithm. *SIAM Review*, 31(1):1–19, March 1989.
8. Madhurima Ghose and Edward Rothberg. A parallel implementation of the multiple minimum degree ordering heuristic. Technical report, Old Dominion University, Norfolk, VA, 1994.

9. Anshul Gupta. Fast and effective algorithms for graph partitioning and sparse matrix reordering. Technical Report RC 20496 (90799), IBM T. J. Watson Research Center, Yorktown Heights, NY, July 10, 1996. Available on the WWW at the *IBM Research CyberJournal* site at *http://www.research.ibm.com:8080/*.

10. Anshul Gupta. WGPP: Watson graph partitioning (and sparse matrix ordering) package: Users manual. Technical Report RC 20453 (90427), IBM T. J. Watson Research Center, Yorktown Heights, NY, May 6, 1996.

11. Anshul Gupta. Graph partitioning based sparse matrix ordering algorithms for finite-element and optimization problems. In *Proceedings of the Second SIAM Conference on Sparse Matrices*, October 1996.

12. Anshul Gupta, George Karypis, and Vipin Kumar. Highly scalable parallel algorithms for sparse matrix factorization. Technical Report 94-63, Department of Computer Science, University of Minnesota, Minneapolis, MN, 1994. To appear in *IEEE Transactions on Parallel and Distributed Systems*, 1997. Postscript file available via anonymous FTP from the site *ftp://ftp.cs.umn.edu/users/kumar*.

13. Anshul Gupta and Vipin Kumar. Parallel algorithms for forward and back substitution in direct solution of sparse linear systems. In *Supercomputing '95 Proceedings*, December 1995.

14. Bruce Hendrickson and Robert Leland. A multilevel algorithm for partitioning graphs. In *Supercomputing '95 Proceedings*, 1995. Also available a Technical Report SAND93-1301, Sandia National Laboratories, Albuquerque, NM.

15. Bruce Hendrickson and Edward Rothberg. Improving the runtime and quality of nested dissection ordering. Technical Report SAND96-0868J, Sandia National Laboratories, Albuquerque, NM, 1996.

16. George Karypis, Anshul Gupta, and Vipin Kumar. Parallel formulation of interior point algorithms. Technical Report 94-20, Department of Computer Science, University of Minnesota, Minneapolis, MN, April 1994. A short version appears in *Supercomputing '94 Proceedings*.

17. George Karypis and Vipin Kumar. METIS: Unstructured graph partitioning and sparse matrix ordering system. Technical report, Department of Computer Science, University of Minnesota, 1995.

18. George Karypis and Vipin Kumar. Multilevel k-way partitioning scheme for irregular graphs. Technical Report TR 95-064, Department of Computer Science, University of Minnesota, 1995.

19. George Karypis and Vipin Kumar. Parallel multilevel graph partitioning. Technical Report TR 95-036, Department of Computer Science, University of Minnesota, 1995.

20. B. W. Kernighan and S. Lin. An efficient heuristic procedure for partitioning graphs. *The Bell System Technical Journal*, 1970.

21. Vipin Kumar, Ananth Grama, Anshul Gupta, and George Karypis. *Introduction to Parallel Computing: Design and Analysis of Algorithms*. Benjamin/Cummings, Redwood City, CA, 1994.

22. Alex Pothen and C.-J. Fan. Computing the block triangular form of a sparse matrix. *ACM Transactions on Mathematical Software*, 1990.

The Design, Implementation, and Evaluation of a Banded Linear Solver for Distributed-Memory Parallel Computers

Anshul Gupta[1], Fred G. Gustavson[1], Mahesh Joshi[2], and Sivan Toledo[1]

[1] IBM T.J. Watson Research Center,
P.O. Box 218, Yorktown Heights, NY 10598, U.S.A.
[2] Department of Computer Science, University of Minnesota,
200 Union Street SE, Minneapolis, MN 55455.

Abstract. This paper describes the design, implementation, and evaluation of a parallel algorithm for the Cholesky factorization of banded matrices. The algorithm is part of IBM's Parallel Engineering and Scientific Subroutine Library version 1.2 and is compatible with ScaLAPACK's banded solver. Analysis, as well as experiments on an IBM SP2 distributed-memory parallel computer, show that the algorithm efficiently factors banded matrices with wide bandwidth. For example, a 31-node SP2 factors a large matrix more than 16 times faster than a single node would factor it using the best sequential algorithm, and more than 20 times faster than a single node would using LAPACK's DPBTRF. The algorithm uses novel ideas in the area of distributed dense matrix computations that include the use of a dynamic schedule for a blocked systolic-like algorithm and the separation of the input and output data layouts from the layout the algorithm uses internally. The algorithm also uses known techniques such as blocking to improve its communication-to-computation ratio and its data-cache behavior.

1 Introduction

This paper describes the design, implementation, and evaluation of a solver for banded positive-definite symmetric linear systems with a reasonably wide band. It is based on ideas that were first described by Agarwal et al. [1]. The interface of the solver is identical to the interface of ScaLAPACK's new banded linear solver [4], which is designed for and restriced to narrow-band matrices. The two solvers therefore complement each other. The paper focuses on the Cholesky factorization of the matrix. The companion banded triangular solve subroutine is not discussed, since its design is completely different. The solver is now part of Version 1.2 of IBM's Parallel Engineering and Scientific Subroutine Library (PESSL). Our analysis shows that this solver is highly scalable and this is confirmed by the performance results on an IBM SP2 distributed-memory parallel computer. An additional important contribution of this paper is that it makes a strong case for run-time scheduling and data distribution for parallel algorithms with high computation to data ratios. Redistributing the data at run time allows the user to lay out the data using a simple data distribution, and at the same time enables the algorithm to work with the most appropriate distribution.

Our performance results indicate that our approach represents a viable approach to the design of numerical algorithms for regular problems with high computation to data ratios. The separation of the scheduler module and the actual numerical computation module is also attractive from the software engineering point of view. One can be easily modified or replaced without affecting the other. The fact that our algorithm performs well even on full (not banded) problems leads us to beleive that our methodology is general enough to be applicable to a variety of dense matrix algorithms with minor modifications.

Our goal was to design a solver that achieves high single-node performance through blocking the Cholesky factorization, that minimizes communication and work, and that enables the use of large numbers of processors. Achieving this goal required a number of design innovations and departures from current practices. We realized several things early in the project: a) that using a conventional block-cyclic layout coupled with an "owner-computes" scheduling rule would prevent us from achieving our goals, b) that since the algorithm performed matrix-matrix operations that take hundreds or thousands of cycles to complete as its primitive building blocks, we could delay scheduling decisions to runtime with a negligible performance penalty, and c) that for moderate to large bandwidths, the time spent on floating-points arithmetic would be large compare to the time required to reshape the data layout of the matrix. Consequently, we decided to reshape the matrix prior to the factorization so that the algorithm could work with a more appropriate data layout.

Our solver breaks the input matrix into blocks whose size depends on the number of processors and the bandwidth of the matrix. It then computes a static schedule that determines which processor works on which block and in what order. The matrix is then reshaped according to the requirements of the schedule, factored, and then the factor is put back together in the input data layout. The reshaping of the matrix as well as the runtime computation of an irregular schedule for a regular problem represent departures from current practices in the design of parallel algorithms. Our performance results, which are reported in Section 6, show that the solver is efficient and suggest that current practices should be reexamined.

The remainder of the paper is organized as follows. Section 2 describes the integration of the solver into PESSL, a ScaLAPACK-compatible subroutine library and our overall implementation strategy. Section 3 presents an overview of the factorization algorithm. The details of the algorithm, together with a complete analysis of the assignment of processors to block operations are discussed in Section 4. Section 5 explains how the matrix is reshaped. Section 6 presents experimental results that show that the solver performs well and that substantiate our main claims. Section 7 presents our conclusions from this research.

2 Library Considerations and Implementation Issues

This section describes the input and output formats that the solver uses, as well as the overall implementation strategy.

The interface of the solver is compatible with ScaLAPACK's parallel band solver [4, 5]. There are three user-callable subroutines, a factorization subroutine that computes the factorization $LL^T = A$, a triangular-solve subroutine that given the factorization solves the linear system $AX = LL^TX = B$, and a combined factor and solve subroutine. Either the lower or the upper triangle of A is represented in the input data structure.If the lower part of A is supplied, then the factorization routine returns L. If the upper part is stored, the factorization returns L^T. We only consider the case where the lower part is represented.

The solver assumes that the lower parts of the input band matrix and the output factor that overwrites it are stored in packed format in a global array. Columns of the matrix occupy columns of the array, with diagonal matrix elements stored in the first row of the array. The global array is distributed in a one-dimensional block data layout in which a contiguous group of columns is stored on each processor. This data layout is the one-dimensional distributed analog of the lower packed format for storing symmetric band matrices on uniprocessors and shared-memory multiprocessors. Lower packed format is used by numerical linear algebra libraries such as LAPACK [3] and IBM's ESSL [8].

The algorithm is implemented using the single program multiple data (SPMD) model with explicit message passing between processors. The solver uses ScaLA-PACK's own message passing library, the Basic Linear Algebra Communication Subroutines (BLACS), as much as possible, in order to maintain compatability and interoperability with ScaLAPACK. We also used the Message Passing Interface (MPI) in cases where we found it desirable to use nonblocking sends and receives,

Operations on blocks are performed by calls to the sequential level-3 Basic Linear Algebra Subroutines (BLAS) [6] and to a sequential Cholesky factorization subroutine from either LAPACK or the IBM Engineering and Scientific Subroutine Library (ESSL).

3 An Overview of the Factorization Algorithm

The block Cholesky factorization of an n-by-n matrix A with can be summarized as follows, assuming that the factor L overwrites the lower part of A. Each nonzero r-by-r block A_{ij} of the matrix undergoes the transformation

$$\tilde{A}_{ij} = A_{ij} - \sum_{k=0}^{j-1} L_{ik}L_{jk}^T \tag{1}$$

(we use zero-based indices throughout the paper). Diagonal blocks are subsequently factored,

$$\tilde{A}_{ii} = L_{ii}L_{ii}^T . \tag{2}$$

A nondiagonal block L_{ij} of the factor is computed by solving a triangular linear system

$$L_{ij} L_{jj}^T = \tilde{A}_{ij} . \tag{3}$$

We refer to this operation as scaling. When the matrix is banded, blocks that are identically zero are ignored, so the transformation formula (1) changes to

$$\tilde{A}_{ij} = A_{ij} - \sum_{k=\max(0,i-m_r)}^{j-1} L_{ik} L_{jk}^T , \tag{4}$$

where m_r is the block half-bandwidth. We implement equation (4) by $j-\max(0, i-m_r)$ multiply-subtract operations of the form

$$\tilde{A}_{ij} \leftarrow \tilde{A}_{ij} - L_{ik} L_{jk}^T . \tag{5}$$

The blocks A_{ij}, $\tilde{A}_{ij}$, and L_{ij} can all occupy the same location in memory, which we informally denote by L_{ij}. Equations (2), (3), and (4) can be combined into a single expression that we refer to as the *Cholesky formula* for block (i,j),

$$L_{ij} = (A_{ij} - \sum_{k=\max(0,i-m_r)}^{j-1} L_{ik} L_{jk}^T) L_{jj}^{-T} , \tag{6}$$

where the matrices A_{ij}, L_{ij}, L_{ik}, and L_{jk} are square of size r except in the last block row, and where L_{jj}^{-T} denotes $(L_{jj}^{-1})^T$. Note that the last operation to be performed in the formula, the multiplication by L_{jj}^{-T}, requires the solution of a triangular system of linear equations if $i > j$, and the Cholesky factorization of an r-by-r block if $i = j$. In the factorization algorithm, Cholesky formula L_{ij} for block (i,j) is computed in $j + 1 - \max(0, i - m_r)$ consecutive block operations of types (2), (3), and (5).

The equations impose a partial order on the scheduling of the block operations of Equation (6) in the algorithm, because a multiply-subtract operation cannot be performed until the two blocks of L that are involved have been computed, and the final scaling or factorization applied to a block cannot proceed until all multiply-subtract operations have been completed. Our solver uses a systolic schedule. By a systolic schedule we mean a schedule in which all the operations on a block are performed in consecutive time steps, and in which all the arguments of an operation arrive at a block exactly when they are needed. In each time step a processor performs (at most) one block operation (2), (3), or (5), as well as sending and receiving up to two blocks.

Our solver assigns the block operations in Equations (2), (3), and (5) to processors. The assignment works in two levels. In the first level, all the block operations in the Cholesky formula (6) for block (i,j) are assigned to a single *process*, also denoted by (i,j). In the second and more specific level, a set of Cholesky formulas, or processes, is assigned to a single physical processor. We denote the processor to which process (i,j) is assigned by $P(i,j)$. A processor computes a single Cholesky formula assigned to it in consecutive time steps. The processor that computes a Cholesky formula stores the corresponding block in its local memory for the entire factorization algorithm. A processor executes one formula after another until all the blocks it was assigned have been factored.

More specifically, the algorithm uses the following schedule. A block formula L_{ij} starts its computation when its first block operand(s) arrives, except for L_{00} which starts in the first time step with no operands. After a block is computed by the processor assigned to it, it immediately starts to move. A nondiagonal block L_{jk} of the Cholesky factor moves one column to the right in every systolic time step. This block participates in a multiply-subtract operation $\tilde{A}_{jl} \leftarrow \tilde{A}_{jl} - L_{jk}L_{lk}^T$ with the block (j, l) that it is passing through, where $k < l \leq j$. After block (j, k) passes through diagonal block (j, j), it starts moving down column j, again participating in the multiply-subtract operation $\tilde{A}_{ij} \leftarrow \tilde{A}_{ij} - L_{ik}L_{jk}^T$ with every block (i, j) it is passing through, where $j < i \leq k + m$. As can be seen, each nondiagonal block (except for blocks in the last m_r block rows) updates exactly m_r blocks using Equation (5). A diagonal block L_{jj} is factored, as in Equation (2), immediately after all the updates (4) have been applied to it. It then starts moving down column j. It participates in a triangular solve (3) in every subsequent systolic time step, in rows $i = j + 1$ through $i = j + m_r$. It stops moving when it reaches the last nonzero block in column j.

The solver factors the matrix in five major phases. The first phase determines, based on the half-bandwidth m and the number p of processors, the largest block size r that still permits the algorithm to efficiently use p processors. The second phase computes a schedule in which each processor is assigned a set of Cholesky formulas and the order in which these formulas will be computed. The third phase partitions the input matrix into r-by-r blocks and sends each block to the processor which is assigned to compute it. The fourth phase executes the schedule. The fifth phase reshapes the matrix again so that the factor computed in phase four overwrites the input matrix in its original data layout. We remark that phases three and five are performed only in order to provide the user with a simple and convenient input and output format: they are not part of the factorization itself.

4 A Detailed Analysis of the Algorithm

In this section we formally define and analyze a systolic factorization schedule and the assignment of actual processors to Cholesky formulas. The section proves that the schedule is correct and it establishes the number of physical processors required to simulate the systolic algorithm. We show that the number of systolic processes that are simultaneously active is $\lceil (m_r + 1)(m_r + 2)/6 \rceil$ and that the same number of processors are capable of simulating the systolic algorithm. In addition, we show that the systolic factorization algorithm almost always balances the amount of local storage required on each processor.

The algorithm partitions the band matrix into an n_r-by-n_r block band matrix, with blocks of size r-by-r and block half-bandwidth m_r. Only blocks in the lower part of the matrix are stored and used. There are $(m_r + 1)n_r - m_r(m_r + 1)/2$

nonzero blocks in the matrix, $m_r + 1$ in a block column (except for the last m_r block columns). The original and block half bandwidths are related by the equation $m = rm_r - l_r$. The last block will be an upper triangular matrix with the first l_r diagonals of the upper triangular matrix equal to zero.

The systolic algorithm works in discrete time steps. Each step takes a single time unit, which is the time it takes a single processor to perform one block multiply-subtract operation (a GEMM level-3 BLAS), send two blocks and receive two blocks, where all the blocks are square of order r. In some time steps, instead of a block multiply-subtract operation, a processor may need to solve an r-by-r triangular linear system with r right hand sides (a TRSM level-3 BLAS), factor an r-by-r block (implemented by LAPACK's POTRF or ESSL's POF), or perform a multiply-subtract that updates an r-by-r symmetric matrix (a SYRK level-3 BLAS). We assume that these operations take less time than a multiply-subtract operation. The assumption is justified for the values of r in which we are interested, since operation counts in these operations are one half, one third, and one half, respectively, of the operation count of a multiply-subtract (GEMM). (When r is very small, divides and square roots can dominate the running time of these operations, so operation counts do not provide good estimates of the running times.)

The systolic schedule multiplies L_{ik} by L_{jk}^T and subtracts the product from A_{ij} in time step $t = i + j + k$. The operation is performed by processor $P(i,j)$, which stores A_{ij} locally. This processor receives at the beginning of the time step the block L_{ik} from processor $P(i, j-1)$ and the block L_{jk}^T from processor $P(i-1, j)$. The final operation in Equation (6), either the solution of a triangular linear system or the factorization of a block, is performed at time step $i + 2j$.

We now give a detailed description of our algorithm. We specify the algorithm for a diagonal blocks A_{ii} first, followed by the code for a nondiagonal block A_{ij}. Comments are preceded by a percent sign.

For $k = \max(0, i - m_r)$ to $i - 1$
    % Iteration k of the loop is performed during time step $i + i + k$
    Receive L_{ik} from $P(i, i-1)$
    If $k > \max(0, i - m_r)$ then send L_{ik} to $P(i, i+1)$
    Update $A_{ii} \leftarrow A_{ii} - L_{ik}L_{ik}^T$ by calling SYRK
End for
Factor $A_{ii} = L_{ii}L_{ii}^T$ during time step $3i$ (by calling LAPACK's POTRF or ESSL's POF)
If $i < n_r - 1$ then send L_{ii} to $P(i, i+1)$ during time step $3i$

Next, we give the code for a nondiagonal block A_{ij}. Note that for the last block in a columns, that is, when $i = j + m_r$, the "for" loop is empty and the block A_{ij} is upper triangular with zeros in the first l_r diagonals.

For $k = \max(0, i - m_r)$ to $j - 1$
    % Iteration k of the loop is performed during time step $i + j + k$
    Receive L_{ik} from $P(i, j-1)$
    Receive L_{jk} from $P(i-1, j)$

	j = 0	j = 1	j = 2	j = 3	j = 4	j = 5	j = 6	j = 7	j = 8	j = 9
i = 0	0:0(0)									
i = 1	1:1(2)	2:3(1)								
i = 2	2:2(4)	3:4(3)	4:6(0)							
i = 3	3:3(2)	4:5(4)	5:7(2)	6:9(1)						
i = 4	4:4(1)	5:6(3)	6:8(4)	7:10(3)	8:12(0)					
i = 5		7:7(0)	8:9(2)	9:11(4)	10:13(2)	11:15(1)				
i = 6			10:10(1)	11:12(3)	12:14(4)	13:16(3)	14:18(0)			
i = 7				13:13(0)	14:15(2)	15:17(4)	16:19(2)	17:21(1)		
i = 8					16:16(1)	17:18(3)	18:20(4)	19:22(3)	20:24(0)	
i = 9						19:19(0)	20:21(2)	21:23(4)	22:25(2)	23:27(1)

Fig. 1. The systolic schedule and the assignment of processors for the case $n_r = 10$, $m_r = 4$. The figure shows for each block the time steps in which the block's formula starts and ends its activity, separated by a colon, and the physical processor assigned to the formula, in parenthesis.

> Send L_{ik} to $P(i, j + 1)$
> If $k > \max(0, i - m_r)$ then send L_{jk} to $P(i + 1, j)$
> Update $A_{ij} = A_{ij} - L_{ik} L_{jk}^T$ by calling GEMM
> End for
> Receive L_{jj} from $P(i - 1, j)$ during time step $i + 2j$
> If $i < j + m_r$ then send L_{jj} to $P(i + 1, j)$ during time step $i + 2j$
> Compute $L_{ij} = A_{ij} L_{jj}^{-T}$ by calling TRSM during time step $i + 2j$
> Send L_{ij} to $P(i, j + 1)$ during time step $i + 2j$

The following theorem proves that the timing indicated in the code is correct.

Theorem 1. *The timing indicated in the algorithm is correct. That is, the block L_{ik} that is supposed to be received by processor $P(i, j)$ during time $t = i + j + k$ from processor $P(i, j - 1)$ is always sent by $P(i, j - 1)$ during time $t - 1$, and the block L_{jk} that is supposed to be received by processor $P(i, j)$ during time $t = i + j + k$ from processor $P(i - 1, j)$ is always sent by $P(i - 1, j)$ during time $t - 1$.*

The proof, which uses induction and is fairly straightforward, is omitted from this abstract.

It is easy to see that a block (i, j) becomes active during time step $2i + j - m_r$ (or at time step $i + j$ if $j < m_r$) and it is completed during time step $i + 2j$. Therefore, the Cholesky formula for block (i, j) is active during

$$(i + 2j) - (2i + j - m_r) + 1 = (j - i) + m_r + 1 = -d + m_r + 1$$

time steps, where d denotes the diagonal $i - j$ of block (i, j) (except in the first $m_r - 1$ columns where formulas can be active for fewer time steps than that). Successive formulas along a diagonal start and end their activities three time steps apart. Figure 1 shows an example of the schedule.

We now prove a main result of this section, namely, that the number of active processes is at most $\lceil (m_r + 1)(m_r + 2)/6 \rceil$.

Theorem 2. *There are at most $\lceil (m_r + 1)(m_r + 2)/6 \rceil$ processes active at any time step.*

Proof. Let $m_r = 3q + z$, where $0 \leq z \leq 2$. We treat the three cases $z = 0$, $z = 1$, and $z = 2$ separately.

We start with the case $m_r = 3q + 1$. We prove the theorem by exhibiting an assignment of

$$\left\lceil \frac{(m_r + 1)(m_r + 2)}{6} \right\rceil = \frac{(m_r + 1)(m_r + 2)}{6} = \frac{(3q + 2)(q + 1)}{2}$$

processors to all the Cholesky formulas in the schedule. We assign exactly $q + 1$ processors to each pair of diagonals d and $m_r - d$ for each value of d between 0 and $\lfloor m_r/2 \rfloor$, as well as $(q + 1)/2$ processors to diagonal $m_r/2$ if m_r is even. (Note that if m_r is even, then there is an odd number of diagonals and q is odd.) When m_r is odd, the total number of processors in the assignment is $(q + 1)(m_r + 1)/2$. When m_r is even, the total number of processors in the assignment is $(q + 1)(m_r/2) + (q + 1)/2 = (q + 1)(m_r + 1)/2$. To see that the assignment of $(q + 1)$ processors per pair of diagonals is necessary, note that blocks on diagonal d require $m_r - d + 1$ time steps and that blocks on diagonal $m_r - d$ require $m_r - (m_r - d) + 1 = d + 1$ time steps. A block from diagonal d and a block from diagonal $m_r - d$ therefore require $m_r + 2 = 3(q + 1)$ time steps together. We now show that assigning $q + 1$ processors for a pair of diagonals is sufficient. Assign a single processor to block $(j + d, j)$ on diagonal d and to block $(j + m_r - q, j + d - q)$ on diagonal $m_r - d$. Since block $(j + d, j)$ completes at time step $(j + d) + 2j = 3j + d$ and since block $(j + m_r - q, j + d - q)$ starts at time step $2(j + m_r - q) + (j + d - q) - m_r = 3j + d + (m_r - 3q) = 3j + 1$, this single processor can execute both formulas. Since this processor spends $3(q + 1)$ steps on both, and since blocks along a diagonal start 3 time steps apart, we can also assign the same processor to blocks $(j + d + w(q + 1), j + w(q + 1))$ and their "mates" on diagonal $m_r - d$ for any integer w. We therefore need q additional processors to cover these two diagonals. The same holds for all pairs of diagonals. If there is an odd number of diagonals, the middle diagonal requires $(m_r + 2)/2 = (3q + 3)/2$ time steps per block. Therefore, the processor that is assigned to the block in column j in this diagonal can also compute the blocks in columns $j + w(q + 1)/2$ for any integer w. Hence $(q + 1)/2$ processors can cover the middle diagonal. This concludes the proof for the case $m_r = 3q + 1$.

We omit the other cases, which can be reduced to the previous case. The full paper [7] details the proof for all cases.

Table 1 shows an example of a complete schedule for the case $m_r = 3q + 1 = 4$, $n_r = 10$, and $p = 5$. In the first and last $6q = 6$ time steps, some processors are idle. The total number of idle time steps in the example is $6qp = 30$, so the inefficiency is $30/140 = 21\%$. It can be shows that in general the number of idle time steps is about $6qp \approx 2m_r$, and that they occur in the first and last $6q$ or so steps, with equal numbers of idle steps in the beginning and end of the factorization. We omit further details.

π	0	1	2	3	4	5	6	7	8	9	10	11	12	13
0	(000)	—	—	—	(220	221	222)	(511)	(440	441	442	443	444)	(733)
1	—	—	(110	111)	(400)	—	(330	331	332	333)	(622)	(551	552	553
2	—	(100)	—	(300)	—	(320	321	322)	(521	522)	(541	542	543	544)
3	—	—	—	(210	211)	(410	411)	(430	431	432	433)	(632	633)	(652
4	—	—	(200)	—	(310	311)	(420	421	422)	(531	532	533)	(642	643

π	14	15	16	17	18	19	20	21	22	23	24	25	26	27
0	(662	663	664	665	666)	(955)	(884	885	886	887	888)	—	—	
1	554	555)	(844)	(773	774	775	776	777)	—	(995	996	997	998	999)
2	(743	744)	(763	764	765	766)	(965	966)	(985	986	987	988)	—	—
3	653	654	655)	(854	855)	(874	875	876	877)	—	—	—	—	—
4	644)	753)	754	755)	(864	865	866)	(975	976	977)	—	—	—	—

Table 1. A complete schedule for the case $m_r = 3q + 1 = 4$, $n_r = 10$, and $p = 5$. An ijk entry in location (π, t) in the table indicates that processor π is computing during time step t the kth block operation of Cholesky formula (i, j) in Equation (6). Individual Cholesky formulas are enclosed in parenthesis.

5 Reshaping the Data Layout

Even though reshaping a distributed array is a conceptually simple operation in which each array element is sent from its source to its destination, it can take a significant amount of time if not done carefully. Specifically, complex address calculations must be avoided whenever possible, memory-to-memory copying must be done efficiently to minimize cache and TLB (translation lookaside buffer) misses, and interprocessor communication must often be done in large blocks to reduce the effects of communication latency and of frequent processor interrupts. The design of the reshaping module of the solver, which is based on the design of similar subroutines in a parallel out-of-core library [9], aims to avoid these problems.

The main idea behind the reshaping module is the computation of the intersection of two data distributions. Consider reshaping a distributed array from a source distribution D_s to a target distribution D_t. Each distribution decomposes the array into contiguous regions (not necessarily axis-parallel), each of which is simultaneously stored on a single processor. The term contiguous region is used in a loose sense to mean that it and its subsets can be efficiently copied to and from a contiguous buffer in memory (i.e. packed and unpacked). The intersection of the distributions is the decomposition of the global array into a set of maximal contiguous regions such that each region is the intersection of a single set of D_s and a single set of D_t. The reader can visualize the intersection by considering the array with the regions of D_s bounded by red lines and the regions of D_t bounded by green lines. In Figure 2, place the two diagrams with the heavy lines colored red and green one on top of the other. The intersection is the collection of contiguous regions bounded by lines of any color, including lines of mixed colors (See Figure 2). The property of the intersection which is of interest is that each of its maximal regions is stored on a single processor in the

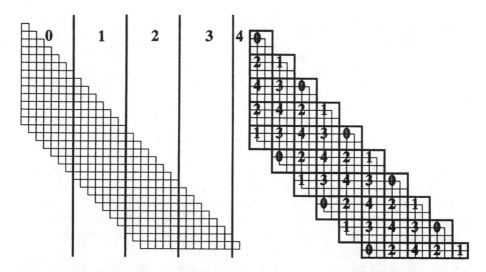

Fig. 2. Distributed layouts of a symmetric banded matrix with 29 rows and columns and a half bandwidth of 12. Only the lower part of the matrix is stored. The figure on the left shows the input and output layout on 5 processors. The heavy lines that show the data distribution are called red red in the text. It is a ScaLAPACK-compatible block distribution of the columns of the matrix, where each processor (except the last) stores a group of 7 columns. The numbers represent the processor that owns each group of columns. The columns are stored locally on every processor packed into an array with at least 13 rows. The figure on the right shows the layout that the systolic factorization algorithm uses in this case. The heavy lines are called green in the text. The matrix is layed out in 3-by-3 blocks, and the numbers show which processor is assigned, and therefore stores, each block. This same layout is shown in Figure 1.

source distribution and on a single processor in the target distribution.

In our case, one distribution is always a column block distribution of a banded matrix and the other is always an apparently irregular distribution of square blocks. The intersection is therefore a decomposition of the matrix into rectangular blocks of varying sizes, which are easy to enumerate. Blocks that contain only zero diagonals of the banded matrix are ignored by the reshaping routines, as well as by the rest of the solver.

6 Performance of the Solver

The experiments were performed on an IBM SP2 parallel computer [2]. The machine was configured with so-called thin nodes with 128 Mbytes of main memory ran AIX version 4.1.3. Thin nodes have a 66.7 MHz POWER2 processor, 64 Kbytes 4-way set associative level-1 data-cache, no level-2 cache, and a 64-bit-wide main memory bus. They have smaller data paths between the cache and the floating-point units than all other POWER2-based SP2 nodes. In all the experiments the message passing layer used the network interface in user-space mode and did not use interrupts. For block operations on the nodes we used IB-M's Engineering and Scientific Subroutine Library (ESSL) version 2.2. For some comparisons we used IBM's Parallel Engineering and Scientific Subroutine Library (PESSL) version 1.1, which based on and compatible with ScaLAPACK, a public domain linear algebra package for linear algebra computations.[3] We used POWER2-specific versions of all the libraries.

The interaction between our algorithm and the architectural balance of the SP2 is best explained with a few examples. In typical time step of the schedule, a node receives two r-by-r matrices, multiplies two r-by-r matrices, and sends two. A single node can multiply two 512-by-512 matrices in less than 1.26 seconds, giving a rate of more than 213 million operations per second (this is a measured rate. The peak rate is 266 Mflops). Sending and receiving the four matrices would take less than 0.21 seconds at the peak 41 Mbytes per second rate. Even if the effective rate is only half of that, and if no overlapping of communication and computation occurs, the communication time represents less than 25% of the time it takes to complete a time step. If the blocks size is only 256-by-256, however, the matrix multiplication takes only 0.16 seconds at the same rate and communication takes more than 0.05 at a rate of 41 Mbytes/sec. At half the communication rate, communication time represents about 40% of the time step. We conclude that while communication costs do not overwhelm the running time when the block size is larger than about 200, they represent a significant overhead even for operations on fairly large dense submatrices.

Table 2 shows that the performance of our factorization algorithm on the SP2 is excellent compared to the performance of other distributed dense matrix computations in PESSL. The performance of our algorithm is also good relative to the performance of the corresponding sequential factorization subroutine in ESSL. The sequential subroutine factored matrices of order $n = 25,000$ and half-bandwidths m ranging from 50 to 400 on a single thin SP2 node at rates of 146–182 Mflops. (The corresponding sequential factorization algorithm in LAPACK on a single thin node is between 1.2 times slower for $m = 400$, to 3.1 times slower for $m = 50$.)

Two important performance trends emerge from Table 2. First, the table shows that larger block sizes usually yield better performance, because the computation-to-communication ratio increases. Second, the table shows that for a given block size, performance improves with the number of processors, because the bandwidth of the matrix increases. When the block bandwidth increases, the fraction of the systolic steps that involve a matrix multiply-subtract increases. When the bandwidth is small, on the other hand, there are relatively more block operations that requires less arithmetic operations than a multiply-subtract, such as scaling. Processors that perform such operations remains idle for part of the systolic time step, waiting for other processors to complete multiply-subtract operations. Table 2 also shows that the overhead of reshaping the data usually accounts for less than 20% of the total factorization time. In many cases the

---

[3] PESSL also contains routines for Fourier transforms and related computations that are not part of ScaLAPACK.

p	n	$m+1$	r	T_t	T_f	T_r	MF/p	n	$m+1$	r	T_t	T_f	T_r	MF/p
4	34100	300	100	25.4	19.5	5.9	30	17000	600	200	32.9	27.3	5.6	45
5	32000	400	100	26.2	20.5	5.7	39	16000	800	200	33.0	27.4	5.5	60
7	35800	500	100	30.2	24.1	6.0	42	17800	1000	200	37.6	31.6	6.0	65
10	42600	600	100	35.1	29.3	5.8	43	21200	1200	200	44.1	38.1	6.0	67
12	43800	700	100	36.1	30.2	5.8	49	21800	1400	200	44.9	39.2	5.7	76
15	48000	800	100	39.5	33.3	6.2	51	24000	1600	200	49.9	43.3	6.6	78
19	54000	900	100	44.1	37.6	6.4	52	27000	1800	200	55.3	49.3	6.0	80
22	56300	1000	100	46.0	39.6	6.4	55	28000	2000	200	57.2	51.2	6.0	85
26	60500	1100	100	49.3	42.2	7.0	56	30200	2200	200	62.0	55.4	6.6	86
31	66100	1200	100	54.5	47.5	6.8	56	33000	2400	200	68.5	61.3	7.1	85

p	n	$m+1$	r	T_t	T_f	T_r	MF/p	n	$m+1$	r	T_t	T_f	T_r	MF/p
4	8400	1200	400	46.9	41.6	5.3	58	5400	1800	600	55.1	49.3	5.8	62
5	8000	1600	400	46.4	41.0	5.4	76	4800	2400	600	48.2	43.3	4.9	76
7	8800	2000	400	51.8	46.0	5.8	82	5400	3000	600	55.4	50.5	4.9	79
10	10400	2400	400	61.2	55.6	5.6	83	6600	3600	600	71.5	65.5	6.0	76
12	10800	2800	400	64.4	58.1	6.2	91	7200	4200	600	78.7	72.1	6.5	82
15	12000	3200	400	71.7	64.9	6.8	94	7800	4800	600	86.2	80.0	6.2	82
19	13200	3600	400	78.9	71.9	7.0	93	9000	5400	600	101	93.7	7.0	82
22	14000	4000	400	84.4	77.1	7.3	98	9000	6000	600	102	94.2	7.4	81
26	14800	4400	400	88.8	81.2	7.6	100	9600	6600	600	108	101	7.2	81
31	16400	4800	400	99.9	91.4	8.5	98	10800	7200	600	124	115	8.5	81

Table 2. The performance of the factorization algorithm on an SP2 with thin nodes. The tables show the performance using four block sizes r, 100, 200, 400, and 600. The total amount of storage per processor is kept approximately constant in all the experiments, at about 20 million bytes per processor for the matrix itself. The total running time is denoted by T_t, the factorization time is denoted by T_f, the reshaping time is denoted by T_r, and the number millions of floating-point operations per second per processor is denoted by MF/p.

overhead is smaller than 10%, even on thin nodes.

7 Conclusions

This paper describes the design, proof of correctness, implementation, and evaluation of a band Cholesky factorization algorithm for distributed parallel computers. Both the analysis and the experiments indicate that the algorithm delivers excellent performance on wide-band matrices, especially with a large number of processors. The algorithm uses several novel ideas in the area of distributed dense matrix computations, including the use of a dynamic schedule that is based on a systolic algorithm and the separation of the input and output data layouts from the layout that the algorithm uses internally. The algorithm also uses known techniques such as blocking to improve its communication-to-computation ratio and minimize the number of cache misses.

Acknowledgments. Thanks to Ramesh Agarwal, Sunder Athreya, Susanne M. Balle, Robert Blackmore, John Lemek, Clint Whaley, and Mohammad Zubair for their help with this project.

References

1. Ramesh Agarwal, Fred Gustavson, Mahesh Joshi, and Mohammad Zubair. A scalable parallel block algorithm for band Cholesky factorization. In *Proceedings of the 7th SIAM Conference on Parallel Processing for Scientific Computing*, pages 430–435, San Francisco, February 1995.
2. T. Agerwala, J. L. Martin, J. H. Mirza, D. C. Sadler, D. M. Dias, and M. Snir. SP2 system architecture. *IBM Systems Journal*, 34(2):152–184, 1995.
3. E. Anderson, Z. Bai, C. Bischof, J. Demmel, J. Dongarra, J. Du Croz, A. Greenbaum, S. Hammarling, A. McKenney, S. Ostrouchov, and D. Sorensen. *LAPACK User's Guide*. SIAM, Philadelphia, PA, 2nd edition, 1994. Also available online from http://www.netlib.org.
4. Anonymous. ScaLAPACK's user guide. Technical report, University of Tennessee, 1996. Draft.
5. J. Choi, J. Dongarra, R. Pozo, and D. Walker. ScaLAPACK: A scalable linear algebra for distributed memory concurrent computers. In *Proceedings of the 4th Symposium on the Frontiers of Massively Parallel Computation*, pages 120–127, 1992. Also available as University of Tennessee Technical Report CS-92-181.
6. Jack J. Dongarra, Jeremy Du Cruz, Sven Hammarling, and Ian Duff. A set of level 3 basic linear algebra subprograms. *ACM Transactions on Mathematical Software*, 16(1):1–17, 1990.
7. Anshul Gupta, Fred G. Gustavson, Mahesh Joshi, and Sivan Toledo. The design, implementation, and evaluation of a banded linear solver for distributed-memory parallel computers. Technical Report RC20481, IBM T.J. Watson Research Center, Yorktown Heights, NY, June 1996. Available online from from the IBM Research CyberJournal at http://www.watson.ibm.com:8080.
8. IBM Corporation. *Engineering and Scientific Subroutine Library, Version 2 Release 2: Guide and Reference*, 2nd edition, 1994. Publication number SC23-0526-01.
9. Sivan Toledo and Fred G. Gustavson. The design and implementation of SOLAR, a portable library for scalable out-of-core linear algebra computations. In *Proceedings of the 4th Annual Workshop on I/O in Parallel and Distributed Systems*, pages 28–40, Philadelphia, May 1996.

A New Parallel Algorithm for Tridiagonal Symmetric Positive Definite Systems of Equations

Fred G. Gustavson and Anshul Gupta

Mathematical Sciences Department
IBM Research Division
T.J. Watson Research Center
Yorktown Heights, NY 10598

Abstract. A new parallel algorithm for solving positive definite symmetric tridiagonal systems of linear equations is presented. It generalizes to the band and block tridiagonal cases. It is called the middle p-way BABE (burn at both ends) algorithm. It has a redundancy of two: The best serial algorithm requires 3N multiply-adds, 2N multiplies and N divides. This algorithm requires 6N/p multiple-adds, 5N/p multiplies and N/p divides. There is only one global communication step which is an all-to-all concatenation of six double words from each process.

1 Introduction

This paper considers a new algorithm for solving, in parallel, a tridiagonal positive definite symmetric system of linear equations. A tridiagonal matrix is a band matrix with a band width of one. Because our algorithm generalizes to the band case and is also most closely related to published band algorithms, we refer to the band literature. The algorithm is a modification of the Lawrie, Sameh [10], and Dongarra, Sameh [6] algorithm and is related to the Johnsson [9], Dongarra, Johnsson [5] and Hajj, Skelboe [8] algorithms. These algorithms are concerned with banded systems in the diagonally dominant and positive definite symmetric cases. The algorithm is also related to algorithms of Sun, Zhang, and Ni [11]. A survey paper by Arbenz and Gander [2] discusses some of the above and other parallel algorithms for banded linear systems. Because of space limitations we cannot adequately cover the large literature on this subject.

We call the new algorithm the middle p-way BABE (Burn at Both Ends) [4] algorithm. Given a tridiagonal system, $Tx = r$, of order N we break it into p systems of order $n = N/p$.

Each of the p systems is mainly described by its positive definite symmetric tridiagonal matrix $T_i, 0 \leq i < p$. For each middle $T_i, 1 \leq i < p-1$ we remove its first and last equations. This gives rise to another positive definite symmetric tridiagonal matrix M_i. For the first and last system, $i = 0$ and $i = p-1$, we remove only one equation namely the last and first respectively. The middle BABE algorithm is a divide and conquer type algorithm. It has three phases:

1. Simultaneously each process i "BABE factors and forward eliminates" $M_i x = r$ and computes its part of the Schur complement of the reduced system.
2. Perform a global MPI all to all concatenation of each part of the reduced system so that each process has the entire reduced system. Then, using a generalization of the BABE technique, each process solves for its two solution components.
3. Simultaneously each process "BABE back solves" for its remaining solution values.

We show that the redundancy for this process is two; i.e. based on operation count the parallel algorithm performs twice as much work as the serial algorithm. For positive definite band systems the redundancy is four [2]. The global all to all concatenation consists of $\log_2 p$ individual point to point communications with a total bandwidth of $6p$ double words. It is possible to interleave computation (cyclic reduction) with these $\log_2 p$ individual stages and thus reduce the computation cost of solving the reduced system to $0(\log_2 p)$. We suggest that it is better to let the highly optimized MPI global communication routine handle these $\log_2 p$ individual communications and to not break the $\log_2 p$ phases up into communication and computation steps. This is especially true for our reduced tridiagonal system as the computation requirement to solve the reduced system is $4p$ multiply-adds, $4p$ multiplies and $2p$ divides.

The paper consists of three sections. In section 2 we show that the middle BABE algorithm can be viewed a serial LDL^T factorization of PTP^T where P is a permutation matrix. For the specific P, we show the reduced system is also positive definite symmetric tridiagonal. Section 2 can be viewed a proof of correctness of the algorithm. In section 3 we give full details about the three main phases of the algorithm. The cost of solving $Tx = r$ is $3N$ multiply-adds, $2N$ multiplies and N divides. Section 3 concludes with showing that additional operations needed by the algorithm are $3N$ multiply-adds and $3N$ multiplies plus the cost of solving the reduced system. This cost is $2K$ multiply-adds, K multiplies and K divides where $K = 2p - 2$. When $N >> p$, this establishes that the redundancy is two. Because of space limitations we cannot go into implementation details. We mention that we have introduced the BABE technique to introduce a two way parallelism into the individual process computations. For RISC processors we call this functional parallelism and we refer the reader to [1] for some details. The accuracy of BABE type algorithms have been studied and the literature indicates that these BABE algorithms have the same or better accuracy than standard LDL^T factorization and solution algorithms. See [12, 3, 7] for details.

2   A Serial View and Proof of Correctness

If T is a positive definite symmetric tridiagonal matrix, then PTP^T is a positive definite symmetric matrix for an arbitrary permutation P. We use this fact to claim that our parallel algorithm constitutes a serial factorization, LDL^T of PTP^T, for a specific P. In fact, for the specific P that we use the reduced system will itself be a positive definite symmetric tridiagonal matrix. Let p be the number of processes and $\sum_{i=0}^{p-1} n_i = N$. Each process $i, 0 \leq i < p$ will be given the $i-th$ set of n_i equations that make up the global system $Tx = r$. From now on we set $n = N/p$ and drop the dependence on i. Let T_i be tridiagonal matrix associated with the $i-th$ set of n equations. For $0 < i < p-1$, define $M_i = T_i(2 : n-1, 2 : n-1)$ as the middle rows and columns of T_i. The first and last equations associated with T_i will become part of the reduced system (Schur complement). Removing these two rows is crucial to producing an algorithm that has only one global communication step. For process 0 define $M_0 = T_0(1 : n-1, 1 : n-1)$ and for process $p-1$ define $M_{p-1} = T_{p-1}(2 : n, 2 : n)$. The size of the reduced system will be $2p-2$. For process 0, $n-1$ equations of M_0 come first and its last equation is the first reduced equation. For process $i, 0 < i < p-1$ its $n-2$ equations of M_i follow the $n-2$ equations of process $i-1$ ($n-1$ equations if $i = 1$) and its first and last equation follows the two reduced equations of process $i-1$ (one equation if $i = 1$). For process $i = p-1$ its last $n-1$ equations follow the $n-2$ equations of process $p-2$ and its first equation is the last equation (equation N of the global system) of the reduced system. The specific P is made up of p sub-permutations $P_i, 0 \leq i < p$. The definition of each P_i can be deduced from the above description. Hopefully, Figure 1 will make this clear. Here $p = 4, N = 40$ and $n = 10$. $P =$(0:8, 11:18, 21:28, 31:39, 9, 10, 19, 20, 29, 30, 31). In the figure the very light shading represents the diagonal and the light shading represents the off (upper) diagonal and the dark shading lower diagonal. The lower diagonal is the transpose of the upper diagonal and for tridiagonal is equal to the off diagonal. We have added the redundant lower diagonal to the figure for clarity of discussion. We have used the labels A, B, C, D to denote the individual elements of processes 0 to 3. Now consider the reduced system (equations 34:39 of Figure 1). Each of the $p = 4$ processes are able to compute their part of the Schur complement independently (in parallel).

There are $p = 4$ rectangular matrices in columns 34 to 39. Their transposes are shown in rows 34 to 39. Each rectangular matrix has two nonzero elements which are the first and last off diagonal elements of T_i. The part of Schur complement for the middle processes (B and C) is a 2 by 2 positive definite symmetric matrix and a 1 by 1 positive matrix for the end processes (A and D). Let $S = M_i^{-1}$. To compute the part of the Schur complement we need only compute $S_{11}, S_{1j} = (S_{j1})$ and S_{jj} where j is the order of M_i. Only the middle systems (B and C) produce a single fill to the reduced systems off diagonal elements. Process $i, 0 \leq i < p-1$, off diagonal connecting original last element (element n), which is not part of T_i, (off diagonal elements A, B, C of the reduced system) are not changed in the Schur complement.

3 Algorithm Details and its Operation Count

Consider any middle process $i, 0 < i < p-1$. We want to solve

$$c_0 e_1 x_0 \;+\; Tx \;+\; c_n e_n x_{n+1} \;=\; r$$

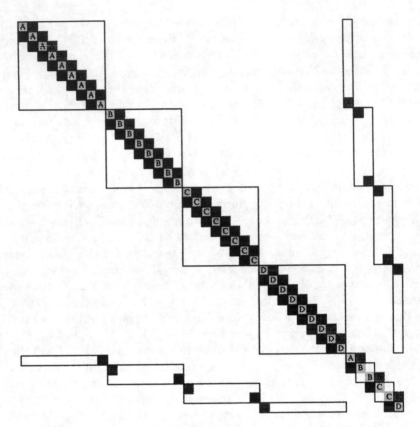

Fig. 1. The reordered matrix PTP^T, where $N = 40$ and $p = 4$.

$$\tag{1}$$

where $x = (x_1, x_2, \cdots, x_n)^t, r = (r_1, r_2, \cdots, r_n)^t, T$ is a symmetric positive definite tridiagonal matrix of order n and e_j is the $j - th$ unit vector and t denotes transpose. (We drop the dependence on i as doing so will not effect the generality of the description and will aid in notational clarity.) Matrix T is described by two vectors $b = (b_1, b_2, \cdots, b_n)^t$ and $c = (c_1, c_2, \cdots, c_{n-1})^t$ where b is the main diagonal and c is the off-diagonal. We want to transform the first and last equations of (1) into two reduced equations. We shall do this by working the middle $n - 2$ equations of (1). We have

$$c_1 e_1 x_1 \quad + \quad M x(2 : n-1) \quad + \quad c_{n-1} e_n x_n = r(2 : n-1)$$

$$\tag{2}$$

where e_j is the $j - th$ unit vector of order $n - 2$. Here M is a symmetric positive definite tridiagonal matrix of order $n - 2$.

Equations (2) are the same as Equations (1) except in (2) we have omitted the first and last equations of Equations (1).

We write the n equations (1) in Tableau form :

$x_0\ x_1\ x_2$			$x_{n-1}\ x_n$	x_{n+1}	RHS
$c_0\ b_1\ c_1$					r_1
$c_1\ b_2\ c_2$					r_2
	$\cdots$				
		$\cdots$			
		$c_{n-2}\ b_{n-1}\ c_{n-1}$			r_{n-1}
			$c_{n-1}\ b_n$	c_n	r_n

$$(3)$$

Let $S = M^{-1}$ which is also a positive definite symmetric matrix. Multiply equations (2) by S to get the following Tableau form:

$x_0\ x_1$		$x_2\ x_{n-1}\ x_n$		x_{n+1}	RHS
$c_0\ b_1$		c_1			r_1
$c_1\ S_{1,1}$		1.0	$S_{1,n-2}c_{n-1}$		y_2
	$\cdots$				$\cdots$
		$\cdots$			$\cdots$
$c_1 S_{n-2,1}$		1.0	$S_{n-2,n-2}c_{n-1}$		y_{n-1}
		$c_{n-1}\ b_n$		c_n	r_n

$$(4)$$

In (4), $y(2 : n - 1) = Sr(2 : n - 1)$. Now use equation two of Tableau (4) to eliminate variable x_2 of equation one and equation $n - 1$ to eliminate variable x_{n-1} of equation n. We get the two reduced equations in Tableau below :

x_0	x_1	x_n	x_{n+1}	RHS
c_0^t	bb_1	bb_2	0.0	rr_1
0.0	bb_2^t	bb_n	c_n	rr_n

$$(5,6)$$

where $bb_1 = b_1 - c_1^t S_{1,1} c_1, bb_2 = -c_1 S_{1,n-2} c_{n-1} = -c_1^t S_{n-2,1}^t c_{n-1}^t$, $bb_n = b_n - c_{n-1}^t S_{n-2,n-2} c_{n-1}, rr_1 = r_1 - c_1 y_1$, and $rr_n = r_n - c_{n-1}^t y_{n-1}$.
Equations (5,6) are the reduced equations and the values are scalars. However, the results are also true for matrices and therefore we give these results where c_0, bb_1, bb_2, bb_n, and c_n are block matrices of order m and the vectors $x_0, x_1, x_n, x_{n+1}, rr_1$, and rr_n are of order m. Each component of $x(2 : n - 1)$ is an m vector. For tridiagonal, $m = 1$.
For process $0, c_0 = 0$ and we choose M to be the first $n - 1$ rows and columns of T. For process $p - 1, c_n = 0$ and we choose M to be the last $n - 1$ rows and columns of T. In each case there is only one reduced equation; equation n for process 0 and equation 1 for process $p - 1$. We get for process 0

x_n	x_{n+1}	RHS
bb_n	c_n	rr_n

$$(7)$$

where $bb_n = b_n - c_{n-1}^t S_{n-1,n-1} c_{n-1}$ and $rr_n = r_n - c_{n-1}^t y_{n-1}$ and for process $i = p - 1$ we get

$$\frac{x_0 \quad x_1 \quad \text{RHS}}{bb_1 \quad c_1 \quad rr_1}$$

$$(8)$$

where $bb_1 = b_1 - c_1^t S_{1,1} c_1$ and $rr_1 = r_1 - c_1 y_1$.

Now note that process $i-1$ has its $c_n = c_0^t$ for process i. This follows from symmetry. It is clear that (7), ((5), (6) for $i = 1, p-2$), and (8) form a $2p-2$ symmetric system of equations. Each process puts its six coefficients $bb_1, bb_2, rr_1; bb_n, c_n, rr_n$ in a send buffer. (process 0 does not store coefficients 1,2,3 and process $p-1$ does not store coefficients 4,5,6).

There are three main steps to the algorithm. Main step 1 partially "transforms" Equation (3) into Equations (4-8).

Now we discuss how we handle each middle tridiagonal system (2). Each middle process starts solving three tridiagonal systems $Mx = r$, $Mv = c_1 e_1$ and $Mw = c_{n-1} e_{n-2}$ where e_1 and e_{n-2} are the first and last $(n-2th)$ unit vectors. Process 0 does not solve $Mv = c_0 e_1$ as $c_0 = $ zero. Process $p-1$ does not solve $Mw = c_n e_{n-1}$ as $c_n = $ zero. The factorization starts at both ends using the BABE technique. $Mx = r$ and $Mv = c_1 e_1$ goes top to middle and $Mx = r$ and $Mw = c_n e_{n-1}$ goes bottom to middle. The factorization for $Mx = r$ stops at the middle and coefficients $a1, a2; b1, b2; r1, r2$ are computed for the later back solution computation. Let $j = (n-2)/2$. Then j and $j+1$ represent the middle two equations. Let a be a vector of length $n-2$. Initialize $a_1 = c_1$ and $a_{n-2} = c_{n-1}$ and $a_i = 0$ otherwise. Vector a will contain information about v and w as well as be the scaling vector (spike) for later back solution. Forward elimination zeros below the diagonal and backward elimination zeros above the diagonal. Upon reaching the middle Tableau (2) looks like

x_1 x_2 x_3 $\cdots$ x_j x_{j+1} $\cdots$ x_{n-2} x_{n-1} x_n	RHS
a_1 b_2 c_2 $\qquad\qquad\qquad\qquad$ 0	r_2
a_2 0 b_3 c_3 $\qquad\qquad\qquad$ 0	r_3
$\cdots\cdots 0$ $\cdots$ $\qquad\qquad\qquad$ $\cdots$	
a_j $\qquad$ 0 b_j c_j $\qquad\qquad$ 0	r_{j+1}
0 $\qquad\qquad$ c_j b_{j+1} 0 $\qquad$ a_{j+1}	r_{j+2}
$\cdots$ $\qquad\qquad\qquad$ $\cdots$ 0 $\cdots$ $\cdots$	
0 $\qquad\qquad$ c_{n-3} b_{n-2} 0 $\qquad$ a_{n-3}	r_{n-2}
0 $\qquad\qquad\qquad$ c_{n-2} b_{n-1} a_{n-2}	r_{n-1}

$$(9)$$

Note that we can now solve three sets of 2 by 2 equations: process the middle two equations, j and $j+1$ to solve for $a1, a2; b1, b2;$ and $r1, r2$.

$$\begin{vmatrix} b_j & c_j \\ c_j & b_{j+1} \end{vmatrix} \begin{vmatrix} a1 & b1 & r1 \\ a2 & b2 & r2 \end{vmatrix} = \begin{vmatrix} a_j & 0 & r_{j+1} \\ 0 & a_{j+1} & r_{j+2} \end{vmatrix}$$

After doing so, the middle equations of Tableau (9) become:

x_1 x_2 x_3 $\cdots$ x_j x_{j+1} $\cdots$ x_{n-2} x_{n-1} x_n	RHS
$a1$ $\qquad\qquad$ 1 0 $\qquad\qquad$ $b1$	$r1$
$a2$ $\qquad\qquad$ 0 1 $\qquad\qquad$ $b2$	$r2$

Later on, when we know x_1 and x_n we can use these six values $a1, a2; b1, b2;$ and $r1, r2;$ to compute x_j, x_{j+1}. Then we use the rest of the above Tableau (9) to compute, via back and forward solution, the solution x_i for $1 < i < j$ and $j + 1 < i < n$.

The algorithm needs to compute v_1, w_{n-2} and $v_{n-2} = w_1$ where v_1 and v_{n-2}, w_1 and w_{n-2} are the first and last components of v and w. $v_{n-2} = w_1$ as M is symmetric and v and w are the first and last columns of $S = M^{-1}$. It turns out that $w_1 = a2a_{j+1}$ and that v_1 and w_{n-2} can be found via backward and forward substitution : Set $v1 = a1$ and $wn = b2$. Also, set $y2 = r1$ and $ynm1 = r2$. The following back solve code computes from Tableau (9) the values $v1, wn, y2,$ and $ynm1$.

```
C    Here n is n-2. Then, j=no2 = n/2.
C    forward solve equation j=no2 + 2 when n is odd
     if (2*no2.lt.n) then
        j = no2
        ynm1= r(np1-j)-c(n-j)*ynm1
        wn= a(np1-j)-c(n-j)*wn
        ynm1=ynm1*b(np1-j)
        wn=wn*b(np1-j)
     endif
C    back-solve remaining equations
     do j=no2-1,1,-1
        y2= r(j)-c(j)*y2
        v1= a(j)-c(j)*v1
        y2=y2*b(j)
        v1=v1*b(j)
        ynm1= r(np1-j)-c(n-j)*ynm1
        wn= a(np1-j)-c(n-j)*wn
        ynm1=ynm1*b(np1-j)
        wn=wn*b(np1-j)
     enddo
```

We claim that the values $y_2, y_{n-1}, S_{1,1}, S_{1,n-2}, S_{n-2,n-2}$ of Tableau (4) $= y2, ynm1, v1, w1, wn$. Once these values are found we can compute five of the six coefficients of the reduced equations (5,6). On process 0, equation (7) is used and on process $p - 1$, equation (8) is used. Overall, the p processes generate coefficients for $2p - 2$ equations.

We now describe main step 2. All processes, 0 to $p - 1$, participate in an all to all concatenation so that each process receives the entire $2p - 2$ positive definite symmetric system. Each process then solves for two variables that it needs for doing the back solution; process i computes variables x_{2i}, x_{2i+1} for $0 < i < p$. Process 0 solves for x_1 and process $p - 1$ solves for x_{2p-2}. Note that to solve a tridiagonal system for all variables requires factorization, forward and back solution. Here, because we only want up to two solution components, we can "half" this computation by avoiding the back solution phase. We apply a generalization of the BABE procedure and meet at variable k instead of at the specific value $k = n/2$. Reference [3] states that this procedure has good numerical properties. Therefore, each process does "half" of the computation and all processes are load balanced.

We now describe main step 3. Note that knowing x_1, x_n on process i allows us to compute x_i for $1 < i < n$ as follows. When step 2 completes these two variables (one on processes 0 and $p-1$) are on their process. Recall that the Tableau (10) allows us to compute x_j and x_{j+1} once x_1 and x_n are known. When x_j and x_{j+1} are known we use the rest of Tableau (9) to compute via back and forward solution the solution x_i for $1 < i < j$ and $j + 1 < i < n$.

There is *only* one global communication step in the entire algorithm. In main step 1, each process expresses the solution vector x in terms of r and the spike vector a which contains part of vector v, components 1 to $n02$ and part of vector w, components $n02 + 1$ to n. The Tableau's of main step 1 are further used to compute Schur complement which in this case involves finding three corner elements of $S = M^{-1}$. Also, two RHS values of the reduced equations (Schur complement RHS) are obtained using the Tableau. Once the pieces of Schur complement system is known the entire system is communicated to all processes. Then each process *only* computes its two solution values! Then, main step 3 begins where the scalars x_1, x_n act as scaling coefficients for the spike vector a in the final parallel back solution process (see Tableau (9)).

In main step 1 we do "BABE factorization and forward elimination" of $Mx = r$. The operation count is $2n$ multiply-adds, n multiplies and n divides. An additional n multiplies is done to compute the spike vector a. The value w_1 is a by-product of this computation. Next "BABE back solution" is done to compute the four scalars v_1, w_n, y_2, y_{n-1}. The cost is $2n$ multiply-adds and $2n$ multiplies. Thus the cost of phase 1 is $4n$ multiply-adds, $4n$ multiplies and n divides. In main step 3 we do $2n$ multiply-adds and n multiplies. The total operation count is $6n$ multiply-adds, $5n$ multiples and n divides. We note that serial tridiagonal solution requires $3n$ multiply-adds, $2n$ multiplies and n divides. Based on operation count alone this shows that the redundancy is 2. However, because pipeline latencies effect the performance of tridiagonal solution, the operation count is *not* a good indication of performance. Additionally a divide operation on RISC processors tends to be out of proportion large relative to multiply-add cost; e.g. on RS/6000 divide cost is 19 cycles and multiply-add cost is 1 cycle! Neglecting pipeline latency, but using this weighted operation count, brings the redundancy factor down from 2 to $30/24 = 1.2$. In closing, we mention that the BABE technique reduces the effect of pipeline latency and thus it was used in our implementation of the parallel tridiagonal solver.

4   Summary

A new parallel algorithm, called the middle p-way BABE algorithm, for solving symmetric positive definite systems of equations is presented. A proof of correctness and full algorithm details are presented. It generalizes to the the band and block tridiagonal cases but only sketchy details are given. With easy modifications the algorithm can be applied to diagonally dominant systems of equations. The redundancy of the algorithm (based on operation count) is two. However, on RS/6000, the redundancy is 1.2 if one uses a weighted operation count. Such measures are not too reliable, especially for tridiagonal because of pipeline latencies in the factorization. We introduced the BABE factorization to reduce the effect of this latency. Compared to standard factorizations BABE factorizations are comparable or better in terms of accuracy.

Acknowledgement The first author thanks M. Zubair for stimulating his interest in this problem during 1992 by showing him [13]. He also says thank you very much to Jerzy Wasniewski for being a driving force behind the successful PARA conferences.

References

1. Ramesh C. Agarwal, Fred G. Gustavson, and M. Zubair. Exploiting functional parallelism of POWER2 to design high-performance numerical algorithms. *IBM Journal of Research and Development*, 38(5):563–576, 1994.
2. Peter Arbenz and Walter Gander. A survey of direct parallel algorithms for banded linear systems. Technical Report 221, Institute of Scientific Computing, ETH Zurich, 1994.
3. I. Babuska. Numerical stability in problems of linear algebra. *SIAM Journal on Numerical Analysis*, 9(1):53–77, 1972.
4. Jack J. Dongarra et al. *Linpack*. SIAM, Philadelphia,PA, 1979.
5. Jack J. Dongarra and S. L. Johnsson. Solving banded systems on a parallel processor. *Parallel Computing*, 5:219–246, 1987.
6. Jack J. Dongarra and A. H. Sameh. On some parallel banded system solvers. *Parallel Computing*, 1:223–235, 1984.
7. K. Vince Fernando, Beresford N. Parlett, and Inderjit S. Dhillon. A way to find the most redundant equation in a tridiagonal system. Technical report, Computer Science Department, University of California, Berkeley, 1995.
8. Ibrahim N. Hajj and Stig Skelboe. A multilevel parallel solver for block tridiagonal and banded linear systems. *Parallel Computing*, 15:21–45, 1990.
9. S. L. Johnsson. Solving narrow banded systems on ensemble architectures. *ACM Transactions on Mathematical Software*, 11:271–288, 1985.
10. D. H. Lawrie and A. H. Sameh. The computation and communication complexity of a parallel banded system solver. *ACM Transactions on Mathematical Software*, 10:185–195, 1984.
11. Xian-He Sun, Hong Zhang, and Lionel M. Ni. Efficient tridiagonal solvers on multicomputers. *IEEE Transactions on Computers*, 41(3), 1992.
12. Henk A. van der Vorst. Large tridiagonal and block tridiagonal linear systems on vector and parallel computers. *Parallel Computing*, 5:45–54, 1987.
13. M. Zubair. A coarse grained tridiagonal solver for distributed memory parallel machines. Technical report, Computer Science Department, Old Dominion University, Norfolk, VA, 1993.

An Environment for the Parallel Solution of Coupled Problems on Structured Grids*

Matthias G. Hackenberg[1], Wolfgang Joppich[1] and Slobodan Mijalković[2]

[1] SCAI – Institute for Algorithms and Scientific Computing
GMD – German National Research Center for Information Technology
53754 Sankt Augustin, Germany
[2] Faculty of Electronic Engineering, University of Niš
Beogradska 14, 18000 Niš, Yugoslavia

Abstract. The solution of coupled problems is one of the grand challenges in scientific computing. Many problems can only be solved partially. Not all dependencies can be taken into account, e.g. only subproblems can be solved completely, even when optimal numerical methods (multigrid methods) are used. Therefore, parallel simulation is inevitable.

Coupled simulation of problems with different numerical complexities is possible now. An environment has been developed, which is an extension of L_iSS (environment for the parallel multigrid solution of partial differential equations on general 2D domains). Block-structured grids with overlap are the basis for parallelization. In order to achieve acceptable load-balance each subproblem is distributed to all processors. Components of the multigrid algorithm can be selected separately for each subproblem, thus allowing optimal methods for each problem. In addition, a large number of coupling strategies can be used, varying from coupling after each time step to coupling after each relaxation.

This strategy can be used in a real life example from VLSI wafer fabrication, where oxidation and diffusion are simulated coupled and in parallel.

1   Introduction

1.1   Coupled Problems

Coupled problems arise when different processes on connected domains have to be simulated. Examples are the shaking of an aircraft wing combined with the flow around the wing, the reaction of a ship's hull together with the movement of the load and the flow of water, or the movement of cardiac valves and their effect on the blood. Usually only single processes are considered, e.g. aerodynamics or structural mechanics of fixed wings. The coupled simulation of all processes is very difficult.

A very interesting problem is coupled diffusion and oxidation in process simulation [3]. In VLSI device fabrication the method of local oxidation is used

---

* This work was supported within the GRISSLi project by the German Federal Ministry for Research and Education under contract no. 01IS512C.

for the creation of isolation layers and doping masks. After masking the silicon (Si) with silicon-nitride (Si_3N_4) and implanting dopant material (boron, arsenic, phosphorus,...), silicon-dioxide (SiO_2) develops under oxidizing conditions, i.e. the presence of oxygen or water vapor at high temperatures. The initial geometry and the geometry after a certain time $t > 0$ can be found in Fig. 1.

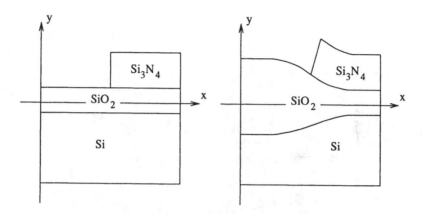

Figure 1. Initial geometry and geometry after a certain time $t > 0$

The growing oxide behaves like a highly viscous flow well described by the Navier-Stokes equations. The diffusion of dopant material is modeled with the help of diffusion equations. As the geometries are time-dependent two additional equations used for the creation of new grids have to be solved on each subdomain. Seven partial differential equations have to be solved in the silicon-dioxide whereas only three equations have to be considered in the silicon. Obviously, this imbalance is the major factor requiring special care when developing strategies for the parallel solution.

1.2 Parallel Multigrid Methods

Usually, optimal multigrid methods [3] are used for the solution of partial differential equations. L_iSS [4] developed by GMD is an environment for the parallel multigrid solution of partial differential equations on general 2D domains. The block-structures can be composed of several arbitrarily shaped logically rectangular blocks. The general approach is grid partitioning [1], i.e. each block of the block-structured grid is enlarged by an overlap and is mapped to one processor of the parallel computer. The parallelization is simplified by the use of a communication library for block-structured grids based on the PARMACS. In the near future the PARMACS are likely to be replaced by MPI.

Communication of blocks is done with the help of the overlaps as shown in Fig. 2. The dashed lines indicate two identical grid lines. The alteration rights along common block boundaries are distributed uniquely according to the block numbers assigned by the user. The block with higher number is only updated by the neighboring block within the overlap area. Vice versa, data is transferred to the overlap of the block with lower update priority. In addition to that, those inner points of the latter block at the common block boundary marked by the dashed line in Fig. 2 are updated by the block with higher priority. Adhering to these principles efficient parallel multigrid algorithms can be obtained [1].

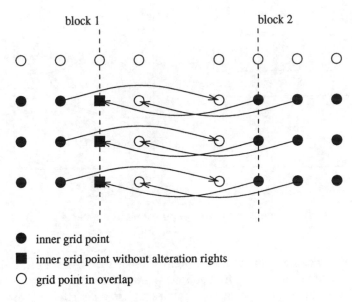

Figure 2. Communication of blocks

2   Parallel Coupled Solution

2.1   Creation of New Structures

There are two possible approaches to parallel coupled solution. On the one hand the whole problem is solved by all processors. This means that the grid covering the domain of the global problem is partitioned according to the number of grid points. Each part is mapped to one processor. Considering the special application some processors simulate the silicon-dioxide and others perform the computation in the area of the silicon, i.e. some processors solve seven equations with seven unknowns and others only three equations with three unknowns. As those processors which have to solve seven equations have to perform considerably more work than those with only three equations the main disadvantage of this approach is quite evident: load imbalance. Partitioning according to both number of grid points and number of unknowns would be an obvious way to a more balanced distribution. However, the correct ratio of grid points and number of unknowns cannot be evaluated in general. The inability to predict the cost of all future computations can be considered as the main reason for this.

On the other hand each subproblem is spread over all processors, i.e each subproblem is solved by all processors one after another. In order to do so, each subdomain is partitioned according to the number of grid points leading to as many parts as processors available for the computation. As the number of equations is constant within each subproblem and the number of grid points of each part of the subproblem is almost identical, the computational work of all processors is very similar. Thus load balance can be preserved. In order to spread the subproblems over all processors new block-structures have to be created. This principle is shown in Fig. 3.

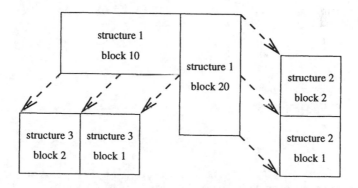

Figure 3. Creation of new structures

Allowing different algorithms for each subproblem is another advantage of spreading each subproblem over all processors. Apart from the number of unknowns and equations, for example, a special number of levels of the multigrid cycle can be used for each subproblem. In addition to that, the stepping in time is simplified, especially for subproblems requiring different time step sizes. Therefore, optimal algorithms can be formed for the computation on each structure.

2.2 Communication of Structures

Naturally, the new structures have to communicate in order to solve the global problem. Similar to blocks, the structures are enlarged by an overlap. Unlike the grid functions at block boundaries, grid functions at the common boundary of two structures are in general not unique. Due to the type of interface conditions used, grid function values at interface points have to be stored separately in each structure. Therefore, the communication strategy of blocks cannot be used for structures. The modified update sequence shown in Fig. 4 has to be used. Grid points on identical positions are marked by the dashed line. Grid function values in the overlap of structures can be used for the formulation of coupling conditions at common interfaces of structures.

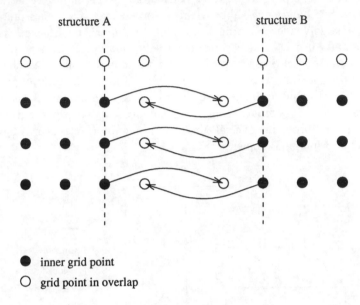

Figure 4. Communication of structures

The information received from neighboring structures can be used in two ways for the modification of interface conditions. First, grid functions received from neighboring structures can be used in a Jacobi-like manner. After completing the computation on all structures, communication between structures is started. The currently calculated values are used for the modification of the boundary equations. This principle can be found in Fig. 5. Communication after each cycle combined with V-cycles using three levels on two structures is shown. This strategy is referred to as "parallel emulation of communication".

Second, the grid functions of neighboring structures can be used in a Gauß-Seidel-like way. Each piece of information is used for the modification of the boundary conditions as soon as possible. The example of communication after each cycle can be found in Fig. 6. Three levels are used on each structure. This communication strategy is named "sequential communication".

Experiments with model problems show that sequential communication is generally better than parallel emulation of communication [2]. This result is similar to the convergence properties of Jacobi-type relaxations compared to iterations of Gauß-Seidel-type.

2.3 Switching of Structures

Usually, the partial solutions of the coupled problems depend on each other. Within iterative processes the current approximation of the solution of one sub-

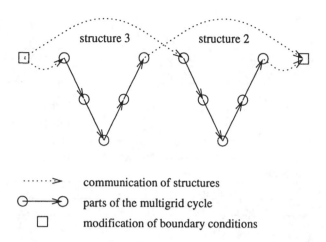

Figure 5. Parallel emulation of communication of structures

problem can be used as initial value for the computational solution of the other subproblem. As each subproblem is treated as a new structure on all processors, structures have to be handled one after another. Switching from one structure to another can take place at certain points within the multigrid algorithm. Several possibilities are shown in Table 1.

Obviously, the tighter the coupling the more similar the algorithms have to be. For example, coupling after each relaxation within the multigrid algorithm requires identical algorithms on all structures. However, when coupling after each time step, a single grid method for one subproblem can be combined with iterative solution by e.g. multigrid using several levels on the other structure.

A full multigrid cycle (FMG cycle) consisting of two V-cycles on each of three levels is shown as an example. Possible points for switching between structures can be found in Fig. 7. The numbers indicate the different coupling strategies.

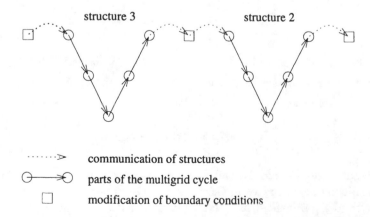

Figure 6. Sequential communication of structures

Table 1. Possible switching points

strategy	activation of next structure after ...
0	No communication of structures
1	each time step
2	the last FMG interpolation, then after the last cycle
3	each FMG interpolation, on the finest grid after the last cycle
4	the last FMG interpolation, then after each cycle
5	each FMG interpolation, on the finest grid after each cycle
6	each cycle
7	each smoothing-step
8	each relaxation

Sequential communication and parallel emulation of communication can be combined with each strategy. Strategy number i combined with parallel emulation of communication is called strategy 10i. The component of the algorithm shown between two switching points is repeated for each structure.

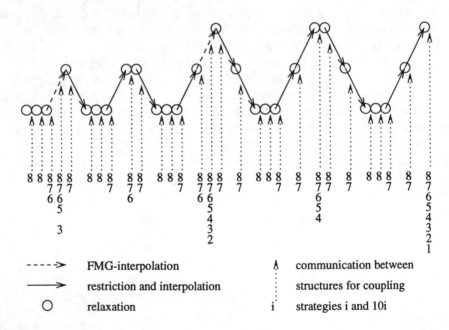

Figure 7. Switching points

As presented in [2] in more detail for model problems, frequent coupling leads to empirical convergence rates identical to rates obtained by computation with boundary conditions instead of interface conditions. The boundary conditions used are obtained by passing the exact solution of the model problems to the interface conditions. The tighter the coupling the better the convergence.

3 Example

Four processors of the parallel machine SP2 are used for the computation of the real life example of coupled oxidation and diffusion. The block structures of the final geometry after oxidation and diffusion for 60 minutes can be found in Fig. 8. The numbers within the blocks identify the processor numbers, those at the boundaries mark the codes for the complex boundary conditions. Each code identifies different complex conditions for each grid function.

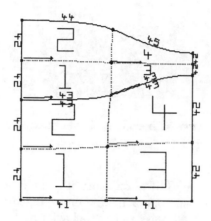

Figure 8. Block structures

Hitherto, the silicon-nitride is not simulated itself. At the moment it is modeled by a special system of boundary conditions (code 45 in Fig. 8). The free oxide surface is described by another system of boundary conditions (code 44). Boundary condition code 43 is used for the interface between the structures. The oxidation rate is implemented in the corresponding equations. Symmetry conditions are specified as code 42. The situation within the silicon far from the interface is described by boundary conditions with code 41.

The final concentration of boron in silicon-dioxide and in silicon after 60 minutes can be seen in Fig. 9. The jump of the concentration at the common boundary of oxide and silicon is caused by the segregation condition used as interface condition applied to the equations describing the diffusion of dopant. Sequential communication after each cycle of the multigrid iteration is chosen as coupling strategy for the computation.

4 Conclusion

An environment for the parallel solution of coupled problems has been developed. It is designed for applications on general 2D block-structured grids. Coupled

Figure 9. Concentration of boron in silicon-dióxide and in silicon after 60 minutes

oxidation and diffusion of VLSI wafer fabrication is an example demonstrating the capability of the chosen approach. The future extension of this approach to 3D-problems is encouraged by the positive behavior both with respect to parallel and numerical results.

References

1. McBryan, O. A., Frederickson, P. O., Linden, J., Schüller, A., Solchenbach, K., Stüben, K., Thole, C.-A., Trottenberg, U., *Multigrid Methods on Parallel Computers – A Survey of Recent Developments*, Impact Comput. in Sci. Engrg. **3** (1991) 1–75
2. Hackenberg, M. G.: Untersuchungen zur parallelen Lösung gekoppelter Modellprobleme auf strukturierten Gittern, Diplomarbeit, Universität zu Köln, 1996.
3. Joppich, W., Mijalković, S.: Multigrid Methods for Process Simulation, Computational Microelectronics, Springer Verlag, Wien, New York, 1993.
4. Ritzdorf, H., Schüller, A., Steckel, B., Stüben K.: L_iSS - An environment for the parallel multigrid solution of partial differential equations on general 2D domains, Parallel Computing **20** (1994) 1559–1570

PARSMI, a Parallel Revised Simplex Algorithm Incorporating Minor Iterations and Devex Pricing

J. A. J. Hall and K. I. M. McKinnon

Department of Mathematics and Statistics, University of Edinburgh

Abstract. When solving linear programming problems using the revised simplex method, two common variants are the incorporation of minor iterations of the standard simplex method applied to a small subset of the variables and the use of Devex pricing. Although the extra work per iteration which is required when updating Devex weights removes the advantage of using minor iterations in a serial computation, the extra work parallelises readily. An asynchronous parallel algorithm PARSMI is presented in which computational components of the revised simplex method with Devex pricing are either overlapped or parallelism is exploited within them. Minor iterations are used to achieve good load balance and tackle problems caused by limited candidate persistence. Initial computational results for an six-processor implementation on a Cray T3D indicate that the algorithm has a significantly higher iteration rate than an efficient sequential implementation.

1 Introduction

The revised simplex method for solving linear programming (LP) is one of the most widely used numerical methods. Although barrier methods are often the preferred method for solving single large LP problems, the revised simplex method is the most efficient solution procedure when families of related LP problems have to be solved. Despite its importance, relatively little progress has been reported on parallel methods based on the revised simplex algorithm.

Special techniques such as minor iterations or edge weight based pricing strategies are used by serial implementations of the revised simplex method in order to achieve faster solution times. The major computational steps in the revised simplex method with minor iterations and the Devex pricing strategy are described in Section 2. These steps are also used in the parallel algorithm, PARSMI, which is described in Section 3. PARSMI exploits parallelism within the computational steps which parallelise naturally and overlaps the remaining computational steps.

Promising computational results obtained using a prototype implementation of PARSMI on six processors of a Cray T3D are given in Section 4. These results illustrate the behaviour a major part of the algorithm. The extension of the implementation to more processors is discussed in Section 5.

1.1 Background

A linear programming problem has standard form

$$\begin{aligned}
\text{minimize } & f = c^T x \\
\text{subject to } & Ax = b \\
& x \geq 0
\end{aligned} \tag{1}$$

$$\text{where } x \in I\!R^n \quad \text{and} \quad b \in I\!R^m.$$

At any stage in the simplex method the indices of the variables are partitioned into set $\mathcal{B}$ of m basic variables and set $\mathcal{N}$ of $n - m$ nonbasic variables. The nonbasic variables are given the value zero and the values of the basic variables are determined by the equations in (1). The components of c and columns of A corresponding to the index set $\mathcal{B}$ are the basic costs c_B and basis matrix B. Those corresponding to the set $\mathcal{N}$ are the non-basic costs c_N and matrix N.

A feasible point x corresponding to such a partition of the indices is a vertex of the feasible region of (1). At this vertex, the simplex method first determines the direction of an edge of the feasible region along which the objective function decreases and then finds the step to the next vertex of the feasible region along this edge. This corresponds to increasing a nonbasic variable from zero until one of the basic variables is reduced to zero. Repeating this process determines a path along edges of the feasible region which terminates at a vertex for which there is no edge along which the objective decreases.

The two main variants of the simplex method correspond to different means of calculating the data required to determine the step to the new vertex. The first variant is the standard simplex method in which the directions of all edges at the current vertex and the gradient of the objective in each of these directions are maintained in a rectangular tableau. In the revised simplex method, the vector of gradients and the direction of the chosen edge are determined by solving systems involving the matrix B, using a factored form of its inverse, and forming matrix-vector products using the matrix N. For large sparse problems the standard simplex method is completely uncompetitive in speed compared with the revised simplex method. It is also much less stable numerically.

A simple parallel implementation of the revised simplex method using a sparse LU decomposition is described by Shu and Wu in [7]. However they only parallelise the sparse matrix-vector products and, as a result, report little or no speed-up over their serial implementation for general problems. Bixby and Martin discuss parallelising the CPLEX dual simplex implementation in [2] where, once again, parallelism was only successfully exploited in the sparse matrix-vector products, with minimal speed-up on general problems.

Only recently have parallel algorithms been developed which give worthwhile speed-up relative to an efficient serial implementation. Hall and McKinnon have developed an algorithm, ParLP, which is described in [5] and achieve a speed-up of up to 5 on general problems. Wunderling described an algorithm in [8] which is completely parallel for 2 processors.

2 The serial revised simplex method

It is possilbe incorporate good features of the standard into the revised simplex method by using minor iterations. Within minor iterations the standard simplex method is applied to the LP sub-problem corresponding to a small subset of the variables. This is a central feature of PARSMI. There are various strategies for weighting each reduced cost. They aim to predict the likely cost reduction in an iteration and so reduce the number of iterations . Devex pricing described by Harris in [6] is such a strategy and is used in PARSMI. It is commonly the default in efficient sequential implementations. The major computational steps of the revised simplex method with minor iterations and Devex pricing are illustrated in Figure 1.

CHUZC: Scan $\hat{c}_N$ for a set Q of good candidates to enter the basis.
FTRAN: Form $\hat{a}_j = B^{-1}a_j$, $\forall j \in Q$, where a_j is column j of A.
Loop {minor iterations}
    CHUZC_MI: Scan $\hat{c}_Q$ for a good candidate q to enter the basis.
    CHUZR: Scan the ratios $\hat{b}_i/\hat{a}_{iq}$ for the row p of a good candidate to
        leave the basis, where $\hat{b} = B^{-1}b$ (ratio test). Let $\alpha = \hat{b}_p/\hat{a}_{pq}$.
    UPDATE_MI: Update $Q := Q\backslash\{q\}$.
        Update $\hat{b} := \hat{b} - \alpha\hat{a}_q$.
        Update the columns $\hat{a}_Q$ and reduced costs $\hat{c}_Q$.
        Update the Devex weights for the candidates in Q.
End loop {minor iterations}
For {each basis change} do
    BTRAN: Form $\pi^T = e_p^T B^{-1}$.
    PRICE: Form pivotal row $\hat{a}_p^T = \pi^T N$.
        Update reduced costs $\hat{c}_N := \hat{c}_N - \hat{c}_q\hat{a}_p^T$ and Devex weights.
    If {growth in factors} then INVERT: Form factored inverse of B.
    else UPDATE: Update the inverse of B corresponding to the basis change.
End do

Fig. 1. The revised simplex method with minor iterations and Devex pricing

When a single nonbasic variable x_j increases by unit amount, the basic variables must change by $-B^{-1}a_j$ to as to maintain equation $Ax = b$. (Here a_j is column j of A.) Thus the corresponding change in the objective function value is the *reduced cost* $\hat{c}_j = c_j - c_B^T B^{-1}a_j$. The traditional 'Dantzig' criterion for determining the quality of nonbasic variables is just the size of the reduced cost. However, if the vector $\hat{a}_j = B^{-1}a_j$ is large relative to $\hat{c}_j$ it is likely that only a small increase in x_j will be possible before one of the basic variables is reduced to zero. Thus edge weight based pricing strategies, such as Devex, weight the reduced cost by dividing it by (a measure of) the length of $\hat{a}_j$. The Devex strategy maintains weights $s_j \approx 1 + \|\hat{a}_j\|_R^2$, where $\|.\|_R$ is the 2-norm taken over a set R. This set R is initially empty, with corresponding unit initial weights. Following each subsequent basis change, one index is either added to or removed from the set R and periodically R is reset to be empty. The details of this are described by Harris in [6] and the computation required in order to update the weights is outlined below.

2.1 The revised simplex method with minor iterations and Devex pricing

At the beginning of a major iteration in Figure 1, it is assumed that the vector of reduced costs $\hat{c}_N^T = c_N^T - c_B^T B^{-1} N$ and the vector $\hat{b} = B^{-1} b$ of current values of the basic variables are known and that a factored inverse of the basis matrix B is available. The first operation is CHUZC which scans the weighted reduced costs to determine a set Q of good candidates to enter the basis. The inner loop then applies the standard simplex method to the LP problem corresponding to the candidates in Q so requires the corresponding columns $\hat{a}_j = B^{-1} a_j$ of the standard simplex tableau. These columns are formed by passing forwards through the factors of B^{-1}, an operation known as FTRAN. The matrix formed by the columns $\hat{a}_j$, $j \in Q$ and the corresponding vector of reduced costs for the candidates $j \in Q$ are conveniently denoted by $\hat{a}_Q$ and $\hat{c}_Q$.

In each minor iteration, CHUZC_MI scans the weighted reduced costs of the candidates in Q and selects one, q say, to enter the basis. The vector $\hat{a}_q$ is referred to as the *pivotal column*. The variable to leave the basis is determined by CHUZR which scans the ratios $\hat{b}_i / \hat{a}_{iq}$. In the discussion below, p is used to denote the index of the row in which the leaving variable occurred, referred to as the *pivotal row*, and p' denotes the index of the leaving variable. The value $\alpha = \hat{b}_p / \hat{a}_{pq}$ is the new value of x_q. The vector of new values of the basic variables is given by

$$\hat{b} := \hat{b} - \alpha \hat{a}_q \tag{2}$$

and clearly has a zero value in component p. Once the indices q and p' have been interchanged between the sets $\mathcal{B}$ and $\mathcal{N}$, a *basis change* is said to have occurred. The standard simplex tableau corresponding to the new basis is obtained by updating the previous tableau in an operation known as UPDATE_MI. The values of the basic variables are updated according to (2). The matrix $\hat{a}_Q$ and reduced costs $\hat{c}_Q$ are updated by a Gauss-Jordan elimination step and the Devex weights are updated using row p of the updated tableau.

Since the variable to enter the basis is removed from the set Q, the number of minor iterations which are performed is limited by the initial cardinality of Q. However, due to the changes in the reduced costs, Q may consist of purely unattractive candidates before it becomes empty. In this case the work done in forming and updating the corresponding standard simplex columns is wasted. The number of minor iterations which are actually performed depends on the extent to which candidates remain attractive. This is an example of the property of *candidate persistence* which is highly problem-dependent and has a major influence on the effectiveness of the parallel algorithms described by Hall and McKinnon in [5] and Wunderling in [8].

Before the next major iteration can be performed it is necessary to update the reduced costs and Devex weights and obtain a factored inverse of the new basis. Updating the reduced costs and Devex weights following each basis change requires the corresponding pivotal row $\hat{a}_p^T = e_p^T B^{-1} N$ of the (full) standard simplex tableau. This is obtained in two steps. First the vector $\pi^T = e_p^T B^{-1}$ is formed by passing backwards through the factors of B^{-1}, an operation known as BTRAN, and then the vector $\hat{a}_p^T = \pi^T N$ of values in the pivotal row is formed. This sparse matrix-vector product with N is referred to as PRICE. Once the reduced costs and Devex weights have been updated, the factored inverse of the new basis is obtained by updating the current factored inverse (the UPDATE operation). Note that, eventually it will be either more efficient, or necessary for numerical stability, to find a new factored inverse using the INVERT operation.

2.2 FTRAN and BTRAN

In order to describe the parallel algorithm PARSMI, it is necessary to consider the FTRAN and BTRAN operations in further detail. Let the basis matrix whose inverse is factored by INVERT be denoted by B_0 and the basis matrix after a further U basis changes be denoted by $B_U = B_0 E_U$ so $B_U = E_U^{-1} B_0^{-1}$. Let $\mathcal{P}$ be the set of rows in which basis changes have occurred. Note that for large LP problems $P = |\mathcal{P}| \ll m$. The most convenient representation of the matrix E_U^{-1} is as a product of Gauss-Jordan elimination matrices. In FTRAN the vector $\hat{a}_j = B_U^{-1} a_j$ is formed by the operations $\tilde{a}_q = B_0^{-1} a_q$ and $\hat{a}_q = E_U^{-1} \tilde{a}_q$. In BTRAN the vector $\pi^T = e_p^T B_U^{-1}$ is formed by the operations $\tilde{\pi}^T = e_p^T E_U^{-1}$ and $\pi^T = \tilde{\pi}^T B_0^{-1}$. Note that forming $\tilde{\pi}$ only requires rows $i \in \mathcal{P}$ of the elimination matrices.

3 PARSMI, a parallel revised simplex algorithm

The concept of PARSMI is to dedicate some processors to performing minor iterations in parallel, while simultaneously using other processors to select and form the vectors $\tilde{a}_{q'}$ for new good candidates to be added to $\mathcal{Q}$. Other processors are used to form the π vectors and perform the PRICE operations required to bring the reduced costs and Devex weights up-to-date.

One processor, referred to as the MI_CT processor, is dedicated to controlling the minor iterations and performing the computation which uses the data in rows $i \in \mathcal{P}$ of the elimination matrices which represent E_U^{-1}. When it receives the vector $\tilde{a}_{q'}$ for a good candidate, q' is added to $\mathcal{Q}$ and its current reduced cost is calculated (and subsequently updated) using just the data in rows $i \in \mathcal{P}$ of the elimination matrices. Only when a candidate is actually chosen as the variable x_q to enter the basis is the full vector $\hat{a}_q = E_U^{-1} \tilde{a}_q$ actually computed. This operation, together with CHUZR and operation (2), called *update RHS*, can be distributed by rows on the minor iteration (MI) processors.

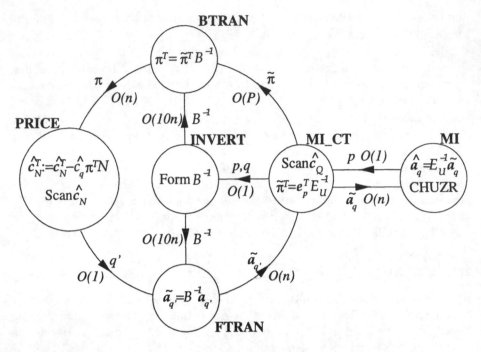

Fig. 2. Six-processor implementation of PARSMI

Once a basis change has been determined, the vector $\tilde{\pi}^T = e_p^T E_U^{-1}$ is formed on the MI_CT processor and one of several BTRAN processors completes the calculation of $\pi^T = \tilde{\pi}^T B_0^{-1}$. The subsequent calculation of $\pi^T N$, update of the reduced costs and Devex weights and the CHUZC operation are distributed over a number of PRICE processors and yields new good candidates for which vectors $\tilde{a}_{q'}$ are formed on the FTRAN processors. One or more INVERT processors are dedicated to keeping the basis B_0 as up-to-date as possible in order to reduce the operations required to form $\hat{a}_q$ from $\tilde{a}_q$.

3.1 Implementation

The algorithm could be implemented on a shared memory multiprocessor or a distributed memory machine with high ratio of communication speed to computation speed. The minimum number of processors required to implement the algorithm is six, and such an implementation is currently being developed on the Edinburgh Cray T3D. The implementation shares its fundamental computational components with a highly efficient sequential implementation of the revised simplex method. This is competitive with commercial simplex solvers and is used to assess the performance of the parallel implementation. Most message passing, in particular all short messages, is done via the Cray-specific shared memory SHMEM [1] subroutine library, with the remaining message-passing being done via the Parallel Virtual Machine (PVM) subroutine library [4].

On a distributed memory machine there are two benefits of splitting the FTRAN and the BTRAN operations. Firstly it avoids the cost of forming $\hat{a}_j$ for candidates which are discarded without entering the basis. Secondly, the data required to form $\hat{a}_q$ from $\tilde{a}_q$ is already available on the processor on which it is used.

The major computation and communication requirements of the six-processor implementation of PARSMI are illustrated in Figure 2. An alternative view of the algorithm is provided by the Gantt chart in Figure 3. Each box represents the time spent performing a particular operation on a given processor, with SI corresponding to sending a new factored inverse, C being CHUZC, F being FTRAN and R being CHUZR. The unlabelled box following CHUZR corresponds to updating the RHS.

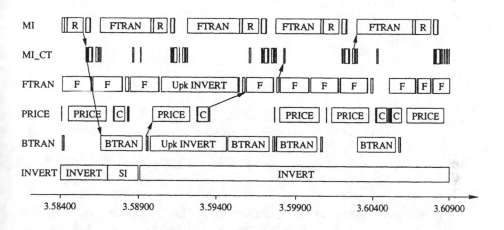

Fig. 3. Gantt chart of processor activity for 25FV47

4 Computational results

The results presented in this section are for the six processor implementation on the Edinburgh Cray T3D. They demonstrate the effectiveness of a key part of the algorithm and motivate the discussion of implementations using a larger number of processors in Section 5. The current implementation is only for feasible starting bases and only permits a few tens of iterations to be performed. However this is long enough to get a good assessment of the steady state time per iteration when started from a feasible basis. Thus the decrease in the time per iteration of the six-processor implementation over that for a comparable serial implementation may be determined. The effect on the total number of iterations due to using out-of-date reduced costs cannot yet be assessed.

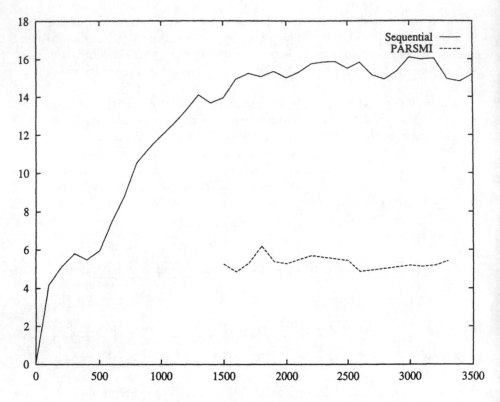

Fig. 4. Time per iteration for 25FV47

A number of experiments were performed using the classic Netlib [3] test problem 25FV47 for which the serial revised simplex implementation obtains feasibility after 1500 iterations. Every hundredth basis from this point was stored and the time per iteration (averaged over 5 iterations) at each of these bases was recorded. This is represented by the solid line on the graph illustrated in Figure 4 and is approximately $15ms$ from the point at which feasibility is obtained. The six-processor implementation of PARSMI was started from each of the stored feasible bases and run until a steady state average time per iteration was reached. This time per iteration is represented by the dotted line on the graph illustrated in Figure 4 and is approximately $5ms$, corresponding to a speed-up of about 3. Given that six processors are being used the efficiency is about 50%. Analysis of the Gantt chart in Figure 3 shows that the bottleneck in the parallel implementation is the time taken to complete FTRAN and then perform CHUZR and update RHS. For the serial implementation these have average computation times of $1.8ms$, $1.0ms$ and $0.3ms$ respectively, a total of $3.1ms$, an iteration speed which, if achieved by the parallel implementation would correspond to a best possible speed-up of 5. With the parallel implementation the average time taken to complete FTRAN is $2.7ms$ so the total time for the bottleneck operations is $4.0ms$, an iteration speed which would correspond to a speed-up of 4. There remains scope for improvement in the efficiency of the current parallel implementation which will allow the best possible speed-up to be approached.

The reason for the greater average time taken to complete FTRAN within the parallel implementation is readily explained. The serial implementation reinverts the basis every 25 iterations leading to an average number of 12 updates which must be applied. Within the parallel implementation approximately 12 basis changes occur during INVERT thus the number of updates which must be applied when completing FTRAN ranges between 12 and 24: the average of 18 corresponds to the 50% increase in the average time.

5 Conclusions

The computational results given in the previous section for the (partial) six-processor implementation of PARSMI demonstrate the effectiveness of part of the algorithm: the increase in the iteration speed approaches the best that can be achieved. The bottleneck in the parallel implementation which limits the decrease in the time per iteration is the computation required to complete of FTRAN and then perform CHUZR and update RHS. The computation required by each of these operations could be distributed over two or more processors, without significant duplication of work, allowing the best possible time per iteration to be reduced by a corresponding factor. Note that as the time per iteration decreases, there is a consequent increase in the number of basis changes by which the factored inverse is out-of-date when INVERT is completed, and hence the number of updates which must be applied also increases. Thus the time per iteration cannot be scaled down indefinitely without additional processors being employed to increase the frequency with which fresh factored inverses are available.

However, a more critical limitation of the iteration speed once the minor iterations are parallelised is likely to be the rate at which the pivotal columns of candidates to enter the basis can be generated. The average time taken to perform FTRAN (with B_0^{-1}) is $1.8ms$ so, even if all candidates were attractive, this would represent a limit on the time per iteration unless more than one processor were devoted to FTRAN.

If the BTRAN and PRICE processors cannot produce candidates with sufficient frequency, candidates will be chosen with respect to increasingly out-of-date reduced costs. As a result, a decreasing proportion of these will prove to be attractive when their reduced cost is brought up-to-date. This will not only decrease the frequency of basis changes but is likely to lead to a significant increase in the number of iterations required to solve the problem since the candidate which actually enters the basis may be far from being the best with respect to the Devex criterion. Progress may even temporarily stall even though the current basis is not optimal because the most up-to-date reduced cost available indicate no candidates remain. Parallelising PRICE and overlapping more than one BTRAN would therefore be attractive.

For a given time per iteration t, the number of basis changes by which the reduced cost of a candidate is out-of-date when it enters the basis is at least T/t, where T is the total time required to perform a BTRAN, PRICE and start FTRAN (form $\tilde{a}_q = B_0^{-1} a_q$). The sequence of arrows in Figure 3 show the communication and computation which is required between a basis change being determined and a candidate, chosen with respect to the reduced costs for that basis, being ready to enter a future basis. In this example the reduced cost is three basis changes out-of-date. Even if parallelism is fully exploited within a single PRICE operation, T is at least the serial time to perform BTRAN and start FTRAN, unless parallelism is exploited within BTRAN and FTRAN a task which is beyond the scope of this project.

With a larger number of processors available, the optimal distribution of the processors to activities will vary both during the solution procedure (as PRICE becomes relatively cheaper than FTRAN and BTRAN) and from one LP problem to another. LPs with a large ratio of columns to rows will benefit from a larger number of processors being devoted to PRICE operation. Problems with proportionally more rows, and those for which the vectors $\hat{a}_j$ are relatively dense will benefit from relatively more processors performing minor iterations. If strategies can be developed to prevent the number of iteration required to solve a problem increasing significantly, the potential fast iteration speed and adaptability of the algorithm to the full range of LP problems can be expected to lead to a parallel implementation which gives a significantly improvement over the sequential implementation of the revised simplex method.

References

1. R. Barriuso and A. Knies. *SHMEM User's guide for Fortran.* Cray Research inc.
2. R. E. Bixby and A. Martin. Parallelizing the dual simplex method. Technical Report SC-95-45, Konrad-Zuse-Zentrum für Informationstechnik Berlin, 1995.
3. D. M. Gay. Electronic mail distribution of linear programming test problems. *Mathematical Programming Society COAL Newsletter*, 13:10–12, 1985.
4. A. Geist, A. Beguelin, J. Dongarra, W. Jiang, R. Manchek, and V. Sunderam. *PVM: Parallel Virtual Machine - A User's Guide and Tutorial for Networked Parallel Computing.* MIT Press.
5. J. A. J. Hall and K. I. M. McKinnon. An asynchronous parallel revised simplex method algorithm. Technical Report MS 95-50, Department of Mathematics and Statistics, University of Edinburgh, 1995.
6. P. M. J. Harris. Pivot selection methods of the Devex LP code. *Math. Prog.*, 5:1–28, 1973.
7. W. Shu and M. Wu. Sparse implementation of revised simplex algorithms on parallel computers. In *Proceedings of* 6th *SIAM Conference on Parallel Processing for Scientific Computing*, pages 501–509, 1993.
8. R. Wunderling. Parallelizing the simplex algorithm. ILAY Workshop on Linear Algebra in Optimzation, Albi, April 1996.

Parallel Optimization
of Interplanetary Trajectories*

Beidi HAMMA

Parallel Algorithms Team, CERFACS,
European Centre for Research and Advanced Training in Scientific Computing
42 av. Coriolis, F-31057 Toulouse-Cedex, France. e-mail: hamma@cerfacs.fr

Abstract. This paper describes the software developed at CNES within the context of the VESTA Franco-Soviet program, for the optimization of interplanetary trajectories. The sequentiel software, intended for the determination of multiple flyby trajectories and their optimization, was developed previously in collaboration with CNES and LAO of Paul Sabatier University, Toulouse, France. In this software, an impulse manoeuvre is performed at each flyby. The possibility to add an intermediate manoeuvre between flybys and the use of the swing-by aim to reduce the total trajectory costs. The optimization routines are parallelized.

Keywords: Parallel Optimization, Interplanetary Trajectories, Swing-by, Multiple-Flyby, Lambert Problem.

NOTATIONS

(P_i)	$i = 1, N$, Celestial bodies passed on flyby
(T_i)	$i = 1, N$, Dates of flyby
$(\mathbf{X_i}, \mathbf{W_i})$	Positions and speeds of P_i on date T_i
$\mathbf{V_i}^+$	Absolute speed of the probe on departure from P_i
$\mathbf{V_i}^-$	Absolute speed of the probe on arrival at P_i
$\mathbf{v_i}^+$	Relative speed of the probe on departure from P_i
$\mathbf{v_i}^-$	Relative speed of the probe on arrival at P_i
$\Delta\mathbf{V_i}$	Increment on flyby over P_i
$\mathbf{X}_{Mi}$	Position of the intermediate manoeuvre
T_{Mi}	Dates of intermediate manoeuvre
$\mathbf{v}_{Mi}^-$	Probe speed before the intermediate manoeuvre
$\mathbf{v}_{Mi}^+$	Probe speed after the intermediate manoeuvre
$\Delta\mathbf{v}_{Mi}$	Increment of the intermediate manoeuvre

N.B. $\mathbf{V}$ (in bold style) denotes a vector V, since the springer style llncs does not allow an arrow over v.

---

* This work was initiated and funded by CNES (French National Space Centre), Toulouse, contract 873- CNES-87-4937 between CNES and a previous structure of LAO, Paul Sabatier University. The work, except for the parallelization of the algorithms, has been carried out in collaboration with the Laboratoire d'Approximation et d'Optimisation (LAO), Université Paul Sabatier, Toulouse, France.

1 Introduction

This paper is concerned with a work done within the context of the VESTA franco-soviet program on Optimization of Interplanetary Trajectories. The project aim is to rapidly determine a catalogue of possible trajectories in accordance with given ballistic constraints. That is, the problem to solve is:
- define a model which finds flyby trajectories for visiting comets and big asteroids between Mars and Jupiter
- optimize the selected trajectories with respect to a cost function which is the total impulses trajectory from departure (the earth) to the end of the mission.
For reducing the cost of a given trajectory we use two strategies. The first is to use a model with special features, which gives a low initial cost, and the second is to apply suitable optimization algorithms to reduce this initial cost.
Our low initial cost model is characterized by the use of "swing-by" for reducing maoeuvring costs, and the use of intermediate manoeuvres, between each two flybys, to avoid limitations in manoeuvring.
For optimizing the intial costs we developed global optimization codes based on Multistart and a mixture of Simulated Annealing.
The parallel experiments with the optimization codes are performed on a transputer system. The codes are written in OCCAM2. The whole code is simulated in pseudo-parallel.
The VESTA project is concerned with the problem of sending a probe which should be able to visit big (enough) asteroides and comets between Mars and Jupiter. For that, a general software program HOPTIMA has been developed at CNES, Toulouse. The program HOPTIMA includes both the determination of intial trajectories allowing the visits and the process for optimizing those initial trajectories. Our main contribution was in the last stage. Historically, trajectory design for an interplanetary mission required the solution of a deterministic boundary value problem. Given a launch date and an arrival date, the heliocentric Lambert Problem is solved giving the hyperbolic excess velocity vectors at each planet.
Mission analysis often requires the determination of multiple flyby trajectories for which the trajectory of the spacecraft with respect to some central body also involves close flybys of some satellites of the central body. The actually flown trajectory for any space mission usually contains one or more relatively small thrusting manoeuvres, which are typically modelled as instantaneous changes in velocity. These ΔV's are of two types: *deterministic* or *statistical*. A deterministic ΔV is designed into the trajectory to propulsely alter the flight path. A statistical ΔV corrects random error sources on the trajectory such as manoeuvre execution errors or orbit determination errors. In our case, we are concerned with a procedure which minimizes only the total deterministic ΔV. Statistical ΔV's are not considered.
Within the VESTA Franco-Soviet program, it was planned to launch a probe which is composed of two parts: the first will stabilise around the "red planet", while the other will be in charge of flybying over a maximum of asteroides possible for a mission not exceeding 5 years. One of the main phases of the project

was the choice and the calculation of adequate trajectories. For that, we have to take into account several criteria like the transfer times between two bodies (which should not be long) and the size of bodies (the bigger are the better). For enabling scientific studies, the relative flyby speeds on the bodies should not to be very high. Finally, the probe can not board more than a certain quantity of ergol (fuel). As can be seen, the choice is not so easy, especially if we add that the number of asteroides known is more than 11 449. Hence, there is a need for optimizing the trajectory. The optimization will deal with the consommation (fuel). In fact, for getting the maximum scientific data during the mission, we need to board many scientific material; in order to reduce the total weight we will try to minimize the required amount of fuel, and then try to consume fuel as less as possible so that the probe can travel as far as possible. This is one of the reasons why Swing-by (a kind of gravitational tremplin) is used to modify the probe direction without making costly manoeuvres.

2 Interplanetary Flyby Trajectories

2.1 Determination of Trajectories

The model retained in CNES (see [CNES 1], [CNES 2]) is a model in which the method of trajectory calculation used is focused on speed rather than on accuracy: the "patched conics" method. In this model, we assume that at any stage the probe is attracted by only one body at the time: either the sun (most of time) or a planet (in the neighborhood of the concerned planet). Then, a trajectory is a succession of heliocentric ellipse arcs (with the sun as center) and planetocentric hyperbolas at the moment of flyby over the planet. The continuity in speed where the arcs meet is ensured by means of impulse manoeuvres. The sphere of influence of the various planets are assimilated at an infinite distance from the planet i.e. the speeds at infinity for planetrocentric arcs are given by the relative speeds for the heliocentric arcs. All the speeds are then calculated either by solving the Lambert equation or by extrapoling Keplerian orbits in the case of intermediate manoeuvres.

2.2 Improved trajectories

In order to get a low cost for the trajectories defined above, two techniques are used:

Swing-by. The use of gravity assistance in the neighborhood of a planet can help to obtain good parameters (speed, angles) for the continuation without a need of expensive manoeuvres. In our case, for instance, swing-by is done over Mars.

Intermediate Manoeuvres. When manoeuvres are done only during flybys of planets or bodies, there is a limitation: thrust, angles, ... And this has a significative cost in terms of impulse to correct the probe trajectory. In order to avoid this limitation, a part of the required manoeuvre can be done before the flyby or after the flyby. To give an idea of the importance of intermediate manoeuvres, let us take a real-life analogical example: driving a car in town. In order to full stop at traffic lights, usualy one has to start braking many meters (hundreds) in advance, to adjust the speed before reaching the traffic lights (instead of a "catastrophic" brake just in front of the lights!).

It is easy to show that those intermediate manoeuvres contribute in reducing the trajectory cost.

3 Trajectory Optimization

Once a model of determining trajectories is accepted, we have to define , how to optimize an initial trajectory given by this model.

The optimization problem to solve is the following:

Given a (initial) trajectory of a probe, visiting N bodies P_i, $i = 1, N$, at dates T_i, $i = 1, N$, optimize this trajectory subject to:

   - manoeuvres that are impulses

   - manoeuvres that are done at each flyby, with the possibility to add inter-mediate manoeuvres between flybys

   - the cost function which is the total sum of the modules of ΔV resulting from manoeuvres.

The optimization constraints are:

   - departure speed from first boby may not exceed some limit V_{Max}

   - a minimum altitude r_{im} is required for flybying some planets.

Hence, as stated in the introduction, for determining the multiple flyby trajectory, we have to solve a finite number of two-body problems (see for instance [CNES 3]).

3.1 Formulation without intermediate manoeuvres

This is the basic case. We want to find the trajectory for flybying the N (selected) bodies P_i, $i = 1, N$, at dates T_i, $i = 1, N$, using impulse manoeuvres during the flyby of each body. This conducts to the resolution of a sequence of Lambert problems (i.e. find N times the Lambert transfer from body P_i (at date T_i) to body P_{i+1} (at date T_{i+1})). The variables are the N dates T_i, $i = 1, N$. The cost (objective) function can be expressed as the total sum of the modules of all the increments:

$$J = \sum_{i=1}^{N-1} \|\Delta v_i\|$$

where the v_i are both relative and absolute speeds.

3.2 Formulation with intermediate manoeuvres

In this case we take into account intermediate manoeuvres. Intermediate manoeuvres can be seen as manoeuvres on fictious non-mass bodies (i.e. with negligible mass) M_i, $i = 1, N$. So the problem to solve is: *extrapole the trajectory* of the real body P_i (flybyed at date T_i) to the fictious non-mass body M_i (at date T_{Mi} using the relative departure speed V_i^+ of the probe from P_i. This allows us to calculate the position of the intermediate manoeuvre just by making an orbit propagation. Then we have to find the Lambert transfer from M_i (at date T_{Mi}) to (real) body P_{i+1} (at date T_{i+1}) in the same way as in the previous section. This makes a total of $5N - 4$ variables namely:

- the N flyby dates
- the $N - 1$ manoeuvring (intermediate) dates
- the $3(N-1)$ components of the relative speed on departure from the bodies.

Hence the cost J can be expressed as the total sum of the modules of increments (included those during the intermediate manoeuvres):

$$ J = \sum_{i=1}^{N-1} ||\Delta v_i|| + \sum_{i=1}^{N-1} ||\Delta v_{Mi}|| $$

4 HOPTIMA : Trajectories Optimization Software

4.1 Computation of the objective function and the constraints

Increments are made when necessary for completing the trajectory.

The increments are of three types: first-body increment, Swing-by increment and negligible body increment.

The first body increment is due to the manoeuvre done at the first body for taking into account the departure speed V_1^+ constraint which has to be less than a fixed limit V_{MAX}.

Swing-by increments are preformed during planet flybys.

Negligible bodies increments are increments due to manoeuvres during flyby of asteroides, comets, fictious bodies (for intermediate manoeuvres) which are considered as negligible (in term of mass) bodies.

a) **First body increment** : ΔV_1.

Basically we have to treat the following constraint $V_1^+ \leq V_{MAX}$. A way to do this is to correct the probe speed as follows:

$$
\begin{cases}
\Delta \mathbf{V_1} = 0 & \text{if } V_1^+ \leq V_{MAX} \\
\Delta \mathbf{V_1} = \frac{V_1^+ - V_{1MAX}}{V_1^+} . \mathbf{V_1^+} & \text{if } V_1^+ > V_{MAX}
\end{cases}
$$

b) **Planet (swing-by) increment** : ΔV_i.

The swing-by will help to reduce the manoeuvre cost. We also have to take into account the minimum altitude r_{im} for flybying the planet. This is done as follows: let us recall that the maximum deviation angle of a swing-by cone is $\theta_{iM} = 2 \arcsin((1)/(1 + (r_{im}v^2)/(i)))$, where $v = min\ (V_i^+, V_i^-)$. Then the necessary angle is given by $\theta_i = \arccos((\mathbf{V_i^-}.\mathbf{V_i^+})/(V_i^- V_i^+))$

So, the optimal swing-by is performed either :

– before the flyby, whenever $V_i^+ > V_i^-$. We have to reduce the arrival speed (to equal its module to the module of the departure speed during the flyby) and to give the needed rotation.

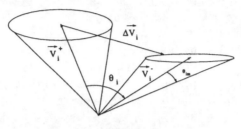

Manoeuvre before flyby

– or after the flyby, if $V_i^+ \leq V_i^-$. In this case, we have to increase the arrival speed and add the required surplus rotation .

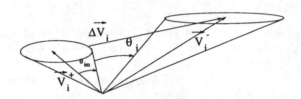

Manoeuvre after flyby

Depending on the value of θ_{iM}, the increment is:

$$\begin{cases} \Delta V_i = & |V_i^- - V_i^+| & if \ \theta_i \leq \theta_{iM} \\ \Delta V_i = \sqrt{(V_i^-)^2 + (V_i^+)^2 - 2V_i^- V_i^+ \cos(\theta_i - \theta_{iM})} & if \ \theta_i > \theta_{iM} \end{cases}$$

c) Negligible body increment : ΔV_i or ΔV_{iM}.
Here we are concerned with non-mass bodies numbers j: for real bodies $j = i$ (comets, asteroids) and for fictitious bodies $j = iM$ (positions of intermediate manoeuvres).

$$||\Delta V_j|| = ||\mathbf{V_j^+} - \mathbf{V_j^-}|| = ||\mathbf{v_j^+} - \mathbf{v_j^-}||$$

Gradient of J. The calculation of ∇J involves finding the differential of speeds. Earlier work by Bonneau et al. (see [CNES 3] for instance) showed how to calculate ∇J and integrate the computation in the part of trajectory determination to make it directly available for the optimization routines.

4.2 Parallel Optimization Routines

We performed the optimization process in the following manner:

Procedure PAROPT

1. **Initialization**: from the initial trajectory draw k initial close trajectories.
2. **Perform in parallel** a local minimization from each of the k initial trajectories.
3. **Run from each termination point** (given by the parallel process) the *heating-cycles Simulated Annealing* (cf. [5]).

It is clear that this process is fully parallelizable on k processors.

Heating-cycles is an algorithm introduced by Bonnemoy and Hamma ([5]) in order to ameliorate simulated annealing algorithm. The principle is to increase the temperature in the simulated annealing after classical stopping tests are satisfied. Otherwise the number of total iterations, at this stage of the algorithm, is very high and the acceptance probability, which is the key, is almost zero (ecall that this probability depends in decreasing sense on the iterations). Then there will be no more accepted non-descent points (which are supposed to enable the algorithm to escape from local minimum states).

We used the PAROPT procedure in order to find good minimums (hoping to find the global). As one can see, this procedure is a kind of Multistart algorithm for global optimization. We have used Simulated Annealing at the end of the local optimization processes because the kind of problem we are treating has many local minimums close to each other. Thus, Simulated Annealing which can be summarized as *Move - Evaluate - Accept*, is well suited to try to escape from local minimums. After the third step is used, there are two possibilities: either one at least of the k processors finds a better point than step 2, or one at least of processors confirms one of the attraction regions found in step 2 by giving more accuracy on the optimum value (both variable and function value). For local optimization, we developed routines based on BFGS quasi-Newton algorithms.

5 Numerical Results

In this section, we give some numerical results. The general software HOPTIMA was first implemented in Fortran 77 on Cyber CDC with NOS/VE system. The optimization part was then parallelized on a transputer system (Hathil of Abo Akademi in Finland) using the OCCAM2 language. We ran HOPTIMA on the transputer system. The following results are obtained by running in parallel the optimization part on 8, 16 or 32 processors. We present the best results obtained for some intial trajectories.

The parallelization decreases the execution time in a significant way since most of the time during the optimization is consumed by function evaluation. And since the parallel programs are using the subroutines developed at CNES (e.g. Lambert Transfert problem ....), the desired data for almost all processors can be calculated in one step in each subroutine. This is due to the fact that for a range of the variables (small perturbation on dates, velocities, ...), many of the subroutines can switch from one calculation to another by using previous results.

For all results, the initial weight = 2000 kg and the Braking Dv = 0.100 km/s. The initial weight is (suposed to be) the total mass of a penetrator (500 kg), the fuel and the carrying capacity. That is, at the end of the mission the remaining mass corresponds to the initial mass minus the mass of the pentrator and minus the mass of the fuel necessary for making all the ΔV's. A positive remaining mass means that the mission is successful and that the corresponding mass can be used as carrying capacity while a negative mass means a deficit.

Body	Initial Trajectory		Optimal Trajectory	
	Flyby date	Minimum Flyby Altitude	Flyby date	Intermediate Manoeuvre
Earth	1996-11-29	110 km	1996-12-01 at 23H	1997-03-17 at 09H
Mars	1997-06-28	10 km	1997-07-04 at 07H	1997-12-14 at 13H
Hestia	1998-05-26		1998-05-22 at 00H	1999-05-26 at 15H
1932RB	2000-04-04		2000-04-05 at 03H	2000-07-20 at 22H
Veseli	2000-11-04		2000-11-05 at 06H	2001-01-18 at 22H
Sapho	2001-04-10		2001-04-10 at 02H	
	Initial Cost: $J = 10.325\, km/s$		Optimal Cost: $J = 1.743\, km/s$	
	Initial Remaining Mass: $M = -2.2\, kg$		Optimal Remaining Mass: $M = +716.3\, kg$	

RESULT N. 1

The table shows that after optimization we will have in our disposal a mass of 716 kg (instead of being debtor of 2.2kg) which can be used for additional scientific materials for the mission.

	Initial Trajectory		Optimal Trajectory	
Body	Flyby date	Minimum Flyby Altitude	Flyby date	Intermediate Manoeuvre
Earth	1994-11-01	110 km	1994-10-30 at 01H	1997-03-17 at 09H
Mars	1995-08-10	10 km	1995-08-24 at 19H	1997-12-14 at 13H
Ast-4152	1996-12-10		1996-12-13 at 19H	1999-05-26 at 15H
Mars	1997-08-15	10 km	1997-08-19 at 18H	2000-07-20 at 22H
Ast-2087	1998-05-05		1998-05-10 at 22H	2001-01-18 at 22H
Ast-1	1999-03-03		1999-03-03	
	Initial Cost: $J = 10.109\ km/s$		Optimal Cost: $J = 0.947\ km/s$	
	Initial Remaining Mass: $M = -1.8\ kg$		Optimal Remaining Mass: $M = +1039.0\ kg$	

RESULT N. 2

In the above table the optimization procedure gives a mass of 1039.0 kg (instead of a debit of 1.8 kg).

6 Summary and directions for future work

We have presented a parallel computation of optimal multiple flyby trajectories. This parallel process is characterized by:

1. determination of suitable trajectories by using spatial techniques developed at CNES. For this, we retained a process with swing-by and possible intermediate manoeuvres.
2. parallelization of the optimization process by using a Multistart technique for global optimization. The local optimization routines are based on BFGS quasi-newton algorithms. The trajectory determination includes both the cost function computation and gradients, and takes into account all constraints (ballistic) for optimization routines.
3. an improvement of the optimization process by the use of heating-cycles Simulated Annealing developed especialy for this purpose.

As future work, a profound study of all CNES routines involved in the function evaluation (there are about 60, developed by different departments for internal use) will help to point out all parallelizable parts. Then, using a recent version of parallel simulated annealing, based on [9] for instance, will probably reduce the execution time considerably (up to half an hour, for the moment).

Acknowledgements

The author gratefully acknowledge:

- Professor J.-B. Hiriart-Urruty from LAO, University Paul Sabatier, Toulouse, who was the advisor of this work for the optimization part

- CNES researchers who initiated this work and helped a lot. In chronological order they are: Michel Rouze, F. Bonneau, C. Aubert, G. Lassalle-Balier and R. Epenoy.

- A. Torn and S. Viitanen of the computer science department of Abo Akademy, Finland, for helping with experiments on transputer system.

7 References

[1] [CNES 1] F. Bonneau, P. Dorio, HOPTIMA : optimisation de trajectoires interplanétaires avec manoeuvres intermédiaires, CT/DTI/MS/MN/283, CNES Toulouse June 1986

[2] [CNES 2] F. Bonneau, P. Dorio, L. Fontaine, *Optimisation de trajectoires interplanétaires avec comtraintes supplémantaires, application au projet VESTA*, CT/DTI/MS/MN/321, CNES Toulouse, Dec. 1986

[3] [CNES 3] F. Bonneau, L. Fontaine, *Test d'optimalité et animation de trajectoires interplanétaires*, CT/DTI/MS/MN/127, CNES Toulouse, 1986

[4] [CNES 4] F. Bonneau, P. Dorio, *VESTA détermination de trajectoires créneau 1994*, CT/DTI/MS/MN/127, CNES Toulouse, May 1986

[5] Bonnemoy C. and B. Hamma,*La méthode du recuit simulé: opimisation globale dans $\mathbb{R}^n$*, Rairo Afcet APII, vol 25, N.5, 1991, pp. 477-496.

[6] D'Amario L. A., Byrnes D.V. and R. H. Stanford,*A new method for optimizing Mutliple-Flyby Interplanetary Trajectories*, J. of Guidance and Control, Vol.4. N.6, Nov.-Dec. 1981.

[7] B. Hamma, *Etude de Problèmes d'optimisation de trajectoires interplanétaires*, Final Report, marché CNES-873-4937, July 1991.

[8] B. Hamma, *La méthode du Recuit Simulé: Problèmes de convergence et de parallélisation dans le cas de l'optimisation continue*, Tech. Rep. N.5, LACO, Dept. Maths, Université de Limoges, 1994.

[9] B. Hamma, Torn A. and S. Viitanen, *Parallel Continuous Simulated Annealing for Global Optimization*, submitted to Optimization Methods and Software, 1996.

Filter Model of Reduced-Rank Noise Reduction

Per Christian Hansen[1] and Søren Holdt Jensen[2]

[1] UNI•C, Building 304, Technical University of Denmark, DK-2800 Lyngby, Denmark
[2] CPK, Aalborg University, Fredrik Bajers Vej 7, DK-9220 Aalborg Øst, Denmark

Abstract. The key step in reduced-rank noise reduction algorithms is to approximate a matrix by another one with lower rank, typically by truncating a singular value decomposition (SVD). We give an explicit and closed-form derivation of the filter properties of the rank reduction operation and interpret this operation in the frequency domain by showing that the reduced-rank output signal is identical to that from a filter-bank whose analysis and synthesis filters are determined by the SVD. Our analysis includes the important general case in which pre- and dewhitening is used.

1 Introduction

Reduced-rank noise reduction, in which a signal matrix is approximated by another one with lower rank, may be viewed as an "energy decomposition" filtering out parts of the corresponding spectrum with low energy [1]. This techniques is the underlying principle in the noise reduction algorithms proposed in [2, 3]. The key idea is to form a Hankel (or Toeplitz) matrix from the input signal, compute the singular value decomposition (SVD) of the matrix, discard small singular values to obtain a matrix with reduced rank, and finally construct the output signal from this generally unstructured matrix by arithmetic averaging along its antidiagonals (or diagonals). This algorithm, which we shall refer to as the truncated SVD (TSVD) algorithm, has been used successfully for reduction of white noise in speech signals [2].

Prewhitening of the signal—corresponding to multiplying the signal matrix with a certain matrix—is sometimes used if the noise cannot be considered as white. The prewhitening operation can be included as an integral part of the algorithm, which then requires the computation of the quotient SVD (QSVD) of the signal-prewhitener matrix pair. This is the truncated QSVD (TQSVD) algorithm [3], in which the reduced-rank matrix is obtained by discarding small quotient singular values.

In [4], we have derived a filter model of the TSVD algorithm and the TQSVD algorithm, respectively. This work was based on ideas presented in [5], where it is concluded that the TSVD algorithm corresponds to subtracting from the input signal information contained in eigen-residuals corresponding to the smallest singular values. Here we present further contributions along the line described in [4]. To be specific, we study the relation between the finite impulse response (FIR) filters in the TSVD/TQSVD filter model and the eigenfilters of the autocorrelation matrix of the signal under consideration.

2 Notation

Given a vector $\mathbf{x} = (x_1, x_2, \ldots, x_N)^T$ of length N, we define the following two Hankel matrices of dimension $m \times n$ and $(N + n - 1) \times n$, respectively, where $m + n - 1 = N$:

$$
\mathcal{H}(\mathbf{x}) \equiv
\begin{pmatrix}
x_1 & x_2 & \cdots & x_n \\
x_2 & x_3 & \cdots & x_{n+1} \\
\vdots & \vdots & \vdots & \vdots \\
x_m & x_{m+1} & \cdots & x_N
\end{pmatrix},
$$

$$
\mathcal{H}_p(\mathbf{x}) \equiv
\begin{pmatrix}
0 & 0 & \cdots & x_1 \\
\vdots & \vdots & \vdots & \vdots \\
0 & x_1 & \cdots & x_{n-1} \\
x_1 & x_2 & \cdots & x_n \\
x_2 & x_3 & \cdots & x_{n+1} \\
\vdots & \vdots & \vdots & \vdots \\
x_m & x_{m+1} & \cdots & x_N \\
x_{m+1} & x_{m+2} & \cdots & 0 \\
\vdots & \vdots & \vdots & \vdots \\
x_N & 0 & \cdots & 0
\end{pmatrix}.
$$

The subscript p in $\mathcal{H}_p(\mathbf{x})$ comes from the fact that this matrix is obtained from $\mathcal{H}(\mathbf{x})$ by *pre-* and *post-augmentation* with triangular Hankel matrices.

Next, let $\mathbf{J}$ denote the symmetric matrix derived from the identity matrix by rearranging its columns in reverse order. We now define two Toeplitz matrices derived from $\mathcal{H}(\mathbf{x})$ and $\mathcal{H}_p(\mathbf{x})$ by reversing their columns, i.e., $\mathcal{T}(\mathbf{x}) \equiv \mathcal{H}(\mathbf{x})\,\mathbf{J}$ and $\mathcal{T}_p(\mathbf{x}) \equiv \mathcal{H}_p(\mathbf{x})\,\mathbf{J}$. Premultiplication of an n-vector $\mathbf{y}$ with either $\mathcal{H}(\mathbf{x})$ or $\mathcal{H}_p(\mathbf{x})$ corresponds to filtering $\mathbf{x}$ with a FIR filter whose coefficients are the elements of the vector $\mathbf{y}$. The only difference between the two operations lies in the way the two ends of the output signal vector are computed; the vector $\mathcal{H}(\mathbf{x})\mathbf{y}$ has length m, while the vector $\mathcal{H}_p(\mathbf{x})\mathbf{y}$ has length N. Similarly, premultiplication of $\mathbf{y}$ with either $\mathcal{T}(\mathbf{x})$ or $\mathcal{T}_p(\mathbf{x})$ corresponds to filtering $\mathbf{y}$ with a FIR filter whose coefficients are the elements of $\mathbf{x}$ in reverse order.

Finally, we define the averaging operator $\mathcal{A}$ that transforms a given $m \times n$ matrix $\mathbf{M}$ into a signal vector $\mathbf{s} \equiv \mathcal{A}(\mathbf{M})$ of length N by arithmetic averaging along the $N = m + n - 1$ antidiagonals of $\mathbf{M}$:

$$
s_i = \frac{1}{\beta - \alpha + 1} \sum_{k=\alpha}^{\beta} \mathbf{M}_{i-k+2,k}, \quad i = 1, \ldots, N,
$$

where $\alpha = \max(1, i - m + 2)$ and $\beta = \min(n, i + 1)$.

For the special case where $\mathbf{M}$ is a rank-one matrix, i.e., where we can write $\mathbf{M} = \mathbf{u}\mathbf{v}^T$, the averaging operation can be expressed in the following simple form: $\mathcal{A}(\mathbf{M}) = \mathcal{A}(\mathbf{u}\mathbf{v}^T) = \mathbf{D}\mathcal{T}_p(\mathbf{u})\mathbf{v}$, where $\mathbf{D}$ is an $N \times N$ diagonal matrix given by $\mathbf{D} = \mathrm{diag}(1, 2^{-1}, \ldots, \mu^{-1}, \mu^{-1} \ldots, \mu^{-1}, \ldots, 2^{-1}, 1)$, with $\mu = \min(m, n)$. The first $\mu - 1$ and last $\mu - 1$ diagonal elements account for the fact that the corresponding antidiagonals of $\mathbf{M}$ do not have full length μ.

3   TSVD algorithm

Given an input signal vector s_{in} of length N, consisting of a pure signal plus additive *white* noise, we first form the associated $m \times n$ Hankel matrix

$$\mathbf{H} = \mathcal{H}(s_{\mathrm{in}}).$$

The next step is to compute the SVD of $\mathbf{H}$:

$$\mathbf{H} = \sum_{i=1}^{n} \sigma_i \mathbf{u}_i \mathbf{v}_i^T,$$

where the left and right singular vectors $\mathbf{u}_i$ and $\mathbf{v}_i$ are orthonormal, and the singular values σ_i are non-negative and appear in non-decreasing order, $\sigma_1 \geq \cdots \geq \sigma_n \geq 0$.

The third step is to approximate $\mathbf{H}$ by a rank-k matrix $\mathbf{H}_k$ with $k \leq n$. There are several possibilities here, and in a unified notation we can write $\mathbf{H}_k$ as

$$\mathbf{H}_k = \sum_{i=1}^{k} w_i \sigma_i \mathbf{u}_i \mathbf{v}_i^T, \qquad k \leq n.$$

The least squares approximation, which is closest to $\mathcal{H}(s_{\mathrm{in}})$ in the 2-norm and Frobenius norm, is obtained with $w_i = 1$, $i = 1, \ldots, k$. The minimum variance approximation [6], which is the best estimate of the pure-signal matrix that can be obtained by making linear combinations of the noisy data in the matrix $\mathcal{H}(s_{\mathrm{in}})$, is obtained with $w_i = 1 - \sigma_{\mathrm{noise}}^2 / \sigma_i^2$, $i = 1, \ldots, k$, where $\sigma_{\mathrm{noise}}^2$ is the white noise variance.

The final step is to compute the output signal vector s_{out} of length N from $\mathbf{H}_k$. This is done by arithmetic averaging along the antidiagonals of $\mathbf{H}_k$, i.e.,

$$s_{\mathrm{out}} = \mathcal{A}(\mathbf{H}_k).$$

To derive the filter model of the TSVD algorithm, we use the definitions and equations from the previous section together with the identity $\mathbf{H}\mathbf{v}_i = \sigma_i \mathbf{u}_i$, and we obtain

$$s_{\mathrm{out}} = \mathbf{D} \sum_{i=1}^{k} w_i \, \mathcal{H}_p(\mathcal{H}(s_{\mathrm{in}})\mathbf{v}_i) \, (\mathbf{J}\mathbf{v}_i).$$

This equation defines the precise relation between the input vector s_{in} and the output vector s_{out}.

We see that the output signal essentially consists of a weighted sum of k intermediate signals s_i given by $s_i = \mathcal{H}_p(\mathcal{H}(s_{\mathrm{in}})\mathbf{v}_i) \, (\mathbf{J}\mathbf{v}_i)$, $i = 1, \ldots, k$, of which

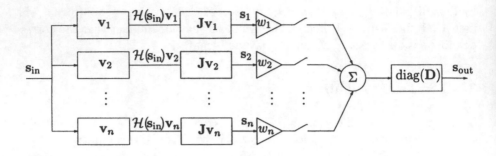

Fig. 1. Filter model of the TSVD algorithm.

$\mathcal{H}(s_{in})v_i$ is a signal obtained by passing s_{in} through a FIR filter with filter coefficients v_i, and s_i is a signal obtained by passing $\mathcal{H}(s_{in})v_i$ through a FIR filter with filter coefficients Jv_i, i.e., the coefficients of the first filter in reverse order. It is well known that this results in a zero-phase filtered version of s_{in}. The weights simply represent k amplifiers with gain w_i and the matrix D represents an N-point window whose elements are the diagonal elements of D.

From the above discussion it is evident that the FIR filters v_i and Jv_i constitute an analysis bank and a synthesis bank, respectively. Figure 1 shows the filter model of the TSVD algorithm. For completeness we have included all n filters corresponding to the n SVD components of H, plus a switch in each filter branch. The TSVD output signal s_{out} is then obtained by closing the first k switches corresponding to the largest k singular values used in (3). We note that exact reconstruction of s_{in} is obtained with $k = n$ and all $w_i = 1$.

There is a direct relation between the FIR filters v_i and the eigenfilters whose coefficients are the elements of the eigenvectors of the autocorrelation matrix of the signal under consideration. Let $C = \mathcal{H}(s_{in})^T\mathcal{H}(s_{in})$ be a scaled estimate of the $n \times n$ autocorrelation matrix for s_{in}. Then the eigenvalues and eigenvectors of C are related to the SVD of $\mathcal{H}(s_{in})$ as

$$C = \mathcal{H}(s_{in})^T\mathcal{H}(s_{in}) = \sum_{i=1}^{n} v_i\, \sigma_i^2\, v_i^T.$$

Since the right-hand side is an eigen-decomposition, we see that the FIR filters characterized by v_i are approximations to the eigenfilters associated with s_{in}.

4 TQSVD algorithm

If the noise in the input signal s_{in} is *not white*, then it is common to apply pre- and dewhitening in the TSVD algorithm. If e denotes the pure-noise component of s_{in}, and if we compute the QR-factorization of the Hankel matrix associated with e, $\mathcal{H}(e) = QR$, then $\mathcal{H}(e)^T\mathcal{H}(e) = R^TR$ is a estimate of the noise correlation matrix, and the prewhitened matrix that we work with is $\tilde{H} = HR^{-1}$.

We omit the case where $\mathbf{R}$ is singular. After the truncation has been applied to $\tilde{\mathbf{H}}$, yielding the rank-k matrix $\tilde{\mathbf{H}}_k$, the dewhitened output signal is computed as $\mathbf{s}_{\text{out}} = \mathcal{A}(\tilde{\mathbf{H}}_k \mathbf{R})$.

The TQSVD algorithm, which is based on the QSVD of the matrix pair $(\mathcal{H}(\mathbf{s}_{\text{in}}), \mathcal{H}(\mathbf{e}))$, incorporates the pre- and dewhitening as an integral part of the algorithm. The QSVD is given by

$$\mathcal{H}(\mathbf{s}_{\text{in}}) = \sum_{i=1}^{n} \delta_i \hat{\mathbf{u}}_i \mathbf{x}_i^T, \qquad \mathcal{H}(\mathbf{e}) = \sum_{i=1}^{n} \mu_i \hat{\mathbf{v}}_i \mathbf{x}_i^T,$$

where $\hat{\mathbf{u}}_i$ and $\hat{\mathbf{v}}_i$ are orthonormal vectors, the vectors $\mathbf{x}_i$ are linearly independent, and $\boldsymbol{\Delta} = \text{diag}(\delta_1, \ldots, \delta_n)$ and $\mathbf{M} = \text{diag}(\mu_1, \ldots, \mu_n)$ with $\delta_i^2 + \mu_i^2 = 1$ for $i = 1, \ldots, n$. The quotient singular values are δ_i / μ_i, and they satisfy $\delta_1 / \mu_1 \geq \cdots \geq \delta_n / \mu_n \geq 0$. In the white-noise case, the QSVD yields the ordinary SVD of $\mathbf{H} = \mathcal{H}(\mathbf{s}_{\text{in}})$.

In the TQSVD algorithm, the rank-k matrix approximation that corresponds to $\mathbf{H}_k$ is given by

$$\mathbf{Z}_k = \sum_{i=1}^{k} w_i \delta_i \hat{\mathbf{u}}_i \mathbf{x}_i^T,$$

and the TQSVD output signal is $\mathbf{s}_{\text{out}} = \mathcal{A}(\mathbf{Z}_k)$. The weights w_i are computed by the same formulas as in the TSVD algorithm, but with the singular values σ_i replaced by the quotient singular values δ_i / μ_i and σ_{noise}^2 being the noise variance of the prewhitened signal.

Also here, we obtain an expression for $\mathbf{s}_{\text{out}} = \mathcal{A}(\mathbf{Z}_k)$ that leads to a filter model. The main difference is that we need an additional set of vectors $\boldsymbol{\theta}_1, \ldots, \boldsymbol{\theta}_n$ defined such that $(\boldsymbol{\theta}_1, \ldots, \boldsymbol{\theta}_n)^T = (\mathbf{x}_1, \ldots, \mathbf{x}_n)^{-1}$. We note that the two sets of vectors are biorthonormal, i.e., $\boldsymbol{\theta}_i^T \mathbf{x}_j = 1$ for $i = j$ and zero otherwise. To derive the filter model of the TQSVD algorithm, we use the identity $\mathbf{H}\boldsymbol{\theta}_i = \delta_i \hat{\mathbf{u}}_i$ and obtain

$$\mathbf{s}_{\text{out}} = \mathbf{D} \sum_{i=1}^{k} w_i \, \mathcal{H}_p(\mathcal{H}(\mathbf{s}_{\text{in}})\boldsymbol{\theta}_i)\,(\mathbf{J}\mathbf{x}_i).$$

Figure 2 shows the FIR filter model of the TQSVD algorithm.

The filters that arise in the filter model have filter coefficients $\boldsymbol{\theta}_i$ and $\mathbf{J}\mathbf{x}_i$, and they are not zero-phase filters as in the white-noise case. Perfect reconstruction is still obtained when $k = n$ and all $w_i = 1$.

As in the white-noise case, the TQSVD FIR filters are related to eigenfilters, but not as directly as in the TSVD algorithm. Given the QR-factorization $\mathcal{H}(\mathbf{e}) = \mathbf{Q}\mathbf{R}$ associated with the pure-noise matrix, the symmetric matrix

$$\left(\mathbf{R}^{-1}\right)^T \mathbf{H}^T \mathbf{H} \mathbf{R}^{-1} = \left(\mathbf{H}\mathbf{R}^{-1}\right)^T \left(\mathbf{H}\mathbf{R}^{-1}\right)$$

is an estimate of the autocorrelation matrix for the *prewhitened* signal when prewhitened by $\mathbf{R}^{-1}$. The eigen-decomposition of this matrix is given by

$$\left(\mathbf{H}\mathbf{R}^{-1}\right)^T \left(\mathbf{H}\mathbf{R}^{-1}\right) = \sum_{i=1}^{n} (\mu_i^{-1}\mathbf{R}\boldsymbol{\theta}_i)(\delta_i/\mu_i)^2(\mu_i^{-1}\mathbf{R}\boldsymbol{\theta}_i)^T,$$

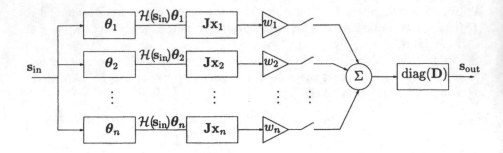

Fig. 2. Filter model of the TQSVD algorithm.

showing that $\lambda_i = (\delta_i/\mu_i)^2$ and $z_i = \mu_i^{-1}R\theta_i$ are the eigenvalues and orthonormal eigenvectors, respectively, of the prewhitened autocorrelation matrix. Hence, the vectors θ_i that determine the analysis filters of the TQSVD filter model are equal to $\mu_i R^{-1}$ times the eigenvectors z_i of the prewhitened autocorrelation matrix:

$$\theta_i = \mu_i R^{-1} z_i, \qquad i = 1, \ldots, n. \tag{1}$$

The synthesis filters, in turn, are determined by the vectors x_i as

$$x_i = \mu_i^{-1} R^T \left(\mu_i^{-1} R \theta_i \right) = \mu_i^{-1} R^T z_i, \qquad i = 1, \ldots, n. \tag{2}$$

In other words, the vectors x_i that determine the synthesis filters of the TQSVD filter model are equal to $\mu_i^{-1} R^T$ times the eigenvectors z_i of the prewhitened autocorrelation matrix.

Equations (1) and (2) are important because they allow us to compute the resulting analysis and synthesis filters even if prewhitening with R^{-1} is explicitly used.

5 Numerical Example

In order to illustrate our filter-bank interpretation, we apply the *TQSVD* algorithm to a realistic example in speech enhancement.

We consider a segment s_{in} of voiced speech in additive colored noise. The speech signal was sampled at 8 kHz and filtered by a first order FIR preemphasis filter:

$$P(z) = 1 - \beta z^{-1}, \quad \beta = 0.95.$$

The noise was generated such that its power spectral density (PSD) was

$$S(\omega) = \frac{1}{(1 + a^2) - 2a \cos(\omega T_s)}, \quad a = 0.5, \quad T_s = 125 \mu s$$

and scaled such that the *SNR* of the signal s_{in} was 15 dB. Figure 3 shows the linear predictive coding (LPC) model spectrum of the pure speech segment (top) and the PSD of the noise (bottom). The segment length N was set to 160 samples corresponding to 20 ms, which is the typical block length in speech processing because this length is short enough for the segment to be nearly stationary. The size of both $\mathbf{H}$ and $\mathbf{N}$ was set to 141×20.

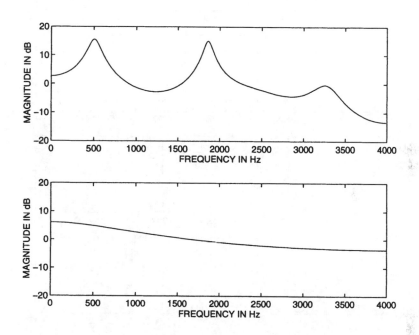

Fig. 3. The LPC model spectrum of the pure preemphasized speech segment (top) and the PSD of the noise (bottom).

Table 1. The 20 quotient singular values of the matrix pair $(\mathbf{H}, \mathbf{N})$.

i	δ_i/μ_i	δ_{5+i}/μ_{5+i}	δ_{10+i}/μ_{10+i}	δ_{15+i}/μ_{15+i}
1	12.42	4.58	4.03	2.44
2	11.73	4.50	3.81	2.11
3	10.33	4.35	3.79	1.63
4	9.99	4.16	3.38	1.51
5	4.90	4.08	2.85	1.47

Table 1 lists the 20 quotient singular values δ_i/μ_i of $(\mathbf{H}, \mathbf{N})$, and Fig. 4 shows the frequency response of the first 15 combined analysis/synthesis filters (cf. Fig. 2) associated with the QSVD of $(\mathbf{H}, \mathbf{N})$. It is clearly seen that the

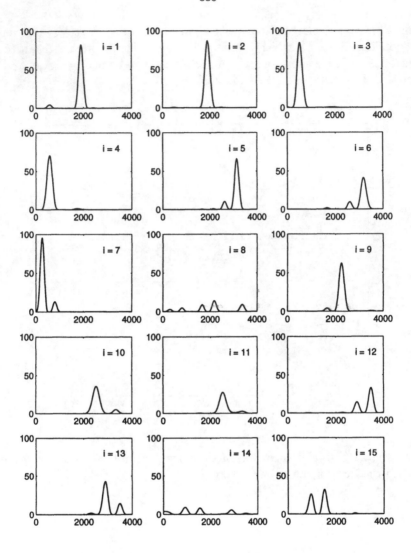

Fig. 4. Frequency responses of the first 15 analysis/synthesis filters.

first 6 analysis/synthesis filters have band-pass characteristics and capture the 3 formants (i.e., the maxima in the LPC model spectrum) of the speech segment. These three formants, and in particular the two with lowest frequency, are so "peaked" that they can be modeled with good approximation by pure sinusoids— hence the shape of the corresponding FIR filters.

The remaining 14 filters (of which the first 9 are shown in Fig. 4) capture those spectral components of the signal that lie in between the formants (including the noise), and these components are much less important when reconstructing the signal. In this way, we achieve a good noise reduction of voiced speech sounds in the TQSVD method.

6 Conclusion

Reduced-rank noise reduction is interpreted as a filter bank, consisting of analysis/synthesis FIR filters whose filter coefficients are derived from the SVD of the signal matrix. The number of filter branches equals the rank of the approximating matrix.

In the white-noise case (no prewhitening), the analysis filter coefficients are the right singular vectors of the signal matrix. The synthesis filter coefficients are the same, except that they appear in reverse order.

When pre- and dewhitening is used, then the analysis and synthesis filters are determined by the right quotient singular vectors of the signal-prewhitener matrix pair. These coefficients, in turn, are related to the right singular vectors of the prewhitened signal.

Traditionally, reduced-rank techniques have been analyzed and interpreted in linear algebra terms, in particular signal and noise subspaces. Our analysis gives new insight: it interprets reduced-rank techniques as filter operations and therefore allows an analysis in the frequency domain. This insight can, for example, provide an aid to choosing an appropriate prewhitener.

Acknowledgements

Søren Holdt Jensen was supported in part by the European Community Research Program HCM, contract no. ERBCHBI-CT92-0182. Per Christian Hansen was supported in part by grant 9500764 from the Danish Natural Science Research Council.

References

1. L. L. Scharf and D. W. Tufts, "Rank Reduction for Modeling Stationary Signals," *IEEE Trans. Acoust., Speech, Signal Processing*, vol. 35, pp. 350–355, March 1987.

2. M. Dendrinos, S. Bakamidis, and G. Carayannis, "Speech Enhancement from Noise: A Regenerative Approach," *Speech Communication*, vol. 10, pp. 45–57, Feb. 1991.

3. S. H. Jensen, P. C. Hansen, S. D. Hansen, and J. A. Sørensen, "Reduction of Broad-Band Noise in Speech by Truncated QSVD," *IEEE Trans. Speech, Audio Processing*, vol. 3, pp. 439–448, Nov. 1995.

4. S. H. Jensen and P. C. Hansen, "Reduced-rank noise reduction: A filter-bank interpretation," to appear in *Proc. VIII European Signal Processing Conference (EUSIPCO-96)*, Trieste, Italy, Sept. 1996.

5. I. Dologlou and G. Carayannis, "Physical Interpretation of Signal Reconstruction from Reduced Rank Matrices", *IEEE Trans. Signal Processing*, vol. 39, pp. 1681–1682, July 1991.

6. B. De Moor, "The Singular Value Decomposition and Long and Short Spaces of Noisy Matrices," *IEEE Trans. Signal Processing*, vol. 41, pp. 2826–2838, Sept. 1993.

On the Crossover Points for Dynamic Load Balancing of Long-Chained Molecules

David F. Hegarty, M.T. Kechadi and K.A. Dawson

Advanced Computational Research Group,
University College Dublin, Belfield, Dublin 4, Ireland

Abstract. We presented a parallel simulation environment for polymers, DNA and protein molecular chain simulations. The system is intended to help automatically parallelise computer simulations of complex polymers, DNA and protein molecules. In this paper we look at a load balancing algorithm for a class of computational problems where the topology of the problem is important, as well as just the dynamic problem structure. We study the algorithm and find the crossover points when it is beneficial to apply the algorithm. Both a theoretical model and experimental data from an implementation on a Cray T3D are used.

1 Introduction

Understanding the behaviour of molecular systems such as DNA, proteins, fluid flow dynamics or lattice gas simulations, and determining their microscopic structure are central problems in the field of physical chemistry. However, determining this structure is a hard task, and it is too difficult to build analytical models for such complex structures.

Computer simulation of these problems allows the estimation of average properties since sufficient high-quality experimental data for accurate structure determination is often too difficult and expensive to obtain [6]. Large heterogeneous distributed computing systems, composed of a variety of high-performance architectures are becoming readily available to researchers and experimentalists [8], but to fully exploit these new systems software must be provided to manage the complexity of the underlying physical architecture for the user.

Our objective is to develop a software environment to mask the heterogeneity of computation and communication resources from the user. The approach chosen is detailed in [12], and is based on the three following criteria: 1) the system has to be easy to use, 2) it must exploit the heterogeneity of the hardware, and 3) it has to achieve high performance via parallelism. These three criteria are addressed using three components: a graphical user interface, a virtual machine, and a runtime system.

The problems encountered in simulations of polymers, DNA and proteins are difficult to parallelise efficiently, since they are a priori unpredictable, and so completely dynamic in nature. Domain decomposition methods are frequently used with these problems due to their suitability for parallel implementation [4, 5, 13]. Techniques such as dynamic load balancing must be applied to dynamically adapt the decomposition in order to continuously keep up with the changing nature of the problem. The correct application of dynamic load balancing is often critical to the performance of the parallel simulation. To resolve the problem several load-balancing techniques have been proposed for SIMD [2, 11] and MIMD [9, 14, 16, 17, 19].

Most of these methods are not well suited to problems where a fixed topology has to be taken into account in addition to the instantaneous problem structure. In this paper we present the Positional Scan Load Balancing (PSLB) algorithm for this class of applications, where the locality of the decomposition must be preserved. In the rest of the paper these problems are termed topological problems. Applying the PSLB algorithm incurs a certain overhead, which must be outweighed by the benefits gained. We study the algorithm and find the crossover points when it is beneficial to apply the algorithm. Both a theoretical model and experimental data obtained from an implementation on a Cray T3D are used.

2 Problem Decomposition

Domain decomposition is used to solve the problem of mapping the application to the appropriate machine[10]. The problem space is broken into several sub-domains, which in a parallel environment are processed concurrently. We distinguish two levels of description for the decomposition of topological applications. The first is a spatial domain decomposition, where the problem is decomposed into tasks based on its fixed (or initial) structure. The second level takes place in the calculation domain, where the work due to the interactions between tasks is dynamically mapped to the processors.

The problem is represented as a set of nodes $V = \{(v_i, n_i)\}, n_i \in \mathbf{N}$, and edges between these nodes. Each node v_i has a corresponding weight n_i which represents the amount of work contained within that node. To represent topological problems two edge sets are used. Edges represent calculations due to a pair of work nodes interacting. The first set E^s is the set of fixed static edges. Since these edges are fixed in time an initial static decomposition may be used. The second set consists of the dynamic edges, which depend on time, E_t^d.

Spatial decomposition is based on a mapping of the work nodes V to processing nodes, while the second level is obtained from a mapping of the dynamic edge set to the processing nodes. In the algorithm description the edges in E_t^d are represented by work units w_j^i contained within the work nodes. For our purposes we assume we have an initial mapping ν of work nodes V to the set of processing nodes $P = \{p_i\}$. To preserve the decomposition locality all the work units are assigned an order according to the topology.

Let N_i be the total number of work units assigned to the processing node p_i, $N_i = \sum_{k, \nu(v_k)=i} n_k$. Then the **work unit index** of w_j^i, the jth work unit within the processing node v_i, is given by $I_j^i = j + \sum_{k<i} N_i$. This definition is used to calculate destination processors for the migration of work by the load balancing algorithm in the next section.

3 PSLB Algorithm

The Positional Scan Load Balancing (PSLB) algorithm is a dynamic load balancing algorithm preserving the decomposition locality. The technique is based on the parallel prefix operation, or **scan**.

Let p be the number of processing nodes and let each p_i have a power π_i, which is the number of work units per unit time it can calculate. Π is the total power of the processing system, with $\Pi = \sum_{i=1}^{p} \pi_i$. Then the normalised relative power γ_i for p_i is $\gamma_i = \pi_i/\Pi$.

The algorithm is given in figure 1. It first calculates the scan $S = (+, L)$ where $L = (N_1, N_2, \cdots, N_q)$. Then the total work W is calculated as the value S_p of the scan in the last processor added to its own work load Np. Let k_0 be the last k such that $\lambda_k \leq I_j^i/W$. Then the destination processing node for w_j^i is k_0 and the index within that node is $\gamma_{k_0}(I_j^i - \lambda_{k_0}W)$.

Figure 2(a) shows how the work is distributed among processors with three different processing powers, according to the initial mapping given. To equalise the processing time of each processing node, the algorithm distributes the work according to the relative power of these processing nodes. Figure 2(b) shows that the algorithm assigns the most work to the two last processors since they have the most processing power. The redistribution completely maintains the original order of the work, represented in the figure by a different shade of grey for each processing node's original work.

```
forall processors i in parallel
  1 S_i ⟵ ∑_{k<i} N_k
  2 W ⟵ S_p + N_p
  3 calculate λ_j = ∑_{k<j} γ_k
  4 for j = 1 to N_i
    4.1 find last k s.t. λ_k ≤ (j + S_i)/W
    4.2 ℓ ⟵ (i + S_i)γ_k − λ_k γ_k W
    4.3 migrate(w_ℓ^k ⟵ w_j^i)
    end for
end forall
```

Fig. 1. The Positional Scan Load Balancing algorithm.

The **migrate** routine has two functions. The first is the straightforward communication of work units between processors to balance the load. This should not be carried out within the loop, but rather exchanged in a single communication after the loop. The second accommodates the migration of work nodes between processors to reduce the communication cost by identifying work node pairs with high communication traffic.

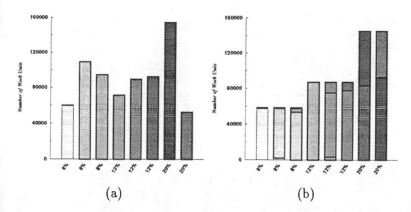

(a)   (b)

Fig. 2. (a) Distribution of work units from 8,192 bead polymer on 8 processors before load balancing, with each bar being a separate processing node, and the percentage labels being the percentage of the total power of the system for that processing node (b) Distribution of work after load balancing using the PSLB algorithm.

4 Analysis and Results

In this section the performance of the PSLB algorithm is analysed using a theoretical model and experimental results. Both the simulation application and the PSLB implementation were written as SPMD applications using the MPI (Message Passing Interface) library [7], running on a 384 processor Cray T3D in the Edinburgh Parallel Computer Centre at the University of Edinburgh.

4.1 Simulation Model

A polymer simulation model is used as a platform to provide a basic scientific simulation for analysing the algorithm. A polymer is represented using the freely-jointed bead-spring model[1]. All beads interact with pairwise Van der Waals forces, modelled by the Lennard-Jones potential, given for any two beads i, j by $V_{i,j} = \frac{A}{r_{ij}^{12}} - \frac{B}{r_{ij}^{6}}$. A and B are constants and r_{ij} is the distance between the two beads i and j. Nearest neighbouring beads along the chain are connected by springs, giving an additional harmonic interaction with the potential given by $I_{i,j} = -\frac{1}{2}kr_{ij}^{2}$. These interactions correspond to the static edges E^{s}.

The Lennard-Jones interaction between two beads is neglected when r_{ij} is greater than some cutoff distance giving rise to the dynamic edges E_{t}^{d}. Each bead varies in the number of interactions it must calculate, leading to load imbalance. Figure 3 shows the number of work units assigned to each of 128 processing nodes on the Cray T3D for a typical polymer configuration. We can see a considerable load imbalance since the work could be completed in nearly half the time if each processor carried out the same amount of work.

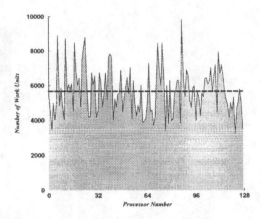

Fig. 3. Load imbalance for 128 processors for a 16384 bead homo-polymer with the average value given as a dashed line.

4.2 BSP Model Analysis

An analysis of the algorithm and polymer simulation has been carried out using the BSP (Bulk Synchronous Parallel) model [3, 15, 18]. In this model a computation is broken into a set of super–steps, consisting of calculations on local data followed by a communication and synchronisation between processors at the end. The cost of a super-step is parameterised by the number of processors p, the speed s of these processors, the global performance g of the communication network, and the synchronisation overhead l. A maximum cost is assigned for each super–step S_i of the algorithm as a function of these parameters as

$$C_i = w + l + g \max(h_s, h_r) \tag{1}$$

where w is the number of local operations carried out in S_i and h_s, h_r are the maximum amounts of data sent and received respectively.

Let A_1, A_2, A_3 be the size in words of a work unit, a calculation result and a bead, respectively. Let W_1 be the number of operations needed to determine the destination processor and index of a work unit. Let W_2 be the number of operations required to calculate an interaction and W_3 the cost of up updating a local position. α is the maximum proportion of the total number of beads, N, allowed on a single processor, $0 \leq \alpha \leq 1$. Then the cost of a simulation step (without load balancing) is written according to equation (1) as

$$C_{\text{sm}} = \alpha N((N - 1)W_2 + W_3) + A_3(N - 1)g + l. \tag{2}$$

The cost of the load balancing step is

$$C_{\text{dw}} = \alpha N(N - 1)W_1 + (2 + \alpha N(N - 1)A_1(p - 1)/p)g + 2l, \tag{3}$$

and the maximum cost of sending the results back to the originator is

$$C_{\text{rs}} = (A_2 \alpha N(N - 1)(p - 1)/p)g + l. \tag{4}$$

These cost expressions allow the prediction of performance on any parallel architecture, by substituting the appropriate values of the BSP model parameters g, l and p. The complexity of both the simulation and the PSLB algorithm is $\mathcal{O}(N^2)$ since load balancing takes place at the interaction level, whose number is bounded by $(N^2 - N)$.

Changing the work unit decomposition results in a change of the effective value of α. Let the value of α after load balancing be $\hat{\alpha}$. Then the total cost of the simulation including the PSLB algorithm is

$$C_{tc} = \hat{\alpha} N (W_3 + (N - 1)(W_1 + W_2)) +$$
$$\left[(N - 1)(\hat{\alpha} N \left(\frac{p - 1}{p} \right) (A_1 + A_2) + A_3) + 2 \right] g + 4l. \tag{5}$$

Load balancing is carried out at every time step since the interaction decomposition is dynamic. However this is not necessarily beneficial due to the overhead of the PSLB algorithm. There is therefore some crossover point before which the balancing communication and calculation overhead do not make it worthwhile to load balance at every time-step. After load balancing, each processor is assigned W/p work units. Therefore $\hat{\alpha} = 1/p$. The crossover point for any parameter can be found by solving the equation

$$C_{sm} = C_{tc}, \tag{6}$$

for that parameter. In section 4.3 the crossover point for W_2 is found experimentally, holding the other parameters constant.

4.3 PSLB Performance

Figure 4(a) shows the time taken by the PSLB algorithm for different numbers of processors on the T3D. It shows the two main phases of the algorithm - the migration of the work to be calculated, and the return of the results. The execution time of the algorithm decreases as the number of processors is increased, since the average amount of data transferred per processor reduces (see table 1).

The PSLB algorithm is sender initiated, i.e. it is the processing node that is assigned a work unit that determines where it should go. It would therefore be well suited to a one-sided communication interface implementation. However MPI does not currently support one sided communications. Therefore the sender must tell the receiver how much work is going to be received before actually sending the work. This explains the difference between the time for the migration of the work and the return of the results.

Figure 4(b) shows the time taken for the different parts of PSLB: the execution of the scan, the calculation of the work unit indices, the sending of information to the receiver to allow the work to be communicated, and the communication of the work. Only 4,8,16 and 32 processors are shown for clarity. The time to send the work dominates, accounting for approximately 90% of the total time. This indicates that the section of the algorithm which is unbalanced - the overhead due to the calculation of work indices - is not excessive.

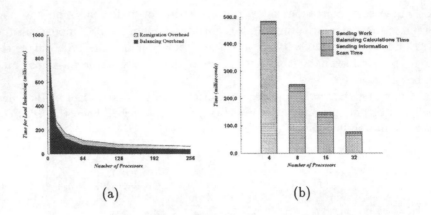

(a)   (b)

Fig. 4. (a)Time taken for PSLB algorithm on from 2 to 256 processors, with both the time taken to migrate the work to an even distribution and the time taken to redistribute the results to the originators shown for a homo-polymer with 8,192 beads. (b)Breakdown of the work migration phase into its component parts.

Number of Processors	Average Data Sent (KB)	Time (ms)	Average Bandwidth (KB/s)
4	192.0	837.3	229.3
8	144.0	432.3	333.1
16	75.9	236.8	320.5
32	49.0	120.1	408.0

Table 1. Showing the amount of data sent by processors for a 8,192 bead polymer.

4.4 PSLB Overhead

Figure 5(a) shows the time taken without the PSLB overhead for the calculation of interactions with and without load balancing. The time taken to calculate the simple Lennard-Jones interaction is at the left of this graph. We see here that no appreciable gain is made by employing the PSLB algorithm. However as soon as we use more complex interactions, the time taken to complete the calculations becomes significantly shorter.

The crossover point for W_2 (the complexity of an interaction) is shown in figure 5(b) on 256 processors including all overhead. The crossover points for different numbers of processors are given in table 2. These decrease with increasing number of processors, showing that the point at which it becomes beneficial to apply load balancing involves less and less calculations. This is due to the amount of data sent per processor, as discussed under figure 4(a).

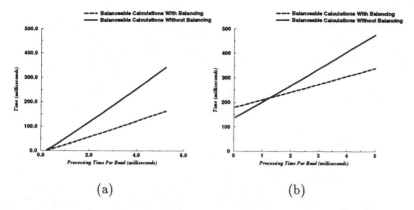

(a)   (b)

Fig. 5. (a) Times for interaction calculations with and without load balancing. (b) Times for interaction calculations including all overhead.

Number of Processors	4	8	16	32	64	128	256
Crossover Point	5.40	3.46	2.82	1.52	1.21	1.09	1.25

Table 2. Point for beneficial load balancing, in processing time per bead.

Figure 6(a) shows the speedup of an 8,192 bead polymer simulation algorithm with PSLB for up to 256 processors. Close to 100% efficiency is obtained for less than 256 processors, at which point it drops to 80%. This drop in efficiency accounts for the increase in the last crossover point in table 2.

The figures discussed previously were for fixed problem size scaling, i.e. how the performance varied with differing numbers of processors with constant problem size. Figure 6(b) shows the total time taken to complete an iteration of the program on 256 processors scaling the problem size. The data were obtained just after the crossover point to show the gain as the problem size is increased. As reported in figure 5, increasing the problem complexity would result in further gains. The scaling with size is very closely matched by a quadratic curve because the problem being studied is an N^2 one, as shown in section 4.2. We can see that the total time taken scales more slowly with load balancing than without.

5   Conclusion

Computational chemistry, and in particular biomolecular modelling including protein folding, takes up a considerable proportion of supercomputer time. Often these simulations can be hampered by the problem of load imbalance giving us a strong motivation to work in this area. The system proposed allows the monitoring of, and provides the means to rectify this problem through the use of dynamic load balancing and scheduling techniques. We have detailed a synchronous load balancing scheme which considers the topology of the problem

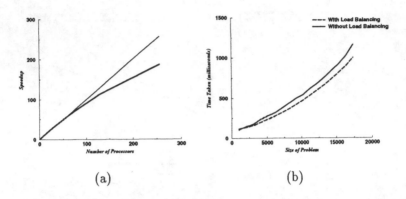

(a)   (b)

Fig. 6. (a)Speedup obtained from 2 to 256 processors. (b)Scaled problem size performance with 256 processors.

when re-balancing the workload. A theoretical model is provided to express the cost and analyse the complexity of the algorithm. The benefits and scalability of PSLB are examined with experimental results on a massively parallel machine.

We have quantified the overhead of the PSLB algorithm and have found the crossover points beyond which load balancing is recommended. The load balancing phase is managed by the runtime system. A combination of two different decomposition schemes is used which gives PSLB the flexibility to balance the workload at two levels of granularity.

Our results were obtained on a T3D, a tightly coupled parallel machine with all processors having the same processing power, leading to the homogeneous version of the algorithm being implemented. We are currently evaluating the general version of PSLB. This implementation runs on a workstation cluster.

6 Acknowledgements

This work was partly carried out during a research visit to the Edinburgh Parallel Computer Centre in the University of Edinburgh, Scotland, funded by the Human Capital Mobility grant number ERB-CHGE-CT92-0005.

References

1. A. Baumgartner. Statics and dynamics of the freely jointed polymer chain with lennard-jones interaction. *Journal of Chemical Physics*, 72(2):871–879, Jan. 1980.
2. E.S. Biagioni. *Scan Directed Load Balancing.* PhD thesis, University of North Carolina at Chapel Hill, 1991.
3. R. H. Bisseling and W. F. McColl. Scientific computing on bulk synchronous parallel architectures. Technical Report 836, Dept. of Mathematics, University

of Utrecht, December 1993. Short version in Proc. 13th IFIP World Computer Congress. Volume I (1994), B. Pehrson and I. Simon, Eds., Elsevier, pp. 509–514.

4. W.J. Camp, S.J. Plimpton, B.A. Hendrickson, and R.W. Leland. Massively Parallel Methods for Engineering and Science Problems. *Communication of the ACM*, 37(4):31–41, April 1994.

5. C. Che Chen, J.P. Singh, W.B. Poland, and R.B. Altman. Parallel Protein Structure Determination from Uncertain Data. In *Proc. of Supercomputing'94*, 1994.

6. E.A. Colbourn. *Computer Simulation of Polymers*. Longman, 1994.

7. Message Passing Interface Forum. *MPI: A Message-Passing Interface Standard*. University of Tennessee, Knoxville, Tennessee, May 1994.
   URL: http://www.mcs.anl.gov/Projects/mpi/standard.html.

8. R.F. Freund and H.J. Siegel. Heterogeneous Processing. *IEEE Computer*, (6):13–17, June 1993.

9. M. Furuichi, K. Taki, and N. Ichiyoshi. A Multi-level Load Balancing Scheme for OR-Parallel Exhaustive Search Programs on the Multi-PSI. In *Proc. of the 2nd ACM SIGPLAN Symposium on Principles and Practice of Parallel Programming*, pages 50–59, 1990.

10. David F. Hegarty, M. Tahar Kechadi, and K.A. Dawson. Dynamic Domain Decomposition and Load Balancing for Parallel Simulations of Long-Chained Molecules. In *Proc. of PARA'95*, volume 1041 of *Lecture Notes in Computer Science*, pages 303–312, Copenhagen, Denmark, August 1995. Springer-Verlag.

11. G. Karypis and V. Kumar. Unstructured Tree Search on SIMD Parallel Computers. Technical Report 92-21, Dept. of Computer Science, University of Minnesota Minneapolis, MN 55455, April 1992.

12. T. Kechadi, P. Kiernan, D. Hegarty, and K. Dawson. Parallel Simulation Environment for Polymers, DNA and Protein Molecular Chains. In *Proc. of the International Conference and Exhibition on High-Performance Computing and Networking (HPCN'96)*, Brussels, Belgium, April 1996.

13. D. E. Keyes and W. D. Gropp. A Comparison of Domain Decomposition Techniques for Elliptic Partial Differential Equations and their Parallel Implementation. *SIAM Journal of Scientific and Statistical Computing*, 8:166–202, March 1987.

14. M.H. Willebeek-Le Mair and A.P. Reeves. Strategies for Dynamic Load Balancing on Highly Parallel Computers. *IEEE Transactions on Parallel and Distributed Systems*, 4(9), September 1993.

15. W. F. McColl. Bsp programming. In G. E. Blelloch, K.M. Chandy, and S. Jagannathan, editors, *Specification of Parallel Algorithms. Proc. DIMACS Workshop, Princeton, May 9–11, 1994*, volume 18 of *DIMACS Series in Discrete Mathematics and Theoretical Computer Science*, pages 21–35. American Mathematical Society, 1994.

16. S. Patil and P. Banerjee. A Parallel Branch and Bound Algorithm for Test Generation. *IEEE Transactions on Computer Aided Design*, 9(9), March 1990.

17. W. Shu and L.V. Kale. A Dynamic Scheduling Strategy for the Chare-Kernel System. In *Proc. of Supercomputing'89*, pages 389–398, 1989.

18. L. G. Valiant. A bridging model for parallel computation. *Communications of the ACM*, 33(8):103–111, 1990.

19. B.W. Wah and Y.W. Eva Ma. MANIP - A Multicomputer Architecture for Solving Combinatorial Extremum-Search Problems. *IEEE Transactions on Computers*, C-33(5), May 1984.

Parallel Computation of Spectral Portrait of Large Matrices

V. Heuveline and M. Sadkane

IRISA-INRIA. Campus Universitaire de Beaulieu. 35042 Rennes Cedex, France

Abstract. This paper presents a parallel version of a Davidson type method for the computation of spectral portrait of large non hermitian matrices. Performance results obtained on the machine Paragon are reported.

1 Introduction

The notion of *pseudospectra* is useful for the eigenvalue analysis of non-normal matrices and operators [11]. It generalizes the notion of eigenvalues of a matrix in the sense that, instead of only representing an eigenvalue by its computed approximation, one may consider a neighborhood of it which may be defined by some tolerance threshold. Applications of pseudospectra are widespread. See [12] for a review of some important applications that have been carried out recently.

Let $A \in \mathbb{C}^{n \times n}$ and denote $\|.\|$ the euclidean norm or its induced matrix norm. For each $\epsilon \geq 0$, the ϵ-*pseudospectrum* of A is the set of eigenvalues of all perturbed matrices $A + \Delta$ with $\|\Delta\| \leq \epsilon$:

$$\Lambda_\epsilon(A) = \{z \in \mathbb{C} : z \text{ is an eigenvalue of } A + \Delta; \|\Delta\| \leq \epsilon\} \tag{1}$$

$$= \{z \in \mathbb{C} : \|(A - zI)^{-1}\| \geq \frac{1}{\epsilon}\}. \tag{2}$$

The above definition includes the convention $\|(A - zI)^{-1}\| = \infty$ if $z \in \Lambda_0(A)$. Thus the 0-pseudospectrum is nothing but the set of eigenvalues of A. It is clear that if A is normal, then $\Lambda_\epsilon(A)$ is the union of closed balls $\{z \in \mathbb{C} : |z - \lambda| \leq \epsilon\}$ centered at the eigenvalue λ of A and of radius ϵ, so that the eigenvalues alone can explain the behavior of A. If A is non-normal, its ϵ-pseudospectrum can be much larger than its set of eigenvalues. In this case the eigenvalues can only give asymptotic information on the behavior of A.

The spectral portrait of the matrix A [5] is the representation of the function

$$sp_A : z \rightarrow \|(A - zI)^{-1}\| \tag{3}$$

by means of level curves. Its computation amounts to evaluating the function :

$$sp_A : z \rightarrow \frac{1}{\sigma_{min}(zI - A)}, \quad \text{for } z \in \mathbb{C} \tag{4}$$

where $\sigma_{min}(zI - A)$ stands for the smallest singular value of the matrix $zI - A$.

For large matrices, it is clear that the Singular Value Decomposition [6] cannot be used since it induces high computational cost and storage. However, noticing that $\sigma_{min}^2(A - zI) = \lambda_{min}((A - zI)^H(A - zI))$ where λ_{min} denotes the smallest eigenvalue, one may use an efficient sparse symmetric eigenvalue solver for computing the smallest eigenvalue. In [2], the Davidson method was successfully used in this context, and allowed to plot the spectral portrait for large matrices on some rectangular domains of the complex plane. Theoretical as well as numerical aspect of this method are given in [3, 8]. The use of an eigenvalue solver adapted to large matrices, such as Davidson's method, reduces the complexity required for computing $\sigma_{min}(A - zI)$ but the total CPU time for the spectral portrait is still high.

In this paper, we consider the parallelization of this method. The goal is to reduce the computational time and to be able to deal with larger matrices than reasonably feasible on a sequential platform. In section 2, we describe an adaptation of Davidson's method [8] which allows to compute the smallest singular value, and emphasize the necessary details for drawing the spectral portrait of a given matrix, using this algorithm. In its generalized form, Davidson's method involves the choice of a preconditioner which may greatly influence the performance of the algorithm. In section 3, the selection of two parallelizable preconditioners is discussed as well as the parallelization of the whole method. Finally, in section 4, we give some numerical tests and discuss the performances of both approaches on our target machine Paragon.

2 Computation of the spectral portrait

To compute the spectral portrait of a given matrix $A \in \mathbb{C}^{n \times n}$, we need the smallest singular value σ_{min} of $A - zI$ at any point $z \in \mathbb{C}$. We thus consider the modification of Davidson's method proposed in [8]. This method uses the fact that $\sigma_{min}^2(A_z) = \lambda_{min}(A_z^H A_z)$ with $A_z = A - zI$, i.e. it computes the smallest eigenpair of the hermitian matrix $A_z^H A_z$. Since the smallest eigenvalues of $A_z^H A_z$ can become clustered, we choose to use a block version of this algorithm which allows the computation of several eigenpairs simultaneously and, therefore, ensures the convergence of the smallest one (See Algorithm 1). $Precond_{A_z}$ stands for a preconditioner whose choice is discussed in subsection 3.4. Note that the explicit computation of $A_z^H A_z$ is not required. All that it is required is the product of A_z and of its transpose by vectors, so that the structure of A_z can be exploited. At step k, the basis V_{k+1} is obtained from V_k by incorporating the vectors $t_{k,i}$, $i = 1, \ldots, l$ after orthonormalization. See [3, 8] for a detailed description of the algorithm and convergence proof.

In practice we do not need to compute the spectral portrait on the whole complex plane, but only in the neighborhood of the eigenvalues of interest. In the example discussed in section 4, we are interested, for stability reasons, in the rightmost eigenvalues. Therefore we discretize a rectangular domain where these eigenvalues are located and then compute the smallest singular value $\sigma_{min}(A - zI)$ for each z on the grid of the discretization. For two closed points z and z'

of the complex plane the matrices A_z and $A_{z'}$ have closed singular values and often closed singular vectors. To improve performances we therefore scan the grid columnwise in order to encounter neighboring points, and, for a given point of the grid, we start the Davidson algorithm with the singular subspace computed on its predecessor (See Algorithm 2).

Algorithm 1 : Davidson(A_z,m,l)
Choose l the block size and m the maximum size of the basis;
Choose an initial orthonormal matrix $V_1 = [v_1, \cdots, v_l] \in \mathbb{C}^{n \times l}$;
For $k = 1, 2, \cdots$

    1: *Compute the matrix $U_k = A_z V_k$;*
    2: *Compute the matrix $W_k = A_z^H U_k$;*
    3: *Compute the Rayleigh matrix $H_k = V_k^H W_k$;*
    4: *Compute the l smallest eigenpairs $(\nu_{k,i}^2, y_{k,i})_{1 \leq i \leq l}$ of H_k;*
    5: *Compute the Ritz vectors $x_{k,i} = V_k y_{k,i}$ $i = 1, \cdots, l$;*
    6: *Compute the residuals $r_{k,i} = W_k y_{k,i} - \nu_{k,i}^2 x_{k,i}$ for $i = 1, \cdots, l$;*
    If *convergence* **then** **exit**
    7: *Compute the new directions $t_{k,i} = Precond_{A_z}(r_{k,i})$ $i = 1, \cdots, l$;*
    8: **If** $dim(V_k) \leq m - l$
    then $V_{k+1} = orthonormalize([V_k, t_{k,1}, \cdots, t_{k,l}])$
    else $V_{k+1} = orthonormalize([x_{k,1}, \cdots, x_{k,l}, t_{k,1}, \cdots, t_{k,l}])$

endfor

Algorithm 2 : Draw_portrait($A, (x_1, y_1), (x_2, y_2), nx, ny$)
(x_1, y_1) (resp. (x_2, y_2)) denotes the bottom left (resp. upper right) point of the grid, and nx (resp. ny) is the number of points in the horizontal (resp. vertical) direction;

1: $h_x = \frac{x_2 - x_1}{nx - 1}$, $h_y = \frac{y_2 - y_1}{ny - 1}$, $z = x_1 + iy_1$;
For $j = 1 : nx$
    For $i = 1 : ny$
        2: *Compute $\sigma_{min}(A - zI)$ with Algorithm (1);*
        (Start with the previous computed singular subspace if available)
        3: $z = z + ih_y$ *(next line);*
    endfor
    4: $z = z + h_x$ *(next column);*
    5: $h_y = -h_y$ *(change the sweep direction along the column);*
endfor

---

3 Parallel implementation

In this section, we describe the parallelization of Algorithm 2 on a distributed memory platform. Our implementation has been developed on Paragon which will be first described. The direct parallelization of the two main loops of algorithm (2) is not possible for large matrices because of the restrictive available memory on each node. We therefore consider the parallelization of the main operations in the Davidson Algorithm 1, viz. the matrix-by-vector multiplications, the orthogonalization processes and the preconditioning step.

3.1 Intel Paragon

The Paragon is a massively parallel distributed memory MIMD (DM-MIMD) supercomputer produced by the Supercomputer Systems Division of Intel Corporation. It interconnects, on a high-performance network, up to several hundred processing nodes. An active backplane with routine chips connected in two-dimensional grid provides input ports to the network at each processing nodes and to nodes implementing the I/O subsystems. Its peak hardware bandwidth is of 175 MB/s in each directions. The Paragon on which we carried out our experiments, consists of a meshgrid of $16 \times 4 = 64$ nodes. 8 nodes are devoted to the system management (service and I/O nodes) while 56 remain for application purposes (code, data, message-buffers). Each node comprises two i860 XP microprocessor chips. One chip is entirely devoted to communications. The i860 XP is a 50 MHz, 64-bit microprocessor with a double precision peak performance of 75 MFlops. An amount of 16 MB of RAM is also available on each node but approximately 8 to 9 are consumed by OSF such that 7 MB are really available for the users. This restriction has some consequences on the implementation strategy of the proposed algorithm and will be further discussed in the next subsection. The Task-to-Task communication latency is about $50\mu s$ while the bandwidth is approximately 92 MB/s under OSF.

For the data exchanges between the processors, we make use of the Paragon **nx** message-passing library. This library includes not only a subset of basic commands but also optimized reduction and broadcast operations. Our code is written in Fortran 77 and, to achieve performances we have used, whenever possible, asynchronous communication commands.

3.2 Matrix-by-vector multiplication

The matrix-by-vector multiplications are commonly the most CPU-time consuming operations in iterative solvers.

In Algorithm 1, we encounter both, the matrix-by-vector multiplication and the multiplication of its transpose with a vector. To attain good performances, one could store twice the matrix, under its usual and transpose form, and apply a unique efficient matrix-by-vector multiplication procedure on them. However, due to the restricted available memory ($\sim 7Mb$) on each node, we cannot consider this approach. Both operations, the matrix-by-vector multiplication and the multiplication of its transpose with a vector, are therefore separately implemented, using the same storage of the considered matrix.

Due to these memory size constraints, the basis V_k in Algorithm 1 cannot be entirely stored on one node for large matrices. It is, therefore, not possible to compute independently the points of the spectral portrait on a single node. This coarse *granularity* approach is therefore rejected in our context.

Let us now describe the data repartition used to store a sparse matrix on several nodes and its related matrix-by-vector multiplication scheme.

The main goal of the data repartition is to allow as much as possible local computation on each nodes and whenever possible to overlap the communications with computations. Since the efficiency of such a scheme greatly depends on the structure of the matrix, it is not possible, for a given scheme, to attain optimal performances for all matrices.

In our implementation, the matrices and vectors are distributed row-wise among the processors. Such a scheme, used in a straightforward way, may lead to large variations of the data and hence the CPU-time requirements on each processor. To ensure a favorable *load balancing*, any permutation of the rows as well as the number of rows on each node may be chosen for a given matrix.

On each node, the matrix components are themselves stored row-wise. They are separated in two parts : *local* and *exterior* parts. We assume that the local part contains the components for which no communication is needed for the matrix-by-vector multiplication, i.e. the needed components of the vector are already available on the same processor. We denote A_{lock} the local part of A on the node number k and A_{extk}^j the j-th exterior part of the matrix A on the node number k.

$$A = \begin{bmatrix} A_{loc0} & A_{ext0}^1 & \cdots & \cdots & A_{ext0}^{p-1} \\ A_{ext1}^0 & A_{loc1} & \cdots & \cdots & A_{ext1}^{p-1} \\ & \vdots & & & \vdots \\ & \vdots & & & \vdots \\ A_{ext(p-1)}^0 & A_{ext(p-1)}^1 & \cdots & \cdots & A_{loc(p-1)} \end{bmatrix} \quad v = \begin{bmatrix} v_{loc}^0 \\ v_{loc}^1 \\ \vdots \\ \vdots \\ v_{loc}^{p-1} \end{bmatrix} \begin{array}{l} \to Node\ 0 \\ \to Node\ 1 \\ \to \vdots \\ \to \vdots \\ \to Node\ (p-1) \end{array} \quad (5)$$

A corresponding matrix-by-vector multiplication algorithm can then be written as follows:

Algorithm 3 : Matrix-by-vector multiplication Av
We use the same notation as in (5).
k is the considered processor and p is the number of processors.

Step 1: <u>*Send*</u> *the needed components of v_{loc}^k to other processors.*
Step 2: *Compute $w^k = A_{lock} v_{loc}^k$.*
Step 3: <u>*Receive*</u> *the needed components of v_{loc}^j for $j \neq k$ from other processors.*
Step 4: *Compute $w^k = w^k + \sum_{j \neq k} A_{extk}^j \cdot v_{loc}^j$.*

Since we use asynchronous commands, the communication step 1 overlaps the computational step 2. The multiplication of the transpose of the same matrix with a vector involves a slightly different algorithm. Here the overlap between communication and computation occurs between step 2 and step 3.

Algorithm 4 : Matrix transpose-by-vector multiplication $A^T v$:
We use the same notation as in algorithm (3).

Step 1: *Compute $y^j = A_{extk}^{jT} v_{loc}^k$ for $j \neq k$.*
Step 2: <u>*Send*</u> *y^j for $j \neq k$ to node number j.*
Step 3: *Compute $w^k = A_{lock}^T v_{loc}^k$.*
Step 4: <u>*Receive*</u> *$y^{j'}$ with $j' \neq k$ from other processors.*
step 5: *Compute $w^k = w^k + \sum_{j' \neq k} y^{j'}$.*

3.3 Parallelization of the orthogonalization step

The step 8 in Algorithm (1) involves orthogonalization processes in two different situations: first, the orthogonalization of a (small) block of vectors $T_{k,l} = [t_{k,1}, \cdots, t_{k,l}]$ against an orthogonal basis V_k (see Algorithm 8) and second, an orthogonalization of the whole obtained basis $V_{k+1} = [V_k, T_{k,l}]$ (see Algorithms 5,6,7). We briefly describe the strategy adopted in our parallel environment to perform these operations.

It is well known that the Classical Gram-Schmidt method and its modified counterpart may be used to orthogonalize a block of vectors. Both methods basically include *Dot-product* and *SAXPY* operations. Since the components of the vectors are distributed row-wise among the processors, any *Dot-product* necessitates communications while the *SAXPY* operations may be locally executed.

In the following, we denote $V = [v_1, v_2, \cdots, v_l] \in \mathbb{R}^{n \times l}$ with $v_i \in \mathbb{R}^n$ $i = 1, \cdots, l$, with $v_{i,loc}^k$ the local components on the node k of the vector v_i.

The Classical Gram-Schmidt (CGS) and Modified Gram-Schmidt (MGS) methods may then be written in the following way:

Algorithm 5 : CGS(V)
k *is the node number.*
$do\ j = 1 : l$
$\quad do\ i = 1 : j - 1$
$\quad\quad \alpha(i) = v_{i,loc}^{kT} . v_{j,loc}^k$
$\quad enddo$
$\quad Globalsum(\alpha(i), i = 1 : j - 1)$
$\quad do\ i = 1 : j - 1$
$\quad\quad v_{j,loc}^k = v_{j,loc}^k - \alpha(i) . v_{i,loc}^k$
$\quad enddo$
$enddo.$

Algorithm 6 : MGS(V)
k *is the node number.*
$do\ j = 1 : l$
$\quad do\ i = 1 : j - 1$
$\quad\quad \alpha = v_{i,loc}^{kT} . v_{j,loc}^k$
$\quad\quad Globalsum(\alpha)$
$\quad\quad v_{j,loc}^k = v_{j,loc}^k - \alpha . v_{i,loc}^k$
$\quad enddo$
$enddo.$

The Globalsum call performs the componentwise sum of the considered data on each node. The result is itself available on each node. This operation is a *synchronizing-call*, i.e. if any node makes this call, it blocks until all other nodes perform the same call. On Paragon this call is optimized since it considers several ways of exchanging the data and select the most appropriate size and shape of the problem. However *synchronization overhead* must be paid for each call of Globalsum. Consequently CGS will offer better performances than MGS since l calls of Globalsum suffices in CGS whereas MGS needs $\frac{l(l-1)}{2}$ Globalsum calls.

Nevertheless, it is well known that the MGS method is numerically more stable than the standard CGS formulation. At first glance, we could then assert that, as usual in numerical analysis, the numerical robustness has to be paid .

As a matter of fact, this is not entirely true in our situation, since for the second step of the orthogonalization process, the whole basis that should be orthogonalized is available. A simple change of loop order allows to gain the numerical robustness of MGS with computational performances of CGS. The method described below is strictly equivalent to MGS and only needs l Globalsum calls. We denote it by IMGS : Intermediate Modified Gram-Schmidt. It corresponds to the implemented version.

Algorithm 7 : IMGS(V)	Algorithm 8 : ORTHOG(V,T)
k is the node number.	k is the node number.
do $j = 1 : l$	do $i = 1 : l$
$\quad$ do $i = j + 1 : l$	$\quad$ do $j = 1 : m$
$\quad\quad \alpha(i) = v_{j,loc}^{kT} . v_{i,loc}^{k}$	$\quad\quad \alpha(j) = t_{j,loc}^{kT} . v_{i,loc}^{k}$
$\quad$ enddo	$\quad$ enddo
$\quad Globalsum(\alpha(i), i = k+1 : l)$	$\quad Globalsum(\alpha(j), j = 1 : m)$
$\quad$ do $i = k+1 : l$	$\quad$ do $j = 1 : m$
$\quad\quad v_{i,loc}^{k} = v_{i,loc}^{k} - \alpha(i) . v_{j,loc}^{k}$	$\quad\quad t_{j,loc}^{k} = v_{i,loc}^{k} - \alpha(i) . t_{k,loc}^{k}$
$\quad$ enddo	$\quad$ enddo
enddo.	enddo.

In the first step of the orthogonalization process we consider the orthogonalization against V of the block of vectors $T = [t_1, t_2, \cdots, t_l] \in \mathbb{C}^{n \times l}$ where $t_i \in \mathbb{C}^n$, $i \in 1, \cdots, l$, and $t_{i,loc}^{k}$ denotes the local components on the node k of the vector t_i. (See Algorithm (8)). Since the size of the basis V is larger than the blok size T in Davidson's algorithm, the proposed algorithm is not optimal but preserves the numerical stability inherent to the MGS algorithm.

3.4 Parallelization of the preconditioner

In the preconditioning step 7 of Algorithm (1), one usually solves the linear systems

$$C_k t_{k,i} = r_{k,i} \quad i = 1, \cdots, l \qquad (6)$$

where C_k is an approximation of $(A_z^H A_z)^{-1}$ [2, 8].

The main difficulty of this step is that $(A_z^H A_z)^{-1}$ may completely loose the sparsity pattern of A_z and therefore, for large matrices, this stage may become the bottleneck of the algorithm. Several preconditioners have been investigated to partly circumvent these difficulties. These preconditioners include the incomplete LU factorization of A_z [2, 8], the incomplete QR factorization of A_z and the incomplete Cholesky factorization of $(A_z^H A_z)$ using the data structure of A_z [8]. However, the major drawback for these preconditioners is that the considered factorization as well as the necessary resolutions of the corresponding triangular linear systems, are intrinsically sequential operations.

We therefore consider in step 7 of Algorithm 1 two preconditioners based on a Conjugate Gradient (CG) solver:

$$\textbf{Precond1}: \quad [(I - x_{k,i}x_{k,i}^H)(A_z^H A_z - \sigma_{k,i}^2 I)(I - x_{k,i}x_{k,i}^H)] t_{k,i} = r_{k,i} \qquad (7)$$

$$\textbf{Precond2}: \qquad\qquad\qquad\qquad (A_z^H A_z)t_{k,i} = r_{k,i} \qquad (8)$$

for $i = 1, \cdots, l$. **Precond1** is the preconditioner used in the Jacobi-Davidson method [9]. In **Precond2** we try to solve iteratively the exact linear system $(A_z^H A_z)t_{k,i} = r_{k,i}$. This approach is intended to approximate the one adapted in [2] where a direct linear system solver was applied to (6) with an approximation of $A_z^H A_z$.

In order to obtain an approximation of the solution of both systems (7) and (8) we apply a fixed number k_{pmax} of iteration of the (CG) algorithm (see

Table 1). The main operations encountered in the (CG) iterations are again the matrix-vector or transpose of matrix-vector multiplications, the *Dot-product* and *SAXPY* operations. The implementation of each of this operation has already been described in subsections (3.1) and (3.2) respectively.

4 Numerical tests

We have illustrated the numerical behavior, as well as the performances of David-son's method with the two different preconditioners **Precond1** (7) and **Pre-cond2** (8) for drawing the spectral portrait of large matrices on several test problems [7]. But because of space limitation we report the result on only one matrix: PDE2961 taken from [1] .

The eigenvalues of interest are the rightmost ones. We therefore use an Arnoldi type method to estimate the domain where these rightmost eigenval-ues are located and then draw the spectral portrait in this domain (see Figures 1,7,8).

In the following, (x_1, y_1) (resp. (x_2, y_2)) denotes the bottom left (resp. upper right) point of the grid, n_x (resp. n_y) is the number of points in the horizontal (resp. vertical) direction of the grid. For the Davidson algorithm parameters, we use the same parameters m,l as in algorithm (1) (See Table 1). *tol* is the tolerance for which the condition

$$\|(A_z^H A_z)\tilde{x} - \tilde{\nu}^2 \tilde{x}\| \leq tol.\|A_z^H A_z\|_F \tag{9}$$

is satisfied, where $(\tilde{\nu}^2, \tilde{x})$ is the smallest computed eigenpair and $\|.\|_F$ is the Frobenius norm.

During our experimentations, we have to cope with the lack of memory that large matrices necessitate and the performance degradation due to memory swaps. For this reason, we run our experiments on sufficient number of nodes in order to fit the whole data on them without performance overhead. Let r be the smallest number of such nodes, the speed-up sp_q with respect to q processors is therefore calculated as follows:

$$sp_q = r \frac{T_q}{T_r} \tag{10}$$

where T_q is the computational time on q nodes. The number r is indicated for each speed-up curve along the y-axis.

PDE2961	Grid	(x_1, y_1)	(x_2, y_2)	(n_x, n_y)
		$(9.5, -0.8)$	$(10.0, 0.8)$	$(60; 60)$
	Davidson	$m = 100$	$l = 3$	$tol = 1.00E - 08$
	Preconditioner	$k_{pmax} = 10$	$tol_p = 1.00E - 10$	

Table 1: Test parameters for PDE2961

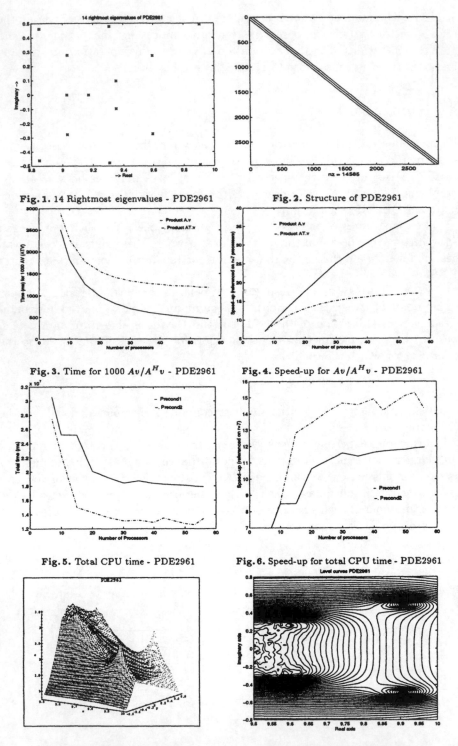

Fig. 1. 14 Rightmost eigenvalues - PDE2961

Fig. 2. Structure of PDE2961

Fig. 3. Time for 1000 $Av/A^H v$ - PDE2961

Fig. 4. Speed-up for $Av/A^H v$ - PDE2961

Fig. 5. Total CPU time - PDE2961

Fig. 6. Speed-up for total CPU time - PDE2961

Fig. 7. Spectral portrait of PDE2961

Fig. 8. Level curves for PDE2961

The performances of both the matrix-by-vector and the matrix transpose -by-vector multiplications are given in Figures 3 and 4 for a number of nodes varying between 7 and 56. For the computation of the whole spectral portrait, the number of restarts is approximately 59000 for **Precond1** and 58000 for **Precond2**. On 7 processors, 76 % (resp. 70 %) of the CPU time is necessary for the preconditioning step **Precond1** (resp. **Precond2**) and 70 % (resp. 55 %) on 56 processors. The performances of the whole algorithm in its both versions, **Precond1** and **Precond2**, are summarized in Figures 5 and 6.

Other test matrices taken from [1] have also been studied (See [7] for more details). In general, **Precond1** behaved better numerically than **Precond2**, but necessitated expensive *Dot-product* operations in our distributed environment.

References

1. Bai, Z., Barret, R., Day, D., Demmel, J., Dongarra, J.:   Test matrix collection (non-hermitian eigenvalue problems). Manuscript, (1995)
2. Carpraux, J.F., Erhel, J., Sadkane, M.: Spectral portrait for non hermitian large matrices. *Computing*, **53** (1994) 301–310
3. Crouzeix, M., Philippe, B., Sadkane, M.: The Davidson method. *SIAM J. Sci. Comput.* **15** (1994) 62–76
4. Davidson, E.R.: The iterative calculation of a few of the lowest eigenvalues and corresponding eigenvectors of large real-symmetric matrices. *Comp. Phys.* **17** (1975) 87-94
5. Godunov, S.K.: Spectral portrait of matrices and criteria of spectrum dichotomy. In L. Athanassova and J. Herzberger, editors, *in the third international IMACS-CAMM symposium on Computer arithmetic and enclosure methods*. Oldenburg, North-Holland,(1991)
6. Golub, G.H., Van Loan, C.F.: *Matrix Computations*. The Johns Hopkins University Press, Baltimore, (1989)
7. Heuveline, V., Sadkane, M.: Parallel computation of spectral portrait of large matrices. Technical Report 1037, INRIA-IRISA, (1996)
8. Philippe, B., Sadkane, M.: Computation of the fundamental singular subspace of a large matrix. *To appear in Lin. Alg. Applic.*
9. Sleijpen, G.L.G., Van der Vorst, H.A.: A Jacobi-Davidson iteration method for linear eigenvalue problems. *SIAM J. Matrix Anal. Appl.* **17** (1996) 401-425
10. Toumazou, V., Marques, O.A.:   Spectral portrait computation by a Lanczos method. Technical Report TR/PA/95/05, CERFACS, (1995)
11. Trefethen, L.N.: Pseudospectra of matrices. In D.F. Griffiths and G.A. Watson, editors, *in 14th Dundee Biennal Conference on Numerical Analysis*, (1991)
12. Tretethen, L.N.: Pseudospectra of linear operators. In Berlin Akademie-Verlag, editor, *ICIAM'95: Proceedings of the third international congress on industrial and applied mathematics*, (1995)

Computer Parallel Modular Algebra

Sergey A. Inutin

Department of Information Technologies, Surgut State University, Energetikov 14, Surgut, 626400, Russia

Abstract. The computer algebra of parallel modular operations is described. The base set of the algebra is a finite dimension metric space of modular integer vectors. Two metrics are introduced. An orthogonal normal basis is employed to reconstruct the value of the integer number corresponding to the vector. An analog of the inner product is used to advance beyond the additive range, and the vector product is defined in two ways. The algebra could serve as the basis for parallel computer arithmetic of unbounded digit integers.

1 Introduction

Parallel algorithms are a good source of speed increase for electronic computational devices [1]. The paper describes a computer algebra of parallel modular operations.

2 Algebra Description

To define the parallel computer algebra first the base set is described. The finite dimensional metric space of V - vectors of modular components is defined in the following way.

The elements of the space are the vectors of modular components

$$\overline{a} = \left(\alpha_1,...,\alpha_n\right) = \left(\left(\overline{a}\cdot\overline{e}_1\right),...,\left(\overline{a}\cdot\overline{e}_n\right)\right),$$

where $\overline{a} \in p_1'\times...\times p_n'$, $\alpha_i = |A|_{p_i} \subset p_i' = \{0,...,p_{i-1}\}$, $\left(\overline{a}\cdot\overline{e}_i\right)$ is the inner product of the vectors $\overline{a}$ and $\overline{e}_i$, i = 1,...,n . The components of the vectors are the residues of $\mathrm{mod}\ p_i$ of a number $A \in N$, and $\{\overline{a}\} \leftrightarrow \{A\}$.

The vector space is linear, i.e.

$$\overline{c} = \mu\cdot\overline{a} + v\cdot\overline{b} = \left(....,\left|\mu\cdot\alpha_i + v\cdot\beta_i\right|_{p_i},...\right) \in \prod_{i=1}^{n} p_i' \ .$$

The inner product defined as

$$\left(\overline{a}\cdot\overline{b}\right) = \sum_{i=1}^{n}\alpha_i\cdot\beta_i \ ,$$

has the following properties:

$$\left(\overline{a}\cdot\overline{a}\right) > 0, \quad \left(\overline{a}\cdot\overline{a}\right) = 0 \Rightarrow \overline{a} = \left(0,\dots,0\right);$$
$$\left(\overline{a}\cdot\overline{b}\right) = \left(\overline{b}\cdot\overline{a}\right);$$
$$\left(\left(\overline{a}+\overline{c}\right)\cdot\overline{b}\right) = \left(\overline{a}\cdot\overline{b}\right) + \left(\overline{c}\cdot\overline{b}\right).$$

The characteristic of a vector is defined as its modulus:

$$\left(\overline{a}\cdot\overline{a}\right) = \sum_{i=1}^{n} \alpha_i^2 = \left|\overline{a}\right|^2.$$

The inner product defines a metric but not a norm because $\left(\lambda\cdot\overline{a}\cdot\overline{a}\right) \le \lambda\cdot\left(\overline{a}\cdot\overline{a}\right)$.

The residue distance between vectors $\overline{a}$, $\overline{b}$, analogous to the Hemming distance is defined as

$$\omega\left(\overline{a}\cdot\overline{b}\right) = \sum_{i=1}^{n} \delta\left(\left|\alpha_i - \beta_i\right|_{p_i}\right),$$

where

$$\delta\left(\left|\alpha_i - \beta_i\right|_{p_i}\right) = \begin{cases} 1, & \text{if } \alpha_i \ne \beta_i \\ 0, & \text{if } \alpha_i = \beta_i \end{cases}, i = 1,\dots,n.$$

The residue distance is a metric, indeed:

1. $\omega\left(\overline{a},\overline{b}\right) \ge 0$,

2. $\omega\left(\overline{a},\overline{b}\right) = \omega\left(\overline{b},\overline{a}\right)$,

3. $\omega\left(\overline{a},\overline{b}\right) \le \omega\left(\overline{a},\overline{c}\right) + \omega\left(\overline{c},\overline{b}\right)$.

The known fact is that the modular coding between the sets $[0, P-1] \subset N$ and $\left\{\left(\alpha_i,\dots,\alpha_n\right)\right\}$, where $P = \prod_{i=1}^{n} p_i$, establishes a biection mapping.

Equivalence classes on the set of modular vectors are introduced by fixing the modulus of a modular vector. Note that the minimum increment of a vector by a unit in the sense of the Lee distance causes the change of the vector modulus squared by an odd integer. Indeed,

$$\left|\overline{a}\right|^2 = \sum_{i=1}^{n}\left(\alpha_i^2 + 2\alpha_i + 1\right), \text{ where } \alpha_i \in \left\{0,\dots,p_i - 1\right\}.$$

The normal vectors in the modular vector space are the basis vectors only since $\left|\overline{e_i}\right| = 1$ for them only. The modulus of a vector is equal zero for the null vector only, and the distance from the null vector to a vector is equal one for the basis vectors only.

The orthogonal normal basis of the n-dimensional space with

$$P = \prod_{i=1}^{n} p_i$$

consists of the vectors $\left\{\left(\overline{m_i P / p_i}\right)\right\}$.

Proposition. The following statements are true:

1. $\left(\left(\overline{m_i P / p_i}\right) \cdot \left(\overline{m_j P / p_j}\right)\right) = 0$ for all $i \neq j$;

2. $\left(\left(\overline{m_i P / p_i}\right) \cdot \left(\overline{m_i P / p_i}\right)\right) = 1$ for all i;

3. $p_i\left(\overline{m_i P / p_i}\right) + \left(\overline{p_i}\right) = p_i\left(\overline{1}\right)$, where

$m_i = \left|P / p_i\right|_{p_i}^{-1}$.

A number $A \in N$ and the vector $\overline{a}$ are interrelated through the following:

$$A = \sum_{i=1}^{n}\left(\overline{a} \cdot \left(\overline{m_i P / p_i}\right)\right) \cdot m_i P / p_i - rP,$$

$$A = \left(\overline{a} \cdot \overline{M}\right) - rP,$$

where r - is the characteristic of A depending on the choice of the basis,

$\overline{M} = \left(\ldots, m_i P / p_i, \ldots\right)$.

The modular algebra is defined by the following signature:

$$\left(V, +, -, \cdot, \times, /, \div\right),$$

where $V = p_1' \times \ldots \times p_n'$ is the basis set (its properties are described above);

+ and - are the additive operations of component by component addition and subtraction by the system moduli;

· is the component by component multiplication by the system moduli of the first type;

× is the second type multiplication;

/ is the component by component analog of the exact division or the first type multiplication by an inverse element [2];

÷ is the analog of the exact division or the second type multiplication by an inverse element.

To define the above mentioned operations we need a set of n numbers $\left\{p_i\right\}$ (in general case, of mutually prime ones) which are the moduli of the modular system.

Lets consider additive and multiplicative binary vector operations. The additive operations are defined in the following way:

$$\overline{a} \pm \overline{b} = \left(\ldots, \left|\alpha_i \pm \beta_i\right|_{p_i}, \ldots\right).$$

Vector multiplication in the modular algebra could be defined in two ways. The modular multiplication of the first type is a component by component operation. The result of this operation is the vector

$$\overline{a} \cdot \overline{b} = \left(\ldots, \left| \alpha_i \cdot \beta_i \right|_{p_i}, \ldots \right).$$

Particularly, $\left(\left(\overline{m_i P / p_i} \right) \cdot \left(\overline{m_j P / p_j} \right) \right) = 0$ for all $i \neq j$.

Vector multiplication of the second type $\overline{a} \times \overline{b}$ is defined in the standard way. The product is the vector orthogonal to the vectors-multipliers; its modulus is the product of the multipliers' moduli times the sine of the angle between them.

Remark 1. If n=3, then for the orthogonal vectors $\left(p_i - 1, 0, 0 \right)$ and $\left(0, p_j - 1, 0 \right)$ the product is orthogonal to the vectors - multipliers, and its modulus is equal to

$$\begin{vmatrix} p_i - 1 & 0 \\ 0 & p_j - 1 \end{vmatrix} = \left| \left(p_i - 1 \right) \cdot \left(p_j - 1 \right) \right|_{p_i p_j} = p_i p_j - 1.$$

The above mentioned result is consistent with the way the nonprime bases of modular system are introduced.

Remark 2. The moduli of the product vectors computed using the different multiplication definitions are not equal.

$$\sum_{i=1}^{n} \left| \alpha_i \cdot \beta_i \right|_{p_i}^2 \neq \sum_{i=1}^{n} \alpha_i^2 \cdot \sum_{i=1}^{n} \beta_i^2 - \left(\sum_{i=1}^{n} \alpha_i \cdot \beta_i \right)^2,$$

and, particularly, for n=2,

$$\left| \alpha_1 \cdot \beta_1 \right|_{p_1}^2 + \left| \alpha_2 \cdot \beta_2 \right|_{p_2}^2 \neq \left(\alpha_2 \cdot \beta_1 - \alpha_1 \cdot \beta_2 \right)^2$$

The operation of the formal division of the first type is defined accordingly:

$$\overline{a} / \overline{b} = \overline{a} \cdot \left| \overline{b} \right|_P^{-1} = \left(\ldots, \left| \alpha_i \left| \beta_i \right|_{p_i}^{-1} \right|_{p_i}, \ldots \right),$$

where vector $\left| \overline{b} \right|_P^{-1}$ is the one for which $\left\| \overline{b} \right|_P^{-1} \cdot \overline{b} \right|_P = 1$. The inverse vector is defined for every vector in V.

The operation of the formal division of the second type is defined in the following way:

$$\frac{\overline{a}}{\overline{b}} = \overline{a} \times \overline{b}^{-1} = \left(\ldots, \left| \frac{\alpha_i}{\sqrt{\sum_{i=1}^{n} \beta_i^2}} \right|_{p_i}, \ldots \right),$$

where vector $\bar{b}^{-1}$ is the vector orthogonal to the vector $\bar{b}$ for which the following holds:

$$\bar{b}^{-1} = \left(..., \left| \frac{1}{\sqrt{\sum\limits_{i=1}^{n} \beta_i^2}} \right|_{p_i} ,... \right), \text{ and } \bar{b}^{-1} \times \bar{b} = \bar{1}.$$

Since there exist quadratic non-residues taken by a prime p_i, the inverse vector in the sense of the second type multiplication exists not for every vector in V. Consequently, not for every vector of the space the division of the second type is defined.

References

1. Koliada, A.A.   Modular structures of the conveyer handling of digital information. Minsk, Universitetskoie, 1992.
2. Inutin, S.A. A Method of an inverse element computation in a finite field. *Scientific Works of the Surgut State University*, Surgut, S-S regional publishing house,1: 102-107, 1995.
3. Munro, I. The computational complexity of algebraic and numerical problems. *American Elseyier*, 7,1980.

Parallel Operating System for MPP System : Design and Implementation

Hae Jin Kim, Jae Kyung Lee, Kee Wook Rim

System S/W Section, ETRI, 161 Kajong-Dong, Yusong-Gu, Taejon 305-350, KOREA

Abstract. This paper presents the design and implementation of the microkernel based parallel operating system named MISIX for the highly parallel processing system called SPAX. First, we discuss the design specifications and practical design issues of MISIX. Then, we present the design and implementation of MISIX. Finally, we summerize the lessons we have learned throughout the design and implementations.

1 Introduction

Traditional bus-based symmetric multiprocessors(SMPs) are limited in the number of processors, memory, and I/O bandwidth that they can support. As client and server environments and downsizing technologies are prevailing, the demand for the development of high performance server has been accelerated. In result, powerful, modular, and scalable computing systems can be built using relatively inexpensive computing nodes coupled with high-speed interconnection networks. Such systems can take the form of loosely-coupled systems, built out of workstations[1], massively-parallel processing systems[2], or perhaps as a collection of small SMPs interconnected through a low-latency high-bandwidth network. With these kinds of high-speed and large-scale parallel computers, there is no doubt that *it is crucial to design and implement powerful parallel operating system* which works on top of them. Recently, some brand-new multicomputer systems such as Cray System T3D[3], Unisys OPUS[4], and IBM SP2[5] was released with their parallel operating systems. This paper present the development of the MISIX, microkernel based parallel operating system for highly parallel processing system called SPAX(Scalable Parallel Architecture computer based on X-bar network).

2 System Architecture

The SPAX[6] consists of *multiple clusters* which communicate with one another *via 2-level interconnection networks* (Xcent-Net). The Xcent-Net[7] provides dual paths, so that it can achieve high availability in presence of faulty path. Each cluster can have eight nodes, which can be classified into three categories: the Processing Node(PN), the Input/Output Node(ION), and the communication Connection Node(CCN). Basically, each node has the same basic hardware components called Main Processing Engine(MPE). As some special devices are attached, each node can be personalized into the PN's, the ION's and the CCN's.

The MPE implements *the concept of generalized node* from the hardware point of view. The MPE consists of three distinct parts: the CPU, which contains four Intel P6 microprocessors, and 256 KB or 512 KB of cache memory; the 1 GB of local shared memory; the XNIF, the network interface to the Xcent-Net; and the PCI slots. The system architecture shown in Figure 1 requires parallel operating system which exploits the massively-parallel characteristics of the underlying hardware system.

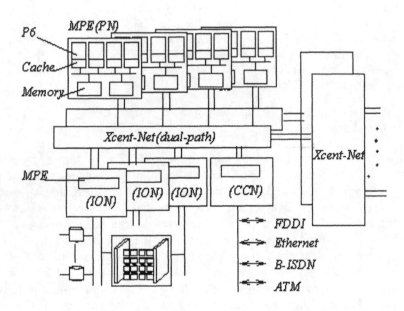

Fig. 1. SPAX system architecture

3 MISIX Requirements for SPAX

The requirements of the MISIX, the parallel operating system for the SPAX are as follows:

Performance. It should support *excellent scalability and high performance I/O throughput*, so that the target system can reach the designated 20 GIPS in overall system performance and the 10,000 tpmC(TPC-C) in database processing power.

Availability. It has to enhance system *reliability, fault-resilient capability, and serviceability*, so that it can satisfy maximum 99.99

Single System Image and Open System. It should support the single system image throughout the whole system and standard UNIX interfaces to users.

Parallel processing. It should provide *parallel programming model*, multithread model, and user-driven dynamic load balancing mechanism, so that it can support applications run in parallel.

4 Design Issues for MISIX

There exist several important issues in designing the MISIX, parallel operating system[8]. Belows are the list of them.

Microkernel and System Servers. Since the dawn of microcomputing, users and developers have argued with one other to defend the honor of their chosen operating systems. A microkernel[9] is a tiny operating system core that provides the foundation for modular, portable extensions. Recently, microkernels are *the central design element of new operating systems.* Every next-generation operating system will have one. However, there's plenty of disagreement about *how to organize operating system services relative to the microkernel.* Questions include how to overcome the performance degradation due to the message passing, how to design device drivers to get best performance while abstracting their functions from the underlying hardware, whether to run nonkernel operations in kernel or user space, and whether to keep existing legacy code or to throw everything away and start from scratch.

Single System Image (SSI). The fact that the computer is actually built out of multiple computing nodes should be invisible to the user. Therefore, the SSI issue is *how to provide the image of a single large UNIX operating system to the user and applications* while managing the separate, physically distributed system resources such as memory, disk, and other devices across node and clusters. This is one of the key technical issues of multicomputer operating system because SSI is directly coupled with ease of administration, load balancing which enhance overall system response time in times of heavy loading, and preserving standard APIs with existing applications.

Distribution of Resources. This issue is *how to exploit the parallelism and physically dispersed hardware in order to benefit users.* The operating system must be designed to interface with mass storage devices or computer networks and to enable scaling of input/output subsystem in order to meet the system's computing power.

High Availability (HA). The HA is related with *the mechanism to avoid a system crash in the presense of a single point of failure.* All mechanisms for fault detection, fault handling, and failover must be provided. The SSI and HA features are hard to be co-existent in one system by their nature.

5 The Design of MISIX Parallel Operating System

5.1 Goals

The prime goals we have pursued are to provide the microkernel based, scalable, open parallel operating system which has the single system image and fault-tolerant features. This goals are hard challenging. In order to meet the basic goals, we established our design direction : we do not follow the temptation of designing the MISIX from scratch; on the other hand, we adapt the state-of-the-art technologies which are widely accepted as the most stable and prevailing products, so that we can focus our strengths to develop advanced features of the MISIX.

5.2 MISIX Configurations

We have designed the MISIX operating system to have *highly flexible and configurable structure* to meet the various operating requirements. As such, each node in the system can be configured to be either a specialized node or a generalized node. *The specialized node configuration* makes a node act as one of the special nodes such as PN's, ION's and CCN's. The specialized node configuration of one cluster environment shown at Figure 2 distributes its functionalities across the different nodes throughout the whole system, depending on the roles of each node. This configuration is naturally suitable to support shared-disk model. In this configuration, the MISIX provides excellent scalability, and enhances system availability by placing important servers duplicated on the other nodes. *In case of generalized configuration*, each node can have almost the same role. Thus, it can be treated as just a general node. All component of the MISIX but some specific centralized servers are loaded in all the nodes. On the contrary to the specialized configuration, this configuration is well applicable to the shared-nothing model. When it comes to the parallel DBMS, MISIX guarantees the flexibility to support both the shared-disk model and the shared- nothing model. As mentioned above, the specialized configuration is related with the shared disk model while generalized configuration with the shared nothing model. Basically, we assume that the MISIX has the specialized configuration, which implies the shared disk model.

5.3 MISIX Microkernel

To provide the microkernel, we decided to *adapt the CHORUS microkernel*[10] developed by CHORUS System Inc., the world-wide leading provider of commercial microkernel-based operating system. The microkernel provides hardware resource management, process scheduling and process placement, virtual memory management, and message passing based interprocess communication and internode mechanism. With the original Chorus microkernel, we have made some modifications in a large portion of the source codes, so that it could properly operate on all kinds of nodes : PN's, ION's, and CCN's. The modification includes

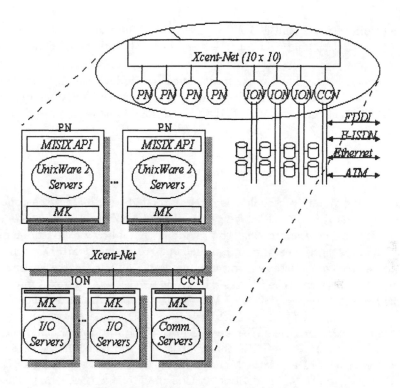

Fig. 2. Specialized node configuration

the kernel initialization, interprocess communication, the interrupt processing, the time of day clock, and the platform specific portions.

Kernel Initialization. The kernel initialization includes all the initializations regarding microkernel : kernel virtual address space management; Intel P6 internal resources such as cache and local Advanced Programmable Interrupt Controller(APIC); and MPE local resources such as the Orion PCI Bridge(OPB), the Orion Memory Controller(OMC), and the PCI. It also handles the activation of non-boot processors.

Interprocess Communication(IPC). The MISIX IPC contains both the interprocess communication among servers in a node through the local shared memory, and the *distributed interprocess communication among servers in different nodes through XNIF,* The IPC mechanism provides three types of messages : URGENT, CONTROL, and DATA. It also supports three ways of message transmission : poin-to-point, multicast, and the broadcast.

Interrupt Processing. It contains the assignment of the interrupt priority and the interrupt vector number, interrupt masking and un-masking, and the design of interrupt service routine. It also activates both the local APIC interrupts such as timer interrupt, inter-processor interrupt, and spurious interrupt which are coming from Intel P6, and I/O APIC interrupts from the over-heat, the system failures, the DUART interrupt, the XNIF interrupt, and the RTC interrupt.

Time Management. The time is maintained by the local APIC timer, and the timer is initialized to 100 Hz. The time-of-the-day clock(TODC) is maintained by DS1287 RTC chip, and the TODC in each node is synchronized by every minute.

5.4 MISIX System Servers

The operating system personality is directly related with maintaining open system and standard UNIX APIs. We *personalized UnixWare 2 as a variety of system servers on top of microkernel* to provide open application environments and make an advantage of maximum parallelism. We designed MISIX to interface with mass storage devices or computer networks and to enable scaling of input/output subsystem. The PN has a lot of system servers on top of microkernel such as Process Manager(PM), File Manager(FM), Path Name Manager(PNM), Coherency Manager(CM), DeadLock Detector(DLD), Device Port Server(DPS), Pipe Manager(PIPE), Stream Manager(STM), Fault Manager(FTM), and other servers. These servers gives UNIX personalities to users. The ION also has I/O related servers such as Disk Cache Manager(DCM), Fast Back-up Manager(FBM), Disk Driver Actor(DDA), and Back-up Console Manager(BCONM). The CCN has system servers like STREAMS Driver Actor(SDA), Boot Manager(BM), and Configuration Manager(CFGM). Consequently, *MISIX can be entitled as the multicomputer version of UnixWare2/MK.*

Single System Image. MISIX distributes functions over multiple nodes while maintaining a single system image. The single system image provides *facilities for transparent remote execution and remote file access.* MISIX provides name- space definitions for files, devices, and IPC mechanisms to maintain a single system image. Load balancing scheme and a shared-memory model across homogeneous processors is included to provide maximum use of distributed resources while maintaining SSI transparency and application semantics.

MISIX FM/DDA. We separated the DDA from the FM, and the SDA from the STM to get the best performance while abstracting their functions from the underlying hardware. Then, we developed the MISIX FM/DDA which can operate on the shared disk system environments using the following steps :
Stage 1 : We designed and implemented the Device Numbering Scheme(DNS) for SPAX disk devices. Because the FM/DDA in a node is limited to access its local disk only, a FM in the local node can't access either the DDA in the remote node directly nor the DDA through the FM in the remote node indirectly. Therefore, we implemented *indirect access scheme from the local FM to the remote DDA* shown at Figure 3. Please notice the root file system is in a left-side node.

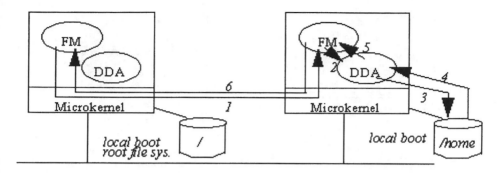

Fig. 3. Accessing the remote DDA through the remote FM

Stage 2 : We implemented *direct access mechanism from the local FM to the remote DDA* using the similar concept of remote mount, and thus removed the FM in the remote node. This remote node in the Figure 4 became very similar to the ION node in SPAX.

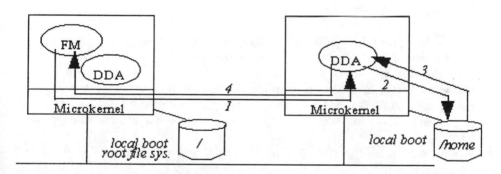

Fig. 4. Accessing the remote DDA directly

Stage 3: We implemented the *shared disk version of FM/DDA* by making the root file system on the remote node. Please notice that the location of the root file system changed to the remote node and the local FM can communicate with the remote DDA directly.

MISIX STM/SDA. We separated the SDA from the STM[11] to minimize the total number of the message traffics. The STM runs in PN, but the separated SDA runs in CCN. We introduced the *stub driver* to interface the the STM and the SDA, and implemented message handling interface according to the Inter-Actor Protocol(IAP).

MISIX API. It includes existing UnixWare 2 API, extended API for single system image, and extended API for direct microkernel invocation from application

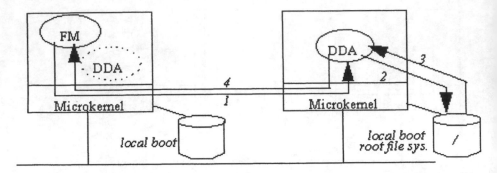

Fig. 5. Disk sharing by making root file system on the remote node

programs. The SSI API provides a few commands and libraries which supports migration, colocation and so on. The MK API gives various system call primitives based on the MISIX microkernel.

6   Implementation

For the implementation of the MISIX, we established the *testbed system environment* which simulates only one cluster of the SPAX system using four Compaq Proliant 4000 systems which were interconnected via commercial Ethernet. A Compaq system has exact resemblance with the one node in SPAX because it has four Intel P5 processors and incorporates SMP features. The prototype implementations have been done through the three steps in the testbed system.

First. We designed and implemented the emulator program[12] and the kernel message passing routines which controls the hardware module that is responsible for the message passing through Xcent-Net.

Second. We implemented the MISIX microkernel by adapting the kernel initialization, the IPC, the interrupt and timer processing.

Finally. We implemented the shared disk version of FM/DDA and separated the SDA from the STM.

In this prototype implementation, we found and resolved many kinds of bugs in MISIX. Those are some unexpected hang-ups in the middle of the boot, the synchronization failures between the master node and the slave node, the problems in the SSI and the multiprocessor capabilities.

7   Lessons from MISIX Implementation

From the development of MISIX, we learned some valuable lessons as follows.

First, there were significant performance degradations when we tried to execute the SMP application which was developed to be run in the single SMP platform using System V IPC mechanism upon the MISIX environment which does not provide the Distributed Shared Memory(DSM) features. The MISIX does not support the DSM facility. Instead, the MISIX provides very high-speed Message Passing Interface(MPI)[13] based on high bandwidth interconnection network.

Second, we decided the direction of MISIX development as the way to avoid the temptations of developing it from scratch. We could focus our *research activities on key technologies* by *keeping existing legacy code* instead of throwing everything away and starting from scratch.

Third, we learned that it is very important for a multicomputer operating system to have *flexible configuration* in related with supporting parallel DBMS. The MISIX can be easily configured for the parallel DBMS of shared nothing model as well as of shared disk model.

Fourth, we recognized that the *Single System Image* is the key feature of a distributed operating system.

Finally, the experience with the prototype leaded us to ensure that *there will be no significant performance differences between local call and remote call* because the Xcent-Net, our interconnection network provides very high transmission bandwidth for the message transfer. In summary, we implemented the MISIX operating system by using the kernel-level system severs instead of employing the user-level middleware.

8   Conclusions

We have described *the design and prototype implementation of the parallel operating system* for massively parallel processing computer which *requires the very advanced technologies and tremendous amount of works.* Throughout the implementation of prototype parallel operating system which supports single system image in the hierarchically clustered environments, we have experienced the possibility of using MPP system in the area of commercial enterprise computing such as on- line transaction processing. Besides, we have acquired the *proponent opinions for sustaining microkernel based parallel, multicomputer operating system approach* because it supports the single system image, high scalability and high level of parallelism. The High Availability (HA) features[14] are not yet fully implemented and those will be implemented in the future. At that time, MISIX will provide high availability by preventing the crash of the whole system using fault tolerant software when there occurs a failure. All mechanisms for fault detection, fault handling, and finally switching and takeover will be provided. Besides the HA, the dynamic load balancing scheme and the solution for providing a single IP address should be found in the near future.

Acknowledgements. We would like to acknowledge Joo Man Kim, Sung In Jung, Sun Hee Ahn, Sun Ja Kim, Young Jin Nam, Nak Joo Jeong, Jeong Nyeo Kim and Il Yeon Cho in the System Software Section for their contributions to the project.

References

1. Anderson, E., Culler, E., Patterson A.: A Case for NOW(Networks of Workstations). IEEE Micro, February (1995)
2. Unisys Corp.: Scalable Parallel Processor(SPP) Hardware Overview. White Paper 4125 3030-000, April (1995)
3. Oed, W.: The Cray Research Massively Parallel Processor System CRAY T3D. Technical Report, November (1993)
4. Unisys Corp.: Unisys SVR4/MK Operating System with Single System Image. White Paper 4125 2990-000, April (1995)
5. Snir, M., Hochschild, P., Frue, D., Gildea, K.: The Communication Software and Parallel Environment of the IBM SP2. IBM Systems Journal, Vol. 34, No. 2, (1995) 205-221
6. Yang-Woo K., Se-Wong O., Jin-Won P.: Design Issues and System Architecture of TICOM-IV, A Highly Parallel Commercial Computer. The 3rd Euromicro Workshop on Parallel and Distributed Processing (1995) 219-226
7. Sang-Man, M., Sang-Seok, S.: A Message Transfer Protocol for a Packet-Switched Interconnection Network. Proceedings of ITC-CSCC'96, Seoul, Korea, July (1996) 625-628
8. Schroder-Preikschat, W.: The Logical Design of Parallel Operating Systems. Prentice Hall, Englewood Cliffs, NJ (1994)
9. Peter D. Varhol .: Small Kernels Hit it Big. Special BYTE Report:Advanced Operating Systems, January (1994)
10. Rozier, M., et al: Overview of the CHORUS Distributed Operating Systems. Technical Report CS/TR-90-25.1, Chorus Systems (1991)
11. Young-Jin N., Joo-Man K.: The Design and Implementation of the Terminal Device Driver Server on top of a Microkernel for SPAX. Lecture Notes in Computer Science **1067** (1996) 1009-1010
12. Sung-In J., Jae-Kyung L., Hae-Jin K.: XNIFsoft: The Design and Implementation of Emulator for Crossbar Network Router in Cluster Systems. Proc. of PDPTA '96 (1996)
13. Message Passing Interface Forum: MPI: A Message-Passing Interface Standard. The MPI Forum, The University of Tennessee, Knoxville, Tennessee, May (1994)
14. Sunil, K., Douglas, S., Francois, A., Jim, L.: Fault Tolerance in a Distributed CHORUS/MiX System. USENIX Technical Conference, San Diego, January (1996) 219-228

New Graph Model and Its Routing Algorithm for Rearrangeable Networks *

Yanggon Kim[1] and Tse-yun Feng[2] **

[1] Virginia Commonwealth University, Richmond, VA 23284-2014, USA
[2] The Pennsylvania State University, University Park PA 16802, USA

Abstract. This paper introduces a new simple graphical representation, called an *N-leaf Dual Complete Binary Tree (N-leaf DCB-tree)*, for $(2\log_2 N - 1)$-stage networks. The N-leaf DCB-tree representation method has hierarchical and recursive characteristics. Due to the hierarchical property, it provides an analytical and systematic model for a multistage interconnection network with respect to a permutation realizability. Based on the recursive property of DCB-tree structure, we present a universal necessary and sufficient condition of a conflict-free connection pattern for N one-to-one simultaneous connection paths in multistage interconnection networks. Depending on the class of a given $(2\log_2 N - 1)$-stage network, this universal condition can be changed. Also this N-leaf DCB-tree model is shown to be very useful in comparing the permutation capabilities of various network topologies. The N-leaf DCB-tree model can convert a routing problem to a *conflict-free assignment problem* for N binary numbers where each binary number is a $(\log_2 N - 1)$-bit string. Converting to a conflict-free assignment problem may make it easier to attack the rearrangeablity of a $(2\log_2 N - 1)$-stage network in a class of Shuffle-Exchange equivalent netwrks (*E-class*).

1 Introduction

A class of interconnection networks, called permutation networks, can be represented by various theoretical models. A good model with good fundamental theorems is a useful tool to solve many problems related to permutation networks. Many good mathematical models for permutation networks have been developed for the special purpose : algebraic notation to verify the relationship among various networks [1], a set of algebraic tools to prove the equivalencein permutation capability for different networks [2], a functional f ormula to describe permutation networks and study the equivalence of networks [3], a simple directed graph to analyze and design the MINs [4], an orthogonal graph to define and analyze MINs [5], and a conflict graph to determine the optimal number of passes required to realize a permutation [6]. The conflict graph, in particular, is

---

* This work was partially supported by the National Science Foundation under Grant CDA-9320642.
** Presently at NSF/CISE/CDA, Arlington, VA 22230, on leave from The Pennsylvania State University.

useful to the conflict resolution problem, which can be formulated as the classical graph coloring problem in which one has to color the nodes of a graph with a minimum number of colors, so that no two adjacent nodes receive the same color. This problem has been shown to be NP-hard[7].

However, most of the proposed graphical models have been defined and represented only for the $(\log_2 N)$-stage cube-type networks. For $(2\log_2 N - 1)$-stage networks, we will introduce a new, simple graphical representation called *N-leaf Dual Complete Binary Tree(N-leaf DCB-tree)*, and will show its hierarchical properties for $(2\log_2 N - 1)$-stage networks. This graphical representation can be used for a MIN with the arbitrary number of stages. This DCB-tree model can transform the conflict-free routing problem into the *conflict-free assignment problem* on N $(n - 1)$-bit binary numbers where $n = \log_2 N$.

Furthermore, by using the hierarchical and systematic property of a tree structure, a universal necessary and sufficient condition for the conflict-free network connection patterns will be shown. This condition can apply to any $(2\log_2 N - 1)$-stage network. That is, this condition is applicable on a network in a class of Shuffle-Exchange equivalent networks ($E-class$) as well as in a class of Beneš-equivalent networks ($R-class$). The main contribution of N-leaf DCB-tree representation is to reduce a routing problem to a conflict-free assignment problem. According to a given permutation, the number of possible conflict-free network connection patterns in a $(2\log_2 N - 1)$-stage network varies.

2 An N-leaf Dual Complete Binary Tree(N-leaf DCB-tree)

2.1 Preliminary

An one-to-one connection path for a pair of input/output terminals in a *concatenated* $(2\log_2 N - 1)$-stage interconnection network can be determined by an $(n-1)$-bit binary number (T^i) which is a substring of its $(2n-1)$-bit self-routing tag. Let an overlapped $(2\log_2 N - 1)$-stage network be named by a *concatenated* $(2\log_2 N - 1)$-*stage network* and denoted by a $\Delta \oplus \Delta'$, which is constructed by merging two $(log_2 N)$-stage cube-type networks with overlapped center stages. Here, Δ and Δ' are called as a *left subnetwork* and a *right subnetwork*, respectively. To specify a path for a given pair of input/output terminals, T^i should be assigned as a binary number ranged between $\underbrace{000\cdots00}_{(n-1)\ bits}$ and $\underbrace{111\cdots11}_{(n-1)\ bits}$ inclusive.

Definition 1. *A conflict-free assignment of T*
Let a set $T = \{T^i \mid T^i = t_1^i t_2^i \cdots t_n^i,$ for $0 \le i \le (N-1)$ and $n = \log_2 N\}$ where N is a power of 2. If we assign an n-bit binary number to each T^i satisfying $T^i \ne T^j$ for $i \ne j$ where $0 \le i, j \le (N-1)$, then it is called a *conflict-free assignment*.

For example, let a set T have four 2-bit binary codes such as T^0, T^1, T^2, and T^3. If they are assigned as $T^0 = 10$, $T^1 = 11$, $T^2 = 00$, and $T^3 = 01$,

INPUT SIDE   OUTPUT SIDE

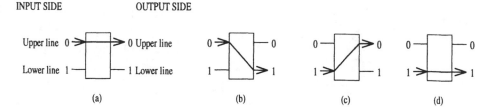

Fig. 1. Four connection patterns of 2×2 SE by s_β^α : (a) s_0^0 (b) s_1^0 (c) s_0^1 (d) s_1^1

then $T = \{00, 01, 10, 11\}$ is called a *conflict-free assignment*. Depending on the assignments of T^i for all i where $0 \leq i \leq (N-1)$, the corresponding connection pattern for a given permutation, P, is either conflict-free or not.

Since a 2×2 SE used in a multistage interconnection network has 2 input lines and 2 output lines, each SE can be set into one of the four connection patterns. If we consider an input line position as well as an output line position together, each path has four kinds of SE connection patterns. Those four possible connection patterns of an SE are defined as follows:

Definition 2. *Connection patterns of an SE for a path :* S_β^α
A connection pattern of an SE is denoted by s_β^α where α and β stand for an input line position and an output line position of an SE, respectively. If a path comes in through an upper input line of an SE, then $\alpha = 0$. Otherwise, $\alpha = 1$. For the output line position, the value of β is assigned in the same way as is the value of α.

Let α_i be an input line position of a 2×2 SE at the ith stage and β_i be an output position of a 2×2 SE at the ith stage where $1 \leq i \leq n$. With left-to-right routing direction, Fig. 1 shows four connection patterns of an SE, which are s_0^0, s_1^0, s_0^1, and s_1^1 as shown in Fig. 1(a), (b), (c), and (d), respectively.

2.2 The Structure of N-leaf DCB-tree

The structure of an N-leaf DCB-tree consists of two complete binary trees, T_{up} and T_{down}, merged back-to-back and both binary trees sharing N-leaf nodes, in which leaf nodes are labeled by v_0 to v_{N-1} as shown in Fig. 2. A left subnetwork, Δ, relates to an upper complete binary tree, T_{up}, and a right subnetwork, Δ', closely relates to a lower complete binary tree, T_{down}. Although a permutation (P) changes, an upper complete binary tree, T_{up}, in an N-leaf DCB-tree is not changed. The edge-connection pattern of T_{up} only depends upon a left subnetwork topology of a $(2 \log_2 N - 1)$-stage network. Due to the unique path property of cube-type networks, there are 2^k reachable input terminals. Then, each input terminal of those 2^k reachable input terminals can be connected to 2^{k-1} SEs at the kth-stage through a unique path. Because of the link pattern of cube-type networks, those input terminals have the same set of reachable SEs at the kth-stage. According to this SE-connectivity of cube-type networks, all

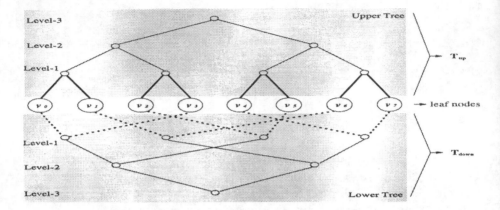

Fig. 2. An example of 8-leaf Dual Complete Binary Tree

of the kth-stage SEs would be decomposed into $N/2^k$ sets and all of the input terminals wo uld be grouped into $N/2^k$ sets also with respect to the decomposed sets of SEs at the kth-stage.

For examples, if a left subnetwork is Banyan, Baseline, Baseline^{-1}, or Flip, then $s_0, s_1, \cdots, s_{k-1}$ of input terminal labels should be don't care bits (x's). In other words, the labels of input terminals in a set can be denoted by one of $\underbrace{00\cdots0}_{(n-k)}\underbrace{xx\cdots x}_{k}$ to $\underbrace{11\cdots1}_{(n-k)}\underbrace{xx\cdots x}_{k}$. Therefore, for all of $N/2^k$ level-k internal nodes, N leaf-nodes are decomposed into $N/2^k$ sets disjointedly. Each set consists of 2^k input terminals whose labels satisfy $(n-k)$ bits in common, such as $\underbrace{00\cdots0}_{k bits}$ to $\underbrace{11\cdots1}_{k bits}$ exclusive of don't care bits.

In order to generalize this decomposition rule, we can generate the following formula:

A set of input terminals, IT_j, according to SE-decomposition at the kth stage

$$
\begin{cases}
\text{If } \Delta \in \{\text{Banyan, Baseline, Baseline}^{-1}, \text{Flip}\}, \text{ then} \\
\quad IT_j = \{S^i \mid S^i = \underbrace{s_l s_{l-1} \cdots s_k}_{n-k} \underbrace{xx\cdots x}_{k} \text{ and } s_l s_{l-1}\cdots s_k = j\} \\
\text{If } \Delta \in \{\text{Data Manipulator}\}, \text{ then} \\
\quad IT_j = \{S^i \mid \underbrace{xx\cdots x}_{k-1} \underbrace{s_{l-k+1} s_{l-k} \cdots s_1}_{n-k} x \text{ and } s_{l-k+1} s_{l-k} \cdots s_1 = j\} \\
\text{If } \Delta \in \{\text{Omega}\}, \text{ then} \\
\quad IT_j = \{S^i \mid \underbrace{xx\cdots x}_{k} \underbrace{s_{l-k} s_{l-k-1} \cdots s_0}_{n-k} \text{ and } s_{l-k} s_{l-k-1} \cdots s_0 = j\}
\end{cases}
\tag{1}
$$

where $0 \le j \le (N/2^k - 1)$.

From Equation 1 for all of the level-k internal nodes in Tup, the following

I. $\forall i, j$ such that $i \neq j$ and $0 \leq i, j \leq (N/2^k - 1)$: $IT_i \cap IT_j = \phi$ (2)

II. $\bigcup\limits_{j=0}^{N/2^k - 1} = \{S^0, S^1, \cdots, S^{N-1}\}$ (3)

(4)

By applying this decomposition rule from level-1 internal nodes to level-n internal node, we can build an upper tree, T_{up} recursively.

Similarly, we can construct a lower complete binary tree, T_{down}. Due to the SE-connectivity of cube-type networks, the $(2n - k)^{th}$-stage SEs and output terminals can be decomposed into $N/2^k$ disjointed sets. The SE-decomposition formula as follows:

A set of output terminals, OT_j, according to SE-decomposition at the $(2n - k)^{th}$ stage

$$
\begin{cases}
\text{If } \Delta' \in \{\text{Data Manipulator, Baseline, Baseline}^{-1}, \text{Omega}\}, \text{ then} \\
OT_j = \{D^i \mid D^i = \underbrace{d_l d_{l-1} \cdots d_k}_{n-k} \underbrace{\text{xx} \cdots \text{x}}_{k} \text{ and } d_l d_{l-1} \cdots d_k = j\} \\
\text{If } \Delta' \in \{\text{Banyan}\}, \text{ then} \\
OT_j = \{D^i \mid D^i = \underbrace{\text{xx} \cdots \text{x}}_{k-1} \underbrace{d_{l-k+1} d_{l-k} \cdots d_1}_{n-k} \text{x} \text{ and } d_{l-k+1} d_{l-k} \cdots d_1 = j\} \quad (5) \\
\text{If } \Delta' \in \{\text{Flip}\}, \text{ then} \\
OT_j = \{D^i \mid D^i = \underbrace{\text{xx} \cdots \text{x}}_{k} \underbrace{d_{l-k} d_{l-k-1} \cdots d_0}_{n-k} \text{ and } d_{l-k} d_{l-k-1} \cdots d_0 = j\}
\end{cases}
$$

where $0 \leq j \leq (N/2^k - 1)$.
From Equation 5 for all of the level-k internal nodes in T_{down}, the following equations are given:

I. $\forall i, j$ such that $i \neq j$ and $0 \leq i, j \leq (N/2^k - 1)$: $OT_i \cap OT_j = \phi$ (6)

II. $\bigcup\limits_{j=0}^{N/2^k - 1} = \{D^0, D^1, \cdots, D^{N-1}\}$ (7)

Therefore, for all of $N/2^k$ level-k internal nodes in T_{down}, N leaf-nodes are decomposed into $N/2^k$ sets disjointedly. Each set consists of 2^k output terminals whose labels satisfy $(n - k)$ bits in common such as $\underbrace{00 \cdots 0}_{k-bits}$ to $\underbrace{11 \cdots 1}_{k-bits}$ exclusive of don't care bits.

2.3 Construction of N-leaf DCB tree

We are concerned with the N-leaf DCB-tree constructing procedure described by using the formal and rigorous notations. For the sequence of columns of a given permutation to correspond with the sequence of leaf nodes, v_i is labeled by (S^i, D^i), in which $S^i = i$, i.e. each node is labeled by a pair of the ith column in Equation ??. An N-leaf DCB-tree related P is built as follows:

An N-leaf DCB-tree for a $(2 \log_2 N - 1)$-stage network, $\Delta \oplus \Delta'$
Begin

1. *Labeling of N-leaf nodes* :
   For each leaf node, v_i, assign a pair of input/output terminals, (S^i, D^i) where $S^i = i$ and $0 \le i \le (N-1)$.

2. *Upper complete binary tree, T_{up}* :

   Case 1 : $\Delta \in$ {Banyan, Baseline, Baseline^{-1}, Flip}
   For each internal node at level-k, $1 \le k \le l$,
   it has 2^k-leaf nodes satisfying $S^i = s^i_l s^i_{l-1} \cdots s^i_k xx \cdots x$,
   i.e., $s^i_l s^i_{l-1} \cdots s^i_k \in S^i$ to be identical.

   Case 2 : Δ is Data Manipulator
   For each internal node at level-k, $1 \le k \le l$,
   it has 2^k-leaf nodes satisfying $S^i = xx \cdots x s^i_{l-k+1} s^i_{l-k} \cdots s^i_1 x$,
   i.e., $s^i_{l-k+1} s^i_{l-k} \cdots s^i_1 \in S^i$ to be identical.

   Case 3 : Δ is an Omega network
   For each internal node at level-k, $1 \le k \le l$,
   it has 2^k-leaf nodes satisfying $S^i = xx \cdots x s^i_{l-k} s^i_{l-k-1} \cdots s^i_0$,
   i.e., $s^i_{l-k} s^i_{l-k-1} \cdots s^i_0 \in S^i$ to be identical.

3. *Lower complete binary tree, T_{down}* :

   Case 1 : $\Delta' \in$ {Data Manipulator, Baseline, Baseline^{-1}, Omega}
   For each internal node at level-k, $1 \le k \le l$,
   it has 2^k-leaf nodes satisfying $D^i = d^i_l d^i_{l-1} \cdots d^i_k xx \cdots x$,
   i.e., $d^i_l d^i_{l-1} \cdots d^i_k \in D^i$ to be identical.

   Case 2 : Δ' is Banyan
   For each internal node at level-k, $1 \le k \le l$,
   it has 2^k-leaf nodes satisfying $D^i = xx \cdots x d^i_{l-k+1} d^i_{l-k} \cdots d^i_1 x$,
   i.e., $d^i_{l-k+1} d^i_{l-k} \cdots d^i_1 \in D^i$ to be identical.

   Case 3 : Δ' is a Flip network
   For each internal node at level-k, $1 \le k \le l$,
   it has 2^k-leaf nodes satisfying $D^i = xx \cdots x d^i_{l-k} d^i_{l-k-1} \cdots d^i_0$,
   i.e., $d^i_{l-k} d^i_{l-k-1} \cdots d^i_0 \in D^i$ to be identical.

4. *Create a root node at level-n for T_{up} and T_{down}* :
   For T_{up}, T_{down}, create a root node at level-n and connect it to two internal nodes at level-l.

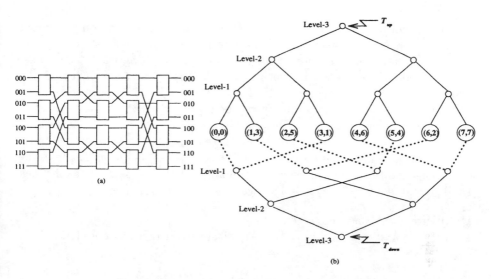

Fig. 3. (a) 8×8 5-stage Baseline $\oplus$ Banyan network (b) its corresponding 8-leaf DCB-tree

End

Fig. 3(a) illustrates a 5-stage network, $\Delta \oplus \Delta'$ =Baseline $\oplus$ Banyan and (b) shows the corresponding DCB-tree for a given permutation, $P = \begin{pmatrix} 0\ 1\ 2\ 3\ 4\ 5\ 6\ 7 \\ 0\ 3\ 5\ 1\ 6\ 4\ 2\ 7 \end{pmatrix}$. In this 8-leaf DCB-tree, by Equation 1, any two leaf nodes connecting to a level-1 internal node in T_{up} have the identical binary strings except for the last bits of S^i, such as 00x, 01x,10x, or 11x. Similarly, by Equation 5, a level-1 internal node in T_{down} connects two leaf nodes whose two leftmost bits of D^i are identical, such as 00x, 001, 10x or 11x.

According to the recursive property, the level-2 internal node in T_{up} has two level-1 internal nodes as its children, and its four leaf nodes are 0xx or 1xx for S^i. In like manner, the level-2 internal node in T_{down} has four leaf nodes whose D^is are 0xx or 1xx. Therefore, for a given $(2 \log_2 N - 1)$-stage network, only the edge-connection pattern between level-1 internal nodes and leaf nodes in T_{down} drawn by the dotted lines in Fig. 3(a), can be changed as a permutation is changed. The other edge-connection patterns are fixed.

3 N-leaf DCB-tree in N simultaneous connections

To reduce an N simultaneous routing problem in a $(2 \log_2 N - 1)$-stage interconnection network to a conflict-free assignment problem of an N-leaf DCB-tree for a given permutation, three following reduction rules should be satisfied. First, a given permutation should be mapped to an N-leaf DCB-tree for a given $(2 \log_2 N - 1)$-stage interconnection network. Second, all the possible alternative one-to-one connection paths for a given pair of input/output terminals should be specifiable and distinguishable. Finally, there should exist the necessary and sufficient condition for the N-leaf DCB-tree of a given permutation to detect a conflict on an N simultaneous routing connection pattern.

In order to denote a selected connection path of (S^i, D^i), let T^i be assigned to a leaf node of a DCB-tree in which $T^i = t_1^i t_2^i \cdots t_{n-1}^i$ is a l-bit binary variable and $(S^i, D^i) = (i, D^i)$. Thus, if all of the T^is are given, its network connection pattern for a given permutation can be produced in a $(2 \log_2 N - 1)$-stage interconnection network. Next, we propose the necessary and sufficient condition for a conflict-free routing problem of a given permutation to reduce to the conflict-free assignment problem in a DCB-tree without its proof. The following definitions and theorems will lead to our main results on reduction to an N-leaf DCB-tree model for $(2 \log_2 N - 1)$-stage interconnection networks.

Definition 3. *A substring of T^i*
Let $T^i(a : b) = t_a^i t_{a+1}^i \cdots t_b^i$ be a $(b-a+1)$-bit substring of a T^i where $1 \leq a, b \leq n$ and $a \leq b$.

Definition 4. $T(up, k, m)$, $T(down, k, m)$, *and* $T'(down, k, m)$

- $T(up, k, m) = \{T^i(1 : k) \mid T^i \in$ leaf nodes of a level-k internal node of $T_{up}\}$ be a set of the leftmost k-bit substrings of T^is which are leaf nodes of a level-k internal node in an upper binary complete tree, T_{up}, of an N-leaf DCB-tree.

- $T(down, k, m) = \{T^i(1 : k) \mid T^i \in$ leaf nodes of a level-k internal node of $T_{down}\}$ be a set of the leftmost k-bit substrings of T^is which are leaf nodes of a level-k internal node in a lower binary complete tree, T_{up}, of an N-leaf DCB-tree.

- $T'(down, k, m) = \{T^i(n-k : n-1) \mid T^i \in$ leaf nodes of a level-k internal node of $T_{down}\}$ be a set of the rightmost k-bit substrings of T^is which are leaf nodes of a level-k internal node in a lower binary complete tree, T_{down}, of an N-leaf DCB-tree.

where m is the index of a level-k internal node and $1 \leq m \leq \frac{N}{2^k}$.

For a given permutation, P, to be realizable without conflict in a network, the N-leaf DCB-tree should satisfy the following necessary and sufficient condition depending on the class of the $(2 \log_2 N - 1)$-stage interconnection network, such as *E-class* or *R-class*. These theorems reduce a routing problem to a conflict-free assignment problem. The proofs of the following theorems are given in [8].

Theorem 5 (Conflict-free connection patterns for R-class networks). *Suppose* $\Delta \oplus \Delta' \in$ *R-class. If and only if for all k, m, $1 \leq k \leq (n-1)$ and $1 \leq m \leq \frac{N}{2^k}$, $T(up, k, m)$ and $T(down, k, m)$ satisfy the conflict-free assignment conditions, its connection pattern for a given permutation, P, of a $\Delta \oplus \Delta' \in$ R-class is realized without conflict.*

Theorem 6 (Conflict-free connection patterns for an E-class network). *If and only if for all $1 \leq k \leq (n-1)$ and $1 \leq m \leq \frac{N}{2^k}$, $T(up, k, m)$ and $T'(down, k, m)$ satisfy the conflict-free assignment conditions, its connection pattern for a given permutation, P, is realized without conflict.*

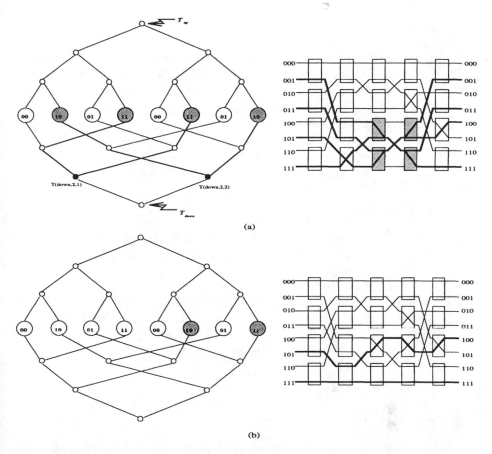

(a)

(b)

Fig. 4. (a) an example of an 8-leaf DCB-tree for Baseline ⊕ Banyan with conflict assignments (b) its 8-leaf DCB-tree with conflict-free assignments and its connection pattern.

As shown in Fig. 4(a), this network connection pattern for $P = \begin{pmatrix} 0\ 1\ 2\ 3\ 4\ 5\ 6\ 7 \\ 0\ 3\ 5\ 1\ 6\ 4\ 2\ 7 \end{pmatrix}$, has conflicts at SE-3 and SE-4 of the stage-3 in 8×8 5-stage Baseline ⊕ Banyan, in which two input lines of SE are sharing one output line of SE. Because of Baseline ⊕ Banyan ∈ R-*class*, Theorem 5 should be valid on this 8-leaf DCB-tree for a conflict-free connection pattern.

However, in terms of two level-2 internal nodes of T_{down} in Fig. 4(a), $T(down, 2, 1)$ and $T(down, 2, 2)$ are not assigned without conflicts, i.e., $T(down, 2, 1) = \{T^0 = 00, T^3 = 11, T^2 = 01, T^5 = 11\}$ and $T(down, 2, 2) = \{T^1 = 10, T^6 = 01, T^4 = 00, T^7 = 10\}$. Due to $T^3 = T^5$ and $T^1 = T^7$, two paths for $(1,3)$ and $(7,7)$ cause conflict as well as two paths for $(3,1)$ and $(5,4)$. Resulting from these conflicts, the lower input line of SE-2 at stage-4 is shared with two paths of $(1,3)$ and $(7,7)$. In the same way, two paths of $(3,1)$ and $(5,4)$ should share the upper input line of SE-3 at stage-4. These two SEs are drawn by the shaded boxes of the 8-leaf

DCB-tree shown in Fig. 4(a). Thus, if we re-assign T^i for $0 \leq i \leq 7$ satisfying the conflict-free necessary and sufficient condition in Theorem 5, we can generate the 8-DCB-tree with a conflict-free assignment and its network connection pattern as shown in Fig. 4(b). As can be seen in the 8-leaf DCB-tree of Fig. 4(b), T^5 and T^7 are re-assigned as 10 and 11 respectively. Re-assigning these two numbers results in the conflict-free assignments for all the internal nodes, i.e. it satisfies the necessary and sufficient condition for conflict-free connection pattern. To verify this condition, we can generate a connection pattern shown in Fig. 4(b) with only T^i values. Here, the re-assignments of T^5 and T^7 establish their alternative paths and they result in avoiding the previous conflicts at stage-4.

4 Conclusion

In this paper, the novel graphical representation, called an N-leaf DCB-tree, has been introduced and its hierarchical and recursive characteristics have been discussed. Then, the universal necessary and sufficient condition for a conflict-free connection pattern in a $(2 \log_2 N - 1)$-stage interconnection network for a given permutation has been proposed. Based on the tree structure of a N-leaf DCB-tree, a routing problem and a rearrangeability problem were reduced to a conflict-free assignment problem for N $(n - 1)$-bit binary numbers. To reduce a routing problem to a conflict-free assignment problem may make it easier to attack the rearrangeablity of a $(2 \log_2 N - 1)$-stage network in a E-class which has not been solved yet.

References

1. Wu, C.L., Feng, T.: On a Class of Multistage Interconnection Networks. IEEE Trans. Comput. **29** (1980) 694–702
2. Parker, D.S.: Notes on Shuffle/Exchange-Type Switching Networks. IEEE Trans. Comput. **29** (1980) 213–222
3. Pradhan, D.K., Kodandapani, K.L.: A Uniform Representation of Single and Multistage Interconnection Networks Used in SIMD Machine. IEEE Trans. Comput. **29** (1980) 777–790
4. Agrawal, D.P.: Graph Theoretical Analysis and Design of Multistage Interconnection Networks. IEEE Trans. Comput. **32** (1983) 637–648
5. Scherson, I.D.: Orthogonal Graphs and the Analysis and Construction of a Class of Multistage Interconnection Networks. Proc. of Int. Conf. on Parallel Processing (1990) 380–387
6. Das, N, Bhattacharya, B.B., Dattagupta, J.: Isomorphism of Conflict Graphs in Multistage Interconnection Networks and Its Application to Optimal Routing. IEEE Trans. Comput. **42** (1993) 665–677
7. Bernhard, P.J., Rosenkrantz, D.J.: An Efficient Method for Representing and Transmitting Message Patterns on Multiprocessor Interconnection Networks. IEEE Trans. Parallel and Distributed System **11** (1991) 72–85
8. Kim, Y., Feng, T.: A New Tag Schem and Its Tree Representation For A Shuffle-Exchange Network. Proc. of Int. Conf. on Parallel Processing (1994) 109–112

Modified Dorn's Algorithm with Improved Speed-Up

Ayse Kiper

Department of Computer Engineering, Middle East Technical University
06531 Ankara-TURKEY

Fax : (+90) 312 210 1259, Tel : (+90) 312 210 5594

ayse@rorqual.cc.metu.edu.tr

Abstract. The polynomial evaluation is one of the most common standard problems and has attracted the attention of researches for many years. In this study a brief up-to-date survey of existing parallel methods is given and the possibility of improving the parallelisation of Dorn's algorithm [1] (which is the parallel implementation of the well known Horner's method) is studied. In a previous work [5] the decoupling algorithm [6] for solving bidiagonal systems has been simplified, modified and applied to parallelise Horner's method. This paper is the extension and the generalisation of the approach followed in [5] for Dorn's method by reformulating as a set of independent matrix equations with special bidiagonal coefficient matrices. These independent equations are solved in parallel leading to improvement in the speed-up of the algorithm. Performance parameters the speed-up and the efficiency expressions are given and a comparative analysis of the algorithm with the methods studied in [1] and [5] are presented in a table in terms of the total number of unit time steps, the number of processors used and the degree of polynomials.

1. Introduction

The polynomial evaluation is one of the most common standard problems and has attracted the attention of researchers. An investigation of parallel algorithms for polynomial evaluation has been given by Munro and Paterson [8]. The paper includes the summary and discussion of the algorithms proposed by Estrin [2], Dorn [1] and analysis of a model of arithmetic computation permitting parallelism. The method of binary splitting based on divide and conquer strategy is due to Estrin [2]. Method of Muraoka [9] is sometimes called folding method and is based on a splitting by golden ratio. A fundamental result for a lower bound on time required to evaluate a polynomial under unbounded parallelism is given by Kosaraju [4]. Hyfil and Kung [3] presents a class of algorithms called reduction algorithms. Lakshmivarahan and Dhall [7] have extensively studied and given an up-to-date survey of parallel polynomial algorithms in their book.

Horner's rule is essentially a sequential method to calculate the value of a polynomial

$$p(x) = a_0 + a_1 x + a_2 x^2 + \ldots + a_n x^n \qquad (1)$$

in 2n (n additions and n multiplications) sequential steps which requires the successive computations

$$\left.\begin{array}{l} b_n = a_n \\ b_i = a_i + b_{i+1}x \end{array}\right\} \quad i = \text{n-1, n-2, ...,1,0} \qquad (2)$$

Then $p(x) = b_0$.

Dorn [1] represents kth-order Horner's method which allows parallel implementation. Dorn's formulation for a polynomial of degree n and k (>1) number of processors is

$$p(x) = p_0(x^k) + xp_1(x^k) + x^2 p_2(x^k) + ... + x^{k-1} p_{k-1}(x^k) \qquad (3)$$

where

$$p_i(x^k) = a_i + a_{i+k}x^k + a_{i+2k}x^{2k} + ... + a_{i+\lfloor n/k \rfloor k}x^{\lfloor n/k \rfloor k},$$
$$i = 0, 1, 2, ..., k-1 \qquad (4)$$

Dorn's parallel algorithm of (4) suggests a three stage process:

(i) Compute $x, x^2, ..., x^k$ using k/2 number of processors in $\lceil \log k \rceil$ unit times,

(ii) Compute $p_i(x^k)$ for $i = 0, 1, 2, ..., k-1$ using Horner's method in parallel with k processors in at most $2\lfloor n/k \rfloor$ unit time steps,

(iii) Multiply $p_i(x^k)$ by x^i for $i = 0, 1, 2, ..., k-1$ in one step and add them up by fan-in algorithm in $\lceil \log k \rceil$ steps.

Then the total time to compute p(x) of degree n using k processors is

$$T(n,k) = 2\lceil \log k \rceil + 2\lfloor n/k \rfloor + 1 \qquad (5)$$

In this paper the decoupling algorithm (hereafter will be called DECAL) presented by Kowalik and Kumar [6] will be simplified, modified and used to propose a further parallelisation of Dorn's method.

2. Matrix Equation Formulation of Dorn's Method

Evaluation of k number of subpolynomials $p_i(x^k)$ of Equation (4) by Horner's method represents recurrence system with r = (n+1)/k

$$b_r^{(i)} = a_r^{(i)} = a_{i+\lfloor n/k \rfloor k}$$

$$b_{r-1}^{(i)} = a_{r-1}^{(i)} + x^k b_r^{(i)} = a_{i+(\lfloor n/k \rfloor - 1)} + x^k b_r^{(i)}$$

$$\begin{array}{c} . \\ . \end{array} \qquad (6)$$

$$b_1^{(i)} = a_1^{(i)} + x^k b_2^{(i)} = a_i + x^k b_2^{(i)}$$

$p_i(x^k)$ for i =0,1,2,.....,k-1 are evaluated independently in parallel as

$$p_i(x^k) = b_1^{(i)} \text{ , for } i = 0,1,2,.....,k-1 \tag{7}$$

Equation (7) for each i can be represented as a matrix equation of the form

$$\tilde{b}^{(i)} = A\tilde{b}^{(i)} + \tilde{a}^{(i)} \tag{8}$$

where

$$A = \begin{bmatrix} 0 & & & & & \\ x^k & 0 & & & & \\ & x^k & 0 & & & \\ & & & \cdot & \cdot & \\ & & & & \cdot & \cdot \\ & & & & \cdot & \cdot \\ & & & & x^k & 0 \end{bmatrix} \quad \tilde{b}^{(i)} = \begin{bmatrix} b_r^{(i)} \\ b_{r-1}^{(i)} \\ \cdot \\ \cdot \\ \cdot \\ b_2^{(i)} \\ b_1^{(i)} \end{bmatrix} \quad \tilde{a}^{(i)} = \begin{bmatrix} a_r^{(i)} \\ a_{r-1}^{(i)} \\ \cdot \\ \cdot \\ \cdot \\ a_2^{(i)} \\ a_1^{(i)} \end{bmatrix} \tag{9}$$

Equation (8) yields the matrix form

$$D\tilde{b}^{(i)} = \tilde{a}^{(i)} \text{ , i = 0,1,2,....,k-1} \tag{10}$$

where D is an rxr matrix equation of the structure

$$D = \begin{bmatrix} 1 & & & & & \\ -x^k & 1 & & & & \\ & -x^k & 1 & & & \\ & & & \cdot & \cdot & \\ & & & & \cdot & \cdot \\ & & & & \cdot & \cdot \\ & & & & -x^k & 1 \end{bmatrix} \tag{11}$$

Following the method of [5] , further parallelisation is possible for the solution of each matrix equation of (10) using DECAL.

3. Decoupling Algorithm (DECAL)

The method of Kowalik and Kumar [6] allows parallel solution of (10) for each i using p number of processors much smaller than the number of equations. Method has been described for a first order, linear recurrence system and has been extended to solve a single set of tridiagonal equations.

The solution of the bidiagonal system $E x = r$

$$
\begin{bmatrix}
1 & & & & & \\
e_2 & 1 & & & & \\
 & e_3 & 1 & & & \\
 & & . & . & & \\
 & & & . & . & \\
 & & & & . & . \\
 & & & & e_n & 1
\end{bmatrix}
\begin{bmatrix}
x_1 \\ x_2 \\ x_3 \\ . \\ . \\ . \\ x_n
\end{bmatrix}
=
\begin{bmatrix}
r_1 \\ r_2 \\ r_3 \\ . \\ . \\ . \\ r_n
\end{bmatrix}
\tag{12}
$$

is completed in three stages.

(i)   Elimination Stage. The matrix system (12) is restructured. If $s = n/p$ then

Processor i , $1 \le i \le p$

$$f_{(i-1)s+1} = e_{(i-1)s+1}, \text{ not computed for i=1}$$

for $j = (i-1)s+2$ to is do

    begin

        $f_j = -e_j f_{j-1}$, not computed for $i = 1$

        $r_j = r_j - e_j r_{j-1}$

    end

The new coefficient matrix is almost diagonalised but new non zero elements have been created and is in the form of Equation (7).

$$
\begin{bmatrix}
1 & & & & & & \\
& 1 & & & & & \\
& f_{s+1} & 1 & & & & \\
& & & \ddots & & & \\
& f_{2s} & & 1 & & & \\
& & & & \ddots & & \\
& & & & \ddots & & \\
& & & f_{n-(s-1)} & & 1 & \\
& & & & & & \ddots \\
& & & f_n & & & 1
\end{bmatrix}
\begin{bmatrix}
x_1 \\ \cdot \\ x_s \\ x_{s+1} \\ \cdot \\ x_{2s} \\ \cdot \\ \cdot \\ \cdot \\ x_{n-(s-1)} \\ \cdot \\ x_n
\end{bmatrix}
=
\begin{bmatrix}
r_1 \\ \cdot \\ r'_s \\ r'_{s+1} \\ \cdot \\ r'_{2s} \\ \cdot \\ \cdot \\ \cdot \\ r_{n-(s-1)} \\ \cdot \\ r_n
\end{bmatrix}
$$

$$(13)$$

(ii) Decoupling Stage. Is the decoupling of the system by finding the values of $x_s, x_{2s}, \ldots, x_{n-ks}, x_n$. The solution process is sequential and obtained by

$$x_s = r'_s$$
$$f_j x_{j-s} + x_j = r'_j, j = 2s, 3s, \ldots, ps$$

(iii) Solution Stage. The decoupled system now can be solved in parallel on $(p-1)$ processors as

  Processor i, $2 \le i \le p$

   for $j = (i-1)s+1$ to $is-1$ do

    $$x_j = r'_j - f_j x_{(i-1)s}$$

4. Dorn's Method Using DECAL

The first two stages of the three stage algorithm DECAL is satisfactory to evaluate a polynomial. Matrix D in Equation (10) is a special case of matrix E of Equation (12) such that all subdiagonal elements $d_{i+1,i}$ (i–1,2,...,i-1) are equal to x^k. Assuming

that p $(p_1, p_2, \ldots, p_p)$ number of processors is available for each subpolynomial, than the matrix D of Equation (10) can be partitioned into subsystems of size $m = (n+1)/kp$. The elimination stage of DECAL is performed in parallel on p processors for each subpolynomial as

$$d_{i,i-1} \qquad\qquad on\ p_1$$
$$d_{i+m,(i-1)+m} \qquad on\ p_2$$
$$\left.\begin{array}{l} d_{i,i-1} \qquad\qquad on\ p_1 \\ d_{i+m,(i-1)+m} \quad on\ p_2 \\ \quad . \qquad\qquad . \\ \quad . \qquad\qquad . \\ \quad . \qquad\qquad . \\ d_{i+(p-1),(i-1)+(p-1)m} \quad on\ p_n \end{array}\right\} i = 2,3,....,m \qquad (14)$$

The resulting system is given by Equation (15)

$$\begin{bmatrix} 1 & & & & & & & \\ & . & & & & & & \\ & & 1 & & & & & \\ & & -x^k & 1 & & & & \\ & & -x^{2k} & . & & & & \\ & & . & & . & & & \\ & & -x^m & & 1 & & & \\ & & & & . & & & \\ & & & & & . & & \\ & & & & & . & & \\ & & & & -x^k & 1 & & \\ & & & & -x^{2k} & . & & \\ & & & & . & & . & \\ & & & & -x^m & & 1 \end{bmatrix} \begin{bmatrix} b_r^{(i)} \\ b_{r-1}^{(i)} \\ . \\ b_{r-(m-1)}^{(i)} \\ b_{r-m}^{(i)} \\ b_{r-(m+1)}^{(i)} \\ . \\ b_{r-(2m-1)}^{(i)} \\ b_{r-2m}^{(i)} \\ . \\ . \\ b_{m+1}^{(i)} \\ b_m^{(i)} \\ b_{m-1}^{(i)} \\ . \\ b_1^{(i)} \end{bmatrix} = \begin{bmatrix} a_r^{(i)} \\ a_{r-1}^{\prime(i)} \\ . \\ a_{r-(m-1)}^{\prime(i)} \\ a_{r-m}^{(i)} \\ a_{r-(m+1)}^{\prime(i)} \\ . \\ a_{r-(2m-1)}^{\prime(i)} \\ a_{r-2m}^{(i)} \\ . \\ . \\ a_{m+1}^{\prime(i)} \\ a_m^{(i)} \\ a_{m-1}^{\prime(i)} \\ . \\ a_1^{\prime(i)} \end{bmatrix} \qquad (15)$$

with new created non-zero elements

$(x^k)^2, (x^k)^3, ..., (x^k)^m$ which are computed in

$$T_{el_1} = \lceil \log k \rceil + \lceil \log m \rceil \qquad (16)$$

steps using the first k/2, then m/2 number of processors. For k-subpolynomials kp number of processors in parallel completes the elimination process. The new right-hand-side vector a elements are created for i=0,1,2,...,k-1 and j= 2,3,...,m-1 as

$$\left.\begin{array}{l} a'^{(i)}_{r-1} = a^{(i)}_{r-1} + x^k a^{(i)}_r \\[2mm] a'^{(i)}_{r-j} = a^{(i)}_{r-j} + x^k a'^{(i)}_{r-(j-1)} \end{array}\right\} on\ \ p_1$$

$$\left.\begin{array}{l} a'^{(i)}_{r-(m+1)} = a^{(i)}_{r-(m+1)} + x^k a^{(i)}_{r-m} \\[2mm] a'^{(i)}_{r-(m+j)} = a^{(i)}_{r-(m+j)} + x^k a'^{(i)}_{r-(m+j-1)} \end{array}\right\} on\ \ p_2$$

$$.$$
$$.$$
$$.$$

$$\qquad\qquad\qquad\qquad\qquad\qquad\qquad\qquad\qquad\qquad (17)$$

$$\left.\begin{array}{l} a'^{(i)}_{r-(r-m+1)} = a^{(i)}_{r-(r-m+1)} + x^k a^{(i)}_{r-(r-m)} \\[2mm] a'^{(i)}_{r-(r-m+j)} = a^{(i)}_{r-(r-m+j)} + x^k a'^{(i)}_{r-(r-m+j-1)} \end{array}\right\} on\ \ p_p$$

The right-hand-side vector elements corresponding to the first equation of each subblock of Equation (15) does not alter during the elimination process. Therefore using kp number of processors, only kp(m-1) new right-hand-side vector elements are created and required total number of parallel operations is

$$T_{el_2} = 2(m-1) \qquad\qquad\qquad\qquad\qquad\qquad\qquad (18)$$

Each elimination requires two sequential arithmetic operations consisting of one multiplication and one addition. The total parallel steps for elimination stage is

$$T_{el} = T_{el_1} + T_{el_2} = \lceil \log k \rceil + \lceil \log m \rceil + 2(m-1) \qquad\qquad (19)$$

In Stage (ii) , each of the k subpolynomials presented in the matrix form (15) is partitioned into p number of subsystems which can be solved independently. The decoupling operation is done by determining the last unknown element of subblocks except the first which does not need any computation. From Equation (15)

$$b^{(i)}_{r-(m-1)} = a'^{(i)}_{r-(m-1)} \qquad\qquad\qquad\qquad\qquad\qquad (20)$$

others are computed using the recursion

$$b^{(i)}_{r-(mj-1)} = a'^{(i)}_{r-(mj-1)} + x^{km} b^{(i)}_{r-[m(j-1)-1]}\ \ j = 2,3,.......,p \qquad (21)$$

Equation (21) has a sequential character, but can be parallelised by reformulating as a multiplication of series of (p-1) number of 2x2 matrices and the initial vector as

$$\begin{bmatrix} b^{(i)}_1 \\ 1 \end{bmatrix} = \begin{bmatrix} x^{km} & a'^{(i)}_1 \\ 0 & 1 \end{bmatrix} \begin{bmatrix} x^{km} & a'^{(i)}_{m+1} \\ 0 & 1 \end{bmatrix} .... \begin{bmatrix} x^{km} & a'^{(i)}_{r-(m-1)} \\ 0 & 1 \end{bmatrix} \begin{bmatrix} b^{(i)}_{r-(m-1)} \\ 1 \end{bmatrix} \qquad (22)$$

for i = 0,1,2,...,k-1 in parallel.

(p-1) matrices can be multiplied up by recursive doubling in $\lceil \log(p-1) \rceil$ steps. Resulting matrices obtained by multiplications are of the same structure as the initial ones, so that they are suitable for parallel evaluation. Each matrix multiplication step requires two sequential steps in parallel. Computation of Stage (ii) is completed in

$$T = 2\lceil \log(p-1) \rceil + 2 \tag{23}$$

parallel steps using k (p-1) processors for k subpolynomials.

The last Stage (iii) is the evaluation of

$$p(x) = b_1^{(0)} + x b_1^{(1)} + x^2 b_1^{(2)} + \ldots + x^{k-1} b_1^{(k-1)} \tag{24}$$

Multiplications of $b_1^{(i)}$ by x^i for i = 0,1,2,......,k-1 is performed in one step.

Additions in Equation (24) is completed by fan-in algorithm in $\lceil \log k \rceil$ steps. The total number of parallel steps to evaluate a polynomial of degree n using Dorn's method with DECAL is

$$T(n, kp) = 2\lceil \log k \rceil + 2\lceil \log(p-1) \rceil + \lceil \log m \rceil + 2m + 1 \tag{25}$$

5. Performance of the Method

The speed-up and efficiency of the Dorn's algorithm using DECAL can be used to analyse and compare the method with the well known polynomial evaluation methods. The speed-up of the method over the sequential Horner's method and the efficiency can be given successively as

$$S_{kp} = \frac{2n}{2\lceil \log k \rceil + 2\lceil \log(p-1) \rceil + \lceil \log m \rceil + 2m + 1} \tag{26}$$

and

$$E_{kp} = S_{kp} / kp \tag{27}$$

6. Results and Discussion

Dorn's algorithm with DECAL can be summarised in three steps:

(i) Compute $(x^k), (x^k)^2, (x^k)^3, \ldots, (x^k)^m$ in

$$T = \lceil \log k \rceil + \lceil \log m \rceil \quad T = \lceil \log k \rceil + \lceil \log m \rceil$$

steps using k/2 and m/2 number of processors successively.

(ii) Compute $p_i(x^k)$ for i = 0,1,...,k-1 in parallel and each individually parallelised using DECAL so that it is completed

$$T = 2(m-1) + 2\lceil \log(p-1)\rceil + 2$$

steps using kp number of processors.

(iii) Multiply $p_i(x^k)$ by x^i for i = 0,1,2,...,k-1 in one step and add them up by fan-in algorithm in $\lceil \log k\rceil + 1$ steps using k/2 number of processors.

Comparison of the two algorithms , namely Dorn's and Dorn's with DECAL shows that difference in parallel evaluation steps T originates from the computation of subpolynomials $p_i(x^k)$. This yields the comparison of the partial expressions of T's for these algorithms

for Dorn's : $2\lfloor n / k\rfloor$

for Dorn's with DECAL : $\lceil \log m\rceil + 2\lceil \log(p-1)\rceil + 2m$

n	k	* Dorn's alg.	*** Dorn's alg. with DECAL			** Horner's alg. with
			p=2	p=4	p=8	DECAL
15	2	17	13	12	-	19
	4	11	10	-	-	14
31	2	33	22	17	14	36
	4	19	15	14	-	23
	8	13	12	-	-	16
63	2	65	39	26	19	68
	4	35	24	19	16	40
	8	21	17	16	-	25
	16	15	14	-	-	18
255	2	257	137	77	45	263
	4	131	74	45	30	138
	8	69	43	30	23	75
	16	39	28	23	20	44

Table 1. Some representative values of T(n,p) for the algorithms of Dorn's, Dorn's with DECAL and Horner's with DECAL. Where p represents the number of processors used.

* $T(n,k) = 2\lceil \log k\rceil + 2\lfloor n / k\rfloor + 1$, [7]

** $T(n,k) = \lceil \log m\rceil + 2\lceil \log(k-1)\rceil + 2m$. [5]

*** $T(n, pk) = 2\lceil \log k\rceil + 2\lceil \log(p-1)\rceil + \lceil \log m\rceil + 2m + 1$

where

n : degree of polynomial
k : number of subpolynomials for (*) and (***) and number of subblocks for (**)
p : number of subblocks for each subpolynomial for (***)
m : size of subblocks fore (**) and (***)

Some representative numerical values of parallel steps T(n,p) are given in Table.1 for various values of polynomial degree n and number of processors p used to compare the three methods namely Dorn's, Dorn's with DECAL and Horner's with DECAL. Values in Table .1 of Dorn's and Dorn's with DECAL algorithms for the same number of subpolynomials k show that the decoupling of subpolynomials introduces additional decrease in T with the cost of increase in the total number of processors used . The number of processors used in Dorn's and Dorn's with DECAL is k and kp successively. If the comparison is done with respect to the number of processors then Dorn's and Dorn's with DECAL algorithms are comparable. It had also been shown [5] that Dorn's and Horner's with DECAL were comparable with reference to the number of processors. It should be also noted that the implementation of DECAL is simple. It suggests that if a faster parallel algorithm is developed to solve Equation (21) , the speed-up of Dorn's with DECAL can be improved.

References

[1]   W. S. Dorn, "Generalisation of Horner's Rule for Polynomial Evaluation", *IBM J. Res. Dev.*, 6 , pp. 239-245, 1962.

[2]   G. Estrin, " Organisation of a Computer System - The Fixed Plus Variable Structure Computer", *Proc. Western Joint Computer Conf.*, AFIPS Press, Montvale, NJ, pp. 33-40, 1960.

[3]   L Hyafil and H. T. Kung, "The Complexity of Parallel Evaluation of Linear Recurrences", *J. ACM*, 24, pp. 513-521, 1977.

[4]   S. R. Kosaraju, "Parallel Evaluation of Division-Free Arithmetic Expressions", *Proc. Symp. Theory of Computing*, pp.231-239, 1986.

[5]   A. Kiper, "Parallel Polynomial Evaluation by Decoupling Algorithm", *Parallel Algorithms and Applications*, 9, pp.145-152, 1996.

[6]   J. S. Kowalik and S. P. Kumar, "Parallel Algorithms for Recurrences and Tridiagonal Equations", *Parallel MIMD Computation: HEP Supercomputer and its Applications*, (ed. J. S. Kowalik) , MIT Press, pp.295-307, 1985.

[7]   S. Lakshmivarahan and S. K. Dhall, *Analysis and Design of Parallel Algorithms*, McGraw-Hill, 1990.

[8]   I. Munro and M. Paterson, "Optimal Algorithms for Parallel Polynomial Evaluation", *J. Comp. Syst. Sci.*, 7, pp. 189-198, 1973.

[9]   Y. Muraoka, "Parallelism Exposure and Exploitation in Programs", *TR 424*, Dept. of Comp. Sci. , Univ. of Illinois, Urbana-Champaign, IL, 1971.

Comparison of Two Short-Range Molecular Dynamics Algorithms for Parallel Computing

J. Kitowski[12], K. Boryczko[1] and J. Mościński[12]

[1] Institute of Computer Science, AGH, al. Mickiewicza 30, 30-059 Cracow, Poland
[2] Academic Computer Centre CYFRONET, ul. Nawojki 11, 30-950 Cracow, Poland
kito@uci.agh.edu.pl

Abstract. In the paper we compare timing results for short-range 12/6 Lennard-Jones molecular dynamics simulation for two kinds of data structures resulting from two different kinds of 3-D computational box. For parallel and distributed computing the domain decomposition method and message-passing paradigm were applied. The programs were run on parallel computers like HP/Convex SPP1200 and Intel Paragon XP/S as well as on a network of workstations.

1 Introduction

Rapid development in area of numerical simulation tools and substantial increase of computational power of modern computers as well as cost-effective use of network of computers have made possible complicated large-scale simulations. One of the fast developing approaches to material science is application of particle simulation, mainly Molecular Dynamics (MD) method to solving the problems out of reach of classical models based on flow continuity of matter, momenta and energy. Some early results of such simulations (e.g. [1]-[4]) have been followed by other works (e.g. [5]-[10]) showing needs for enlargement of simulation scale.

In the last years many parallel MD algorithms have been proposed. The most popular approach is to use spatial decomposition with coarse grained cells. Some of the algorithms adopt also Verlet neighbour tables. Rather seldom data-parallel model is applied, however substantial development has been observed in that field also (e.g.[11]). A good survey of the methods can be found in [12].

The purpose of the paper is to verify parallel efficiency of 3-D short range molecular algorithms using two kinds of multiprocessor computers and a network of workstations. One of the algorithms, *Pipe Method*, is treated as a reference algorithm, while the second, *MDMEGA*, is taken for the study from CCP5 Library (at Daresbury Laboratory) and adopted for efficient use on HP/Convex SPP1200 multiprocessor. Its future use covers 3-D simulations of microhydrodynamics effects. Efficiency of the both algorithms is compared with others known from the literature.

2 Characteristcs of the algorithms

Two kinds of data structures resulting from two different kinds of 3-D computational box are applied in the algorithms. Both use short-range interactions defined by 12/6 Lennard-Jones pair potential and leap-frog algorithm for solving Newtonian equations of motion; simulation of microcanonical ensemble is adopted.

One of the computational box is a long cylinder, so the algorithm is called the *Pipe* Method. It was originally developed for vector computers [13, 14].

Periodic boundary conditions are introduced along the cylinder axis only (in z-direction). On the base of the cutoff radius, R_C, the integer cutoff number n_C is introduced, where n_C is a number of neighbours interacting potentially with a given particle. Particles in the cylinder are sorted due to their Z coordinates and the index vector is set up to returning to original particles indices. For forces calculation the computational cylinder is stepwise shifted in respect to its copy to subsequent neighbouring particles from 1 to n_C.

For parallel computing domain decomposition is adopted – the whole cylindrical computational box is composed of domains allocated on different nodes loosely coupled in z-direction. Although the decomposition can be done with the fixed and with the scalable problem sizes [15] we discuss the algorithm (and the results) for the last case only. The initial number of particles for each node, N_K, is fixed – increase of the number of nodes, K, results in increase of the overall number of particles in simulation and greater statistical ensemble, since $N = K \times N_K$. For parallel processing the algorithm is coded in C-language with PVM or NX2 calls.

The second computational box is a classical 3-D box with periodical boundary conditions applied in every direction. The program *MDMEGA* was developed in Fortran 77 in Daresbury Laboratory [16] for Intel iPSC/860 parallel computer. Forces acting on particles are obtained from the parallel link-cell approach. For the purpose of the study and for future applications the program was re-written in C with PVM calls. The PVM use can easily be replaced by MPI standard emerging more interest at present.

3 Computers and environments

Main feature of three NUMA multiprocessor computers applied in the study are summarized in Table 1.

HP/Convex SPP1200 system consists of hypernodes, each of 8 processing elements connected with 5x5 high-speed crossbar. The elements are HP PA7200 processors. The hypernodes are coupled together with 1D torus (CTI rings). Cache coherency is offered by the system. Two-level organization of the interconnection network can result in difference in access time. ConvexPVM and Convex MPI are the proprietary versions of PVM and MPI environments. Partitions of processors (subcomplexes) can be built up from processors of any hypernode.

In Paragon XP/S system all nodes are uniformly integrated in the 2D mesh interconnection network. Each node consists of two i860 (50MHz) processors, from which one is used for computing and one for communication. The pipelined nature of wormhole routing makes network latency almost independent of the

System name	Nodes	Inter-connection	System: version and environment	Options
HP/Convex SPP1200/XA	32	two-layer – crossbar/ 1D torus	SPP-UX: 3.2.129, C (6.4), ConvexPVM (3.3.10)	-O2
Intel Paragon XP/S	98	2D mesh	OSF/1: 1.0.4, C, NX2 (R4.5) 1.3.3, F77 NX2 (R5.0) 1.3.3, C (R5.0) MPI	-O, -Knoieee for *PIPE* -O4 -Knoieee for *MDMEGA* -O4 -Knoieee for *MDMEGA*
IBM RS6K/320 SUN Sparc2	2 2	Ethernet network	AIX 3.2.4 SunOS 4.1 C, PVM (3.3.8)	-O

Table 1. Computer systems and environments

distance between sending and receiving nodes. The well established parallel computing methodology is message-passing with proprietary NX2 environment. MPI mesage-passing environment is also available.

The network computer was established of four workstations of similar performance ($2\times$IBM RS6000/320 and $2\times$SUN SPARCStation2), connected together with Ethernet LAN and constituting a cluster with public domain PVM. A simple load-balancing procedure (for the *PIPE* method) has been applied to get similar wall-clock execution time from each node [17].

4 Results

In Fig. 1 comparison of *PIPE* algorithm with *MDMEGA* one is presented. τ represents the wall-clock execution time per MD timestep and particle. Due to shortage of system memory on a cluster of workstations *MDMEGA* was run for smaller number of particles than using Paragon and SPP1200 computers. Scalability of the both algorithms is close to linear, however for the *Pipe* algorithm parallel efficiency is higher due to smaller communication overhead (see Fig. 2.). This is especially valid for distributed computing with the cluster, for which communication to computation ratio is lower than for multiprocessors (Fig.1).

On Paragon machine we also compare results from NX2 and MPI environments (Fig.2). MPI has been found less efficient than NX2, however the difference can be neglected in that case. The results for the rest of figures concern NX2 enviroment only. The main advantage of MPI are its standarized features and availability on different platforms. Some preliminary results for HP/Convex SPP1200/XA show that MPI (ConvexMPI) is more efficient than ConvexPVM, especially when asynchronous communication with MPI is concerned.

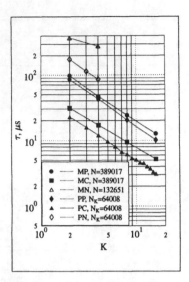

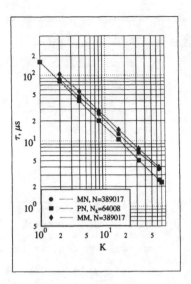

Fig. 1. Execution time, τ, for different number of computing nodes, K, for MDMEGA (Mx) and PIPE (Px) algorithms. Computers: Paragon (yP), SPP1200 (yC), network of workstations (yN). N, N_K – numbers of particles.

Fig. 2. Execution time, τ, for different number of computing nodes, K, for MDMEGA (MN for NX2 and MM for MPI) and PIPE (PN for NX2) algorithms for Paragon computer. N, N_K – numbers of particles.

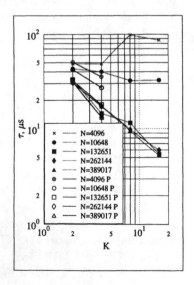

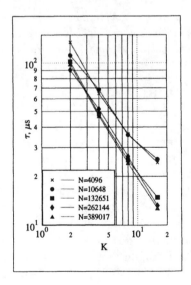

Fig. 3. Execution time, τ, for different number of computing nodes, K, for MDMEGA algorithm and SPP1200 computer. Two different subcomplexes. N – number of particles.

Fig. 4. Execution time, τ, for different number of computing nodes, K, for MDMEGA algorithm and Paragon XP/S computer. N – number of particles.

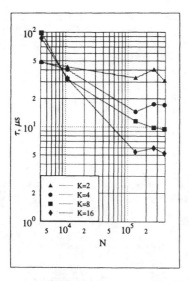

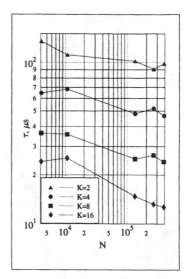

Fig. 5. Execution time, τ, for different number of particles, N, for MDMEGA algorithm and SPP1200 computer. K – number of computing nodes.

Fig. 6. Execution time, τ, for different number of particles, N, for MD-MEGA algorithm and Paragon XP/S computer. K – number of nodes.

In Fig. 3 we present performance of *MDMEGA* algorithm on two different subcomplexes of SPP1200 computer. One subcomplex, mentioned by P at the end of the number of particles, was defined with 7 processors from the same hypernode, thus the interprocessor communication ratio was the highest in that case. The second subcomplex applying up to 16 proceessors was constructed with 4 processors from 4 different hypernodes. In spite of two level communication no substantial degradation of performance was observed in that case. For small number of particles however ($N = 4096 - 10648$), parallel efficiency turn out to be not acceptable in particular for the subcomplex of 16 processors. On Paragon (Fig. 4) for small number of particles degradation of performance less evident. This is due to lower node performance in comparison with SPP1200 and regular communication pattern.

In Figs. 5 and 6 dependency of τ on number of particles, N, is shown for different number of computing nodes, K. For small number of particles the communication overhead is dominant, since increase of N results in decrease of τ. For higher N, when computing load dominates communication, τ is almost independent of N. Its value depends however linearly on the number of processing nodes – twofold increase of K results in similar ratio of τ decrease – showning order $o(N)$ of the algorithm. For Paragon this feature is less clearly observed due to better balance between communication and computation.

Computer	Nodes, K	$N(\times 10^6)$	N_K	τ, μs/timestep/particle
Paragon XP/S (*PIPE*, NX2)	70	4.5	64008	2.3
Paragon XP/S (*PIPE*, NX2)	70	13.4	192024	2.3
Paragon XP/S (*MDMEGA*,NX2)	64	0.39	—	3.7
SPP1200/XA (*PIPE*)	16	0.90	64008	3.0
SPP1200/XA (*MDMEGA*)	16	0.39	—	5.1
CM5E (*BU-TMC* [12])	256	43.9	171500	0.45
CM5 (*SPaSM* [12])	1024	50	48825	0.28
CM5 (*SPaSM* [12])	1024	600	585937	0.40

Table 2. Results for different number of particles for the *PIPE* and *MDMEGA* algorithms compared with other MD results from literature

5 Conclusions

We have shown influence of the computational box on timing results for two different structures - elongaled and cubic ones. The both algorithms turn out to be efficient in parallel implementation for moderate number of particles and computing nodes (see Table 2), although for greater problems simulated with much larger number of nodes other algorithms shown better performance. The *Pipe* Method has slightly better parallel characteristics than *MDMEGA* algorithm. We believe that the last algorithm could be used for efficient simulation of larger ensembles of particles.

6 Acknowledgments

We are very grateful to Dr. Jeremy Cook from University of Bergen for supporting us with the computer time on Paragon. The work was supported by Polish Scientific Committee (KBN) Contracts PB 2 P302 073 05 and 8 S503 006 07.

References

1. Rapaport, D.: Microscale hydrodynamics: Discrete particle simulation of evolving flow patterns, *Phys.Rev.*, **A36**, 7 (1987) 3288.
2. Koplik, J., Banavar, J.R., Willemsen, J.F.: Molecular dynamics of Poiseuille flow and moving contact lines, *Phys. Rev. Lett.*, **60** (1988) 1282.
3. Rapaport, D.: Molecular dynamics study of Rayleigh-Benard convection, *Phys.Rev.Let.*, **60**, 24 (1988) 2480.
4. Puhl, A., Monsour, M.M., Mareshal, M.: Quantitive comparison of molecular dynamics with hydrodynamics in Rayleigh–Benard convection, *Phys. Rev.*, **A40** (1989) 1999.
5. Cui, S.T., Evans, D.J.: Molecular dynamics simulation of two dimensional flow past a plate, *Mol. Simul.*, **9** (1992) 179.
6. Rapaport, D.: Unpredictable convection in a small box: molecular dynamics experiments flow patterns, *Phys.Rev.*, **A46**, 4 (1992) 1971.

7. Lomdahl, P.S., Beazley, D.M., Giles, R., Gronbech-Jensen, N.: Multimillion particle molecular dynamics on the CM5, *Int. J. of Modern Physics C* (1993) 1075.
8. Dzwinel, W., Alda, W., Kitowski, J., Mościński, J., Wcisło, Yuen, D.A.: Macroscale simulation using molecular dynamics method, *Mol.Simul.*, **15** (1995) 343.
9. Dzwinel, W., Alda, W., Kitowski, J., Mościński, J., Yuen, D.A.: An examination of long-rod penetration in microscale using particles, *J.Metal Proc.Technol.*, **60** (1996) 415.
10. Alda, W., Bubak, M., Dzwinel, W., Kitowski, J., Mościński, J., Pogoda, M., Yuen, D.A.: Fluid dynamics simulation with particles, to be presented at *Int. Conference Physics Computing'96*, Sept. 17-21, Kraków, Poland, 1996.
11. Nielsen, O.H.: Data-parallel molecular dynamics with neighbor lists, in: J. Dongarra, K. Madsen and J. Waśniewski (eds.) *Proc. of Second Int. Workshop, PARA'95*, August 21-24, Lyngby, 1995, Lecture Notes in Computer Science, vol. 1041, Springer-Verlag, 1995, p. 443.
12. Beazley, D.M., Lomdahl, P.S., Gronbech-Jensen, N., Giles, R., Tamayo, P.: Parallel algorithms for short-range molecular dynamics, *World Scientific's Annual Reviews in Computational Physics*, **3** (1995), also Los Alamos Report, 1995.
13. Mościński, J., Kitowski, J., Rycerz, Z.A., Jacobs, P.W.M.: A vectorized algorithm on the ETA 10-P for molecular dynamics simulation of large number of particles confined in a long cylinder, *Comput. Phys. Commun.*, **54** (1989) 47.
14. Kitowski, J., Mościński, J.: Microcomputers against Supercomputers ? – On the geometric partition of the computational box for vectorized MD algorithms, *Mol. Simul.*, **8** (1992) 305.
15. Kitowski, J.: Distributed and parallel computing of short-range molecular dynamics, in: J. Dongarra, K. Madsen and J. Waśniewski (eds.) *Proc. of Second Int. Workshop, PARA'95*, August 21-24, Lyngby, 1995, Lecture Notes in Computer Science, vol. 1041, Springer-Verlag, 1995, p. 345.
16. Smith, W.: Daresbury Laboratory CCP5 program for molecular dynamics simulation of Lennard-Jones atoms with a spherical cutoff. *CCP5 Program Library*, Daresbury Lab. Warrington (1992).
17. Boryczko, K., Kitowski, J. Mościński, J.: Load balancing procedure for distributed short-range molecular dynamics, in: J. Dongarra and J. Waśniewski (eds.) *Proc. of First Int. Workshop, PARA'94*, August 21-24, Lyngby, 1995, Lecture Notes in Computer Science, vol. 879, Springer-Verlag, 1994, p. 100.

Parallelising Large Applications

Jesper Larsen, Thomas Christensen, Lars Frellesen

Math-Tech ApS, Admiralgade 22, 1066 Copenhagen K, Denmark**

Abstract. The paper presents experience obtained with the parallelisation of two large applications, one from the petroleum industry - oil
reservoir simulation, and one from environmental management - a three
dimensional hydrodynamic code for coastal waters including a eutrophication module.
We give a brief presentation of the two applications and a description of
the parallelisation strategies. Performance tests are reported.

1 Introduction

Parallel computation is now established as a means of obtaining super computer
performance for large applications. However the current state of compiler and
library technology for parallel computers still does not allow a trivial port from
a sequential machine to a parallel machine. When embarking on the task of
producing a parallel application out of a sequential application many considerations have to be taken. In the present paper we present the considerations we
have taken in parallelising two applications, one from the petroleum industry
- oil reservoir simulation, and one from environmental management - a three
dimensional hydrodynamic code for coastal waters including a eutrophication
module.

The oil reservoir application is based on a finite volume discretisation of the
governing equations with an implicit time discretisation yielding a system of non-
linear equations which are solved by Newtons method. In each Newton iteration
a large sparse system of equations has to be solved. The parallel TFQMR-method
with a block-ILU preconditioner is used as the linear equation solver together
with a domain decomposition for the rest of the code.

The environmental application is based on a staggered grid in space and time
yielding an ADI solution method, the core of which is a tridiagonal solver. An
explicit solver for the advection of various quantities (e.g. salinity and temperature) takes a substantial part of the cpu-time. For a cluster of workstations a
functional decomposition of the code is used, whereas for more powerful parallel
environments a domain decomposition is preferred.

The paper is organised as follows. First we give a brief presentation of the oil
reservoir application followed by a description of the parallel version. Next a brief
presentation of the environmental model is given followed by a description of the
parallelisation strategy for this application. Finally we present some scalability
tests with a discussion of the performance of the two parallel codes.

** The support given by the Danish Agency for Development of Trade and Industry is
gratefully acknowledged

2 COSI

The reservoir simulator COSI was originally developed by the Danish National Laboratory RISØ, The Technical University of Denmark and the consulting engineering company COWIconsult. Since 1990 COWIconsult has been responsible for the development of the functionality of the simulator.

In the EUREKA project PARSIM, COSI has been parallelised in a collaboration between COWIconsult, Math-Tech and the Institute of Advanced Scientific Computing, Liverpool University (IASC).

COSI is a three-dimensional, three-phase, isothermal, numerical reservoir simulation model capable of running in black oil, wet gas or fully compositional mode. The simulator is especially good for modelling low permeable and dual porosity reservoirs.

The simulator is fully implicit, both in pressure and saturation. Space integration is performed by an integral finite volume technique, which enables an arbitrary combination of grid cell types, such as Cartesian, cylindrical and irregular grid cells. The pressure of the oil phase and the component masses are chosen as primary variables. The non-linear equation system is solved by the Newton-Raphson technique.

3 Parallel COSI

The parallelisation strategy for COSI has been described in detail in [3]. It falls into two main steps

- parallelising the linear equation solver
- parallelising the rest of the code.

A library code from IASC [1] based on the transpose free quasi minimum residual (TFQMR) method with a block incomplete LU decomposition as preconditioner [2] has been used for the parallelisation of the linear equation solver. For the rest of the code a domain decomposition of the reservoir compatible with data structure used by the linear equation solver has been applied.

4 MIKE 3

MIKE 3 is a commercial code developed the Danish Hydraulic Institute for the solution of three dimensional flows. MIKE 3 can be applied to:

- Oceanographic studies
- Coastal circulation studies
- Water pollution studies
- Heat and salt recirculation studies

MIKE 3 is composed of three fundamental modules: The hydrodynamic (HD) module, the turbulence module and the transport or advection-dispersion (AD) module. Various features such as free surface description, laminar flow description and density variations are optionally invoked within the three fundamental modules.

On top of the hydrodynamic part of the code various modules for calculating environmental parameters are included in the package. These are a water quality module and a eutrophication module that traces various biological substances. The structure of the code is shown in Fig. 1.

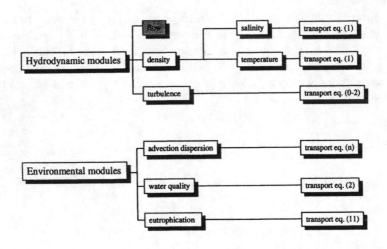

Fig. 1. The MIKE 3 code structure

The governing equations are discretized using a finite difference technique. The resulting difference equations are solved using and Alternating Direction Implicit (ADI) method. The transport equations are all solved using an explicit method. In a moderate sized example the distribution of cpu-time is as follows:

- basic hydrodynamics: 23 %
- density (salinity transport only): 5 %
- turbulence(zero equation model): 2.5 %
- eutrophication: 68 %.

It is seen that the transport equation takes up about three quarter of the computational time in this example. In typical applications this is a lower bound.

5 Parallelising MIKE 3

Various options exist for the parallelisation of MIKE3. In Fig. 2 we have shown four different possibilities.

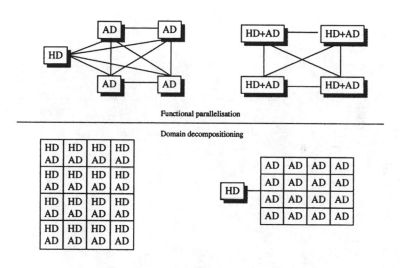

Fig. 2. The parallelisation strategies for MIKE 3

Using a functional decomposition each of the transport modules (AD-modules) is run on its own processor. Since the transport equations are solved explicitly, no inter-processor communication is needed. The hydrodynamics (the HD module) is either run on a separate processor and the flow field is distributed for the AD calculation after each time step, or the HD module is run (redundantly) on all processors limiting the amount of communication. The latter is the preferred solution for the functional decomposition.

Using domain decomposition both the HD computation and the AD computation can be parallelised. However, it is difficult to parallelise the hydrodynamics module due to the ADI algorithm. Hence in an initial implementation only the AD computation is decomposed. This limits the number of processors that can be usefully applied for the parallel implementation.

Comparing the functional decomposition with the parallel decomposition we remark that the number of parallel AD calculations in a typical run is more than twelve. Hence a functional decomposition can utilise as many processors as a domain decomposition and is preferred compared to the latter.

5000 cells

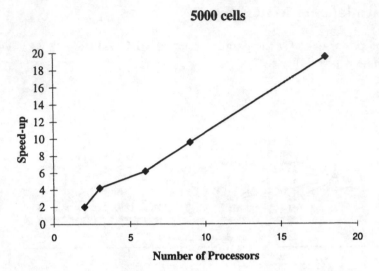

10000 cells

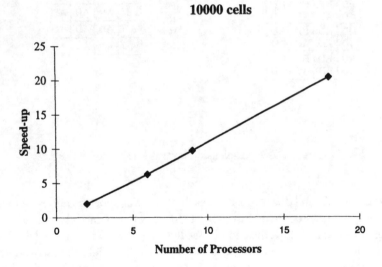

Fig. 3. Speed-up results for the parallel COSI

If more processors are available the HD module should be parallelised as well in order to get a good speed-up. But in doing so alternative algorithms to the ADI method should be investigated.

6   Scalability Results

In Fig. 3 we show the speed-up obtained with two moderately sized problems run on an IBM SP2 computer at UNI-C. We have not shown results for a single processor. In fact we have a sequential code with a different preconditioner which is much faster that the parallel code on a single processor. So although our results show excellent scalability, we are still looking for a better parallel linear equation solver.

In Fig. 4 we show preliminary scalability results for the environmental application run on a cluster of PC's. The parallel part, i.e. the computation of the AD module, scales well. Only a minor increase in the sequential part, i.e. computation of the HD module, is observed. It is evident that the HD module should be parallelised as well, if more processors should be used.

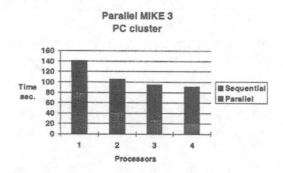

Fig. 4. Scalability results for the parallel MIKE3 on a cluster of PC's

In Fig. 5 similar scalability results on an IBM SP2 are shown.

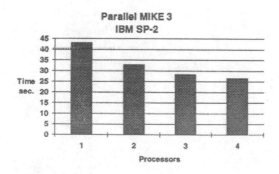

Fig. 5. Scalability results for the parallel MIKE3 on an IBM SP2

7 Concluding Remarks

Concludingly we remark that

- all parts of code should be parallelised
- use of library routines speeds up parallelisation
- alternative algorithms should be considered when parallelising existing applications.

References

1. C. Addison. A preliminary investigation into sparse iterative solvers for oil reservoir simulation. *Proceedings of HPCN'94, Munich, Germany*, April 1994.
2. R. W. Freund. A Transpose-Free Quasi-Minimal Residual Algorithm for non-Hermitian Linear Systems. *SIAM J. Sci. Comput,* 14: No. 2, 470-482, 1993.
3. J. Larsen, L. Frellesen, J. Jansson, F. If, C. Addison, A. Sunderland and T. Oliver. Parallel Oil Reservoir Simulation. *Proceedings of PARA95, Lyngby, Denmark*, August 1995.

High Performance Fortran Interfacing to ScaLAPACK

Paulo A. R. Lorenzo[1], Andreas Müller[1], Yoshimichi Murakami[1,2]
& Brian J. N. Wylie[1]
wylie@cscs.ch

[1] Centro Svizzero di Calcolo Scientifico (CSCS/SCSC),
CH-6928 Manno, Switzerland
[2] CSCS Summer Intern 1995, during leave from Kanazawa University, Japan

Abstract. The ScaLAPACK numerical library for MIMD distributed-memory parallel computers comprises highly efficient and robust parallel dense linear algebra routines, implemented using explicit message passing. High Performance Fortran (HPF) was developed as an alternative to the message-passing paradigm. It extends Fortran 90 with directives to automatically distribute data and to parallelize loops, such that all required inter-processor communication is generated by the compiler. While HPF can ease parallelization of many applications, it still does not make sense to re-program existing libraries like ScaLAPACK. Rather, programmers should have the opportunity to use them from within HPF programs.

HPF interfaces to routines in the ScaLAPACK library are presented which are simplified considerably through exploitation of Fortran 90 array features. Substantial performance benefits from interfacing to efficient ScaLAPACK routines are also demonstrated via a comparison with equivalent HPF-coded functions. Finally, standard ScaLAPACK optimizations, tuning block sizes and processor topology/mapping, are found to be equally effective from HPF.

1 Introduction

A large number of high-quality numerical libraries — such as EISPACK, LINPACK, LAPACK, FFTPACK, and SPARSKIT — have been developed to solve commonly occuring problems such as the solution of systems of linear equations, performing fast Fourier transformations, and so on. Such libraries have become *de facto* standards and are used in a vast spectrum of applications. The importance of such libraries should not be underestimated: they free the developer from programming common routines and allow efficient and portable code development.

With the advent of parallel computers, the existing sequential libraries — which were not designed with parallelism in mind — needed to be extended. ScaLAPACK (Scalable LAPACK) [1], is likely to become a *de facto* standard in the near future. It is based on block-cyclic data distributions of dense full and trapezoidal matrices [2], and portable communication routines from BLACS [3, 4] and MPI [5].

At the same time, due to the increasing complexity of computer architectures, tools which reduce the user's tasks of programming on these machines become more and more necessary. Such a tool is the High Performance Fortran (HPF) language [6], which realizes a data-parallel programming paradigm. While HPF can ease parallelization of many applications, it still does not make sense for the programmer to re-program library routines. Rather, the user should be offered the opportunity to use existing library routines, such as those in ScaLAPACK.

HPF has a clean and straightforward syntax. The HPF programmer does not worry about internal parallelization details, inter-processor communication handles or distributed data bounds. The philosophy of HPF is to strengthen the compiler, letting the programmer focus on the application. ScaLAPACK routines, on the other hand, require the programmer to provide a complete description of data distribution and communication handles.

To make the ScaLAPACK library available to the HPF programmer, we propose HPF interfaces to the ScaLAPACK routines. These wrappers will enable the programmer to conveniently define distributed arrays using HPF directives and call ScaLAPACK routines from within HPF applications. Invisible to the programmer, the interface routines bridge the gap between two different programming models — the data-parallel model and the message passing paradigm. Since the interface routines are written in HPF, portability is thus guaranteed to any machine where an HPF compiler is installed.

A brief overview of HPF and ScaLAPACK in Section 2 leads into Section 3 which describes the design and implementation of the interface routines. Section 4 presents some simple experiments made to compare the performance of applications that call ScaLAPACK routines through the interface to 'plain' HPF implementations, which are summarized in the concluding section.

2 Overview of HPF and ScaLAPACK

HPF is an extension of Fortran 90 for MIMD parallel machines including both distributed-memory and shared-memory computers with non-uniform memory access times. The language was designed to allow 'easy' parallelization of existing sequential codes as well as the writing of efficient new code. It provides the user with the illusion of a single thread of control as well as a global name space: the programmer just has to specify how data should be distributed and the rest is the task of the compiler.

The data-parallel programming paradigm realized with HPF provides directives to specify how array sections should be distributed and to inform the compiler how to parallelize the program. Blockwise and cyclic distributions may also be combined in a blocked-cyclic distribution. In addition to containing all the functionality of Fortran 90, HPF supplies a FORALL statement to extend Fortran 90 array operations, automatic conversion of one data distribution to another, ways to align different distributions and otherwise give the compiler more information to parallelize code better, and a large number of standard library functions.

On the other hand, HPF in itself does not remove the need for commonly-used library routines. Additionally, HPF can have performance bottlenecks when inherent parallelism is not recognized. Therefore, it is therefore natural and necessary to supplement HPF with other software.

ScaLAPACK is a public-domain library now ported to many parallel computers. It consists of over 400 FORTRAN 77 subroutines for solving some of the most common problems in numerical linear algebra on vector and parallel computers. ScaLAPACK has as its basic communication layer the Basic Linear Algebra Communication Subprograms (BLACS) — message-passing routines that communicate matrices among processors arranged in a virtual two-dimensional processor topology [3].

Like LAPACK (Linear Algebra PACKage), ScaLAPACK is based on block-partitioned data sets in order to minimize the frequency of data movement between different levels of the memory hierarchy. This concept is extended to parallel computers by using block-cyclic data distribution. It provides a simple, yet general-purpose, way of distributing a block-partitioned matrix on distributed-memory concurrent computers.

The block-cyclic data distribution algorithm gives high individual node performance based on level-2 and level-3 BLAS operations [7], while also ensuring good load balancing. ScaLAPACK therefore offers the application developer, highly efficient robust parallel dense linear algebra routines for vector-vector, matrix-vector, and matrix-matrix operations, matrix factorizations and eigenvalue solvers.

3 HPF/ScaLAPACK Interface

3.1 Design

The purpose of the HPF interface to ScaLAPACK is to provide high-level access to ScaLAPACK routines. Interface routines completely wrap the respective ScaLAPACK routines and appear as Fortran 90 library routines to the programmer. The interface thereby unburdens the programmer of needing to know the internal mapping of the distributed data. In addition, the number of subroutine arguments is reduced considerably, since array sections can be handled transparently by Fortran 90 and HPF, i.e., there is no need to pass a set of lower and upper array bounds when the routine should operate on sections of the arrays passed as arguments to the ScaLAPACK routine.

Within the interface, a BLACS routine is called to initialize the communication grid required by ScaLAPACK. Then, for every distributed array, a descriptor is set up using Fortran 90 and HPF standard library functions. The appropriate ScaLAPACK routine is then called, before ultimately deleting the communication grid, again by calling a BLACS routine.

HPF provides a virtual, single-threaded model of execution, whereas the implementation of the ScaLAPACK routines is based on the multi-threaded SPMD model of execution. Therefore, these routines have to be declared EXTRINSIC (F77_LOCAL). Such routines are always called by *all* processors simultaneously. If distributed arrays are passed, every processor receives a pointer to that part of the array stored locally, rather than a copy of the entire array. That is exactly what the ScaLAPACK routines expect.

In the execution of an HPF program, a fork-like procedure is performed to generate the parallel processes. That means that the initialization routines of the communication library which is used by BLACS cannot be called from the running HPF program, because this would again generate processes. To avoid the problems that arise from this, BLACS and the HPF communication layer have to be based on the same communication library, such as MPI or PVM.

3.2 Implementation

To explain the implementation of the HPF interface, the double-precision real matrix-matrix multiplication routine of ScaLAPACK, PDGEMM, is used as an example:

```
pdgemm (transa, transb, m, n, k, alpha,
        A, ia, ja, DESCA,
        B, ib, jb, DESCB,
        beta,
        C, ic, jc, DESCC)
```

PDGEMM performs the following operation:

$$C_{(ic:ic+m-1,jc:jc+n-1)} =$$
$$\alpha \times A_{(ia:ia+\eta-1,ja:ja+\theta-1)} B_{(ib:ib+\mu-1,jb:jb+\kappa-1)} + \beta \times C_{(ic:ic+m-1,jc:jc+n-1)}$$

where A is a $(m \times k)$, B is a $(k \times n)$, and C is a $(m \times n)$ distributed matrix of double-precision real values. α and β are double-precision real scalars. The character arguments *transa* and *transb* indicate whether the matrices A and B are transposed, in this case:

$$transa = \text{'N (no transpose)} \implies \eta = m, \ \theta = k$$
$$transa = \text{'T' or 'C' (transpose)} \implies \eta = k, \ \theta = m$$
$$transb = \text{'N' (no transpose)} \implies \mu = k, \ \kappa = n$$
$$transb = \text{'T' or 'C' (transpose)} \implies \mu = n, \ \kappa = k$$

ia, ja, ib, jb, ic and *jc* are the lower and upper bounds of the sections of the matrices to be multiplied. *DESCA, DESCB* and *DESCC* are the array descriptors of the corresponding distributed array, each consisting of 8 integer elements with the following significance:

DESCA(1) : number of rows in the distributed matrix A.
DESCA(2) : number of columns in the distributed matrix A.
DESCA(3) : blocking factor used to distribute the rows of matrix A.
DESCA(4) : blocking factor used to distribute the columns of matrix A.
DESCA(5) : process row over which the first row of matrix A is distributed.
DESCA(6) : process column over which the first column of matrix A is distributed.
DESCA(7) : BLACS context handle, indicating global context of the matrix operation.
DESCA(8) : leading dimension of local array storing local blocks of distributed matrix A.

The argument list of the HPF interface to PDGEMM, HPF_PDGEMM becomes:

```
hpf_pdgemm (transa, transb, alpha, A, B, beta, C)
```

As can be readily seen, the use of the HPF interface is much simpler. The HPF programmer just needs to to include the INTERFACE block for the desired routines and link their program with the BLACS, ScaLAPACK and HPF interface libraries.

The core part of the interface is the routine SET_DESC which calls BLACS for the initialization of the communication grid, and initializes the descriptor of each distributed data. Due to its modularity, the interface is easily debugged and extremely flexible.

The main routines called in SET_DESC are the following: SIZE, HPF_DISTRIBUTION, BLACS_GET and BLACS_GRIDINIT. SIZE is the Fortran 90/HPF intrinsic function which returns the size of an array, and it is used to set the first and second components of the ScaLAPACK descriptor. HPF_DISTRIBUTION is an HPF intrinsic routine which returns the information regarding the distribution of the ultimate align-target associated with a variable. The third and fourth components of the descriptor are set based on that information. BLACS_GET and BLACS_GRIDINIT are routines in the BLACS library which set-up the communication grid and initialize the communication handle ICONTEXT. ICONTEXT is passed to ScaLAPACK as the seventh component of the descriptor. As the communication handles are possibly different on different processors, ICONTEXT is a distributed array with each processor owning one element. The fifth and sixth elements of the descriptor are set to zero, and the eighth element is set based on the actual distribution among the processors. The setting of these three elements of the descriptor is highly dependent on the compiler implementation.

There are over 400 routines in ScaLAPACK and to individually implement all of the corresponding interface routines would both take a long time and be prone to subtle errors. Fortunately, some sets of interfaces provide the same functionality, only differing in the data types. This means that interface routines can be efficiently generated using a macro processor, such as GNU m4.

Further implementation details, including example program listings, are available in Technical Report CSCS-TR-96-13 (May 1996).

4 Experiments

To verify the correctness and evaluate their performance, the two equivalent implementation approaches to solve a given numerical problem are investigated:

1. Code programmed entirely in HPF (called "Plain HPF," hereafter); and,
2. HPF code which calls ScaLAPACK routines through the new interfaces.

Following each approach, two test programs were implemented with which to test the corresponding interface routines:

(a) Matrix-matrix multiplication; and,
(b) LU decomposition (splitting a matrix into upper- and lower-triangular parts).

The experiments presented were performed on a non-dedicated SUN SPARC-server 1000 with 8 processors and 448 Mbytes of memory. Execution times presented are the average of a few measurements, and considered to be typical execution times: variations of up to 10% are common, and occasionally much slower executions are observed (due to contention with other jobs for CPU or other system resources). The HPF compiler used for the tests came from the Portland Group Inc. (PGI), pghpf release 2.1-1 [8], and was used with maximum optimization (-O4). Plain HPF codes were run with PGI's (default) RPM environment — a proprietary low-overhead transport mechanism — whereas with ScaLAPACK, MPI(CH) was explicitly used. ScaLAPACK version 1.0 (with appropriate BLACS library based on MPI) was obtained from Netlib.

4.1 Preamble

Different algorithms were investigated to determine the most appropriate Plain HPF version for comparison with ScaLAPACK equivalents. For double-precision matrix-matrix multiplication of a square Householder matrix to itself, a straightforward triply-nested loop algorithm was found to give equivalent performance to a simple call to the Fortran 90/HPF intrinsic routine, MATMUL. Maximizing simplicity, the latter was chosen as a reference code for comparison against the HPF interface to ScaLAPACK. The LU decomposition code which will also provide a reference is based on the Gaussian Elimination benchmark in HPF by Hon W. Yau (NPAC, reference kernel 0022). Optimizations were not investigated, as the performance of the HPF code (which improves with successive compiler releases) is provided only as a general reference point.

The overhead of using the HPF interface routines was also investigated and found to be minimal (always less than 0.1 second and generally much less) for all cases investigated, independent of the problem size or of the number of processors. This leads us to conclude that the EXTRINSIC (F77_LOCAL) interface works efficiently: there is no copying of data in the call of the HPF interface to ScaLAPACK.

4.2 Tuning the block size

The interface should bridge HPF to ScaLAPACK in a way which is transparent for the HPF programmer: there should be no undue restrictions. If the kind of distribution which has been chosen doesn't match ScaLAPACK, expensive re-mapping will be unavoidable. On the other hand, different block sizes, which are known to significantly influence the performance of ScaLAPACK [2], can be freely investigated.

A quick check for an appropriate block size for the matrix-matrix multiplication test program was undertaken. Figure 1 shows execution times for a 1000×1000 square matrix size problem on 6 processors varying the block size: a selection of block sizes up to 40 were examined, and the execution time found to vary by less than 5% for blocks of size 6 and larger. Notably, the default block size of 1 is a factor of two or more worse, and should be avoided.*

Based on these results, a block size of 26 was chosen to be the common value in all subsequent tests, standardizing the other performance experiments. Arbitrarily, the same block size of 26 was also used for LU decomposition runs too.

4.3 Scalability

Several scalability measurements were performed using the matrix-matrix multiplication and LU decomposition test programs. Figure 2 shows the execution time for LU decomposition using HPF/ScaLAPACK and Plain HPF on 8 processors, varying the size of the square matrix from 100×100 up to 3000×3000 — this size of problem was found to roughly correspond to an execution time of 20 minutes, and considered to be as large as would be likely to run on such platforms. Plain HPF versions have more restricted memory limitations, and the execution times are approximately six times longer.

---

* Similar variations were also observed for larger block sizes in several instances where the partitioning of arrays onto processors were particularly unfavorable.

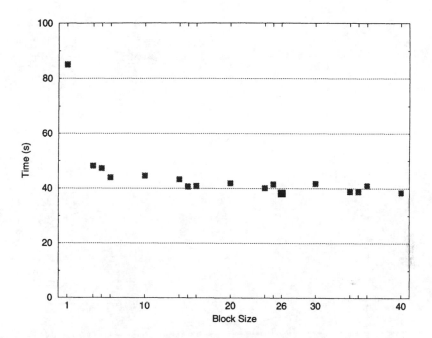

Fig. 1. HPF/ScaLAPACK execution times for double-precision matrix-matrix multiplication of 1000×1000 matrices on 6 processors varying the block size.

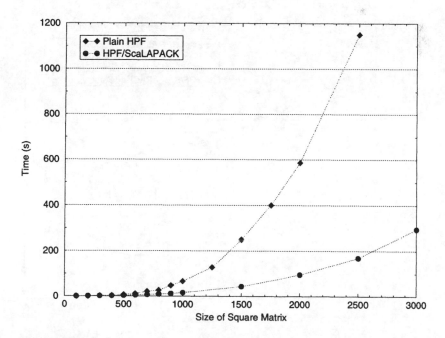

Fig. 2. HPF/ScaLAPACK and Plain HPF execution times for double-precision LU decomposition on 8 processors varying the size of the square matrix (block size 26).

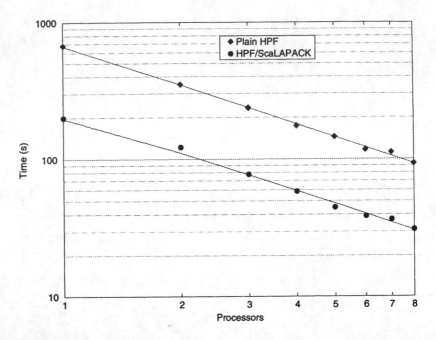

Fig. 3. HPF/ScaLAPACK and Plain HPF double-precision matrix-matrix multiplication execution times for a square matrix of size 1000 (block size 26) varying the number of processors.

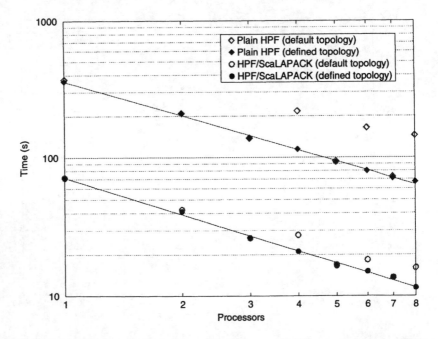

Fig. 4. HPF/ScaLAPACK and Plain HPF double-precision LU decomposition execution times for a square matrix of size 1000 (block size 26) varying the number of processors, both for the default processor topology and a defined linear processor mapping.

Figures 3 and 4 show the execution times of HPF/ScaLAPACK and Plain HPF versions of matrix-matrix multiplication and LU decomposition for a constant size (1000×1000) square matrix on varying numbers of processors. Best-fit lines from regression analysis are also included, with the Plain HPF and HPF/ScaLAPACK versions showing similar respectible speed-ups. HPF/ScaLAPACK gives rather better performance than the Plain HPF versions considered: the difference is a factor of 3 for matrix-matrix multiplication and almost a factor of 6 for LU decomposition.

4.4 Processor topology

HPF programmers have the option of specifying the processor topology they would like to have. Using the PROCESSORS directive, one or more rectilinear processor arrangements can be declared, specifying for each one its name, its rank (number of dimensions), and the extent in each dimension. Figure 4 also shows performance results for LU decomposition using the default and a defined grid. One version defined a one-dimensional arrangement of processors, known to be most appropriate for LU decomposition, whereas the other used the default, letting the compiler decide which arrangement of processors to have.

The stepped shape of the default version is caused by the fact that the PGI HPF compiler automatically chooses a rectangular grid for even numbers of processors (e.g., 4×2), instead of the more favorable vector arrangement (8×1): since ScaLAPACK's two-dimensional distribution was specified as a directive, it is not too unreasonable that the compiler has prefered compact two-dimensional processor grids over linear alternatives. This default behavior of the compiler is good for some problems, indifferent for others (such as for the matrix-matrix multiplication, Figure 3), and sub-optimal for others (such as the LU decomposition, Figure 4, where even on 8 processors the difference in performance amounts to 40%). However, the use of the HPF PROCESSORS directive gives programmers the possibility to inform the compiler of the most appropriate processor topology when this is advantageous.

5 Conclusions

HPF proposes an straightforward programming style that unburdens the programmer of much of the parallelization complexity, providing high-level structures that exploit the parallelism of the application. Although the advantages of programming in HPF are clear, it still does not make sense to re-program routines already available in libraries. Libraries such as ScaLAPACK that are robust and widely-used in both academic and commercial applications should be readily accessible to HPF application developers.

The design and implementation of an HPF interface to ScaLAPACK has been presented. This interface library allows HPF programmers to use ScaLAPACK routines as if they were ordinary HPF routines. Long lists of ScaLAPACK routine arguments are reduced to a minimum, with only the essential arguments remaining, such as the input/output matrices and vectors. HPF programmers can therefore conveniently take full advantage of ScaLAPACK.

The HPF interface is designed to be portable, using only Fortran 90/HPF standard routines, and the modular implementation makes it possible to automatically generate interfaces to ScaLAPACK routines working with different data types, which eases maintenance. Further work is necessary to verify portability and to improve the robustness of the interface, with additional checks for unusual, but still valid, input data.

Measurements show that the overhead of the HPF interface, over direct use of ScaLAPACK routines, is negligible. Comparing the performance of ScaLAPACK routines accessed through the HPF interface with equivalent implementations written entirely in HPF clearly demonstrated the superior HPF/ScaLAPACK performance. While only small numbers of processors were investigated here, performance and scalability matched that of ScaLAPACK itself, therefore it is expected that much larger parallel computers should also be efficiently exploited.

ScaLAPACK's prescribed data mapping is not optimal in all cases, but even so, it is very efficient and optimized for a broad range of situations. Just as ordinary ScaLAPACK programmers have control over block sizes and the processor topology/mapping, should they desire to investigate these options, so do HPF programmers. The advantages of pursuing such optimizations have also been demonstrated from HPF.

Acknowledgments This work was initiated during the 1995 CSCS Summer Student Internship Program (SSIP'95), as a project proposed and advised by Prof. Jack J. Dongarra (UTK/ORNL) and financially supported by NEC Corporation. It was subsequently completed when the latest HPF compiler from the Portland Group Inc. became available. A number of people provided additional guidance and valuable assistance throughout the duration of this work: thanks go to Denney Cole, Paul Kinney and Douglas Miles of PGI, Yoshiki Seo (NEC C&CRL), and in particular to colleagues, Vaibhav Deshpande, Norio Masuda and William Sawyer.

References

1. J. Choi, J. J. Dongarra, R. Pozo, and D. W. Walker, "ScaLAPACK: A scalable linear algebra library for distributed memory concurrent computers," in *Proc. 4th Symp. Frontiers of Massively Parallel Computation (Frontiers '92)*, IEEE Computer Society Press, 1992.
2. J. Choi, J. J. Dongarra, L. S. Ostrouchov, A. P. Petitet, D. W. Walker, and R. C. Whaley, "The design and implementation of the ScaLAPACK LU, QR, and Cholesky factorization routines," Tech. Rep. ORNL/TM-12470, Oak Ridge National Laboratory, September 1994.
3. V. Deshpande, W. Sawyer, and D. W. Walker, "An MPI implementation of the BLACS," in *Proc. 2nd MPI Developers Conf., (MPIDC'96, Notre Dame, IN, USA)*, pp. 195–198, IEEE Comp. Soc. Press, July 1996. [ISBN: 0-8186-7533-0] Also available as CSCS-TR-96-11.
4. J. J. Dongarra and R. C. Whaley, "A user's guide to the BLACS v1.0," Technical Report CS-95-281, LAPACK Working Note 94, University of Tennessee, 1995.
5. Message Passing Interface Forum, "MPI: A Message-Passing Interface Standard," *International Journal of Supercomputer Applications*, vol. 8, no. 3&4, pp. 157–416, 1994.
6. High Performance Fortran Forum, "High Performance Fortran Language Specification: Version 1.0," *Scientific Programming*, vol. 2, no. 1&2, 1993.
7. J. J. Dongarra, J. Du Croz, S. Hammarling, and I. Duff, "A set of level 3 basic linear algebra subprograms," *ACM Transactions on Mathematical Software*, vol. 16, pp. 1–17, Mar. 1990.
8. The Portland Group, Inc. (PGI), 9150 SW Pioneer Court (Suite H), Wilsonville, Oregon 97070, USA, *pghpf User's Guide and Reference Manual*, 2.1 ed., May 1996.

Partitioning an Array onto a Mesh of Processors

Fredrik Manne and Tor Sørevik

Department of Informatics, University of Bergen,
N-5020 Bergen, Norway
email: {fredrikm,tors}@ii.uib.no

Abstract. Achieving an even load balance with a low communication overhead is a fundamental task in parallel computing. In this paper we consider the problem of partitioning an array into a number of blocks such that the maximum amount of work in any block is as low as possible. We review different proposed schemes for this problem and the complexity of their communication pattern. We present new approximation algorithms for computing a well balanced generalized block distribution as well as an algorithm for computing an optimal semi-generalized block distribution. The various algorithms are tested and compared on a number of different matrices.

1 Introduction

A basic task in parallel computing is the partitioning and subsequent distribution of data to processors. The problem one faces in this operation is how to balance two often contradictory aims; finding an equal distribution of the computational work and at the same time minimizing the imposed communication.

In the data parallel model this can be modeled as a graph partitioning problem where the vertices represents data and the edges indicate that results obtained from processing one data unit will be needed for further processing of the other. Finding an optimal solution is know to be NP-hard [8], and hence impossible to solve to optimum for large instances.

In settings where locality is of importance the partitioning and resulting mapping should as far as possible be done such that adjacent nodes are mapped to the same processor. Thus the dataset should be partitioned into connected components. If the data is stored in an array these components might for reasons of both efficiency and simplicity be restricted to be rectangular blocks of the array. Several high-performance computing languages include the possibility for the user to specify such a partitioning and distribution of data onto a logical set of processors. The compiler then maps the data onto the physical processors and determines the communication pattern. An example of one such scheme is the block distribution found in languages such as Vienna Fortran [4] and HPF [10].

In general the block distribution will result in equal size blocks and therefore cannot adapt to load imbalance that might be present. Consider ocean modeling where the presence of land gives irregular areas for which no computations are needed. As demonstrated in [2] a block distribution that takes this into account reduces the time spent on a parallel computation. More general partitioning schemes that have been proposed for these kinds of problems include the generalized and semi-generalized block distribution [5, 15, 16, 17].

In this paper we discuss a number of different partitioning schemes. In particular, we describe an efficient iterative algorithm that computes a well balanced generalized block distribution. We also show how an optimal semi-generalized block distribution can be found. The performance of these algorithms are compared with other orderings such as the uniform block distribution and the binary recursive decomposition [1, 3]. The algorithms presented extend earlier work for one dimensional arrays [13, 14].

The paper is organized as follows: Section 2 gives formal definitions of the different partitioning schemes and relate these to each other. Section 3 presents new algorithms for computing the different distributions and Section 4 reports on the performance of these. Finally, in Section 5 we conclude and point to areas of further work.

2 Structured Distributions

In this section we define and relate the different types of distributions and discuss what kind of communication pattern they impose. One measurement of the communication complexity is the maximum number of neighbors a block can have. In this paper we only consider arrays of dimension two. All results, however, may easily be extended to arrays of higher dimensions.

Let $A \in \Re^{m \times n}$ and let p and q be integers such that $1 \leq p \leq m$ and $1 \leq q \leq n$. Let $R = \{r_o, r_1, ..., r_p\}$ be integers such that $1 = r_0 \leq r_1 \leq ... \leq r_p = m + 1$. Then R defines a *partitioning* of $[1..m]$ into p consecutive intervals $[r_i, ..., r_{i+1} - 1]$, $0 \leq i < p$. We denote this interval by $[r_i, ..., r_{i+1}]$. A partitioning of $[1..m]$ into p intervals and of $[1..n]$ into q intervals defines a partitioning of A into $p \times q$ blocks.

We now define the different types of distributions of A. The distributions are given in increasing order of complexity.

2.1 Non Recursive Distributions

The most simple distribution we consider is the uniform distribution:

Definition 1 Uniform Block Distribution. The interval $[1..m]$ is partitioned into p consecutive intervals of size $\lceil \frac{m}{p} \rceil$ with the possible exception of the last interval. Similarly $[1..n]$ is partitioned into q intervals of size $\lceil \frac{n}{q} \rceil$.

The uniform distribution divides A into $p \times q$ equally sized blocks. See Figure 1a for an example. In certain applications the amount of data that needs to be communicated is proportional to the perimeter of each block. In this setting the uniform distribution minimizes the time needed for communication. If the work associated with each element of A is equal it also gives a perfectly balanced workload. But being fixed a priori it has no possibility to adapt to load imbalance if the computational work varies throughout A. This might be mitigated by moving any of the $p + q - 2$ interior delimiters. By doing so we allow more flexibility in the size of the blocks while at the same time keeping the regular communication pattern of the uniform block distribution.

Definition 2 Generalized Block Distribution (GBD). The interval $[1..m]$ is partitioned into p consecutive intervals without restrictions on the size of each interval. Similarly $[1..n]$ is partitioned into q intervals.

See Figure 1b for an example of the GBD. The GBD was discussed by Fox et. al. [7] and implemented as part of Superb environment [17] and later in Vienna Fortran [6]. It is also a candidate to be included as part of the ongoing HPF2 effort [11]. For examples of how the GBD can be used in areas such as sparse-matrix and particle-in-cell computations see [5] and [12].

While the GBD has the same structured communication pattern as the uniform distribution the blocks sizes vary. The time spent on communication is therefore likely to be higher than with the uniform distribution.

In some cases it is only necessary to have a structured distribution in one dimension. If this is the horizontal direction we may relax the partitioning conditions in the vertical direction. One would thus allow for an individual partitioning of the columns in each row segment given by the horizontal distribution.

Definition 3 Semi-Generalized Block Distribution (SBD). The interval $[1..m]$ is partitioned into p consecutive intervals $[r_i, r_{i+1}]$, $1 \leq i \leq p$ without restrictions on the size of $r_{i+1} - r_i$. For each horizontal interval $[r_i, r_{i+1}]$ the interval $[1..n]$ is partitioned into q intervals.

See Figure 1c for an example of the SBD. Ujaldon et. al. proposed a partitioning scheme called Multiple recursive decomposition which results in a SBD [15, 16]. It was designed for solving problems from sparse linear algebra on parallel computers.

The SBD has a possibility to adapt better to load imbalance than the GBD. But since a block may have as many as $2q + 2$ neighbors the communication pattern becomes less structured.

In a parallel environment the time spent on a computation is determined by the processor taking the longest time. To estimate the time needed to process each block we define a non-negative cost function ϕ on contiguous blocks of A. The assumptions on ϕ are that if a and b are blocks of A such that $a \subseteq b$ then $\phi_a \leq \phi_b$, and $\phi_a = 0$ if and only if a is the empty block. For most reasonable functions ϕ we expect that if the value of a (or b) is known then the value of b (or a) can be computed in $O(|b| - |a|)$ time. An example of ϕ might be the number of non-zero elements or the sum of the absolute values of the elements in a block.

We now get the following optimization problem:

Partition A in such a way that $\max_{i=1:p, j=1:q} \phi_{i,j}$ is minimized.

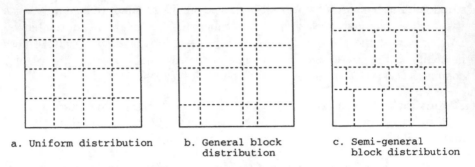

a. Uniform distribution   b. General block   c. Semi-general
                           distribution   block distribution

Fig. 1. Examples of the different distributions with $p = q = 4$

The relationship between the optimal values for each of the different distributions is captured in the following:

Theorem 4. *Let $A, p, q,$ and ϕ be defined as above. Let further S be the set of all SBDs on A, $\mathcal{R}$ the set of all GBDs, and $\mathcal{U}$ the uniform distribution. Then the following is true:*

$$\min_{\forall S} \max_{i=1:p, j=1:q} \phi_{i,j} \leq \min_{\forall \mathcal{R}} \max_{i=1:p, j=1:q} \phi_{i,j} \leq \max_{(i,j) \in \mathcal{U}} \phi_{i,j} \qquad (1)$$

Proof: The result follows trivially from the fact that $\mathcal{U} \in \mathcal{R} \subseteq S$. $\square$

As is evident from Theorem 4 the more unstructured the distribution is the more even we can get the load balance, but as discussed above, the more complex and time consuming the communication becomes.

2.2 Recursive Distributions

It is possible to generalize Definitions 2 and 3 to make the distributions recursive. The partitioning would then recursively be applied to each block for a number of d levels. Thus A is partitioned into $p^d \times q^d$ blocks. The relationships given by Theorem 4 still holds true for the recursive distributions. Note also that for both the GBD and the SBD the minimum cost of the recursive distribution is lower than the minimum cost of the non-recursive version partitioned into $p^d \times q^d$ blocks.

The maximum number of neighbors any block can have above or below and to the right or left is q^{d-1} and p^{d-1} for the recursive GBD and q^d and p^{d-1} for the recursive SBD. Thus the recursive orderings give more complicated communication patterns than their non-recursive counterparts.

For $p = q = 2$ the recursive SBD gives the well known binary recursive decomposition [1]. Figure 2 shows examples of the recursive GBD and the binary recursive decomposition.

Less restricted block distributions than the ones presented here may lead to a better load balance but are likely to give more irregular communication patterns that would be difficult to implement efficiently.

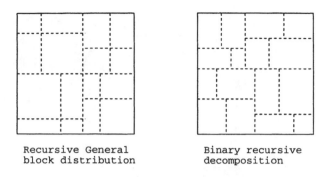

Recursive General
block distribution

Binary recursive
decomposition

Fig. 2. Recursive distributions with $p = q = d = 2$

3   Algorithms

In this section we describe an efficient iterative algorithm for computing a well balanced GBD. The solution obtained is shown to be a local optimum. We also show how an optimal SBD can be found. For completeness we describe a number of proposed approximation algorithms.

All presented algorithms extend previous results on partitioning a one dimensional array. We therefore start by recapturing this problem and its solution.

The problem is identical to the generalized partitioning problem discussed so far but with $n = q = 1$. Thus we are partitioning a vector of length m into p consecutive intervals. The current fastest algorithm for solving this problem is based on dynamic programming and runs in time $O(p(m - p))$ [14]. This is based on the same assumptions on ϕ as for the general problem. Thus the cost function can be computed in time $O(1)$ when the size of an interval changes by one. Every function evaluation in the algorithm is of this type. Thus if the time to calculate the cost function is c the time complexity becomes $O(p(m - p)c)$.

3.1   The Generalized Block Distribution

Consider first the problem of partitioning $[1..n]$ into q intervals when the partitioning of $[1..m]$ into p intervals has been fixed. The placement of vertical delimiters i and $i + 1$ then defines p blocks of cost $\phi_{i,j}$, $1 \leq j \leq p$. We define a new cost function $\theta_i = \max_{1 \leq j \leq p} \phi_{i,j}$. The function θ has the same monotone properties as ϕ. Thus we can reduce the placement of the vertical delimiters to solving the one-dimensional case with cost function θ. Using the dynamic programming algorithm we can now find an optimal placement of the vertical delimiters. This is also true if the vertical delimiters are fixed and an optimal placement of the horizontal ones is desired.

We suggest an algorithm where this step is applied iteratively: The delimiters are fixed in one direction and placed optimally in the other. This is then repeated while alternating which delimiters are fixed until no decrease in the maximum cost is obtained. At this stage a local optimum has been reached, where moving

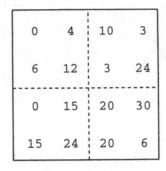

Fig. 3. A local optimum which is not globally optimal.

any set of either horizontally or vertically delimiters will not decrease the overall cost of the partition.

However, as the example in Figure 3 shows, the solution obtained might not be globally optimal. Here $p = q = 2$ and the cost function is the sum of the elements. The indicated solution of cost 76 cannot be improved by moving any one of the delimiters, whereas the optimal solution has the lower right hand block of size one and is of cost 70.

With the assumptions made on ϕ in Section 2 the time complexity of calculating θ when the size of an interval changes by one is $O(m)$. Thus the time complexity of one iteration of algorithm becomes $O(qm(n - q) + pn(m - p))$.

For most natural cost functions it is possible to improve the time complexity by collapsing parts of the array in each iteration. The assumption we make is that the contribution to the cost function from one column (or row) of a block can be reduced to one number. Given a partition of the rows we can then collapse each column segment to one number. This reduces the size of the matrix from $m \times n$ to $p \times n$ and the time complexity becomes $O(mn + pq(n - q) + pq(m - p))$ where the mn term comes from the collapsing step.

We note that if the contribution from a single row or column segment $[i, j]$ can be calculated as $\phi(1..j) - \phi(1..i)$ then the rows can be collapsed in $O(pn)$ time and the columns in $O(qm)$. This is done by pre-computing $\phi(1..i)$ for every value of i for every row and column. The expense for this speedup is that we need an $O(mn)$ pre-computational step and extra storage to hold the values from this step.

A more simple approximation algorithm for computing a GBD was presented in [12]. In this algorithm the rows and columns are collapsed and partitioned separately giving a time complexity of $O(p(m - p) + q(n - q) + mn)$.

3.2 The Semi-Generalized Block Distribution

In this section we show that the dynamic programming algorithm can be extended to compute an optimal SBD.

Consider the row interval $[r_i, r_{i+1}]$ in a SBD. Let γ be the value of an optimal q-partition of the column segments in this interval. Since ϕ is monotone it follows that γ is also monotone and $\phi = 0$ if and only if $\gamma = 0$. Thus we can use the dynamic programming algorithm to compute an optimal p-partition of $[1..m]$ using γ as cost function resulting in an optimal SBD.

The time complexity of finding the optimal q-partition of $[r_i, r_{i+1}]$ is $O(q(n - q)(r_{i+1} - r_i))$. The function γ needs to be evaluated $O(p(m - p))$ times and therefore the overall time complexity becomes $O(pqm(m - p)(n - q))$.

This result can be improved if it is possible to collapse the columns of A. Recall that in the dynamic programming algorithms the function value of an interval is always obtained after one of the delimiters of the interval has been moved exactly one place. Thus if we performed a collapsing of the columns the previous time we evaluated γ we can update this in time $O(n)$ to the value needed in the current evaluation. This reduces the time complexity of evaluating γ to $O(n + q(n - q))$ and the overall time complexity becomes $O(p(m - p)(n + q(n - q))) = O(pq(m - p)(n - q))$.

An approximation algorithm for computing a SBD can be obtained by first determining an optimal partition on the collapsed rows of A. The columns of each row segment are then collapsed and optimally q partitioned. The time complexity of this algorithm is $O(mn + p(m - p) + pq(n - q))$.

For completeness we also describe the Multiple recursive decomposition [16]. Let $p_1, p_2, ..., p_k$ be the prime factorization of p in descending order of magnitude. The rows are first partitioned into p_1 intervals such as to minimize the cost of the most expensive interval. Each interval is then further partitioned into p_2 intervals and so on until p intervals have been obtained. This process is then repeated for each row segment using the prime factors of q. We note that if p and q are powers of 2 and it is possible to collapse the columns and rows of A, the time complexity of this algorithm is $O(m \log p + pn \log q + mn)$.

4 Numerical experiments

We have implemented the algorithms of Section 3, and performed a number of experiments in order to investigate how well they partition an array with respect to load balance. The cost function we have used is the sum of the elements. For each data set we report the results from the following algorithms:

- **SBD** The optimal SBD distribution as described in Section 3.2.
- **ASD** The approximation algorithm described in Section 3.2.
- **MRD** The Multiple recursive decomposition as described in Section 3.2.
- **GBD** The iterative algorithm from Section 3.1 for computing a GBD.
- **GBA** The simple approximation algorithm mentioned in Section 3.1.
- **UD** The uniform distribution.
- **BRD** The binary recursive decomposition [1].

To illustrate the behavior of the algorithms we tried them on 3 different types of test matrices. The function $rand()$ generates numbers from a uniform random sequence of non-negative integers less than $2^{15} - 1$.

1. **Skewed matrix:** The weights of the elements are skewed to the bottom right of the array. The matrix elements are: $a_{ij} = (rand() \bmod 7)(i+j)$
2. **Peak matrix:** The matrix has a distinct peak, randomly chosen at (r, c). The matrix elements are: $a_{ij} = (rand() \bmod 127)/((|r-i|)(|c-j|) + 1.0)$
3. **Diagonal dominant matrix:** The matrix elements are: $a_{ij} = (rand() \bmod 127)/((|i-j|) + 2.0)$

For each type of matrix we have performed three series of experiments. First we keep the size of the matrix and the value of q fixed, while p is increased. We then increase both p and q while keeping the size of matrix fixed and finally we keep both p and q fixed while changing the size of the matrix. The results from the tests are presented in Table 1 through 3. For all test matrices the results are normalized relative to $(\sum_{i,j} a_{i,j})/(pq)$ which is the theoretical lower bound for any distribution.

The algorithm for computing the GBD distribution has been initialized in three different ways: (i) Starting with the delimiters as far left as possible, (ii) as far right as possible, and (iii) the Uniform distribution. The results obtained differ only marginally from each other. We therefore only report the best result for each test case. The number of iterations needed before a converged solution is obtained varies from 3 to 25 with an average of 6. The number of iterations is the sum of the number of vertical and horizontal partitionings performed. There does not appear to be any correlation between the number of iterations and the goodness of the obtained solution.

Problem size				Semi-general dist.			General dist.			
m	n	p	q	SBD	ASD	MRD	GBD	GBA	UD	BRD
256	256	2	2	1.01	1.01	1.01	1.06	1.06	1.50	1.00
256	256	4	2	1.01	1.01	1.01	1.06	1.11	1.64	
256	256	8	2	1.02	1.02	1.02	1.07	1.17	1.66	
256	256	16	2	1.04	1.04	1.06	1.09	1.24	1.72	
256	256	32	2	1.06	1.07	1.12	1.13	1.20	1.78	
256	256	64	2	1.12	1.12	1.25	1.18	1.28	1.75	
256	256	4	4	1.01	1.02	1.01	1.09	1.15	1.73	1.02
256	256	8	8	1.04	1.04	1.05	1.11	1.23	1.88	1.04
256	256	16	16	1.06	1.07	1.08	1.18	1.32	1.93	1.10
256	256	32	32	1.16	1.17	1.23	1.32	1.55	2.15	1.23
256	256	64	64	1.30	1.30	1.66	1.65	1.92	2.50	1.50
256	32	16	16	1.27	1.35	1.41	1.35	1.44	2.40	1.42
256	64	16	16	1.16	1.19	1.24	1.27	1.42	2.19	1.26
256	128	16	16	1.10	1.11	1.13	1.20	1.39	1.95	1.12
256	256	16	16	1.07	1.07	1.12	1.18	1.32	1.93	1.10

Table 1. Results from test using skewed matrices.

As expected the SBDs give a better load balance than the GBD.

Problem size				Semi-general dist.		MRD	General dist.			BRD
m	n	p	q	SBD	ASD		GBD	GBA	UD	
256	256	2	2	1.01	1.01	1.02	1.01	1.04	1.50	1.01
256	256	4	2	1.02	1.01	1.03	1.11	1.14	2.25	
256	256	8	2	1.02	1.07	1.02	1.18	1.33	3.16	
256	256	16	2	1.04	1.07	1.07	1.24	1.57	4.50	
256	256	32	2	1.10	1.10	1.27	1.31	1.37	6.44	
256	256	64	2	1.19	1.21	1.53	1.26	1.37	4.87	
256	256	4	4	1.02	1.06	1.02	1.23	1.40	4.41	1.04
256	256	8	8	1.04	1.07	1.05	1.35	2.27	6.94	1.06
256	256	16	16	1.12	1.21	1.34	1.65	3.15	16.00	1.48
256	256	32	32	1.39	1.58	1.70	1.75	3.33	23.43	2.19
256	256	64	64	9.93	9.93	9.93	9.93	9.93	25.16	9.93
256	32	16	16	1.40	1.63	1.96	1.68	2.11	9.32	1.63
256	64	16	16	1.27	1.54	1.54	1.67	2.16	8.73	1.41
256	128	16	16	1.16	1.26	1.33	1.48	3.34	9.27	1.32
256	256	16	16	1.12	1.27	1.55	1.55	2.73	14.05	1.24

Table 2. Results from tests using peak matrices.

Problem size				Semi-general dist.		MRD	General dist.			BRD
m	n	p	q	SBD	ASD		GBD	GBA	UD	
256	256	2	2	1.00	1.01	1.00	1.66	1.67	1.69	1.00
256	256	4	2	1.01	1.01	1.02	1.68	1.76	1.68	
256	256	8	2	1.03	1.04	1.03	1.68	1.80	1.79	
256	256	16	2	1.04	1.04	1.06	1.68	1.80	1.79	
256	256	32	2	1.09	1.09	1.11	1.75	1.87	1.86	
256	256	64	2	1.12	1.17	1.18	1.81	1.98	1.92	
256	256	4	4	1.02	1.02	1.02	2.02	2.91	2.74	1.03
256	256	8	8	1.06	1.06	1.09	3.22	4.98	4.31	1.07
256	256	16	16	1.14	1.17	1.26	5.02	8.15	6.47	1.20
256	256	32	32	1.50	1.70	1.87	7.63	13.07	9.40	1.61
256	256	64	64	3.93	4.72	4.94	11.08	18.71	14.34	3.02
256	32	16	16	1.81	2.59	2.59	2.68	5.33	9.17	1.68
256	64	16	16	1.37	1.51	1.86	3.51	5.47	7.74	1.65
256	128	16	16	1.21	1.22	1.44	4.08	6.69	6.71	1.41
256	256	16	16	1.13	1.13	1.30	5.26	8.03	6.58	1.25

Table 3. Results from tests using diagonal dominant matrices.

For almost every test problem the optimal SBD is fairly close to the lower bound given by the average cost. The most noticeable exception is the peak matrix of size 256×256 with $p = q = 64$. In this case the most expensive block consists of one single matrix element for the optimal SBD as well as for the iterative GBD. Thus the load imbalance we see here is inherit in the problem.

The ASD distribution is never far from the lower bound given by the optimal SBD. Compared with the lower time-complexity of the ASD this might make it a good choice. It also gives better load balance than the MRD distribution. While the binary recursive decomposition places in between ASD and MRD.

The GBD distribution outperforms the GBA distribution. This must, however, be compared with the higher time complexity of computing the GBD distribution.

For both the skewed and the peak matrices the GBD distribution is fairly close to the optimal SBD. From Theorem 4 it follows that for these matrices the presented GBD must be close to optimal. For the diagonally dominant matrices the difference is larger. However, we believe this to be a feature inherit in the definition of the GBD and not a consequence of our algorithm.

In general we see that it becomes harder to obtain a well balanced distribution as the ratios $\frac{m}{p}$ and $\frac{n}{q}$ become smaller.

It follows from the test results for both the SBD and the GBD that the more time one is willing to spend on obtaining a good distribution the better the load balance becomes.

As expected the results confirm that the uniform distribution is not suitable for matrices with non-uniform load.

5 Conclusion

We have presented an efficient iterative algorithm that computes a well balanced GBD. We have also developed an algorithm that computes an optimal SBD. These were tried on a number of test problems and compared with other approximation algorithms. This showed that the SBDs in general gave a more even load balance than the GBDs. This must, however, be compared with the more complicated communication pattern given by the SBD.

When choosing a distribution one must first determine what kind of communication needs one has. Based on this and the criticality of achieving an even load balance one can decide which type of distribution to use. Then depending on how much time one is willing to spend on calculating a distribution one can decide which algorithm to use.

An advantage of the GBD is that it is easy to specify, only requiring two vectors of length q and p whereas the SBD requires $p + p * q$ data elements.

In a resent development it has been shown that computing the optimal GBD is NP-hard for certain cost functions [9].

As a continuation of this work we are currently implementing several sparse matrix algorithms on a parallel computer. The object is to investigate how well the different partitioning schemes behave on real-world problems where both load balance and communication influence the overall time.

References

1. M. J. BERGER AND S. H. BOKHARI, *A partitioning strategy for nonuniform problems on multiprocessors*, IEEE Trans. Comput., C-36 (1987), pp. 570–580.

2. R. BLECK, S. DEAN, M. O'KEEFE, AND A. SAWDEY, *A comparison of data-parallel and message-passing versions of the Miami Isopycnic Coordinate Ocean Model (MICOM)*, Parallel Comput., 21 (1995), pp. 1695–1720.
3. S. H. BOKHARI, T. W. CROCKETT, AND D. M. NICOL, *Parametric binary dissection*, Tech. Rep. ICASE Report No. 93-39, Nasa Langley Research Center, 1993.
4. B. CHAPMAN, P. MEHROTRA, AND H. ZIMA, *Programming in Vienna Fortran*, Sci. Prog., 1 (1992), pp. 31–50.
5. ———, *High performance Fortran languages: Advanced applications and their implementation*, Future Generation Computer Systems, (1995), pp. 401–407.
6. ———, *Extending HPF for advanced data parallel applications*, IEEE Trans. Par. Dist. Syst., (Fall 1994), pp. 59–70.
7. G. FOX, M. JOHNSON, G. LYZENGA, S. OTTO, J. SALMON, AND D. WALKER, *Solving Problems on Concurrent Processors*, vol. 1, Prentice-Hall, Englewood Cliffs, NJ, 1988.
8. M. R. GAREY AND D. S. JOHNSON, *Computers and Intractability*, Freeman, 1979.
9. M. GRIGNI AND F. MANNE, *On the complexity of the generalized block distribution*. To appear in the proceedings of 1996 Workshop on Irregular Problems, 1996.
10. HIGH PERFORMANCE FORTRAN FORUM, *High performance language specification. Version 1.0*, Sci. Prog., 1–2 (1993), pp. 1–170.
11. *High Performance Fortran Forum Home Page*. http://www.crpc.rice.edu/HPFF/home.html.
12. F. MANNE, *Load Balancing in Parallel Sparse Matrix Computations*, PhD thesis, University of Bergen, Norway, 1993.
13. F. MANNE AND T. SØREVIK, *Optimal partitioning of sequences*, J. Alg., 19 (1995), pp. 235–249.
14. B. OLSTAD AND F. MANNE, *Efficient partitioning of sequences*, IEEE Trans. Comput., 44 (1995), pp. 1322–1326.
15. M. UJALDON, S. D. SHARMA, J. SALTZ, AND E. ZAPATA, *Run-time techniques for parallelizing sparse matrix problems*, in Proceedings of 1995 Workshop on Irregular Problems, 1995.
16. M. UJALDON, E. L. ZAPATA, B. M. CHAPMAN, AND H. P. ZIMA, *Vienna-Fortran/HPF extensions for sparse and irregular problems and their compilation*. Submitted to IEEE Trans. Par. Dist. Syst.
17. H. ZIMA, H. BAST, AND M. GERNDT, *Superb: A tool for semi-automatic MIMD/SIMD parallelization*, Parallel Comput., (1986), pp. 1–18.

P_ARPACK: An Efficient Portable Large Scale Eigenvalue Package for Distributed Memory Parallel Architectures

K. J. Maschhoff and D. C. Sorensen

Rice University

Abstract. P_ARPACK is a parallel version of the ARPACK software. ARPACK is a package of Fortran 77 subroutines which implement the Implicitly Restarted Arnoldi Method used for solving large sparse eigenvalue problems. A parallel implementation of ARPACK is presented which is portable across a wide range of distributed memory platforms and requires minimal changes to the serial code. The communication layers used for message passing are the Basic Linear Algebra Communication Subprograms (BLACS) developed for the ScaLAPACK project and Message Passing Interface(MPI).

1 Introduction

ARPACK is a collection of Fortran77 subroutines designed to solve large scale eigenvalue problems. ARPACK stands for ARnoldi PACKage. ARPACK software is capable of solving large scale non Hermitian (standard and generalized) eigenvalue problems from a wide range of application areas. The software is designed to compute a few, say k, eigenvalues with user specified features such as those of largest real part or largest magnitude using only $n \cdot \mathcal{O}(k) + \mathcal{O}(k^2)$ storage. A set of Schur basis vectors for the desired k dimensional eigen-space is computed which is numerically orthogonal to working precision. Eigenvectors are also available upon request. Parallel ARPACK (P_ARPACK) is provided as an extension to the current ARPACK library and is targeted for distributed memory message passing systems. The message passing layers currently supported are BLACS and MPI.

The Arnoldi process is a technique for approximating a few eigenvalues and corresponding eigenvectors of a general $n \times n$ matrix. It is most appropriate for large structured matrices $\mathbf{A}$ where structured means that a matrix-vector product $\mathbf{w} \leftarrow \mathbf{A v}$ requires $\mathcal{O}(n)$ rather than the usual $\mathcal{O}(n^2)$ floating point operations (Flops). This software is based upon an algorithmic variant of the Arnoldi process called the Implicitly Restarted Arnoldi Method (IRAM). When the matrix $\mathbf{A}$ is symmetric it reduces to a variant of the Lanczos process called the Implicitly Restarted Lanczos Method (IRLM). These variants may be viewed as a synthesis of the Arnoldi/Lanczos process with the Implicitly Shifted QR technique that is suitable for large scale problems. For many standard problems, a matrix factorization is not required. Only the action of the matrix on a vector is needed.

The important features of ARPACK and P_ARPACK are:

- A reverse communication interface.
- Ability to return k eigenvalues which satisfy a user specified criterion such as largest real part, largest absolute value, largest algebraic value (symmetric case), etc. For many standard problems, the action of the matrix on a vector $\mathbf{w} \leftarrow \mathbf{Av}$ is all that is needed.
- A fixed pre-determined storage requirement suffices throughout the computation. Usually this is $n \cdot \mathcal{O}(k) + \mathcal{O}(k^2)$ where k is the number of eigenvalues to be computed and n is the order of the matrix. No auxiliary storage or interaction with such devices is required during the course of the computation.
- Sample driver routines are included that may be used as templates to implement various spectral transformations to enhance convergence and to solve the generalized eigenvalue problem.
- Special consideration is given to the generalized problem $\mathbf{Ax} = \mathbf{Mx}\lambda$ for singular or ill-conditioned symmetric positive semi-definite $\mathbf{M}$.
- Eigenvectors and/or Schur vectors may be computed on request. A Schur basis of dimension k is always computed. The Schur basis consists of vectors which are numerically orthogonal to working accuracy. Computed eigenvectors of symmetric matrices are also numerically orthogonal.
- The numerical accuracy of the computed eigenvalues and vectors is user specified. Residual tolerances may be set to the level of working precision. At working precision, the accuracy of the computed eigenvalues and vectors is consistent with the accuracy expected of a dense method such as the implicitly shifted QR iteration.
- Multiple eigenvalues offer no theoretical or computational difficulty other than additional matrix-vector products required to expose the multiple instances. This is made possible through the implementation of deflation techniques similar to those employed to make the implicitly shifted QR algorithm robust and practical. Since a block method is not required, the user does not need to "guess" the correct block size that would be needed to capture multiple eigenvalues.

2 Parallelizing ARPACK

The parallelization paradigm found to be most effective for ARPACK on distributed memory machines was to provide the user with a Single Program Multiple Data (SPMD) template. The reverse communication interface is one of the most important aspects in the design of ARPACK and this feature lends itself to a simplified SPMD parallelization strategy. This approach was used for previous parallel implementations of ARPACK [2] and is simple for the the user to implement. The reverse communication interface feature of ARPACK allows the P_ARPACK codes to be parallelized internally without imposing a fixed parallel decomposition on the matrix or the user supplied matrix-vector product.

Memory and communication management for the matrix-vector product can be optimized independent of P_ARPACK . This feature enables the use of various matrix storage formats as well as calculation of the matrix elements on the fly.

The calling sequence to ARPACK remains unchanged except for the addition of the BLACS context (or MPI communicator). Inclusion of the context (or communicator) is necessary for global communication as well as managing I/O. The addition of the context is new to this implementation and reflects the improvements and standardizations being made in message passing [9, 7].

2.1 Data Distribution of the Arnoldi Factorization

The numerically stable generation of the Arnoldi factorization

$$\mathbf{AV}_k = \mathbf{V}_k\mathbf{H}_k + \mathbf{f}_k\mathbf{e}_k^T$$

where

$\mathbf{A}$, $n \times n$ matrix
$\mathbf{H}_k$, $k \times k$ - projected matrix (Upper Hessenberg)
$\mathbf{V}_k$, $n \times k$ matrix, $k \ll n$ - Set of Arnoldi vectors
$\mathbf{f}_k$, residual vector, length n, $\mathbf{V}_k^T\mathbf{f}_k = 0$

coupled with an implicit restarting mechanism [1] is the basis of the ARPACK codes. The simple parallelization scheme used for P_ARPACK is as follows.

Arnoldi Factorization

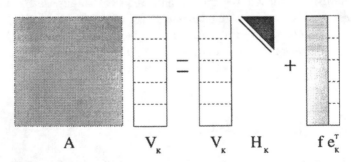

- $\mathbf{H}_k$ replicated on every processor
- $\mathbf{V}_k$ is distributed across a 1-D processor grid. (Blocked by rows)
- $\mathbf{f}_k$ and workspace distributed accordingly

The SPMD code looks essentially like the serial code except that the local block of the set of Arnoldi vectors, $\mathbf{V}_{loc}$, is passed in place of $\mathbf{V}$, and n_{loc}, the dimension of the local block, is passed instead of n.

With this approach there are only two communication points within the construction of the Arnoldi factorization inside P_ARPACK: computation of the 2-norm of the distributed vector f_k and the orthogonalization of f_k to V_k using Classical Gram Schmidt with DGKS correction [5]. Additional communication will typically occur in the user supplied matrix-vector product operation as well. Ideally, this product will only require nearest neighbor communication among the processes. Typically the blocking of $\mathbf{V}$ is commensurate with the parallel decomposition of the matrix $\mathbf{A}$. The user is free, however, to select an appropriate blocking of $\mathbf{V}$ such that an optimal balance between the parallel performance of P_ARPACK and the user supplied matrix-vector product is achieved.

The SPMD parallel code looks very similar to that of the serial code. Assuming a parallel version of the subroutine matvec, an example of the application of the distributed interface is illustrated as the follows:

```
10   continue
     call psnaupd (comm, ido, bmat, nloc, which, ...,
    *                         Vloc , ... lworkl, info)
     if (ido .eq. newprod) then
        call matvec ('A', nloc, workd(ipntr(1)), workd(ipntr(2)))
     else
        return
     endif
     go to 10
```

Where, nloc is the number of rows in the block Vloc of V that has been assigned to this node process.

Typically, the blocking of V is commensurate with the parallel decomposition of the matrix A as well as with the configuration of the distributed memory and interconnection network. Logically, the V matrix be partitioned by blocks

$$\mathbf{V}^T = (\mathbf{V}^{(1)^T}, \mathbf{V}^{(2)^T}, ..., \mathbf{V}^{(nproc)^T})$$

with one block per processor and with $\mathbf{H}$ replicated on each processor. The explicit steps of the process responsible for the j block are shown in Table 1.

Note that the function $gnorm$ at Step 1 is meant to represent the global reduction operation of computing the norm of the distributed vector $\mathbf{f}_k$ from the norms of the local segments $\mathbf{f}_k^{(j)}$ and the function $gsum$ at Step 3 is meant to represent the global sum of the local vectors $\mathbf{h}^{(j)}$ so that the quantity $\mathbf{h} = \sum_{j=1}^{nproc} \mathbf{h}^{(j)}$ is available to each process on completion. These are the only two communication points within this algorithm. The remainder is perfectly parallel. Additional communication will typically occur at Step 2. Here the operation $(\mathbf{A}loc)v$ is meant to indicate that the user supplied matrix-vector product is able to compute the local segment of the matrix-vector product $\mathbf{A}v$ that is consistent with the partition of $\mathbf{V}$. Ideally, this would only involve nearest neighbor communication among the processes.

(1) $\beta_k \leftarrow gnorm(\|f_k^{(*)}\|); \quad v_{k+1}^{(j)} \leftarrow f_k^{(j)} \cdot \frac{1}{\beta_k};$

(2) $w^{(j)} \leftarrow (Aloc)v_{k+1}^{(j)};$

(3) $\begin{pmatrix} h \\ \alpha \end{pmatrix}^{(j)} \leftarrow \begin{pmatrix} V_k^{(j)T} \\ v_{k+1}^{(j)T} \end{pmatrix} w^{(j)}; \begin{pmatrix} h \\ \alpha \end{pmatrix} \leftarrow gsum \left[\begin{pmatrix} h \\ \alpha \end{pmatrix}^{(*)} \right]$

(4) $f_{k+1}^{(j)} \leftarrow w^{(j)} - (V_k, v_{k+1})^{(j)} \begin{pmatrix} h \\ \alpha \end{pmatrix};$

(5) $H_{k+1} \leftarrow \begin{pmatrix} H_k & h \\ \beta_k \ e_k^T & \leftarrow \end{pmatrix};$

(6) $V_{k+1}^{(j)} \leftarrow (V_k, v_{k+1})^{(j)};$

Table 1. The explicit steps of the process responsible for the j block.

Since **H** is replicated on each processor, the parallelization of the implicit restart mechanism described in [1, 2] remains untouched. The only difference is that the local block $\mathbf{V}^{(j)}$ is in place of the full matrix **V**. All operations on the matrix **H** are replicated on each processor. Thus there is no communication overhead but there is a "serial bottleneck" here due to the redundant work. If k is small relative to n this bottleneck is insignificant. However, it becomes a very important latency issue as k grows and will prevent scalability if k grows with n as the problem size increases.

The main benefit of this approach is that the changes to the serial version of ARPACK were very minimal. Since the change of dimension from matrix order n to its local distributed blocksize `nloc` is invoked through the calling sequence of the subroutine `psnaupd`, there is no fundamental algorithmic change to the code. Only eight routines were affected in a minimal way. These routines either required a change in norm calculation to accomodate distributed vectors (Step 1), modification of the distributed dense matrix-vector product (Step 4), or inclusion of the context or communicator for I/O (debugging/tracing). More specifically, the commands are changed from

```
rnorm = sdot (n, resid, 1, workd, 1)
rnorm = sqrt(abs(rnorm))
```

to

```
rnorm = sdot (n, resid, 1, workd, 1)
call sgsum2d(comm,'All',' ',1, 1, rnorm, 1, -1, -1 )
rnorm = sqrt(abs(rnorm))
```

where `sgsum2d` is the BLACS global sum operator. The MPI implementation uses the MPI_ALLREDUCE global operator. Similarly, the computation of the matrix-vector product operation $h \leftarrow V^T w$ requires a change from

```
    call sgemv ('T', n, j, one, v, ldv, workd(ipj), 1,
  *             zero, h(1,j), 1)
```

to

```
    call sgemv ('T', n, j, one, v, ldv, workd(ipj), 1,
  *             zero, h(1,j), 1)
    call sgsum2d( comm, 'All', ' ', j, 1, h(1,j), j,
  *               -1, -1 )
```

Another strategy which was tested was to use Parallel BLAS (PBLAS) [8] software developed for the ScaLAPACK project to achieve parallelization. The function of the PBLAS is to simplify the parallelization of serial codes implemented on top of the BLAS. The ARPACK package is very well suited for testing this method of parallelization since most of the vector and matrix operations are accomplished via BLAS and LAPACK routines.

Unfortunately this approach required additional parameters to be added to the calling sequence (the distributed matrix descriptors) as well as redefining the workspace data structure. Although there is no significant degradation in performance, the additional code modifications, along with the data decomposition requirements, make this approach less favorable. As our parallelization is only across a one dimensional grid, the functionality provided by the PBLAS was more sophisticated than we required. The current implementation of the PBLAS (ScaLAPACK version 1.1) assumes the matrix operands to be distributed in a block-cyclic decomposition scheme.

2.2 Message Passing

One objective for the development and maintenance of a parallel version of the ARPACK [3] package was to construct a parallelization strategy whose implementation required as few changes as possible to the current serial version. The basis for this requirement was not only to maintain a level of numerical and algorithmic consistency between the parallel and serial implementations, but also to investigate the possibility of maintaining the parallel and serial libraries as a single entity.

On many shared memory MIMD architectures, a level of parallelization can be accomplished via compiler options alone without requiring any modifications to the source code. This is rather ideal for the software developer. For example, on the SGI Power Challenge architecture the MIPSpro F77 compiler uses a POWER FORTRAN Accelerator (PFA) preprocessor to automatically uncover the parallelism in the source code. PFA is an optimizing Fortran preprocessor that discovers parallelism in Fortran code and converts those programs to parallel code. A brief discussion of implementation details for ARPACK using PFA preprocessing may be found in [6]. The effectiveness of this preprocessing step is still dependent on how suitable the source code is for parallelization. Since most of the vector and matrix operations for ARPACK are accomplished via BLAS and LAPACK routines, access to efficient parallel versions of these libraries alone will provide a reasonable level of parallelization.

Unfortunately, for distributed memory architectures the software developer is required to do more work. For distributed memory implementations, message passing between processes must be explicitly addressed within the source code and numerical computations must take into account the distribution of data. In addition, for the parallel code to be portable, the communication interface used for message passing must be supported on a wide range of parallel machines and platforms. For /small P_ARPACK, this portability is achieved via the Basic Linear Algebra Communication Subprograms (BLACS) [7] developed for the ScaLAPACK project and Message Passing Interface (MPI) [9].

3 Parallel Performance

To illustrate the potential scalability of Parallel ARPACK on distributed memory architectures some example problems have been run on the Maui HPCC SP2. The results shown in Table 1 attempt to illustrate the potential internal performance of the of the P_ARPACK routines independent of the users implementation of the matrix vector product.

In order to isolate the performance of the ARPACK routines from the performance of the user's matrix-vector product and also to eliminate the effects of a changing problem characteristic as the problem size increases, a test was comprised of replicating the same matrix repeatedly to obtain a block diagonal matrix. This completely contrived situation allows the workload to increase linearly with the number of processors. Since each diagonal block of the matrix is identical, the algorithm should behave as if $nproc$ identical problems are being solved simultaneously (provided an appropriate starting vector is used). For this example we use a starting vector of all "1's". The only obstacles which prevent ideal speedup are the communication costs involved in the global operations and the "serial bottleneck" associated with the replicated operations on the projected matrix H. If neither of these were present then one would expect the execution time to remain constant as the problem size and the number of processors increase.

The matrix used for testing is a diagonal matrix of dimension $100,000$ with uniform random elements between 0 and 1 with four of the diagonal elements separated from the rest of the spectrum by adding an additional 1.01 to these elements. The problem size is then increased linearly with the number of processors by adjoining an additional diagonal block for each additional processor. For these timings we used the non-symmetric P_ARPACK code pdnaupd with the following parameter selections: mode is set to 1, number of Ritz values requested is 4, portion of the spectrum is "LM", and the maximum number of columns of V is 20.

Number of Nodes	Problem Size	Total Time (s)	Efficiency
1	100,000 * 1	40.53	
4	100,000 * 4	40.97	0.98
8	100,000 * 8	42.48	0.95
12	100,000 * 12	42.53	0.95
16	100,000 * 16	42.13	0.96
32	100,000 * 32	46.59	0.87
64	100,000 * 64	54.47	0.74
128	100,000 * 128	57.69	0.70

Table 2. Internal Scalability of P_ARPACK

4 Availability

The codes are available by anonymous ftp from

`ftp.caam.rice.edu`

The ARPACK package is in

`pub/software/ARPACK/arpack96.tar.gz`

Parallel ARPACK (P_ARPACK) is in

`pub/software/ARPACK/parpack96.tar.gz`

Follow the instructions in the README files. ARPACK and P_ARPACK software is also available on Netlib in the directory `scalapack`.

5 Summary

We have presented a parallel implementation of the ARPACK library which is portable across a wide range of distributed memory platforms. The portability of P_ARPACK is achieved by utilization of the BLACS and MPI. We have been quite satisfied with how little effort it takes to port P_ARPACK to a wide variety of parallel platforms. So far we have tested P_ARPACK on a SGI Power Challenge cluster using PVM-BLACS and MPI, on a CRAY T3D using Cray's implementation of the BLACS, on an IDM SP2 using MPL-BLACS and MPI, on a Intel paragon using NX-BLACS and MPI, and on a network of Sun stations using MPI and MPI-BLACS.

6 Research Funding of ARPACK

Financial support for this work was provided in part by the National Science Foundation cooperative agreement CCR-912008, and by the ARPA contract number DAAL03-91-C-0047 (administered by the U.S. Army Research Office).

References

1. D. C. Sorensen, *Implicit application of polynomial filters in a k-step Arnoldi method,* SIAM Journal on Matrix Analysis and Applications, 13(1):357-385, January 1992.
2. D. C. Sorensen, *Implicitly-Restarted Arnoldi/Lanczos Methods for Large Scale Eigenvalue Calculations,* (invited paper), in Parallel Numerical Algorithms: Proceedings of an ICASE/LaRC Workshop, May 23-25, 1994, Hampton, VA, D. E. Keyes, A. Sameh, and V. Venkatakrishnan, eds., Kluwer, 1995 (to appear).
3. R.B. Lehoucq, D.C. Sorensen, P.A. Vu, and C. Yang, *ARPACK: Fortran subroutines for solving large scale eigenvalue problems,* Release 2.1
4. R. B. Lehoucq and D.C. Sorensen *Deflation Techniques for an Implicitly Re-started Arnoldi Iteration* , To appear in SIAM Journal of Matrix Analysis
5. J. Daniel, W.B. Gragg, L. Kaufman, and G.W. Stewart *Reorthogonalization and stable algorithms for updating the Gram-Schmidt QR factorization* , Mathematics of Computation, 30:772-795, 1976
6. M.P. Debicki, P. Jedrzejewski, J. Mielewski, P. Przybyszewski, and M. Mrozowski *Application of the Arnoldi Method to the Solution of Electromagnetic Eigenproblems on the Multiprocessor Power Challenge Architecture,* Technical Report Number 19/95, Department of Electronics, Technical University of Gdansk, Poland.
7. J. J. Dongarra and R. C. Whaley *LAPACK Working Note 94, A User's Guide to the BLACS v1.0,* , June 7, 1995
8. J. Choi, J. Dongarra, S. Ostrouchov, A. Petitet, D. Walker, and R. C. Whaley *LAPACK Working Note 100, A Proposal for a Set of Parallel Basic Linear Algebra Subprograms* , May 1995
9. Message Passing Interface Forum, *MPI: A Message-Passing Interface Standard* , International Journal of Supercomputer Applications and High Performance Computing, 8(3/4), 1994

Parallel Distributed Representation of Sparse Grids Using Process Arrays *

Michael May

Technical University of Munich
Institute of Computer Science
Department of Computer Architecture and Computer Organization
Arcisstr.21,
D - 80290 Munich, Germany
email: maym@informatik.tu-muenchen.de

Abstract. In this paper a new data structure for the distributed representation and efficient handling of sparse grids on parallel architectures is introduced. The new data layout is making use of the message passing paradigm and provides dynamic partioning and load balancing transparent to the application making use of it. This way sequential partial differential equations (PDE) solvers based on sparse grids can be data parallelized with hardly any source code modifications.

Sparse Grids

In every mathematic model that needs adaptive full grids in order to reduce the number of unknowns, *sparse grids*, which were introduced 1990 by [4], can help to further decrease memory and computing requirements while maintaining nearly the same accuracy as on full grids ([3]).

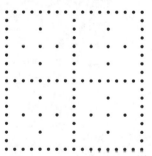

Figure 1: A sparse grid $S_{4,4}^{(2)}$

In figure 1 we see a simple sparse grid on the unit square with mesh width $h = 2^{-4}$ whose structure corresponds to a binary tree of binary trees[2]. Every node of the tree represents a grid line in x-direction whereas the level of a node

---

* This work is being funded by the ministry for education, science, research and technology BMBF within the project PAR-CVD (Parallel simulation of Chemical Vapor Deposition processes)

[2] the boundary grid lines have to handled separately

determines the number of grid points on this grid line. The points on a given grid line can be respresented by individual second order binary trees starting from the grid line nodes. This way sparse grid points can be described by a level and an index information, a fact which can be exploited for hashing.

By summing up we see immediately that sparse grids only contain $(h^{-1} \cdot (\mathrm{ld}(h^{-1}))^{d-1})$ grid points compared to (h^{-d}) for usual full grids, where d denotes the dimension of the problem and h the mesh width on the boundary. On the other hand the interpolation error in the L_2 norm only increases slightly from order (h^2) to $(h_l^2 \cdot (\mathrm{ld}(h^{-1}))^{d-1})$ with sufficiently smooth functions.

Further optimizations in memory and time can be achieved by adapting the sparse grid to the function to be interpolated. This involves coarsening the grid on interpolation points that contribute hardly any weight to the error and the solution vector and - on the other hand - refining the grid on points being crucial to both the error and the solution ([4]).

The just mentioned properties of sparse grids are sufficient to understand the scope of this paper beginning in the next section. For the curious reader however, we refer to [4], [1] and [3] .

Data Structures

In literature the only data structure considered useful to implement sparse grids are systems of binary trees. They represent the hierarchical structure and the low density of sparse grids very well, in contrast to inefficient and expensive arrays.

The problems arising from this data structure layout are obvious: depending on the target function, the binary trees can become arbitrarily deep and hence degenerate into linked lists. This fact hinders both the efficient access and distribution of the data onto the available processors, which are only possible from balanced trees.

For use with sparse grid based PDE solvers we thus need a representation of sparse grids that offers both efficient search operations for arbitrary grid points[3] and easy data partitioning for parallel processing. In the next section we present a module `Sparse_Grid` for the parallel representation of such grids satisfying the demands just mentioned.

The Process Array approach

We propose an new data layout – based on an multikey file structure for accessing disk storage presented by [2] – for the parallel representation of sparse grids and will discuss its suitability for partitioning and load balancing. The implementation and properties (variables) of the data layout are encapsulated inside a module and can only be accessed by means of exported module methods which we will present later. This way the numerical algorithm using the

---

[3] while insertions and deletions during the building and destruction of the sparse grid may be slow

module `Sparse_Grid` does not and need not know anything about the internal representation and distribution of the grid points. Modifications to step from a sequential numerical program to a data parallel one are thus minimal.

An incarnation of `Sparse_Grid` is characterized by the size of the individual dimensions and the number of maximal grid points leading to the notion of a *virtual* array as the sparse grid points are internally stored in hash tables instead of arrays. Instances of `Sparse_Grid` can be initialized and destroyed by the methods

$$\text{SG\_alloc(), SG\_dealloc()}$$

The properties of each individual grid point are the level and index information and a special field to hold additional information like function values or the hierarchical surplus. In order to read and set these properties there are the methods

$$\text{SG\_insertnode(), SG\_deletenode, SG\_findnode(), SG\_set()}$$

In order to initialize, synchronize and re-gather the data distribution and parallel calculation there are three more external methods,

$$\text{SG\_distribute(), SG\_sync(), SG\_collect(),}$$

Everything else, for instance the partitioning and load balancing schemes by means of processor *splits* and processor *merges*, the internals of synchronization, the storing and accessing of grid points in processor local hash tables, is fully transparent to the numerical algorithm. In the subsequent subsections we will give an overview of all hidden features of the module `Sparse_Grid` and discuss its suitability for partitioning and load balancing.

Process Arrays

For notational simplicity we only regard the two-dimensional case, as the adaption to the general case is straight-forward.

For a parallel computation the target area covered by a sparse grid should be divided into subregions of uniform point density. This is achieved by means of a so called *process array*, a two-dimensional array

$$\text{process\_array[0..nx-1][0..ny-1],}$$

defining the correspondence of convex 2-dimensional quads of the grid to associated processors or processes. Each cell of `process_array` holds the identification number of the processor or process responsible for the appropriate subregion. The grid points in these subregions are stored in (local) processor memory in (individual) hash tables.

The partition of sparse grid (that yields the process array) is determined by so called *scales*, vectors

$$\text{xscale[0..nx]} \; and \; \text{yscale[0..ny]}$$

describing the partition of each dimension of the grid respectively. These data structures are replicated on each processor of the parallel machine.

The access (find, insert, delete) to a sparse grid point with level l and index i is as follows (see figure 2):

1. The level and index of the grid point are checked against the x-scale and y-scale to determine the appropriate row x and column y in the process array.
2. Next it is checked whether the pointer process_array[x][y] matches the processor of the computing process. If so, the grid point is accessed in the local processor memory by efficient hashing techniques. A write access can lead to subsequent processor splits or merges.
3. If process_array[x][y] does not match, the access to the grid point is delegated by passing a message to the corresponding processor.

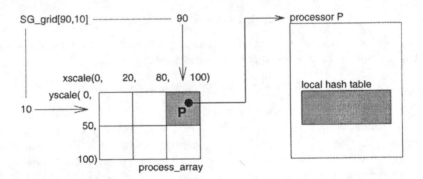

Figure 2: Access to a sparse grid point

Splitting and Merging

Partitioning and dynamic load balancing are achieved by means of processor *splits* and processor *merges*, if individual processor loads[4] reach a maximal (minimal) value by the repeated insertion (deletion) of grid points. Two parameters influence the dynamic correspondence of the processors to the sparse grid sub-regions, the *splitting policy* and the *merging policy*.

1. The **splitting policy** is characterized by three features:
   □ The (maximal) *processor load*, of which any overflow leads to a split.
   □ The *dimension* that is chosen for the next split, i.e. the axis to which the partitioning hyper-plane is orthogonal.[5]
   □ The *location* of the split point, i.e. the point at which the linear scale is partitioned

---

[4] e.g. the number of locally stored sparse grid points
[5] For instance, a very simple splitting policy would choose the dimensions cyclically. More elaborate policies could favor some dimensions and thus lead to higher resolutions.

2. The **merging policy** is determined by the following decisions:
   - ☐ The (minimal) *processor load*, of which any underflow triggers a merge operation.
   - ☐ Which pair of adjacent subregions is a possible *candidate* for a merge.[6]
   - ☐ If there are several candidates, which one wins the *competition*.

These policies of the module enable an equal distribution of the virtual sparse grid array over the processors and re-balance the processor load after each split or merge thus implementing a dynamic partitioning.

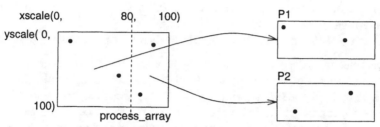

In order to explain the dynamic behaviour of the module `Sparse_Grid` we supply an example of a splitting operation in figure 3 and 4.

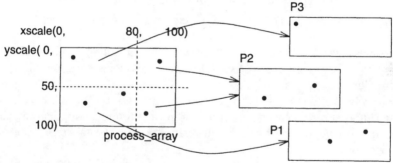

As there is no a-priori information about the number of points contained in a sparse grid, a dynamic distribution using subsequent processor splits from the very beginning can lead to unequally balanced grid points and thus processor loads. In such cases it is more efficient to build up the full sparse grid on all processors, split the grid in subregions and then delete the grid points not needed for the local computation on each processor respectively. A further adaption of the sparse grid to the target function can be balanced by processor splits again.

Hash Tables

For memory and performance reasons the points of the virtual sparse grid are internally stored in hash tables. In the case of sparse grids it is evident from the introduction that information of level and index is sufficient to identify a

---

[6] Examples for algorithms to choose possible merge partners are the well known *buddy* system and the more general *neighbor* system.

sparse grid point. Consequently, the position of a node is determined by the result of a hash function that computes an array-index of the hash table from the parameters level and index.

Synchronization

As already mentioned above, access to non-local grid points is delegated by passing a message to the processor responsible for the appropriate subregion. Messages in the module `Sparse_Grid` are used in a uniform manner:

1. The messages are tagged with a flag indicating the type of action to be performed by the receiving processor, and
2. contain all the data necessary for the indicated operation.

For example, in order to insert a new grid point in the grid region of another processor, a message containing the level, index and additional information is packed, tagged with `SIG_INS` and sent to the appropriate processor. At a given time, the appropriate function on the remote processor, here `SG_insertnode()`, is called with the parameters that were delivered in the message.

The message tags are

$$SIG\_INS, SIG\_DEL, SIG\_GET, SIG\_SPLIT, SIG\_MERGE$$

to signal a nonlocal insertion, deletion, find, split and merge operation respectively.

The messages are handled and buffered in a message queue until they are received by the appropriate process. This message queue is worked off by a message handler inside the module, `SG_sync()`. Only by calling `SG_sync()` messages will be received and corresponding actions performed. By placing calls to the SG message handler on synchronization points in the numerical application, the user has complete control over the strictness and timing of synchronization among the processors.

For instance, in conventional scientific computing, a call to `SG_sync()` should be placed after every time step, i.e. at the end of the time-step loop. To support local time stepping inside the target domains, `SG_sync()` could be called only every, say 50 time steps. Eventually, calling `SG_sync()` after every write access to the sparse grid would implement a very strict synchronization.

Conclusion and Future Work

We have presented a module built on top of a message passing library to distribute and parallelize sparse grids nearly transparently to the numerical application. As a case study we are about to parallelize a sparse grid based Navier-Stokes-solver for use with `Fastest` 3D, a simulation package for chemical vapor deposition processes in the electronic and optoelectronic field.

As the module `Sparse_Grid` is suitable for grids of arbitrary dimensions, it is obvious to implement simulations of d-dimensional problems as well i.e. to handle heat transport and chemical reaction equations.

References

1. Bungartz, H.:
   Dünne Gitter und deren Anwendung bei der adaptiven Lösung der dreidimensionalen Poisson-Gleichung
   Ph.D. Thesis, Technical University of Munich (1992)
2. Nievergelt, J., Hinterberger, H., Sevcik, K. C.:
   The Grid File: An Adaptable, Symmetric Multikey File Structure
   ACM Transactions on Database Systems, Vol. 9, No. 1 (1981)
3. Griebel, M.:
   Parallel Multigrid Methods on Sparse Grids
   SFB-Bericht Nr. 342/30/90 A, Technical University of Munich (1990)
4. Zenger, C.:
   Sparse Grids
   SFB-Bericht Nr. 342/28/90 A, Technical University of Munich (1990)

Optimal Scheduling, Decomposition and Parallelisation

S McKee, [†]

P Oliveira, [‡] C Aldridge[†]
J McDonald [††]

Abstract. This study is concerned with the optimal scheduling of an electricity power system. Using Lagrangian relaxation, the original primal problem may be rewritten in a dual formulation; the problem then admits decomposition into more tractable subproblems. Genetic algorithms are discussed and applied to this problem allowing even greater parallelisation. Finally, the concept of genetically engineered algorithms is introduced, and a project which is underway at the University of Strathclyde is described.

1. Introduction

Power system scheduling involves decisions concerning which units should be run and what their level of output should be. These two decisions are usually referred to as the Unit Commitment and Economic Dispatch problems. Together these manifest themselves as a large-scale mixed integer programming problem. Power systems can vary from a small number of units to several hundred; and the planning period can vary from short term (1 to 7 days) to long term (several weeks up to a year).

This highly combinatorial problem is further complicated by two main stochastic inputs: the demand which the system must satisfy and, in the case of hydro systems, the inflows to the reservoirs. The full extent of this problem is such that it is essential to employ the most efficient algorithm and the most appropriate computer architecture configuration. To this end, a dual formulation has been obtained through Lagrangian relaxation of the original primal problem. This admitted decomposition into more tractable subproblems which has allowed the implementation of the algorithm on the Edinburgh Concurrent Supercomputer (ECS).

The purpose of this paper is to emphasise the potentiality for parallelisation of this complicated problem. With this in mind, and for reasons both of space and clarity of exposition, we shall restrict this discussion to the problem with thermal units only.

[†] Department of Mathematics, University of Strathclyde, Glasgow

[‡] Department of Production and Systems Engineering, Universidade do Minho, Braga

[††] Department of Electronic and Electrical Engineering, University of Strathclyde, Glasgow.

2. Lagrangian Relaxation

Consider the following problem (eg Geoffrion (1970a,b), Luenberger (1984)):

$$\min_{\mathbf{x}} \sum_{i=1}^{q} f_i(\mathbf{x}_i)$$

subject to

$$\sum_{i=1}^{q} \mathbf{h}_i(\mathbf{x}_i) = 0 \quad \text{and} \quad \sum_{i=1}^{q} \mathbf{g}_i(\mathbf{x}_i) \le 0$$

where the components of the n-vector $\mathbf{x}$ are partitioned into q disjoint subvectors, $\mathbf{x} = (\mathbf{x}_1, \ldots, \mathbf{x}_q)$, not necessarily with the same number of components. Associating the dual variable $\boldsymbol{\lambda}$ with the equality constraints and $\boldsymbol{\mu} \ge 0$ with the inequality constraints, the dual function becomes

$$\Phi(\boldsymbol{\lambda}, \boldsymbol{\mu}) = \min_{\mathbf{x}} \sum_{i=1}^{q} \left\{ f_i(\mathbf{x}_i) + \boldsymbol{\lambda}^T \mathbf{h}_i(\mathbf{x}) + \boldsymbol{\mu}^T \mathbf{g}_i(\mathbf{x}_i) \right\}$$

The corresponding dual problem decomposes into q separable problems

$$\min_{\mathbf{x}_i} \left\{ f_i(\mathbf{x}_i) + \boldsymbol{\lambda}^T \mathbf{h}_i(\mathbf{x}_i) + \boldsymbol{\mu}^T \mathbf{g}_i(\mathbf{x}_i) \right\} \quad (i = 1, \ldots, q)$$

which in principle can be solved more efficiently than the original problem.

The unit-commitment-economic dispatch problem has special characteristics which can be exploited in order to construct separable problems. These characteristics (Tong and Shahidiehpour (1989)) can be stated as follows:

(1)   The commitment variables are the only ones to be restricted to integer values; when these are fixed, the problem becomes one of continuous optimisation.

(2)   The demand and reserve constraints are global in the sense that they couple all the generating units; all other constraints are local, representing the different operating restrictions of each unit.

In Lagrangian relaxation (Lauer et al. (1982); Bertsekas et al. (1983); Merlin & Sandrin (1988); Oliveira et al. (1992a)), the second characteristic is used to create a separable problem by relaxing the coupling constraints so that each subproblem involves only one individual unit subject to its own local (operating) restrictions. The local subproblems are parameterized by the Lagrange multipliers, and the master problem maximises a dual function, producing new estimates of the Lagrange multipliers while ensuring that the two global constraints are met.

3. The Mathematical Model

The electricity generating system under consideration includes a nuclear component of the "must-run" type. It is clear that, for optimal scheduling, the nuclear component may be subtracted out. The system to be considered here, therefore, consists of thermal, and hydro and pumped-storage units only. For brevity we shall only treat the thermal units; details of the others may be found in Oliveira et al. (1992a,b). The time interval over which the system operates is divided into T equal time periods.

3.1 The thermal system

Suppose there are N thermal units, and let the power (MW) produced by a unit i during period t be x_i^t. Each unit has an associated integer variable which denotes whether the unit is committed (on) or not (off)

$$\alpha_i^t = \begin{cases} 1 & \text{if unit } i \text{ is on during period } t, \\ 0 & \text{otherwise,} \end{cases} \quad \underline{x}_i \alpha_i^t \le x_i^t \le \overline{x}_i \alpha_i^t,$$

for $i = 1, \ldots, N$ and for $t = 1, \ldots, T$, where $\underline{x}_i$ and $\overline{x}_i$ are respectively the minimum and maximum power output for unit i.

The costs are of four types

(a) Running cost per unit is

$$F_i \alpha_i^t + V_i x_i^t \ (i = 1, \ldots, N; \ t = 1, \ldots, T).$$

Here F_i is the fixed cost and V_i is the slope of the linear approximation to the fuel cost over the range $[\underline{x}_i, \overline{x}_i]$. The running costs are zero for $x_i^t = 0$.

(b) Start-up cost per unit

$$U_i \beta_i^t \ (i = 1, \ldots, N; \ t = 1, \ldots, T).$$

Here U_i is the start-up cost for the unit i, and β_i^t is an integer variable defined as

$$\beta_i^t = \begin{cases} 1 & \text{if unit } i \text{ is started in period } t, \\ 0 & \text{otherwise.} \end{cases}$$

Hence

$$\beta_i^t \ge \alpha_i^t - \alpha_i^{t-1} \ (i = 1, \ldots, N; \ t = 2, \ldots, T).$$

(c) Shut-down cost per unit is

$$D_i \gamma_i^t \ (i = 1, \ldots, N; \ t = 1, \ldots, T),$$

where D_i is the shut-down cost for unit i, and γ_i^t is an integer variable defined as

$$\gamma_i^t = \begin{cases} 1 & \text{if unit } i \text{ is shut down in period } t, \\ 0 & \text{otherwise.} \end{cases}$$

So

$$\gamma_i^t \geq \alpha_i^{t-1} - \alpha_i^t \ (i = 1, \ldots, N; \ t = 2, \ldots, T).$$

(d) The state of the thermal unit τ_i^t, the number of hours the unit has been on or off, is defined as

$$\tau_i^{t+1} = \begin{cases} \tau_i^t + 1 & \text{if } \tau_i^t \geq 1 \quad \text{and} \quad \alpha_i^{t+1} = 1, \\ 1 & \text{if } \tau_i^t \leq -1 \quad \text{and} \quad \alpha_i^{t+1} = 1, \\ \tau_i^t - 1 & \text{if } \tau_i^t \leq -1 \quad \text{and} \quad \alpha_i^{t+1} = 0, \\ -1 & \text{if } \tau_i^t \geq 1 \quad \text{and} \quad \alpha_i^{t+1} = 0, \end{cases}$$

for $i = 1, \ldots, N$ and for $t = 1, \ldots, T$.

Hence

$$\tau_i^t \in \begin{cases} \{1, \ldots, \overline{\tau}_i\} & \text{if } \alpha_i^t = 1, \\ \{\underline{\tau}_i, \ldots, -1\} & \text{if } \alpha_i^t = 0, \end{cases}$$

where $\underline{\tau}_i$ and $\overline{\tau}_i$ are respectively the minimum down and up times of unit i.

3.2 The demand and reserve constraints

In practice power output has to satisfy a stochastic demand. However, time series analysis (or some other technique) may be employed to yield a deterministic demand and this is the approach taken here. Moreover, if one unit breaks down, then it is not possible to start an uncommitted thermal unit immediately, and so a certain amount of reserve, known as spinning reserve, should be available. There are several reserve policies that could be adopted: a constant reserve in every time period; a variable reserve; a reserve enough to cover the loss of the largest thermal unit; or a reserve assessed on the basis of risk analysis (eg Muckstadt and Koenig (1977); Merlin and Sandrin (1983)). In this study, we shall consider satisfying the demand, while allowing just sufficient excess capacity to cover the constant reserve imposed upon the generator by its adherence to a regulatory authority.

Therefore

$$\begin{aligned} \sum_{i=1}^{N} x_i^t &\geq d^t \\ \sum_{i=1}^{N} \overline{x}_i \, \alpha_i^t &\geq d^t + R \end{aligned} \qquad (t = 1, \ldots, T),$$

where both hydro and pump-storage have been omitted for clarity (see eg Oliveira et al. (1992a).

3.3 The mixed integer model

In summary, the scheduling problem can be modelled as the mixed integer
linear programming problem (P) defined by

$$
\min_{\substack{\alpha_i^t,\, x_i^t \\ i=1,\ldots,N \\ t=1,\ldots,T}} \sum_{t=1}^{T} \sum_{i=1}^{N} \left(U_i\,\beta_i^t + F_i\,\alpha_i^t + V_i\,x_i^t + D_i\,\gamma_i^t \right)
$$

subject to the constraints of Sections 3.1 and 3.2.

4. Lagrangian Relaxation in Power Scheduling

The implementation of Lagrangian relaxation for a particular problem can
only be achieved when certain questions have been answered (Fisher (1981),
(1985)), and these questions condition the implementation itself. For instance,
which restrictions should be relaxed from all those constraining the problem?
Clearly the answer to this is problem-specific and must be such as to create
a much simpler problem to solve. In power scheduling, the most suitable restrictions to be relaxed are the global ones, that is, the demand and reserve
constraints. The inclusion of these two constraints gives the Lagrangian problem defined by

$$
\Phi(\lambda_1,\lambda_2) = \min_{\substack{\alpha_i^t,\, x_i^t \\ (i=1,\ldots,N) \\ (t=1,\ldots,T)}} \left\{ \sum_{t=1}^{T} \sum_{i=1}^{N} \left(U_i\,\beta_i^t + F_i\,\alpha_i^t + V_i\,x_i^t + D_i\,\gamma_i^t \right) \right.
$$

$$
+ \sum_{t=1}^{T} \lambda_1^t \left(d^t - \sum_{i=1}^{N} x_i^t \right)
$$

$$
\left. + \sum_{t=1}^{T} \lambda_2^t \left(d^t + R - \sum_{i=1}^{N} \overline{x}_i\,\alpha_i^t \right) \right\}
$$

To emphasise the problem decomposition we write

$$
\Phi(\lambda_1,\lambda_2) = \sum_{i=1}^{N} \Phi_i^1(\lambda_1,\lambda_2) + \sum_{t=1}^{T} (\lambda_1^t\,d^t + \lambda_2^t(d^t + R))
$$

where

$$
\Phi_i^1(\lambda_1,\lambda_2) = \min_{\substack{\alpha_i^t,\, x_i^t \\ (t=1,\ldots,T)}} \sum_{t=1}^{T} \left(U_i\,\beta_i^t + F_i\,\alpha_i^t + V_i\,x_i^t \right.
$$

$$
\left. + D_i\,\gamma_i^t - \lambda_1^t\,x_i^t - \lambda_2^t\,\overline{x}_i\,\alpha_i^t \right), \quad (i=1,\ldots,N)
$$

represents the thermal-units subproblems. Each of these subproblems is locally constrained by the operating characteristics of the individual units. The second term

$$\sum_{t=1}^{T} (\lambda_1^t \, d^t + \lambda_2^t (d^t + R))$$

represents the overall demand and reserve.

5. Parallelisation

Figure 1 displays the information exchange between the master problem and the local subproblems. Clearly, given the values of the Lagrange multipliers, the subproblems can be solved independently of each other. Consequently an algorithmic parallelisation can be implemented such that all subproblems are solved simultaneously; so the subproblems may be regarded as one stage in a two-stage sequential program consisting of this stage and the master problem. This will be efficient provided that the master program is not too time-consuming compared with the subproblems and there are not great imbalances between the computational times of the subproblems.

Thus, the dual problem (D) can be stated as

$$\max_{\lambda_1, \lambda_2} \Phi(\lambda_1, \lambda_2)$$

subject to $\quad\quad\quad\quad\quad \lambda_1 \geq 0, \quad \lambda_2 \geq 0.$

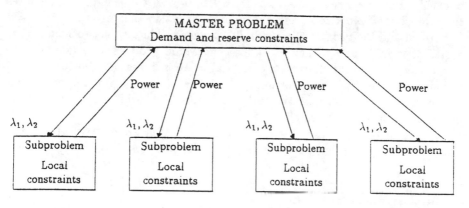

Figure 1

In order to solve the dual problem D, a solution to the subproblems must be sought. Since the dimensionality of each subproblem is quite small this was accomplished by dynamic programming, which can cope with the nonlinearities in the constraints region of the subproblems.

On the other hand the maximisation of Φ with respect to λ_1 and λ_2 implies the successive computation of the multipliers λ_1 and λ_2, and this raises the question as to which method should be chosen to compute those multipliers. There are several methods available, for instance, the subgradient method, and multiplier adjustment methods. It should be noted that $\Phi(\lambda, \lambda_2)$ is a concave function and not differentiable. Fisher ((1981), (1985)) notes the success of the subgradient method on a large number of different applications, pointing out not only its ease of programming but also its robustness. For these reasons the subgradient method was selected. Details of the implementation may be found in Oliveira et al. (1992a).

6. Genetic algorithms

Genetic algorithms (GAs) (see eg Oliveira et al. (1994)) are search and optimisation techniques based loosely on the evolutionary adaption of genes. They differ from conventional optimisation techniques in that they work with a population of candidate solutions which are encoded as strings (or vectors).

The bits making up these strings are usually 0, 1 variables. A fitness function (or objective function) is employed to evaluate the strings. Successive generations are created by applying genetic operators and at each generation the basic GA selects fitter solutions, applies the genetic operators, crossover and mutation, with certain prescribed probability and then tests for convergence. (Convergence in the case of GAs means convergence to a uniform population.)

If at generation N we select, say, the Kth member of the population (string K) according to its fitness F_K then mutation involves randomly choosing a particular bit and changing its value from 1 to 0 (or 0 to 1). Crossover requires the selection of two strings and the random selection of a bit where a cut is performed and the two substrings are interchanged. Thus if the two strings were represented by $\mathbf{x}$ and $\mathbf{y}$ then a random cut would produce $\mathbf{x} = (\mathbf{x}_1, \mathbf{x}_2)$ and $\mathbf{y} = (\mathbf{y}_1, \mathbf{y}_2)$ where the dimension of x_i and y_i were the same. Crossover would produce the new strings $(\mathbf{x}_1, \mathbf{y}_2)$ and $(\mathbf{y}_1, \mathbf{x}_2)$. Generally, the probability of crossover occurring tends to be taken to be relatively high, while the probability of mutation is chosen to be small.

It is clear that GAs are simple and flexible and they are massively parallelisable. They do, however, demand large, often intractably large computational time.

7. Further Parallelisation

There are several ways in which GAs could be implemented to solve the electricity generating problem. We suggest the following. Genetic algorithms are used to generate a population of commitment schedules, that is a population of binary strings of the form $(\alpha_1^1, \alpha_2^1, \ldots, \alpha_N^1, \alpha_1^2, \ldots, \alpha_N^T)$. The next (sequential) step is to evaluate the corresponding schedules using a standard optimisation technique, for instance, linear programming in the case of a linear objective function.

The reserve constraint

$$\sum_{i=1}^{N} \bar{x}_i \, \alpha_i^t \geq d^t + R$$

is evaluated in the integer solution by, for example, checking that each potential string satisfies this constraint. Moreover, minimum up and down times for each unit can be automatically respected in the problem with a subsequent reduction in the feasible solution space (eg see the ideas of Saitoh et al. (1994)).

The linear programming problem is still a global problem due to the presence of the global demand constraint. However the coefficient matrix has the following structure

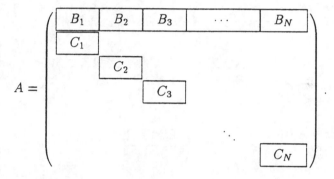

The ideas of the Dantzig-Wolfe (Dantzig and Wolfe (1960)) decomposition method can then be applied to this problem to allow further parallelisation. Figure 2 conveys the structure of the method. This work is near completion and will be reported on in a later paper.

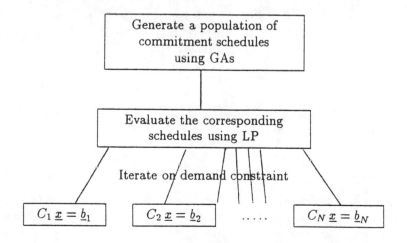

Figure 2

8. Genetically Engineered Algorithms

A large project is underway at Strathclyde, funded partly by the National Grid Company plc, to produce and understand genetically engineered algorithms for electricity scheduling. The idea is to substantially reduce the computation time by underpinning the solution technique by an expert system or systems. Existing knowledge can be elicited from the dispatch operators or from archived schedules. However, scheduling software needs to be run continually, possibly every hour, as new estimates of the demand are calculated. Moreover, the domain seldom changes—new power stations or hydro-dams are only built infrequently. Thus rule induction techniques may be employed to deduce new knowledge which ultimately will be likely to be more reliable than operator knowledge. The project may be summarised schematically.

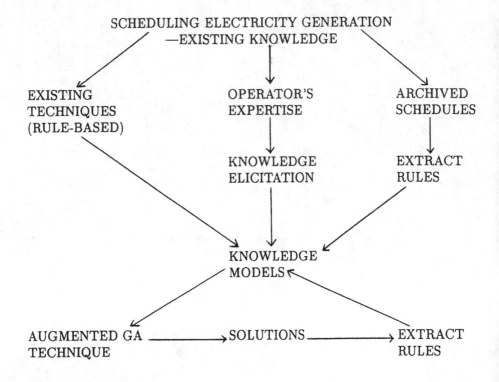

9. Conclusions

This paper contains essentially three ideas. First a Lagrangian formulation with decomposition was described using the subgradient method and dynamic programming. This approach was implemented on the Edinburgh Concurrent Supercomputer which in 1992 consisted of 300 T800 transputers and 100 T414 transputers. All details may be found in Oliveira et al. (1992b). In short it certainly resulted in increased efficiency, but the pumped-storage led to an unbalanced workload. Since pumped-storage is a small part of most systems this may not be too important. The second idea was further improved parallelisation through the use of genetic algorithms. Finally, the concept of genetically engineered algorithms was introduced and a project currently underway at the University of Strathclyde was briefly described.

Acknowledgements

The authors would like to acknowledge funding from the EPSRC and the National Grid Company plc.

References

Bertsekas, D, Lauer, G, Sandell, N and Posbergh, T. Optimal short-term scheduling of large-scale power systems. IEEE Trans AC **28** (1983) 1-11.

Dantzig, G and Wolfe, P. A decomposition principle for linear programs. Ops Res **8** (1960) 101–111.

Fisher, M. The Lagrangian relaxation method for solving integer programming problems. Mgmt Sci **27** (1981) 1-18.

Fisher, M. An applications orientated guide to Lagrangian relaxation. Interfaces, **15** (2) (1985) 10-21.

Geoffrion, A. Elements of large scale mathematical programming–Part I: concepts. Mgmt Sci **16** (1970a) 652-675.

Geoffrion, A. Elements of large scale mathematical programming–Part II: synthesis of algorithms and bibliography. Mgmt Sci **16** (1970b) (676-691).

Lauer, G, Bertsekas, D, Sandell, N and Posbergh, T. Solution of large-scale optimal unit commitment. IEEE Trans PAS **101** (1982) 79-86.

Luenberger, D. Linear and Nonlinear Programming. Addison-Wesley (1984).

Merlin, A and Sandrin, P. A new method for unit commitment at Electricité de France. IEEE Trans PAS **102** (1983) 1218–1225.

Muckstadt, J and Koenig, S. An application of Lagrangian relaxation to scheduling in power-generation systems. Ops Res **25** (1977) 387-403.

Oliveira, P, McKee, S and Coles, C. Lagrangian relaxation and its application to the unit-commitment-economic-dispatch problem. IMA J Maths Appl Bus and Indust. **4** (1992a) 261-272.

Oliveira, P, Blair-Fish, J, McKee, S and Coles, C. Parallel Lagrangian relaxation in power scheduling. Comput Sys Engng **3** (1992b) 609-612.

Oliveira, P, McKee, S and Coles, C. Optimal scheduling of a hydro thermal power generation system. Eur J Op Res **71** (1993) 334-340.

Oliveira, P, McKee, S and Coles, C. Genetic algorithms and optimising large nonlinear systems. In: Artificial Intelligence in Mathematics by J Johnson, S McKee and A Vella, OUP (1994).

Saitoh, H, Inoue, K and Toyoda, J. Genetic algorithm approach to unit commitment. In: Proceedings of the International Conference on Intelligent System Application to Power Systems—A Hertz, A Holden and J Rault (Eds), (1994) 583-589.

Tong, S and Shahidehpour, S. An overview of power generation scheduling in the optimal operation of a large scale power system. Proceedings of the Power Industry Computer Applications (1989).

An Implementation of a Tree-Based N-Body Algorithm on Message-Passing Architectures

Abdullah I. Meajil

The George Washington University
Department of Electrical Engineering and Computer Science
Washington, DC 20052

abdullah@seas.gwu.edu

Abstract

In this paper, a "manager-worker" model for a parallel implementation of hierarchical N-body algorithm is introduced. We describe a load-balanced, extremely simple algorithm for solving the Astrophysics simulation of N-body problem using tree-based data structures and coarse-grained parallel architectures. This algorithm, based on the Barnes-Hut method, first assembles a tree data structure which represents the distribution of bodies, or particles, at all length-scales. A domain decomposition, or an adaptive load balancing technique is used to assign bodies to processors as well as to insure that processors are assigned equal amounts of work. Therefore, the problem of load balancing for parallelized particle simulations implemented on MIMD machines is addressed and a simple dynamic load balancing algorithm, called costzones, is employed. A number of measurements were carried out in order to reveal the behavior of the N-body application with respect to a partitioning technique and load imbalance overhead. We also show that, with using the introduced manager-worker model and the costzones domain decomposition technique, the algorithm is load balanced and that the majority of the time of the algorithm is spent in performing on-processor functions and not in inter-processor communications. We have conducted our study on several MIMD machines such as the 256-processor Cray T3D and the 64-processor Intel Paragon at JPL/ESS-NASA and on the 32-processor Thinking Machines CM-5 at UMC.

Keywords: Astrophysics Simulations; Barnes-Hut Algorithm; Dynamic Load Balancing; MIMD Machines; Performance Analysis

1 Introduction

The study of physical systems by particle simulation is called the "many-body" or the "N-body" problem. N-body simulations have been used to study a wide variety of astrophysical systems during the past years, ranging from small clusters of stars to galaxies and the formation of large-scale structure in the universe. Such studies are conducted in celestial mechanics, plasma physics, and fluid dynamics, as well as in semiconductor device simulation. Particle simulations, or N-body problems, were among the first applications to be implemented when scalar computers first came into widespread use about thirty years ago, and have since played an important role in many science and engineering applications [1]. The advent of parallel computers promises to revolutionize this field, because of the inherent parallelism in all particle algorithms. The basic particle algorithm consists of performing a well defined set of calculations for each particle in the system. Therefore, the classical N-body problem simulates the evolution of a system comprising *n* bodies, or particles, under the influence of forces exerted on each body by the whole system. Notice that, both body and particle terms are being exchangeablely used in this paper.

The particle interactions themselves can be broadly classified into two classes: (1) those involving long-range interactions and (2) those involving short-range interactions. Typical examples from class (1) include the infinite range electromagnetic and gravitational forces, while a typical class (2) example is the elastic impact interaction. Although the algorithmic approaches for these two classes share some commonalties, the differences, especially when it comes to parallelization, are so great that they are usually treated separately [2].

Tree-based algorithms approximate the solution to equation of Newtonian gravity. In these algorithms the particles are sorted into a spatial hierarchy which forms a tree date structure. Each node in the tree then represents a grouping of particles. Data which represents average quantities of these particles (e.g., total mass, center of mass, and high order moments of the mass distribution) are computed and stored at the nodes of the tree. The forces are then computed by having each particle search the tree and pruning subtrees from the search when the average data stored at that node can be used to compute a force on the searching particle below a user supplied accuracy limit. For a fixed level of accuracy this algorithm scales as $n \log n$. However, since the tree search for any one particle is not known in advance and the tree is unstructured, this presents problems for distributed memory parallel implementations of this algorithm (because minimizing any off processor accesses of data is needed). On the other hand, the problem does posses a highly parallel component: each particle searches the tree structure completely independently of all other particles in the system.

In this paper we introduce a *"manager-worker"* model, in which a manager processing node starts up several worker processing nodes then gathers and interprets their results, for parallelizing the Barnes-Hut algorithm. With this model, the manager creates the tree where all spatial information about all particles are inserted. Then the manager broadcasts the tree to all worker processing nodes. Each worker processing node does its work without access to data that is being held by other nodes. Each worker node, then, sends its updated particles to the manager node in order to create an updated tree which is to be used in the next time-step.

This paper is organized as follows. Next section overviews the mathematical basis for the N-body problem as well as some of methods to solve the N-body simulation in the literature. Section 3 presents the sequential Barnes-Hut method. A simple parallel algorithm of Barnes-Hut method used "manger-worker" model is introduced in section 4. Section 5 discusses the data structures for this parallel implementation as well as the target high performance parallel computing platforms for this study. Section 6 presents the performance measurements analysis. Finally, conclusions for this paper results are in section 7.

2 N-body Algorithms on Literature

2.1 Mathematical Basis

Because the N-body problem can not be done in closed form, the calculation must be done numerically. That is, at each time t, the gravitational forces of each mass on each of the others may be computed by Newton's laws. Newton's law of gravity states that the force between any pair of particles is proportional to the product of their masses divided by the square of the distance between them. When there are many particles, the acceleration of each particle is given by the sum of the accelerations (as computed by Newton's law) induced by all the other particles. By many iterations of this method of calculation, the position of each particle after an arbitrary length of time may be found.

The general N-body problem may be stated as the following set of differential equations [10]:

$$dx_i/dt = \upsilon_i \tag{1}$$

$$m_i \times d\upsilon_i/dt = \Sigma_{j \neq i} F_{ij}$$

(2)

In astrophysical simulations, the force term, F_{ii} is the Newtonian gravity:

$$F_{ii} = Gm_i m_i r_{ii} / |r_{ii}|^3 \tag{3}$$

This is simply a large set of differential equations that integrated numerically using a simple algorithm.

The gravitational force is "long-range", meaning that there is no cutoff, beyond which the force may be considered negligible. In principle, it is necessary to evaluate the entire sum on the right-hand side of equation (2) at each time-step dt of the time integration. Naively, this requires $O(n^2)$ operations at each time-step.

2.2 N-body Methods

There are different algorithms to solve this problem. In the simple method, all pair-wise forces are computed directly which involves the calculation of all $n(n-1)/2$ forces. The time complexity is $O(n^2)$ where n is the number of particles. Therefore, this method allows an accurate description of the dynamically evolution but at a price that grows rapidly for increasing number of particles.

In the Miller method, the solution technique represents the problem in a position-velocity space, and transforms the force field using a fast Fourier Transform into a form where it can be applied in linear time. This method is of complexity $O(n \log n)$. Although the calculations can be performed more quickly, the result will be of less accuracy.

The Aaerseth method involves in keeping track, for each particle, of the sets of "nearby" particles and "faraway" particles. The faraway particles may be integrated with large time-steps than the nearby particles. When the particles are uniformly distributed, the time complexity is $O(n^{1.5})$, otherwise $O(n^2)$.

In the Hockney technique, the universe is divided into cells. First the particle-to-particle interactions are computed, and then the cell-to-cell interactions are handled. This technique has complexity $O(n^{4/3})$ for a uniform distribution. There is a variation of Hockney method, of complexity $O(n \log n)$, where the cell-to-cell interactions are computed by a Fast Fourier Transform.

A new way of realizing a tree-based force calculation with logarithmic growth, $O(n \log n)$, of force terms per particle that avoids the tree-tangling complications has been presented by Barnes-Hut [3]. This algorithm allows rigorous upper bounds for errors that arise from neglecting internal lump structure, and also offers a well-defined procedure for estimating more typical, average errors. The Barnes-Hut algorithm, which is the most widely used and promising hierarchical N-body, uses a tree-structured hierarchical subdivision of space into cells, each of which is recursively divided into eight subcells whenever more that one particle is found to occupy the same cell. The potential method involves a number of operations that grow only as $(n \log n)$. This calculations can be performed more quickly with the time complexity $O(n \log n)$.

3 The Barnes-Hut Method

3.1 Basic Principles

Because interactions occur between each pair of particles in an N-body simulation, the computational work scales asymptotically as n^2. Much effort has been expended to reduce the computational complexity of such simulations, while retaining acceptable accuracy. One approach is Barnes-Hut method. The Barnes-Hut (BH) algorithm [3] is one of a number of algorithms [1-5] which use a multipole expansion and a hierarchical data structure to reduce the complexity of computing long-range interactions like gravity. The multipole expansion allows one to treat a collection of bodies as a point mass (perhaps with quadruple and higher moments) located at the center of mass. In Figure 1, the force on point x_i may be evaluated approximately as:

$$F_i = \Sigma_i \; Gm_i m_j r_{ij} / |r_{ij}|^3 \text{, or}$$

$$F_i \approx Gm_i MR_{cm} / |R_{cm}|^3 \tag{4}$$

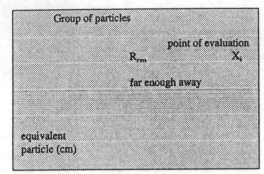

Figure 1: *Approximation of particles by a single equivalent particles.*

The quality of the approximation in equation (4) is a decreasing function of the ratio: $b/|R_{cm}|$, where b is the radius of the collection of bodies. In the BH algorithm, multipole moments are computed for cubical cells for an oct-tree of variable depth. The tree is constructed at each time-step with the following properties: (*i*) the root cell encloses all the bodies; (*ii*) no terminal cell contains more than 1 body; and (*iii*) any cell with 1 body is a terminal cell.

To compute the force on a body, one traverses the tree staring at the root. Any time a cell with a sufficiently small value of $b/|R_{cm}|$ is encountered, the multipole approximation is utilized. Thus, distance cells, which comprise many individual bodies, may be approximated in unit time. The resulting algorithm, when applied to all bodies, requires $O(n \log n)$ operations to evaluate forces on all n bodies [10].

3.2 The Barnes-Hut Algorithm

In the BH method [3], the force-computation phase within a time-step of an N-body calculation is expanded into the following phases:

(**I**) **Building the tree:** The current positions of the bodies, or particles, are first used to determine the dimensions of the root cell of the tree (since particles move between time-steps, the bounding box that contains the entire distribution may change and has to be recomputed every time-step). The tree is then built by adding particles one by one into the initially empty root cell, and subdividing a cell into its four children as soon as it contains more than a single particle (the termination condition for the Barnes-Hut method). The result is a tree whose internal nodes are cells and whose leaves are individual particles. Empty cells resulting from a cell subdivision are ignored. The tree (and the Barnes-Hut algorithm) is therefor adaptive in that it extends to more levels in regions that have high particle densities. Since n particles are being loaded into an initially empty tree, and since the expected height of the tree when the ith particles is inserted is $\log (i)$, the expected computational complexity of this phase is

$$\sum_{i=1}^{n} \log(i) = \log(n!) = O(n \log n).$$

(**II**) **Computing cell centers of mass:** Once the tree is built, the next step is to compute the centers of mass of all internal cells, since cells are approximated by their centers of mass when permitted by the hierarchical force-computation algorithm (see step 3 below). An upward pass is made through the tree staring at the leaves, to compute the center of mass of internal cells in the tree, and hence the expected computational complexity of this phase is $O(n)$.

(**III**) **Computing forces:** The tree is traversed once per particle to compute the net force acting on that particle. The force-computation algorithm for a particle starts at the root of the tree and conducts the following test recursively for every cell it visits: If the cell's center of mass is far enough away from the particle, the entire subtree under that cell is approximated by a single particle at the cell's center of mass, and the force this center of mass exerts on the particle is computed; if, however, the center of mass is not far enough away from the particle, the cell must be "opened" and each of its subcells visited. A cell is determined to be far enough away if the following condition is satisfied:

$$l / d < \theta \tag{5}$$

where l is the length of a side of the cell, d is the distance of the particle from the center of mass of the cell, and θ is a user-defined particle traversed more levels of those parts of the tree which represent space that is physically close to it, and groups other particles at a hierarchy of length scales.

The complexity of the force-computation phase is determined by the number of particle-particle or particle-cell interactions examined. Since forces are computed on particles one by one, and there are n particles, the expected complexity is of the order of n multiplied by the average number of particles/cells that a particle interacts with. This average number of instructions per particle depends on both n as well as θ. Therefore, when θ goes to 0, a particle computes interactions with all other particles, in which case the complexity of force computation is $O(n^2)$. However, for particle values of θ $(0.3 < \theta < 1.2)$, the number of interactions per particle is of the order of the depth of the tree, or $O(\log n)$, in terms on n and $O(1/\theta^2)$ in terms of θ. The complexity of the force-computation phase is, therefore, $O(1/\theta^2 n \log n)$. It is the only phase whose complexity depends on θ.

(IV) Updating particle properties: Finally, the force acting on a particle is used to update such particle properties as acceleration, velocity and position. This phase does a constant amount of work per particle, so its computational complexity is $O(n)$.

To this end, the overall expected complexity of the sequential Barnes-Hut application, therefore, is $O(n \log n) + O(1/\theta^2 n \log n) + O(n)$ or $O(1/\theta^2 n \log n)$ asymptotically, which is the same as the complexity of the force-computation phase.

Partitioning and the domain decomposition: This phase is added in our parallel implementation in order to expedite the computation and minimize the load imbalance. The technique used for the domain decomposition is called costzones which was originally implemented on shared-memory parallel architectures [7]. This technique will be discussed in the next section.

4 Manager-Worker Model for Our Barnes-Hut Parallel Algorithm

Typically, in the sequential algorithm, the program first initializes the particle positions and velocities, and then iterates over a fixed number of time-steps to compute the inter-particle forces. Over 95% of the sequential execution time is spent in computing the inter-particle forces, this percentage increasing with problem size [6]. This routine consists of a few tens of lines of code, so it is essential to parallelize and tune it in order to obtain the maximum possible performance and computation speedup. However, it is evident that the tree building stage takes a relatively small fraction of the time in this case. If the number of particles per processor stays very large, where in the most realistic machines, tree building stage is not likely to take much time relative to the rest of the time-step computation. This is the reason why the tree building stage has not been parallelized.

In any message-passing parallel architecture, each node is essentially a separate computer, which has a CPU unit (processor) and a distinct memory space. Therefore, nodes can run distinct programs; and they can team up to work on the same problem and exchange data by passing messages. The most common programming model used with parallel computers is the "Single Program, Multiple Data (SPMD)" model. Because each node is an independent computer, we can use other programming model, the "*manager-worker*" model, in which one "manager" program starts up several "worker" programs on other nodes, then it gathers and interprets their results.

This programming model is used on our parallel tree-based N-body problem application. With this model, the manager node creates the tree where all spatial information about all particles is inhabitant. Then, the "manager" node sends the tree to each node which has a subset of particles to work on in order to compute the essential forces that interact with its particles. This step is done at every "worker" node in the machine. Therefore, each node does its work without access to data that is being held by other nodes. In this case, each node operates completely independently. So, each "worker" node would update information of all its particles which are a subset of all particles in the universe. The "worker" node, finally, sends its updated particles to the "manager" node in order to create an updated tree that would be used in the next time-step. Therefore, developing a parallel implementation of the Barnes-Hut algorithm proceeds in three steps:

(I) Domain decomposition: Since the goal of the load balancing algorithm is to track fluctuations in the particle density, then any procedure should converge at least as fast as these fluctuations develop. Therefore, to provide load balancing, it is clearly not good enough to partition the domain space statically among processors, since this will lead to very poor load balancing. Thus, domain decomposition technique is needed to divide the total workload equally amongst the processors. The total work is a summation of all works associated with particles. Therefore, the costzones domain decomposition technique in [7] would be adopted and implemented in our message-passing parallel implementation. Costzones technique was originally implemented for shared-memory parallel machines such as Stanford DASH system. This technique is very simple and does not have much computational overhead associated with it. It superior to other domain decomposition techniques such as the Orthogonal Recursive Bisection (ORB) [8], in terms of simplicity and computational overhead, and as well as the performance of load balancing.

The costzones technique takes advantage of another key into the hierarchical methods for classical N-body problems, which is that they already have a representation of the spatial distribution encoded in the tree data structure they use. Therefore, it can partition the tree rather than the partition the space directly. The cost of (or work associated with) every particle, which is the total amount of interactions between the particle and all others, as counted in the previous time-step, is stored with the particle. Every internal cell holds the sum of the costs of all cells (either leaf, particle, or internal) that are contained within it plus its own cost; these cell costs are computed during the upward pass through the tree that computes the cell centers of mass, or multipole expansions.

The total cost in the domain is divided among processors so that every processor has a contiguous, equal range or zone of costs (hence the name of costzones). Which costzone a particle belongs to is determined by the total cost up to that particle in an inorder traversal of the tree. In the costzones technique, processors descend the tree in parallel, picking up the particles that belong in their costzone. The partitioning technique requires only a few lines of code, has negligible runtime overhead and yields a very good balance. Therefore, a processor performs only a partial traversal of the tree.

(II) Build the tree: After building the BH tree in the "manager" node, the "manager" node will broadcast the BH tree to each "worker" node in the partition. Therefore, an identical copy of the BH tree would be available in each "worker" node. Therefore, this phase consumes a very little time of execution.

(III) Evaluate the forces: Proceeding with the force-evaluation algorithm exactly as in the sequential case, i.e., traverse the tree once for each particle, or body. Since each node has a subset of the bodies and a whole copy of the BH tree, there is no more need for inter-processor communications. In fact, the original sequential code for force evaluation may be used completely unchanged. The parallel algorithm produces results identical to the one in the serial algorithm, except for a very small amount of round-off error which results from the non-associatively of floating-point operations.

5 Implementation on Message-Passing Machines

5.1 Principle Data Structures

Conceptually, the main data structure in the application is the Barnes-Hut tree. Since the tree changes every time-step, it is implemented in the program with two arrays: an array of bodies that are leaves of the tree, and an array of internal cells in the tree. Among other information, every cell has pointers to its children, and it is these pointers that maintain the current structure of the tree. These is also a separate array of pointers to bodies and one of pointers to cells. These arrays are used by the processors to determine which bodies and cells they own: Every processor owns an equal contiguous chunk of pointers in these arrays, each chunk sized to be larger than the maximum number of the program is linearly proportional to the number of bodies for both uniform distributions (balanced tree) and non-uniform ones.

We have written the code, to run under distributed memory-parallel platforms, entirely in ANSI C and augmented with the appropriate NX (Paragon), CMMD (CM-5), or PVM (T3D) communication calls. This code has been ported to several message-passing parallel systems such as: the 256-processor Cray T3D and the 64-processor Intel Paragon at the Jet Propulsion Laboratory / NASA, and the 32-processor Thinking Machines CM-5 at UMC. Notably, the code also compiled and runs with no modifications, but the explicit system calls and their associates, on sequential platforms that supply an ANSI C compilers. In this paper, the results have been obtained for the N-body gravitational code on platforms of the above machines. The reported results have been produced by running the code for 10 time-steps and the user-defined parameter θ for errors in the accelerations was set to 0.50. The particle distribution is generated by using the RAND() function that generates a simulation for the universe. While this procedure may be questionable from a physical point of view, it ensures that the scaling behavior (as a function of n) will not be contaminated by differences arising from different spatial distributions of bodies.

5.2 Message-Passing Parallel Systems

The JPL/ESS Intel Paragon and Cray T3D systems and the Thinking Machines CM-5 at UMC were used to conduct this study. The Intel Paragon has a total of 64 nodes organized into a 16x4 mesh of which 54 are compute nodes and 8 are service nodes. Each node, an Intel GP node, is essentially a separate computer, with one compute and one communication $i860$ processors. Each of the 56 compute nodes has 32 Mbytes of memory. The programs can be developed in C or FORTRAN which are supported by NX library routines for communication and synchronization purposes. The JPL Intel Paragon is operated as application development platform, with interactive access to all of the compute nodes.

The Cray T3D is MIMD system with physically distributed but globally addressed memory. The JPL Cray T3D has a Cray Y-MP as its host system and currently consists of 256 processors each with 2 MWords (16 Mbytes) of DRAM memory. Each processor is a 64-bit DEC Alpha microprocessor with a frequency of 150 MHz capable of achieving 150 MFLOPS. The system is space-shared into partitions where the numbers of processors are powers of two. A node consists of two processors sharing a network support logic. All processors are connected by a bi-directional 3-D Torus system interconnect network. This topology ensures short connection paths and high bisectional bandwidth. The system software includes FORTRAN, C, and C++ compilers as well as tools for application performance analysis and parallel code debugging. The PVM is currently supported as are some lower level Cray libraries for passing data and messages among processors.

The Thinking Machines CM-5 at UMC is a multi-user, timeshared massively parallel system comprising 32 processing nodes, each with its own memory, supervised by a control processor, known as partition manager (PM). A CM-5 processing node (PE) consists of a RISC processor, a Network Interface chip, 4 memory units of 8 Mbytes RAM, 4 Vector Unit arithmetic accelerators connected through a 64-bit M-bus. The RISC processor (a SPARC chip with a clock rate of 32 MHz) acts as a control and scalar processing resource for the Vector Unit. PM and PEs in a CM-5 are connected to two communication networks, organized in a fat tree architecture: the data network, used for bulk data transfers in which each item has a single source and destination, and the control network, used for operations that involve all the nodes at once, such as synchronization operations, broadcasting and combing. The CM-5 offers two high level languages for data parallel programming: CM-Fortran and C*. For the message passing programming model CM-Fortran, C* and the standard C, C++, and FORTRAN 77 languages are available. All the languages are supported for the message passing primitives by the CMMD library.

6 Performance Measurements

Performance problems in asynchronous massively parallel programs are often the result of unforeseen and complex asynchronous interactions between autonomous processors. These parallel programs are run on massively parallel computers, which is informally defined as a collection of tens or even hundreds of processors, that cooperate in performing a single task. In a multiple-instruction, multiple-data (MIMD) computer, processors execute independent serial programs as processes that communicate with processes on other processors when needed; on other words, processes on individual processors communicate by explicitly sending and receiving messages.

Our extensive series of performance measurements were carried out on the 256-processor Cray T3D and the 64-processor Intel Paragon at JPL/NASA and on the 32-processor Thinking Machines CM-5 at UMC in order to reveal the behavior of the N-body simulation with respect to scalability and parallelization overheads including partitioning and load imbalance overhead. These were performed for different input sizes as well as for different number of processors. The extensive series of timings on the three parallel platforms are shown in Figures 2, 3, and 4 for T3D, Paragon, and CM-5 machines, respectively. The times reported are the total time for the entire experiments, where each experiment runs for 10 time-steps.

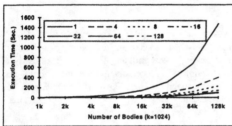

Figure 2: *Running time for N-body on T3D.*

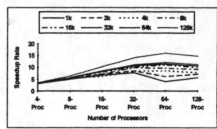

Figure 5: *Scalability on Cray T3D.*

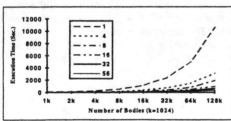

Figure 3: *Running time for N-body on Paragon.*

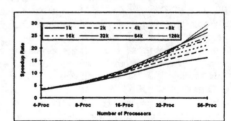

Figure 6: *Scalability on Paragon.*

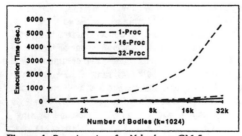

Figure 4: *Running time for N-body on CM-5.*

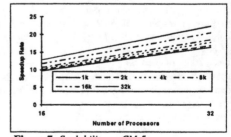

Figure 7: *Scalability on CM-5.*

Figures 5, 6, and 7 summarize the scalability measurements. In these figures, N-body scales nicely with the increasing number of processors, particularly when large data sets are used. This is consistent with the intuition driven from the way the parallel program works. In the used parallel program, building the tree was done sequentially at the manager node, recall the manager-worker model. This sequential part requires traversing the tree only once at each time-step. On the other hand, computing the forces at each particle was parallelized, which requires N traverses of the tree in the sequential case. With such N:1 growth ratio in the parallel to the sequential parts, near-linear speedup is expected. However, due to the rising communications cost from the manager to the workers, as the size of the machine grows and processors become more distant from one another, a gradual drop in efficiency is observed.

The overhead may be separated into components which combine additively to give the total parallelization overhead, i.e., $f = f_{comm} + f_{redun} + f_{imbal}$. The communication overhead, f_{comm}, arises from the time required to exchange data between processors. The redundancy overhead, f_{redun}, arises from the additional work required by the parallel implementation over and above that required by the serial implementation, i. e., the load balancing routine, costzones. Finally, f_{imbal}, measures the difference in speed between the average and slowest processor. A detailed study of overhead performance budget and scaleability for message-passing parallel architectures is presented in [9].

Figures 8 reports the communication overhead measurements for different input size sets on Intel Paragon system. As the number of processors increase, a corresponding increase of communications overhead takes place. This is a side effect due to the use of the manager-worker model, as distance variability from the manager increases with the increased number of workers. However, as the input data size increases, most of this overhead is amortized nicely.

The load balancing scheme, costzones, currently attempts to load balance only the work involved in force evaluation since this phase consumes most of the execution time. Therefore, the domain decomposition and tree construction work, which scales closely with the particle number, causes the load imbalance.

The load imbalance overhead, f_{imbal}, is shown in Figures 9, 10, and 11 for T3D, Paragon and CM-5, respectively. The load imbalance overhead for our N-body parallel implementation is less than 1% on Intel Paragon when $n > 8k$ bodies and on the average 4% on CM-5 of the total execution time. These Figures also demonstrate that the overhead f_{imbal} decreases as more bodies are added on a larger machine. Nevertheless, for large number of bodies, the overhead f_{imbal} is quite very small. This means that, one can expect improved performance from newer, larger machines. Besides, the cost of costzones partitioning grows very slowly with number of bodies or number of processors. Thus, the costzones scheme is not only much simpler to implement, but also results in better performance, particularly on larger machines. For more detailed study on load imbalance and the cost of partitioning technique, refer to [6].

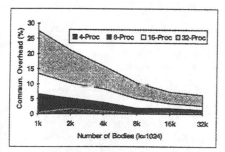

Figure 8: *Communication overhead on T3D.*

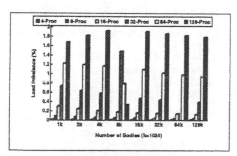

Figure 9: *Load imbalance on Cray T3D.*

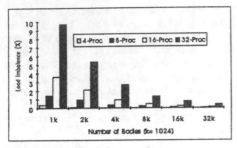

Figure 10: *Load imbalance on Paragon.*

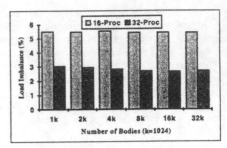

Figure 11: *Load imbalance on CM-5.*

7 Conclusions

We have described a simple load-balanced algorithm for implementing a gravitational N-body tree-based code on a message-passing parallel computer architectures. This parallel implementation of Barnes-Hut method is based on the manager-worker model. A simple mechanism for dynamically balancing the loads encountered in $O(N \log N)$ N-body simulations has been presented and shown to perform satisfactorily. The dynamic load balancing algorithm, costzones, presented here is local in the sense that only those processors directly affected participate in the decision making process. This domain decommission technique, in spite of being simple, let to a very balanced execution across the processors and resulted in a code that is easily portable onto any message-passing parallel system without much regard to machine topology.

As for practical applications of the present work, the N-body algorithm discussed in this paper has been implemented on several distributed memory parallel machines including the Cray T3D and the Intel Paragon systems at JPL/NASA and the CM-5 at UMC. In all cases the scalability and parallelization overheads have shown satisfactory results including the dynamic load balancing which has significantly increased the program performance.

References

[1] R. W. Hockney and J. W. Eastwood, *Computer Simulation Using Particles*, Adam Hilger, Bristol 1988.

[2] J. Katzenelson, *"Computational Structure of the N-Body Problem,"* SIAM J. Sci. Stat. Comp., v.10, no.4, pp.787-815, 1989.

[3] J. Barnes and P. Hut, *"A Hierarchical O(NlogN) Force-Calculation Algorithm,"* Nature, v. 324, pp. 446-449, 86.

[4] A. W Appel, *"An Efficient Program for Many-Body Simulation,"* SIAM J. Sci. Stat. Comput., v. 6, pp. 85-93, 85.

[5] L. Greengard and V. Rokhlin, *"A Fast Algorithm for Particle Simulations,"* J. Comp. Phys., v.73, pp. 325-348, 87.

[6] A.I. Meajil, *"A Load Balancing of O(NlogN) N-Body Algorithm on Message-Passing Architectures,"* 9th Intl. Conf. on Parallel and Distributed Computing Systems, Sep 25-27, 1996, Dijon, France.

[7] J. P. Singh and C. Holt, *"A Parallel Fast Multipole Method,"* Proc. Supercomputing'93, pp. 54-65, November 93.

[8] G. Fox, *Numerical Algorithms for Modern Parallel Computer Architectures*, pp. 37-62, Springer-Verlag, 1988.

[9] T. El-Ghazawi, T. Sterling, A. Meajil, and A. Ozkaya, *"Overhead and Salability Measurements on the Cray T3D and Intel Paragon Systems,"* Research Report to CESDIS, Goddard Space Flight Center, NASA, June 1995.

[10] J. K. Salmon, *Parallel Hierarchical N-body Methods*, Ph.D. thesis, California Institute of Technology, Dec. 90.

A Quantitative Approach for Architecture-Invariant Parallel Workload Characterization

Abdullah I. Meajil[†], Tarek El-Ghazawi[†], and Thomas Sterling[‡]

† Electrical Engineering and Computer Science Department
The George Washington University
‡ Center of Excellence in Space Data & Information Sciences
NASA/Goddard Space Flight Center

abdullah@seas.gwu.edu

Abstract

Experimental design of parallel computers calls for quantifiable methods to compare and evaluate the requirements of different workloads within an application domain. Such metrics can help establish the basis for scientific design of parallel computers driven by application needs, to optimize performance to cost. In this work, a parallelism-based framework is presented for representing and comparing workloads, based on the way they would exercise parallel machines. This method is architecture-invariant and can be used effectively for the comparison of workloads and assessing resource requirements. Our workload characterization is derived from parallel instruction centroid and parallel workload similarity. The centroid is a workload approximation which captures the type and amount of parallel work generated by the workload on the average. The centroid is an efficient measure which aggregates average parallelism, instruction mix, and critical path length. When captured with abstracted information about communication requirements, the result is a powerful tool in understanding the requirements of workloads and their potential performance on target parallel machines. The parallel workload similarity is based on measuring the normalized Euclidean distance (ned), which provides an efficient means of comparing workloads, between workload centroids. This provides the basis for quantifiable analysis of workloads to make informed decisions on the composition of parallel benchmark suites. It is shown that this workload characterization method outperforms comparable ones in accuracy, as well as in time and space requirements. Analysis of the NAS Parallel Benchmark workloads and their performance is presented to demonstrate some of the applications and insight provided by this framework. The parallel-instruction workload model is used to study the similarities among the NAS Parallel Benchmark workloads in a quantitative manner. The results confirm that workloads in NPB represent a wide range of non-redundant benchmarks with different characteristics.

Keywords: Instruction-Level Parallelism; Parallel Computer Architecture; Parallel Workload Characterization; Performance Evaluation; Workload Similarity

1 Introduction

The design of parallel architectures should be based on the requirements of real-life production workloads, in order to maximize performance to cost. One essential ingredient in this is to develop scientific basis for the design and analysis of parallel benchmark suites. It was observed that benchmarking should also become a more scientific activity [1]. Selection of test workloads from an application domain must be determined by specific metrics that delineate the salient equivalencies and distinctions among candidate test codes. Previous efforts have addressed the problem of characterizing and measuring specific aspects of parallel workloads. Depending on the irrespective objectives, these efforts have separately quantified attributes such as the total number of operations, average degree of parallelism, and instruction mixes [2-10]. We point here that equally important is how these factors interrelate to make up the workload as presented to the system on a cycle by cycle basis. Therefore, still needed is the means to characterize parallel workloads based on how they are expected to exercise parallel architectures. In this study, we propose an application characterization based on the dynamic parallel instruction sequence in workloads. With this in place, the similarity among each pair of workloads is measured using the normalized Euclidean distance, a computationally efficient technique for pattern matching.

Since measuring parallel instructions is of interest to this study, we consider efforts that examined instruction-level parallelism. Researchers have measured instruction-level parallelism to try different parallel compilation concepts and study their effect on parallelism. Most of these studies measured the limits of (average) parallelism under ideal conditions, such as the oracle model where parallelism is only limited by true flow dependencies. Then, they examined the drop in parallelism when specific architectural or compilation implementation concepts were introduced into the model [2-6].

In this work we only consider workload characterization based on parallel instructions, which encompasses information on parallelism, instruction mix, and amount and type of work on a cycle-by-cycle basis. Bradley and Larson [11] have considered parallel workload characterization using parallel instructions. Their technique compares the differences between workloads based on executed parallel instructions. Executed parallelism is the parallelism exploited as a result of interaction between hardware and software. This technique is, therefore, an architecture-dependent technique due to its dependency on the specific details of the underlying architecture. In their study, a subset of the Perfect™ Benchmarks has been chosen to run on the Cray Y-MP. Then a multidimensional matrix that represents the workload parallelism profile was constructed. The Frobenius matrix norm is then used to quantify the difference between the two workload parallelism matrices. In addition to requiring a lot of space and time, this method is restricted to comparing identical executed parallel instructions only.

The technique proposed here, in contrast, uses the vector-space model [12] to provide a single point (but multidimensional) representation of parallel workloads and measures the degree of similarity between them. The similarity is derived from the spatial proximity between workload points in that space and therefore provides a collective measure of similarity based on all instructions. In specifics, each workload in a benchmark suite is approximated by a parallel-instruction centroid under this model. The difference between two workloads are quantified using appropriately normalized Euclidean distance (*ned*) between the two centroids.

Architecture-independence of our parallel-instruction vector-space model is derived from using the oracle abstract architecture model [3, 7]. In order to simulate such an oracle, two major modules are used: an interpreter and a scheduler [7, 13]. The interpreter accepts the assembly instructions generated from high-level code and execute it. The stream produced is passed to the scheduler which places each instruction at the earliest possible level for execution, based on the dependencies between the current instruction and the previously scheduled ones. The "two pass" nature of this process gives us the oracle. Or loosely speaking is to assume that an *Oracle* is present to guide us at every conditional jump, telling us which way the jump will go each reference, and resolving all ambiguous memory references [3]. Therefore, the oracle model is an idealistic model that considers only true flow dependencies. The parallel instructions (*PIs*) are generated by scheduling sequential instructions that are traced from a RISC processor execution onto the oracle model. The traced instructions are packed into parallel instructions while respecting all flow dependencies between instructions. To compare our technique with the parallelism-matrix one, we consider an extended version of the parallelism-matrix technique which is made architecture-independent by replacing the Cray Y-MP simulator with the oracle model.

In this paper we present the concept of parallel-instruction vector space model and a parallel-instruction workload similarity measurement technique. We compare this technique to the parallelism-matrix method [11]. It will be shown here that our method is not only machine-independent, but also better represents the degree of similarity between workloads. Further, the technique is very cost efficient when compared with similar methods. We also show that the parallel-instruction vector space model provides a useful framework for the design and analysis of benchmarks, as well as for performance prediction. This is demonstrated by analyzing some of the NAS Parallel Benchmark workloads [14] using this model. The NAS Parallel Benchmark (NPB) suite is rooted in the problems of computational fluid dynamics (CFD) and computational aerosicences. It consists of eight benchmark problems each of which is focusing on some important aspect of highly parallel supercomputing, for aerophysics applications [15]. This paper is organized as follows: next section presents our parallel-instruction vector space model. Section 3 presents an overview of related previous work; also, the comparison between that technique and ours will be discussed in this section. Experimental measurements for NAS Parallel Benchmarks will be presented in section 4. Finally, conclusions are given in section 5.

2 The Parallel-Instruction Vector-Space Model

Our Parallel-Instruction Vector Space Model is represented here provides for an effective workload representation (characterization), as will be shown. Effectiveness, in this regard, refers to the fidelity of the representation and the associated space and time costs. In this framework, each parallel instruction can be represented by a vector in a multidimensional space, where each coordinate corresponds to a different instruction/operation type (*I-type*) or a different basic operation (ADD, LOAD, FMUL, ...). The position of each parallel instruction in the space is determined by the magnitude of the *I-types* in that vector.

2.1 Definitions

Architecture-Invariant Characterization: Machine invariance in this context refers to having a characterization model and methodology that produces workload characteristics that depend only on the application and are valid for any target parallel machine. To produce workload characterizations that are machine invariant, we consider extracting pure workload characteristics by scheduling the workload to run on an ideal architecture. The oracle model is basically a data-flow model, which is described below in terms of modern parallel architectures.

Parallel Workload Instant and Parallel Work: The workload instant for a parallel computer system is defined here as the types and multiplicity of operations presented for execution by an idealistic system (oracle model), in one cycle. A workload instant is, therefore, represented as a vector quantity (parallel instruction) where each dimension represents an operation type and the associated magnitude represents the multiplicity of that operation in the parallel instruction. Parallel workload of an application is the sequence of instances (parallel instructions) generated from that application.

Workload Centroid: The centroid is a parallel instruction in which each component corresponds to the average occurrence of the corresponding operation type over all parallel instructions in the workload. Centroid, therefore, can be thought of as the point mass for the parallel workload body.

Workload Similarity: Two workloads exhibited by two applications are, thus, considered identical if they present the machine with the same sequence of parallel instructions. In this case both workloads are said to be exercising the machine resources in the same fashion.

2.2 A Vector-Space Model for Workload

Consider three types of operation (*I-types*) such as arithmetic operations (ALU), floating-point operations (FP), and memory access operations (MEM), then the parallel-instruction vector can be represented in a three-dimensional space as a triplet:
$$PI = (MEM, FP, ALU).$$
If a parallel instruction in a workload is given by
$$PI_i = (4, 7, 2),$$

then this ith parallel instruction in the workload has 4 memory operations, 7 floating-point operations, and 2 arithmetic operations. The total operations in this parallel instruction would be 13 operations that can be run simultaneously. In general, parallel instructions are represented as t-vectors of the form

$$PI_i = (a_{i1}, a_{i2}, \dots, a_{it})$$ (1)

where the coefficient a_{ik} represents the count of operations of type k in parallel instruction PI_i.

Comparing workloads based on sequence of parallel instructions could be quite complex and prohibitive, for realistic workloads. This is because the comparison requires examining each parallel instruction from one workload against all parallel instructions in the other workload, which has very high computational and storage requirements. This has led us to propose the concept of centroid for workload representation and comparison, which is a cost-effective means to represent workloads.

The centroid is a parallel instruction in which each component corresponds to the average occurrence of the corresponding operation type over all parallel instructions in the workload. Given a set of n parallel instructions constituting a certain workload, the corresponding centroid vector is

$$C = (CI_1, CI_2, \dots, CI_t)$$ (2)

where: $$CI_k = 1/n \sum_{i=1}^{n} a_{ik}$$ (3).

In addition to simplifying the analysis, centroids have the quality of providing an easy way to grasp the workload characteristics and the corresponding resource requirements. This is because the centroid couples instruction-level parallelism and instruction mix information to represent the types and multiplicity of operations that the machine is required to perform, on the average, in one cycle. This also represents the functional units types and average number of them needed in the target machine in order to sustain a performance rate close to the theoretical rate. Due to their simplicity and physical significance, as discussed above, centroids are used in the rest of this work as the basis for workloads representation and comparisons.

2.3 Workload Comparison Using the Vector-Space Model

Measuring similarity based on centroids mandates the selection of a similarity metric which can generate easy to understand numbers. Thus, a good metric should have the following characteristics:
- (i) Generates normalized values between 0 and 1.
- (ii) "0" represents one extreme (e.g. dissimilar), while "1" represents the other extreme (e.g. similar).
- (iii) Scales appropriately between these two extremes as the similarity between the compared workloads changes.

This leads us to select the normalized Euclidean distance (ned) between two centroids, representing two different workloads, as follows. Let point u be the t-tuple $(a_1, a_2, \dots, a_t)$ and point v be the t-tuple $(b_1, b_2, \dots, b_t)$; then the Euclidean distance (d), from Pythagoras' theorem, is

$$d(u,v) = \sqrt{(a_1 - b_1)^2 + (a_2 - b_2)^2 + \dots + (a_t - b_t)^2}$$ (4).

In order to conform with the aforementioned metric characteristics, the Euclidean distance between any two workloads in a benchmark suite needs to be normalized. To do so, normalization by the distance between the origin and the point made out of the maximum coordinates in the two workloads is used. Let WL_r and WL_s be two workloads in a benchmark suite, where each can be characterized by a t-centroid vector (t instruction types) as follows:

$$WL_r = (CI_{r1}, CI_{r2}, \ldots, CI_{rt}), \text{ and}$$

$$WL_s = (CI_{s1}, CI_{s2}, \ldots, CI_{st}).$$

And let CI_{ik} represent the centroid magnitude of the kth instruction type in workload i. The maximum centroid-vector in this workloads can be represented as follows.

$$C_{max}(WL_r, WL_s)=(max(CI_{r1}, CI_{s1}), max(CI_{r2}, CI_{s2}), \ldots, max(CI_{rt}, CI_{st})) \qquad (5).$$

Then, the normalized Euclidean distance (*ned*) between the two workloads can be measured as:

$$ned(WL_r, WL_s)=d(WL_r, WL_s) / d(C_{max}, null\text{-}vector) \qquad (6)$$

where null-vector (origin) is a t-vector in which each element equals to 0; hence, *null-vector* = $(0,0,\ldots,0)$. In this case, 1 represents identical workloads while 0 represents orthogonal workloads that use different operations and thus, would exercise different aspects of the target machine.

3 Comparison with Previous Related Work

3.1 The Parallelism-Matrix Technique

This technique represents an executed-parallelism workload profile in a multidimensional matrix (n-matrix). Each dimension in this n-matrix represents a different instruction type in a workload. In this method "work" has been defined to be the total number of operations of interest a workload can have [11]. When there is only one instruction type of interest, work is considered to be the total operations of that type in a workload. Therefore, a natural extension to the simple post-mortem average is a histogram $W = <W_0, \ldots, W_t>$, where W_i is the number of clock periods during which i operations of interest type were completed simultaneously. The sum

$$t = \sum_{i=0}^{t} W_i \qquad (7)$$

is the number of clock periods consumed by the entire workload, and the weighted sum

$$w = \sum_{i=0}^{t} iW_i \qquad (8)$$

is the total amount of work performed by the workload. To facilitate comparisons between workloads that have different execution times, each entry in the histogram is divided by t, the total execution time in clock periods, to produce a normalized histogram called the *parallelism vector* $P = <P_0, \ldots, P_t>$, where $P_i = W_i/t$. By construction, each entry P_i has a value between 0 and 1 that indicates the fraction of time during which i units of work were completed in parallel.

In a similar way, executed parallelism matrices of arbitrary dimension can be constructed with one dimension for each of the different kinds of work that are of interest. Some other possibilities for "work" include logical operations, integer operations, and I/O operations. Depending on how work is defined, various parallelism profiles from an executed-parallelism workload matrix can be obtained.

The parallelism profiles for two workloads, thus far, can be compared by comparing the parallelism matrices for each workload using the Frobenius matrix norm to quantify the difference. If A is the two-dimensional $m \times n$ parallelism matrix for workload 1 and B is the $m \times n$ parallelism matrix for workload 2, then the difference in

$$Diff(A, B) = \| A - B \|_F$$

$$= \sqrt{\sum_{i=0}^{m} \sum_{j=0}^{n} |a_{ij} - b_{ij}|^2} \tag{9}.$$

Intuitively, the Frobenius norm represents the "distance" between two matrices, just as the Euclidean formula is used to measure the distance between two points. This distance may range from 0.00, for two workloads with identical executed parallelism distributions, to $\sqrt{2}$ in the case where each matrix has only one non-zero element (with value 1.00) in a different location. Thus, the numbers produced by this method do not scale in way that can provide an intuition understanding to the degree of similarity. Further, should two workloads be 100% dissimilar, they can still produce different numbers. Additional problems that relate to the processing and memory cost of this method are addressed in the next subsection.

3.2 Comparison Study with Our Model

Similarity measurements obtained by parallelism-matrix technique have many shortcomings. In the parallelism-matrix model, the similarity value of two workloads that have no identical parallel instructions would differ from one case of dissimilar workloads to another. However, if there are some identical parallel instructions in the workloads, then the similarity value becomes more meaningful. For example, under the parallelism-matrix approach, a parallel instruction with 49% loads and 51% adds is considered completely dissimilar to one that has 51% loads and 49% adds. Thus, the parallelism-matrix technique lacks the ability to compare realistic workloads when they lack identical parallel instructions. In the parallel-instruction vector space technique, however, the more continuos representation space provides always meaningful similarity values through the distance metric. When two workloads are quite different, the distance values are high. On the other hand, when there are some differences in the workloads, the distance value changes proportionally between 0 and 1.

In general, the parallel-instruction vector space method presents more detailed information. For each workload centroid, each attribute represents an arithmetic mean of a type of instruction in the workload. By comparing this centroid to other workload centroid, each matching attribute will be compared. This comparison tells in which direction these two workloads are different.

The parallel-instruction vector space method is also more efficient in time and space. After producing parallel instructions, both techniques make two steps in order to measure the workload similarity. The first step is workload representation, and the second is workload comparison. The parallelism-matrix technique represents a workload in a t-dimensional matrix where each dimension represents an instruction type. The maximum magnitude of a dimension is $n + 1$, where n represents the maximum instruction type occurrences in any parallel instruction in that workload. Therefore, the parallel matrix technique needs as much storage as the size of the matrix, i.e. has a storage complexity of $O(n^t)$. On the other hand, the parallel-instruction vector space model represents a workload by a centroid of length t, $O(t)$ storage. The time to produce the workload representation, in the parallelism-matrix technique, is the parallel-instruction counts (p) times the parallel-instruction length (t), or $O(p \cdot t)$. This is because all parallel instructions have to be generated first, before constructing and filling the matrix. However, in the parallel-instruction vector space model, the computational complexity is $O(t)$. This is due to the fact that the workload centroid is calculated on-the-fly.

In the comparison step (measuring similarity), the parallelism-matrix technique compares every element of one matrix with the corresponding element in the other matrix. Therefore, the computational complexity of this technique is $O(n^t)$. In the parallel-instruction vector space model, however, the computational complexity is $O(t)$. This is due to the fact that the workload centroid has t types of instructions. Thus, our parallel-instruction vector space model outperforms the parallelism-matrix technique for measuring the workload similarity in all essential aspects.

4 NAS Parallel Benchmark Experimental Workload Comparisons

In order to demonstrate the potential utility of this model and verify the underlying concepts with real-life applications we study the NAS Parallel Benchmark (NPB) suite [14] using our model. The NPB represents an implementation independent problem set, representative of Computational Aerosciences workload computations. The NPB workloads have been implemented on nearly every parallel platform and results have been reported by the vendors. Also, it is now widely cited in the high performance computing world as a reliable measure of sustained performance.

We start by representing the workloads in this suite using parallel instruction centroids. Then we characterize the similarity among different workload pairs using the normalized Euclidean distance (*ned*) approach. We also demonstrate that in real-life workloads, such as those of NPB, parallelism smoothability is high enough to use average degree of parallelism to represent parallel activities, as in the centroids.

4.1 A NAS Parallel Benchmark Overview

The NPB suite consists of two major components: five parallel kernel benchmarks and three simulated computational fluid dynamics (CFD) application benchmarks. The kernels are relatively compact problems that can be implemented fairly quickly and provide insight into the general levels of performance that can be expected for particular types of numerical computations. The simulated computational fluid dynamic (CFD) applications usually required more effort to implement, but they are more indicative of the types of actual data movement and computation required in state-of-the-art CFD application codes. This benchmark suite successfully addresses many of the problems associated with benchmarking parallel machines. They mimic the computation and data movement characteristics of large-scale computational fluid dynamic applications, numeric-scientific workloads, which frequently use floating-point arithmetic and often operate on arrays. These benchmarks are specified only algorithmically. Figure 1 gives an overview of the NPB kernels. A detailed description of these NPB problems is given in [14].

o *embar* is an "embarrassingly parallel" kernel, which evaluates an integral by means of pseudorandom trails. It is a typical of many Monte Carlo applications. Two-dimensional statistics accumulated from a large number of Gaussian pseudorandom numbers, generated according to a scheme that well suited for parallel computation.

o *mgrid* executes four iterations of the V-cycle multigrid algorithm to obtain an approximate solution to the discrete Poisson problem $\nabla^2 u = v$ on a $256 \times 256 \times 256$ grid with periodic boundary conditions.

o *cgm* is a conjugate gradient method that computes an approximation to the smallest eigenvalue of a large, sparse, symmetric positive definite matrix of order 14,000 with a random pattern of nonzeros. This problem is typical of unstructured grid computations and it uses sparse matrix-vector multiplication.

o *fftpde* uses FFT's On $256 \times 256 \times 128$ complex array to solve a three-dimensional partial differential equation. This benchmark represents the essence of many "spectral" codes or eddies turbulence simulations.

o *buk* is a large integer sort. This kernel performs a sorting operation that is important in "particle method" codes. It is similar to "particle in cell" physics applications, where particles are assigned to cells and may drift out. This problem is unique in that floating-point arithmetic is not involved.

Figure 1: *An overview of the NAS Parallel Benchmark kernels.*

4.2 Experimental Measurement Tools and Process

In order to explore the inherent parallelism in workloads, instructions traced are scheduled for the oracle model architecture. This model presents the most ideal machine that have unlimited processors and memory, and does not incur any overhead. The Sequential Instruction Trace Analyzer (SITA) is a tool that measures the amount of parallelism which theoretically exists in a given workload [7, 13]. SITA takes a dynamic trace generated by *spy* tool [15] from a sequential execution of a conventional program, and schedules the instructions according to how they could be executed on an idealized architecture while respecting all relevant dependencies between instructions.

The analysis process of a SPARC workload or benchmark takes four steps. First, a SPARC executable file is created, using the desired optimization level. The program is statically linked to eliminate the spurious instructions used in linking a program to the libraries. Secondly, the pre-analyzer (*sitapa*) is run with *spy* and executable to extract a list of basic blocks and frequencies of the workload, which is then read by the control-dependence analyzer (*sitadep*) to produce an annotated list, as the third step. This annotations include control-dependency relationships between the blocks and destination frequencies. Finally, the scheduler (*sitarun*) is run with the annotated list as input, and generally with *spy* and executable. The scheduler produces output indicating the parallelism available for the given input trace under the given oracle model. There are 69 basic instruction operations in SPARC. These instructions mainly fall into five basic categories: load/store, arithmetic/logic/shift, floating-point, control transfer, and read/write control register operate. Therefore, each parallel instruction presented by a vector of length five [16].

4.3 Workload Centroids for NAS Parallel Benchmarks

Parallel-instruction centroid vectors can reveal differences in workload behavior that can not be distinguished by averages of parallelism degrees as shown in Table 1. Therefore, the parallelism behavior of two workloads can be efficiently compared by using the aforementioned parallel-instruction vector space model and the similarity function, expression (6), to quantify the similarity between these workloads.

Table 1: *Centroids for the NAS Parallel Benchmarks.*

	MemOps	ALUops	Fpops	TransferOps	OthersOps
embar	116.27	79.013	53.37	20.613	7.785
mgrid	32.217	18.22	9.08	0.714	0.108
cgm	128.622	69.803	34.034	7.355	0.12
fftpde	195.68	122.11	74.58	27.37	2.37
buk	2.424	1.55	0.626	0.448	0.18

Smoothability [7] is a metric designed to capture the parallelism profile variability around the average degree of parallelism. The interest in smoothability stems from the fact that the centroid is based upon the average degree of parallelism for each type of operation. Therefore, for centroids to well represent workloads, those workloads should have relatively high smoothability (close to 1). In this section we show that typical real-life applications, such as those represented by NPB, have high smoothability. Our results indicate that the parallelism obtained has a relatively smooth temporal profile which exhibits a high degree of uniformity in the parallelism except for the *cgm* benchmark whose smoothability is 68%. In all cases, but the *cgm* benchmark, the smoothability is better than 83%. Most importantly, in the context of this study, the smooth temporal behavior supports the fidelity of representing practical workloads using parallel instruction centroids

4.4 Similarity Measurements for NAS Parallel Benchmarks

Table 2 quantifies the normalized Euclidean distance (*ned*) between each pair of benchmarks in the NAS Parallel Benchmark suite, using expression (6). Again, note that the similarity in parallelism is not a transitive relation. We first compare *fftpde* and *mgrid*. The relatively high *ned* value, 0.84, illustrates that these two workloads have different parallelism behaviors. In the case of *embar* and *cgm*, however, the relatively small *ned* value, 0.18, illustrates that these two workloads have relatively similar parallelism properties. Although the two workloads come from different application areas, each workload is expected to exercise target machines with a similar mix of parallelism. However, the communications requirements for these workloads, *embar* and *cgm*, are extremely different.

Table 2: *ned values for the NAS Parallel Benchmarks.*

	embar	mgrid	cgm	fftpde	buk
embar	0.0000				
mgrid	0.7554	0.0000			
cgm	0.1799	0.7468	0.0000		
fftpde	0.3819	0.8452	0.3949	0.0000	
buk	0.9806	0.9227	0.9804	0.9879	0.0000

5 Conclusions

This paper introduced a parallelism-based methodology for an easy to understand representation of workloads. The method is architecture-independent and can be used effectively for the comparison of workloads and assessing resource requirements. A method for comparing workloads based on the notion of the centroid of parallel instructions was introduced. The normalized Euclidean distance (*ned*) to provide an efficient means of comparing the workloads. The notion of the centroid was coupled with the normalized Euclidean distance (*ned*) among pairs of workloads. This has provided the basis for quantifiable analysis of workloads to make informed decisions on the composition of parallel benchmark suites and the design of new parallel computers. Analysis of existing benchmarks is also provided for by this model in which the centroid sheds light on the hardware resource requirements and how benchmark is exercising the target machines. Furthermore, the distance among workload pairs allows identifying possible correlations.

A study between a comparable method, the parallelism-matrix technique, and our parallel-instruction vector space model was also presented. It was shown that the parallelism-matrix technique depends only on identical rather than similar parallel instructions. Our parallel-instruction vector space model, however, takes all parallel instructions into account when representing workloads and their similarities. Therefore, unlike the parallelism-matrix technique, *ned* scales with the degree of dissimilarity between workloads in a representative fashion. Furthermore, *ned* has meaningful boundary values, where 0 means identical workloads and 1 means completely different workloads. As far the storage requirements, the parallelism-matrix technique requires much more memory space than our parallel-instruction vector space model. In addition, when two workloads are compared, the computational cost in the parallelism-matrix technique is extremely expensive while our parallel-instruction vector space model requires much less computational time. Hence, the parallel-instruction vector space model does not only provide more accurate, but also more cost-effective parallelism-based representation of workloads.

The parallel-instruction workload model was used to study the similarities among the NAS Parallel Benchmark workloads in a quantitative manner. The results confirm that workloads in NPB represent a wide range of non-redundant applications with different characteristics.

References

[1] R. Hockney, *The Science of Computer Benchmarking*, SIAM Publisher, Philadelphia , PA., 1996.

[2] G.S. Tjaden and M.J. Flynn, *"Detection and Parallel Execution of Independent Instructions,"* IEEE Transactions on Computers, vol. C-19, no. 10, pp. 889-895, October 1970.

[3] A. Nicolau and J.A. Fisher, *"Measuring the Parallelism Available for Very Long Instruction Word Architectures,"* IEEE Transactions on Computers, vol. 33, no. 11, pp. 968-976, Nov. 1984.

[4] M. Butler, T. Yeh, Y. Patt, M. Alsup, H. Scales, and M. Shebanow, *"Single Instruction Stream Parallelism Is Greater Than Two,"* Proc. of the 8*th* Annual Symp. on Comp. Arch., pp. 276-286, May 1991.

[5] Kumar, *"Measuring Parallelism in Computation-Intensive Scientific/Engineering Applications,"* IEEE Transactions on Computers, vol. C-37, no. 9, pp. 1088-1098, Sep. 1988.

[6] T. Sterling, T. El-Ghazawi, A. Meajil and A. Ozkaya, *"NASA Science Workload Characterization for Scalable Parallel Computer Architectures,"* Supercomputing 93, technical program, 1993, pp. 78.

[7] K.B. Theobald, G.R. Gao, and L.J. Hendren, *"On the Limits of Program Parallelism and its Smoothability,"* Proc. of the 25th Annual International Symp. on Micro-Architecture (MICRO-25), pp. 10-19, Portland, Oregon, December 1992.

[8] T. Conte and W. Hwu, *"Benchmark Characterization,"* IEEE Computer, pp. 48-56, January 1991.

[9] M. Calzarossa and G. Serazzi, *"Workload Characterization for Supercomputer,"* Performance Evaluation of Supercomputers, J.L. Martin (editor), pp. 283-315, North-Holland, 1988.

[10] J. Martin, *"Performance Evaluation of Supercomputers and Their Applications,"* Parallel Systems and Computation, G. Paul and G. Almasi (Editors), pp. 221-235, North-Holland, 1988.

[11] D. Bradley and J. Larson, *"A Parallelism-Based Analytic Approach to Performance Evaluation Using Application Programs,"* Proceedings of the IEEE, vol., 81, no. 8, pp. 1126-1135, August 1993.

[12] A.I. Meajil, *"An Architecture-Independent Workload Characterization Model for Parallel Computer Architectures,"* Technical Report No. GWU-IIST 96-12, Department of Electrical Engineering and Computer Science, George Washington University, July 1996.

[13] K.B. Theobald, G.R. Gao, and L.J. Hendren, *"Speculative Execution and Branch Prediction on Parallel Machines,"* ACAPS Technical Memo 57, McGill University, December 21, 1992.

[14] D. Bailey et al. *"The NAS Parallel Benchmarks,"* RNR Technical Report RNR-94-007, March 1994, NASA Ames Research Center, Moffett Field, CA.

[15] G. Irlam, *The Spa package*, version 1.0, October 1991.

[16] The SPARC Architecture Manual, Version 8, SPARC Int'l, Inc., Menlo Park, CA, 1991.

Parallel Mapping of Program Graphs into Parallel Computers by Self-Organization Algorithm *

O.G.Monakhov, O.Y.Chunikhin

Computing Centre, Sibirian Division of Russian Academy of Science,
Pr. Lavrentiev 6, Novosibirsk, 630090, Russia,
e-mail: monakhov@ssd.sscc.ru

Abstract. The optimization problem of mapping program graphs to parallel distributed memory computers is considered. An algorithm which is based on the self-organizing neural networks is proposed. We tried to apply the ability of Kohonen neural networks to compute a neighbourhood preserving mapping to complicated topologies of program and processor graphs. The goal of the algorithm is to produce an allocation with the minimal communication cost and the computational load balance of processors.

1 Introduction

There is considered a problem of mapping of the information graph of a complex algorithm into the graph of the interprocessor network of the parallel computer system (CS), which consists of elementary processors (PE) with local memory, interconnected by a programmable communication network. In this paper the parallel algorithm is proposed for optimal or suboptimal solution of the mapping problem, the objective function for mapping is developed and experimental results are presented.

2 Optimal Mapping Problem

Let a model of a parallel program be the graph $G_p = (M, E_p)$, where M is a set of modules(processes), $|M| = n$, E_p is a set of edges, representing information connections between modules. Let t_i be defined as weight of the module $i \in M$, representing the execution time (or the number of computational steps) of the module i. Let a_{ij} be defined as weight of the edge $(i, j) \in E_p$, representing the number of information units passed from module i to module j.

A model of multiprocessor system with distributed memory is an undirected graph $G_s = (P, E_s)$ representing the network topology (structure)of the system, where P is a set of PE, and edges E_s represent interconnection links between PE, $|P|$=m.

---

* This work is supported by RFBR project N96-01-01632

The distance between nodes i and j of the graph G_s is denoted as d_{ij}. It represents cost of data transfer between processors i and j. Neighborhood of the node i with radius ρ is the set $D_\rho(i) = \{j \in M : 1/a_{ij} < \rho\} \cup \{i\}$.

Let $a_{ij} = 0$ if there is no link, but if there is no link between i and j in G_s we can let d_{ij} be equal to different values:

1)$d_{ij} = \infty$ -data transfer is impossible

2)d_{ij} - the shortest distance between nodes i and j in the graph or the information is transferred in the shortest way.

Let $\varphi : M \rightarrow P$ is the mapping of an information graph of the parallel program G_p into the structure G_s of CS. Let the mapping φ be represented by vector $X = \{x_{ij} : i \in M, j \in P\}$, where $x_{ij} = 1$, if $\varphi(i) = j$ and $x_{ij} = 0$, if $\varphi(i) \neq$ j.

Let the quality of the mapping parallel program graph into structure of CS for the given vector X is described by functional:$F(X) = F_E(X) + F_C(X)$, where $F_E(X)$ represents computational cost (the overall module execution time of parallel program on system or the load balancing of PE for given X) and $F_C(X)$ represents interprocessor communication cost. The optimal mapping problem of parallel program graph into CS structure consists in optimization of the functional $F(X)$ by means of the parallel algorithm.

3 Cost Functional of Mapping

Now let us describe the objective cost function which represents the computational load balance of PE and communication cost for the given vector X.

$$F(x) = w_1 \sum_{\alpha=1}^{m} (\sum_{i=1}^{n} t_i * x_{i\alpha} - M_t)^2 + w_2 \sum_{i=1}^{n} \sum_{j=1}^{n} \sum_{\alpha=1}^{m} \sum_{\beta=1}^{m} a_{ij} d_{\alpha\beta} x_{i\alpha} x_{j\beta},$$

where $M_t = \frac{1}{m} \sum_{j=1}^{n} t_j$ - ideal average processor load,
w_1 and w_2 weights of the first and the second terms of the functional.

The first term in this expression describes deviation of PE_k load, the second one describes the quality of using the interprocessor links.

The function may be presented as follows:

$$F(X) = \sum_{i=1}^{n} \sum_{\alpha=1}^{m} x_{i\alpha}[I_{i\alpha} + \sum_{j=1}^{n} \sum_{\beta=1}^{m} T_{i\alpha j\beta} x_{j\beta}],$$

where

$$T_{i\alpha j\beta} = w_1 t_i t_j \delta_{\alpha\beta} + w_2 d_{\alpha\beta} a_{ij},$$

$$I_{i\alpha} = -2w_1 M_t t_i,$$

$\delta_{\alpha\beta} = 1$, if $\alpha = \beta$ and $\delta_{\alpha\beta} = 0$, if $\alpha \neq \beta$ - delta-function.

4 Parallel Algorithm for Mapping Optimization

The algorithm is based on using the ability of the Kohonen type of neural networks to compute neighborhood preserving mapping. A neuron is corresponded to each process, whose state is described by a vector $x_i (i \in M)$. The length of the vector is equal to the number of processors in the CS. Its components are denoted $x_{i\alpha} (\alpha \in P)$. The intuitive idea is that the self-organization process results in a good topology correspondent between program and system graphs and at the same time the algorithm is constructed so that the objective function has to decrease on each step.

On each step of the algorithm we choose a neuron (i.e process) by chance, calculate the objective function supposing that this process is working on a concrete processor, choose a processor which minimizes the function, and move some neighborhood of the neuron(the distance depends on the interprocessor links activity) towards the winner processor (i.e. we increase corresponding components of the neurons state vectors and decrease the other ones to save their sum equal to 1). The radius of moved neighborhood and step size are decreased with number of iterations.

Mapping algorithm:

1. Initialize all state vectors x_i i.e. $\forall i \in M, \forall \alpha \in P : x_{i\alpha} := 1/m$

2. For $t := 1$ to $t_{\max}$ do
   3. wpr:=(random element from M)
   4. For each $k \in P$ do

$$5. \forall i \in M, i \neq \text{wpr}, \forall \alpha \in P : Y_{i\alpha}^k := X_{i\alpha}, \forall \alpha \in P : Y_{\text{wpr},\alpha}^k := \delta_{\alpha k}$$

$$6. F_k := F(Y^k)$$

7. end for
   8. Let $wp : F_{wp} := \min\{F_k : k \in P\}$

$$9. \forall i \in D_\rho(wpr), \forall \alpha \in P : x_{i\alpha} := x_{i\alpha} + \epsilon(t)\delta_{k\alpha}$$

$$10. \forall i \in D_\rho(wpr) : x_i := x_i / \sum_\alpha x_{i\alpha}$$

11. end for

Where $\epsilon(t)$ is monotonically decreasing function meaning step size;$\rho(t)$ is also monotonically decreasing function - radius of moved neighborhood.

On steps 5 and 6 we compute the objective function supposing that process wpr is working on the processor k in other equal conditions.

The step 2 may look as follows :

$$2. \text{ While } \exists i \in M, \forall \alpha \in P | X_i - e_\alpha | > d$$

where e_α is unit vector with a unit in the position α, d is the threshold, i.e. the algorithm finishes its work if all x_i are close enough to unit vectors.

Parallel algorithm works a on computer system topology "ring". It is used to calculate the set of F_k. We could see that calculating the function $F(X)$ consists of multiplying the matrix T on the vector X, addition the vector I and multiplying final vector on X again.

Let the computer system consists of r processors $PE_1 \ldots PE_r$. Let us break P into r parts $P_1 \ldots P_r : P_1 \cup \ldots \cup P_r = P, P_i \cap P_j = \emptyset$, $i \neq$j. This program do the partition so that difference between numbers of elements in P_i and P_j is not more than 1. Then we break vector $x_{i\alpha}$ into corresponding parts, so PE_i stores components $x_{i\alpha} : i \in M, \alpha \in P_i$ and all elements a_{ij} and $d_{\alpha\beta}$. Steps 4-7 are paralleled so that PE_i calculates $F_k : k \in P_i$. We could see that calculation of F_k consist in multiplication of matrix T and vector X, addition vector I and than multiplication the obtained vector and vector X again. In order to do it blocks $\{x_{i\alpha} : i \in M, \alpha \in P_i\}$ are transferred in the ring, so during one cycle all blocks visit each processor and after two cycles all processors calculate their F_k.

5 Experimental Results

Experiments were conducted in a computer system MVS-100 based on eight Intel860 processors. The algorithm finds out suboptimal mapping and demonstrates linear acceleration if each P_i consists of more than 3-4 elements.

For example there are results of solving the problem for two variants of input graphs:

1. Program graph is a grid 4x4, system graph is a grid 2x2 - very simple example, the program always finds out optimal result. Here is a time table:

Number of processors	1	2	4
Time per iteration	0.03	0.02	0.03

Average number of iterations is 250.

2. Program graph is a grid 4x4, system graph is a grid 3x3. In 11 experiments the algorithm found out the optimal solution for 4 times and close to optimal one for 5 time.

Number of processors	1	2	4	6	8
Time per iteration	0.35	0.18	0.13	0.13	0.12

Average number of iterations is 400.

PARA'96 Conference Dinner Talk
1996 August 20

Peter Naur

DIKU - Department of Computer Science
University of Copenhagen
Universitetsparken 1
DK 2100 Copenhagen Ø
Denmark

Ladies and Gentlemen,

I feel honoured and privileged to have been given this opportunity to address the participants of the PARA96 Workshop. I shall make use of this to share with you some of my concerns with regard to the field of computing. I shall do this with the confidence that as participant at this workshop you will be able to appreciate and perhaps even share my concerns.

This confidence derives more specifically from the appearance of the word 'applied' in the title of your workshop. But I have to add that, as I see matters, having this word in your title is both a good and a bad sign.

What is good is that you are working on applying computing. I consider this to be highly important.

The bad thing is that it is necessary to say this. Why is that not a matter of course? In other words, how could there be computing which is not applied? Or at least, how could computing which is not applied be worthy of scientific attention?

And this brings in the key issue, the matter of *scientific validity*. This is what I want to go into. This may appear to be unnecessary in a computing context. Do we not have computer science, and related, special departments such as cognitive science? Do these not take care of the scientific validity?

My answer is **No**, they do not, and worse, these very designations confuse the issue of scientific validity. Where they are least damaging is in naming academic departments. But where they are really harmful is in naming fields of human concern and activity. When, for example, they suggest that work in what is called computer science must automatically be scientifically valid.

I wish to maintain that the talk of sciences, as fields of human activity and insight, is misguided and harmful. Scientific validity is not a matter of fields of inquiry.

What is it then, scientific validity? My answer is that science, together with scholarship – all that which in some languages, for example Danish, has one collective name, videnskab – my answer is that science and scholarship are centered around *descriptions of aspects of the world*. Not any descriptions, but descriptions having certain characteristics, the most important one being that scientific descriptions are *coherent* within themselves and with one another. Coherence in this context of course includes the old requirement that your description, or theory, must agree with the observations.

You may notice that in focussing on descriptions I implicitly have rejected the common idea that science is a matter of logic, of truth, or of true theories. The relation of a description to that which it describes is not one of truth or correctness, but rather a matter of *relative adequacy*. Moreover, that which we want to describe can only be a selected, limited aspect of the world. There can be no complete description of any part of the world, and certainly not of the whole world.

Any description consists of certain elements brought together to form a connected structure of a kind. Thus our attention is turned towards the elements used in scientific descriptions. What are they, or what should they be?

A commonly accepted answer to this is to say that they should be mathematical, or at least that the best scientific descriptions are mathematical. I consider this to be totally misguided. For one thing, what exactly is meant by a description being mathematical is unclear. Surely any description will have to include certain parts that are formulated in ordinary language. For another, some of the greatest scientific discoveries have been expressed by means of elements that have no mathematics in them. A good example is the structure of DNA as discovered by Watson and Crick. They described the molecular structure by means of metal rods forming a three-dimensional grid.

What I want to bring out is, first, that the elements used to form scientific descriptions may be of any kind. They may be static, such as texts and figures, and they may be dynamic, such as processes executed in a computer. Second, in any scientific description the choice of the form of description is a highly important and entirely non-trivial matter. I believe it is obvious that if Watson and Crick had not chosen to work with their metal rods they would not have made their discovery. Upon consideration of great developments in physics one will find that they in several important cases have been closely related to a novel choice of descriptive elements. Quantum mechanics is one outstanding case.

With this notion, a valid scientific/scholarly contribution is a novel description of some aspect of the world that achieves high decree of coherence with the best other descriptions that have been produced at the time. The novelty may reside in what aspect is described, or in the form of the description and the choice of descriptive elements, or in both.

It follows that each scientific effort may try to develop either the aspect of the world being described or the form and elements used in the description of it. This is reflected in the fact that some of the specializations of scientific activity are primarily concerned with an aspect of the world, for example biology and geology, while some are concerned with particular elements and forms of description, the prominent cases being mathematics and computing.

Or at least, that is how I view the scene. And now we come to what I consider scientific perversions.

Already there is a long tradition among mathematicians, that mathematics used to describe the world is inferior, that mathematics really should be pursued just for its own sake, for the beauty of it. I regret to say that this attitude appears to be quite dominant among academic mathematicians. As a result I believe that most present day mathematics has become totally insignificant to anything else, it has become just a game for the insiders.

And now to some perversions specifically related to computing, as I see it. They are of several sorts. One is that what is offered by computing is taken dogmatically to be a proper form of description of some aspect of the world, without the question being raised of the propriety of this form for the particular aspect. The dominating case of this kind is the description of the human mental activity as pursued in what is called artificial intelligence. It is expressed in talking about "the human information processor". The trouble is that this sort of description fits the human mental activity very badly. This is ignored in artificial intelligence, however, and accordingly the classical insight into the human mental activity, as established a hundred years ago by William James, is ignored. The situation is that computing in this way has effected a decay of much of psychology.

Another sort of perversion of computing is brought out by the designation 'theoretical computer science'. What is published under this designation is work on data structures and data processes developed without regard to any application. Such work is quite analogous to pure mathematics, pursued merely for the beauty of it. The use of the word 'theoretical' lends a certain air of significance to the activity. This is a misuse of the word 'theoretical', however. As this word has been used over the past several centuries, a theory is not independent of aspects of the world, but is a description which covers the common aspects of a range of phenomena, by the generality of the description. This makes sense in fields that are oriented towards particular aspects of the world. In geology one may for example have a theory of the interior of the Earth. But in relation to computing such a notion of theory does not apply.

The temptation to play with computing techniques, as in theoretical computer science, rather than to develop the techniques in the context of concrete descriptions, is undoubtedly very great. Part of the temptation lies in the fact that it is much easier just to play with programs than to make sure that they achieve adequate description of some aspect of the world.

Even in the context of a workshop such as your present one this is quite visible. In the abstracts of the 52 papers to the workshop one finds a specific application area mentioned in only about one fourth of them. Some of the other abstracts mention that the program being discussed has important applications, without giving specifics.

The situation in mathematics is illustrative. Here the pure variety has been defended with the claim that its results will later prove to be important. I consider this to be a wholly irresponsible claim. The situation to-day is that around 200 thousand theorems are published in the mathematical literature every year, and that no-one, absolutely no-one, can tell even which ones are mathematically important, if any. Much less can anybody tell how they might help to achieve better descriptions of the world.

Bad consequences of what I consider scientific perversions of computing are glaringly visible in the way computing comes into our lives all the time. Let me mention just a few examples. Misconceptions about the human mental activity are reflected in the design of the computer interfaces presented to the human operators. It is simply an everyday fact that many of these are so poor that the operators are in a state of perpetual fight with them, resulting in waste of time and errors. Typically I go to the railway station and order reservations for a simple trip to Paris. The operators spend the best part of an hour with their terminal to produce the ticket. When I examine the ticket I find that the dates of the reservations do not match. There is first a night train to Cologne. But the date of the reservation for the continuation from Cologne has been made the same as that of first part of the trip, ignoring that a night has passed. Or I go to the public library to find the score of Haydn's symphony no. 95. I try to press the buttons of the terminal. The screen tells me, there is no such author as Haydn. Or take the stuff shown on your screen when you examine the email sent to you. There you are presented with a jungle of irrelevant figures and dates and whatnot, giving you a losing fight to find what might be relevant to you. I have given up email, I will have none of it.

Another area of scientific perversions related to computing flourishes in academic teaching of computer programming. Here there is a stress on program development methods and formal program specifications that supposedly will help the students to produce perfect programs. This is misguided in two ways. First, it ignores that people differ profoundly. A method and notation that one person finds useful, another person may find worthless. Second, stressing the logic properties of programs turns the attention away from the important relation of programs to that which they describe, which is not a question of logic or form.

Let me end these remarks by a return to the issue of science and scholarship. Perhaps you will counter my reflections by questioning my claim that science and scholarship are matters of coherent descriptions of the world. Perhaps you will say that I am mistaken, that science is a question of logic and truth, as so many people have thought. To this I will say, on this view science is a sort of game, a game played by the scientists. They like it and are happy to play it. But how will you explain that science is important to humanity at large? To most people logic and truth are of very minor interest and concern.

On the view that science is a matter of coherent description, the matter is very different, the fact being that description is a vital issue in the development of human civilization. Description is the carrier of exogenetic heredity, the transfer from one generation to the next of the experience gained about the world. Thus science, taken to be concerned with coherent descriptions of the world, is obviously of vital importance to humanity.

But genuine science sells badly, or rather, it does not sell at all; it is the very antithesis of salesmanship. Thus it is not to be wondered that in our overcommercialized society, science worthy of the name is a losing proposition.

Applied computing is a remaining stronghold of genuine science. Keep it up!

References

1. Naur, P.: 1995, "Datalogi som Videnskab", Rapport nr. 95/94, University of Copenhagen, Department of Computer Science.
2. Naur, P.: 1995, "Knowing and the Mystique of Logic and Rules", Kluwer Academic Publishers, Dordrecht, The Netherlands.

Compiler Optimizations for Red-Black HPF Codes

C. Alexander Nelson

Digital Equipment Corporation, nelson@hpc.pko.dec.com

Abstract. A case study is presented of the compiler optimizations that have been used to generate efficient code for a red-black relaxation problem coded in HPF and targeted at distributed memory. High Performance Fortran (HPF) gives the programmer the ability to express the parallelism at a high level without entering into the low-level details of message-passing and synchronization thereby reducing the time and effort required for parallel program development. Because the HPF compiler is responsible for scheduling the parallel operations on the physical machines, HPF opens up a vast area of optimizations which the compiler must perform in order to generate efficient code. These are optimizations which would otherwise have been performed by the programmer at a lower level (using explicit message passing). Some timings from the Digital Fortran 90 compiler showing the effect of these optimizations are presented.

1 Language constructs: brief

The HPF program fragment in Figure 1 shows three different ways of stating the same algorithm in Fortran 90. Of course the *do-loop* syntax does not necessarily imply that the computations can be done in parallel even though that information is sometimes discoverable by the compiler. The *forall* and *vector* assignment syntax not only express the parallelism inherent in the computation but also have the advantage of clarity.

The HPF directive states the the arrays A and B should have their elements distributed across the processors available for a given dimension by dividing the elements of that dimension into equal parts and giving the first part to the first processor and similarly for the rest of the processors. Figure 2 shows pictorially the distribution given to the arrays A and B in the code in Figure 1. The numbers inset into the four regions indicate which processor owns which submatrix of the distributed array.

2 Nearest Neighbor Optimizations

2.1 Owner Computes Rule

Many compilers targeted to distributed memory implement the assignment statement using the owner computes rule. This rule states that all of the computations

```
subroutine sub
!hpf$ processors P(2,2)

integer, parameter                :: n=16
integer, dimension(n,n)           :: A, B
!hpf$ distribute(block,block) onto P :: A
!hpf$ align with A                :: B
integer                           :: i,j

! DO-LOOPS
do i=2,n-1
 do j= 1,n
  A(i,j) = B(i-1,j) + B(i+1,j)
 end do
end do

...
! FORALL
forall(i=2:n-1,j=1:n) A(i,j) = B(i-1,j) + B(i+1,j)

...
! VECTOR ASSIGNMENT
A(2:n-1,:) = B(1:n-2,:) + B(2:n,:)

end subroutine sub
```

Fig. 1. An HPF Program Fragment

must happen on the processor that owns the value that is being assigned to. The storing processor is known as the "owner". This rule has the consequence of requiring all of the values that participate in a store to be moved to the owning processor using interprocessor communication. After these values have been moved, the owner now has all the values it needs to do the computation(s). These computations now happen without interprocessor communication.

On distributed memory any interprocessor communication is expensive, and given a fixed byte count sending fewer long messages is preferable to sending many short messages. Using these performance assumptions, the naive approach to implementing assignments has several problems. If the right-hand-side has many computations that could be done without interprocessor communication, these computations should occur first thereby reducing the amount of data that must be sent to other processors. Then for those values that must be sent, as few messages as possible containing many values should be used to transfer the

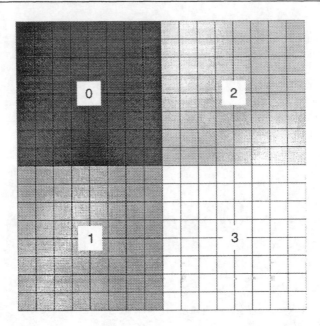

Fig. 2. Block-Block Two-dimensional Array Onto a 2x2 Processor Grid

values (as opposed to many messages that move just one value). In fact the Digital Fortran 90 compiler does not strictly adhere to the owner computes rule for these reasons.

2.2 The Problem

The naive approach is even worse for a class of problems known as nearest-neighbor codes. Figure 3 shows a two-dimensional *forall* statement. This type of code is often referred to as a nearest-neighbor code because an element of an array is updated using values that are near (in the sense of array coordinates) to the element being updated. Block distributions preserve this nearness by keeping neighboring elements on the same or immediately adjacent processors. The *forall* from Figure 3 (whose arrays are pictured in Figure 2), shows that processor 0 needs the first row from processor 1 and the first column from processor 2 to calculate the new values to be stored on processor 0. In fact most of the values that processor 0 needs are already on processor 0. Only a very few of the values needed by processor 0 to do its computation are not available locally. However, because no reference on the right-hand-side is exactly aligned with the left-hand-side reference, naive application of the owner computes rule requires temporaries for each of the four references on the right-hand-side. These temporaries will be locally aligned with the left-hand-side reference and consist mostly of values that

```
subroutine sub
!hpf$ processors P(4)
integer, parameter               :: n=16
integer, dimension(n,n)          :: A, B
!hpf$ distribute(block,block) onto P :: A
!hpf$ align with A               :: B
integer                          :: i


forall(i=2:n-1,j=2:n-1) &
A(i,j) = B(i+1,j) + B(i-1,j) + B(i,j-1) + B(i,j+1)

end subroutine sub
```

Fig. 3. 2-Dimensional Nearest-Neighbor Code Fragment

were already available locally. For example, on processor 0 a temporary that lives on processor 0 will be created to hold the values from the B(i+1,j) reference that corresponds with the 8x8 submatrix of A(i,j) locations that will be stored to. Seven of the eight rows of this temporary will be filled in with values from the local (to processor 0) submatrix of B. The last row will be filled in with the eight edge values from the submatrix of B owned by processor 1.

2.3 Smaller Temporary Arrays

Since only the values on processor 1 that "border" processor 0 need to be moved, a temporary to hold those values can be made much smaller. In this example the local temporaries can be of size 1x8 instead of size 8x8. This approach ameliorates the problem of allocating storage to hold copies of values that are already available locally and significantly reduces the memory requirements of the application.

2.4 Overallocate and Move Shadow Edges

If, however, storage is overallocated for the submatrix of the array A on processor 0, then the border values from processor 1 can be moved directly into the subsection of A on processor 0. Imagine that processor 0 has enough storage allocated for A to store a 10x10 submatrix instead of just the 8x8 submatrix for the values that it owns. These extra edges are known as "shadow edges" or "ghost cells" and will hold copies of the near edges of the submatrices owned by neighboring processors. This has the advantage (over using a temporary) of improving cache usage when the local computation is performed because there will be no references to a temporary array and all references will be to contiguous memory. For example when A(8,1) is updated it needs the value B(9,1) which, though owned by processor 1, has been stored locally in the shadow edge of the local submatrix of B.

2.5 Move Edges as Chunks

In fact instead of moving each value from the edges of the submatrix owned by processor 1 to its shadow edge on processor 0, the entire edge can be moved in one send. This reduces the number of send-receives which will increase application performance. Using block move code to move the shadow edge from its submatrix into the message buffer that moves values between processors rather than using element by element copying is also desirable for cache reasons. Similarly, a block move is also used to move the edge values out of the message buffer and into the shadow edge on the receiving processor.

2.6 Move edges for each array not for each array reference

It is often the case in nearest-neighbor codes that the array references on the right-hand-side of the assignment are different sections of the same array. Recognizing that only one move of shadow edges for a given array is necessary removes the need to move the same values more than once. For example in Figure 4, the

---

```
subroutine sub(n)
integer, dimension(n)    :: B
!hpf$ distribute(block)   :: B
integer                  :: i

forall(i=2:n-2) B(i) = B(i+1) + B(i+2)

...
end subroutine sub
```

Fig. 4. One edge Nearest-Neighbor Code Fragment

---

array B is referenced twice, but its shadow edges only need to be moved once. So moving the shadow edges required by the B(i+2) reference is sufficient; no further sending of data is required by the B(i+1) reference. Here the size of the shadow edge is two instead of one as in the previous examples. A naive application of owner computes would have degraded performance by moving the two different but overlapping sections of B; i.e. the B(i+1) edge values would have been moved twice.

2.7 Move only the necessary edges for the statement

In fact some codes are simple enough that not all of the edges need to be moved. Referring again to the code fragment in Figure 4 there are no references to B(i-1) so only half the shadow edges need filling in this example; there is no need to fill up all of the shadow edges. Contrast this situation to the code in Figure 1 where the array B needed to have its shadow edges moved both directions in the first dimension.

2.8 Corners are moved for free

Careful consideration of how the shadow edges are moved for the two dimensional case shows that corners are moved for free. Consider Figure 1 again. Suppose the assignment statement had been

```
A(i,j) = B(i+1,j) + B(i+1,j+1)
```

In this case, processor 0 would have needed values from not only processors 1 and 2, but also the corner value from processor 3. Figure 5 shows that the value from processor 3 is moved to processor 0 as a by-product of moving the shadow edges of processors 1 and 2. A similar analysis shows that if top and bottom edges are moved before left and right edges, that the corner value would move from processor 3 to processor 2 and from there to processor 1. Thus the order of filling shadow edges does not affect the resulting values in those shadow edges.

In this trivially small example one might attempt to overlap the sending of the edges up with the sending of edges to the left. Remember that for larger processor grids most processors will not be edge processors. The non-edge processors will send their edge values to the appropriate shadow edges to the left and then right and then up and then down in the most general case. So in general it is not possible to overlap sending edge values left with sending edge values up.

2.9 Avoid moving edges that are up to date

As a final optimization, a compiler can look across statements and determine that the shadow edges for a particular array do not need to be moved because the shadow edges were fetched by a previous statement and those previously moved values are still valid. This situation occurs in many nearest neighbor programs including shallow water codes and red-black relaxation codes.

In red-black codes a detailed analysis is required to recognize when shadow values are still valid. A three dimensional red-black relaxation code is shown in Figure 6. This code consists of eight *forall* statements; the first four statements update the red values and the second four the black values. A careful inspection of the second *forall* shows that the shadow edges for U were fetched by the first *forall* and that the values fetched out of the shadow edges of U by the second *forall* do not use values that were updated by the first *forall*. Thus, though some of the shadow edges that were fetched by the first *forall* were updated by the first *forall*, the shadow values needed by the second *forall* were not among these updated

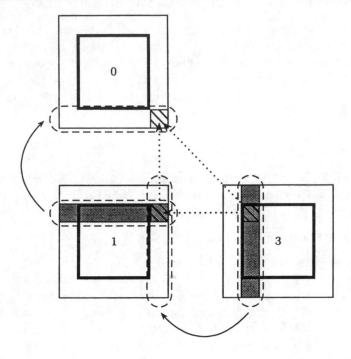

Fig. 5. Moving the Corner Values

values. After applying this analysis to the entire code fragment, it becomes clear that fetching the shadow edges of U is only necessary before updating the red points and just before updating the black points (after doing the red points). A compiler must use sophisticated section analysis to determine that these are the only two shadow fetches that are needed. Removing the unnecessary updates of the shadow edges for U for six of the eight nearest-neighbor assignments significantly improves the performance of this application.

3   Some timing results

The timing results in Figure 7 and Figure 8 demonstrate the performance improvement of the previously explained optimizations for a three dimensional red-black relaxation and a shallow water code. These two codes had their problem sizes adjusted to make the execution times be under two minutes. Actual timings would depend on the problem size under consideration and the amount of tolerance allowed before convergence is assumed. An NProcs value of 0 means the application was not compiled for parallel execution. A Nprocs value of "n" means the application was compiled for parallel execution on "n" processors. In general, a slight improvement is seen moving from 0 to 1 processors as a more

```
do nrel=1,iter
! RED-BLACK RELAXATION: THE RED POINTS (Fetch U's shadow edges)
 forall (i = 2:nx-1:2, j=2:ny-1:2, k=2:nz-1:2)
   u(i, j, k) = factor*(hsq*f(i, j, k) +  u(i-1,j,k) + u(i+1,j,k) &
   + u(i,j-1,k) + u(i,j+1,k) + u(i,j,k-1) + u(i,j,k+1))
 end forall

  ! Compiler avoids refetching shadows for U until the black points
 forall (i = 3:nx-1:2, j=3:ny-1:2, k=2:nz-1:2)
   u(i, j, k) = factor*(hsq*f(i, j, k) +  u(i-1,j,k) + u(i+1,j,k) &
   + u(i,j-1,k) + u(i,j+1,k) + u(i,j,k-1) + u(i,j,k+1))
 end forall

 forall (i = 3:nx-1:2, j=2:ny-1:2, k=3:nz-1:2)
   u(i, j, k) = factor*(hsq*f(i, j, k) + u(i-1,j,k) + u(i+1,j,k) &
   + u(i,j-1,k) + u(i,j+1,k) + u(i,j,k-1) + u(i,j,k+1))
 end forall

 forall (i = 2:nx-1:2, j=3:ny-1:2, k=3:nz-1:2)
   u(i, j, k) = factor*(hsq*f(i, j, k) + u(i-1,j,k) + u(i+1,j,k) &
   + u(i,j-1,k) + u(i,j+1,k) + u(i,j,k-1) + u(i,j,k+1))
 end forall

! RED-BLACK RELAXATION: THE BLACK POINTS  (Refetch shadow edges for U)
 forall (i = 3:nx-1:2, j=2:ny-1:2, k=2:nz-1:2)
   u(i, j, k) = factor*(hsq*f(i, j, k) + u(i-1,j,k) + u(i+1,j,k) &
   + u(i,j-1,k) + u(i,j+1,k) + u(i,j,k-1) + u(i,j,k+1))
 end forall

  ! Compiler avoids refetching shadows for U until the black points
 forall (i = 2:nx-1:2, j=3:ny-1:2, k=2:nz-1:2)
   u(i, j, k) = factor*(hsq*f(i, j, k) +  u(i-1,j,k) + u(i+1,j,k) &
   + u(i,j-1,k) + u(i,j+1,k) + u(i,j,k-1) + u(i,j,k+1))
 end forall

 forall (i = 2:nx-1:2, j=2:ny-1:2, k=3:nz-1:2)
   u(i, j, k) = factor*(hsq*f(i, j, k) +  u(i-1,j,k) + u(i+1,j,k) &
   + u(i,j-1,k) + u(i,j+1,k) + u(i,j,k-1) + u(i,j,k+1))
 end forall

 forall (i = 3:nx-1:2, j=3:ny-1:2, k=3:nz-1:2)
   u(i, j, k) = factor*(hsq*f(i, j, k) + u(i-1,j,k) + u(i+1,j,k) &
   + u(i,j-1,k) + u(i,j+1,k) + u(i,j,k-1) + u(i,j,k+1))
 end forall
enddo
```

Fig. 6. 3D Relaxation Code Fragment

aggressive use of dependence analysis is used when generating parallel code. Of course no communication is generated for a one element processor array. There is usually a performance degradation in moving to two processors. Having only two processors means that the computation per processor will at best be halved, but communication is so expensive that for the small problem sizes used in these timings the overall execution time is increased. In these examples, the nearest neighbor optimizations removed enough of the communication that the improvement in computation speed gained by distributing the data was enough to cover the added cost of communication. Superlinear speedups can be achieved for small problem sizes if all of the data on one processor can fit into the cache. All timings were done on a workstation farm of Digital Alphastation 3000 Model 700 machines interconnected via Digital's Gigaswitch/FDDI using the code generated by the Digital Fortran 90 compiler.

NProcs	No NN Opts	With NN Opts
0	60.1	NA
1	55.8	NA
2	99.0	26.1
4	95.8	19.1
8	87.0	14.8

Fig. 7. Timings (in seconds) for Red-Black Code

NProcs	No NN Opts	With NN Opts
0	31.4	NA
1	29.1	NA
2	50.7	12.8
4	25.0	5.8
8	12.4	2.6

Fig. 8. Timings (in seconds) for Shallow Water Code

4 Conclusion

Compiling parallel algorithms coded in HPF can lead to extremely efficient code for certain classes of problems. HPF compilers must learn to recognize and optimize these problems to generate high quality code. The number of problems that can be handled efficiently by HPF compilers will continue to increase as the compilers mature and their use becomes ever more prevalent in high performance computing.

Parallelization of a Local Area Ocean Model

Luka Onesti and Tor Sørevik

Para//ab, University of Bergen

Abstract. In this paper we describe the parallelization of a local area ocean circulation model using HPF. We find that this particular problem, leading to a typical data parallel algorithm is easily expressed in HPF. The problems which occurs is mainly due to restrictions in the current HPF-compilers.

We have run our program on 3 different platforms, using different HPF-compilers. and will report on the status of these compilers as measured by our problem.

1 Introduction

Parallel computing has proved to offer the computational power it claimed a decade ago, and today all computers on the TOP500 list are parallel computers. In spite of the enormous performance gains offered by parallel computers there is still some reluctance among application programmers to use these. The reason for this is the huge effort needed to port large application programs to parallel platforms. In particular this is true when using the message passing programming paradigm for distributed memory computers.

High performance Fortran (HPF) [Hig] [KLS+94] is an attempt to meet the need for easier parallel programming, removing the burden of figuring out the communication pattern of data from the programmer to the compiler. HPF is built upon FORTRAN 90 (F90) [ofSC91] by adding directives and a few new language features and intrinsics to F90. Being close to sequential F90 programming an additional advantage of HPF/F90 would be the need to maintain only one code for sequential as well as parallel programming.

While HPF in theory offer an easy way to parallel programming., life is somewhat different i practice. In particular, the present stage the HPF-compilers available to us has certain shortcomings. In part they do not implement the full HPF (or F90) standard, nor do they agree on what not to implement, which make portability difficult.

Being a much richer language than F77, F90 gives, in many cases, multiple choices for how to express complex statements. Although being semantic equivalent the different syntactical variants sometimes gave huge and unexpected performance differences. Moreover, we found that although an HPF code is portable across a variety of platforms, it is not necessarily performance portable. We contribute these obstacles to the shortcomings of current HPF (and F90) compilers, and expect these to improve in the future.

In this paper we describe the parallelization, using HPF, of a local area ocean model, developed jointly between the Univ. of Bergen and the Norwegian Marine Research Institute [BES96].

After a brief description of the model in section 2, we introduce and discuss the F90 and HPF features necessary for parallel implementation of this application in section 3. In section 4 we discuss the data mapping. In section 5 we present our numerical result and discuss them.

2 The Ocean Model

The governing equations of the model is the standard equations for continuity, momentum and conservations, with the modifying hydrostatic assumption that the weight of the fluid identically balances the pressure, and the Boussinesq approximation that density differences are neglected unless the differences are multiplied by gravity. Motions induced by small scale processes are parametrized by horizontal and vertical eddy viscosity/diffusivity terms.

Even with these assumptions the equations constitute a complicated coupled set of non-linear time dependent PDEs in three spatial dimensions.

The model is a local area model as opposed to a global model. Thus the vertical boundaries are in part fixed (over land) and open (over sea). Fluxes at the open boundaries are taken from a global model. The horizontal boundaries impose von Neuman boundary conditions on the vertical velocity.

The spatial variables is discretized by central finite differences, using a staggered C-grid. To take care of the irregularity imposed by bottom topography the σ-coordinate system is used. This is a linear transformation of the depth such that all gridpoints, with the same vertical coordinate, have the same relative position between the surface and the bottom. In each horizontal layer a cartezian grid with rectangular shape was used. The irregular shape of the coastline implies that we include some onshore area by this discretization. All variables are defined here too, but have, of course, no meaningful physical interpretation. Computation over these gridpoints can easily be omitted by masking.

With this spatial discretization all data is stored in regular 2 or 3-dimensional arrays, corresponding to their physical displacement.

To advance the solution in time an operator splitting technique is used. The effect of each term in the governing equations is computed separately and added up to give the total contribution. Gravity waves are computed by an spatial split implicit method in each horizontal layer. Thus one have to solve multiple, independent tridiagonal linear system in each timestep. All other terms are propagated in time by explicit timestepping. At boundary points data for central differences is not available, thus these points are taken care of by special code segments.

3 The HPF implementation

3.1 Platforms used

Available to us for this project were three different HPF-compilers, running on different hardware platforms. Below we give a brief description of the compilers and their related hardware.

The bulk of the porting and the initial testing was done on Digital's proprietary F90/HPF compiler, version 4.0. This is an integrated compiler where the user determine by a switch whether he wants to compile for sequential execution (=F90) or parallel execution (=HPF). Digital has been a forerunner in developing a full F90-compiler, and that part of the compiler appear to have reached a mature state. This software was run on a cluster of 8 DEC workstations, each equiped with a 233 Mhz alpha processor, connected with an fddi-switch, DEC's own Gigaswitch. The programs were run under Digital's PSE (Parallel Software Environment), which enables you to use the cluster with the same ease as any "all-in-one-box" MPP system.

We've also used Portland Group (PGI)'s portable compiler. This compiler compiles HPF code into f77 + message passing. The message passing layered used can be adapted to the hardware and software platform available. We have run version 2.0 of the compiler on a 100 processors Intel Paragon using Intel's nx as the message passing layer.

The third compiler we have used is IBM's proprietary HPF compiler version 1.0 running on an IBM SP2. This compiler is, from the user perspective, quite similar to DEC's. The HPF parts are the parallel extension to the F90-compiler.

3.2 Language features used

F90 provides an array syntax for matrix and vector operations. In addition to its beauty and convenience, this notation also provides compiler writers with good opportunities for optimization and parallelization. Whenever possible this notation has been used in the present code.

Other features, new to FORTRAN with the F90-standard, used in this code are: MODULE used to implement global variables in place of COMMON, INTERFACE used to assist the compiler with interprocedural analysis, WHERE used to mask out array elements, not needed to be updated in array statements. We also experimented with dynamic memory allocation, but this lead to a serious performance degradation, and was for that reason abandoned. The FORALL statement of HPF was also used.

A complicating feature of F90 is that it does not allow nested WHERE. This forced us to rewrite the code, throwing complicated, concatenated, logical statement into one WHERE construct. This did not only make the code pretty ugly, but turned out to be a tough case for the compiler. It tended to perform badly in these cases. An alternative to WHERE is to use FORALL with a masking statement, but on the DEC-compiler this had even worse performance characteristics. The reason seems to be that when a masking array is found within a FORALL statement the compiler by default choose to replicate the masking array on all processors. If the masking array is aligned with the other distributed arrays this impose a enormous data traffic and we experience slow down when using more than one processor.

Communication imposed when distributing data to multiple processors is only nearest neighbor. Consequently the data was distributed BLOCK-wise. A TEMPLATE was used to defined the data distribution and all 2- or 3- dimensional arrays was align with this. Global arrays, defined in MODULE, were given their distribution here. When the MODULE was incorporated in a program segment with the USE statement, the data distribution automaticly followed. This strategy works fine on the DEC-platform, but although the MODULE feature is supposed to be supported in the last version of the Intel HPF compiler, we were not able to use it as indicated above. On the Paragon we therefore arranged the global arrays in traditional common statement and collected these in an INCLUDE-file, which also contained the HPF-attribute statements and the template declaration as well.

Local arrays were always aligned with a global array. When a local array was passed to a subroutine through a dummy argument, we intended to use the INHERIT statement to make sure the already defined data distribution was inherit. This approach gave, however, bad performance on both platforms. We found the most efficient implementation to be using the descriptive ALIGN directive, in which we declared the dimension of the arrays explicitly and then aligned them to a template and passed it to the subroutine using module (or common in the case of the Intel Paragon). We believe the difference to be that the runtime calculation of indexes is no longer necessary.

We also tried using ASSUMED-sized array. This works, but slow, and was for that reason abandoned in the final version.

We also took advantage of some of the new array inquiry function of F90, like MAXVAL, ANY and MAX. Using these was necessary to get good parallel performance on multiple processors. The standard F77-code with an IF-test within a set of nested DO-loops, was always executed sequential.

Features that is not yet implemented or does not yet work properly is the INDEPENDENT statement of HPF. This is the directive needed to tell the compiler that F77 DO-loops are parallel. Without this directive one are forced to change all code into F90-array syntax and/or using FORALL. We also had problems with the PURE-attribute. Either it didn't work (on the PGI-compiler) or it was not used to parallelize the code.

3.3 Code rewriting

The procedure we followed in the porting procedure was the obvious one. First rewriting the code to F90 and testing this version for correctness and efficiency on a sequential platform, and than introducing the HPF-compiler directives for parallelization. The great advantage of this is the incremental nature of this step by step procedure. As expected the bulk of the work is related to doing the translation to F90. The sometimes unexpected performance behavior also created some headache.

Some times substantial rewriting was needed to better express the true data parallel nature of the problem. In particular code like:

```
DO I = 1, N
   TMP1 = ....
   TMP2 = ....
   CALL FOO (A(I), TMP1, TMP2)
END DO
```

was expanded to

```
FORALL (I = 1:N)
       TMP1(I) = ...
       TMP2(I) = ...
END FORALL
CALL FOO (A, TMP1, TMP2)
```

where FOO was rewritten to operate on arrays instead of scalars. In particular this was important for solving multiple tridiagonal systems in parallel.

Note that this rewriting expands temporary variables to arrays instead of scalars and thereby increasing the memory requirements. We experience that the compilers we used also introduced this kind of temporary arrays in their intermediate code.

4 Data distribution

The scheme for computing the vertical mixing leads to $N_x \times N_y$ independent tridiagonal linear equation systems in each horizontal grid point. For this part of the code to be solved efficiently in parallel, all vertical gridpoints in the same horizontal cell have to remain at the same processor. This gives a straightforward parallelization of the vertical mixing.

Except for the gravity waves all timestepping are done by a simple explicit scheme, independently for the different vertical layers. Thus a simple block partition in one or two of the x- and y- direction imposes a simple and regular nearest neighbor communication, which has proved efficient in many similar cases.

On a $N_x \times N_y \times N_z$ grid the amount of data needed to be sent in each step, if divided in the y-direction is:

$$Comm_{1d} = 2\alpha + 2\beta N_x N_z \tag{1}$$

similar if divided in the x-direction. A 2d- partition on $P \times P$ processors gives

$$Comm_{2d} = 4\alpha + 2\beta(N_x N_z + N_y N_z)/P \tag{2}$$

Here α is the latency and β represents the inverse of the bandwidth for a give system.

As is evident from these formulas a 2d partitions is always best for a large number of processors. For less than 4 processors a 1d partion parallel to the axis with fewest gridpoint is the best. Exactly where the crossover point is depends on the system (α and β) and the problem size.

The simple model above does not take into account the cost of data access. In FORTRAN data in multidimensional arrays is stored in column major order. Thus to send a column of a matrix, one need to retrieve and store data which is stored consecutively in memory. While accessing data in an array segment imposes stride in memory. For this reason we do expect a higher crossover point between 1-d and 2-d distribution, than predicted by the formula above.

The only nonobvious data distribution occurs when solving the multiple tridiagonal systems of equations for the implicit timestepping of the gravity waves. The operations are by themselves "embarassingly parallel", but the data needed to construct the coefficient matrix of each individual tridiagonal system is distributed across the processors. The coefficients for each tridiagonal system is, as usually, stored in 4 vectors The coefficients for all the systems is kept in 4 2-d arrays. To solve one specific tridiagonal system for splitting in "X-direction" one has to traverse the corresponding columns of these matrices. This can be done without communication if the coefficient matrices are distributed (*,BLOCK). When solving in "Y-direction" one need to traverse rows of the matrices, consequently the matrices should be distributed (BLOCK,*). These matrices are constructed using data involved in other parts of the computation which might, as described above, be distributed in either of these ways. The important thing is, however, that for the reminding computation to execute efficiently all these arrays have to be aligned. Thus at least one of the two sets of coefficients are not aligned with the data it is constructed from, and thereby imposing communication.

5 Numerical Experiments

All experiments reported here are based on the full application code or part of it. The grid size we used was $125 \times 113 \times 12$. The work associated with a gridpoint depends on whether or not it is over land or sea. When distributing the gridpoints the current HPF standard provides no mechanism for taking the differences in workload of the gridpoints into account.

5.1 Syntax dependent Performance

In the explicit timestepping procedure a number of array assignment is used. Typically we need to handle boundary points separately. Moreover for efficiency reasons we will like to mask out "on-shore" gridpoints. F90 provides different ways of expressing this and we tried some of these.

All these variants have the option of replacing the masking with an appropriate placed multiplication with zero. This does, however, require careful implementation to avoid division by zero. On the Digital platform we found that the WHERE construct got serious performance degradation when the masking statement become complicated. The FORALL statement was for all practical purposes impossible to use on more than one processor, due to the need for replicating masking arrays. Wherever possible we therefore replaced masking with multiplication with zero in Figure 1 This was not possible everywhere, hence variant 3 outperformed the two other constructs.

In table 2 we summarize how the different flavors of the code perform sequential on the DEC alpha and on the i860 processor of the Paragon

On the DEC-cluster all versions suffer a severe loss in performance as soon as we move to parallel execution, (See Figure ??). In particular this is true for the FORALL-version. This is due to the (unnecessary) replication of the masking array. The relatively poor performance of the interconnecting network is also a part of the explanation.

All versions show a very good speed-up on the Paragon up to 32 nodes, (See Figure ??). Only the communication free, tridiagonal solver scales beyond this. With a 1-d data distribution and the current resolution this is as expected. We believe the superlinear speed-up when moving from 1 to 2 processor is due to memory access pattern.

On the SP2 the WHERE-construct shows decent speed-up, while the FORALL-construct should be avoid.

5.2 Which Blocking to choose?

As discussed in section 4 the choice between 1-dimensional and 2-dimensional distribution is not obvious for moderate number of processor. Moreover when

1. FORALL with masking as in:

```
FORALL (I = 2:IM,J = 2:JM-1,K = 1:KB-1,DUM(I,J).NE.O.)
       XFLUXUP(I,J,K) = AREATU1(I,J,K)*FIELD(I-1,J,K)
       XFLUXLW(I,J,K) = ....
       .........
END FORALL
```

2. Array-syntax and the WHERE-statement as in the included example. In this case the masking array must have the same shape as the arrays it applies to. Thus one either have to expand DUM with a third dimension on the expenses of extra storage or wrapping the WHERE-construct in a DO-loop keeping the third index of XFLUXUP, XFLUXLW, ... fixed.

```
DO K = 1, KB-1
WHERE (DUM(2:IM,2:JM-1).NE.O.)
       XFLUXUP(2:IM,2:JM-1,K) = AREATU1(2:IM,2:JM-1,K)) &
              *FIELD(1:IM-1,2:JM-1,K))
       XFLUXLW(2:IM,2:JM-1,K) = ....
       .........
END WHERE
END DO
```

3. FORALL without masking, but call to a pure function as in:

```
PURE FUNCTION SUPERBEEF_X(...,SCALAR1, SCALAR2, DUM,...)
...................
FORALL (I = 2:IM,J = 2:JM-1,K=1:KB-1)
       XFLUX(I,J,K)=SUPERBEEF_X(...,FIELD(I,J,K),&
                    FIELD(I-1,J,K),DUM(I,J),...)
```

In this case the masking is done with a traditional IF-statement inside SUPER-BEEF_X which only takes scalar variables. Here each processor runs a copy of the sequential routine SUPERBEEF_X with the data associated with the grid point it is responsible for.

Fig. 1. Different syntax for expressing masking on arrays

	DEC's HPF alpha cluster	PGI's HPF Paragon	IBM's HPF SP-2
Forall+pure function	.52	Doesn't work	4.83
Forall with masking	1.58	6.27	2.73
Where with 2d masking	1.57	10.02	1.82
Where with 3d masking	-	10.22	1.50

Fig. 2. *Single node performance in seconds of the different version of the code-segment displayed in Figure 1*

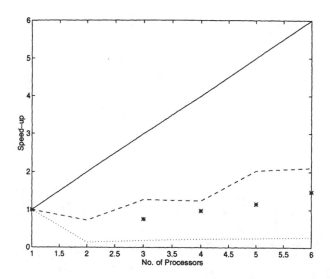

Fig. 3. *The straight line represent linear speed-up. The '*' is FORALL with PURE-function, the dashed line is WHERE and dotted line is the FORALL-version*

the prime factorization of the processor number contains more than 2 primes there is multiple choices for 2-dimensional distribution. This choice could be left to the compiler, or made explicit by the programmer by using the following form of the distribution directive:

`!HPF$ DISTRIBUTE A (BLOCK,BLOCK) ONTO (P,Q)`
where (P,Q) are given parameters. To give an idea of how this effect the performance, we have run the entire code on 36 processor of the Intel Paragon with all possible BLOCK distribution. Parameters one expect to influence the performance is load imbalance and ratio between computation an communication. In our example these can be measured as the maximal number of grid point on one processor ($\lceil 125/P \rceil \times \lceil 113/Q \rceil \times 12$) and the amount of data to be communicated. This is proportional to the vertical surfaces of the "box" for individual processor. This should be (($\lceil 125/P \rceil + \lceil 113/Q \rceil) \times 2 \times 12$) The result is given in figure 6.

As expected (*,BLOCK) is better than (BLOCK,*), since passing a column instead of row means accessing consecutive data in memory. This effect seems to dominate the communication performance. The amount of data to be communicated and the load imbalance for computation indicated seems to have only minor effect on the efficiency.

When ONTO is not used the compiler does not have the number of processor available at compile time, thus some index calculation have to be deferred to runtime. Whether this explains the non-optimal performance of (BLOCK,BLOCK,*) without ONTO or the compiler simply choose a non-optimal factorization we don't know.

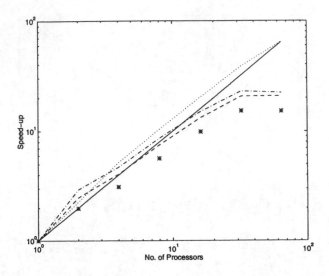

Fig. 4. *The straight line represent linear speed-up. The dash-dotted line is the forall-version, * is WHERE with 2d-masking and dash-dotted line is WHERE with 3d masking. The dotted line is the speed-up of the multiple tridiagonal solver.*

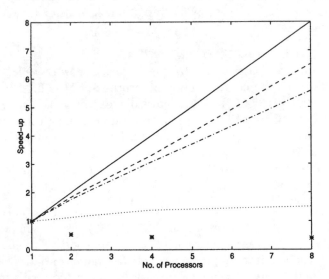

Fig. 5. *The straight line represent linear speed-up. The '*' is using FORALL with a PURE-function, The dashed line is WHERE with 2d-masking. The dotted line is the FORALL-version with masking. The dash-dotted line is WHERE with 3d-masking*

distribution	Max gridpoint	Data to send	Time in sec.
(*,BLOCK,*)	6000	3000	3.39
(BLOCK,*,*)	5425	2712	17.83
(BLOCK,BLOCK,*) ONTO (18,2)	4788	1536	7.03
(BLOCK,BLOCK,*) ONTO (12,3)	5016	1176	8.54
(BLOCK,BLOCK,*) ONTO (9,4)	4872	1032	7.01
(BLOCK,BLOCK,*) ONTO (6,6)	4536	960	3.99
(BLOCK,BLOCK,*) ONTO (4,9)	4992	1080	7.11
(BLOCK,BLOCK,*) ONTO (3,12)	5040	1248	8.65
(BLOCK,BLOCK,*) ONTO (2,18)	5292	1680	6.29
(BLOCK,BLOCK,*)			7.09

Fig. 6. *The elapsed time in seconds pr. timestep as a function of the distribution*

5.3 Scalability and Data distribution

On the basis of the above experiment we have picked the most efficient versions of the different subroutines and run the full scale application program on different systems, with different number of processor and with the (BLOCK,BLOCK,*) and (*,BLOCK,*) distribution. The result is presented in figure 7 below.

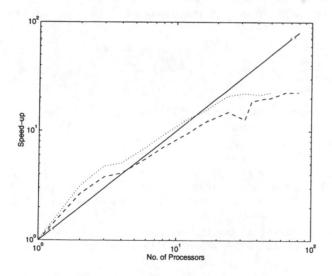

Fig. 7. *The Straight line represent linear speed-up. The dashed line is the BLOCK,BLOCK distribution while the dotted line is the (*,BLOCK) distribution*

As expected the 1-d distribution performs best on low to moderate number of processors, but it doesn't scale well. The 2-d distribution scales somewhat better.

6   Conclusions

This exercise has convinced us that HPF is a much easier road to parallel programming than message passing. The incremental way of porting large application code makes the debugging much easier. The advantage of having only one code to maintain is also huge. It does make meaningful to do almost all the developing work on your (sequential) local workstation, and only when finished move your code to the central (parallel) HPC-platform.

The reasons for the problems we occurred appears to be immature compilers. Bugs and missing features cost us lot time and frustrations. The quality of the compilers are however, rapidly increasing and we have seen great improvements over the months we worked on this project.

Also confusing and frustrating is the inconsistent performance behavior of the different compilers. Since performance is a key issue it is important that the code not only works when moved to another platform, but still runs efficiently. This is not the case now. Choosing the most efficient HPF-programming style for the platform you do the development work on, is by no means any guarantee that you have an efficient code when moving to another platform.

We do believe these differences will diminish as the vendors put more effort into optimizing their compilers. Some of the differences are due to the underlying sequential Fortran compilers and although FORTRAN 90 has been around for some years, not all vendors have put enough effort into optimizing their F90-compiler.

References

[BES96]   J. Berntsen, T. O. Espelid, and M. D. Skogen. Description of a σ-coordinate ocean model. Technical Report ??, Department of Mathematics, University of Bergen, 1996.

[Hig]   High Performance Fortran Forum Home Page.
http://www.crpc.rice.edu/HPFF/home.html.

[KLS+94]   C. H. Koelbel, D. B. Loveman, R. S. Schreiber, G. L. Steele, and M. E. Zosel. *The High Performance Fortran Handbook*. The MIT Press, Cambridge Massachusetts, 1994.

[ofSC91]   International Organization for Standardization and International Electrotechnical Commission. Fortran 90, iso/iec 1539: 1991], 1991.

Parallel Solution of Sparse Problems
by Using a Sequence of Large Dense Blocks

Tz. Ostromsky[1], S. Salvini[2], J. Wasniewski[3] and Z. Zlatev[4]

[1] Central Laboratory for Parallel Information Processing,
Bulgarian Academy of Sciences,
Acad. G.Bonchev str., bl. 25-A, 1113 Sofia, Bulgaria;
e-mail: ceco@iscbg.acad.bg
[2] The Numerical Algorithms Group Ltd (NAG),
Wilkinson House, Jordan Hill Road, Oxford, UK, OX2 8DR
e-mail: stef@nag.ac.uk
[3] Danish Computing Centre for Research and Education (UNI-C), DTU, Bldg. 304,
DK-2800 Lyngby, Denmark;
e-mail: jerzy.wasniewski@uni-c.dk
[4] National Environmental Research Institute,
Frederiksborgvej 399, P. O. Box 358, DK-4000 Roskilde, Denmark
e-mail: luzz@sun2.dmu.dk

Abstract. Linear least squares problems arise in many important fields of science and engineering (such as econometry, geodesy, statistics, structural analysis, fluid dynamics, etc.). Applications from all these fields are to be handled numerically in the solution of various models used in the development of new industrial products. At the National Environmental Research Institute (Denmark), linear least squares problems have been applied in the efforts to determine optimal values of some of the parameters involved in air pollution models.

Large and sparse least squares problems can be handled by treating a sequence of dense blocks. Standard parallel subroutines, which perform orthogonal decomposition of dense matrices (as, for example, subroutines from LAPACK, SCALAPACK or NAG), can be used to handle successively the dense blocks. A preliminary reordering procedure, LORA, is to be applied at the beginning of the computational process. The size of the blocks can be adjusted to the particular computer used. Results obtained on a Silicon Graphics POWER CHALLENGE are given.

The same method can also be used in the solution of large and sparse systems of linear algebraic equations.

1   Using dense blocks in the solution of sparse problems

Dense matrix computations are becoming more and more popular when sparse matrices are handled on large modern high-speed computers. The first work where dense matrix computations were advocated when sparse matrices are to be handled is perhaps [2]. The **multifrontal method** is often used in the new sparse codes (see [3], [4], as well as the references given there). In this method one tries to carry out the computations by frontal matrices (which are either dense or are treated as dense); see, for example, [10], [12]. Another method, which is based on the use of a sequence of large dense blocks, will be described in this paper. All floating-point computations are performed by dense matrix kernels applied to factorize large dense blocks. These kernels calculate an orthogonal decomposition of each dense block by calling, many times, BLAS routines. **The dense kernels are performing the most time-consuming part of the computational work.** Therefore it should be expected to obtain good results on any high-speed computer for which high quality software for dense matrices is available. The efficiency of this approach in the case where the high-speed computer is a CRAY C92A has been demonstrated in [9]. The performance on a Silicon Graphics POWER CHALLENGE computer will be discussed in this paper.

2   General discussion of the algorithm

Consider the linear least squares problem: $Ax = b - r$ with $A^T r = 0$ where $A \in \mathbf{R}^{m \times n}$, $x \in \mathbf{R}^{n \times 1}$, $b \in \mathbf{R}^{m \times 1}$, $r \in \mathbf{R}^{m \times 1}$, $m \geq n$ and $rank(A) = n$. Assume that matrix A is large and sparse. The main objective is to compute an upper triangular matrix $R \in \mathbf{R}^{n \times n}$ by using orthogonal transformations in a sequence of large dense blocks. The algorithm from [9] will be used here. The major steps of this algorithm are shortly described in Table 1. The subroutines based on this algorithm are united in package PARASPAR-L.

Step 2 - Step 4, from the algorithm given in Table 1, are carried out in a loop. Step 1 - Step 5 are not discussed in detail here; these steps are fully described in [9].

Step	Task	Description of the actions taken during the task
1	LORA	Reorder the non-zeros of the mattrix to an upper triangular form with as many non-zeros under the diagonal blocks as possible.
2	Scatter	Form a large block. Scatter its non-zeros in a two dimensional array. Fill the empty locations with zeros.
3	Compute	Call routines which perform some **dense** orthogonal decomposition of the block determined in Step 2.
4	Gather	Gather the non-zeros of the decomposed block (without the non-zeros of Q) in the sparse arrays. Optionally, use dropping. If the remaining active sub-matrix is still sufficiently sparse, then **GO TO STEP 2.**
5	Last block	Form the last block. Switch to **dense** matrix technique.

Table 1
Major steps in the algorithm based on the use of large dense blocks.

3   Preliminary reordering by using LORA

LORA stands for "locally optimized reordering algorithm". It has been studied in [7], [8] and [11]. **LORA** reorders the matrix to a block upper triangular form with rectangular diagonal blocks (Fig. 1) with an additional requirement to put as many zeros as possible under the diagonal blocks. There are three types of blocks: (i) dense blocks in the separator (the black boxes in Fig. 1), (ii) zero blocks under the separator (the white boxes in Fig. 1), (iii) sparse blocks over the separator (the shaded boxes in Fig. 1; some of these blocks may contain only zeros).

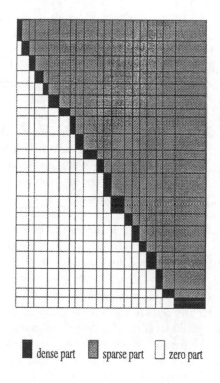

dense part ■   sparse part ▨   zero part ☐

Figure 1
Sparsity pattern of a rectangular matrix obtained by LORA.

4 Creating large dense blocks

The most straight forward way to create dense blocks is by using directly the dense blocks created by LORA (the separator in Fig. 1). One can take a block-row corresponding to the first dense block from the separator in Fig. 1. Assume that this block-row contains r_1 rows. The non-zeros in these r_1 rows are stored in a two-dimensional array containing r_1 rows and c_1 columns, where c_1 is the union of the sparsity pattern of the r_1 rows in the first block-row. Assume that the first dense block from the separator in Fig. 1 has q_1 columns ($rank(A) = n$ implies $q_1 \leq min(r_1, c_1)$). Then q_1 orthogonal stages have to be performed by the dense matrix kernel chosen when the first dense block is treated (non-zeros under the diagonal of column i, $i = 1, 2, \ldots, q_1$, are produced during the orthogonal stage i by some kind of orthogonal transformations).

The creation of the next dense blocks is more complicated (because the fact that there are some unfinished rows after the treatment of the previous dense blocks must be taken into account). The process is fully described in [9]. The important fact is that three parameters are to be taken into account when any dense block i is handled by dense subroutines: (a) the number of rows r_i, (b) the number of columns c_i and (c)the number of stages q_i that are to be performed.

The dense blocks created by using block-rows induced by the dense blocks from the separator in Fig. 1 are in general small. Therefore one must use block-rows obtained by combining several dense blocks from the separator in Fig. 1. If this is done, then the situation becomes more complicated (see [9]). However, the three parameters r_i, c_i and q_i are again important when the dense block i, $i = 1, 2, \ldots, I$ (I being the number of large dense blocks) is to be treated.

The three parameters can be used in the efforts to get dense blocks of appropriate size; not too small (because the dense kernels will not be very efficient if this is so), but not too large either (because the number of arithmetic operations grows rather quickly when these parameters become large). It is difficult to use c_i, because this parameter is determined at the end of the process in which a dense block is produced. Parameter q_i can be used rather efficiently. Assume that we are creating a block-row by using s successive dense blocks from separator in Fig. 1; say, $d_{i_1,i_1}, d_{i_2,i_2}, \ldots, d_{i_s,i_s}$. Assume also that the next dense block is $d_{i_{s+1},i_{s+1}}$. Then q_i is determined by the positions of the first non-zeros in d_{i_1,i_1} and $d_{i_{s+1},i_{s+1}}$.

The use of parameter q_i to put some lower limit of the number of orthogonal stages that are to be carried out in each dense block is often sufficient; one can, for example, require $q_i \geq 100$ (this means that the chosen values of q_i will be very close to 100, because the dense blocks from the separator in Fig. 1 are small). However, for very long matrices (m much greater than n), the use only of q_i may lead to the creation of very large dense blocks. Such a situation occurred in some of our experiments with (100000×10000) the matrices. Therefore, it is worthwhile to use additionally r_i. This parameter is also easily available (determined of the row index of the first row in d_{i_1,i_1} and the row index of the first row in $d_{i_{s+1},i_{s+1}}$). Parameter r_i is used in an attempt to limit the number of rows in the dense block under consideration; more precisely, the attempts to get a dense block with q_i orthogonal stages are continuing, by adding more blocks from the separator in Fig. 1, until the number of rows r_i remains less than some limit.

5 Treating the large dense blocks

Let us reiterate here that the i'th dense block contains r_i rows and c_i columns, while q_i stages are to be performed. Four algorithms have been tested. Different dense kernels are used during the treatment of the large dense blocks in these four algorithms.

5.1 ALG1: Direct use of LAPACK subroutines

The LAPACK subroutine ([1]) DGEQRF is called to produce zeros under the diagonal elements of first q_i columns of the $(r_i \times c_i)$ dense matrix by Householder reflections. After that the LAPACK subroutine DORMQR is called to modify the last $c_i - q_i$ columns of the i'th dense block-row by using the orthogonal matrix Q_i obtained during the call of DGEQRF. The speed-up was rather low (even when tests in which only the LAPACK routines were called). The reason for this (according the specialists from Silicon Graphics) was that fact that the parallel tasks in the LAPACK routines are too small for this machine.

5.2 ALG2: Use of Silicon Graphics versions of DGEQRF

A new version of DGEQRF, received from Silicon Graphics, has been tested. The speed-ups in the cases where q_i were greater than 300 was quite satisfactory. However, the speed-up was rather poor on 8 processors when $q_i \leq 300$. We understand that the updated subroutine does not try to exploit all 8 processors in this situation.

5.3 ALG3: Use of Salvini's version of DGEQRF

S. Salvini prepared a fast version of DGEQRF ([13]). This version has been attached and tested in PARASPAR-L. The results were always better than those produced by the previous two algorithms.

5.4 ALG4: Use of Salvini's version of DGEQRF/DORMQR

In the previous three algorithms the original LAPACK subroutine DORMQR has been used with different versions of DGEQRF. It is natural to try to improve also the performance of DORMQR. This has been done by applying a subroutine prepared by S. Salvini ([13]) which performs the work done by both DGEQRF and DORMQR. In the previous three algorithms we had only to link the proper subroutine DGEQRF. In ALG4 we had to modify the code so that the new subroutine could be called.

The results in the next section show clearly that ALG4 performs considerably better than the other three algorithms, and in some cases much better.

5.5 Common discussion of the kernels used in ALG3 and ALG4

A short discussion of the main ideas used in the development of the dense kernels used in ALG3 and ALG4 is given below; more details can be found in [13].

New linear algebra routines were used (i) for the QR factorization of the matrix formed by the first q_i columns of the dense block i and (ii) for the orthogonal transformation of the remaining $c_i - q_i$ columns. The first routine (applied in ALG3) replaced the LAPACK QR factorization routine DGEQRF, though maintaining an identical interface; the second routine (used in ALG4) replaced the combined effects of DGEQRF and DORMQR, The new routines produced considerable gains in parallelism and efficiency over the LAPACK routines:

- **Parallelism was introduced at the level of the routines** themselves, rather than at the level of the underlying BLAS. Hence parallel thread creation and destruction take place only once on entry and exit, respectively, from the routine, thus reducing the associated overheads.
- In DGEQRF, Level-2 BLAS are used to compute the block reflectors. Their parallel efficiency is so low that to all effects the computation of block-reflectors can be viewed as a quasi-scalar bottleneck. As the number of processors increases, this effect cannot be neglected, and it can seriously affect performance, particularly for the matrix sizes of interest in this work. The solution was to overlap the computation of the block reflectors with the updating of the trailing sub-matrix, thus removing the bottleneck.
- Serial BLAS or BLAS-like routines are used on each processor to carry out block operations: some of these were optimized for the highly rectangular sub-matrices used during the computation.
- Static load balancing was employed, using only one synchronization barrier per block iteration.
- A variable block-size strategy was used.

6 Moving non-zeros from dense arrays to sparse arrays

Consider the i'th dense block. Assume that the orthogonal decomposition of this block has been calculated (by some dense kernels). After that the non-zeros must be gathered in the sparse arrays. This can cause difficulties, because the number of non-zeros per row is in general changed during the orthogonal decomposition. Therefore some operations that are traditionally used in sparse techniques for general matrices (performing copies of rows at the end of sparse arrays or even garbage collections) must be used; such operations are discussed in [6] and [15]. This extra work can be reduced by (i) dropping small elements and (ii) avoiding the storage of Q_i in the sparse arrays.

Dropping has two effects: (i) the numbers of copies and/or garbage collections is often reduced and (ii) the sizes of some of the following blocks are sometimes reduced (when c_i is reduced for some values of i). The second effect is more important than the first one. If dropping is used, then one should try to regain the accuracy lost by using R in **a preconditioned conjugate gradients (PCG) method**. The preconditioned system is $Cz = d$, where $C = (R^T)^{-1}A^T A R^{-1}$, $z = Rx$ and $d = (R^T)^{-1}A^T b$. The PCG method is applicable, because C is symmetric and positive definite. C is never formed explicitly; one works the whole time with A and R. Q, which is normally rather dense, is neither stored nor used in the iterative process (see more details in [15]).

Dropping is very successful for some matrices but it should not be used if the matrix is very ill-conditioned. Direct methods may work better in the latter case. The storage of Q can be avoided also in this situation. One can calculate $c = Q^T b$ during the decomposition and then x can be found from $Rx = c$. The storage of Q can be avoided even in the case where c has not been calculated;

by solving $R^T R x = A^T b$. This may lead to less accurate results (comparing with the case where $R x = c = Q^T b$ is solved).

7 Numerical results

Most of the numerical results have been obtained on Silicon Graphics POWER CHALLENGE. The code is now running on other computers. All times given in the tables are **the total computing times** needed to solve the problems chosen (i.e. these times are the sums of the computing time needed to perform the preliminary reordering by **LORA**, the scattering procedure, the factorization carried out by dense kernels, the gathering procedure, the solution of the system $R x = c = Q^T b$ and the iterative process when the PCG method is used).

7.1 Matrices used in the runs

Rectangular matrices from the Harwell-Boeing set ([5]) have been used in [9]. Larger sparse matrices (created by a sparse matrix generator from [15], [16]) will be used in this paper. It is desirable to be able to check the accuracy of the computed solution. Therefore vector b has been created as Ax with a vector x with components $x_i = 1$ for $i = 1, 2, \ldots, n$.

7.2 Comparison of the four algorithms

Large matrices, created by the sparse matrix generator from [15] with $m = 100000$ and with 10 different values of n, $n = 10000(10000)100000$, have been run with the four algorithms sketched in Section 5. Some of the results are given in Table 2 (obtained with drop-tolerance $TOL = 0.0$), Table 3 (obtained with $TOL = 0.015625$) and Table 4 (obtained with $TOL = 0.5$).

Three major conclusions can be drawn by comparing the computing times shown in Table 2 - Table 4 (it should be mentioned, however, that many other results have also been taken into account: (i) ALG4 performs best for all drop-tolerances, (ii) the use of a positive drop-tolerance leads to a considerable improvement of the results for all four algorithms and (iii) the use of a very large drop-tolerance does not necessarily lead to improvement of the computing time (even if the PCG method converges, which will not be the case when the matrices are ill-conditioned; see Section 6).

7.3 Parallel runs

The best algorithm has been run on 1, 2, 4 and 8 processors. The drop-tolerance used was $TOL = 0.015625$. The density of the matrices tested was varied by using one of the five parameters in the sparse matrix generator from [15]; parameter r (the number of non-zeros in these matrices is $NZ = mr + 110$). ALG4 was used with $r = 5(5)100$. The matrices used are with 20000 rows and 10000 columns. Some of the results are given in Table 5.

Number of columns	ALG1	ALG2	ALG3	ALG4
10000	186.4	148.0	94.5	54.8
20000	262.9	219.6	151.8	97.5
30000	176.5	139.8	102.5	54.5
40000	406.6	339.5	270.3	196.3

Table 2

Computing times achieved when matrices with 100000 rows and different
numbers of columns are run by four algorithms on POWER CHALLENGE
using 8 processors. The drop-tolerance is $TOL = 0.0$.

Number of columns	ALG1	ALG2	ALG3	ALG4
10000	96.1	73.8	51.5	25.5
20000	109.7	80.6	55.9	28.8
30000	69.3	58.1	44.2	22.1
40000	73.4	62.7	46.8	22.4

Table 3

Computing times achieved when matrices with 100000 rows and different
numbers of columns are run by four algorithms on POWER CHALLENGE
using 8 processors. The drop-tolerance is $TOL = 0.015625$.

Number of columns	ALG1	ALG2	ALG3	ALG4
10000	44.4	35.4	26.8	15.2
20000	71.4	54.5	38.1	21.7
30000	77.2	62.3	49.0	27.8
40000	106.1	90.5	71.3	38.6

Table 4

Computing times achieved when matrices with 100000 rows and different
numbers of columns are run by four algorithms on POWER CHALLENGE
using 8 processors. The drop-tolerance is $TOL = 0.5$.

Parameter r	1 processor	2 processors	4 processors	8 processors
10	63.7	35.5 (1.8)	22.6 (2.8)	16.5 (3.9)
20	65.7	36.7 (1.8)	23.9 (2.7)	17.3 (3.8)
30	67.9	38.5 (1.8)	24.9 (2.7)	18.6 (3.7)
40	70.1	40.5 (1.7)	26.4 (2.7)	19.6 (3.6)

Table 5

Computing times achieved when matrices with 20000 rows and 10000 columns
are run on POWER CHALLENGE using different numbers of processors.
ALG4 has been used with a drop-tolerance is $TOL = 0.015625$.

The speed-up is slowly decreasing when the density of non-zeros grows (the
speed-up on 8 processors is 2.8 when $r = 100$). There are several reasons for
this. Only Step 3 (see Table 1) is optimized on this stage; this is the most

time-consuming step; see [9]. The other steps are at present carried out in a sequential mode. Some attempts to improve the efficiency of these parts (and, thus, the efficiency of the whole code) will also be carried out in the future. This will be an important issue for computers with many processors. The number of iterations is increasing when r grows. This indicates that some efforts to optimize the PCG method are also needed.

8 Conclusions and plans for further improvements

The most important task which has to be done in the future is to check systematically the efficiency of the code on different high-speed computers; especially computers with shared memory and computers with distributed memory (on many computers with distributed memory **MPI**, [14], could be applied). The work on this task has been started. Some preliminary results have been obtained on IBM SP2.

Acknowledgments

This research was partially supported by the Bulgarian Ministry of Education (grant I-505/95), NATO (OUTR.CGR 960312) and the Danish Natural Sciences Research Council (SNF).

Dr. Guodong Zhang from the Application Group at Silicon Graphics sent us the last updated version of the LAPACK routine DGEQRF tunned on POWER CHALLENGE (this subroutine was used in ALG2).

References

1. Anderson, E., Bai, Z., Bischof, C., Demmel, J., Dongarra, J, Du Croz, J., Greenbaum, A, Hammarling, S., McKenney, A., Ostrouchov, S., and Sorensen, D.i "LAPACK: Users' guide", SIAM, Philadelphia, 1992.
2. Duff, I. S., "Full matrix techniques in sparse Gaussian elimination", In: "Numerical Analysis Proceedings, Dundee 1981" (G. Watson, ed.), pp. 331-348. Springer, Berlin, 1981.
3. Duff, I. S., "A review of frontal methods for solving linear systems", Computer Physics Communications; to appear.
4. Duff, I. S., "Sparse numerical linear algebra: direct methods and preconditioning", Report No. RAL-TR-96-047, Department for Computation and Information, Atlas Centre, Rutherford Appleton Laboratory, Oxon OX11 0QX, England, 1996.
5. Duff, I. S., Grimes, R. G. and Lewis, J, G, "Sparse matrix test problems", ACM Trans. Math. Software, 15 (1989), 1-14.
6. Duff, I, S. and Reid, J. K., "Some design features of a sparse matrix code", ACM Trans. Math. Software, 5 (1979), 18-35.
7. Duin, A. C. N. van, Hansen P. C., Ostromsky, Tz, Wijsoff, H. and Zlatev, Z., "Improving the numerical stability and the performance of a parallel sparse solver", Comp. Math. Appl., Vol. 30, No. 12 (1995), 81-96.

8. Gallivan, K., Hansen P. C., Ostromsky Tz., and Zlatev, Z., "A locally optimized reordering algorithm and its application to a parallel sparse linear system solver", Computing, 54 (1995), 39-67.
9. Hansen, P. C., Ostromsky, Tz., Sameh, A. and Zlatev, Z., "Solving sparse linear least squares problems on some supercomputers by using a sequence of large dense blocks", Internal report, 1995.
10. Matstoms, P., "The multifrontal solution of sparse linear least squares problems", Thesis, Department of Mathematics, Linköping University, Linköping, Sweden, 1991.
11. Ostromsky, Tz., Hansen, P. C. and Zlatev, Z., "A parallel sparse QR-factorization algorithm", In: "Applied Parallel Computing in Physics, Chemistry and Engineering Science" (J. Dongarra, K. Madsen and J. Waśniewski, eds), pp. 462-472, Springer-Verlag, Berlin, 1996.
12. Puglisi:, C., "QR factorization of large sparse overdetermined and square matrices using a multifrontal method in a multiprocessor environment", PhD Thesis. Report No. TH/PA/93/33, CERFACS, Toulouse, 1993.
13. Salvini, S., "Linear algebra subprograms on shared memory computers: Beyond LAPACK", Springer-Verlag, Berlin, to appear.
14. Walker, D. W. and Dongarra, J. J., "MPI: A standard message-passing interface", SIAM News, Vol. 29, No. 1, 1-9.
15. Zlatev, Z., "Computational methods for general sparse matrices", Kluwer Academic Publishers, Dordrecht-Toronto-London, 1991.
16. Zlatev, Z., Wasniewski, J. and Schaumburg, K., "A testing scheme for subroutines solving large linear systems", Comput. Chem., 5 (1981), 91-100.

The Parallel Surrogate Constraint Approach to the Linear Feasibility Problem

Hakan Özaktaş, Mustafa Akgül, and Mustafa Ç. Pınar*

Department of Industrial Engineering
Bilkent University
06533 Bilkent, Ankara, Turkey

Abstract. The linear feasibility problem arises in several areas of applied mathematics and medical science, in several forms of image reconstruction problems. The surrogate constraint algorithm of Yang and Murty for the linear feasibility problem is implemented and analyzed. The sequential approach considers projections one at a time. In the parallel approach, several projections are made simultaneously and their convex combination is taken to be used at the next iteration. The sequential method is compared with the parallel method for varied numbers of processors. Two improvement schemes for the parallel method are proposed and tested.

Key Words. Linear and convex feasibility, projection methods, distributed computing, parallel algorithms.

Subject Classifications (AMS). 90C25, 90C26, 90C60.

1   Introduction to Projection Methods for the Convex Feasibility Problem

In this paper some experimental results with the surrogate constraint algorithm described in [Yang & Murty 92] are given. The algorithm has both sequential and parallel versions. Emphasis is put on the parallel implementation.

The problem is to find a feasible point with respect to a set of linear inequalities explicitly defined as $Ax \leq b$, where $x \in R^n$ and A is an m by n matrix. It is assumed that the polyhedron defined by these inequalities is nonempty.

An iteration of the algorithm is carried out by projecting the current point onto a surrogate constraint—defined by a set of violated constraints at the current iteration. The algorithm may be seen as an extended version of the classical methods like Cimmino's algorithm and the relaxation algorithm for linear inequalities.

Before going into the sequential and parallel algorithms, an overview of several projection algorithms to the feasibility problem will be given. Such algorithms are useful when the problem is to determine a feasible point in a nonempty set defined by the intersection of many convex sets.

* The authors are indebted to K. Madsen for providing financial support to this project.

A projection of a point $x^k \in R^n$ onto a convex set Ω gives a point (if there are any) $x \in \Omega$ which has the minimal Euclidean distance to x^k [Hir.-Urr. & Lemar. 93]. However, projections are usually defined more generally as the nearest points contained in the convex bodies with respect to appropriate distance definitions.

The method of successive orthogonal projections of Gubin, Polyak and Raik works by projecting the current point onto a convex set at each iteration until a point which is contained in the intersection of these convex sets is found. At a typical iteration of the SOP algorithm (actually in many of these algorithms) an overrelaxed or an underrelaxed step is taken in the computed projection direction. Hence an iterative step becomes:

$$x^{k+1} = x^k + \lambda_k(P_{C_i}(x^k) - x^k) \qquad 0 < \lambda_k < 2 \tag{1}$$

where P_{C_i} is the projection operator onto the closed convex set C_i and λ_k is the so called relaxation parameter. When $\lambda_k = 1$, the next point generated is the exact orthogonal projection of the current point. On the other hand, when $\lambda_k > 1$, one has longer steps, which is the case of overrelaxation and when $\lambda_k < 1$, one has shorter steps, which is the case of underrelaxation [Censor & Zenios 95].

One has to compute the projection of x^k onto C_i for which $x^k \notin C_i$. This projection requires the solution of the following problem:

$$P_{C_i}(x^k) = \min_{x \in C_i} ||x^k - x|| \tag{2}$$

In this case, the minimization is made over the Euclidean distance, and the nearest point of C_i to x^k is found.

For the cases where $C_i = \{x \in R^n : f_i(x) \leq 0\}$, $f_i(x)$ convex and differentiable, the moving direction $P_{C_i}(x^k) - x^k$ is given by $\nabla f_i(x^{k+1})$. However, to determine the direction and the next point, one needs to solve the minimization problem defined above and in a way, to find the point x^{k+1}, in advance. The cyclic subgradients projection method overcomes this difficulty by picking up an alternate moving direction, $\nabla f_i(x^k)$ (or a suitable subgradient at x^k, for the nondifferentiable case) [Censor & Lent 82], [Censor & Zenios 95].

When an explicitly defined linear feasibility problem is considered, both SOP and CSP methods are equivalent to the relaxation procedure of Agmon, Motzkin and Schoenberg [Censor & Zenios 95]. In that case, the intersection of many halfspaces define the convex set. At an iteration point x^k, the routine proceeds by considering a violated inequality $a^i x^k > b_i$ and calculating the next point as:

$$x^{k+1} = x^k - \lambda_k \frac{a^i x^k - b_i}{||a^i||^2} a^i \tag{3}$$

where $0 < \lambda_k < 2$. Clearly the gradient of $f_i(x) = a^i x - b_i$, is a^i regardless of x, hence there is no essential difference between these routines.

Cimmino's algorithm considers all violated inequalities at a time. Projections are made separately and their convex combination is taken. The routine is quite slow when the number of constraints is large. The interest in the surrogate constraint algorithm stems from this fact. In the Yang-Murty algorithm instead of making projections for all violated constraints, block projections are made. However, the paper by Yang and Murty reports computational results only with the sequential surrogate constraint method. Our point of departure is to investigate the performance of the parallel version of this algorithm. Implementation of a similar but somewhat different parallel algorithm is described in [Gar.-Pal. & Gon.-Cas. 96] which will also be discussed in Section 3.

2 The Sequential Surrogate Constraint Method

The matrix A and the RHS vector b are split into p blocks so that one has $A^t x \leq b^t$, $t = 1, ..., p$. Each block consists of m_t rows. During the iterations, projections are made not to individual half spaces defined by violated inequalities, but to sets defined by surrogate constraints. The surrogate constraint for each block is $\pi^t A^t x \leq \pi^t b^t$, where π^t is an appropriate weight vector.

The sequential surrogate constraint algorithm of Yang and Murty is given as follows:

Step 0. Generate or read a feasible problem, with $A \in R^{m \times n}$, $b \in R^m$ with previously known values of n, m, p, $m_1, \ldots, m_p$. Initially, let $k = 0$ and $t = 1$. Fix a value of λ so that $0 < \lambda < 2$.

Step 1. Check, if $A^t x^k \leq b^t$. If so, then let $x^{k+1} = x^k$, otherwise let

$$x^{k+1} = x^k - \lambda \frac{(\pi^{t,k} A^t x^k - \pi^{t,k} b^t)(\pi^{t,k} A^t)}{||\pi^{t,k} A^t||^2} \tag{4}$$

where $\pi_i^{t,k} > 0$, if constraint i is violated and $\pi_i^{t,k} = 0$ otherwise, and $\sum_{i=1}^{m_t} \pi_i^{t,k} = 1$ is required for convenience. Update the value of the counter of violated inequalities.

Step 2. If $t < p$, let $k = k + 1$, $t = t + 1$ and go to *Step 1*. If $t = p$ and if the total number of violated constraints in the major iteration is zero then stop, the current solution is feasible. Otherwise, assign zero as the new value of the counter of violated constraints, let $t = 1$, $k = k + 1$ and go to *Step 1*.

Note that one can make the feasibility check also by setting $x_{\text{old}} = x^k$ at the end of each major iteration, i.e. when $t = p$. If at the end of the following major iteration, the current x is same with x_{old}, then this solution is feasible.

This algorithm is finitely convergent, if the feasibility check in *Step 1* is adjusted to allow for a certain degree of tolerance. Hence, one needs to make the comparison between $a^i x^k$ and $b_i + \varepsilon$ where ε is a small positive value [Yang & Murty 92].

During our implementation, A has been generated as a sparse matrix without any structure. The blocks are divided almost evenly, since the matrix doesn't have a special structure. Implementation results of the sequential algorithm is given in Table 1. Details related to problem generation, weight selection rules and matrix storage are given in Section 5.

p //	2 //	4 //	8 //	16 //
$m,\ n$				
500, 1000	15.4(7.2)	27.8(6.2)	51.8(5.6)	104.6(5.6)
2000, 1000	119(59)	203(50)	365.4(44.8)	651.8(39.8)
5000, 2500	106.6(52.8)	191.8(47.4)	367(45)	681.6(43.6)

Table 1. Average number of iterations (numbers in the parentheses represent the major iterations) of an implementation of the sequential algorithm. Matrices are randomly generated with a sparsity value of 0.02. m and n are the row and column sizes respectively and p represents the number of blocks. Five test problems for each size have been solved.

3   The Parallel Surrogate Method

The parallel algorithm of Yang and Murty uses the same structure. The problem is divided into blocks evenly and surrogate constraints are considered for block projections. But in this case, the projections are made simultaneously and a convex combination of these is taken. Thus, a single combined step is taken at a major iteration when compared to p distinct movements in the sequential algorithm.

Hence, the algorithm will be modified as follows:

Step 0. Generate or read a feasible problem. Let $k = 0$ and fix λ so that $0 < \lambda < 2$.

Step 1. For $t = 1, ..., p$,
check, if $A^t x^k \le b^t$. If so, then let $P_t(x^k) = x^k$, otherwise let

$$P_t(x^k) = x^k - \frac{(\pi^{t,k} A^t x^k - \pi^{t,k} b^t)(\pi^{t,k} A^t)}{||\pi^{t,k} A^t||^2} \tag{5}$$

where $\pi^{t,k}$, is the same as that of the sequential algorithm. When the entire matrix is processed, let $P(x^k) = \sum_{t=1}^{p} \tau_t P_t(x^k)$ where $\sum_{t=1}^{p} \tau_t = 1$, $\tau_t \ge 0$, and $\tau_t > 0$ for all blocks which violate feasibility. The next point is generated as:

$$x^{k+1} = x^k + \lambda(P(x^k) - x^k) \tag{6}$$

Update the total number of violated constraints, in all blocks.

Step 2. If the total number of violated constraints in the major iteration is zero then stop, the current solution is feasible. Otherwise, assign zero to the number of violated constraints, let $k = k + 1$ and go to *Step 1.*

Our experimental results reveal that the parallel algorithm as given in this pure form is much slower than the sequential algorithm. Comparing the results in Table 1 and Table 2 (assuming that an iteration of the sequential routine is more or less equivalent to one major iteration of the parallel routine) indicates

p // m, n	2 //	4 //	8 //	16 //
500, 1000	27.6	69.4	209.4	507
2000, 1000	267.4	1422.8	3393.8	6921.4
5000, 2500	1087.6	3274.8	6524.2	13069.2

Table 2. Average number of major iterations of an implementation of the parallel algorithm. p represents the number of blocks and hence the processors. Same test problems with the sequential experiments given in Table 1, are used.

that the two algorithms have incomparable performances. However, it is still desirable to benefit from the effects of parallelization. Clearly, one should be able to obtain some better results when it is possible to distribute the feasibility checks of the blocks to distinct machines.

An examination of the two algorithms reveals the problem of the parallel routine and suggests a partial remedy. The sequential routine takes several steps (the number being equal to the number of blocks which have infeasibility) which accumulate, whereas the parallel routine provides a single step (which is a convex combination of the steps generated from infeasible blocks) during a major iteration. The overall effect of this fact is much worse, as seen from the test results.

Let us recall the sequential moving direction (with an appropriate magnitude based on the Euclidean distance to the surrogate hyperplane) for a certain infeasible block t:

$$- d_t = - \frac{(\pi^{t,k} A^t x^k - \pi^{t,k} b^t)(\pi^{t,k} A^t)}{||\pi^{t,k} A^t||^2} \tag{7}$$

The next test point is calculated as, $x^{k+1} = x^k - \lambda d_t$.

The parallel algorithm also uses the direction given in (7) for the corresponding block. However, instead of accumulating these steps, a convex combination of these, hence a shorter step is taken. The movement in the parallel routine, given in (5) and (6) can be rewritten as (assuming that $d_t^k = 0$ when block t is feasible):

$$x^{k+1} = x^k + \lambda \left(\sum_{t=1}^p \tau_t P_t(x^k) - x^k \right), \qquad \left(\sum_{t=1}^p \tau_t = 1 \right)$$

$$= x^k + \lambda \left(\sum_{t=1}^p \tau_t (x^k - d_t^k) - x^k \right)$$

$$= x^k - \lambda \left(\sum_{t=1}^p \tau_t d_t^k \right)$$

$$x^{k+1} = x^k - \lambda \bar{d}^k, \qquad \left(\bar{d}^k = \sum_{t=1}^{p} \tau_t d_t^k \right) \qquad (8)$$

One can suggest the usage of longer steps having magnitudes comparable to those obtained in the sequential algorithm. An idea is to implement the algorithm by increasing the step size by multiplying it with p at each major iteration.

Using this idea somewhat improves the performance during implementation. However, it will yield unnecessarily long steps which slows down the algorithm, in the cases where some of the blocks reach feasibility immediately and some do not. Therefore, choosing the number of violated blocks at a given iteration as the multiplicative parameter will be a better strategy to approach the trajectory obtained from the sequential algorithm. Thus, one can obtain the next point alternatively as:

$$x^{k+1} = x^k - (\lambda \bar{p}^k) \bar{d}^k, \qquad (9)$$

where $\bar{p}^k$ is the number of blocks which have infeasibility at the kth iteration.

The results obtained with this adjusted step sizing rule is given in Table 3. It can be seen that, for relatively small values of p, the results are better than the pure application of the parallel algorithm. Furthermore, some of these results are better than those obtained by the sequential algorithm, assuming that one major iteration of a parallel routine is more or less equivalent to a normal iteration of the sequential routine. However, utilizing purely (9) might yield us

p //	2 //	4 //	8 //	16 //
$m,\ n$				
500, 1000	7.2	7	6.6	7
2000, 1000	84	152.2	–	–
5000, 2500	86.6	979.6	–	1164.6

Table 3. Average number of major iterations of an implementation of the parallel algorithm with new step sizing policy. Same test problems with the previous experiments given in Table 1, and Table 2 are used. '–' represents a typical case where the routine does not converge to a feasible point.

a nonconvergent routine, especially when p is relatively large. We were not able to obtain the complete test results for $p = 8$ and $p = 16$. The reason is that, the generated steps might still become too long, in some cases. Thus a regulatory mechanism of the step size is required to use (9) successfully.

An improved step for a similar parallel projection algorithm outlined in [Gar.-Pal 93] has been given in [Gar.-Pal. & Gon.-Cas. 96] recently. This paper discusses the slow performance of the parallel Cimmino algorithm (without any adjustment) and proposes an accaleration procedure. The partitioning mechanism and projective subroutines in the resulting algorithm are different than that of the Yang-Murty algorithm, however the overall algorithm is similar. The pure parallel algorithm uses the following combined step:

$$x^k - \lambda \left(\sum_{t=1}^{p} \tau_t d_t^k \right) \tag{10}$$

which is identical with (8). The improved step yields the following test point:

$$x^{k+1} = x^k - \frac{1}{2} \lambda_k \frac{\sum_{i=1}^{p} ||d_t^k||^2}{||\sum_{t=1}^{p} d_t^k||^2} \sum_{t=1}^{p} d_t^k \tag{11}$$

It has been established in [Gar.-Pal. & Gon.-Cas. 96] that the algorithm with adjusted step sizing policy performs better under some assumptions, both theoretically and practically.

We have used the García Palomares-Gonzáles Castaño step given by (11) in the parallel surrogate algorithm of Yang-Murty, for the same test problems. The results are given in Table 4. It can be seen that this step improves the parallel version of the Yang-Murty algorithm, however it seems that using the step given in (9) performs better when p is relatively small.

Another interesting observation is that for a given problem size the number of iterations of the parallel algorithm is almost constant as the number of blocks increases. This is in contrast to the results of Table 2 where the number of major iterations increases directly for a given problem as more blocks are used. Based on this limited evidence we can conclude that the new step provides a significant stabilizing effect on the parallel surrogate constraint algorithm.

p // $m,\ n$	2 //	4 //	8 //	16 //
500, 1000	24.2	24.2	26.6	27.8
2000, 1000	193	200.2	193.4	197.8
5000, 2500	612.8	643	646.6	669

Table 4. Average number of major iterations of an implementation of the parallel algorithm with the García Palomares-Gonzáles Castaño step. Same test problems with the previous experiments given in Table 1, Table 2 and Table 3 are used.

4 Practical Considerations for the Parallel Algorithm

In the parallel routine the number of submatrix blocks should be equal to the number of processors. Each processor deals exactly with a single block, therefore it is quite natural to divide the matrix into p submatrices evenly or almost evenly, so that each submatrix has $\lceil \frac{m}{p} \rceil$ rows. Each processor checks its constraints (which are equal or almost equal in size) for feasibility and computes the projection if necessary. When a processor finishes its task, it waits for the others to finish as well.

During the subroutine operations (feasibility checks and projection calculations for distinct blocks) each processor works independently and no message passing within the machines is required. When all processes are finished the projection data is transferred to one of the processors, where the new point is calculated according to (6). The calculation of the cumulative projection and the new point, is the sequential part of the algorithm. After this step, the new point is broadcasted to all processors and the procedure is repeated until a feasible point is found.

The initialization phase is carried out on each processor independently to avoid further communication within the blocks. If the problem is to be randomly generated, the initial seed is broadcasted to all machines. If data has to be read from the files, this is done by all machines.

A distributed implementation of the algorithm is being developed using PVM 3.3.11 on several Sparc workstations. The algorithm is being governed by a main C routine, which makes calls to a C subroutine (for parallel block operations) and to several PVM functions suitable to C programs (see [Geist et al. 94]). These results will be reported elsewhere.

5 Implementation Issues

It is assumed that the matrix underlying the feasibility problem is sparse. This matrix is stored both in the rowwise format and the columnwise format, to ease the access. When computing $A^t x^k$ or $a^i x^k$, the rowwise format is used and when computing $\pi^{t,k} A^t$ the columnwise format is used. The widely known standard formats are used almost without any changes [Pissa. 84], but to ease the control over the storage arrays, we have added one more cell to each, which marks the end of the array.

Sample test problems have been generated as follows: First of all, a matrix with a given size and sparsity percentage is generated so that the nonzero elements are distributed uniformly, so that each nonzero value is distributed in between -5.0 and 5.0. The random distribution has been done in two ways. In the first, an exact number of nonzero entries is fixed and they are distributed uniformly (with parameters depending on the problem size) to rows of the matrix so that each row has at least one nonzero element. Then, their column numbers are generated uniformly. In the second, a simulated sequence of a Poisson process is obtained and points are generated with exponential interarrival times with parameter equal to the sparsity of the matrix. The exponential density is truncated between 1 and n and the generated interarrival time is rounded to the nearest integer and the next nonzero entry position is found by adding this value to the column number of the previously placed nonzero entry. When one row is finished, the process is continued in the next row. This is somewhat better than the first, since the nonzero entries are generated sequentially and one does not require a reordering when storing the data in rowwise format. Here we utilize the property stating that in a Poisson process, given that n arrivals have occurred within a time interval $(0,t)$, the distribution of the arrival times $S_1, \ldots, S_n$ have the same distribution as the order statistics of n independent random variables distributed on this interval $(0,t)$ (see for example [Ross 83]). We assume that the idea is applicable to a discrete interval since it is quite a long interval, and that the truncation of the exponential distribution does not have a significant effect on the uniformity of the nonzeros over the matrix.

After generating the random matrix, a random x is generated, so that each of its elements are in the interval $(-4.5, 4.5)$. Following this, Ax is computed and the b vector is generated so that $b_i = A_i x + u_i$ where u_i is a discrete uniform random variable between 0 and 1. In this way a feasible polyhedron is created. Our aim is to try to keep this polyhedron somewhat small and distant to the initial point of the algorithm, so that trivial convergence in a few steps will not occur.

An important issue is the selection of the weight vector $\pi^{t,k}$. Weights may be distributed equally among all violated constraints or they can be assigned, proportional to the amount of violations. Or a suitable combination of the two approaches may also be used. In our tabulated results we have used the hybrid approach (which has also been used in [Yang & Murty 92]):

$$\pi_i^{t,k} = \frac{0.2(A_i^t x^k - b_i)}{\sum_{h:A_h^t x^k > b_h^t}(A_h^t x^k - b_h^t)} + \frac{0.8}{\text{number of violated constraints}} \tag{12}$$

For convenience it is assumed that $\sum \pi_i^{t,k} = 1$. The relaxation parameter λ has been taken to be 1.7 and the tolerance limit for feasibility, ε is 10^{-9}.

6 Conclusions and Future Work

In our test problems it has been observed that the parallel surrogate constraint method without any adjustment is quite poor when compared to the sequential surrogate constraint method. The reason is that the magnitudes of the steps obtained in the parallel algorithm remain quite small with respect to the accumulated sequential steps.

In order to compensate for this loss, we have magnified the parallel step length by multiplying it with the number of infeasible blocks at that time. In this way we have obtained some favorable results. We have also used a longer step which has been recently used in a similar (but not the same) parallel routine by [Gar.-Pal. & Gon.-Cas. 96]. The results are also improved when compared to the pure version of the parallel algorithm. This approach also seems to stabilize the number of iterations when more blocks are used to partition the problem.

It seems that increasing the step simply by multiplying it with the current number of infeasible blocks, yields quite an improvement when the number of blocks is relatively small. However, this idea should be used with some sort of regulation over the step given by (9), since the idea gives unnecessarily long steps which may prevent convergence, especially when the number of processors is relatively large.

The issue that is of interest at this point is to obtain an adjusted parallel algorithm and establish convergence. This can be possibly done by using the improved step and adapting a control mechanism. We are also foreseeing the completion of the PVM implementation of the parallel algorithm.

References

[Censor & Lent 82] Censor, Y., Lent, A.: Cyclic Subgradient Projections. *Mathematical Programming*, 24:233–235, 1982.

[Censor & Zenios 95] Censor, Y., Zenios, S. A.: *Parallel Optimization: Theory, Algorithms and Applications* (to be published by Oxford University Press). October 18, 1995.

[Gar.-Pal. 93] García Palomares, U. M.: Parallel Projected Aggregation Methods for Solving the Convex Feasibility Problem. *SIAM Journal on Optimization*, 3–4:882–900, November 1993.

[Gar.-Pal. & Gon.-Cas. 96] García Palomares, U. M., González Castaño, F. J.: Acceleration technique for solving convex (linear) systems via projection methods. Technical Report OP960614, Universidade de Vigo, ESCOLA TÉCNICA SUPERIOR DE ENXEÑEIROS DE TELECOMUNICACIÓN, Lagoas Marcosende 36200 Vigo, España, 1996.

[Geist et al. 94] Geist, A., Beguelin, A., Dongarra, J., Jiang, W., Manchek, R., Sunderam, V.: *PVM: Parallel Virtual Machine. A User's Guide and Tutorial for Networked Parallel Computing.* The MIT Press., Cambridge, Massachusetts, 1994.

[Hir.-Urr. & Lemar. 93] Hiriart-Urruty, Jean-Baptiste and Lemaréchal, Claude: *Convex Analysis and Minimization Algorithms.* Springer-Verlag, Berlin, 1993.

[Pissa. 84] Pissanetzky, Sergio: *Sparse Matrix Technology.* Academic Press Inc., London, 1984.

[Ross 83] Ross, Sheldon M.: *Stochastic Processes.* John Wiley & Sons Inc., 1983.

[Yang & Murty 92] Yang, K., Murty, K.G.: New Iterative Methods for Linear Inequalities. *Journal of Optimization Theory and Applications*, 72:163–185, January 1992.

A Parallel GRASP for MAX-SAT Problems [*]

P. M. Pardalos[1], L.Pitsoulis[1], and M.G.C. Resende[2]

[1] Center for Applied Optimization, Department of Industrial and Systems
Engineering, University of Florida, Gainesville, FL 32611-6595 USA
[2] AT&T Research, Florham Park, NJ 07932 USA

Abstract. The weighted maximum satisfiability (MAX-SAT) problem
is central in mathematical logic, computing theory, and many industrial applications. In this paper, we present a parallel greedy randomized
adaptive search procedure (GRASP) for solving MAX-SAT problems.
Experimental results indicate that almost linear speedup is achieved.
Key words: Maximum Satisfiability, Parallel Search, Heuristics, GRASP,
Parallel Computing.

1 Introduction and Problem Definition

The Satisfiability Problem (SAT) is a central problem in artificial intelligence,
mathematical logic, computer vision, VLSI design, databases, automated reasoning, computer-aided design and manufacturing. In addition, SAT is a core
problem in computational complexity, and it was the first problem shown to be
NP-complete [2]. Since most known NP-complete problems have natural reductions to SAT [5], the study of efficient (sequential and parallel) exact algorithms
and heuristics for SAT can lead to general approaches for solving combinatorial
optimization problems.

In SAT problems we seek to find an assignment of the variables that satisfy
a logic formula or the maximum number of clauses in the formula. More specifically, let $x_1, x_2, \ldots, x_n$ denote n Boolean variables, which can take on only the
values true or false (1 or 0). Define clause i (for $i = 1, \ldots, n$) to be

$$\mathcal{C}_i = \bigvee_{j=1}^{n_i} l_{ij},$$

where the literals $l_{ij} \in \{x_i, \bar{x}_i \mid i = 1, \ldots, n\}$. In addition, for each clause $\mathcal{C}_i$, there
is an associated nonnegative weight w_i. In the weighted *Maximum Satisfiability
Problem* (MAX-SAT), one has to determine the assignment of truth values to
the n variables that maximizes the sum of the weights of the satisfied clauses.
The classical Satisfiability Problem (SAT) is a special case of the MAX-SAT in
which all clauses have unit weight and one wants to decide if there is a truth
assignment of total weight m.

---

[*] Invited paper, PARA96 – Workshop on Applied Parallel Computing in Industrial
Problems and Optimization, Lyngby, Denmark (August 18–21, 1996)

We can easily transform a SAT problem from the space of true-false variables into an optimization problem of discrete 0-1 variables. Let $y_j = 1$ if Boolean variable x_j is true and $y_j = 0$ otherwise. Furthermore, the continuous variable $z_i = 1$ if clause C_i is satisfied and $z_i = 0$, otherwise. Then, the weighted MAX-SAT has the following mixed integer linear programming formulation:

$$\max \ F(y, z) = \sum_{i=1}^{m} w_i z_i$$

subject to

$$\sum_{j \in I_i^+} y_j + \sum_{j \in I_i^-} (1 - y_j) \geq z_i, \quad i = 1, \ldots, m,$$

$$y_j \in \{0, 1\}, \quad j = 1, \ldots, n,$$

$$0 \leq z_i \leq 1, \quad i = 1, \ldots, m,$$

where I_i^+ (resp. I_i^-) denotes the set of variables appearing unnegated (resp. negated) in clause C_i.

Active research during the past decades has produced a variety of exact algorithms and heuristics for SAT problems [7]. Many of these algorithms have been implemented and tested on parallel computers. Efficient parallel implementations can significantly increase the size of the problems that can be solved. For recent advances on parallel processing of discrete optimization problems see the survey article [1] and the books [4, 11].

One successful heuristic for solving large SAT and MAX-SAT problems is GRASP [12, 13]. A greedy randomized adaptive search procedure (GRASP) is a randomized heuristic for combinatorial optimization [3]. In this paper, we describe a parallel implementation of GRASP for solving the weighted MAX-SAT problem. GRASP is an iterative process, with each GRASP iteration consisting of two phases, a construction phase and a local search phase. The best overall solution is kept as the result.

In the construction phase, a feasible solution is iteratively constructed, one element at a time. At each construction iteration, the choice of the next element to be added is determined by ordering all elements in a candidate list with respect to a greedy function. This function measures the (myopic) benefit of selecting each element. The heuristic is adaptive because the benefits associated with every element are updated at each iteration of the construction phase to reflect the changes brought on by the selection of the previous element. The probabilistic component of a GRASP is characterized by randomly choosing one of the best candidates in the list, but not necessarily the top candidate. This choice technique allows for different solutions to be obtained at each GRASP iteration, but does not necessarily compromise the power of the adaptive greedy component of the method.

As is the case for many deterministic methods, the solutions generated by a GRASP construction are not guaranteed to be locally optimal with respect to simple neighborhood definitions. Hence, it is usually beneficial to apply a local

```
procedure grasp(RCLSize,MaxIter,RandomSeed)
1    InputInstance();
2    InitializeDataStructures();
3    BestSolutionFound = 0;
4    do k = 1,..., MaxIter →
5        ConstructGreedyRandomizedSoln(RCLSize,RandomSeed);
6        LocalSearch(BestSolutionFound);
7        UpdateSolution(BestSolutionFound);
8    od;
9    return(BestSolutionFound)
end grasp;
```

Fig. 1. A generic GRASP pseudo-code

```
procedure ConstructGreedyRandomizedSoln(RCLSize,RandomSeed,x)
1    do k = 1,..., n →
2        MakeRCL(RCLSize);
3        s = SelectIndex(RandomSeed);
4        AssignVariable(s, x);
5        AdaptGreedyFunction(s);
6    od;
end ConstructGreedyRandomizedSoln;
```

Fig. 2. GRASP construction phase pseudo-code

search to attempt to improve each constructed solution. Through the use of customized data structures and careful implementation, an efficient construction phase can be created which produces good initial solutions for efficient local search. The result is that often many GRASP solutions are generated in the same amount of time required for the local optimization procedure to converge from a single random start. Furthermore, the best of these GRASP solutions is generally significantly better than the solution obtained from a random starting point.

2 GRASP for the weighted MAX-SAT

As outlined in Section 1, a GRASP possesses four basic components: a greedy function, an adaptive search strategy, a probabilistic selection procedure, and a local search technique. These components are interlinked, forming an iterative method that constructs a feasible solution, one element at a time, guided by an adaptive greedy function, and then searches the neighborhood of the constructed

```
procedure AdaptGreedyFunction(s)
1    if s > 0 →
2        for j ∈ Γ_s^+ →
3            for k ∈ L_j (k ≠ j) →
4                if x_k is unnegated in clause j →
5                    Γ_k^+ = Γ_k^+ - {j}; γ_k^+ = γ_k^+ - w_j;
6                fi;
7                if x_k is negated in clause j →
8                    Γ_k^- = Γ_k^- - {j}; γ_k^- = γ_k^- - w_j;
9                fi;
10           rof;
11       rof;
12       Γ_s^+ = ∅; Γ_s^- = ∅;
13       γ_s^+ = 0; γ_s^- = 0;
14   fi;
15   if s < 0 →
16       for j ∈ Γ_{-s}^- →
17           for k ∈ L_j (k ≠ j) →
18               if x_k is unnegated in clause j →
19                   Γ_k^+ = Γ_k^+ - {j}; γ_k^+ = γ_k^+ - w_j;
20               fi;
21               if x_k is negated in clause j →
22                   Γ_k^- = Γ_k^- - {j}; γ_k^- = γ_k^- - w_j;
23               fi;
24           rof;
25       rof;
26       Γ_{-s}^+ = ∅; Γ_{-s}^+ = ∅;
27       γ_{-s}^+ = 0; γ_{-s}^- = 0;
28   fi;
29   return
end AdaptGreedyFunction;
```

Fig. 3. AdaptGreedyFunction pseudo-code

solution for a locally optimal solution. Figure 1 shows a GRASP in pseudo-code. Lines 1 and 2 of the pseudo-code input the problem instance and initialize the data structures. The best solution found so far is initialized in line 3. The GRASP iterations are carried out in lines 4 through 8. Each GRASP iteration has a construction phase (line 5) and a local search phase (line 6). If necessary, the solution is updated in line 7. The GRASP returns the best solution found.

Next, we describe the GRASP (based on [12, 13]) for the weighted MAX-SAT. To accomplish this, we outline in detail the ingredients of the GRASP, i.e. the construction and local search phases. To describe the construction phase, one

```
procedure LocalSearch(x,BestSolutionFound)
1    BestSolutionFound = C(x);
2    GenerateGains(x,G,0);
3    G_k = max{G_i | i = 1,...,n} ;
4    for G_k ≠ 0 →
5        Flip value of x_k;
6        GenerateGains(x,G,k);
7    rof;
8    BestSolutionFound = C(x);
9    return;
end LocalSearch;
```

Fig. 4. The local search procedure in pseudo-code

needs to provide a candidate definition (for the restricted candidate list), provide an adaptive greedy function, and specify the candidate restriction mechanism. For the local search phase, one must define the neighborhood and specify a local search algorithm.

2.1 Construction phase

The construction phase of a GRASP builds a solution, around whose neighborhood a local search is carried out in the local phase, producing a locally optimal solution. This construction phase solution is built, one element at a time, guided by a greedy function and randomization. Figure 2 describes in pseudo-code a GRASP construction phase. Since in the MAX-SAT problem there are n variables to be assigned, each construction phase consists of n iterations. In MakeRCL the restricted candidate list of assignments is set up. The index of the next variable to be assigned is chosen in SelectIndex. The variable selected is assigned a truth value in AssignVariable. In AdaptGreedyFunction the greedy function that guides the construction phase is changed to reflect the assignment just made. To describe these steps in detail, we need some definitions. Let $N = \{1, 2, \ldots, n\}$ and $M = \{1, 2, \ldots, m\}$ be sets of indices for the set of variables and clauses, respectively. Solutions are constructed by setting one variable at a time to either 1 (**true**) or 0 (**false**). Therefore, to define a restricted candidate list, we have 2 potential candidates for each yet-unassigned variable: assign the variable to 1 or assign the variable to 0.

We now define the adaptive greedy function. The idea behind the greedy function is to maximize the total weight of yet-unsatisfied clauses that become satisfied after the assignment of each construction phase iteration. For $i \in N$, let Γ_i^+ be the set of unassigned clauses that would become satisfied if variable x_i were to be set to **true**. Likewise, let Γ_i^- be the set of unassigned clauses that would become satisfied if variable x_i were to be set to false. Define

$$\gamma_i^+ = \sum_{j \in \Gamma_i^+} w_j \text{ and } \gamma_i^- = \sum_{j \in \Gamma_i^-} w_j.$$

The greedy choice is to select the variable x_k with the largest γ_k^+ or γ_k^- value. If $\gamma_k^+ > \gamma_k^-$, then the assignment $x_k = 1$ is made, else $x_k = 0$. Note that with every assignment made, the sets Γ_i^+ and Γ_i^- change for all i such that x_i is not assigned a truth value, to reflect the new assignment. This consequently changes the values of γ_i^+ and γ_i^-, characterizing the adaptive component of the heuristic.

Next, we discuss restriction mechanisms for the restricted candidate list (RCL). The RCL is set up in MakeRCL of the pseudo-code of Figure 2. We consider two forms of restriction: value restriction and cardinality restriction.

Value restriction imposes a parameter based *achievement level*, that a candidate has to satisfy to be included in the RCL. In this way we ensure that a random selection will be made among the best candidates in any given assignment. Let $\gamma^* = \max\{\gamma_i^+, \gamma_i^- \mid x_i$ yet unassigned$\}$. Let α $(0 \leq \alpha \leq 1)$ be the restricted candidate parameter. We say a candidate $x_i = $ true is a *potential candidate* for the RCL if $\gamma_i^+ \geq \alpha \cdot \gamma^*$. Likewise, a candidate $x_i = $ false is a potential candidate if $\gamma_i^- \geq \alpha \cdot \gamma^*$. If no cardinality restriction is applied, all potential candidates are included in the RCL.

Cardinality restriction limits the size of the RCL to at most maxrcl elements. Two schemes for qualifying potential candidates are obvious to implement. In the first scheme, the best (at most maxrcl) potential candidates (as ranked by the greedy function) are selected. Another scheme is to choose the first (at most maxrcl) candidates in the order they qualify as potential candidates. The order in which candidates are tested can determine the RCL if this second scheme is used. Many ordering schemes can be used. We suggest two orderings. In the first, one examines the least indexed candidates first and proceeds examining candidates with indices in increasing order. In the other, one begins examining the candidate with the smallest index that is greater than the index of the last candidate to be examined during the previous construction phase iteration.

Once the RCL is set up, a candidate from the list must be selected and made part of the solution being constructed. SelectIndex selects at random the index s from the RCL. In AssignVariable, the assignment is made, i.e. $x_s = $ true if $s > 0$ or $x_s = $ false if $s < 0$.

The greedy function is changed in AdaptGreedyFunction to reflect the assignment made in AssignVariable. This requires that some of the sets Γ_i^+, Γ_i^-, as well as the γ_i^+ and γ_i^-, be updated. There are two cases, as described in Figure 3. If the variable just assigned was set to true then Γ^+, Γ^-, γ^+ and γ^- are updated in lines 5, 8, 12, and 13. If the variable just assigned was set to false then Γ^+, Γ^-, γ^+ and γ^- are updated in lines 19, 22, 26, and 27.

2.2 Local search phase

In general, most heuristics for combinatorial optimization problems terminate at a solution which may not be locally optimal. The GRASP construction phase

```
procedure ParallelGRASP(n,dat)
1           GenerateRandomNumberSeeds(s_1,...,s_N);
2           do i = 1,...,N →
3               GRASP(n,dat,s_i,val_i,x_i);
4           od;
5           FindBestSolutions(x_i,val_i,max,x*);
6           return(max,x*);
end ParallelGRASP(n,dat);
```

Fig. 5. Parallel GRASP for MAX-SAT

described in Subsection 2.1 computes a feasible truth assignment that is not necessarily locally optimal with respect some neighborhood structure. Consequently, local search can be applied with the objective of finding a locally optimal solution that may be better than the constructed solution. To define the local search procedure, some preliminary definitions have to be made. Given a truth assignment $x \in \{0,1\}^n$, define the *1-flip neighborhood* $N(x)$ to be the set of all vectors $y \in \{0,1\}^n$ such that, if x is interpreted as a vertex of the n-dimensional unit hypercube, then its neighborhood consists of the n vertices adjacent to x. If we denote by $C(x)$ the total weight of the clauses satisfied by the truth assignment x, then the truth assignment x is a *local maximum* if and only if $C(x) \geq C(y)$, for all $y \in N(x)$. Starting with a truth assignment x, the local search finds the local maximum y in $N(x)$. If $y \neq x$, it sets $x = y$. This process is repeated until no further improvement is possible.

Given an initial solution x define G_i to be the gain in total weight resulting from flipping variable x_i in x, for all i. Let $G_k = \max\{G_i \mid i \in N\}$. If $G_k = 0$ then x is the local maximum and local search ends. Otherwise, the truth assignment resulting from flipping x_k in x, is a local maximum, and hence we only need to update the G_i values such that the variable x_i occurs in a clause in which variable x_k occurs (since the remaining G_i values do not change in the new truth assignment). Upon updating the G_i values we repeat the same process, until $G_k = 0$ where the local search procedure is terminated. The procedure is described in the pseudo-code in Figure 4. Given a truth assignment x and an index k that corresponds to the variable x_k that is flipped, procedure GenerateGains is used to update the G_i values returned in an array G. Note that, in line 2, we pass $k = 0$ to the procedure, since initially all the G_i values must be generated (by convention variable x_0 occurs in all clauses). In lines 4 through 7, the procedure finds a local maximum. The value of the local maximum is saved in line 8.

3 Parallel Implementation

The GRASP heuristic has an inherent parallel nature, which results in an effective parallel implementation [9, 10]. Each GRASP iteration can be regarded as a search in some region of the feasible space, not requiring any information from previous iterations. Therefore, we can perform any number of iterations (searches) in parallel, as long as we make sure that no two searches are performed in the same region. The region upon which GRASP performs a search is chosen randomly, so by using different random number seeds for each iteration performed in parallel we avoid overlapping searches.

Given N processors that operate in parallel, we distribute to each processor its own problem input data, a random number seed, and the GRASP procedure. Each processor then applies the GRASP procedure to the input data using its random number seed, and when it completes the specified number of iterations it returns the best solution found. The best solution among all N processors is then identified and used as the solution of the problem. It is readily seen that the cost of processor interaction is completely absent since each GRASP procedure operates independently, resulting in no communication between the processors. This in turn, results in an almost linear speedup.

Based on the above discussion, the implementation of GRASP to solve the MAX-SAT in parallel is presented in pseudo code in Figure 5. In line 1 the random numbers seeds $s_1, \ldots, s_N$ are generated for each of the N processors, while in lines 2 to 4 the GRASP procedure is distributed into the processors. Each processor i takes as input the problem size n, the input data data, and its random number seed s_i, and returns its best solution found in val_i, with the corresponding truth assignment x_i. In line 5 the best solution among all val_i is found together with the corresponding truth assignment vector x^*.

The code for the GRASP parallel procedure is written in Fortran (f77), and was implemented in a Parallel Virtual Machine (PVM) framework [6]. PVM utilizes a network of Unix workstations preferably sharing the same filesystem, by treating each workstation as a different processor enabling parallel execution. The advantages of PVM are its portability, the ease of implementing existing codes, and the fact that even parallel machines in a network could be used. The main disadvantage is that network communication could cause slow or problematic execution since the communication between the processors is based on the network status. But for inherently parallel algorithms with minimal amount of communication required, PVM presents and ideal framework for implementation and testing purposes.

4 Computational Results

In this section, computational experience regarding the parallel implementation of GRASP for solving MAX-SAT instances is reported. The purpose of this experiment is not to demonstrate the overall performance of GRASP for solving MAX-SAT instances, which is reported in [13], but rather to show the efficiency of the parallel implementation of the heuristic in terms of speedup and solution quality. A sample of ten test problems were used for calculating the average speedup of the parallel implementation, which were derived from the SAT instance class jnh of the 2nd DIMACS Implementation Challenge [8]. These

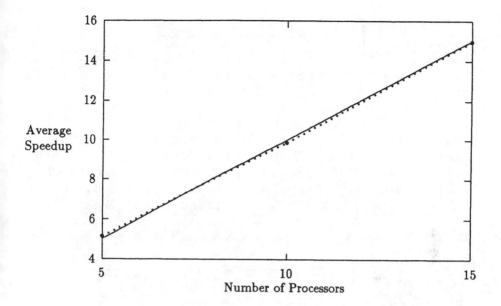

Fig. 6. Average Speedup

problems were converted to MAX-SAT problems by randomly assigning clause weights between 1 and 1000, while their size ranges from 800 to 900 clauses. Furthermore the optimal solution for each instance is known from [13].

The parallel implementation was executed on 15 SUN-SPARC 10 workstations, sharing the same file system, and communication was performed using PVM calls. For each instance we run GRASP in 1, 5, 10 and 15 processors, with maximum number of iterations 1000, 200, 100 and 66 respectively. The amount of CPU time required to perform the specified number of iterations, and the best solution found was recorded.

The computational results are shown in Tables 1 and 2. In Table 1 we can see that that the parallel GRASP with 15 processors always produces better solution than the the serial (1 processor) except in one case. On the average the solutions obtained from the 1, 5, 10 and 15 processors are $0.864 \times 10^{-3}, 0.811 \times 10^{-3}, 0.705 \times 10^{-3}$ and 0.58×10^{-3} percent from the optimal solutions. The solution quality increases on the average, as the number of available processors increases. Moreover, in Table 2 we can see clearly that the speedup of the parallel implementation is almost linear, as illustrated in figure 6 where the average speedup for 5, 10 and 15 processors is shown.

Problem	Optimal	1-proc.		5-proc.		10-proc.		15-proc.	
name		soln	error-%	soln	error-%	soln	error-%	soln	error-%
jnh201	394238	394154	0.21	394238	0.0	394171	0.17	394238	0.0
jnh202	394170	393680	1.24	393708	1.17	393706	1.17	393883	0.72
jnh203	393881	393446	1.10	393289	1.50	393695	0.47	393695	0.47
jnh205	394063	393890	0.43	393958	0.26	394060	0.07	394060	0.07
jnh207	394238	394030	0.52	393929	0.78	393813	1.07	394090	0.37
jnh208	394159	393893	0.67	393585	1.45	393622	1.36	393483	1.71
jnh209	394238	393959	0.70	393805	1.10	393884	0.89	393985	0.64
jnh210	394238	393950	0.73	394238	0.0	394238	0.0	394238	0.0
jnh301	444854	444403	1.01	444577	0.62	444577	0.62	444577	0.62
jnh302	444459	443555	2.03	443911	1.23	443911	1.23	443911	1.23

Table 1. Solutions for the jnh problem class (error-% is expressed in $\times 10^{-3}$).

Problem	1-proc.	5-proc.		10-proc.		15-proc.	
name	Time	Time	Speedup	Time	Speedup	Time	Speedup
jnh201	310.4	62.8	4.94	30.5	10.2	22.2	14
jnh202	312.2	59.8	5.22	31.2	10	23.4	13.3
jnh203	351.2	72.3	4.85	35.2	9.97	23.2	15.13
jnh205	327.8	63.4	5.17	32.1	10.2	22.5	14.56
jnh207	304.7	56.7	5.37	29.6	10.3	19.8	15.38
jnh208	355.2	65.6	5.41	33.2	10.69	21.0	16.9
jnh209	339	60.5	5.6	33.6	10.1	21.6	15.69
jnh210	318.5	57.6	5.52	32.5	9.8	20.8	15.31
jnh301	414.5	85.3	4.85	45.2	9.17	28.3	14.64
jnh302	398.7	88.6	4.5	48.2	8.27	27.0	14.7

Table 2. CPU time and speedup for the jnh problem class

References

1. G.Y. Ananth, V. Kumar, and P.M. Pardalos. Parallel processing of discrete optimization problems. In *Encyclopedia of Microcomputers*, volume 13, pages 129–147. Marcel Dekker Inc., New York, 1993.
2. S.A. Cook. The complexity of theorem-proving procedures. In *Proceedings of the Third annual ACM Symposium on Theory of Computing*, pages 151–158, 1971.
3. T.A. Feo and M.G.C. Resende. Greedy randomized adaptive search procedures. *Journal of Global Optimization*, 6:109–133, 1995.
4. A. Ferreira and P.M. Pardalos, editors. *Solving Combinatorial Optimization Problems in Parallel: Methods and Techniques*, volume 1054 of *Lecture Notes in Computer Science*. Springer-Verlag, 1996.
5. M.R. Garey and D.S. Johnson. *Computers and intractability: A guide to the theory of NP-completeness*. W.H. Freeman and Company, New York, 1979.

6. A. Geist, A. Beguelin, J. Dongarra, W. Jiang, R. Mancheck, and V. Sunderam. *PVM: Parallel Virtual Machine A Users Guide and Tutorial for Networked Parallel Computing*. Scientific and Engineering Computation. MIT Press, Massachusetts Institute of Technology, 1994.

7. J. Gu. Parallel algorithms for satisfiability (SAT) problems. In P.M. Pardalos, M.G.C. Resende, and K.G. Ramakrishnan, editors, *Parallel Processing of Discrete Optimization Problems*, volume 22 of *DIMACS Series in Discrete Mathematics and Theoretical Computer Science*, pages 105–161. American Mathematical Society, 1995.

8. D.S. Johnson and M.A. Trick, editors. *Cliques, coloring, and Satisfiability: Second DIMACS Implementation Challenge*. DIMACS Series in Discrete Mathematics and Theoretical Computer Science. American Mathematical Society, 1996.

9. P.M. Pardalos, L. Pitsoulis, T. Mavridou, and M.G.C. Resende. Parallel search for combinatorial optimization: Genetic algorithms, simulated annealing and GRASP. *Lecture Notes in Computer Science*, 980:317–331, 1995.

10. P.M. Pardalos, L.S. Pitsoulis, and M.G.C. Resende. A parallel GRASP implementation for the quadratic assignment problem. In A. Ferreira and J. Rolim, editors, *Parallel Algorithms for Irregularly Structured Problems – Irregular'94*, pages 111–130. Kluwer Academic Publishers, 1995.

11. P.M. Pardalos, M.G.C. Resende, and K.G. Ramakrishnan, editors. *Parallel Processing of Discrete Optimization Problems*, volume 22 of *DIMACS Series in Discrete Mathematics and Theoretical Computer Science*. American Mathematical Society, 1995.

12. M.G.C. Resende and T.A. Feo. A GRASP for Satisfiability. In D.S. Johnson and M.A. Trick, editors, *Cliques, coloring, and Satisfiability: Second DIMACS Implementation Challenge*, DIMACS Series in Discrete Mathematics and Theoretical Computer Science. American Mathematical Society, 1996.

13. M.G.C. Resende, L. Pitsoulis, and P.M. Pardalos. Approximate solution of weighted max-sat problems using grasp. In DingZu Du, Jun Gu, and Panos M. Pardalos, editors, *Satisfiability Problem: Theory and Applications*, DIMACS Series in Discrete Mathematics and Theoretical Computer Science. American Mathematical Society, 1996.

Applications of HPCN in Manufacturing Industry

John W. Perram
Department of Information Technology
Odense University
Lindø Center for Applied Mathematics
Forskerparken 10
DK-5230 Odense, Denmark

Abstract. In this work, we will consider industrial applications which illustrate two aspects of HPCN. These applications are AMROSE, a system for automatically welding complex curved structures on the basis of a CAD model, and SMART PAINTER, an automatic task curve planning system for spray painting robots on the basis of a CAD description of the surface to be painted and a mathematical model of the painting process. Both applications, of which AMROSE is a working prototype welding the world's largest container ships, are characterized by computationally intensive numerical simulation, complex optimization, control and sensor integration operating in a heterogeneous multi-processor network. In the body of the paper we discuss briefly the software architectures of both applications, the numerical algorithms used in the simulations and the distributed hardware system. We conclude with some speculations of how applying HPCN in manufacturing industry will lead to integrating the production process in the global information society.

* Presented at PARA96. The involvement of all members of the AMROSE and SMART PAINTER teams is gratefully acknowledged. This work received significant financial support from Odense Steel Shipyard, the Danish Research Academy, the Academy for the Technical Sciences and the Research Councils' special program in Information Technology

1 Autonomous Multiple Robot Operation in Structured Environments

AMROSE is a system capable of automatically generating and executing robot programs for welding complex curved steel structures, of the type shown in Figure 1, typically ship sections, from a CAD model of the work cell. The various

Fig. 1. Complex ship section of the type welded by the AMROSE system at Odense Steel Shipyard

components of the software system are shown in Figure 2. Starting at the left, the product model can be regarded as the fusion of geometric data from the CAD database with knowledge about the welding process. The output can be regarded as a complete digital description of the work piece and the tasks which have to be performed on it.

The task, which is abstracted to moving the robot tool center along an arbitrary piecewise 3D twisted curve according to the process rules, is sent to the task scheduler, which breaks down the task into simpler subtasks and optimally schedule them for the equipment available. For welding, this module uses a simulated annealing strategy, but for painting, as we shall see, this module will be a massive distributed optimization system.

The task plan is sent on to both the motion planner and the tool manager. The role of the motion planner is to make a roadmap for the robot to move between the various enclosed spaces in the structure using graph optimization algorithms. The output is a set of intermediate goals for the robot which guide

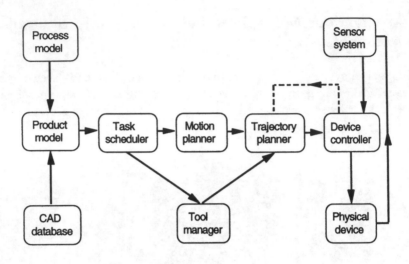

Fig. 2. Architecture for the AMROSE robot trajectory planning system

it through the maze of steel plates.

The role of the tool manager is to take control of the robot when it arrives in the vicinity of a new task and ensure that the tool motion fulfils the requirements of the product model. Algorithms used here are interpolation and continuous optimization [1]. The output here is a set of constraints on the motion of the tool center during task performance.

The trajectory planner is the most computationally intensive part of the software system [3]. It generates the detailed motion of the robot by numerically solving the equations of motion for a mathematical model of the robot subject to metaphysical forces [2] attracting it to the next intermediate goal, repelling it from obstacles (including itself) and subjecting it to extra constraints during task performance. Some idea of how to design these metaphysical force fields is given in [4]. This module uses constraint dynamics algorithms [5] originally developed in molecular dynamics [6]. This module, which runs on a powerful Silicon Graphics work station, generates configurations for the robot at 100 Hz in better than real time. The output is a robot program which is sent on the the device controller.

In the current prototype, the device controller is a proprietary controller for a Motoman K3 industrial robot customized for the shipyard application. Work is currently underway to replace it with an open system using inexpensive motor positioning cards embedded in a pentium PC running Windows NT. The controller takes input from the positioning and welding task sensor systems and computes on-line corrections to the robot program to take account of small deviations of the actual work cell and task curves from the process model.

The physical device is a 9 degree of freedom composite robot consisting of a 3D cartesian gantry which positions a 6 degree of freedom Motoman K4 industrial robot hanging from it. However, one of the advantages of the AMROSE method is that the equations of motion are independent of the robot structure or the number of links. The system can be quickly adapted to different robots, such as the snake-like robot for painting in enclosed spaces shown in Figure 3.

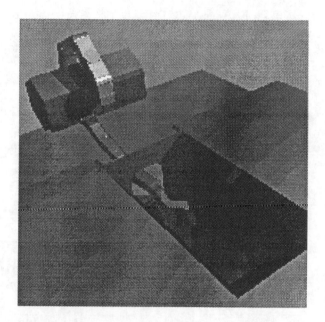

Fig. 3. A snake-like robot for performing tasks in confined spaces

From the above description, the reader can discern that AMROSE is both a distributed modular software system and a multiprocessor hardware system running a number of tightly synchronized, intensive numerical calculations. The software design is Object-Oriented and the code is written in C++.

2 SMART PAINTER

Because of the modular nature of the AMROSE software system, the process model for welding can be replaced by models for other processes. An obvious application is spray painting. The problem here is that the process has hitherto been poorly understood so that current robot painting installations are programmed manually by former manual painters using their imperfect knowledge. Moreover, the programming process is a very slow and expensive one which means that robot painting is confined to mass production. For example, MAERSK Container Industry in Tinglev, Denmark manufactures and paints 17,000 more or less identical dry goods and refrigerated containers per year, more or less sufficient to maintain the size (250,000) of the container inventory of the MAERSK shipping line.

To use robots for painting short production runs requires a different strategy based on a mathematical model of the painting process. The recent article [7] describes how to compute the coverage on a surface from the motion of a calibrated spray gun.

An analytic relation between the 3D paint flux distribution function in the spray cone and the coverage profile for a linear strip painted by a spray gun following the normal vector of a flat plate can be inverted to compute the flux distribution function by solving Abel's integral equation. The calculated distribution function function can then be used to predict the coverage profile for more complicated motions of the spray gun. This is confirmed in Figure 4, which shows a measured coverage profile for a vertically painted strip from which the flux distribution function has been extracted and used to predict the coverage for a strip painted with a gun tilted with respect to the direction of movement of the paint gun.

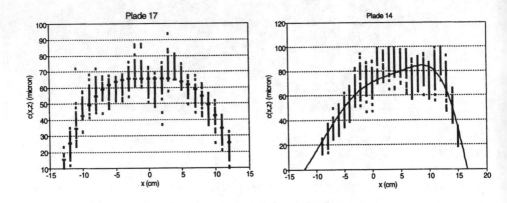

Fig. 4. Showing the fitted experimental coverage profile and the predicted and measured coverage functions painted with a tilted spray

This model will be used to compute the paint coverage profile on a given surface for a given motion of the robot. This can then be used in both a graphical simulator of the painting process so that the user can investigate the consequences of various strategies for overlaying strips, as well as painting in corners and near edges. It can also be used in an optimization tool for automatically generating a trajectory (almost certainly suboptimal) for painting a given surface subject to constraints on minimal local coverage. As the phase space of the paint spray is 12-dimensional, the reader can appreciate that this optimization will be a massive HPCN application in itself.

From the previous section, the reader will have discerned that the state of the art in robot controllers leaves something to be desired, especially if we wish to achieve collaboration with the trajectory planning system. A possible approach is model-based control where a complete mathematical model of the robot is used to compute the motor forces required to make the physical device reproduce the planned motion. If the various parameters affecting the motion were precisely known, we could be confident that our predictions would be achieved. By observing deviations between the actual and planned motion, a smart controller would learn about the current values of these partially known parameters. The advantage of such a controller is that it would largely consist of high level software running on conventional mass-produced microprocessors rather that special purpose electronics.

3 Perspectives for Manufacturing Industry

The ability of robot systems to plan autonomously opens science-fiction like possibilities for manufacturing industry. If we to get robots to execute a design in a physical product, then the way is open to automate custom production of high value added products. Moreover, the designer need no longer take into account the details of the production process. As 3D modelling software becomes sufficiently user friendly, this means that the customer can be integrated into the design and production processes. The possibilities for interactive marketing provided by World Wide Web will then enable manufacturing industry to be integrated in the global information society.

References

1. R. Larsen, J.W. Perram and H.G. Petersen, *Device Independent Tool Center Control*, Int. J. Robotics Res., submitted for publication.
2. O. Khatib, *Real Time Obstacle Avoidance for Manipulators and Mobile Robots*, Int. J, Robotics Res., 5:90-98, (1986).
3. L. Overgaard, H.G. Petersen and J.W. Perram, *A general Algorithm for Dynamic Control of Multi-Link Robots*, Int. J. Robotics Res. 14:281-294, (1995).
4. J.W. Perram and Y. Demazeau, A Multi-Agent Architecture for Distributed Constrained Optimization and Control, to be presented at ICMAS'96, Kyoto.
5. S.W. de Leeuw, J.W. Perram and H.G. Petersen, *Hamilton's equations for constrained dynamical systems*, J. Stat. Phys., 61:1203-1222, (1990).
6. J.P. Ryckaert, G. Ciccotti and H.J.C. Berendsen, *Numerical Integration of the Cartesian Equations of Motion of a System with Constraints: Molecular Dynamics of n-Alkanes*, J. Comp. Phys., 23:327-341, (1977).
7. P. Hertling, L. Høg, R. Larsen, J.W. Perram and H.G. Petersen, Task Curve Planning for Painting Robots: Process Modelling and Calibration, IEEE Trans. Robotics and Automation, 12:324-330, (1996).

An Effective Model to Decompose Linear Programs for Parallel Solution*

Ali Pınar and Cevdet Aykanat

Computer Engineering Department

Bilkent University, Ankara, Turkey

Abstract. Although inherent parallelism in the solution of block angular Linear Programming (LP) problems has been exploited in many research works, the literature that addresses decomposing constraint matrices into block angular form for parallel solution is very rare and recent. We have previously proposed hypergraph models, which reduced the problem to the hypergraph partitioning problem. However, the quality of the results reported were limited due to the hypergraph partitioning tools we have used. Very recently, multilevel graph partitioning heuristics have been proposed leading to very successful graph partitioning tools; Chaco and Metis. In this paper, we propose an effective graph model to decompose matrices into block angular form, which reduces the problem to the well-known graph partitioning by vertex separator problem. We have experimented the validity of our proposed model with various LP problems selected from NETLIB and other sources. The results are very attractive both in terms of solution quality and running times.

1   Introduction

Coarse grain parallelism inherent in the solution of block angular *Linear Programming* (LP) problems has been exploited in recent research works [5, 10]. However, these approaches suffer from *inscalability* and *load imbalance*, since they exploit only the existing block angular structure of the constraint matrix. This work focuses on the problem of decomposing irregularly sparse constraint matrices of large LP problems to obtain block angular structure for scalable parallelization. The objective in the decomposition is to minimize the size of the master problem—the sequential component of the overall parallel scheme—while maintaining computational balance among subproblem solutions.

The literature that addresses this problem is extremely rare and very recent. Ferris and Horn [3] model the constraint matrix as a bipartite graph, and use graph partitioning heuristics for decomposition. However, this model is not suitable for the existing graph partitioning heuristics and tools. In our previous work [12], we have proposed two hypergraph models which reduce the decomposition problem to the well-known hypergraph partitioning problem.

Very recently, multilevel graph partitioning heuristics have been proposed leading to very successful graph partitioning tools; Chaco [6] and Metis [7]. This

---

* This work is partially supported by the Commission of the European Communities, Directorate General for Industry under contract ITDC 204–82166 and The Scientific and Technical Research Council of Turkey under grant no EEEAG 160.

$$A_B^p = \begin{pmatrix} B_1 & & & \\ & B_2 & & \\ & & \ddots & \\ & & & B_k \\ R_1 & R_2 & \dots & R_k \end{pmatrix} \quad A_B^d = \begin{pmatrix} B_1 & & & C_1 \\ & B_2 & & C_2 \\ & & \ddots & \vdots \\ & & & B_k\ C_k \end{pmatrix} \quad A_{DB} = \begin{pmatrix} B_1 & & & C_1 \\ & B_2 & & C_2 \\ & & \ddots & \vdots \\ & & & B_k\ C_k \\ R_1 & R_2 & \dots & R_k\ D \end{pmatrix}$$

Fig. 1. Primal (A_B^p), dual (A_B^d) and doubly-bordered (A_{DB}) block angular forms of an LP constraint matrix A

work proposes a new graph model—Row-Interaction Graph (RIG)—for decomposing the constraint matrices. In RIG, each row is represented by a vertex, and there is an edge between two vertices if there exists at least one column which has nonzeros in both respective rows. This model reduces the decomposition problem into the graph partitioning by vertex separator problem. Vertices in part P_i of a partition correspond to the rows in block B_i, and vertices in the separator correspond to the coupling rows. Hence, minimizing the number of vertices in the separator corresponds to minimizing the size of the master problem.

We have experimented the validity of the proposed graph model with various LP constraint matrices selected from **NETLIB** and other sources. We have used *Metis* tool for multi-way partitioning of sample RIGs by edge separators. Then, we have used various proposed heuristics for refining the edge-based partitions found by *Metis* to partitions by vertex separators. Our results are much better than those of previous methods. We were able to decompose a matrix with 10099 rows, 11098 columns, 39554 nonzeros into 8 blocks with only 517 coupling rows in 1.9 seconds and a matrix with 34774 rows, 31728 columns, 165129 nonzeros into 8 blocks with only 1029 coupling rows in 10.1 seconds. The solution times with $LOQO$[14] are 907.6 seconds and 5383.3 seconds, respectively.

2 Previous Work

2.1 Bipartite Graph Model

Ferris and Horn [3] model the sparsity structure of the constraint matrix as a bipartite graph. In this model (BG), each row and each column is represented by a vertex, and the sets of vertices representing rows and columns form the bipartition. There exists an edge between a row vertex and a column vertex if and only if the respective entry in the constraint matrix is nonzero. This graph is partitioned using Kernighan-Lin [8] heuristic. Then, vertices are removed until no edges remain among different parts. This enables permutation of the matrix into a doubly-bordered form (Fig. 1). Out of the vertices removed, the ones representing columns constitute the row-coupling columns, and the ones representing the rows constitute the column-coupling rows. This doubly-bordered matrix A_{DB} is transformed into a block angular matrix A_B^p by *column splitting*[3].

2.2 Hypergraph Models

In our previous study [12], we have proposed two hypergraph models for the decomposition. A hypergraph $\mathcal{H} = (\mathcal{V}, \mathcal{N})$ is defined as a set of vertices and a set of nets (hypergedges) among these vertices. Each net is a subset of vertices of the hypergraph. In a partition, a net is cut (external), if it has vertices in more than one parts, and uncut (internal), otherwise.

In the first model, namely the *row–net* (RN) model, each row is represented by a net, whereas each column is represented by a vertex. The set of vertices connected to a net corresponds to the set of columns which have a nonzero entry in the row represented by this net [12]. In this model, the decomposition problem reduces to the well-known *hypergraph partitioning* problem. Hypergraph partitioning tries to minimize the number of cut nets, while maintaining balance between the parts. Maintaining balance corresponds to balancing among block sizes in the block angular matrix A_B^p (Fig. 1), and minimizing the number of cut nets corresponds to minimizing the number of coupling rows in A_B^p.

The second model, namely the *column–net* (CN) model, is the dual of the RN model, so partitioning this hypergraph gives dual block angular matrix A_B^d.

3 Graph Partitioning by Vertex Separator

We say that $\Pi^k = (P_1, P_2, \ldots, P_k; S)$ is a k-way vertex separation of $\mathcal{G} = (\mathcal{V}, \mathcal{E})$ if the following conditions hold: each part P_i, for $1 \leq i \leq k$, is a nonempty subset of $\mathcal{V}$; all parts and the vertex separator $S \subset \mathcal{V}$ are mutually disjoint; union of k parts and the separator is equal to $\mathcal{V}$; and there does not exist an edge between two parts P_i and P_j for any $i \neq j$. We also restrict our separator definition as follows: each vertex in the separator S is adjacent to vertices of at least two different parts. Balance criterion for part sizes is defined as: $(W_{max} - W_{avg})/W_{max} \leq \epsilon$ where W_{max} is the size of the part with maximum size, W_{avg} is the average part size, and ϵ is a predetermined imbalance ratio.

Using these definitions, the problem of partitioning by vertex separators can be stated as: *"finding a balanced vertex partition with desired number of parts which minimizes the cardinality of the set S"*.

4 Row Interaction Graph

In this section, we present a new graph model, namely the *row interaction graph* (RIG), for the decomposition. In RIG, each row is represented by a vertex, and there exists an edge between two vertices if and only if there exists at least one column which has nonzeros in both respective rows. So formally:

Definition 1 *A graph $\mathcal{G} = (\mathcal{V}, \mathcal{E})$ is a RIG representation of a sparse matrix $A = (a_{ij})$ iff the following conditions are satisfied.*

- $\mathcal{V} = \{r_1, r_2, \ldots, r_i, \ldots, r_M\}$, where r_i represent the ith row of matrix A.
- $e = (r_i, r_j) \in \mathcal{E} \iff \exists k\ 1 \leq k \leq N \ni a_{ik} \neq 0$ and $a_{jk} \neq 0$

A sample sparse matrix A, and the associated RIG are presented in Fig. 2. In this graph, edge (r_1, r_5) is because of rows 1 and 5 having a nonzero in column 3. However, there does not exist an edge between r_1 and r_2, because there does not exist a column in which both row 1 and row 2 have a nonzero.

A k-way vertex separation $\Pi^k = (P_1, P_2, \ldots, P_k; S)$ of RIG induces a row and column permutation for matrix A transforming it into a block angular form

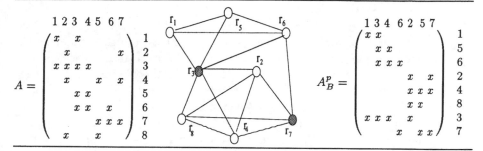

Fig. 2. A sample 8×7 matrix, its associated RIG and its block angular form A_B^p induced by the vertex separation $\Pi^2 = (\{r_1, r_5, r_6\}, \{r_2, r_4, r_8\}; \{r_3, r_7\})$ on the RIG.

A_B^p with k blocks. In a separation Π^k of RIG, vertices in the separator S correspond to the coupling rows of A, and vertices in part P_i correspond to the rows in block B_i. The permutation of the columns is controlled by the rows. Each column is placed in the same block as the rows it shares non-zero(s). By definition of the vertex separator, there are no edges between vertices in different parts, hence there is no column interaction between rows in different blocks, i.e., there are no columns which have nonzeros in two rows at different parts.

Given $\Pi^2 = (\{r_1, r_5, r_6\}, \{r_2, r_4, r_8\}; \{r_3, r_7\})$ as a separation of the RIG in Fig. 2, the associated permuted matrix A_B^p can be obtained as follows. Vertices r_1, r_5, r_6 (r_2, r_4, r_8) are in part P_1 (P_2), so respective rows will be placed in block B_1 (B_2). Rows 3 and 7 will form the coupling block because of the vertices r_3 and r_7 on the separator. Columns 1,3,4,6 (2,5,7) are placed in block B_1 (B_2) since they share nonzeros with rows placed in this block.

RIG model reduces the decomposition problem to a well-known problem, graph partitioning by vertex separators. The problem of graph partitioning by vertex separators has two objectives: (i) minimizing the number of vertices in the separator, (ii) maintaining balance between number of vertices in parts other than the separator. The first objective directly corresponds to minimizing the number of coupling rows, since each vertex in the separator of RIG corresponds to a row in the coupling block of A_B^p. The second objective corresponds to maintaining balance among the block sizes in the block angular matrix A_B^p.

5 Finding Vertex Separators

We have adopted commonly used scheme of finding vertex separators from edge separators. Edges in the edge separator are called the *cut edges*. An edge is cut if it is between two different parts. Each edge can be associated with a weight, and *cutsize* of a partition is the sum of weights of cut edges. In the light of these definitions, problem of graph partitioning by edge separator can be stated as: *finding a balanced partition of vertices of the graph which minimizes the cutsize.*

The set of vertices adjacent to the cut edges is called the *wide separator* [9]. We will call the subgraph induced by the wide separator and the cut edges as the *wide-separator* subgraph $\mathcal{G}_{WS}$. A subset of the vertices in the wide separator can be chosen to form a narrow separator, a feasible separator of smaller cardinality.

5.1 Finding Wide Separators

There are no certain metrics for the "goodness" of a wide separator that will lead to a smaller narrow separator. However, two metrics have gained popularity due to their simplicity and availabilty of appropriate software tools. The first one is minimizing the number of cut edges. Minimizing the number of cut edges can give us a good estimate of a vertex separator, since it finds logical clusters on the graph. The second one is minimizing the number of vertices in $\mathcal{G}_{WS}$. Leiserson and Lewis [9] model the graph with a hypergraph, where there exists a vertex for each vertex in the graph, and there exists a net n_i for each vertex v_i which contains v_i and all vertices adjacent to v_i. With this hypergraph, if a net n_i is on the cut, then the vertex v_i should be on the wide separator. Hence, minimizing the number of cut nets on this hypergraph corresponds to minimizing the number of vertices in $\mathcal{G}_{WS}$.

Although, both metrics are valuable assets for the goodness of a wide separator, they do not guarantee a narrow separator of smaller cardinality.

5.2 Edge Weightening for Better Wide Separators

We propose a heuristic model for finding a better wide separator. Our basic observation is that all edges are not of equal importance for the goodness of a wide separator. Edges incident to a vertex with high degree are less important, since this vertex has a higher probability to be moved to the separator. Here, degree $deg(u)$ of a vertex $u \in \mathcal{V}$ refers to the number of edges incident to u in RIG. So, we can assign weights to the edges inversely proportional to the degrees of its end-vertices. We propose the following weight function:

$$weight((u, v)) = \frac{1}{max(deg(u), deg(v))}$$

Minimizing the cutsize of this edge-weighted RIG is expected to yield good wide separators for refining to narrow separators.

5.3 From Wide Separators to Narrow Separators

This part of the problem is equivalent to finding a minimum vertex cover on $\mathcal{G}_{WS}$. This problem can be solved optimally in polynomial time for two way partitions, by finding maximum matchings on bipartite graphs [13]. However, we need to resort to heuristics for the solution of this problem for multi-way partitions.

We have experimented the greedy heuristics, *maximum-inclusion* (MI) and, *minimum-removal* (MR) proposed in[9]. In this work, we also propose a new heuristic, namely, *one-max-inclusion* (OMI) heuristic which is presented in Figure 3. Our heuristic is similar to MI with the following enhancement: OMI starts with including the vertices adjacent to a vertex of degree 1 to the vertex cover (narrow separator), since this does not destroy our chance to find an optimal solution. When there are no vertices of degree 1, we take a greedy decision similar to that of MI and include the vertex with the highest degree to the separator. Then, we again seek for vertices with degree 1, and repeat this process until all edges are adjacent to a vertex in the separator.

INPUT: Wide Separator subgraph: $\mathcal{G}_{WS} = (\mathcal{V}_{WS}, \mathcal{E}_{WS})$
OUTPUT: Narrow Separator: $S \subset \mathcal{V}_{WS}$

    repeat
        for each $v \in \mathcal{V}$ $deg(v) = 1$ **do**
            $u \leftarrow$ *only neighbor of* v
            $S \leftarrow S \cup \{u\}$;
            for each $x \in Adj(u)$ **do** $\mathcal{E} \leftarrow \mathcal{E} - \{(u, x)\}$ **endfor**;
        endfor
        if $(\mathcal{E} \neq \emptyset)$ **then**
            $v \leftarrow$ vertex with maximum degree ;
            for each $u \in Adj(v)$ **do** $S \leftarrow S \cup \{u\}$ **endfor**;
            for each $x \in Adj(u)$ **do** $\mathcal{E} \leftarrow \mathcal{E} - \{(u, x)\}$ **endfor**;
    until $\mathcal{E} = \emptyset$

Fig. 3. A greedy heuristic for finding a narrow separator from the wide-separator subgraph of a partition of RIG by edge separator

In our experiments, OMI heuristic, overperformed the other two, MI and MR [11]. We have compared the performance of OMI heuristic with optimal solutions obtained by matchings, for bisections. We have seen that, average difference for 27 different data set after 20 runs was only 0.11%, and the peak difference was only 0.58% for one data set.

6   Experimental Results

We have experimented the validity of the model on various LP matrices selected from the **Netlib** suite [4], Kennigton problems [1], and collection of Gondzio[2]. The properties of these problems are presented in Table 1. In this table, M, N, Nz, and D, columns represent the number of rows, columns, nonzeros, and density of the respective constraint matrices, respectively. Here, D is computed as $Nz/(M \times N)$. In Table 1, Nz/N and Nz/M columns denote average number of nonzeros per row and column, respectively, and $|\mathcal{E}/\mathcal{V}|$ column denotes average vertex degree of the associated RIGs. All experiments have been performed on a *SUN Sparc 5* workstation. We have used Metis [7] for graph partitioning, an FM-variant [2] for hypergraph partitioning, and OMI heuristic implemented in C for finding narrow separators from wide separators. For each experiment, partitioning heuristic has been run 20 times with random seeds. Following tables and figures display the averages of these runs

Figure 4 shows the relative performance of the edge-weighted graph (W-RIG) model and hypergraph(H-RIG) model compared to unweighted graph (U-RIG) model in finding narrow separators for 8-way partitioning. W-RIG and U-RIG models correspond to running Metis on weighted and unweighted RIG, respectively, and then refining the resulting wide-separators to narrow separators with OMI. H-RIG corresponds to running the FM-variant [2] on the hypergraph

---

[2] These problems are available by anonymous ftp from IOWA Optimization Center ftp col.biz.uiowa.edu:pub/testprob/lp/gondzio

Table 1. Properties of the Constraint Matrices and their associated RIGs

Problem Name	Constraint Matrix Properties						RIG Properties	
	M	N	N_z	$D\%$	N_z/N	N_z/M	$|\mathcal{E}|$	$|\mathcal{E}|/|\mathcal{V}|$
80bau3b	2262	9799	21002	0.09	2.14	9.28	10074	8.91
bnl2	2324	3489	13999	0.17	4.01	6.02	13457	11.58
cycle	1903	2857	20720	0.38	7.25	10.89	27714	29.13
czprob	929	3523	10669	0.33	3.03	11.48	7072	15.22
d2q06c	2171	5167	32417	0.29	6.27	14.93	26991	24.87
ganges	1309	1681	6912	0.31	4.11	5.28	7656	11.70
greenbea	2392	5405	30877	0.24	5.71	12.91	33841	28.30
sctap3	1480	2480	8874	0.24	3.58	6.00	7386	9.98
ship12l	1151	5427	16170	0.26	2.98	14.05	10673	18.55
stocfor2	2157	2031	8343	0.19	4.11	3.87	12738	11.81
woodw	1098	8405	37474	0.41	4.46	34.13	20421	37.20
cre-a	3516	4067	14987	0.10	3.69	4.26	51015	10.10
cre-c	3068	3678	13244	0.12	3.60	4.32	49025	13.93
cre-d	8926	69980	242646	0.04	3.47	27.18	285068	16.40
osa-07	1118	23949	143694	0.54	6.00	128.53	273779	15.87
CO9	10789	14851	101578	0.06	6.84	9.41	20748	11.80
CQ9	9278	13778	88897	0.07	6.45	9.58	18905	12.32
GE	10099	11098	39554	0.04	3.56	3.92	181670	40.71
NL	7039	9718	41428	0.06	4.26	5.89	52466	93.86
mod2	34774	31728	165129	0.01	5.20	4.75	119208	22.10
world	34506	32734	164470	0.01	5.02	4.77	106156	22.88

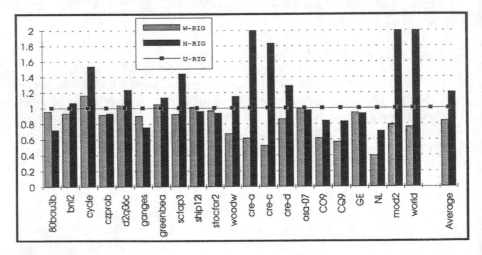

Fig. 4. Narrow separator quality of edge-weighted graph (W-RIG) model and hypergraph (H-RIG) model compared to unweighted graph (U-RIG) model in finding narrow separators of test RIGs for 8-way partitioning. Bars under the baseline indicate that the respective model performs better than the U-RIG model.

representations of the RIGs (as discussed in Section 5.1), and refining the results with OMI. W-RIG model produces 20% better results on the average than U-RIG model. The difference becomes more significant for larger problems. Although H-RIG model is worse than U-RIG model on the average, it produces better results for many of the problems.

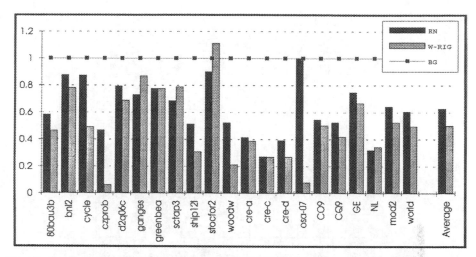

Fig. 5. Quality of edge-weighted RIG model (W-RIG) and RN hypergraph model compared to the BG bipartite graph model for 8-way block angular decomposition of test matrices. Bars under the baseline indicate that the respective model performs better than BG model.

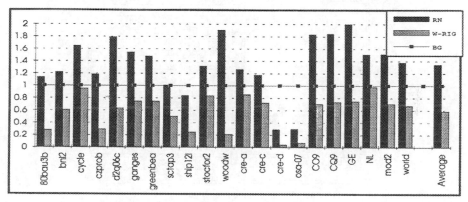

Fig. 6. Execution times of (W-RIG) model and RN model compared to (BG) model for 8-way block angular decomposition of test matrices.

Figures 5 and 6 illustrate quality and execution times of W-RIG model and row-net (RN) hypergraph model compared to the bipartite graph (BG) model for 8-way block angular decomposition of test matrices. W-RIG model overperforms BG model in all problems except for stocfor2. W-RIG results are twice better than BG on the average. The difference becomes drastic for osa-07, czprob, and woodw. The common point in these matrices is the large number of columns with respect to rows. Relative performance of BG deteriorates for matrices with $N \gg M$, since BG treats both rows and columns as decision variables. The difference between run times of BG and W-RIG becomes very significant for 80bau3b, czprob, woodw, cre-d, osa-07, all of which has $N \gg M$. Recall that, we have used the same partitioning tool for both BG and RIG models, hence

Table 2. The effectivity of RIG Model

Problem			k	# Coup. Rows		t_{part}	
		LOQO		abs.	rel.	abs.	rel.
Name	Rows	$t_{sol}(secs)$			%	secs.	%
cycle	1903	110.8	4	64	3.36	0.87	0.79
			8	100	5.25	1.05	0.95
d2q06c	2171	400.0	4	223	10.27	0.96	0.24
			8	293	13.50	1.17	0.29
ganges	1309	21.9	4	68	5.19	0.32	1.46
			8	128	9.78	0.41	1.87
greenbea	2392	166.3	4	125	5.23	1.34	0.81
			8	231	9.66	1.63	0.98
ship12l	1151	20.5	4	49	4.26	0.43	2.10
			8	78	6.78	0.54	2.63
stocfor2	2157	24.8	4	44	2.04	0.53	2.14
			8	120	5.56	0.66	2.66
woodw	1098	80.7	4	68	6.19	0.74	0.92
			8	160	14.57	0.86	1.07
cre-a	3516	40.8	4	112	3.19	1.03	2.52
			8	141	4.01	1.27	3.11
cre-c	3068	40.7	4	102	3.32	0.89	2.19
			8	127	4.14	1.08	2.65
cre-d	8926	6719.9	4	913	10.23	6.12	0.09
			8	1117	12.51	6.73	0.10
osa-07	1118	398.7	4	80	7.16	3.39	0.85
			8	80	7.16	4.05	1.02
CO9	10789	1827.6	4	1099	10.19	4.30	0.24
			8	1363	12.63	4.72	0.26
CQ9	9278	1664.4	4	751	8.09	4.00	0.24
			8	1061	11.44	4.36	0.26
GE	10099	907.6	4	331	3.28	1.71	0.19
			8	517	5.12	1.93	0.21
NL	7039	699.2	4	547	7.77	2.82	0.40
			8	633	8.99	3.22	0.46
mod2	34774	5383.3	4	559	1.61	9.44	0.18
			8	1029	2.96	10.07	0.19
world	34506	25819.7	4	615	1.78	9.24	0.04
			8	1074	3.11	10.02	0.04
Average			4		5.48		0.90
			8		8.06		1.1

the difference is directly due to the effectiveness of the models.

The performances of W-RIG and RN model are quite competitive. W-RIG is better on the average. However, the difference is not too large, and may be due to the partitioning tool used. But a careful observation reveals that the performance of RN model becomes poor for problems with $N \gg M$. This is simply because RN works on too many vertices.

Table 2 shows the overall effectiveness of the proposed model. The number of coupling rows and the percent ratio of the number of coupling rows to the total number of rows, the actual partitioning times and percent ratio of partitioning times to solution times of the problems with $LOQO$ [14] are presented. On the overall average, only 5.48% and 8.06% of the rows are on the coupling block for 4 and 8 block decompositions, respectively. The partitioning times are negligible compared to $LOQO$ solution times (0.9% for 4 blocks, and 1.1% for 8 blocks). Another remarkable point in this table is that partitioning times grow slowly with the problem size, although solution times rapidly increase. This makes decomposition very practical for large problems.

7 Conclusion

We have proposed an effective graph model to decompose LP matrices to block angular form for scalable parallelization. The new model reduced the problem to the well-known graph partitioning by vertex separator problem. The validity of the model has been experimented with various LP matrices, and its performance has been compared with bipartite graph [3] and hypergraph models [12]. The proposed model overperformed the previous two models on the existing graph/hypergraph partitioning tools. The new model is very effective and enables us to decompose a matrix with 10099 rows, 11098 columns, 39554 nonzeros into 8 blocks with only 517 coupling rows in 1.9 seconds and a matrix with 34774 rows, 31728 columns, 165129 nonzeros into 8 blocks with only 1029 coupling rows in 10.1 seconds. The solution times with $LOQO$ are 907.6 seconds for the former and 5383.3 seconds for the latter.

References

1. W. J. Carolan, J. E. Hill, J. L. Kennington, S. Niemi, S. J. Wichmann An Empirical Evaluation of the KORBX Algorithms for Military Airlift Applications *Operations Research* 38(2):240-248, 1990.
2. U. V. Çatalyurek and C. Aykanat, Decomposing Irregularly Sparse Matrices for Parallel Matrix-Vector Multiplication, *Proc. of Irregular 96*,1996, (to appear).
3. M. C. Ferris, and J. D. Horn. Partitioning mathematical programs for parallel solution. Technical report TR1232, Computer Sciences Department, University of Wisconsin Madison, May 1994.
4. D. M. Gay, "Electronic mail distribution of linear programming test problems" *Mathematical Programming Society COAL Newsletter*, 1985.
5. S. K. Gnanendran and J. K. Ho. Load balancing in the parallel optimization of block-angular linear programs. *Mathematical Programming*, 62:41-67, 1993.
6. B. Hendrickson and R. Leland, A Multilevel Algorithm for Partitioning Graphs, Sandia National Laboratories,SAND93-1301, 1993.
7. G. Karypis and V. Kumar, A Fast and High Quality Multilevel Scheme for Partitioning Irregular Graphs, Dept. of Computer Science, Univ. of Minnesota, 1995, TR 95-035.
8. B.W. Kernighan and S. Lin. An efficient heuristic procedure for partitioning graphs. Technical Report 2, The Bell System Technical Journal, Feb. 1970.
9. C. E. Leiserson and J. G. Lewis, Orderings for parallel sparse symmetric factorization, *3rd SIAM Conf. Parallel Processing for Scientific Comp.*, 27-31, 1987
10. D. Medhi. Bundle-based decomposition for large-scale convex optimization: error estimate and application to block-angular linear programs. *Mathematical Programming*, 66:79-101, 1994.
11. A. Pınar, Decomposing Linear Programs for Parallel Solution M.S. Thesis, Bilkent University, July, 1996.
12. A. Pınar, Ü. V. Çatalyurek, C. Aykanat and M. Pınar, Decomposing Linear Programs for Parallel Solution *Lecture Notes in Computer Science*, 1041:473-482, 1996.
13. A. Pothen and C. J. Fan, Computing the Block Triangular Form of a Sparse Matrix, *ACM. Trans. on Math. Software*,16(4):303-324,1990.
14. R. J. Vanderbei, *LOQO* User's Manual. Princeton University, November 1992.

Numerical Libraries on Shared Memory Computers

Stefano Salvini

The Numerical Algorithms Group Ltd, Wilkinson House, Jordan Hill Road, Oxford
OX2 8DR, United Kingdom
E-mail: stef@nag.co.uk

Abstract. This tutorial discusses parallelism on shared memory systems in the context of the development of libraries of numerical subprograms. General issues are first discussed, then the points raised are illustrated by three case studies, corresponding to three different types of parallelism. Conclusions are then drawn to suggest that numerical libraries on shared memory systems will be essential tools for the rapid development of efficient numerical programs.

1 Introduction

The aim of this tutorial is to introduce some of the issues relevant to the development of numerical software on parallel shared memory systems, and in particular to the development, maintentance and portability of libraries of parallel numerical procedures. Three case studies will be presented.

Hardware vendors are dedicating increasing resources to the development of parallel, shared memory architectures (often called SMP or Symmetric Multi-Processors) based on workstation technology, and this class of systems is already having an important impact on the scientific computing community. Indeed, this trend is likely to intensify in the future. SMP systems, particularly those based on workstation, cache-based technology, can provide high computing power at relatively low cost especially in terms of software development costs.

1.1 Software Development and Libraries

In general terms, only considerations of *functional* rather than data parallelism can drive software development on SMP systems. It should also be possible to use a fairly fine granularity, limited by the costs of creating and destroying parallel processes, such as loading the processors' caches, etc

The *fork–join* model of parallelism is typically employed on SMP machines, whereby parallel threads or processes are created on request by a master (serial) thread to carry out some task and destroyed upon completion.

Detailed knowledge of the target system and of its software is not required to great depth in order to achieve substantial efficiency. The use of compiler directives to define parallelism allows the production and maintenance of single source code for both serial and SMP machines.

Optimizing/parallelizing compilers, diagnostic and performance analysis tools are becoming more sophisticated: part of the optimization/parallelization effort could then be left to the system, thus reducing further the workload of the programmer.

A *top-down* modular approach to software development can be employed in many cases, whereby a serial program can be analysed, using profilers or other tools, to identify the most computationally intensive steps. These can be then be encapsulated within modules to be further refined and parallelized.

It is the relative ease of code migration together with the obvious benefits of efficiency, in particular *parallel* efficiency, that highlight the very important roles that software libraries can play for this class of machines. On one side libraries encapsulate the expertise of specialists often accumulated over years of research and development; on the other side it allows the quick prototyping and development of code by providing easy to use, tested and validated modular building blocks, thus allowing the software developer to concentrate on his/her problem rather then on the mechanisms of solution.

For over twenty years, NAG has been recognized as one of the outstanding players in providing Libraries of high quality and efficient numerical routines in several high-level languages and for a variety of platforms [4]. We have also been actively involved in the development of numerical libraries for distributed memory systems through our involvement in several EU-funded projects, and our own products (our PVM and MPI Libraries [4]). We are now enhancing the parallelism for SMP systems in our main libraries, while keeping the user interfaces unchanged. Of course, some degree of parallelism is provided by basic computational kernels such as the BLAS (Basic Linear Algebra Subprograms [2]), usually provided by hardware vendors: however, we will show how this parallelism may be insufficient to tap the true potential of this class of machines.

Retaining the same user interface, while providing highly parallel and efficient routines would allow users to migrate the code with minimum induction and maximum gain of efficiency from serial to SMP systems. This migration is more simple and straightforward than that to distributed memory platforms.

1.2 Mode of Parallelism

Typically, parallelism on a shared memory system consists of the parallel execution of sections of DO loops. Only one level of parallelism can be active at any one time: any lower level parallelism is ignored and execution is carrried out serially. Of course, this also represents the mode of parallelism that can be generated by automatic parallelizing tools or compilers. Other types of parallelism are far more difficult to achieve automatically and would require a far more complex global analysis of a code.

Parallelism can be defined explicitly in either or both of two ways, using system calls or compiler directives. System calls provide a richer and more flexible grammar; however, compiler directives, though more limited in their scope, allow the development of code more easily portable to other SMP systems as well as to serial systems, where the directives are simply ignored.

Within a fork–join region, data can be defined as shared, i.e. accessible by all processor in the participating pool, or local, i.e. accessible only locally by each processor entirely independently from the others. Cache coherence is maintained by the system.

For example, on SGI Power Challenge Platforms [9], basically one native compiler directive is available, the **C\$DOACROSS** directives which requests the concurrent execution of the DO loop immediately following it, in ways and modalities specified in a number of clauses, and specifying the classification of data items specified within the body of the loop as shared or local.

A more complete, and portable, set is given by the PCF directives (**C\$PAR**) which seem to provide a portability base available to most vendors (PCF is the acronym of Parallel Computing Forum, which itself formed the basis for the proposed ANSI-X3H5 standard). These provide a number of constructs which should be sufficient for most cases (semaphores and general locks are not available).

Within a parallel region (a fork–join region), five different constructs can be used:

- Parallel DO: concurrent execution of sections of DO loops (**C\$PAR PDO**).
- Critical Sections: section of code executable only by one process at a time (**C\$PAR CRITICAL SECTION**).
- Serial Sections: sections of code executed (serially) by only one process (**C\$PAR SINGLE PROCESS**).
- Parallel Sections: alternative sections of code executed concurrently (**C\$PAR PARALLEL SECTION**).
- Synchronization barriers (**C\$PAR BARRIER**).

2 Case Studies

This paper reports three cases studies, carried out on the UNI•C SGI Power Challenge, installed at the University of Aarhus, Denmark [7]. This machine includes 16 processors, each comprising a MIPS R8000 CPU and MIPS R8010 FPU, running at a clock speed of 75 MHz, and capable of delivering 4 floating point operations per cycle, hence with a theoretical peak speed of 300 Mflops/sec per processor. Memory is organized in a hierarchical structure of primary and secondary caches and main memory.

The three case studies reported correspond to three different types of parallelism employed:

1. Parallelism consisting of the concurrent execution of sections of DO loops. This is the simplest type of parallelism, and automatic parallelization and optimization tools should, at least in theory, be capable of generating efficient code. BLAS (Basic Linear Algebra Routines) [2] are typical representatives of this type of parallelism. The Level-3 BLAS DGEMM (matrix–matrix multiplication) was chosen for benchmarking. A highly efficient multithread implementation of the Level-3 BLAS is available in the SGI com-

plib_sgimath_mp library to provide a baseline against which the efficiency of the codes generated from Fortran 77 sources could be tested.

2. Routines which do not contain any parallel constructs but rely for their multi-thread efficiency on lower-level, parallel routines. Typical representatives of this class are LAPACK [1] routines, which rely extensively on the parallelism available in the Level-3 BLAS. I will illustrate here the case of the *LU* factorization routine DGETRF.

3. Routines where parallelism is made more complicated by the need for each thread to access and update a globally shared data structure. An adaptive Gauss–Kronrod one-dimensional quadrature [6] routine was chosen for this group.

A number of guidelines were set and adhered to at all times:

- All codes were written in Fortran 77.
- All codes were encapsulated in user-callable routines of identical interface to the existing corresponding serial versions, as available in the NAG Fortran Library [4].
- All parallelism was entirely contained within these routines, i.e. all routines had a simple serial interface.
- All explicit parallelism, i.e. all parallelism defined in the code as opposed to parallelism generated by the parallelization tools, was defined exclusively in terms of PCF (**C$PAR**) directives.
- No SGI-specific system calls or constructs were used.
- All codes, when run in single-thread mode, were required to be at least as efficient as the existing serial codes of identical functionality.

The SGI Power Challenge at UNI•C is a national facility, intensely used by a large community of users spread across Denmark and elsewhere. Therefore, we could not gain single-user access to the system, a pre-requisite for accurate and highly reproducible benchmarks, without affecting the overall availability of the system but had to carry out all tests on a *fully loaded* machine. This implies that all results reported below must be viewed as *lower bounds* on the efficiency achievable, particularly for the cases involving four or more processors.

2.1 Simple DO Loop Parallelism: DGEMM

DGEMM was chosen because of its widespread use and expected ease of parallelization. SGI offers tuned Level-3 BLAS in their **complib_sgimath_mp** library. Benchmarks were carried out using 1, 2, 4 and 8 processors. Three equivalent implementations of DGEMM were benchmarked for each number of processors:

- the SGI version from the **complib_sgimath_mp** library;
- the Fortran model implementation [3] compiled with high optimization, with a wide range of other options set in order to extract as much performance as possible, including source-to-source parallelization and optimization;

– an experimental blocked subroutine (coded in Fortran 77), to illustrate the impact of a block algorithm on efficient cache management.

The SGI version of DGEMM appears very efficient, both in terms of its raw speed and in terms of its scalability. We certainly did not expect the Fortran 77 model implementation to match the SGI version, but the performance achieved fell much shorter than anticipated in some cases: automatic parallelization failed to achieve the efficiency hoped for, even in the relatively simple and regular case of matrix-matrix operations.

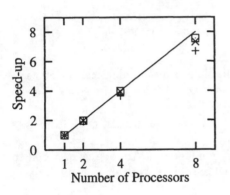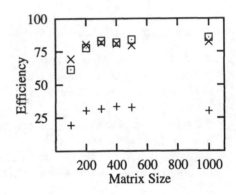

Fig. 1. DGEMM: $C \leftarrow \alpha AB + \beta C$. Speed-up for $m = n = k = 500$ and efficiency for 8 processors, for $m = n = k$.

    □: SGI optimal DGEMM   + : standard Fortran version (F06YAF)
    × : Fortran blocked code

Figure 1 shows some of the results obtained for the case $C \leftarrow \alpha AB + \beta C$. The model implementation, in spite of using the highest level of optimization and of its good scalability, achieved less than half the performance of the SGI version. The experimental version returned a very large increase in performance, exceeding the efficiency of SGI version for the smaller matrices studied, and falling short by less than 5% for the largest matrices. The optimal block-size was found to be of the order of 256. Our blocked version sub-divides the columns of the matrices B and C into equal 'chunks', one for each processor. The lower-level routine used to compute the matrix product of blocks was also written in Fortran.

Figure 2 shows the scalability and efficiency achieved for the case $C \leftarrow \alpha A^T B + \beta C$, using, as before, the SGI version, the Fortran 77 model implementation and the and the experimental Fortran 77 blocked code. With respect to the previous case, the model implementation returns a better efficiency; however, this is considerably degraded for the largest problem studied ($m = n = k = 1000$). This effect is absent in the results from the blocked code, although its efficiency is slightly lower than in the previous case with respect to the SGI version. It

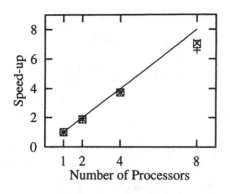 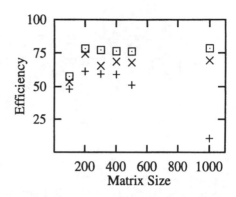

Fig. 2. DGEMM: $C \leftarrow \alpha A^T B + \beta C$. Speed-up for $m = n = k = 500$ and efficiency for 8 processors, for $m = n = k$.
    □: SGI optimal DGEMM   + : standard Fortran version (F06YAF)
    × : Fortran blocked code

seems that the model version fails to achieve a satisfactory utilization of the cache, perhaps because of the Fortran coding, though we expected the optimizing/parallelizing compiler to ease the situation to a better degree.

2.2 Parallelism Encapsulated in Lower Level Routines: LAPACK

The *LU* factorization routine DGETRF is amongst the most heavily and widely used LAPACK subroutines [1]. The code distributed in the LAPACK software employs the *right-looking* version of the blocked algorithm: the matrix is subdivided horizontally and vertically into *blocks* each comprising of b (the block-size) rows/columns, and the following steps are carried out for each diagonal block:

1. Factorize the diagonal and sub-diagonal blocks using partial pivoting (b calls to DSWAP, b calls to DGER, where b is the block-size).
2. Permute the relevant portions of the remaining columns of A ($2b$ calls to DSWAP).
3. Compute the block-row of U (one call to DTRSM).
4. Update the trailing submatrix (one call to DGEMM).

Table 1 reports the performance of the LAPACK routine DGETRF, as implemented in the SGI **complib_sgimath_mp** library, against a specially developed code, briefly described below (marked in the table as 'New'). The same random matrices and timing programs were used for both codes, under identical conditions. The BLAS-based LAPACK approach fails to deliver a good performance for a number of reasons:

1. The fine granularity employed in step 1 above, i.e. the factorization of the diagonal blocks, and its reliance on Level 2 BLAS imply that there is little gain in parallelizing this step. In fact, to all effects, the step is executed in

n_p	1		2		4		8	
n	LAPACK	New	LAPACK	New	LAPACK	New	LAPACK	New
100	9	151	40	154	41	177	40	170
200	151	205	96	282	102	396	105	461
400	202	239	182	366	212	592	208	908
500	210	243	205	398	249	710	248	1024
1000	180	231	266	449	358	891	373	1351

Table 1. *LU* factorization: comparing the speeds, in *Megaflops per second*, of the SGI proprietary version of the LAPACK routine DGETRF (marked 'LAPACK') against the new specially developed code (marked 'New'). Here n_p and n are the number of processors and the order of the matrix, respectively.

single-thread mode, thus providing a serial bottleneck, whose overall effect increases as the number of participating processors increases.

2. Permuting the rows of the matrix requires only $\mathcal{O}(n^2)$ operations, compared with the overall number of operations $\mathcal{O}(n^3)$. However, the permutation of the rows at each stage of the factorization causes considerable cache-thrashing. Its effects are very noticeable indeed: for 1000×1000 matrices, the performance figures for a random matrix, a special matrix which requires always interchange with the bottom row of the active submatrix, and a positive-definite matrix, which requires no row interchanges are 373, 446 and 541 Mflops/sec, respectively, using 8 processors.

3. Each call to the BLAS causes the spawning, and the destruction on exit, of parallel threads. The associated overheads may not be negligible.

4. BLAS do not seem to perform optimally for the type of matrices (highly rectangular), encountered within a typical LAPACK algorithm.

The new code has an identical interface to the LAPACK routine and is equivalent to it. Its enhanced performance was gained in a number of ways:

1. Parallel processes are spawned only once on entry to the routine and destroyed on exit: in other words, parallelism was brought up from the level of the BLAS to the level of the factorization routine.

2. The factorization of individual blocks and updating of the trailing sub-matrices are overlapped to remove the quasi-serial bottleneck.

3. The row permutation (trailing sub-matrix only), updating of the row-block and of the trailing sub-matrix, are carried out on each column block, to improve data locality.

4. All permutations on the left of the pivot columns are carried out after completion of the factorization.

5. Fortran 77 BLAS were developed, particularly suited to the 'thin', elongated submatrices used during the block factorization, to increase performance.

The same approach was used to very good effect for QR and Cholesky factorizations.

Further details on this approach, and the performance obtained for all the factorizations mentioned, can be found in a companion paper to be presented at the PARA'96 Workshop [8].

2.3 More Complex Parallelism: 1-Dimensional Quadrature

The one-dimensional quadrature routine D01AUF is available in both the NAG Fortran 77 Library [4] and NAG Numerical PVM Library [5], a parallel library for distributed memory machines. D01AUF assesses convergence in terms of the *global* error. Hence, whatever parallelism is employed, a global, shared data structure must be used: all participating processors must be able to access and update this data structure. The distributed memory version of D01AUF achieves parallelism via a master–slave paradigm, where global data structures are processed by the 'master', a specified processor, which assigns and distributes parcels of work to the 'slaves', all other processors. The master also collects and processes intermediate results.

A master–slave paradigm was not expected to achieve high efficiency on shared memory machines for a number of reasons.

- In general, shared memory systems consist of fewer processors than distributed memory systems: reducing a more limited pool of resources would reduce efficiency.
- The demand-driven role of the slaves is difficult to implement without resorting to explicit system calls, which we tried to avoid in this work.
- A 'serial' bottleneck would be difficult to avoid in any case: depending on the amount of processing required by the global data structure, considerable degradation of performance may ensue.

Tests showed that a master–slave method would achieve only modest efficiency, and that the coding required would seriously endanger portability. A 'democratic' approach to parallelism was therefore adopted, where no processor is in overall control of the computation. Global data structures are manipulated by each processor, on a first-come first-served basis within a 'critical section' (in PCF parlance), i.e. a section of code which is accessible and executable only by one processor at a time. Particular care was taken in reducing to a bare minimum the number of operations to be carried out in the critical section.

D01AUF is an adaptive routine for the estimation of a one-dimensional integral using any of a set of six Gauss–Kronrod rules with global absolute and/or relative error control. It progressively subdivides the initial range into sub-intervals. Its global data structure consists of a stack which stores for each sub-interval its upper and lower limits, its current integral and (absolute) error estimate. The stack is sorted in descending order of (absolute) error. By popping the top of the stack for next sub-division, a quasi-optimal path is followed to reduce the global error. However, operations on the stack are non-trivial: the ordering of the stack must be maintained, and this may periodically require a global sort.

In the implementation of D01AUF in the Numerical PVM Library version, the master processor maintains the stack; polls, in turn, each slave processor, sending to it a new parcel of work if the slave is found inactive; and collects the results whenever a message reaches it from a slave. Each slave carries out up to a small maximum number of sub-divisions in its sub-range of integration. However, the quasi-optimality of the approach path is necessarily slightly relaxed, allowing for the top of the stack to be popped before the results of the integration over the sub-intervals already sent to the slave for processing have been incorporated into the stack.

An initial approach to develop code for the Power Challenge along these lines proved very inefficient. Therefore, we decided to remove altogether the requirement of using a quasi-optimal sub-division path. This has a number of very important benefits:

- removing at one stroke the bottleneck of sorting a globally shared stack, hence escaping the consequences of Amdahl's law;
- simplifying considerably the code, contrary to the common assumption that parallelism necessarily *increases* code complexity;
- increasing the size of the parcels of work carried out concurrently by each thread;
- simplifying the handling of the global data structure to two sums over its elements and a compression to remove sub-intervals which have converged, which is largely carried out in parallel;
- allowing the use of the same code, with no modifications or special paths, in single-thread mode.

All sub-intervals whose integrals have not converged are sub-divided at the same time in two or more sections, so that every processor carries out the same number of function evaluations. Hence, a satisfactory, albeit static, load balancing can be imposed. The process starts by sub-dividing the initial range in equal parts amongst all participating processors. A sub-integral is deemed to have converged when the sum of its error, normalized to the union of the sub-ranges in the active list of sub-ranges and of the sum of the errors of the converged sub-integrals, is less than the user-defined tolerance. A price has to be paid, in terms of an increase in the number of function evaluations with respect to the serial version of NAG Fortran Library: in all cases studied, this increase has never been found to affect significantly the overall efficiency.

In practice, this approach worked extremely well. First, in all cases studied, the new code in single-thread mode proved at least as fast (in the case of the first integral of Figure 3 twice as fast) as the serial NAG Library code. Then, linear or quasi-linear speed-ups for a range of integrands have been achieved.

Figure 3 shows the speed-ups observed for the two integrals:

$$\int_0^{10} \sum_{k=1}^{10} \cos[1000k\cos(x)]dx \quad \text{and} \quad \int_0^1 \sum_{k=1}^{10} |x^2 + (k+1)x - 2|^{-1/5}dx,$$

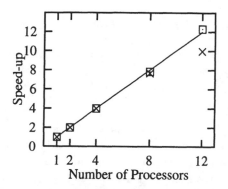

Fig. 3. D01AUF: speed-up for two integrals (see text).
□: first integral   × : second integral.

using the 30/61 Gauss–Kronrod rules for the first, the 10/21 rule for the second integral. Other rules were also used for both integrals, but they are not shown here; as they required more function evaluations, a slightly better speed-up was obtained, though at the cost of extra CPU time.

In single-thread mode, the first integral requires about four seconds, the second about one second. Neither integral involves expensive function evaluations; while the first requires sub-division across the whole range of integration, the second involves sub-divisions concentrated around a discrete number of cusps.

Further, though marginal, increases in performance could be obtained by handling the active list as the union of disjoint lists, one for each of the participating processors; the global step would then be required only to distribute evenly the set of sub-intervals over all participating processors.

3   Conclusions

Shared memory technology is becoming increasingly widespread. It poses new challenges to library developers, but on the other side it provides new opportunities.

- Scalability and efficiency can be achieved, at least in some major areas.
- However, a simple approach to parallelization based on the use of low-level parallel kernels, such as the BLAS, is not guaranteed to provide gains in all cases.
- Algorithms may need to be re-organized, perhaps new ones developed.
- Issues of portability can most likely be successfully tackled. Compiler directives, in particular PCF directives, could allow the maintenance of single-source codes for a spectrum of target systems, inluding *both* serial and shared memory systems.
- The effects of cache and data access and data dependency are very important and need to be addressed. A proper utilization of the cache is paramount for

performance and scalability. In particular, cache thrashing and/or conflicts can degrade performance very markedly.

- Algorithms with explicit parallelism need extensive revision. However, the case of the quadrature routines has shown that this does not necessarily imply an increase of code complexity; the availability of parallelism may actually *reduce* it.

Quite clearly, many of these considerations are beyond the interest and, perhaps, abilities of many code developers. The performance gap between library quality and directly written software can be much more considerable than for serial systems. The expertise required to generate highly efficient code may not be available. But this is just the expertise and 'added-value' which is encapsulated in quality numerical libraries: in our opinion, in all but the most sophisticated instances, the development of numerical software for this class of machines will rely on the availability of quality libraries of routines. This would have a two-fold benefit: it would provide a relatively inexpensive way to high-performance computing particularly in terms of software development costs; but also it would allow an *accessible* route to parallel computing for a much wider community of users than currently.

4 Acknowledgement

This work was partly supported by the Danish Natural Science Research Council through a grant for the EPOS Project (Efficient Parallel Algorithms for Optimization and Simulation).

The author wishes to thank UNI•C for their hospitality and help during his visits there, and in particular Jerzy Wasniewski for his continuous advice.

References

1. E. Anderson, Z. Bai, C. Bischof, J. Demmel, J. Dongarra, J. Du Croz, A. Greenbaum, S. Hammarling, A. McKenney, S. Ostrouchov, and D. Sorensen: LAPACK Users' Guide, Release 2.0. SIAM, Philadelphia, 1995.
2. J.J. Dongarra, J.J. Du Croz, I.S.Duff, S. Hammarling: A Set of Level 3 Basic Linear Algebra Subprograms. ACM Trans. Math, Softw., 16 pp. 1-17, 1990.
3. J.J. Dongarra, J.J. Du Croz, I.S.Duff, S. Hammarling: Algorithm 679: A Set of Level 3 Basic Linear Algebra Subprograms. ACM Trans. Math, Softw., 16 pp. 18-28, 1990.
4. NAG Fortran Library Manual, Mark 17. The Numerical Algorithms Group Ltd, Oxford, 1995.
5. NAG Numerical PVM Library Manual, Release 1. The Numerical Algorithms Group Ltd, Oxford, 1995.
6. R. Piessens, E. De Doncker-Kapenga, E. Überhuber, D. Kahaner: QUADPACK, A Subroutine Package for Automatic Integration. Springer-Verlag, 1983.
7. S.Salvini, J.Waśniewski: Experiences in Developing Numerical Subprograms on a Parallel, Shared Memory Computer. NAG Technical Report TR5/96, Oxford, 1996. Also in UNI•C Technical Report UNIC-96-04, Copenhagen, 1996.

8. S.Salvini, J.Waśniewski: Linear Algebra Subprograms on Shared Memory Computers: Beyond LAPACK. PARA'96 Workshop, 1996.
9. Silicon Graphics Computer Systems, "POWER CHALLENGE Supercomputing Servers", Silicon Graphics Computer Systems, Marketing Dept, Supercomputing Systems Div., 485 Central Avenue, Mountain View, CA 9043, USA, 1994.

Linear Algebra Subprograms on Shared Memory Computers: Beyond LAPACK

Stefano Salvini[1] and Jerzy Waśniewski[2]

[1] The Numerical Algorithms Group Ltd, Wilkinson House, Jordan Hill Road, Oxford
OX2 8DR, United Kingdom
E-mail: stef@nag.co.uk
[2] Computing Centre for Reasearch and Education, UNI•C, DTU, Bldg. 304,
DK-2800 Lyngby, Denmark
E-mail: jerzy.wasniewski@uni-c.dk

Abstract. This paper discusses the implementation of LAPACK routines on a cache-based, shared memory system. The shortcomings of an approach which relies on parallelized BLAS are illustrated in the cases of LU, Cholesky and QR factorizations. An alternative approach to these factorization routines, exploiting explicit parallelism at a higher level in the code, is reported: this provides higher scalability and efficiency in all cases studied. Issues of portability were addressed by using standard Fortran 77 and PCF compiler directives in all codes.

1 Introduction

LAPACK is a collection of routines available in the public domain for the solution of dense and banded linear algebra problems [1]. LAPACK routines have been ported efficiently and reliably to a wide variety of platforms and, because of their quality and range, have become accepted as a *de facto* standard for linear algebra computations.

Blocked algorithms are employed in LAPACK routines to improve data access and re-use and to allow the use of Level-3 BLAS (Basic Linear Algebra Subprograms [2]) to perform the bulk of the computation. The BLAS serve as a portability base for the efficient execution of the LAPACK routines: many hardware vendors offer highly tuned implementations of the BLAS on their systems, which the LAPACK routines can be linked to. This approach includes parallelized implementations of the BLAS for shared memory systems, which allow the essentially sequential higher level LAPACK routines to take advantage of multiple processors on such systems.

This paper discusses how the code for some linear algebra routines – specifically the routines for LU, Cholesky and QR factorization – was rewritten using an alternative approach to achieve a more effective use of parallelism than standard LAPACK on an SGI Power Challenge [5], [3], [4].

2 LAPACK and Shared Memory Systems

It could be expected that parallelized, efficient BLAS would result in high levels of performance for the LAPACK routines. However, the results obtained on the SGI Power Challenge showed only modest levels of parallel performance. For example, for a square matrix of order 1000, on 8 processors, the performances achieved for LU, Cholesky and QR factorizations were 16%, 27% and 23%, respectively, of the theoretical peak speed (see section 4 for more details).

A number of factors, observed for all the routines reported here, contribute to this relative lack of performance. Indeed, some of these factors are *intrinsic* to a cache-based type of architecture and are very likely to be observable on other similar systems.

Overheads are associated with the creation and subsequent destruction of parallel threads: caches needs to be loaded, etc. These costs may be not negligible with respect to the overall computational costs, though their relative importance decreases with increasing problem size. Each call to a BLAS opens a parallel region and closes it on exit. For example, the LU factorization of an $n \times n$ matrix, using a block-size b involves $2\frac{n}{b}$ calls to Level-3 BLAS and n calls to Level-2 BLAS, when row interchanges are ignored.

The factorization of individual blocks (sub-matrices) is carried out mostly by using Level-2 BLAS. The 'density of operations' per memory reference is less favourable than for the updating of the rest of the matrix which relies on Level-3 BLAS; hence the factorization of blocks tends to be carried out in serial or quasi-serial mode, with only very modest gains, if any, in parallelism. This causes a 'quasi-serial' bottleneck whose impact increases with the number of processors. In the example of the LU factorization above, if the current active sub-matrix is $n_A \times n_A$ and the block size is $b \ll n_A$, the number of operations required by the factorization of the current diagonal block is $\sim n_A b^2$, whilst the number of operations required by the updating of the trailing submatrix is $\sim 2n_A^2 b$. For n_p processors, the ratio $\frac{b n_p}{2n_A}$ denotes the fraction of floating-point operations performed in 'quasi-serial' mode. Furthermore, the use of Level-2 BLAS causes the performance of the factorization of the current block to be actually *lower* than that of the Level-3 BLAS used to update the trailing sub-matrix. If $b = 64$ (the LAPACK default), $n_A = 1000$, $n_p = 8$, the factorization of the diagonal block would require at least 25% of the overall CPU time. As the factorization proceeds, that is as the size of the active sub-matrix decreases, this situation worsens. Moreover, increasing the number of processors, can only result in a marginal improvement to the overall performance.

Some cache-thrashing operations are unavoidable without abandoning the BLAS-based approach. For example, rows of matrices need to be interchanged in the LU factorization. Systems with a hierarchical memory structure are ill-suited to this type of operation and considerable penalties are incurred.

3 An Alternative Approach to Parallelism

A different type of approach has proved effective in achieving better levels of performance on shared memory systems.

All code was written using standard Fortran 77. To maximize portability, parallelism was described in terms of PCF compiler directives, which are likely to be available on other platforms.

Explicit parallelism was brought upwards, from the level of the BLAS into the linear algebra routines themselves. The code becomes longer and more complicated: however, *templates* were developed, and different routines could be generated by 'slotting in' the appropriate computational modules. The codes for the QR and LU factorizations reported in Section 4 have the same structure at the highest level where the parallelism is expressed; they differ only in the lower level single-processor routines which they call. As a matter of fact, the LU factorization routine was derived directly from the QR factorization routine. The user interfaces to the new routines were the same as for the corresponding LAPACK routines. Introducing explicit parallelism allowed a number of substantial modifications.

- Parallel thread creation and destruction take place only once on entry and exit, respectively, thus reducing overheads.
- The factorization of blocks is overlapped with the update of the matrix, thus removing at one stroke the quasi-serial bottleneck.
- *Serial* BLAS or BLAS-like routines can be used on each processor to carry out the required block operations.
- Both static and dynamic load balancing are possible and indeed were studied. Static load balancing can be either fixed throughout the computation or adjusted at each block-step: the numbers of column- or row-blocks the processors manipulate are fixed, and only synchronization barriers are required. Dynamic load balancing can be generated using the appropriate parallel DO constructs with dynamic scheduling, or using critical sections.
- The effects of cache-thrashing operations can be kept to a minimum. For example, in the LU factorization, column blocks are manipulated independently from each other in each processor: thus, row interchanges and updates of the block-column are performed one after the other with maximum data re-use. Row interchanges in the column on the *left* of the current pivot block-columns are carried out *after* the computation is completed on each column-block independently, again maximizing data re-use.
- A constant block-size is *not* required. Block-sizes may be varied during the computation for maximum efficiency. Indeed, the most effficient versions for QR and LU factorizations reported below used a simple scheme to vary the block-size.

4 Results

Benchmark results for Cholesky, LU and QR factorizations for 1, 2, 4 and 8 processors and for square matrices of sizes ranging from 100 to 1000 are reported here. All tests were run on the SGI Power Challenge at UNI•C, Denmark. The machine has 16 processors, each running at a clock speed of 75 MHz and capable of delivering 4 floating-point operations per cycle, with a theoretical peak speed of 300 Mflops/sec.

The SGI Power Challenge at UNI•C is a national facility, intensely used by a large community of users spread across Denmark and elsewhere. Therefore, we could not gain single-user access to the system, a pre-requisite for accurate and highly reproducible benchmarks, without affecting the overall availability of the system but had to carry out all tests on a *fully loaded* machine. This implies that all results reported below must be viewed as *lower bounds* on the efficiency achievable, particularly for the cases involving four or more processors.

Considerable gains in scalability and efficiency have been achieved for LU, Cholesky and QR factorization for *all* matrix sizes and *all* numbers of processors studied. The tables below compare the performance of the new routines (marked 'New') against the performance of the LAPACK routines available in the SGI complib_sgimath_mp mathematical library (marked 'LAPACK').

All codes reported below used a static load balancing strategy. Codes employing a dynamic load balancing approach were also developed and tested but their performance was found to be inferior, perhaps because of either or both of two reasons: the overheads associated with dynamic load balancing, which seem rather high on the SGI Power Challenge, and/or memory access competition between the processors.

n_p	1		2		4		8	
n	LAPACK	New	LAPACK	New	LAPACK	New	LAPACK	New
100	86	127	76	195	72	260	69	274
200	131	190	130	323	149	475	131	674
400	172	235	210	423	282	738	281	1161
500	171	243	243	449	334	799	313	1271
1000	174	246	340	497	522	930	559	1594

Table 1. QR factorization performance in Mflops/sec

Table 1 reports the performances obtained using the new QR factorization code, whose interface and specification are wholly equivalent to the LAPACK routine DGEQRF. A static load-balancing approach is used and the block-size is allowed to vary with the number of processors and the number of rows and columns of the active submatrix as the computation progresses. Some new auxiliary routines for the computation of block reflectors were also developed to use Level-3 BLAS-like routines rather than Level-2 BLAS, hence increasing performance. Smaller, though significant, gains were also obtained by writing Fortran 77 BLAS-like routines, tuned to the elongated matrices used in the factorization. For square matrices of order 1000 on 8 processors, the new QR code achieves an efficiency of $> 75\%$ compared to the efficiency of DGEMM for the same matrix sizes (about $2000 - 2100$ Mflops/sec): a higher relative efficiency is achieved for a smaller number of processors. The increase in efficiency is more marked for smaller matrix sizes, which highlights the effects of the quasi-serial bottleneck due to the computation of the block reflectors.

n_p	1		2		4		8	
n	LAPACK	New	LAPACK	New	LAPACK	New	LAPACK	New
100	9	151	40	154	41	177	40	170
200	151	205	96	282	102	396	105	461
400	202	239	182	366	212	592	208	908
500	210	243	205	398	249	710	248	1024
1000	180	231	266	449	358	891	373	1351

Table 2. LU factorization: performance in Mflops/sec

Table 2 shows the results for the LU factorization of dense, square random matrices: for these matrices, random row interchanges take place at each stage of the factorization. In the LAPACK routine DGETRF, this involves accessing rows of the matrix, which results in some cache-thrashing. This effect becomes more noticeable as the problem size increases. Experiments were carried out to estimate the degradation in performance thus caused using positive-definite matrices, requiring no row interchanges, and special matrices which always require interchanges with the bottom row. For $n = 1000$ and 8 processors, DGETRF returned 541 and 446 MFlops/sec for positive-definite and the special matrices described above, respectively. These figures must be compared to the 373 Mflops/sec reported in Table 2. The new code, was not only much faster in all cases than DGETRF, but the effects of row interchanges were very much less marked, almost negligible. As in the case of the QR factorization, a static load balancing strategy with varying block size was used: in fact, the code was directly derived from the QR factorization code by substituting the appropriate lower level numerical kernels.

The results for the Cholesky factorization codes are shown in Table 3. This code did not employ the same template as LU and QR factorizations: a static load balancing, fixed block size strategy was employed. A code based on the QR and LU factorization template with the modifications required by the need to access only either the upper or lower triangle of the matrix, returned performance figures slightly lower than those reported.

n_p	1		2		4		8	
	Upper triangular $A = U^T U$							
n	LAPACK	New	LAPACK	New	LAPACK	New	LAPACK	New
100	66	153	67	138	71	166	72	174
200	115	214	130	270	151	386	160	473
400	162	254	221	396	285	654	326	956
500	175	261	252	414	339	715	387	1111
1000	199	261	348	493	521	916	633	1412
	Lower triangular $A = LL^T$							
n	LAPACK	New	LAPACK	New	LAPACK	New	LAPACK	New
100	65	161	58	130	60	135	56	123
200	120	223	133	267	143	352	142	372
400	178	260	253	401	327	642	365	884
500	193	264	291	425	406	722	420	1076
1000	223	243	349	470	564	886	650	1258

Table 3. Cholesky factorization: performance in Mflops/sec

5 Conclusions

The results obtained confirm on one side the limitations of the parallel BLAS-based LAPACK approach, on the other the validity of the alternative, explicitly parallelized, approach on a cache-based shared memory platform, the SGI Power Challenge. Similar considerations are expected to hold good for other systems in the same class.

All codes were written for easy portability to other systems by using standard Fortran 77 and PCF directives. Future work will be directed to port the existing routines to other platforms and to extend the work carried out to a wider selection of routines, both in linear algebra and in other areas.

6 Acknowledgement

This work was partly supported by the Danish Natural Science Research Council through a grant for the EPOS Project (Efficient Parallel Algorithms for Optimization and Simulation).

References

1. E. Anderson, Z. Bai, C. Bischof, J. Demmel, J. Dongarra, J. Du Croz, A. Greenbaum, S. Hammarling, A. McKenney, S. Ostrouchov, and D. Sorensen: LAPACK Users' Guide, Release 2.0. SIAM, Philadelphia, 1995.
2. J.J. Dongarra, J.J. Du Croz, 1.S.Duff, S. Hammarling: A Set of Level 3 Basic Linear Algebra Subprograms. ACM Trans. Math, Softw., 16 pp. 1-17, 1990.

3. S.Salvini, J.Waśniewski: Experiences in Developing Numerical Subprograms on a Parallel, Shared Memory Computer. NAG Technical Report TR5/96, Oxford, 1996. Also in UNI●C Technical Report UNIC-96-04, Copenhagen, 1996.
4. S.Salvini: Numerical Libraries on Shared Memory Computers, PARA'96 Workshop, 1996.
5. Silicon Graphics Computer Systems, "POWER CHALLENGE Supercomputing Servers", Silicon Graphics Computer Systems, Marketing Dept, Supercomputing Systems Div., 485 Central Avenue, Mountain View, CA 9043, USA, 1994.

Integration of Partitioned Stiff Systems of Ordinary Differential Equations

Stig Skelboe

Department of Computer Science, University of Copenhagen,
Universitetsparken 1, DK-2100 Copenhagen, Denmark
e-mail: stig@diku.dk

Abstract. Partitioned systems of ordinary differential equations are in qualitative terms characterized as monotonically max-norm stable if each sub-system is stable and if the couplings from one sub-system to the others are weak.

Each sub-system of the partitioned system may be discretized *independently* by the backward Euler formula using solution values from the other sub-systems corresponding to the previous time step. The monotone max-norm stability guarantees this discretization to be *stable*. This so-called decoupled implicit Euler method is ideally suited for parallel computers. With one or several sub-systems allocated to each processor, information only has to be exchanged after completion of a step but not during the solution of the nonlinear algebraic equations.

This paper considers strategies and techniques for partitioning a system into a monotonically max-norm stable system. It also presents error bounds to be used in controlling stepsize, relaxation between sub-systems and the validity of the partitioning. Finally a realistic example is presented.

1 Introduction

A previous paper introduced the decoupled implicit Euler method [1] for the parallel integration of partitioned systems of stiff ordinary differential equations. The existence of a global error expansion was proved under very general choice of stepsize, thus permitting the use of Richardson extrapolation. A sufficient condition for stability of the discretization was given. This condition is called monotonic max-norm stability, and it guarantees contractivity.

Define a system of ordinary differential equations,

$$y' = f(t, y), \ y(t_0) = y_0 \text{ and } t \geq t_0 \tag{1}$$

where $y : R \to R^S$, f is Lipschitz continuous in y and $f : R \times R^S \to R^S$. Stable systems of differential equations are considered stiff when the step size of the discretization by an *explicit* integration method is limited by stability of the discretization and not by accuracy. Efficient numerical integration of stiff systems therefore require *implicit* integration methods.

Let the original problem (1) be partitioned as follows,

$$\begin{pmatrix} y_1' \\ y_2' \\ \vdots \\ y_q' \end{pmatrix} = \begin{pmatrix} f_1(t,y) \\ f_2(t,y) \\ \vdots \\ f_q(t,y) \end{pmatrix}, \quad y = \begin{pmatrix} y_1 \\ y_2 \\ \vdots \\ y_q \end{pmatrix}, \quad y(t_0) = \begin{pmatrix} y_{1,0} \\ y_{2,0} \\ \vdots \\ y_{q,0} \end{pmatrix} \tag{2}$$

where $y_r : R \to R^{s_r}$, $f_r : R \times R^S \to R^{s_r}$ and $\sum_{i=1}^{q} s_i = S$. When necessary, the partitioning of y will be stated explicitly like in $f_r(t, y_1, y_2, \ldots, y_q)$.

A linear problem of dimension S, $y' = Ay$ can be partitioned in a similar way by partitioning A into sub-matrices A_{rj} of dimensions $s_r \times s_j$,

$$\begin{pmatrix} y_1' \\ y_2' \\ \vdots \\ y_q' \end{pmatrix} = \begin{pmatrix} A_{11} & A_{12} & \cdots & A_{1q} \\ A_{21} & A_{22} & \cdots & A_{2q} \\ \vdots & \vdots & \cdots & \vdots \\ A_{q1} & A_{q2} & \cdots & A_{qq} \end{pmatrix} \begin{pmatrix} y_1 \\ y_2 \\ \vdots \\ y_q \end{pmatrix}, \quad y(t_0) = \begin{pmatrix} y_{1,0} \\ y_{2,0} \\ \vdots \\ y_{q,0} \end{pmatrix} \tag{3}$$

The following stability condition introduced in [2] plays a crucial role for the stability of decoupled implicit integration methods.

Definition: Monotonic max-norm stability

The partitioned system (2) is said to be monotonically max-norm stable if there exist norms $\| \cdot \|_r$ such that

$$\|u_r - v_r + \lambda [f_r(t,u) - f_r(t,v)]\|_r \geq \|u_r - v_r\|_r + \lambda \sum_{j=1}^{q} a_{rj}(t,u,v)\|u_j - v_j\|_j \tag{4}$$

for all $t \geq t_0$, $\lambda \leq 0$, $u, v \in \Omega_t$ where $\Omega_t \subseteq R^S$ and the following condition holds for the logarithmic max-norm $\mu_\infty(\cdot)$ of the q × q matrix (a_{rj})

$$\mu_\infty\left[(a_{rj}(t,u,v))\right] \leq 0 \tag{5}$$

□

The norms used in (4) can be related as follows using a common norm $\| \cdot \|$

$$d_j\|y_j\|_j \leq \|y_j\| \leq \tilde{d}_j\|y_j\|_j, \quad j = 1, 2, \ldots, q$$

The norms are chosen to minimize $\mu_\infty[(a_{rj})]$ in order to fulfil (5) which states that (a_{rj}) should be diagonally dominant with negative diagonal elements.

The paper [2] includes further results about monotonic max-norm stability including possible choices of (a_{rj}). For the linear problem, (3), the following values of a_{rj} can be used,

$$a_{rr} = \mu_r(A_{rr}), \quad a_{rj} = \frac{\tilde{d}_j}{d_r}\|A_{rj}\|, \quad r \neq j \tag{6}$$

This choice of (a_{rj}) for a linear problem is easily related to the qualitative definition of monotonic max-norm stability, each sub-system is stable ($\mu_r(A_{rr}) < 0$)

and the couplings from one sub-system to the others are weak (diagonal dominance of (a_{rj})). Monotonic max-norm stability admits arbitrarily stiff problems.

The decoupled implicit Euler method is defined as follows where the r'th sub-system is discretized by the backward Euler formula [1]:

$$y_{r,n} = y_{r,n-1} + h_{r,n} f_r(t_{r,n}, \tilde{y}_{1,n}, \ldots, \tilde{y}_{r-1,n}, y_{r,n}, \tilde{y}_{r+1,n}, \ldots, \tilde{y}_{q,n}), \qquad (7)$$

where $n = 1, 2, \ldots$, $t_{r,n} = t_0 + \sum_{j=1}^{n} h_{r,j}$ and the variables $\tilde{y}_{i,n}$ are convex combinations of values in $\{y_{i,k} \mid k \geq 0\}$ for $i \neq r$. The method is called "decoupled" because the algebraic system resulting from the discretization is decoupled into a number of independent algebraic problems. The decoupled implicit Euler formula can be used as the basis of parallel methods where (7) is solved independently and in parallel for different r-values. The method can be used in multirate mode with $h_{r,n} \neq h_{j,n}$ for $r \neq j$, and the multirate formulation can be used in a parallel waveform relaxation method [3].

Theorems 4 and 5 in [2] assure the stability of the discretization by (7) of a monotonically max-norm stable problem and the convergence of waveform relaxation. The stability condition poses no restrictions on the choice of $h_{r,n}$.

The convexity of $\tilde{y}_{i,n}$ is necessary for the stability of the decoupled implicit Euler method. All convex combination coefficients might in general be different, but the convex combinations would typically be either a zero-order extra- or interpolation:

$$\tilde{y}_{i,n} = y_{i,k} \text{ or } y_{i,k+1}, \quad \text{where } t_{r,n} \in (t_{i,k}, t_{i,k+1}]$$

or a first-order interpolation:

$$\tilde{y}_{i,n} = \frac{t_{i,k+1} - t_{r,n}}{t_{i,k+1} - t_{i,k}} y_{i,k} + \frac{t_{r,n} - t_{i,k}}{t_{i,k+1} - t_{i,k}} y_{i,k+1}$$

where $t_{r,n} \in [t_{i,k}, t_{i,k+1}]$.

2 Accuracy

The error of the decoupled implicit Euler formula can be bounded as follows. Consider the Taylor series expansion,

$$\beta_r(t_n) \equiv y_r(t_n) - y_r(t_{n-1}) - h_{r,n} f_r(t_n, y_1(t_n), \ldots, y_r(t_n), \ldots, y_q(t_n))$$
$$= -\frac{h_{r,n}^2}{2} y_r^{(2)}(t_n) + \frac{h_{r,n}^3}{3!} y_r^{(3)}(t_n) - \ldots$$

and the decoupled implicit Euler formula,

$$y_{r,n} - y_r(t_{n-1}) + h_{r,n} f_r(t_n, \tilde{y}_n^r) = 0$$

where $\tilde{y}_n^r = (\tilde{y}_{1,n}, \ldots, \tilde{y}_{r-1,n}, y_{r,n}, \tilde{y}_{r+1,n}, \ldots, \tilde{y}_{q,n})$.

Assume that the monotonic max-norm stability condition (4), (5) is fulfilled with $y_r(t_n), \tilde{y}_n^r \in \Omega_{t_n}$. Then subtraction leads to,

$$\|\beta_r(t_n)\|_r = \|y_r(t_n) - y_{r,n} - h_{r,n}[f_r(t_n, y(t_n)) - f_r(t_n, \tilde{y}_n^r)]\|_r$$
$$\geq [1 - h_{r,n}a_{rr}(t_n, y(t_n), \tilde{y}_n^r)]\|y_r(t_n) - y_{r,n}\|_r$$
$$-h_{r,n}\sum_{j \neq r} a_{rj}(t_n, y(t_n), \tilde{y}_n^r)\|y_j(t_n) - \tilde{y}_{j,n}\|_j$$

The error bound for the decoupled implicit Euler formula is then,

$$\|y_r(t_n) - y_{r,n}\|_r$$
$$\leq [1 - h_{r,n}a_{rr}(t_n)]^{-1}(\|\beta_r(t_n)\|_r + h_{r,n}\sum_{j \neq r} a_{rj}(t_n)\|y_j(t_n) - \tilde{y}_{j,n}\|_j) \qquad (8)$$

The matrix $(a_{ij}(t))$ is defined such that $a_{ij}(t) = \sup_{u,v \in \Omega_t} a_{ij}(t, u, v)$.

A similar bound can be established for the classical implicit Euler formula expressed as follows in a notation analogue to (7),

$$y_{r,n} = y_{r,n-1} + h_{r,n} f_r(t_{r,n}, y_{1,n}, \ldots, y_{r-1,n}, y_{r,n}, y_{r+1,n}, \ldots, y_{q,n})$$

Using the monotone max-norm stability condition, we obtain

$$\beta(t_n) \geq \left\| [I - h_n(a_{ij}(t_n))] \begin{pmatrix} \|y_1(t_n) - y_{1,n}\|_1 \\ \|y_2(t_n) - y_{2,n}\|_2 \\ \vdots \\ \|y_q(t_n) - y_{q,n}\|_q \end{pmatrix} \right\|_\infty$$
$$\geq (1 - h_n\mu_\infty[(a_{ij}(t_n))]) \max_r \|y_r(t_n) - y_{r,n}\|_r$$

where $\beta(t) = \max_r \|\beta_r(t)\|_r$, and the matrix $(a_{ij}(t))$ is defined such that

$$\mu_\infty((a_{ij}(t))) = \sup_{u,v \in \Omega_t} \mu_\infty((a_{ij}(t, u, v)))$$

The analogue bound of (8) for the classical implicit Euler formula is then

$$\|y_r(t_n) - y_{r,n}\|_r \leq [1 - h\mu_\infty((a_{ij}(t_n)))]^{-1}\beta(t_n) \qquad (9)$$

Both (8) and (9) are valid for all values of stepsizes but may be most interesting and useful for values where $\beta(t_n) \approx \|\frac{h_n^2}{2}y''(t_n)\|$.

The main difference between (8) and (9) is the sum term in (8). The local truncation error $\beta(t)$ is $O(h^2)$. The order of the sum term in (8) including $h_{r,n}$ is $O(h^2)$ if $\tilde{y}_{j,n} = y_{j,n-1}$ such that $\|y_j(t_n) - \tilde{y}_{j,n}\|_j$ is $O(h)$. If the off-diagonal elements of (a_{ij}) are very small, the sum term of (8) may be negligible and the error bound of the decoupled implicit Euler formula is almost equal to error bound the classical implicit Euler formula.

If the sum term is not negligible, the value of $\|y_j(t_n) - \tilde{y}_{j,n}\|_j$ can be reduced from $O(h)$ to $O(h^2)$ by using extrapolation for the computation of $\tilde{y}_{j,n}$,

$$\tilde{y}_{j,n} = y_{j,n-1} + \frac{h_n}{h_{n-1}}(y_{j,n-1} - y_{j,n-2}) \tag{10}$$

In the extrapolation formula, the stepsize is assumed to be the same for all sub-systems.

As mentioned in the Introduction, monotonic max-norm stability of the problem no longer guarantees stability of the decoupled implicit Euler formula when extrapolation is used since $\tilde{y}_{j,n}$ computed by (10) is not a convex combination of previously computed values. However, the discretization may still be stable.

An alternative technique for reducing the value of $\|y_j(t_n) - \tilde{y}_{j,n}\|_j$ to $O(h^2)$ is to use relaxation, i.e. solve (7) for $r = 1, 2, ..., q$, using $\tilde{y}_{j,n} = y_{j,n-1}$. The next sweep uses $\tilde{y}_{j,n} = y_{j,n}^{(1)}$ where $y_{j,n}^{(1)}$ is the result of the first sweep, etc. Relaxation does not compromise stability, and besides the monotonic max-norm stability guarantees convergence of the process. However, relaxation is very expensive compared to extrapolation.

When the relaxation process has reached convergence, the decoupled implicit Euler formula is effectively the classical implicit Euler formula, and any standard error estimation technique can be used. When relaxation does not improve the result obtained using the decoupled implicit Euler formula, possibly by employing the extrapolation (10), we have the same opportunity for error estimation. The error bound (8) may be pessimistic, and it is difficult and expensive to estimate the necessary terms.

The organization of the decoupled Euler formula (7) may affect the resulting accuracy. A "Jacobi type" organization as described in connection with relaxation above, is ideal for parallel execution. If parallel execution is not of importance, a "Gauss-Seidel type" organization may provide higher accuracy, especially with careful selection of the sequence of computation.

3 Partitioning criteria

The main objective of the partitioning of a system into sub-systems is to obtain as small sub-systems as possible while retaining the stability properties of the classical implicit Euler discretization. The sub-systems should be small for the sake of computational efficiency.

If possible, the partitioned system should be monotonically max-norm stable to permit flexible multirate discretizations. Except for simple nonlinear problems [2], it is impossible to prove that a partitioning is monotonically max-norm stable. However, this stability condition is only sufficient and not necessary, and therefore a discretization may exhibit the properties of a monotonically max-norm stable partitioning without fulfilling the condition.

The decoupled implicit Euler formula can be considered a generalization of the classical implicit Euler formula so therefore the ultimate check of the quality of a partitioning is to compare the error of the classical implicit Euler formula with the error of the decoupled implicit Euler formula. However, this is expensive in practice, it does not admit dynamic repartitioning and it does not give any hints about how to improve a partitioning.

An obvious approach is to study linearizations of (1) at selected points along the solution trajectory,

$$x' = A_n x \quad \text{where} \quad A_n = \left.\frac{\partial f}{\partial y}\right|_{(t_n, y_n)}$$

A partitioning of the linear problem can be specified by a block diagonal matrix $D = \text{diag}(A_{11}, A_{22}, ..., A_{qq})$ where the dimensions of the diagonal blocks are $s_1, s_2, ..., s_q$ (cf. (3)). The off-diagonal part of A_n is then defined as $B = A_n - D$ and the decoupled implicit Euler formula can be expressed as follows (the "Jacobi" formulation),

$$x_n = x_{n-1} + h_n(Dx_n + B\tilde{x}_n) \tag{11}$$

The "Gauss-Seidel" formulation is obtained by removing the lower triangular part from B and including it in D.

With $\tilde{x}_n = x_{n-1}$, (11) can be expressed as $x_n = M_0 x_{n-1}$ where $M_0 = (I - h_n D)^{-1}(I + h_n B)$.

With $\tilde{x}_n$ being expressed by the extrapolation formula (10), the decoupled implicit Euler formula is

$$x_n = x_{n-1} + h_n(Dx_n + B(x_{n-1} + h_n/h_{n-1}(x_{n-1} - x_{n-2})))$$

This can be reformulated as $(x_n, x_{n-1})^T = M_a (x_{n-1}, x_{n-2})^T$ where

$$M_a = \begin{pmatrix} (M_0 + h_n^2/h_{n-1}(I - h_n D)^{-1}B) & -h_n^2/h_{n-1}(I - h_n D)^{-1}B \\ I & 0 \end{pmatrix}$$

The corresponding matrix for the implicit Euler formula is then $M = (I - h_n A_n)^{-1}$ and both M_0, M_a and M are intended to approximate the matrix exponential $M_e = \exp(h_n A_n)$.

The large eigenvalues (close to one) of M_0, M_a and M determine the accuracy of the integration formula. The small eigenvalues (close to zero) determine the stability properties of the formula. A possible partitioning criterion is therefore that the large eigenvalues of M_0 and M_a are close to the corresponding eigenvalues of M while the small eigenvalues roll off in magnitude as fast or almost as fast as the small eigenvalues of M.

Another possible partitioning criterion is that $\|M_0 - M\|/\|M\|$ is small and that $\|h_n^2/h_{n-1}(I - h_n D)^{-1}B\|/\|M\|$ is small. However, the latter criterion has only proved marginally important in the experiments conducted in following example.

4 Example

The example is the mathematical model of the chemical reactions included in a three-dimensional transport-chemistry model of air pollution. The model is comprised of a system 32 nonlinear ordinary differential equations,

$$y'_i = P_i(y) - L_i(y)y_i, \quad i = 1, 2, ..., 32$$

The nonlinearities are mainly products, i.e. P_i and L_i are typically a sum of terms of the form, $c_{ilm}y_l y_m$ and $d_{il}y_l$, respectively, for $l, m \neq i$.

The chemistry model is replicated for each node of the spatial discretization of the transport part. The numerical solution of a system of 32 ODEs is not very challenging as such, but the replication results in hundreds of thousands or millions of equations, and very efficient numerical solution is crucial. The problem and a selection of solution techniques employed so far are described in [4] and [5].

The system of ODEs is very stiff so therefore implicit integration schemes are required. The resulting nonlinear algebraic problem is the main computational task involved in advancing the numerical solution one time step. A method based on partitioning the system called Euler Backward Iterative method is described in [4]. It can be characterized as a discretization by the implicit Euler formula with block Gauss-Seidel iteration for the solution of the algebraic equations of the discretization. The partitioning in [4] is used as the basis of this example and for the results in [6].

An ideal partitioning would involve "blocks" of size one, i.e. $s_i = 1$ for all i. Such a partitioning is not viable for this problem. The paper [4] identifies a total of 12 out of 32 equations which should be solved in groups of 4, 4, 2 and 2. With a proper ordering of the equations, this corresponds to a partitioning with $s_1 = 4, s_2 = 4, s_3 = 2, s_4 = 2, s_5, s_6, ..., s_{24} = 1$.

In [4] the decoupled Euler formula is relaxed until convergence to obtain the equivalent of the classical implicit Euler formula. The results in [6] are obtained from (7) with a "Gauss-Seidel" scheme and just one sweep of the decoupled implicit Euler formula, i.e. no relaxation. Timing results in [6] show good efficiency of this approach.

Figure 1 shows the errors obtained using the classical implicit Euler formula (solid line) and the decoupled implicit Euler formula with two different partitionings. One partitioning where the 12 equations treated separately in [4] are lumped together into a sub-system of 12 equations (dashed line) corresponding to a partitioning with $s_1 = 12, s_2, s_3, ..., s_{21} = 1$. Another partitioning with a block of 10 equations being a subset of the block of 12, the rest being equal to 1, $s_1 = 10, s_2, s_3, ..., s_{23} = 1$ (dotted line). The extrapolation formula (10) is not used in these examples.

The numerical integration is performed with a fixed time step of $h = 900$ sec. A reference solution is computed using a variable stepsize variable order (max order $= 6$) implementation of the backward differentiation formulas [7] with a bound on the relative local truncation error of 10^{-6}. The errors presented in the

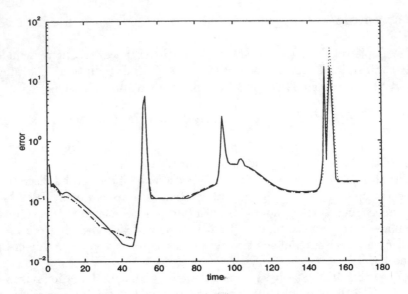

Fig. 1. Integration error for implicit Euler ($s_1 = 32$) – , and decoupled Euler with $s_1 = 12$ - - and $s_1 = 10$ ..... Time unit is 900 sec = 15 min.

figures are the maximum relative errors measured component wise (the values of the components vary widely in magnitude).

Figure 2 shows the difference in integration errors between the implicit Euler formula and the decoupled implicit Euler formula when using the extrapolation formula (10) and not using the extrapolation formula. The difference is typically reduced by an order of magnitude when extrapolation is used. The analogue of Figure 1 presenting the errors of the classical backward Euler formula and the decoupled Euler formula with $s_1 = 12$ and extrapolation does not show any visual difference between the errors of the two methods.

It may be argued that a stepsize leading to a relative error of 0.1 in the smooth parts of the solution and more than 10 in the non-smooth part is far too large. However, from the point of a practical simulation it may be adequate. The very large errors occur in connection with sun rise and sun set where the rate of photo chemical processes changes very abruptly. The errors are probably mainly caused by a simplified modeling of this phenomenon, and the local errors do not propagate to the global error.

When the purpose is to evaluate different integration formulas, the stepsize is appropriate since it tests the methods for both the smooth and the non-smooth parts of the solution.

The linearized analysis presented in the previous section was used in selecting the subset of 10 components for the partitioning with $s_1 = 10$. The initial selection was guided by the values of the off-diagonal elements (the elements of B in a partitioning, cf. (6)). An equation j was chosen as candidate for being taken out of the block of 12 equations if $\|\alpha_{ij}/\alpha_{ii}\| < 1$ for $i = 1, 2, ..., j-1, j+1, ..., 12$

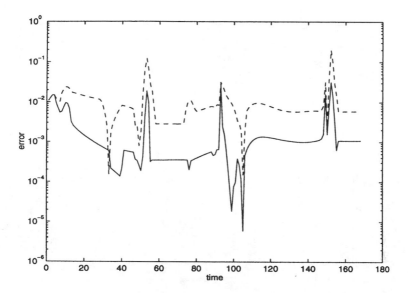

Fig. 2. Integration error difference, implicit Euler ($s_1 = 32$) minus decoupled Euler $s_1 = 12$ using extrapolation –, and implicit Euler minus decoupled Euler $s_1 = 12$ *not* using extrapolation - - . Time unit is 900 sec = 15 min.

(α_{ij} is element ij of A_n). The corresponding M_0 and M_a matrices were then constructed and the eigenvalues computed.

Figure 3 shows an eigenvalue analysis at time step $n = 24$. The accuracy of the partitionings are seen to be good for large eigenvalues with the partitioning with $s_1 = 12$ being slightly better than the partitioning with $s_1 = 10$. The errors of the small eigenvalues roll off just as fast as the magnitude of the eigenvalues. The values of $\|M_0 - M\|/\|M\|$ is 3.236 and 54.877 for $s_1 = 12$ and $s_1 = 10$ respectively, but this difference does not appear to affect the difference in accuracy obtained with the different partitionings, cf. Figure 1.

The only significant difference between the two partitionings is seen in Figure 1 at time step 152. An eigenvalue analysis does not show any difference but $\|M_0 - M\|/\|M\|$ is 0.0014 and 4.44 for $s_1 = 12$ and $s_1 = 10$ respectively. Since the solution has a very strong transient at this point, the magnitude of M_0 probably becomes important.

The initial partitioning given in [4] is probably based on knowledge of the chemical reactions taking place. However, it is also characterized by including the equations with the largest diagonal elements of A_n in the blocks. The approach for partitioning a system is at present to a very large extent based on trial and error supported by some "common sense" guidelines and quick preliminary evaluation by linear analysis.

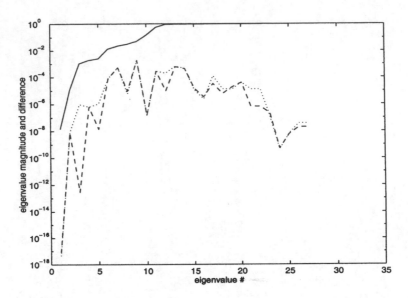

Fig. 3. Linearized system at time step 24 with $h_n = 900$. Sorted eigenvalues of M (implicit Euler) –, difference between eigenvalues of M and M_0 (decoupled Euler) for $s_1 = 12$ - - and $s_1 = 10$ .....

References

1. Skelboe, S., "Methods for parallel integration of stiff systems of ODEs", BIT (1992), vol. 32, pp. 689 – 701.
2. Sand, J. and Skelboe, S., "Stability of backward Euler multirate methods and convergence of waveform relaxation", BIT (1992), vol. 32, pp. 350 – 366.
3. Lelarasmee E., Ruehli A.E., and Sangiovanni-Vincentelli A.L., "The Waveform Relaxation Method for Time-Domain Analysis of Large Scale Integrated Circuits", IEEE Trans. Computer-Aided Design of Integrated Circuits and Systems, 1(1982), pp. 131-145.
4. Hertel, O., Berkowicz, R., Christensen, J. and Hov, Ø., "Test of two numerical schemes for use in atmospheric transport-chemistry models". Atmospheric Environment, 27A (1993), 2591-2611.
5. Gery, M. W., Whitten, G. Z., Killus, J. P. and Dodge, M. C., "A photochemical kinetics mechanism for urban and regional computer modeling". J. Geophys. Res., 94 (1989), 12925-12956.
6. Skelboe, S. and Zlatev, Z., "Exploiting the natural partitioning in the numerical solution of ODE systems arising in atmospheric chemistry". To appear in Springer Lecture Notes in Computer Science, Proceedings of the First Workshop on Numerical Analysis and Applications (WNNA-96), Rousse, Bulgaria, June 24-27, 1996.
7. Skelboe, S., "INTGR for the integration of stiff systems of ordinary differential equations", Report IT 9, March 1977, Institute of Circuit Theory and Telecommunication, Technical University of Denmark.

Direct and Large Eddy Simulations of Thermo-Convective Flows

Jens N. Sørensen, Steen Aa. Kristensen and Lasse K. Christensen

Department of Energy Engineering, Fluid Mechanics Section
Technical University of Denmark, DK-2800 Lyngby

Abstract. Transitional and and turbulent thermo-convective flows in a cylinder heated from below are studied numerically. The governing equations for mass, momentum and energy conservation are solved in vorticity-streamfunction-temperature variables utilizing the Boussinesq approximation and a sub-grid-scale turbulence model. The simulations are performed by a finite-difference code employing a combined 2nd and 4th order accurate spatial discretization. The resulting set of equations is factored into one-dimensional operators that are applied in alternating order. Calculations have been run on the CRAY C92 vector processor as well as on the SP2 parallel machine. The results show good performance, especially on the CRAY C92 where up to 550 Mflops were achieved.

1 Introduction

Since the beginning of this century, the convective flow in a semi-infinite horizontal fluid layer heated from below has been studied as a model example of a system undergoing hydrodynamic instability. From a theoretical as well as a numerical point of view the problem is a particularly simple and accesible case for studying the onset of turbulence and the formation of organized structures. The problem is usually referred to as the Raleigh-Bénard problem and has been largely explored in various regions of parameter space (for a review see e.g. [1]). The picture for other convection geometries, such as e.g. cylindrical containers, is not nearly so complete. A much broader understanding of convection in such geometries would, however, be useful from an application point of view, since the cylindrical geometry is largely employed in e.g. heat exchangers and hot-water tanks. In an effort to improve this situation, we have initiated a numerical study on thermo-convective flows in cylinder geometries of varying aspect ratios. The configuration may be thought of as an idealized hot-water tank.

1.1 The Physical Problem

Consider a closed vertical cylindrical container of radius R and height H, as shown in Fig. 1. At the top and bottom endplates of the cylinder the temperature is kept at constant values T_o and T_1, respectively, where $T_1 > T_o$. The cylinder wall is assumed to be ideally conducting, i.e. $\partial T/\partial r = 0$. Owing to the temperature being higher at the bottom wall the density of the fluid near this will be lower than the one near the top part. Thus, due to bouyancy, the fluid will tend to rise from the lower surface, resulting in natural convection. The motion may, however, depending on the actual circumstances, be counteracted by viscous forces. On the other hand, if the bouyancy force exceeds the visous forces, initial disturbances will grow and convective motion will result. From stability analysis [3] it is shown that the onset of instability is governed by the aspect ratio $\lambda = H/R$ and the Rayleigh number

$$Ra = \frac{gH^3\alpha(T_1 - T_o)}{\kappa\nu},\tag{1}$$

where g is the gravitational acceleration, α is the coefficient of thermal expansion, κ is the diffusivity and ν is the kinematic viscosity. For a semi-infinite domain, that is for $H/R \to 0$, linear stabily analysis shows that the critical, or minimum, value of the Rayleigh number occurs at $Ra_c = 1708$, whereas higher Ra-values are needed for $H/R > 0$ (see e.g. [2]). At first the motion is established as steady rolls, but at increasing values of the Rayleigh number the flow becomes unsteady with an aperiodic behaviour leading ultimately to a fully turbulent flow field.

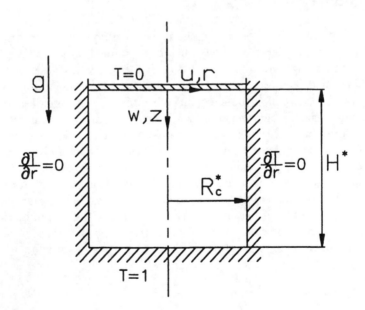

Fig. 1. Geometry of the cylinder. Note that z is directed downwards (in the same direction as g) and that the Origo is located on the top wall.

2 Formulation of the Problem

Consider a fluid with a coefficient of thermal expansion α and density ρ such that

$$\rho = \rho_o \left[1 - \alpha(T - T_o)\right], \tag{2}$$

where T is the temperature and ρ_o is the density when $T = T_o$. Using the Boussinesq approximation, the governing flow equations for the velocity vector V and the temperature can be written

$$\frac{\partial V}{\partial t} + \omega \times V = -\nabla \cdot (\frac{1}{2}\|V\|^2 + g \cdot z + p) - \alpha g(T - T_o) + \nu \nabla^2 V, \tag{3}$$

$$\frac{\partial T}{\partial t} + \nabla \cdot (TV) = \kappa \nabla^2 T, \tag{4}$$

$$\nabla \cdot V = 0, \tag{5}$$

where t is time, p is the pressure and the acceleration of gravity is given as $g = (0, 0, g)$. Furthermore, the vorticity is introduced by the definition

$$\omega = \nabla \times V. \tag{6}$$

As the flow we are going to study is governed by diffusive and heat convective processes, we introduce the characteristic time, velocity, and temperature as, respectively,

$$t_o = R^2/\nu, \qquad V_o = \nu/R, \qquad \Delta T_o = T_1 - T_o.$$

Furthermore, we eliminate the pressure in eq. (3) by applying the curl operator on this and introduce a streamfunction ψ to satisfy the continuity equation (5). This results in the vorticity-streamfunction formulation of the Boussinesq equations, which in a non-dimensionalized and axisymmetric form are written as

$$\frac{\partial \omega}{\partial t} + \frac{\partial(u\omega)}{\partial r} + \frac{\partial(w\omega)}{\partial z} = Gr\frac{\partial T}{\partial r} + \frac{\partial}{\partial r}\left[\frac{1}{r}\frac{\partial(r\omega)}{\partial r}\right] + \frac{\partial^2 \omega}{\partial z^2}, \tag{7}$$

$$\frac{\partial T}{\partial t} + \frac{\partial(uT)}{\partial r} + \frac{\partial(wT)}{\partial z} + \frac{uT}{r} = \frac{1}{Pr}\left[\frac{1}{r}\frac{\partial}{\partial r}(r\frac{\partial T}{\partial r}) + \frac{\partial^2 T}{\partial z^2}\right], \tag{8}$$

$$\frac{\partial}{\partial r}\left(\frac{1}{r}\frac{\partial \psi}{\partial r}\right) + \frac{\partial}{\partial z}\left(\frac{1}{r}\frac{\partial \psi}{\partial z}\right) = \omega, \tag{9}$$

$$u = \frac{1}{r}\frac{\partial \psi}{\partial z}, \qquad w = -\frac{1}{r}\frac{\partial \psi}{\partial r}, \tag{10}$$

where the polar coordinates (r, z) are defined as shown in Fig.1, (u, w) denote the velocity components in the (r, z) directions, respectively, Gr is the Grasshof number and Pr the Prandtl number. These two latter result from the non-dimensionalizing procedure and are given as

$$Gr = \frac{gR^3\alpha(T_1 - T_o)}{\nu^2}, \qquad Pr = \frac{\nu}{\kappa}. \tag{11}$$

It may be noted that the Rayleigh number defined in eq. (1) can be deduced from the relation

$$Ra = Gr \cdot Pr \cdot \lambda^3,$$

where $\lambda = H/R$ is the aspect ratio.

To close the problem the no-slip condition for the velocitiy is imposed on all solid boundaries. Thus in non-dimensionalized variables the boundary conditions read

- Symmetry axis ($r = 0$, $0 \le z \le \lambda$):

$$\psi = u = \omega = \frac{\partial \psi}{\partial r} = 0, \frac{\partial^2 \psi}{\partial r^2} = -w, \frac{\partial T}{\partial r} = 0, \qquad (12)$$

- Cylinder wall ($r = 1$, $0 \le z \le \lambda$):

$$\psi = u = w = \frac{\partial \psi}{\partial r} = 0, \frac{\partial^2 \psi}{\partial r^2} = \omega r, \frac{\partial T}{\partial r} = 0, \qquad (13)$$

- Top wall ($0 \le r \le 1$, $z = 0$):

$$\psi = u = w = \frac{\partial \psi}{\partial z} = 0, \frac{\partial^2 \psi}{\partial z^2} = \omega r, T = 0, \qquad (14)$$

- Bottom wall ($0 \le r \le 1$, $z = \lambda$):

$$\psi = u = w = \frac{\partial \psi}{\partial z} = 0, \frac{\partial^2 \psi}{\partial z^2} = \omega r, T = 1. \qquad (15)$$

Note that the dimensionless variables $r \in [0, 1]$ and $z \in [0, \lambda]$.

2.1 Turbulence Modelling

At high Rayleigh numbers the flow become turbulent which means that it becomes dominated by eddy-structures of varying size. In order to capture the dynamics of these an extremely fine grid has to be employed if Direct Numerical Simulations (DNS) are to be carried out. In principle this limits the practical application of numerical simulations. However, at high Ra-values calculations may be carried by utilizing Large Eddy Simulations (LES). LES is a technique in which only the large-scale motions are computed whereas the smallest turbulent eddies are modelled by a Sub-Grid-Scale (SGS) model. This is accomplished by applying a spatial filter on the governing equations and decomposing the variables in a filtered part and a a fluctuating part:

$$f(r, z) = \bar{f}(r, z) + f'(r, z), \qquad (16)$$

where $f'(r, z)$ denotes the fluctuating and $\bar{f}(r, z)$ the filtered part of the function f. The filtering is performed by folding f by a filter function G, i.e.

$$\bar{f}(r, z) = \int_A G(r - r', z - z') \cdot f(r', z') dA, \qquad (17)$$

where A is the area of the calculation domain. As filter function we apply a box filter, defined as

$$G(r - r', z - z') = \begin{cases} \Delta^{-2} & \text{for } \left\| r - r', z - z' \right\|_\infty < \frac{1}{2}\Delta \\ 0 & \text{for } \left\| r - r', z - z' \right\|_\infty \geq \frac{1}{2}\Delta \end{cases} \quad (18)$$

where $\Delta = (\Delta r \cdot \Delta z)^{\frac{1}{2}}$. Applying the filter on eqs. (7) - (10), we obtain

$$\frac{\partial \bar{\omega}}{\partial t} + \frac{\partial (\bar{u}\bar{\omega})}{\partial r} + \frac{\partial (\bar{w}\bar{\omega})}{\partial z} = Gr \frac{\partial \bar{T}}{\partial r} + \frac{\partial}{\partial r}\left[\frac{1}{r}\frac{\partial (r\bar{\omega})}{\partial r}\right] + \frac{\partial^2 \bar{\omega}}{\partial z^2} + \Omega, \quad (19)$$

$$\frac{\partial \bar{T}}{\partial t} + \frac{\partial (\bar{u}\bar{T})}{\partial r} + \frac{\partial (\bar{w}\bar{T})}{\partial z} + \frac{\bar{u}\bar{T}}{r} = \frac{1}{Pr}\left[\frac{1}{r}\frac{\partial}{\partial r}(r\frac{\partial \bar{T}}{\partial r}) + \frac{\partial^2 \bar{T}}{\partial z^2}\right] + \Theta, \quad (20)$$

$$\frac{\partial}{\partial r}\left(\frac{1}{r}\frac{\partial \bar{\psi}}{\partial r}\right) + \frac{\partial}{\partial z}\left(\frac{1}{r}\frac{\partial \bar{\psi}}{\partial z}\right) = \bar{\omega}, \quad (21)$$

$$\bar{u} = \frac{1}{r}\frac{\partial \bar{\psi}}{\partial z}, \qquad \bar{w} = -\frac{1}{r}\frac{\partial \bar{\psi}}{\partial r}, \quad (22)$$

where

$$\Omega = \frac{\partial}{\partial r}\left[\overline{u\omega} - \bar{u}\bar{\omega}\right] + \frac{\partial}{\partial z}\left[\overline{w\omega} - \bar{w}\bar{\omega}\right], \quad (23)$$

$$\Theta = \frac{\partial}{\partial r}\left[\overline{uT} - \bar{u}\bar{T}\right] + \frac{\partial}{\partial z}\left[\overline{wT} - \bar{w}\bar{T}\right] + \frac{\overline{uT} - \bar{u}\bar{T}}{r}. \quad (24)$$

Except for the terms Ω and Θ, that have to be modelled, it is easily recognized that the filtered equations take the exact same form as the original ones.

To model the additional terms appearing in eq. (19) an eddy-viscosity hypothesis is utilized, that is

$$-[\overline{u\omega} - \bar{u}\bar{\omega}] = \nu_t \frac{\partial \bar{\omega}}{\partial r}, \quad (25)$$

$$-[\overline{w\omega} - \bar{w}\bar{\omega}] = \nu_t \frac{\partial \bar{\omega}}{\partial z}, \quad (26)$$

where the eddy-viscosity is modelled by the SGS model proposed by Lardat and Ta Phuoc [6]

$$\nu_t = (C\Delta)^3 \left[(\frac{\partial \bar{\omega}}{\partial r})^2 + (\frac{\partial \bar{\omega}}{\partial z})^2\right]^{\frac{1}{2}}, \quad (27)$$

where $C = 0.2$ and $\Delta = (\Delta r \cdot \Delta z)^{\frac{1}{2}}$ is the filter width. For the corresponding sub-grid terms of eq. (20) we employ the Boussinesq analogy (see e.g. White [8]) stating that

$$-[\overline{uT} - \bar{u}\bar{T}] = \kappa_t \frac{\partial \bar{T}}{\partial r}, \quad (28)$$

$$-[\overline{wT} - \bar{w}\bar{T}] = \kappa_t \frac{\partial \bar{T}}{\partial z}, \quad (29)$$

where κ_t is the eddy conductivity.

Neither ν_t nor κ_t is a property of the fluid, but they can be formed into a turbulent prandtl number

$$Pr_t = \frac{\nu_t}{\kappa_t},\qquad(30)$$

which, by the so-called Reynolds analogy, can be shown to be of order unity. Thus, we postulate that

$$\kappa_t = \nu_t,\qquad(31)$$

which then closes the system of equations.

3 Numerical Procedure

A fourth-order accurate discretization is used for the velocities u, w (eq.(22)) and the streamfunction ψ (Poisson eq.(21)), based upon 3-point Hermitian formulas,

$$\psi'_{i+1} + 4\psi'_i + \psi'_{i-1} = \frac{3}{h}(\psi_{i+1} - \psi_{i-1}) + O(h^4),\qquad(32)$$

$$\psi''_{i+1} + 10\psi''_i + \psi''_{i-1} = \frac{12}{h^2}(\psi_{i+1} - 2\psi_i + \psi_{i-1}) + O(h^4),\qquad(33)$$

where h denotes the grid spacing, subscript i refers to the point being calculated and the single and double primes denote first and second derivatives respectively. Equations (21)-(22) are solved in combination with eqs. (32)-(33), by making use of the line-relaxation algorithm of Wachspress [9]. Note that the fourth-order accuracy implied in two additional variables, ψ' and ψ''. The advantage is that, for a given accuracy, less grid points are necessary in comparison with lower order methods.

The transport equations (19) for the vorticity and (20) for the temperature are discretized by standard second-order differences for the diffusive terms and the QUICK scheme [4] for the convective terms. The solution is advanced in time by a Crank-Nicolson type formula and the resulting system solved by a succesive-line-relaxation method (ADI). As an example, consider a variable f governed by the equation

$$\frac{\partial f}{\partial t} = L_r(f) + L_z(f) + \phi,\qquad(34)$$

where L_r and L_z denote operators for the derivatives in the r- and z-direction, respectively, and ϕ is a source term. The line-relaxation then proceeds in the two half-steps

Step 1: $\quad f^{n+\frac{1}{2}} - \frac{\Delta t}{2} L_r(f^{n+\frac{1}{2}}) = f^n + \frac{\Delta t}{2}\left[L_z(f^n) + \phi^n\right]$

Step 2: $\quad f^{n+1} - \frac{\Delta t}{2} L_z(f^{n+1}) = f^{n+\frac{1}{2}} + \frac{\Delta t}{2}\left[L_r(f^{n+\frac{1}{2}}) + \phi^n\right],$

such that the complete solution is written as

$$\frac{f^{n+1} - f^n}{\Delta t} = L_r(f^{n+\frac{1}{2}}) + \frac{1}{2}\left[L_z(f^{n+1}) + L_z(f^n)\right] + \phi^n, \tag{35}$$

where Δt is the time increment and superscript n refers to time t^n. The scheme is in principle 2nd order accurate, but as the source term and the boundary conditions are taken at time t^n it becomes formally 1st order accurate. The solution of eq. (21) proceeds in the same manner, but with the time increment replaced by a set of relaxation parameters, λ^k. According to the method of Wachspress [9] k takes the values $1, 2, ..., 2^N$, where N in the present case is put equal to 3. Thus, in total, a time-step is carried out by solving 4 tri-diagonal systems for the 2 momentum euations (19) and (20), and 16 3×3 block-tri-diagonal systems for the Poisson equation system (21), (32) and (33). Both vectorization and parallellization options have been explored for the solution of the systems.

The solution of the complete problem proceeds by first advancing the temperature, eq.(20), in time. Next, the vorticity is advanced in time by introducing the updated temperature field in the right-hand side of eq. (19). Finally, the updated vorticity is employed as a source term on the right-hand side of eq. (21) and the streamfunction, and corresponding velocity field, is calculated from the procedure described above. For a detailed description of the numerical technique, see [7].

4 Results and Discussion

Here results are presented for the early transition of the thermo-convective flow problem depicted in Fig.1. In the work of Charlson and Sani [2] semi-analytical stability analysis was employed to detect the onset of steady rolls. It was found that the onset was governed by the Rayleigh number and the aspect ratio, whereas it was independent of the Prandtl number. Furthermore, the analysis showed that, except for small local oscillation, the critical Rayleigh number increased as function of increasing aspect ratios. Thus, a value of $Ra_c = 1719$ was obtained for $\lambda = 0.125$ whereas $Ra_c = 2262$ for $\lambda = 1.0$.

As a first validation of the developed code the point of instability was calculated and compared to the results of Charlson and Sani for the case $\lambda = 1.0$. The subcritical base solution, which consists of the simple case where no flow motion is present, was employed as initial condition. It is easily verified, by inspection of eqs. (19) - (22), that the base solution takes the form $\psi = u = w = \omega = 0$ and $T = z/\lambda$. To initialize the motion, the base solution was disturbed by perturbing the temperature on the bottom wall by a distribution of the form

$$T_w(r) = 0.98 + 0.02 \cdot \sin(20\pi r), \tag{36}$$

and, by following the value of a monitor point, it was determined if the perturbation persisted or died out as a function of time. The outcome of the study is shown in Fig.2, where the streamfunction in the middle of the calculation domain is chosen as monitor point. The figure shows that the perturbation dies out for $Ra = 2240$ and increases for $Ra = 2247$, whereas neutral stability is

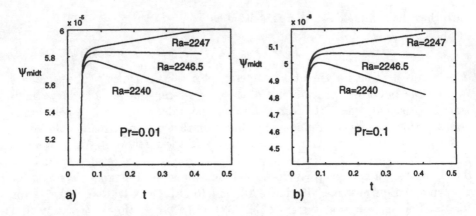

Fig. 2. Time evolution of the monitor point near the critical Rayleigh number; a) Prandtl number 0.01; b) Prandtl number 0.1.

obtained for $Ra = 2246.5$. Furthermore, the results indicate that Ra_c is independent of the value of the Prandtl number. Thus, both for $Pr = 0.01$ (Fig.2a) and $Pr = 0.1$ (Fig.2b) the critical Rayleigh number was found to be equal to 2246.5, a value which is within 1 percent of the analytical result.

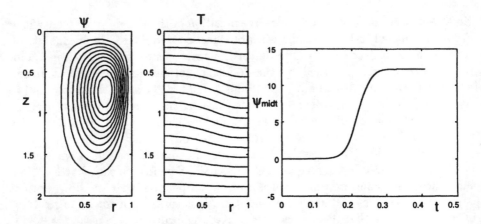

Fig. 3. Steady solution for Rayleigh number 5000; Left: Streamfunction contour; Middle: Iso-temperature field; Right: Time evolution of monitor point.

Fig.3 presents streamlines and the temperature field along with the time evolution of the monitor point for $Ra = 5000$. It is seen that steady state sets in about $t = 0.4$ and that the flow field consists of a single roll, with a temperature gradient decreasing monotonically from the axis to the sidewall of the cylinder.

The shown results are obtained at a spatial discretization $\Delta r = \Delta z = 0.04$ and a time increment $\Delta t = C \cdot \Delta z$, where $C = 0.001$. Thus, $t = 0.4$ corresponds to 10.000 time steps.

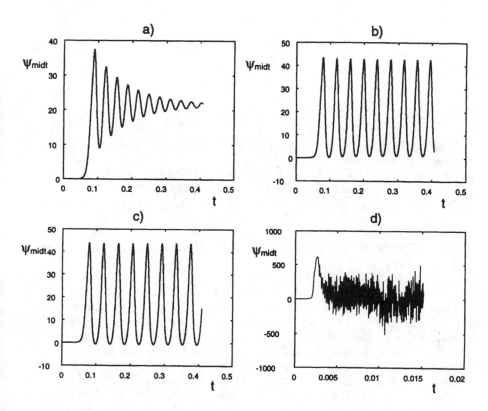

Fig. 4. Time evolution of monitor point; a) Ra = 12.000 (steady); b) Ra = 13.800 (steady, but close to periodic); c) Ra = 13.900 (periodic); d) Ra = 10^6 (chaotic).

In Fig.4 the time evolution of the monitor point is shown at Rayleigh numbers $Ra = 12.000$ (Fig.4a), $Ra = 13.800$ (Fig.4b), $Ra = 13.900$ (Fig.4c) and $Ra = 10^6$ (Fig.4d). The figure demonstrates that the flow is steady for $Ra = 12.000$ and $Ra = 13.800$, whereas it becomes periodic for $Ra = 13.900$ and that chaotic behaviour is accomplished for $Ra = 10^6$. At increasing values of the Rayleigh number it becomes necessary to refine the spatial as well as the temporal discretization, in order to capture the dynamics of the flow. Thus, in order to obtain a grid independent solution in the chaotic regime, a discretization of $\Delta r = \Delta z = 0.01$, with $C = 10^{-4}$, was utilized for the results in Fig.4d.

Finally, in the context of the theme of the present conference, it should be mentioned that the code mainly has been run on work stations and on the CRAY C92 vector processor at UNI-C. The code is highly vectorizable, but the performance depends critically on the number of mesh points used. For the results shown here the performance is about 300 Mflops on the CRAY, but for a discretization of 1000×500 nodepoints up to 550 Mflops was obtained. An early version of the code was tested on the SP2 parallel machine, but no conclusive results on its performance was achieved. However, the ADI technique and the line-relaxation procedures that form the basic algorithm of the code may very well be implemented on different configurations of parallel processors, as has been demonstrated by Johnsson et al. [5]. The potentials of optimizing the algorithm on parallel machines will be further explored in the near future.

References

1. Busse, F.H.: Transition to turbulence in Rayleigh-Bénard convection. Topics in Applied Physics **45** (1981) 97–137
2. Charlson, G.S., Sani, R.L.: Thermoconvective Instability in a bounded cylindrical fluid layer. Int. J. Heat Mass Transfer **13** (1970) 1479–1496
3. Drazin, P.G., Reid, W.H.: Hydrodynamic Stability. Cambridge University Press (1984)
4. Hayase, T., Humphrey, J.A.C., Greif, R.: A consistently formulated QUICK scheme for fast and stable convergence using finite-volume iterative calculation procedures. J. Comp. Physics **98** (1992) 108–118
5. Johnsson, S.L., Saad, Y., Schultz, M.H.: Alternating direction methods on multi-processors. SIAM J. Sci. Statist. Comput. **8(5)** (1987) 686–700
6. Lardat, R., Ta Phuoc Loc : Simulation numérique d'un écoulement turbulent autour d'un profil NACA0012 en incidence. Submitted to La Recherche Aérospatiale (1996)
7. Sørensen, J.N., Ta Phuoc Loc : High-order axisymmetric Navier-Stokes code: Description and evaluation of boundary conditions. Int. J. Num. Meth. Fluids **9** (1989) 1517–1537
8. White, F.: Viscous Fluid Flow. McGraw-Hill (2nd edition) (1991)
9. Wachspress, E.L.: Iterative Soultions of Elliptic Systems. Prentice-Hall (1966)

Multi-tasking Method on Parallel Computers which Combines a Contiguous and a Non-contiguous Processor Partitioning Algorithm

Kuniyasu Suzaki, Hitoshi Tanuma, Satoshi Hirano, Yuuji Ichisugi,
Chris Connelly, and Michiharu Tsukamoto

Electrotechnical Laboratory, MITI, JAPAN
suzaki@etl.go.jp

Abstract. We propose a partitioning algorithm which combines a contiguous and a non-contiguous partitioning algorithm . The proposed partitioning algorithm overcomes the drawbacks caused by contiguous and non-contiguous partitioning algorithms; low processor utilization and long communication path. In this paper we deal with mesh-connected parallel computers and show a partitioning algorithm that combines the Adaptive Scan and the Multi Buddy. To show the performance improved by proposed algorithm, we conducted simulation runs under an assumption of wormhole routing and All-to-All message pattern. From the results we knew that the proposed algorithm improved utilization of a parallel computer and service for each task.

1 Introduction

On parallel computers *multi tasking* of parallel programs is mainly achieved by *space sharing*, because not every application can utilize all processors in a parallel computer. *Partitioning Algorithm* is a method to provide processors of a parallel computer for each task. The processor regions for tasks does not overlap. It enable to run many tasks simultaneously on a parallel computer.

Partitioning algorithms are divided into two categories; *contiguous* and *non-contiguous*. Contiguous partitioning algorithms provide processors which keeps processor connection required by a task. For example, on mesh connected parallel computers a task requires mesh connected parallel processors. A contiguous partitioning algorithm searches a mesh connected idle processors which is same shape of the task. If the idle processors were not found, the search was failed. It caused low processor utilization, when there were much non-connected idle processors.

On the other hand non-contiguous partitioning algorithms can provide processors for each task if there are idle processors more than processors required by the task. The given processors often does not keep the processor connection required by a task. So sometimes processors are scattered with long distance and cross the processors which are given to other tasks. This situation causes long distance of communication path and frequent message collision with other tasks.

To settle these drawbacks, we propose a partitioning algorithm which combines a contiguous and a non-contiguous partitioning algorithm. It enable to achieve high processor utilization and low message conflict. In this paper we deal with mesh-connected parallel computers and propose a partitioning algorithm which combines the contiguous partitioning algorithm Adaptive Scan[1] and the non-contiguous partitioning algorithm Multi Buddy[2]. In order to show the improved performance of proposed algorithm, we conducted simulation runs and compared with the Adaptive Scan and Multi Buddy. On simulation we used a simulator "procsimity"[3] which could simulate some flow controls of message and many message patterns. In this paper we assumed that the flow control was wormhole[4] which was the most popular and All to All message pattern which was know as the weak point for non-contiguous partitioning algorithms.

This paper is organized as follows. In the next section, we introduce partitioning algorithms which are divided into contiguous and non-contiguous one. In section 3, we present the partitioning algorithm which combines a contiguous and a non-contiguous partitioning algorithm. In section 4, we show the performance of proposed partitioning algorithm, which is based on simulation results. Finally, in section 5 we state our conclusions.

2 Partitioning Algorithms

In this section, we introduce contiguous and non-contiguous partitioning algorithms used on mesh-connected parallel computers. The processors of mesh connected parallel computers are denoted by (x, y), which indicates column x and row y and the bottom-left processor of the parallel computer is denoted by $(0,0)$.

2.1 Contiguous Partitioning Algorithms

Contiguous partitioning algorithms can allocate a processor region which keeps the shape of processor connection required by a task. This feature is called "closed partitioning". Under the closed partitioning, messages issued by a task use only the network allocated for the task and there is no message collision between tasks.

Several contiguous partitioning algorithms for mesh-connected parallel computers are proposed. For examples the Frame Slide[5], the 2 Dimensional Buddy[6], the First Fit[7], the Best Fit[7], the Adaptive Scan[1], the Busy List[8], and the Quick Allocation[9]. In this study, we deal with the Adaptive Scan which can achieve high processor utilization in the partitioning algorithms and is easy to implement.

The Adaptive Scan (AS) strategy was proposed by Ding and Bhuyan [1]. It resembles the First Fit (FF) strategy proposed by Zhu [7]. The AS allocates a task to a processor submesh that have the same shape of the task. The AS searches a submesh starting from the bottom left corner $(0, 0)$ of a mesh-connected parallel computer, and then going from left to right and from bottom to top.

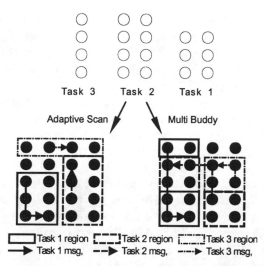

Fig. 1. contiguous and non-contiguous partitioning algorithm

In contrast to the FF, the AS allows the change in the orientation of the incoming tasks. Figure 1 shows an example of the AS strategy, where TASK3 changes its orientation because it can not be allocated with its original orientation.

The AS can achieve high processor utilization in contiguous partitioning algorithms, but it can not overcome non-contiguous partitioning algorithms. For example, if TASK3 requires 2×3 processors, the AS can not allocated TASK3. However non-contiguous partitioning algorithms can allocate, because there is 6 idle processors.

2.2 Non-Contiguous Partitioning Algorithms

Non-contiguous partitioning algorithms can allocate processor region for a task, if there is more idle processors than processors required by the task. It can achieve high processor utilization but it often can not keep the shape of processor connection required by a task. It can not guarantee closed partitioning and some messages make massage collision with messages of other tasks.

Some non-contiguous partitioning algorithms for mesh-connected parallel computers proposed by Liu, Lo, Windisch and Nitzberg. They proposed Random, Page, and Multi Buddy[2]. The three algorithms have same packing ability but distances of communication path are different. In this study, we deal with the Multi Buddy which can achieve short distance of communication path in them.

The Multi Buddy (MB) is based on the 2 Dimensional Buddy(2DB)[6] which is a contiguous partitioning algorithm. The 2DB allocates square processor region whose side length is 2^n and the bottom left corner is $(2^n \times p, 2^n \times q)$. The square regions are called Buddy elements. The Buddy elements for each size are identified by Buddy number which determined by the bottom left corner position[6]. If a task requires processors $W \times H$, the 2DB allocates one of Buddy elements of which size is $2^{\lceil max(log_2 W, log_2 H) \rceil} \times 2^{\lceil max(log_2 W, log_2 H) \rceil}$. If both W and H are not $2^{\lceil max(log_2 W, log_2 H) \rceil}$, the Buddy element includes fragmentation which is a mass of idle processors.

The MB use Buddy elements to manage processor region but it does not cause fragmentation. The MB transform the shape required by a task into the shape composed by Buddy elements. For example in figure 1 TASK1 requires 2×3 processors. The MB transforms the 2×3 processors into one 2×2 Buddy element and two 1×1 Buddy elements.

The MB sometimes allocated non-contiguous region for a task even if a contiguous region for the task existed. The reason is that the MB does not demolish large Buddy element as much as possible even if contiguous region for a task exists. For example in figure 1 TASK1 is allocated separated region in spite that contiguous region exists as the AS allocation. This feature causes long communication path and frequent message collision.

3   Combination of Contiguous and Non-Contiguous one

In order to settle the problems which caused by contiguous and non-contiguous partitioning algorithms, we propose a partitioning algorithm which combines a contiguous and a non-contiguous partitioning algorithm. In this paper we present a partitioning algorithm for mesh connected parallel computers, which combines the Adaptive Scan[1] and the Multi Buddy[2]. We call it "AS&MB".

The AS&MB search a region for a incoming task by the AS at first. If the search by the AS fails, the AS&MB exchanges the searching strategy for the MB. Consequently the AS&MB can allocate a contiguous region for a task if a contiguous region exists on the parallel computer and allocate a non-contiguous region if a contiguous region does not exist.

Figure 2 shows a allocation example by the AS&MB. In the figure TASK1 and TASK2 were allocated by the AS and TASK3 is allocated by the MB. The TASK3 are recognized as $2 \times 4 = (2^1 \times 2^1) \times 2$ and required two $2^1 \times 2^1$ Buddy elements. However there is only one $2^1 \times 2^1$ Buddy element on the parallel computer and then one $2^1 \times 2^1$ Buddy element is destroyed into four $2^0 \times 2^0$ Buddy elements. The AS and MB which compose the AS&MB use different data which indicate processor regions allocated by tasks.

In order to search a free region by the AS, the allocated regions are represented by quadruplet data $< x, y, w, h >$. The x and y indicated the bottom left corner processor of the region and the w and h indicates the width and height of the region. In figure 2 TASK1 and TASK2 is allocated by the AS at the region $< 0, 0, 2, 3 >$ and $< 0, 3, 1, 3 >$ respectively. As TASK3 is allocated by the MB, the allocated region is non-contiguous. The non-contiguous region is composed by five Buddy elements which are contiguous regions;$< 2, 0, 2, 2 >$, $< 0, 4, 1, 1 >$, $< 1, 4, 1, 1 >$, $< 2, 4, 1, 1 >$, and $< 3, 4, 1, 1 >$. Therefore the region data for TASK3 is composed of five contiguous regions.

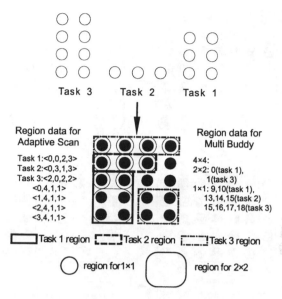

Fig. 2. Partitioning algorithm which combine contiguous and non-contiguous one

In order to search a free region by the MB, the allocated regions are represented by Buddy elements. Therefore the regions allocated by the AS must be represented by Buddy elements. The Buddy element are recognized by the position of the region and the size of the region. The region allocated by the AS is expressed by $< x, y, w, h >$ and the bottom left processor (x,y) is expressed as follows.

$$x = 2^n \times l_n + 2^{n-1} \times l_{n-1} + \cdots + 2^{n-m} \times l_{n-m}$$
$$y = 2^q \times p_q + 2^{q-1} \times p_{q-1} + \cdots + 2^{q-r} \times p_{q-r}$$

The smallest second power 2^{n-m} and 2^{q-r} are the candidates for the size of Buddy element except when the number of x or y is "0". If the number was 0, the candidate for the size would be wild card. Furthermore we must consider the size of the region. The biggest second powers which is included w and h are the candidates; $2^{\lfloor \log_2 w \rfloor}$, and $2^{\lfloor \log_2 h \rfloor}$. The size of Buddy element is the smallest one of $2^{n-m}, 2^{q-r}, 2^{\lfloor \log_2 W \rfloor}$, and $2^{\lfloor \log_2 H \rfloor}$. The number of Buddy element is determined by the size and the position of a Buddy element[2].

Using sample1 in figure3 we explain the translation into Buddy elements. In figure3 a Buddy element is extracted from $< 0, 0, 7, 5 >$ at first. As the bottom left processor is located (0,0), the size of Buddy element depends on the size of region. The size of region is 7×5 and both sides contain 2^2 as the maximum second power. Therefore $2^2 \times 2^2$ Buddy element is extracted form $< 0, 0, 7, 5 >$. The rest region is composed by two rectangle; $< 4, 0, 3, 5 >$ and $< 0, 4, 7, 1 >$. The two candidate regions are overlap at the region $< 4, 4, 3, 1 >$ because we can not determine which rectangle can extract a big Buddy element till we check the candidate regions. As second cut $2^1 \times 2^1$ Buddy element is extracted from $< 4, 0, 3, 5 >$ in same manner and the $< 6, 0, 1, 5 >$ and $< 4, 2, 3, 3 >$ regions are left.

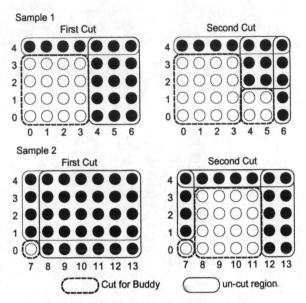

Fig. 3. Transform to Buddy element

If the same size of a task is allocated on different region, not only the number of Buddy element but also the size of Buddy element is different. We explain this case using Sample2 in figure3. At first Buddy element is extracted from $< 7, 0, 7, 5 >$. Since the y position is 0, the x position 7 is a candidate of a size of Buddy element. The value of 7 includes the smallest second power 2^0. Furthermore the size of region is 7×5 and both sides contain 2^2 as the maximum second power. Therefore $2^0 \times 2^0$ Buddy element is extracted form $< 7, 0, 7, 5 >$. The rest region is composed by two rectangle;$< 8, 0, 6, 5 >$ and $< 7, 1, 7, 4 >$ which overlap $< 8, 1, 6, 4 >$. As second cut $2^2 \times 2^2$ Buddy element is extracted from $< 7, 0, 7, 5 >$ in same manner and leave $< 12, 0, 2, 5 >$ and $< 8, 4, 1, 6 >$. At this time the extraction destroies the other candidate. Therefore the other candidates is replaced by $< 7, 1, 1, 4 >$ and $< 7, 4, 1, 7 >$. Here the $< 6, 4, 1, 6 >$ is included in the $< 7, 4, 1, 7 >$ and the $< 6, 4, 1, 6 >$ is eliminated.

4 Simulations

We conducted simulation runs to compare performance of the AS&MB, AS, and MB. We used simulator "procsimity"[3] which is developed by University of Oregon. The procsimity can simulate Store & Forward, Wormhole, and Virtual Cut Through as flow control and estimate the rate of message collision. As message pattern the simulator offers All to All, One to All, FFT, Random, etc.

In this paper we assumed that the flow control was Wormhole(WH)[4] which was the most popular. Under the WH each switch had 1 byte buffer and the flit

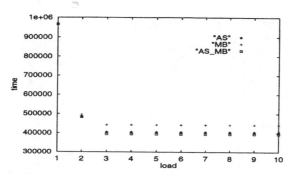

Fig. 4. Makespan(Average number of message:40)

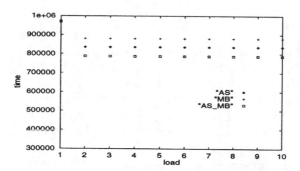

Fig. 5. Makespan(Average number of message:80)

of WH was 1 byte. The delay of sending 1 byte message assumed to be 3 unit time.

The message pattern assumed to be All to All, because it caused much message collision and it was know as the weak point for non-contiguous partitioning algorithms. We assumed number of message followed exponential distribution and average number of message for a task set to 40. Each message included 8 byte data.

All the simulation runs accepted 1000 tasks of which service time followed exponential distribution and average service time set to 1000.0 time units. The simulation stopped when 1000 tasks were completed. The average interarrival time also followed exponential distribution and was changed from 1000.0 to 100.0 time units.

The *load* is defined as the ratio of the average service time $\overline{T}_{service}$ and the average task interarrival time $\overline{T}_{interval}$.

$$Load = \overline{T}_{service}/\overline{T}_{interval} \qquad (1)$$

The load was changed from 1.0 to 10.0, because the average interarrival time was changed form 1000.0 to 100.0 time units. At each load, we investigated makespan and service for each task.

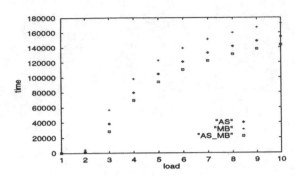

Fig. 6. Average waiting time for allocation

4.1 Makespan

Figure 4 shows the completion time of all tasks, which is called "makespan" in some papers. The x axis indicates load and the y axis indicates makespan. The figure shows results of makespan for the three partitioning algorithms at each load. In the figure all partitioning algorithms showed same makespan at low load. However difference of partitioning algorithms appeared at high load. The order of quickness is the AS&MB, AS, and MB. The deference occurred form packing ability and message delay.

The AS could achieve quick makespan than the MB, although the MB could allocate more tasks than the AS. The reason was that the AS could allocate contiguous region for each tasks. Owing to the contiguous region, the distance of communication path was short and collision of message was little. The slow communication of the MB made makespan long. The AS&MB also was slower message than the AS, but faster than the MB. The message delay by the AS&MB made up with high processor utilization by the AS&MB and the AS&MB could show better performance than the AS. The detail of communication effect shows in the next section.

From this graph we were afraid that the AS overcame the AS&MB at more messages than 40. However the AS&MB could show better performance than the AS. Figure5 show the results when the messages was doubled. Under this condition each message delay was saturated, the effect of contiguous region by the AS was little. The effect of high packing ability by the AS&MB appeared and the AS&MB showed unmistakable better performance than the AS.

4.2 Service for each task

As service for each task we measured two factors which cause delay; the waiting time for allocation and message latency.

If a task is allocated on a processor region, the processor region does not get another task. Therefore one delay factor of of a task is waiting time for finish of allocated tasks. When a task is allocated, the execution time of the task influenced by message delay. The message delay caused by long communication

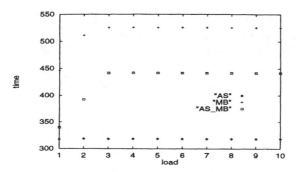

Fig. 7. Average time for latency

path and message collision. The influence of message delay is measured by latency time.

Figure6 shows the average waiting time for allocation. From this figure we knew that there were little waiting time for allocation till load 2 on each partitioning algorithm. However the waiting time increased in proportion to load from load 2 up. The order of short waiting time was the AS&MB,AS, and MB. This waiting time depended on task execution time which was prolonged by late message latency as well as on the packing ability of partitioning algorithms. The packing ability of the MB and AS&MB is higher than the AS. However the latency time of the AS is quicker than the MB and AS&MB, because the AS allocates contiguous processor region. Figure6 showed that packing ability of the AS&MB overcame the prolonged latency time.

Figure7 shows the average time for latency. Form this figure we knew that the latency of tasks allocated by the AS was indifferent from load. This result caused by closed partitioning which prohibited the message collision between tasks. The latency time for the MB and AS&MB are longer than the AS. This tendency resulted in non-contiguous allocation. The latency of the AS&MB is faster than that of the MB, because the AS&MB allocated processor region much closer than the MB. Therefore the distance of communication path by the AS&MB was shorter than the MB and times of message collision of the AS&MB was fewer than that of the MB. The latency for the MB and AS&MB was saturated from load 3 up. This results meant that the packing ability was saturated and task did not allocated anymore. Therefore latency were indifferent form load.

5 Conclusions

We proposed a partitioning algorithm which combine a contiguous and a non-contiguous partitioning algorithm. The algorithm could achieve high processor utilization as non-contiguous one and short distance of communication path. In this paper we dealt with mesh connected parallel computers and proposed the partitioning algorithm which combines the contiguous partitioning algorithm Adaptive Scan and the non-contiguous partitioning algorithm Multi Buddy.

We carried out simulation experiments to compare original partitioning algorithms. On the simulation we assumed wormhole routing and All-to-All message pattern. From the results of simulation we knew that the proposed algorithm could achieve higher utilization of computer (processor and network) and offer better service for each task than the Adaptive Scan and Multi Buddy alone.

This method is not limited to mesh-connected parallel computers. For any parallel computer this method is available and solve the problem which raised from contiguous and non-contiguous partitioning algorithms.

Acknowledgment

The procsimity was made by Dr. Kurt Windisch, Associate Professor Virginia Lo and Dr. Jayne Miller at University of Oregon. I thank them for allowing the use of it. Part of this research was conducted in conjunction with the Real World Computing Project.

References

1. J. Ding and L. N. Bhuyan. An adaptive submesh allocation strategy for two-dimensional mesh connected systems. *Proceedings of International Conference on Parallel Processing*, pages (II)193–200, 1993.
2. W. Liu, V. Lo, K. Windish, and B Nitzberg. Non-contiguous Processor Allocation Algorithms for Distributed Memory Multicomputers. *Supercomputing*, pages 227–236, 1994.
3. K. Windish, V. Lo, and Miller J. ProcSimity: An Experimental Tool for Processor Allocation and Scheduling in Highly Parallel Systems. *Frontiers'95*, pages 414–421, 1995.
4. L.M. Ni and P.K. McKinley. A Survey of Wormhole Routing Techniques in Direct Network. *Computer*, 26(2):62–76, 1993.
5. P. Chuang and N. Tzeng. An efficient submesh allocation strategy for mesh computer systems. *Proceedings of the 11th International Conference on Distributed Computing Systems*, pages 259–263, 1991.
6. K. Li and K. Cheng. A two dimensional buddy system for dynamic resource allocation in a partitionable mesh connected system. *Journal of Parallel and Distributed Computing*, (12):79–83, 1991.
7. Y. Zhu. Efficient processor allocation strategies for mesh-connected parallel computers. *Journal of Parallel and Distributed Computing*, 16:328–337, 1992.
8. D. D. Sharma and D. K. Pradhan. A fast and efficient strategy for submesh allocation in mesh-connected parallel computers. *Procdings of the 5th IEEE Symposium on Parallel and Distributed Processing*, pages 682–689, 1993.
9. S.M. Yoo and H.Y. Youn. An efficient task allocation scheme for two-dimensional mesh-connected systems. *Proceedings of the 15th International Conference on Distributed Computing Systems*, pages 501–508, 1995.

A Highly Parallel Explicitly Restarted Lanczos Algorithm*

M. Szularz[1], J. Weston[1], M. Clint[2] and K. Murphy[2]

[1] School of Information & Software Engineering, University of Ulster, Coleraine
BT52 1SA, Northern Ireland
[2] Department of Computer Science, The Queen's University of Belfast, Belfast BT7
1NN, Northern Ireland

Abstract. The Lanczos algorithm is one of the principal methods for
the computation of a few of the extreme eigenvalues and their corresponding eigenvectors of large, usually sparse, real symmetric matrices.
In this paper a single-vector Lanczos method based on a simple restarting
strategy is proposed. The algorithm finds one eigenpair at a time using a
deflation technique in which each Lanczos vector generated is orthogonalized against all previously converged eigenvectors. The approach taken
yields a fixed k-step restarting scheme which permits the reorthogonalization between the Lanczos vectors to be almost completely eliminated.
The orthogonalization strategy developed falls naturally into the class of
selective orthogonalization strategies as described by Simon.
A 'reverse communication' implementation of the algorithm on an MPP
Connection Machine CM-200 with 8K processors is discussed. Test results using examples from the Harwell-Boeing collection of sparse matrices show the method to be very effective when compared with Sorensen's
state of the art routine taken from the ARPACK library. Advantages of
the algorithm include its guaranteed convergence, the ease with which
it copes with genuinely multiple eigenvalues and fixed storage requirements.
Key words : Lanczos algorithm, restarting, deflation, reorthogonalization, MPP.

1 The Lanczos Algorithm

Assume that $A \in \Re^{n \times n}$ is symmetric and that $q_1 \in \Re^n$ is such that $\| q_1 \|_2 = 1$.
Then the Krylov subspaces generated by A and q_1 are defined to be

$$\mathcal{K}(A, q_1, j) = \mathrm{span}(q_1, Aq_1, \ldots, A^{j-1}q_1)$$

The aim of the Lanczos algorithm, [3], is to generate an orthonormal basis $Q_j = [q_1, \ldots, q_j]$ for $\mathcal{K}(A, q_1, j)$ in such a way that good approximations to the extremal
eigenvalues and corresponding eigenvectors of A may be easily derived. The

* This work was supported by the Engineering and Physical Sciences Research Council
under grants GR/J41857 and GR/J41864 and was carried out using the facilities of
the University of Edinburgh Parallel Computing Centre

vectors in the bases Q_j are referred to as Lanczos vectors. The projection of the original problem into the basis Q_j produces a smaller eigenproblem involving a symmetric tridiagonal matrix $T_j \in \Re^{j \times j}$ which is known as a Lanczos matrix. If $(\theta_i, s_i)_j$ denotes the i-th eigenpair of T_j with $\theta_1 > \theta_2 > \cdots > \theta_j$ then $y_i = Q_j s_i$ is the i-th Ritz vector for the subspace range (Q_j) and θ_i is the corresponding Ritz value. If the ratio $| \lambda_i | / \| A \|_2$ is reasonably close to 1 it turns out that, for surprisingly small j, the Ritz pair $(\theta_i, y_i)_j$ approximates well the i-th eigenpair of A, (λ_i, x_i), where $\lambda_1 \geq \lambda_2 \geq \cdots \geq \lambda_n$.

It is assumed throughout this paper that the $p \ll n$ largest eigenvalues of A, together with their corresponding eigenvectors, are required. If $(\alpha_1, \ldots, \alpha_j)$ and $(\beta_1, \ldots, \beta_{j-1})$ denote the diagonal and off-diagonal elements of T_j, respectively, then the exact arithmetic Lanczos procedure for computing this eigensolution may be described as follows :

Algorithm 1 function$[(\theta_1, y_1), \ldots, (\theta_p, y_p)] = $ Lanczos(A, q_1, p, k)
$\quad \beta_0 \leftarrow 1, q_0 \leftarrow 0$
$\quad$ for $j = 1, 2, \ldots, k$
$\quad\quad \alpha_j \leftarrow q_j^T A q_j$
$\quad\quad q_{j+1} \leftarrow (A - \alpha_j I) q_j - \beta_{j-1} q_{j-1}$
$\quad\quad \beta_j \leftarrow \| q_{j+1} \|_2$
$\quad\quad$ if $\beta_j = 0$ then STOP (eigenvalues of T_j are the j largest eigenvalues
$\quad\quad\quad$ of A)
$\quad\quad q_{j+1} \leftarrow q_{j+1}/\beta_j$
$\quad$ end_for
$\quad$ compute $(\theta_1, y_1), \ldots, (\theta_p, y_p)$.

Thus, the Ritz pair (θ_i, y_i) returned by Algorithm 1 is an approximation to the eigenpair (λ_i, x_i) of A. The accuracy of these approximations can be estimated before y_i is computed, on the basis of the so-called error bounds $| \beta_j \| s_{ji} |$ [3], since

$$| \beta_j \| s_{ji} | = \| A y_i - \theta_i y_i \|_2 \qquad (1)$$

However, a major problem associated with the above algorithm is that k must be chosen to be sufficiently large to guarantee the requested accuracy of the solution. Moreover, the value of k required is not known in advance. Thus, in order to mitigate the attendant storage and computational demands, an explicit restart technique may be introduced. The deflation approach discussed in the following section describes a first step in this direction.

2 The Lanczos Algorithm with Explicit Restart

Let approximated quantities be decorated with the $\hat{}$ symbol. Thus, at the j-th Lanczos step the Ritz pair (θ_i, y_i) is synonymous with $(\hat{\lambda}_i, \hat{x}_i)$. Suppose that the approximated eigenpairs $(\hat{\lambda}_1, \hat{x}_1), \ldots, (\hat{\lambda}_i, \hat{x}_i)$, where $i < p$, are given. Let $\hat{\mathcal{X}}_i = $ span$\{\hat{x}_1, \ldots, \hat{x}_i\}$, and let $\hat{\mathcal{X}}_i^{\perp}$ be its orthogonal complement in $\Re^n$. Hence, if the Lanczos vectors $q_1, \ldots q_k$ are constrained to stay in the subspace $\hat{\mathcal{X}}_i^{\perp}$, the

Lanczos algorithm will converge to the Ritz values of A in the subspace $\hat{\mathcal{X}}_i^\perp$, viz, the desired approximations $\hat{\lambda}_{i+1}, \hat{\lambda}_{i+2}, \ldots$. This observation provides the basis for a simple restart strategy outlined in Algorithm 2 below :

Algorithm 2 $\text{function}[(\hat{\lambda}_1, \hat{x}_1), \ldots, (\hat{\lambda}_p, \hat{x}_p)] = \text{ExpRes}(A, p, k)$

$\quad \hat{\mathcal{X}}_0 = 0$

$\quad$ **for** $i = 1 : p$

$\quad\quad$ choose $q_1 \notin \hat{\mathcal{X}}_{i-1}$

$\quad\quad (\hat{\lambda}_i, \hat{x}_i) \longleftarrow \text{Lanczos}(A, q_1, 1, k)$

$\quad\quad$ **if** $i < p$ **then** $\hat{\mathcal{X}}_i \leftarrow \hat{\mathcal{X}}_{i-1} \oplus \text{span}\{\hat{x}_i\}$

$\quad$ **end_for**

It is necessary to make the assumption that Algorithm 1, as used above, is equipped with a mechanism for projecting each Lanczos vector (including q_1) into $\hat{\mathcal{X}}_i^\perp \neq 0$. This can obviously be achieved by the explicit orthogonalization of each q_j against $\hat{x}_1, \ldots, \hat{x}_i$.

Clearly, the problem of choosing an appropriate value of k for use in Algorithm 2 remains. Nevertheless, since only one eigenpair is sought at a time, it may be expected that this value will be much smaller than that required when Algorithm 1 is used to compute p eigenpairs simultaneously as described in the previous section. However, in order to produce a restarting scheme based on Algorithm 1 where a fixed (and possibly small) k may be used regardless of the accuracy requirements, it is necessary to modify Algorithm 2, thereby yielding Algorithm 3, an exact arithmetic version of which is given below :

Algorithm 3 $\text{function}[(\hat{\lambda}_1, \hat{x}_1), \ldots, (\hat{\lambda}_p, \hat{x}_p)] = \text{ExpRes}(A, p, k, tol)$

$\quad \hat{\mathcal{X}}_0 = 0$

$\quad$ **(1) for** $i = 1 : p$

$\quad\quad$ **(1)** choose $q_1 \notin \hat{\mathcal{X}}_{i-1}$

$\quad\quad$ **(2)** $y_1 \leftarrow q_1$; $\theta_1 \leftarrow y_1^t A y_1$

$\quad\quad$ **(3) while** $\left(\frac{\|A y_i - \theta_i y_i\|_2}{|\theta_i|} > tol \right)$

$\quad\quad\quad$ **(1)** $(\theta_1, y_1) \longleftarrow \text{Lanczos}(A, q_1, 1, k)$

$\quad\quad\quad$ **(2)** $q_1 \leftarrow y_1$

$\quad\quad$ **end_while**

$\quad\quad$ **(4)** $(\hat{\lambda}_i, \hat{x}_i) \leftarrow (\theta_1, y_1)$

$\quad\quad$ **(5) if** $i < p$ **then** $\hat{\mathcal{X}}_i \leftarrow \hat{\mathcal{X}}_{i-1} \oplus \text{span}\{\hat{x}_i\}$

$\quad$ **end_for**

tol is the user supplied tolerance. Again, it is assumed that Algorithm 1, as used above, is equipped with a mechanism for projecting each Lanczos vector (including q_1) into $\hat{\mathcal{X}}_i^\perp \neq 0$. Clearly, according to (1), $\| A y_i - \theta_i y_i \|_2$ may be replaced in the test for convergence by the error bounds $| \beta_j \| s_{ji} | $ [3]. Observe that, in this restarting scheme, $\beta_1 = \| A y_i - \theta_i y_i \|_2$. Thus, the two-norm of the residual

---

[3] in floating point arithmetic relation (1) may, in some circumstances, not even be close to satisfaction. Also, the accuracy of the eigensolution is usually measured by the two-norm of the residual vector, namely, the right-hand side of (1)

vector is provided naturally. Further, since the i-th eigenvalue of A is sought in the subspace $\hat{\mathcal{X}}_{i-1}^{\perp}$, Algorithm 3 is ideal for coping with genuinely multiple eigenvalues of A. It can also be shown formally that Algorithm 3 converges to the p required eigenvalues of A. This follows from the fact that each restart for the i-th eigenpair is achieved using $q_1 \leftarrow y_1$ in the subspace $\hat{\mathcal{X}}_{i-1}^{\perp}$. Hence, within each such subspace, q_1 is identical to the starting vector generated by the Implicitly Restarted Arnoldi Method [10] with 'exact shifts' as used to compute the largest eigenvalue of a symmetric matrix. Sorensen [10] provides a proof of convergence for this method in the case where an arbitrary number of eigenvalues is required. Consider now the use of Algorithm 3 in finite precision arithmetic. In the presence of round-off error the theoretical orthogonality among the Lanczos vectors cannot be guaranteed. This phenomenon seriously complicates the much exploited relationship between the eigenvalues of the Lanczos matrices, T_j, and A, and is discussed more fully in [3], [9]. A straightforward solution to this problem as far as Algorithm 3 is concerned is to include *a complete reorthogonalization* of the Lanczos vectors in each iteration. Thus, Algorithm 1 must be equipped not only with a mechanism for projecting each newly computed Lanczos vector q_{j+1} into $\hat{\mathcal{X}}_i^{\perp} \neq 0$ but also with a mechanism for immediately orthogonalizing it against all previously computed Lanczos vectors $q_1, \ldots, q_j$. Since k is assumed to be small, such a solution is not overly computationally expensive (see also [10] where this approach has been adopted). A variant of Algorithm 3, LAND, which incorporates the above changes has been developed and is discussed further in §5. However, a more efficient solution to the problem with respect to Algorithm 3, within which a finely tuned version of Algorithm 1 is embedded, is now proposed.

3   The Lanczos Algorithm in Finite Precision Arithmetic

If $\mathbf{u}$ is the machine round-off unit then Paige's [5] remarkable result

$$| q_{j+1}^t y_i | \approx \frac{\mathbf{u} \| A \|_2}{| \beta_j \| s_{ji} |} ; i = 1, \ldots, j \qquad (2)$$

establishes that the loss of orthogonality of the Lanczos vectors in floating point arithmetic is not chaotic. Further, Simon [9] shows that the quality of the Lanczos matrix T_j is guaranteed if

$$| q_i^t q_{j+1} | \leq \sqrt{\mathbf{u}} ; i = 1, \ldots, j \qquad (3)$$

It follows from (2) and (3) that the recently computed q_{j+1} may be orthogonalised against all the y_i whose corresponding error bounds satisfy

$$| \beta_j \| s_{ji} | \leq \sqrt{\mathbf{u}} \| A \|_2 \qquad (4)$$

i.e. against the 'converging' Ritz vectors. This approach to the problem of orthogonality is known as *selective reorthogonalization*, see [7], [9].

In contrast, in the variant of Algorithm 3 proposed here, all unwanted, 'converging' Ritz vectors, $y_i : i \neq 1$, are *purged* from the system. Thus, as soon as the error bound associated with a Ritz vector satisfies (4), the Lanczos process is immediately restarted with the current value of y_1, even if $j < k$. Further, if y_1 triggered the restart, all subsequent q_{j+1} computed are not only projected into $\hat{\mathcal{X}}_i^\perp \neq 0$ but are also orthogonalized against the recently computed, 'converging' Ritz vector y_1. This orthogonalization strategy is summarized by Algorithm 4, which then replaces Algorithm 1 as a purpose built function for inclusion in Algorithm 3.

Algorithm 4 function$[(\theta_1, y_1)] = \text{Lanczos}(A, q_1, k, converging)$

    $\beta_0 \leftarrow 1, q_0 \leftarrow 0$

    (1) for $j = 1, 2, \ldots, k$

        (1) if $i \neq 0$ **then** orthogonalize q_j against $\hat{x}_1, \ldots \hat{x}_i$

        (2) $\alpha_j \leftarrow q_j^T A q_j$

        (3) $q_{j+1} \leftarrow (A - \alpha_j I)q_j - \beta_{j-1}q_{j-1}$

        (4) $\beta_j \leftarrow \| q_{j+1} \|_2$

        (5) if $\beta_j = 0$ **then** STOP (eigenvalues of T_j are the j largest eigenvalues of A)

        (6) $q_{j+1} \leftarrow q_{j+1}/\beta_j$

        (7) if $converging = false$ **then**

            (1) compute $s_{j1}, \ldots, s_{jj}$

            (2) if $\min\{| \beta_j \| s_{j1} |, \ldots, | \beta_j \| s_{jj} |\} = | \beta_j \| s_{jr} | \leq \sqrt{\mathbf{u}} \| A \|_2$ **then**

                (1) compute (θ_1, y_1)

                (2) if $r = 1$ **then** $converging \leftarrow true$

                (3) exit

        else

            (3) orthogonalize q_{j+1} against q_1

        end_if

    end_for

    (2) compute (θ_1, y_1).

Observe that p is no longer required as an input parameter and that the new, logical variable *converging* monitors the convergence status of the recently computed Ritz vector y_1 according to (4).

A variant of Algorithm 3 which incorporates Algorithm 4, LAND1, is thus proposed in which **Step (1.3.1)** in Algorithm 3 is replaced by $(\theta_1, y_1) \leftarrow \text{Lanczos}(A, q_1, k, converging)$ and $converging \leftarrow false$ is added to **Step (1.2)**.

4 Implementation

In this section some major implementation issues associated with Algorithm 4 are outlined .

For **Step (2)** bisection, followed by inverse iteration, is a natural choice for computing the eigenvalue $\theta_1(T_k)$ and its corresponding eigenvector s_1. Observe that the adopted restarting strategy provides good lower and upper bounds on θ_1 and that these bounds become increasingly sharp as y_1 converges. Since the bisection algorithm determines the upper bound $\hat{\theta}_1$ on θ_1 (and the lower bound) it follows that $V_k = (\hat{\theta}_1 \, I - T_k)$ is positive definite and, consequently, the LDL^t decomposition of V_k used in the inverse iteration process is stable, see [3]. Hence, the (θ_1, s_1) can be rapidly computed. However, this step must be implemented with great care since even small inaccuracies in the computation of s_1 may impair the stability of the entire algorithm. For this reason the use of Eispack routines such as TINVIT is not recommended.

Steps (1.1),(1.7.3) . These steps may be implemented in the same way. Since nearly perfect orthogonality of q_{j+1} with respect to the converged approximations to $x_1, x_2, \ldots$ is essential in practice, the chosen orthogonalisation technique is *Classical Gram-Schmidt with iterative refinement*, [1], [10]. This technique also has advantages over the *Modified Gram Schmidt* alternative in a shared memory computing environment.

Step (1.7.1) . The bottommost elements of $s_1, \ldots, s_j$ (i.e. of the eigenvectors of the current T_j) may be updated from the previous Lanczos step via the secular equation :

$$\beta_j \sum_{i=1}^{j-1} \frac{s_{ji}^2(T_{j-1})}{\theta_i(T_{j-1}) - \theta} = \alpha_j - \theta \tag{5}$$

see [6]. In such an approach (*direct convergence monitoring*, see [7] for other possibilities) all of the eigenvalues of T_j must first be computed. Using the interlacing properties of the eigenvalues of T_{j-1} and T_j, these may be computed either by solving (5) iteratively with respect to θ, using a Newton-like method, or by the method of bisection. The second option appears to be a better choice since a common bisection routine can be used for steps (2) and (1.7.1). When the eigenvalues $\theta_1, \ldots, \theta_j$ have been computed, the values $| \, s_{jm} \, |$ may be determined using the following formula :

$$| \, s_{jm} \, | = \sqrt{\beta_j \sum_{i=1}^{j-1} \frac{s_{ji}^2(T_{j-1})}{\theta_i(T_{j-1}) - \theta_m(T_j)}} \; ; \quad m = 1, \ldots, j \tag{6}$$

However, without refinement, formula (6) may not be numerically stable (see [8] for details). Nevertheless, as has been pointed out in [7], [9], the accurate evaluation of the right-hand side of (4) is not crucial.

Step (1.7.2) . The value of $\| \, A \, \|_2$ may be approximated by $\max(| \, \theta_1 \, |, | \, \theta_j \, |)$ for T_j.

The Connection Machine CM-200 . Algorithm 4 has been implemented on
the Connection Machine CM-200 as a routine which is called interactively. Thus,
each time that the operation Aq_j has to be performed, control is returned to the
user (*Reverse Communication Strategy*, see [4]). Clearly, Algorithm 3 may be
considered as a driver routine and would normally have to be supplied by the
user who is then free to exploit the structure of the algorithm.

Observe that k is completely independent of p and therefore k can always be
small. Thus, Algorithm 4 lends itself to efficient and highly parallel implementation in the chosen computing environment. To facilitate this all vectors of length
n are declared and stored as distributed CM arrays, whereas all other quantities
are confined to the 'front-end' machine. Hence, all massive *saxpy* type operations
involving the n-vectors are *fine-grain* parallel operations and are consequently
performed on the CM itself. These include the dense matrix-vector products in
the reorthogonalization steps **(1.1)**,**(1.7.3)** and in the computation of the Ritz
vectors in steps **(2)** and **(1.7.2.1)**. In contrast, the remaining computations,
i.e. those involving the matrix T_j, are computationally inexpensive and are performed on the 'front-end' machine, thereby enabling heavy use to be made of
functions from the BLAS library.

The storage and computational requirements of the method are proportional to
$p + k$, rather than to a multiple of p as is the case in, for example [4]. Thus, the
method can be expected to perform well if the number of eigenvalues requested
is relatively large, say ≈ 100.

5 Numerical Results

The performance of the restarting scheme given by Algorithm 3, the driver program, where the value of k was chosen intuitively, has been compared with the
performance of the appropriate driver program calling the symmetric Arnoldi
(Lanczos) routine 'SSAUPD' from the ARPACK library, [4]. The Lanczos routine incorporating Algorithm 4 in a *reverse communication'* manner is called
LAND1. Also, a *full reorthogonalization* variant of LAND1, LAND, has been
implemented. In this variant the entire step **(1.7)** of Algorithm 4 has been replaced by the orthogonalization of q_{j+1} against $q_1, \ldots, q_j$. The performances of
single precision, Connection Machine versions of the three methods, SSAUPD,
LAND and LAND1, have been compared using six sparse matrices selected from
the Harwell-Boeing collection, [2]. The results for p, ranging from 1 to 128, are
shown in Tables 1, 2 and 3. In the header of each Table nz denotes the number
of nonzero elements in the upper triangle of the matrix (for more details see [2]).
Also in the case of the SSAUPD routine k denotes the number of Lanczos steps
after which an 'implicit' restart is made ; the chosen mode is 'exact shifts' (see
[4], [10]).

For reasons of compatibility with the SSAUPD routine the convergence tests in
LAND and LAND1, **Step (1.3)** employ the error bounds $\mid \beta_j \mid\mid s_{ji} \mid$ rather than
the equivalent $\parallel Ay_i - \theta_i y_i \parallel_2$, although the user is free to choose either. The
requested tolerance has been set in all cases to **u**, the relative machine accuracy.

The bottom row of each Table gives the accuracy of the computed solution in the form $c\mathbf{u}$ (or $c\sqrt{\mathbf{u}}$), where c is a natural number. This accuracy has been computed 'independently' as

$$\max(\frac{\| A\hat{x}_i - \hat{\lambda}_i\hat{x}_i \|_2}{| \hat{\lambda}_i |}) \ ; \ i = 1,\ldots,\max(p)$$

In all cases the initial value of q_1 is chosen to be $\frac{1}{\sqrt{n}} \times [1,1,\ldots,1]$ and the products Aq_j are computed using the Connection Machine Scientific Support Library routine 'sparse_matvec_mult'.

	NOS2			NOS3		
	$n = 957$ $(nz = 2,547)$			$n = 960$ $(nz = 8,402)$		
	$\lambda_1 = 1.57283 \times 10^{11}$			$\lambda_1 = 6.89904 \times 10^2$		
	$\lambda_{128} = 1.02958 \times 10^{11}$			$\lambda_{128} = 3.50799 \times 10^2$		
p	SSAUPD $k =$ $\max(16, 2p)$	LAND $k = 64$	LAND1 $k = 64$	SSAUPD $k =$ $\max(16, 2p)$	LAND $k = 32$	LAND1 $k = 32$
1	50.7 (384)	14.8 (832)	5.2 (903)	6.5 (212)	1.3 (96)	0.9 (119)
2	76.4 (1077)	32.3 (1,792)	10.4 (1,756)	8.5 (275)	2.6 (192)	1.7 (228)
4	59.4 (747)	55.2 (3,008)	17.4 (2860)	5.5 (141)	6.3 (452)	3.5 (455)
8	56.9 (573)	83.2 (4,424)	29.2 (4,593)	11.0 (147)	15.2 (1,024)	8.1 (984)
16	73.4 (482)	139.4 (6,976)	48.2 (6,999)	23.1 (150)	33.7 (2,080)	19.6 (2,174)
32	169.0 (423)	229.1 (10,368)	83.3 (10,450)	80.5 (196)	75.0 (3,904)	45.1 (4,275)
64	454.2 (416)	395.3 (15,104)	156.1 (15,472)	363.0 (281)	204.9 (8,032)	120.0 (8,766)
128	741.2 (384)	809.6 (22,912)	327.9 (22,661)	1354.0 (388)	672.3 (17,632)	373.5 (18,571)
	31u	5u	16u	41u	753u	3u

Table 1. Time (in seconds) for matrices NOS2 and NOS3 from the LANPRO collection (figures in brackets show the number of matrix-vector products)

In the case of matrix NOS7 the ARPACK routine failed to find $\hat{\lambda}_{128}$. Also, in a number of examples with simple, diagonal matrices, the ARPACK routine failed to detect multiple eigenvalues.

6 Conclusions

Tables 1 and 2 show that the restart strategies proposed in this paper require significantly more matrix vector products of the form Aq_j than does the ARPACK routine. Nevertheless, their use is to be preferred for the computation of the solution of the majority of examples presented in these two tables. Observe, also, that LAND1 is considerably more efficient than LAND. The reduction in the

	NOS6			NOS7		
	$n = 675$ ($nz = 1,965$)			$n = 729$ ($nz = 2,673$)		
	$\lambda_1 = 7.65060 \times 10^6$, $\lambda_{128} = 6.53867 \times 10^5$			$\lambda_1 = 9.86403 \times 10^6$, $\lambda_{128} = 7.11115 \times 10^2$		
p	SSAUPD $k = \max(16, 2p)$	LAND $k = 32$	LAND1 $k = 32$	SSAUPD $k = \max(16, 2p)$	LAND $k = 32$	LAND1 $k = 32$
1	2.9 (11%)	0.7 (57%)	0.4 (88%)	0.7 (21%)	0.4 (41%)	0.4 (53%)
2	3.9 (14%)	1.5 (56%)	0.8 (85%)	1.4 (18%)	8.1 (42%)	2.5 (83%)
4	3.7 (13%)	2.8 (54%)	1.6 (81%)	4.2 (11%)	24.0 (33%)	3.2 (82%)
8	6.9 (9%)	6.0 (50%)	3.4 (77%)	5.1 (8%)	30.3 (40%)	4.8 (79%)
16	15.5 (4%)	14.4 (43%)	8.1 (69%)	8.5 (4%)	78.9 (34%)	24.1 (65%)
32	59.0 (2%)	38.1 (34%)	21.8 (56%)	18.9 (3%)	107.5 (31%)	48.3 (56%)
64	117.5 (1%)	127.4 (23%)	66.8 (41%)	192.8 (1%)	208.2 (24%)	85.0 (79%)
128	134.3 (1%)	420.5 (13%)	180.0 (28%)	1816.4 ($\approx$0%)	585.9 (15%)	274.5 (28%)
	59u	31u	21u	$11\sqrt{u}$	$4\sqrt{u}$	$17\sqrt{u}$

Table 2. Time (in seconds) for matrices NOS6 and NOS7 from the LANPRO collection (figures in brackets show the percentage of total time taken for the computation of matrix-vector products)

amount of reorthogonalisation required is the contributory factor in this case[4]. For the solutions of the examples given in Table 3 the difference between the two restart strategies is less marked. This may be attributed to the fact that the matrices concerned are much larger than those in the previous tables. Further, for the matrix BCSSTK18, the restart strategies discussed in this paper are much more efficient than the ARPACK routine whereas, in general, the latter is to be preferred in the case of the matrix BCSSTK17. The contributory factor in this case is the denseness of BCSSTK17 which significantly increases the time required for the computation of the matrix vector products. This suggests that the LAND algorithms will tend to perform better on special matrices such as narrow banded symmetric matrices. Moreover, it is expected that these algorithms will be well suited to the following situations :

(i) where p eigenpairs are given and a further r eigenpairs are required
(ii) where p eigenpairs of a matrix A are given and the corresponding p eigenpairs of the slightly perturbed matrix A are required
(iii) where the number of nodes in finite-element approximation has been increased and where the solution for the smaller problem is known.

Although Algorithm 3 seems to cope better than the ARPACK routine with coincident and closely clustered eigenvalues, both algorithms exhibit similar difficulties with clusters of eigenvalues of relatively smaller magnitudes. Thus, for

---

[4] the other factor is that if reorthogonalization can be avoided then a 'purer' Krylov subspace is constructed and, consequently, convergence is faster

	BCSSTK17			BCSSTK18		
	$n = 10,974$ ($nz = 219,812$)			$n = 11,948$ ($nz = 80,519$)		
	$\lambda_1 = 1.29606 \times 10^{10}$, $\lambda_{32} = 4.53540 \times 10^9$			$\lambda_1 = 4.29520 \times 10^{10}$, $\lambda_{32} = 1.93316 \times 10^{10}$		
p	SSAUPD $k = \max(16, 2p)$	LAND $k = 32$	LAND1 $k = 32$	SSAUPD $k = \max(16, 2p)$	LAND $k = 32$	LAND1 $k = 32$
1	9.9 (2)	11.4 (2)	7.4 (1)	8.3 (2)	2.3 (1)	2.9 (1)
2	16.4 (3)	17.1 (3)	16.2 (2)	8.7 (2)	4.6 (2)	6.1 (2)
4	20.5 (4)	28.7 (5)	33.8 (4)	21.3 (5)	11.5 (5)	12.3 (4)
8	34.4 (9)	69.9 (12)	74.3 (9)	36.9 (10)	30.6 (13)	29.4 (8)
16	84.1 (5)	159.5 (27)	150.8 (17)	245.5 (15)	138.8 (43)	116.5 (45)
32	373.1 (4)	825.3 (101)	782.9 (98)	737.6 (7)	393.9 (128)	374.3 (154)
	24u	42u	3u	61u	15u	3u

Table 3. Time (in seconds) for matrices BCSSTK17 and BCSSTK18 from the BCSSTRUC2 collection (figures in brackets show the number of restarts)

matrix NOS7 in Table 2, the accuracy of the solution deteriorates when more than 64 eigenvalues are required.

References

1. Daniel, J., Gragg, W.B., Kaufman, L., and Stewart, G.W., (1976),'Reorthogonalization and stable algorithms for updating the Gram-Schmidt QR factorization',*Math. Comp.*, **30**, 772-795.
2. Duff, I.S., Grimes, R.G., and Lewis, J.G., (1992), 'User's Guide for the Harwell-Boeing Sparse Matrix Collection' (release I), available online ftp `orion.cerfacs.fr`.
3. Golub, G., and Van Loan C.F., (1989), 'Matrix Computations', John Hopkins University Press, London.
4. Lehoucq, R., Sorensen, D.C., and Vu, P.A., (1994), SSAUPD : Fortran subroutines for solving large scale eigenvalue problems, Release 2.1, available from `netlib@ornl.gov` in the `scalapack` directory.
5. Paige, C.C., (1976), 'Error analysis of the Lanczos algorithm for tridiagonalizing a symmetric matrix', *J. Inst. Math. Applic.*, **18**, 341-349.
6. Parlett, B.N., (1980), 'The Symmetric Eigenvalue Problem', Prentice-Hall, Englewood Cliffs, N.J.
7. Parlett, B.N., and Scott, D.S., (1979), 'The Lanczos algorithm with selective orthogonalization', *Math. Comput.*, **33**, 217-238.
8. Rutter, J.D., (1994), 'A Serial Implementation of Cuppen's Divide and Conquer Algorithm for the Symmetric Eigenvalue Problem', LAPACK Working Note 69.
9. Simon, H.D., (1984), 'Analysis of the Symmetric Lanczos Algorithm with Reorthogonalization Methods',*Linear Algebra Appl.*, **61**, 101-131.
10. Sorensen, D.C., (1992), 'Implicit Application of Polynomial Filters in a k-step Arnoldi Method',*SIAM J. Matrix Anal. Appl.*, **13**, 357-385.

Dynamic Systems and Software

Per Grove Thomsen[1]

The Technical University of Denmark, Department of the Mathematical Modelling,
DTU, Bldg. 305, DK-2800 Lyngby, Denmark

Abstract. The present paper reviews some of the experiments that are carried out to study dynamic systems. The dynamic systems are defined mathematically in the form of Ordinary Differential Equations (ODE's) or Differential Algebraic Equations (DAE's).
Experimental methods for studying dynamic systems are applied to identify periodic solutions, stationary points, bifurcations and to make portraits or iterated maps. In such experiments the quality of the results depend heaviky on the properties of the numerical methods that are applied to get the results.
For the two types of systems the appropriate properties are discussed and a new ESDIRK method is presented as a candidate for these studies.

1   Introduction

We will consider dynamic systems as they appear in different application areas like models of mechanical systems, chemical reactions, electrical circuits or classical celestrial mechanics. Common to these problems is that they are studied by looking at the system equations that represent a mathematical model of the system.

The idea of making the study of dynamic systems an experimental activity was first mentioned by Poincaré (1899) who wrote:

> The canonical equations of celestial mechanics do not admit
> (except for some exceptional cases to be discussed seperately)
> any analytical integral besides the energy integral.

Later one of the fathers of modern computing J.von Neumann (1946) made a statement that more explicitly mentions the use of computing:

> Our present analytic methods seem unsuitable for the solution
> of the important problems arising in connection with nonlinear partial
> differential equations
> and in fact with virtually all types of nonlinear problems in pure mathematics.

The sort of dynamic systems we consider belong to the group mentioned by von Neumann and the idea of making experiments of a computational nature to gain insight is a natural consequence of his observation.

Of cause the experiments will involve using numerical methods for the solution of systems of differential equations thus the numerical solution become an integral part of the experiment.

2   Dynamic Systems

We consider systems that can be described by a state vector in N-dimensional vector space

$$u(t) = (u_1(t), u_2(t), \ldots, u_N(t)) \in \Re^N, t \in \Re. \tag{1}$$

The traditional problems that have been treated by textbooks and in the litterature at length are those that assume the system to be determined by a system of ODE's

$$\dot{u}(t) = f(u(t)), u(o) = u_o \in \Re^N, t \in \Re. f : \Re^N \to \Re^N. \tag{2}$$

The unique solution to the system is assumed to exist , imposing smoothness conditions on the function f.

By defining an appropriate semigroup $S(t) : \Re^N \to \Re^N$ we can generate the solution at any point in time as

$$u(t) = S(t)U. \tag{3}$$

We are going to generate the solution by using a numerical method which will generate a sequence of solution points or states. $\{U_n\}_{n=0}^M$ that approximates the exact solution $u(t)$. at discrete points in time. We may assume a constant time interval Δt The solution at $t_n = n\Delta t$ may be written

$$U_n = S_{\Delta t} U_{n-1} = S_{\Delta t}^n U. \tag{4}$$

The goal of the experiment gain insight from the numerical results. This makes it necessary to compare in some sense the results from the numerical computation with the exact solution and in this comparison the concepts of convergence and stability are the tools traditionally used in numerical analysis.

Fixing a particular initial value U and a point point T in time as the point of interest following [5] we will consider convergence in the sense that

$$\limsup_{n\Delta t=T, \Delta t \to 0} \frac{\| u(T) - U_n \|}{\Delta t^r} \leq C \tag{5}$$

$$\limsup_{n\Delta t=T, \Delta t \to 0} \frac{\| S_{Deltat}^n U - S(T)U \|}{\Delta t^r} \leq C \tag{6}$$

for some constant C such that $C\Delta t^r$ is small. The basis for this is the concepts of truncation error of order $r \geq 1$ and the well posedness of the given problem. While these conditions are satisfied by most of the numerical solution methods in the case of systems of ODE's the case of DAE's is far from simple and well understood.

The general DAE' problem may be stated as the solution to an equation

$$F(t, u, \dot{u}, z) = 0 \tag{7}$$

A thorough treatment of the existency of solutions to this general form of problem may be found in [1] here we refer the reader to the excellent review [2] where the results we need are given. For simplicity we only consider the semi-explicit case where the system of DAE's are given in the following form

$$\dot{u} = f(t, u, z) \tag{8}$$
$$0 = g(t, u, z) \tag{9}$$

The question of finding the numerical solution and proving the convergence for this problem is connected to the index, a property of the system connected to the rank of the matrix $D_z(g)$. If it has full rank then we have an index-one problem and the conditions on the convergence of a numerical one-step method can be expressed in a constructive form (see section 4) and such methods are known to exist.

3 Tasks

The types of experiments that are carried out are designed to give different kinds of information about the dynamic system. Searching for periodic solutions may be done by creating iterated maps, singular points can be found using solution of nonlinear systems of algebraic equations. Eigenvalues of local Jacobians may be determined and tests for periodical solutions done using continuation methods that involves solution of the system of ODE's over long time intervals. The tasks will involve for the main part a numerical ODE-solver that should be robust for reasons of correctness of the solution, it should also be efficient because many integrations have to be carried out.

4 Methods

Traditionally studies have been carried out using explicit methods to advance the solution step by step and this type of process works well for non-stiff problems. Hovever when the problem gets stiff and when we are dealing with DAE's this type of method does no longer satisfy requirements of robustness and efficiency in the computations.

Define the stages

$$F(t_n + c_i \Delta t, u_n + \Delta t \sum_{j=1}^{m} a_{ij} U_j, U_i) = 0 \tag{10}$$

$$for U_1, U_2, ..., U_m \in \Re^N \tag{11}$$

$$u_{n+1} = u_n + \Delta t \sum_{i=1}^{m} b_i U_i. \tag{12}$$

The coefficients which characterise the method are traditionally presented in a Butcher tableau of the following form
The implicitness of the method is a necessity in order to satisfy the requirements

$$
\begin{array}{c|cccc}
c_1 & a_{11} & a_{12} & \ldots & a_{1m} \\[4pt]
c_2 & a_{21} & a_{22} & \ldots & a_{2m} \\[4pt]
c_i & a_{i1} & a_{i2} & \ldots & a_{im} \\[4pt]
\hline
u_{n+1} & b_1 & b_2 & \ldots & b_m \\[4pt]
\hline
\tilde{u}_{n+1} & \tilde{b}_1 & \tilde{b}_2 & \ldots & \tilde{b}_m \\
\end{array}
$$

Butcher tableau for a general Implicit Runge Kutta method with m stages..

for the solution of the DAE'type of problems. They may be stated as the property that the internal stages as approximations to the solution must have order greater than one, imposing some conditions on the coefficients in the Butcher tableau. In [3] an account of the necessary requirements that have to be met by a robust method of Runge Kutta type may be found. In the following section we give an example of such a method of the ESDIRK type. It has low order and may be implemented in an efficient way for general purpose use, reference to such an implementation is given in section 6 where the Generic ODE Solving System is presented.

5   ESDIRK-3,4 method

In order for a numerical method to be robust in connection with nonlinear systems for ODE's and DAE's a set of requirements must be fulfilled. Among the family of one-step methods it has been found that the socalled ESDIRK methods with stage order higher than 1 are candidates and we are introducing one with four stages of order three and stage order two for this purpose. The method has the Butcher tableau shown below:

The method has been developed with the purpose of deriving a low order robust method for general use when the accuracy requirements are not too severe and has shown in tests to be competitive for this type of application.

$$
\begin{array}{c|cccc}
0 & 0 \\[4pt]
\frac{5}{6} & \frac{5}{12} & \frac{5}{12} \\[4pt]
\frac{10}{21} & \frac{95}{588} & -\frac{5}{49} & \frac{5}{12} \\[4pt]
1 & \frac{59}{600} & -\frac{31}{75} & \frac{539}{600} & \frac{5}{12} \\[4pt]
\hline
y_{n+1} & \frac{59}{600} & -\frac{31}{75} & \frac{539}{600} & \frac{5}{12} \\[4pt]
\hline
\tilde{y}_{n+1} & \frac{96}{600} & \frac{6}{75} & \frac{4116}{6000} & \frac{18}{132}
\end{array}
$$

Coefficients for the 4-stage ESDIRK method of order 3.

6 GODESS

The GODESS (Generic ODE Solving System) software is aimed at studying properties of numerical methods for ODE's and DAE's. The package is realised by using object oriented programming, a particular method is an instance of an object among many possibilities. The different methods of which there are many to choose from can be specified at the time of execution, in the present version any Runge-Kutta or multistep method can be specified. The methods may be tested in identical environments on the same problems whereby direct comparisons between methods is possible.

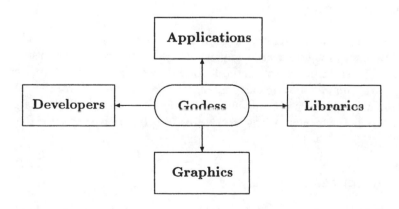

Godess - structure diagram

The diagram represents the way that Godess should be understood and put to work. Godess is a kernel of objects that can be used to solve problems from many applications which may have seperate user interfaces. It delivers results to a graphical system and makes use of one or more libraries of objects and databases with methods and data.

7 Software for Dynamic Systems

References in the litterature on dynamic systems to software for the kind of applications referred to in this paper are quite sparse. Some of the packages are widely used while others are only known to smaller groups. A complete list will be hard to give without forgetting some so here are some of the best known and most recent examples.

NAME	Language	Type	Year
MANPAK	Fortran	package of utility programs	1996
PATH	Fortran	package of utility subroutines	1985
GODESS	C++	Object Oriented class library	1995
AUTO	Fortran	[7]	1986

None of these program products are making use of parallel processing although the repeated solution of initial value problems would lend themselves to parallelization quite easily. Also the iterated maps can be generated in parallel by spreading different starting points over different processors.

8 Conclusion

The study of dynamic systems is an area for development of parallel software products since many tasks are easily distributed over networks or clusters of processors. The state of the art methods for solving ODE's and robust implementations are valuable tools in carrying out the computational work in connection with the numerical experiments involved. It is expected that the near future will see more efficient contributions to the area of software packages for this type of study.

9 References

1. K. E. Brenan, S. L. Campbell, and L. R. Petzold, *Numerical Solution of Initial Value Problems in Differential-Algebraic Equations,* North Holland Publ.Co., New York, NY, 1989.

2. W. C. Rheinboldt, *The Theory and Numerics of Differential-Algebraic Equations* In: Advances in Numerical Analysis,ed. W. Light. Clarendon Press , Oxford 1991.

3. W. C. Rheinboldt, *MANPAK: A Set of Algorithms for Computations in Implicitly defined Manifolds.* , Inst for Comp.Math. and Appl. Univ. of Pittsburgh, Tech Report TR-ICMA-96-198, july 1996.

4. H. Olsson, *Practical Implementation of Runge-Kutta Methods for Initial Value Problems.* Dept. for Computer Science, Lund Institute of Technonogy, nov. 1995 (thesis).

5. W. J. Beyn, *Numerical Methods for Dynamical Systems.* In: Advances in Numerical Analysis,ed. W. Light. Clarendon Press , Oxford 1991.

6. A. Stuart, *Convergence and Stability in the Numerical Approximation of Dynamical Systems* SCCMP, Stanford University, 1996.

7. E. J. Doedel and J. P. Kernevez, *AUTO: Software for continuation and bifurcation problems in ordinary differential equations* Appl.Math. Technical Report, Caltech 1986.

8. Chr. Kaas-Petersen, *Computational Methods for Continuation and Bifurcation of Nonlinear Dissipative Dynamic Systems* DCAMM report No. S28, august 1985. DTH, Lyngby, Denmark.

Parallel Implementations of Classical Optimization Methods

Ole Tingleff

Department of Mathematical Modelling,
The Technical University of Denmark,
DK-2800 Lyngby, Denmark
e-mail: ot @ imm.dtu.dk

Abstract. In this paper we present parallel implementations of 2 classical optimization methods, namely the Quasi-Newton method for non-linear minimization and the Marquardt method for non-linear least squares problems. Both methods are iterative, and we discuss which parts of the computations in an iteration step are suited for parallel computations. We discuss the choice of configuration of the transputer network and present results of test runs of the implementations.

1 Introduction

In optimization we are concerned with the problem of finding an argument vector corresponding to a minimal value of a twice differentiable scalar function of a vector variable. An important special case is the non-linear least squares problem where the function in question is the sum of squares of a number of functions as above, i.e. twice differentiable scalar functions of a vector variable.

Problems of these kinds are very common in practice and often they are quite large or the functions are complicated and take much time to compute. Thus parallel programmes for these problems are very useful.

The parallel programmes described are for a transputer system. One of the advantages of such a system is that it is reconfigurable; we can adapt the hardwire connections of the system to the requirements of the programme. We use a cube structure for the optimization programmes.

We use a communication harness when building the parallel implementations. In such a system there is a (nearly) perfect separation between the parts of the programme concerned with computation and the parts dealing with data transmission. This brings many advantages. See [2] for a discussion of the communication system.

The programming language is occam 2 under the occam toolset, see [1].

2 Classical Methods on Optimization

Consider a scalar, real function of a real vector variable. We assume that the function is twice continuously differentiable and seek an argument vector which gives the least value of the function, i.e.

$$x^\dagger = arg \min_{x} f(x), \ f \in C^2(R^n \to R) \tag{1}$$

The function f is called the criterion function.

An important special case is the problems of non-linear least squares where the criterion function is the sum of squares of a number of twice differentiable functions, i.e.

$$x^\dagger = arg \min_{x} \sum_{1}^{m} (f_j(x))^2, \ f_j \in C^2(R^n \to R) \tag{2}$$

or

$$f(x) = \sum_{1}^{m} (f_j(x))^2$$

Normally, it is very difficult to determine $x^\dagger$ in (1) or (2). Instead of finding $x^\dagger$, the global minimizer, we usually have to make do with one of the local minimizers, x^* which gives minimal value of the criterion function inside a region around x^*, i.e.

$$x^* = arg \min_{x \in D} f(x) \ with \ D = \{x \mid ||x - x^*|| < d\} \tag{3}$$

In this paper we deal with deterministic methods of finding a local minimizer. The methods are iterative; in each iteration step we seek a direction from the present position, in such a way that (at least initially) the value of the criterion function decreases. We determine the length of the step to be taken, either by a socalled linesearch or by other means. Such methods are called descent methods, and a scetch of the algorithm is this:

DESCENT METHOD

```
BEGIN
    k := 0                                  {counter}
    x := x₀                                 {starting position}
    REPEAT
        h := search_direction (x)           {··· from x, downhill}
        IF this is impossible
            stopping_criterion := TRUE
        ELSE
            α := step_length(x, h)          {··from x in direction h}
            x := x + αh                     {new position}
            k := k+1                        {counting}
            stopping_criterion := ···
    UNTIL stopping_criterion OR k > kmax
END         {of Decent Algorithm}
```

Fig. 1   Algorithm for Descent Method

The methods we describe are based on the following equation, Newton's equation:

$$f''(x)h = -f'(x) \tag{4}$$

Here $f'(x)$ is the gradient of the criterion function, the vector of first partial derivatives, and $f''(x)$ is the Hessian matrix containing all the second partial derivatives of f. If we use the solution of (4) as search direction in Fig. 1 and set $\alpha=1$ we get an efficient method. It finds the minimizer of a positive definite quadratic in the first step, and it has quadratic final convergence when used on a convex non-linear function (except in degenerate cases). This method, Newton's method, is not used because in real applications we can't find the second derivatives of the criterion function.

The most widely used method for optimization is the Quasi-Newton method, which is a descent method based on an approximation to Newton's equation (4). The coefficeint matrix B is an approximation to the Hessian matrix $f''(x)$. We update matrix B in each iteration step using the change of the gradient:

QUASI-NEWTON METHOD

```
BEGIN
    k := 0; x := x₀; B := B₀              {innitations}
    REPEAT
        IF f'(x) ≃ 0                       {x stationary}
            stopping_criterion := TRUE
        ELSE
            SOLVE B h = -f'(x)            {··for search-direction h}
            α := step_length (x, h)        {·· from x, along h}
            x := x + αh; k := k+1
            B := update (B)                {··· better approximation}
    UNTIL stopping_criterion OR k > kmax
END          {of Quasi-Newton Algorithm}
```

Fig. 2   Algorithm for Quasi-Newton Method.

In Fig. 2 we normally update the coefficient matrix using the socalled BFGS formula. This consists of a correction term, based on the change of the gradient, and of rank 2. When using this we get convergence towards the Hessian matrix at the solution, $f''(x^*)$ and also ensure that if $\| h \|$ satisfies a certain condition, the matrix B will remain positive definite in the entire iteration, as is the Hessian matrix of convex functions.

For non-linear least squares problems (2) we normally use Marquardt's method also called Damped Gauss-Newton method. We could use a normal quasi-Newton method, but Marquardt's method takes the structure of (2) into account and thus it is more efficient. It is actually a modification of a quasi-Newton method where the fundamental approximation is a linear approximation to each of the function components f_j at the present position.

$$f_j(x + h) \simeq l_j(h) = f_j(x) + h^T f_j'(x) \tag{5}$$

The matrix B is the Hessian matrix of the sum of the squares of l_j. Marquardt introduced a damper, λ ($\lambda \geq 0$) into the coefficient matrix and fixed the step parameter, $\alpha = 1$. Here comes the algorithm:

MARQUARDT'S METHOD

```
BEGIN
    k := 0; x := x₀; λ := λ₀                    {innitations}
    REPEAT
        IF f'(x) ≃ 0                            {x stationary}
            stopping_criterion := TRUE
        ELSE
            SOLVE (B + λI)h = -f'(x)   {··· for search-direction}
            x := x + h; k := k + 1
            λ := update(λ)
    UNTIL stopping_criterion OR k > kmax
END         {of Marquardt Algorithm}
```

Fig. 3   Algorithm for Marquardt's Method.

The damper λ serves 2 purposes. If the B matrix is ill-conditioned or singular, $B + \lambda I$ will be better conditioned and positive definite when λ is big enough. This ensures, that the direction h is downhill. Secondly, if the fundamental approximation (5) is not good enough, a greater value of λ will produce a shorter step h, improving the approximation.

3 Parallel Implementation of the Quasi-Newton Method

In parallel computation the performance decreases when there is too much communication between the processors of the system. For this reason we have chosen to run the main iteration on one of the processors, the Master processor, and use the rest of the processors, the slave processors for parallel computation of selected parts of the algorithm.

In the quasi-Newton algorithm (see Fig. 2) the computationally intensive parts are: 1) determination of the value of the criterion function and its gradient and 2) the solution of the system of equations determining the search direction. The rest of the computations is concerned with inner and outer products of vectors and matrix-vector products, and none of these parallelize well.

The parallel computation of the gradient is suited for a processor farm with tree structure, because the slave processors need not exchange data during the computations. On the other hand when solving the system of linear equations the slaves exchange data during the computations, and a cube structured processor farm is approppriate. Now, it is extremely time consuming (and complicated) to alter the connections of the processor farm during the computations. We had to make a choice, and we chose the cube structure.

In each iteration step, the master transputer computes the value of the criterion function. There is an option for the master to compute some auxiliary terms which help the computation of the gradient components. If, e.g. the function has a term with sin x + cos x, the gradient components have a term cos x - sin x, and there is no need to recompute those. The auxiliary terms are linked to "the rear" of the position vector and broadcast to all the transputers. Each of these determines its share of the gradient components and returns them to the master who collects the entire gradient.

There is a version of the programme where the quasi-Newton system of equations is solved in parallel. The method is a block version of the Gauss-Jordan elimination method, as described in [2]. If the equation system is found to be numerically singular, a constant term is added to all its diagonal coefficients and the iteration step is repeated.

4 Parallel Implementation of Marquardt's Method

When dealing with non-linear least squares problems we consider the vector function $f(x) = [f_j(x)]$ the components of which are the functions whose sum of squares we minimize. The first partial derivatives of f form the Jacobian matrix, $J(x)$ whose rows are the (transposed of the) gradients of each of the functions f_j.

Two parts of the algorithm, Fig. 3 have been parallelized. First the master computes the vector function $f(x)$ and (optionally) some auxiliary terms. After that each transputer determines its share of the rows of the Jacobian matrix, and the complete Jacobian is collected by the master in the same way as in Section 3.

In a version of the programme, Marquardt's system of equations (see Fig. 3) is solved in parallel, exactly as described in section 3. If a system is considered numerically singular, Marquardt's damper λ is increased and the iteration step is repeated.

There is a third possibility for parallel computation in the Marquardt algorithm. If we introduce $f(x)$ and $J(x)$ from above, the central system of equations becomes:

$$(J^T(x)J(x) + \lambda I)h = -J^T(x)f(x)$$

In the present version, both coefficient matrix and right-hand side are computed by the master after parallel computation of $J(x)$. Instead, each transputer could compute its contribution to the coefficient matrix (and the right-hand side) and return it to the master for the final accumulation, i.e.

$$J^T(x)J(x) = \sum_{i=1}^{t} J_i^T(x)J_i(x)$$

where t is the number of transputers. Each term on the right-hand side is a full matrix of the size of the left-hand side. This should parallelize well because there is $O(n^3)$ operations and we only have to transmit $0(n^2)$ data. This is next on the aggenda for the Marquardt programme.

5 The Parallel Derivative Checker

It is our experience that runs with good optimization software are often ruined because the user introduces errors when determining and implementing the derivatives, the gradient vector or the Jacobian matrix. The components of these must be determined by pen and paper or perhaps by an automatic differentiation programme.

For this reason we provide a parallel gradient checker and a parallel Jacobian checker. They compare the user's implementation of the derivatives with forward and backward and also with central difference approximations. The parallel versions of the derivative checkers have not been made in the hope of gaining any speed up, but rather in order to tell the user whether the derivatives have been computed correctly and also transmitted and collected correctly by the communication harness.

6 Conclusion

This paper is meant as a status report on ongoing work.

The non-parallel versions of the derivative checkers are running, as are the parallel versions. The optimization algorithms are running in sequential versions as well as versions with parallel derivative computations and also versions which in addition to this have parallel solution of the systems of linear equations. The programmes are being tested on 3 very simple problems of arbitrary dimension and the results are correct. Still, there are some minor bugs in the newest versions. When they have been found (and the parallel matrix-matrix product has been incorporated) we come to the tuning of the programmes for speed.

We have chosen efficient methods for the non-parallel versions. The equation systems are positive definite so we use an efficient method for such systems, i.e. an LDL^T method with the coefficient matrix packed in an array with only one index. Actually, all matrices in the iteration algorithms are packed in such arrays, and because they are symmetric only one half of the matrices are stored.

The early versions of the parallel programmes have two-dimensional arrays for simplicity and ease of debugging. Arrays of arbitrary size can be transmitted very efficiently on transputer systems provided they only have one index. Thus the matrices have to be packed and repacked, causing some delay; still, this is only $O(n^2)$ allocations, or faster when block moves are used. The innermost loops of an equation solver corresponds to $O(n^3)$ instances each with 2 index calculations. If this is for a one dimensional array (and with clever index calculations) we get a substantial gain in speed; that we know from the sequential implementaion. In the final versions we want packed coefficient arrays everywhere.

Also on the subject of the equation solver: The non-parallel method, the LDL^T does not parallelize too well, so we have chosen the non-symmetric Gauss-Jordan method for the parallel versions. This means that the nominator and denominator of the speedup are execution times for 2 different algorithms. Had we chosen exactly equivalent implementations, one of the methods would be suboptimal. When discussing speedup we need to specify the meaning of the word.

7 References

1. SGS-Thomson: occam 2.1 Toolset, SGS-Thomson (1995)
2. Tingleff, O.: Communication Harnesses for Transputer systems with Tree Structure and Cube Structure. In: Applied Parallel Computing, Dongara, J., Madsen, K. and Wasniewsky, J. (eds.), Springer 1996.

Computer Engineering of Complex Physical and Chemical Systems by Parallel Molecular Dynamics Computations

Søren Toxvaerd

Department of Chemistry, H. C. Ørsted Institute, University of Copenhagen,
DK-2100 Copenhagen Ø, Denmark

Abstract. Parallel Molecular Dynamics (PMD) of complex systems are based on different PVM strategies depending on the nature of the complex system. Two different examples of PMD by PVM will be given. The first example is a simulation, on sp2, of a phase separation in a binary mixture of 343 000 molecules which takes place within millions of time steps. This (physical) system is suitable for a domain decomposition PVM-strategy. The second example is the simulation of an active enzyme in different solvents. The (chemical) system is complex by its nature with many different chemical compounds and suitable for a PVM particle decomposition.

1 Introduction

The time evolution of a complex system can be obtained by integration of the classical equation of motion provided the force field between the center of masses is known. This technique is named Molecular Dynamics (MD) and has been used in physics and chemistry already back in the late fifties and the beginning of the sixties. Of course the "complex" systems simulated at that time were rather simple when looking at today's examples; but still they have played a very important role in understanding the interplay between objects in an ensemble and have contributed significantly to our understanding of complex systems such as the structure and dynamical property of liquids.

Up to this decade a typical simulation was performed by individuals- or at least by small groups which wrote their own master- programs (Fortran) and this master program could typically handle a whole spectrum of accessible systems. The accessibility was set by the capability of a single cpu and its RAM and disk space and a simulation could typically be an investigation of the equilibrium- and dynamical properties of a system of the order thousand particles and followed about a pico second. The limitation was set by the number of particles- or mass centers and the complexity of the force field. By introducing vector facilities and later parallel cpu's this situation changed significantly. It is in fact possible to reformulate the MD programs both for parallel- as well as for vector facilities- or both; but it is probably no longer possible to create a general MD master program and what is perhaps even more important from a scientific point of view, it is not any longer possible for an individual to go the whole way by himself, i.e. to formulate the physical or chemical model, write the program(s) and evaluate the model. The program packages are typically written by teams of scientist. The programs can only handle some limited part of accessible MD systems and they are very difficult to debug partly due to the complexity and partly to the fact that no individual person in the team has a total overlook of the whole code.

When looking at the status of MD on complex system we can now simulate a hundred thousand (simple) particles in up to 10 nanosecond and on the other hand simulate complex biological systems for so long time that its biological functions can be investigated. Futhermore the rapid development in parallel facilities, in performance of the individual cpu's makes it likely that the development will continue. For example we are now able to code MD for vector processing on parallel systems which again give us a factor and means that we can either simulate a system significantly longer or bigger. From an industrial point of view this development is significant. Although MD has been heavily supported not only by universities and by the national science foundations; but also by the industry. Its support, however, has been a kind of long time investment which has not yet paid off fully; but undoubtly will do so in the future.

The talk will concentrate on dealing with the parallel programming of two typical examples from physics and biochemistry and focus on the specific problems and their solutions. The first example is taken from physics and from an industrial point of view its application is in material science. The second example is a biochemical system where the outlook is computer engineering of new and hopefully better enzymes and drugs.

2 A Molecular Dynamics program

Generally speaking a MD program integrates the equation of motions for a given set of mass centers and a specified forcefield between them. Thus the programs naturally consist of four different parts where the first part is the "up start". Here the start positions are specified together with the force field. For a biological enzyme it can be the X-ray crystallographic configuration of the enzyme; but now surrounded by water molecules etc. All masses of the individual atoms in the systems, their chemical bond and the molecular skeleton, given by bondlengths, bond-angles and corresponding forces, must be specified as realisticially as possible and all values are taken from the latest scientific literature. Then at every time step the instant forces must be calculated in the second part of the program. The third part consists of integration of the equation of motions in order to determine the positions at a time δt later, an after looping in the program, time steps for time steps, the fourth part consists of data collection and evaluation. Surprisingly, perhaps the first part: the up start and the third part: the integration of the equation of motions only takes a negligible part of the program and its time-consumption. Almost all the problems- and where most of the cpu time is used, are allocated to calculate the instant force field. It is simply due to the fact that, an ensemble of N particles easily consists of the order of

N^2 forces, since almost all particles interact with all the others. This means that the number of arithmetic operations performed per time steps varies, for a naive MD program, as N^2 with the number of mass centers. This N- dependence of the computer time makes it impossible to simulate even rather small systems; but the dependence can easily be circumvented for some kind of force fields [1]. An example of a forcefield, for which the N^2-dependence can be overcome is what here is named a *physical* MD system. It can be written in PVM applying a domain decomposition between the different nodes in a parallel computer. This system consists of many rather simple and often identical mass centers and with a range of interactions which is small compared to the extension of the whole ensemble of particles. From an industrial point of view it is perhaps misleading to name such an example as belonging to (theoretical) physics since its applications could be many e.g. MD for material properties of polymers. On the other hand it is in contrast to the MD packages for simulating *chemical* and biological materials simply by that the parallel PVM code is totally different. This fact is coming from that we can perform a *domain decomposition* if the range of the interactions are smaller than the system size; but the complex biological materials with an overall active biological structure does not allow for such a decomposition since this active structure might depend on the longer ranged forces. However, also these programs can be written for parallel processing and in PVM because almost all the time is spent on evaluating the instant force field. We can evaluate part of the total sum of forces locally, provided that we know the general behavior of the field and its time- evolution. This strategy is some times named *particle decomposition*; but is not necessarily performed for particles allocated to a "home"-node. The calculations are performed partly on the individual nodes and with a carefully programming the load balance and the data streams can be controlled. The load balance and the necessary data communication between the cpu's, however, vary from model to model and from platform to platform, and this again means that all parallel packages of today are not master programs, but have to be modified in order to ensure an effective performance on a big parallel computer.

But what kind of MD system is interesting from a parallel-MD point of view? The answer is simple. Complex systems which have their interesting properties either because it is caused by many atoms like in the case of a polymer, proteins, enzymes, etc., or systems which only show their characteristic functions within time intervals longer than a nano-second, or both. They are interesting, since they can at the present only be simulated by use of parallel software. This fact is not always acknowledged in the literature; but on the other hand biology, (bio-)chemistry and physics are so rich in examples of systems which have these qualities.

3 Phase separation in complex systems

An example of a system where the behavior is determined by its many-body dynamics which only takes place over a rather long time intervals is the domain decomposition - or phase separation in physical and chemical systems. In fact it has been known for a while that the phases grow "algebraically", i.e. that the mean diameter ,R, of a domain-type increases with time t with an characteristic power- or exponent

$$R(t) \propto t^\alpha \tag{1}$$

and where the exponent α is given by the dominating hydrodynamics in the system, e.g. diffusion along the interfaces [2] and not by the forces between the individual particles. This (three dimensional,(3D)) system can only be evaluated by use of PVM [3]; but we shall, however, for simplicity and illustrative reasons demonstrate it for a 2D system which qualitatively behave identically to a 3D system and where only the value of the exponents might be dimensional dependent.

A system of different objects will separate in sub domains if it is physically-chemically more energetically favorable. This happens in nature if e.g. the temperature is lowered- or if a component is removed or produced behind the concentration of stability. The details are unimportant in this context. What is important is that we know the general physical chemical conditions for separations; but not the hydrodynamics. Furthermore we are not so much interested in the separation of a specific mixture; but rather on the general behavior of phase separation. With all this fact in mind we shall chose a forcefield which is as simple as possible and chemical reactions which contains as little as possible of chemical details; but knowing that the difference between this qualitative master program for phase separation and an real example is a matter of the pre-exponential factor in Eq.(1)-and computer time!

The systems for phase separations consist of many thousand particles, in 3D of 343 000; but for a 2D system 40 000 simple Lennard- Jones particles were found to be sufficient [4]. The immiscibility was ensured by taking different range of interaction between the different species; but all ranges of interactions were much smaller than the size of the container. Let the maximum range of interaction be r_c and the length of the (cubic) box be l. In principle all that has to be done is to perform a domain decomposition and associate particles in a given (cubic) domain to one "home"-node and then perform a list of those particles which are within a distance smaller than r_c from a neighboring domain. These particles interact with particles from the other domain and their positions must be stored in both nodes. Then after all forces between pairs of particles have been calculated, the total force on a given particle can be obtained as the sum of forces from particles in the same "home" domain as the particles plus the forces from the particles in the interface between the different nearby domains.

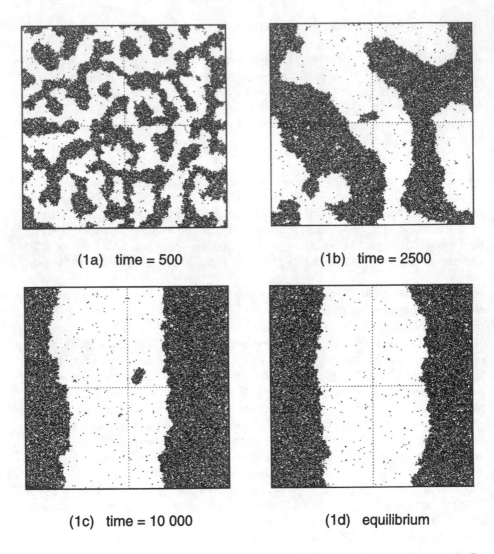

(1a) time = 500

(1b) time = 2500

(1c) time = 10 000

(1d) equilibrium

Fig. 1. Domain evolution in a two dimensional binary mixture of A- and B-(Lennard-Jones)particles. For clarity the figures only show the distribution of A-particles.

Once the force is known the new positions δt later are calculated and the game can go on; but at every time step the positions in the interfaces must be communicated to the neighbor nodes and after the forces between all the pairs have been calculated the forces must be send back. This means, generally speaking that the area of the interface must be small compared to the domain area of for the individual nodes or that

$$r_c \ll l \tag{2}$$

and secondly that the total system cannot be followed longer than what the individual nodes of a system can simulate. An example illustrates the problem: If it is necessary to simulate 1000 000 particles in order to obtain the phenomenon which takes place within 100 nanoseconds and the parallel system consist of 10 nodes we can only and in the very best case evaluate the model for a time interval corresponding to what a single cpu can perform for 100 000 particles. With today's computers (e.g.IBM RISC 6000/590 which is the unit in sp2) we can simulate a time of 1 nanoseconds for 100 000 simple particles within hours ($\approx 10 hours$); but only after a rather complicated programming [1] [3]. This means that a system of 1000 000 only can be followed for 1 nanosecond and only if it is possible to use this optimalisation for the individual cpu's together with the PVM. In other words it means that the system cannot be simulated within days or weeks. In the present case, however, it is in fact possible to combine the different optimalisation strategies and furthermore we noticed that the physics could be obtained for only 343 000 particles within 10 nanosecond, just enough to evaluate the model with the use of 16 nodes on Uni-C's sp2 computer within days and weeks. [3]. Before we show some results for this kind of parallel computations we notice that although the present sp2 system does not support vector processing on the individual cpu's the domain decomposition and with all MD optimalisation allow for this facility. This is not generally known, and the opinion among users of MD is that these programs does not vectorize very efficiently. This is due to the fact that optimalisation consists of making different tabulations of particles within the interaction range resulting in indirect addressing in the inner force-calculation loop. The indirect addressing also means that when applying a vector compiler, the force histograms will be wrong unless the list is sorted in a special way. It is, however, possible to do this by applying a fine domain grid; for details see [5]. The extensive calculations on the sp2-system for the three dimensional system with phase separation [3] determined the dynamics of phase growth, given among other things by the value of the exponent α in Eq. 1. For illustrative reason it is easier to demonstrate the phenomenon in two dimensions where Figure 1. shows the corresponding separation of phases.

The complex MD systems can be extended in many ways, e.g. to include chemical kinetics, [6]. Even such complicated reactions as chemical oscillations [7] can be simulated. Figure 2 shows the concentration of A-particles, [A(t)], as a function of time for the Lotka-Volterra mechanism

$$A + B \overset{k_1}{\rightarrow} 2B$$
$$B + C \overset{k_2}{\rightarrow} 2C$$
$$C + A \overset{k_3}{\rightarrow} 2A$$

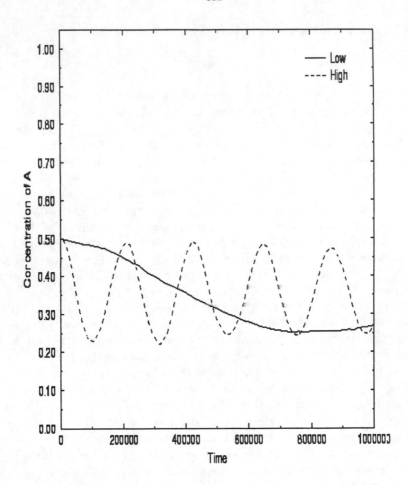

Fig. 2. The concentration of A versus time (the time is given in numbers of timesteps. One step corresponds typically to $\approx 10^{-14}$s in case of condensed simple fluids where the mean collision time is of the order 50 timesteps). "Hi" corresponds to high temperature, i.e. above the critical temperature, while "low" is the concentration below the critical temperature. The number of particles simulated were 16384 in both cases.

where k_1, k_2, and k_3 are the rate constants. As can be seen from the figure the concentration oscillations; for details see [7]. The computations were partly performed on Uni-C's sp2 system.

4 Simulating complex biological systems by parallel systems

During the last decades one of the main use of MD has been simulations of biological systems. These systems are very complicated; they consist typically of a "mother"-molecule of thousands of atoms, mainly carbon, hydrogen and nitrogen atoms. But also other atoms appear and in fact almost the whole periodical table of atoms are used to create us. These mother-molecules are then placed in biological environment like waters and clear enough the MD systems are a drastic reduction of a complex biological system; but the MD system often exhibits the right dynamical (biological) behavior. Still the MD system is big and complex from a computational view point and with a variety of force-fields including long-range forces and it means, as mentioned, that a domain decomposition is no longer effective. This is especially caused by the fact that many chemical forces are long ranged,. e.g. torsion forces and Coulomb forces and thus the domain-interfaces will be relatively big compared to the inner domains. The flexibility of the biological material means that sub domains continuously change its density and compositions so that it is hopeless, in advance to ensure a good load balance. Instead of a domain decomposition one performs what sometimes is named a particle decomposition. It is ,however, not a precise description since one has to distribute the computations of the pair-forces between the cpu's. A guidance is the consumption of cpu-time during a sequential calculation which can be used for a decomposition of the program. There is in principle not a general strategy since the computer time depends on the specific force field, e.g. the degree of electrostatic long-range forces. Nevertheless several MD-packages for simulating complex biological systems have appeared, and no doubt it is only the beginning. The MD packages consist typically of a hundred thousand lines in Fortran and offer a variety of MD facilities for data collections.

One MD package, *GROMACS*, is developed at Professor Berendsen's group at the University of Groningen [8] and is an PVM extension of *GROMOS*. It makes use of *Message Passing* technique, combined with a particle decomposition when the force field histogram is calculated. The programs are available for academics after signing a license agreement. One of us [9] has used the program for simulating an enzyme in an aqueous solution and has obtained a good degree of parallelization. Perhaps the relevant information from a computer calculation point of view is to describe the complexity of the actual biological system in more detail. The system which was simulated by *GROMACS* consists of an enzyme with 269 amino acids and its mass distribution is given by 2331 "atoms" mainly C- and N-atoms; but also H-atoms and a few S- atoms. The molecule, with its flexible skeleton contains bond angle forces between triplets of atoms and torsion forces between fourplets of atoms, Lennard-Jones forces between intra- as well as intermolecular pairs of mass centers. The bond length between neighbour atoms in the enzyme is kept constant by constraint dynamics ("SHAKE"). And the electrostatic (long range) Coulomb forces are also included. The system is sour rounded by 4438 water molecules (iwith three atoms per molecule). The force,

$F_i(t)$, at time t on a mass center no i is obtained as a sum over all mass centres N of the whole system

$$F_i(t) = \sum_{j,\,l,\,k..}^{N} f_{i,j,k..}(t) \tag{3}$$

and the problem, from a load-balance point of view is that this sum cannot easily be divided in advance into terms which take equal computer times, since the structure changes during the computation. Nevertheless one obtained a good degree of performance. The computations were made on a SGI power challenger machine and by using 8 (250 MHZ IP19) processors. By using 8 processors rwe obtained a speed up by a factor 5.

Another parallel MD program, $DL - POLY$, which is also available for academics (after signing a license) is developed at the Daresbury Laboratory by Forester and Smith [10]. It is written (in Fortran) for a distributed memory machine and in PVM. Actually it is tested at the sp2 computer in Daresbury and we have installed the corresponding system at Uni-C. The parallel strategy is slightly different from $GROMACS$'s; it makes use of the big local RAM at the nodes which allow for that the position and velocities can be copied at every node. The advantage of this technique is the efficiency at the individual nodes; but it requires as mentioned an extensive local memory and does have a high communication overhead and thus a need for a fast communication between the nodes.

The general conclusion of our (limited) experience with parallel execution of complex Molecular Dynamics systems is that it is possible to obtain a high degree of parallelism even for strongly non-uniform systems. The development of flexible soft ware system has made it more user-friendly, although we will certainly appreciate the fortcoming development in high performance language.

ACKNOWLEDGMENT

A grant No. 11-0065-1 from the Danish Natural Science Research Council is gratefully acknowledged.

References

1. J. J. Morales and S. Toxvaerd, Comput. Phys. Commun. **71**, 71 (1992).
2. A. D. Bray, Advances in Phys. **43**,357 (1994).
3. M. Laradji, O. G. Mouritsen and S. Toxvaerd, *to appear in* Phys. Rev. Lett. (1996).
4. E. Velasco and S. Toxvaerd, Phys. Rev. Lett. **71**, 388 (1993); Phys. Rev. E. **54**, 605 (1996).
5. R. Everaers and K. Kremer, Comp. Phys. Comm. **81**, 19 (1994).
6. S. Toxvaerd, Phys. Rev. **E 53**, 3710-3716 (1996).
7. K. Geisshirt, E. Præstgaard and S. Toxvaerd, *Submitted to Science*.

8. BIOSON, University of Groningen, Nijenborgh 4, 9747 AG Groningen The Netherlands (fax: 31 50 634800).

9. G.Peters, D. M. F. van Aalten, O. Edholm, S. Toxvaerd and R. Bywater, *To appear in* Biophys. J.

10. T.R. Forrester and W. Smith, CCLRC, Daresbury Laboratory, Daresbory, Warrington WA4 4AD, England.

A Parallelizable and Fast Algorithm for Very Large Generalized Eigenproblems

Henk A. van der Vorst and Gerard L.G. Sleijpen

Mathematical Institute, University of Utrecht,
Budapestlaan 6, Utrecht, the Netherlands

Abstract. We discuss a novel iterative approach for the computation of a number of eigenvalues and eigenvectors of the generalized eigenproblem $Ax = \lambda Bx$. Our method is based on a combination of the Jacobi-Davidson method and the QZ-method. For that reason we refer to the method as JDQZ. The effectiveness of the method is illustrated by a numerical example.

1   Introduction

Iterative techniques for computing eigenvalues and eigenvectors of a large generalized eigenproblem $Ax = \lambda Bx$, usually require that the system is reduced to a standard eigenproblem, for instance $B^{-1}Ax = \lambda x$ (provided that B is nonsingular). This problem can be handled by standard iteration techniques, such as Arnoldi's method [1]. More advanced techniques apply clever implementations of the Arnoldi method with a shift and invert approach:

$$(A - \sigma B)^{-1}Bx = \tilde{\lambda}x,$$

in which σ is chosen close to the desired eigenvalue λ. For an elegant computational approach along these lines, see [7]. Another example of such an approach for problems that we are interested in, see Section 2, is given in [3].

The disadvantage of these conventional approaches is that in each iteration step of the eigensolver a matrix has to be inverted, or, more precisely, that a linear system has to be solved to high accuracy. For very large problems, with sparse matrices A and B, this may pose severe limitations on the order of the matrices for which computation is still feasable.

In our presentation we will describe a new iterative technique for the computation of a number of selected solutions for the eigenproblem

$$(A - \lambda B)x = 0,$$

with A, B general large sparse ($n \times n$) matrices, in which it is avoided to transform the problem by an inversion operation. The technique is based on the Jacobi-Davidson method with reduction to Schur form. With this technique we are able to compute a number of eigenvalues and eigenvectors, with subspaces of restricted dimension, and without inversion of any of the matrices involved. The standard methods of Arnoldi and Lanczos are not well-suited for this.

In the Jacobi-Davidson method a subspace is generated onto which the given eigenproblem is projected. The much smaller projected eigenproblem is solved by standard direct methods, and this leads to approximations for the eigensolutions of the given large problem. This is the 'Davidson' part of the algorithm. Then, a correction equation for a selected eigenpair is set up. The solution of the correction equation defines an orthogonal correction for the current eigenvector approximation; this is the 'Jacobi' part of the algorithm. The correction is used for the expansion of the subspace and the procedure is repeated.

In the new algorithm the large system is projected on a suitable subspace, in a way that may be viewed as a truncated and inexact form of the QR or QZ factorization. The projected system itself is reduced to Schur form by the QR or QZ algorithm. For this reason we refer to the algorithm as JDQR (=Jacobi-Davidson-QR), when $B = I$, and as JDQZ in the general case.

As we will see, the method lends itself perfectly well for parallel execution.

2 An example of very large eigenproblems

Although large eigenproblems arise in many scientific problems, we have been particularly motivated recently by MHD-problems. In particular, we study the dynamics of plasmas in a magnetic field with computational models. The results are applied for further understanding of the stability of Tokomak fusion reactors and of coronal loops, as well as of solar flares.

The interaction of plasma and a magnetic field is governed by essentially the flow equations for fluids combined with the Maxwell equations, and this system has the form

$$
\frac{\partial \rho}{\partial t} = -\nabla \cdot (\rho \mathbf{V})
$$

$$
\rho \frac{\partial \mathbf{V}}{\partial t} = -\rho (\mathbf{V} \cdot \nabla) \mathbf{V} - \nabla p
$$
$$
+ (\nabla \times \mathbf{B}) \times \mathbf{B}
$$

$$
\frac{\partial p}{\partial t} = -(\mathbf{V} \cdot \nabla) p - \gamma p \nabla \cdot \mathbf{V} +
$$
$$
(\gamma - 1) \eta (\nabla \times \mathbf{B})^2
$$

$$
\frac{\partial \mathbf{B}}{\partial t} = \nabla \times (\mathbf{V} \times \mathbf{B}) - \nabla \times (\eta \nabla \times \mathbf{B})
$$

$$
\text{with} \quad \nabla \cdot \mathbf{B} = 0.
$$

The last equation is considered as an initial condition on **B**. One of the approaches, taken in our project, is to consider small perturbations of the unknowns with respect to some known equilibrium, and this leads to a linear system for

688

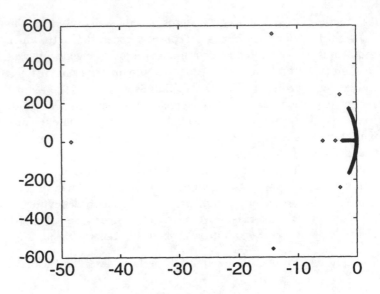

Fig. 1. Eigenvalues for MHD problem

the first order perturbations. The solution for this linearized system is assumed to be of the form $e^{\lambda t}$, and this leads to a large linear generalized eigenproblem. Due to the kind of discretization we use, partly finite differences, and partly Fourier series, this leads to block tridiagonal matrices, with typically $100 - 2,000$ blocks, of size $80 - 320$ each. This amounts to matrices of orders in the range $8,000 - 640,000$.

The matrices have eigenvalues that can be grouped in very large ones (associated with 'Sound waves'), very small ones, and intermediate ones (associated with 'Alfvén' waves), and we are interested in the Alfvén spectrum. In Figure 1 we see the entire spectrum, the middle part of Figure 2 shows the 'Alfvén' spectrum (note the different scales).

The order and the structure of these matrices makes standard direct methods unpractical, and therefore we consider iterative methods. Similar problems have been solved up to orders of a few tens of thousands with the methods of Lanczos (with a code of Cullum et al [3]) and by Shift and Invert Implicitly Restarted Arnoldi method [7]. The applicability of these methods is limited, due to the fact that inversion of matrices is required, and this becomes too expensive for the very large problems that we are interested in. As we will see, inversion of matrices is essential for these methods in order to obtain convergence for the Alfvén spectrum.

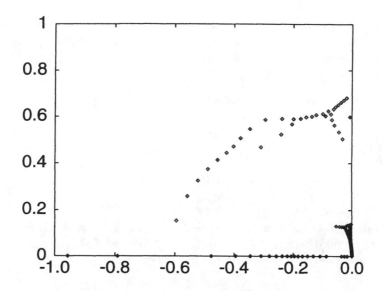

Fig. 2. Alfvén part of spectrum

3   The Jacobi-Davidson method

In this section, we consider the generalized eigenproblem $Ax = \lambda Bx$, and we want to compute a few eigenvalues close to a given target $\tau \in \mathbf{C}$. Suppose that we have a suitable low-dimensional subspace, spanned by $v_1, \ldots, v_m$, and we want to identify good eigenvector approximations in this subspace. Let V_m denote the $n \times m$ matrix with v_i as its i-th column. Let us further assume that the vectors v_i form an orthonormal basis. Then the Petrov-Galerkin approach is to find a vector $y \in \mathbb{R}^m$, and a θ, such that

$$AV_m y - \theta B V_m y \perp \{w_1, \ldots, w_m\},$$

for some convenient set of vectors $\{w_j\}$. We will assume that these vectors are also orthonormal and that they form the columns of a matrix W_m. This leads to an eigenproblem for projected matrices of order m:

$$W_m^* A V_m y - \theta W_m^* B V_m y = 0, \tag{1}$$

where W_m^* denotes the adjoint of W_m. In applications we have that $m \ll n$. The solutions θ are referred to as the (Petrov-)Ritz values (approximations for eigenvalues of $A - \lambda B$), and $V_m y$ is the corresponding (Petrov-)Ritz vector, with respect to the subspace spanned by $v_1, \ldots, v_m$.

The main problem is to find the expansion vectors v_{m+1} and w_{m+1} for the next iteration step in the process for improving the eigen approximations. In [6], it is shown how the expansion vectors can be selected for the standard eigenproblem

$Ax = \lambda x$, and in [4] the same approach is followed for generalized eigenproblems. For this approach it is not necessary to transform the given generalized eigenproblem to a standard one (for instance, by inverting one of the matrices A or B, or a combination).

We will briefly describe this inverse-free Jacobi-Davidson approach, for details we refer to [4]. First we use the QZ algorithm [5] to reduce (1) to a generalized Schur form. With m the dimension of span$\{v_j\}$, this QZ computation delivers orthogonal $(m \times m)$ matrices U_R and U_L, and upper triangular $(m \times m)$ matrices S_A and S_B, such that

$$U_L^* (W_m^* A V_m) U_R = S_A, \tag{2}$$

$$U_L^* (W_m^* B V_m) U_R = S_B. \tag{3}$$

The decomposition is ordered so that the leading diagonal elements of S_A and S_B represent the eigenvalue approximation closest to the target value τ. The approximation for the eigenpair is then taken as

$$(\tilde{q}, \theta) \equiv (V_m U_R(:,1), S_B(1,1)/S_A(1,1)), \tag{4}$$

assuming that $S_A(1,1) \neq 0$ (see [4] for further details). With this approximation for the eigenvector and the eigenvalue, we compute the *residual vector*:

$$r \equiv A\tilde{q} - \theta B\tilde{q}.$$

We also compute an auxiliary vector $\gamma \tilde{z} = A\tilde{q} - \tau B\tilde{q}$, where γ is such that $\|\tilde{z}\|_2 = 1$.
With these two vectors we can define the correction equation for $\tilde{q}$, that is so typical for the Jacobi-Davidson approach: we want to find the correction v for $\tilde{q}$ with

$$\tilde{q}^* v = 0 \quad \text{and} \quad (I - \tilde{z}\tilde{z}^*)(A - \theta B)(I - \tilde{q}\tilde{q}^*)v = -r. \tag{5}$$

In practice, we solve (5) only approximately by a few steps of, for instance, GMRES (sometimes with a suitable preconditioner). Then we orthonormalize the approximation for v with respect to $v_1, \ldots, v_m$, and take the results as our new v_{m+1}. For the expansion vector w_{m+1}, we take the vector $Av - \tau Bv$, orthonormalized with respect to $w_1, \ldots, w_m$.
It can be shown that in this case $\tilde{z} = W_m U_L(:,1)$.

The above sketched algorithm descibes the *Harmonic Petrov value approach* proposed in [4].

4   The Jacobi-Davidson QZ-method

In some circumstances the Jacobi-Davidson method has apparent disadvantages with respect to other iterative schemes, for instance the shift-and invert Arnoldi method [7]. For instance, in many cases we see rapid convergence to one single eigenvalue, and what to do if we want more eigenvalues? For Arnoldi this is not a big problem, since the usually slower convergence towards a particular eigenvalue goes hand in hand with simultaneous convergence towards other eigenvalues. So after a number of steps Arnoldi produces approximations for several eigenvalues. For Jacobi-Davidson the obvious approach would be to restart with a different target, with no guarantee, hoewever, that this leads to a new eigenpair. Also the detection of multiple eigenvalues is a problem, but this problem is shared with the other subspace methods.

A well-known way out of this problem is to use a technique, known as *deflation*. If an eigenvector has converged, then we continue in a subspace spanned by the remaining eigenvectors. A problem is then how to re-use information obtained in a previous Jacobi-Davidson cycle. In [4] an algorithm is proposed by which several eigenpairs can be computed. The algorithm is based on the computation of a partial generalized Schur form for the matrix pair (A, B):

$$AQ_k = Z_k S_k \text{ and } BQ_k = Z_k T_k,$$

in which Q_k and Z_k are n by k orthonormal matrices and S_k, T_k are k by k upper triangular matrices, with $k \ll n$. Note that if (x, λ) is an eigenpair of $S_k x - \lambda T_k x = 0$, then $(Q_k x, \lambda)$ is an eigenpair of $A - \lambda B$.

We now proceed in the following way in order to obtain this partial generalized Schur form for eigenvalues close to a target value τ.

Step I: Given orthonormal subspace bases $v_1, \ldots, v_i$, with matrix V_i, and $w_1, \ldots, w_i$, with matrix W_i, we compute the projected matrices $M_A = W_i^* A V_i$ and $M_B = W_i^* B V_i$. For these $i \times i$ matrices M_A and M_B, we compute the complete generalized Schur form

$$M_A U_R = U_L S_A \text{ and } M_B U_R = U_L S_B, \tag{6}$$

in which U_R and U_L are orthonormal, and S_A and S_B are upper triangular. This can be done with the standard QZ algorithm [5].
Then we orden S_A, and S_B such that the $|S_B(i, i)/S_A(i, i) - \tau|$ form a nondecreasing row for increasing i. The first few diagonal elements of S_A and S_B then represent the eigenapproximations closest to τ, and the first few of the correspondingly reordered columns of V_i represent the subspace of best eigenvector approximations. If memory is limited then this subset can be used for restart, that is the other columns are simply discarded. The remaining subspace is expanded according to the Jacobi-Davidson method.
After convergence of this procedure we have arrived at an eigenpair (q, λ) of $A - \lambda B$: $Aq = \lambda Bq$. The question is how to expand this partial generalized Schur form of dimension 1. This will be shown in step II.

Step II: Suppose we have already a partial Schur form of dimension k, and we want to expand this by an appropriate new column q:

$$A[Q_k, q] = [Z_k, z] \begin{bmatrix} S_k & s \\ & \alpha \end{bmatrix}$$

and

$$B\,[Q_k, q] = [Z_k, z] \begin{bmatrix} T_k & t \\ & \beta \end{bmatrix}.$$

After some standard linear algebra manipulations it follows that

$$(I - Z_k Z_k^*)(\beta A - \alpha B)(I - Q_k Q_k^*)q = 0 \quad \text{and} \quad Q_k^* q = 0,$$

which expresses that the new pair $(q, \lambda \equiv \alpha/\beta)$ is an eigenpair of the matrix pair

$$(I - Z_k Z_k^*)A(I - Q_k Q_k^*), \ (I - Z_k Z_k^*)B(I - Q_k Q_k^*). \tag{7}$$

Approximations for an eigenpair of this matrix pair can be computed by applying the Jacobi-Davidson algorithm (with Schur form reduction, as in step I), with appropriate W and V. Expansion vectors for V and W are computed by solving approximately the Jacobi-Davidson correction equation:

$$(I - \tilde{z}\tilde{z}^*)(I - Z_k Z_k^*)(A - \lambda B)(I - Q_k Q_k^*)(I - \tilde{q}\tilde{q}^*)v = -r, \tag{8}$$

with $Q_k v = 0$, and $\tilde{q}^* v = 0$. This leads to projected matrices M_A and M_B, for which it is easy to prove that:

$$M_A = W^*(I - Z_k Z_k^*)A(I - Q_k Q_k^*)V = W^* A V, \tag{9}$$

and

$$M_B = W^*(I - Z_k Z_k^*)B(I - Q_k Q_k^*)V = W^* B V. \tag{10}$$

For these matrices we construct again, as in step I, the generalized Schur decomposition:

$$M_A U_R = U_L S_A \quad \text{and} \quad M_B U_R = U_L S_B.$$

Upon convergence of the Jacobi-Davidson iteration, we obtain the expansion vector $q = V U_R(:, 1)$ for Q_k, and the expansion vector z for Z_k is computed with the relation $z = W U_L(:, 1)$.

Some notes are appropriate:

1. In our computations, we have used the explicitly deflated matrix $A - \lambda B$ in (8), whereas we could have exploited the much less expensive expressions (9) and (10). We have made this choice for stability reasons. Despite these more expensive computations, it is shown in [4], by numerical experiments, that the entire procedure leads to a very efficient computational process. An explanation for this is that after convergence of some eigenvectors, the pair M_A, M_B will be better conditioned, so that the correction equation (8) in the Jacobi-Davidson step is more easily solved.

2. The correction equation (8) may be solved (approximately) by a preconditioned iterative solver, and it is shown in [4] that the same preconditioner can be used with great efficiency for different eigenpairs. Hence, the costs for constructing accurate preconditioners can be spread over many systems to be solved.

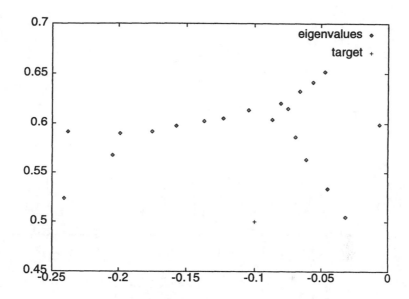

Fig. 3. Computed eigenvalues

5 An example

We will now briefly describe results for the sketched approach for a generalized eigenproblem $Ax = \lambda Bx$, associated to the MHD problems discussed in Section 2. In view of the problems that we actually want to solve, our example is just a 'toy problem' of very modest dimension, $n = 416$. For more information on this particular problem, see for instance [2].

In Figure 1 we see the complete spectrum for this case; Figure 2 shows the so-called Alfvén part of the spectrum. Note the different scales from which it is obvious that the Alfvén spectrum is an interior part of the spectrum, and without Shift-and-Invert it is almost impossible to compute this part with Krylov subspace methods. The 20 eigenvalue approximations, that we have computed with the QZ-variant of our algorithm, are shown in Figure 3. The computations have been carried out in about 15 decimal digits accuracy on a SUN workstation. The target value, indicated in picture 3 by a '+', was $\tau = -.1 + .5i$. The maximum dimension of the subspaces for the Jacobi-Davidson part of the algorithm was fixed to 15. As soon as we arrived at that dimension, the subspace was shrinked, as described in **Step I** above, to 10. An eigenpair was considered to be converged as soon as the norm of the residual for the normalized eigenvector approximation was below 10^{-9}. For the preconditioner in the solution process for the correction equation, we used the exact inverse of $A - \tau B$ (for fixed τ). The correction equations were solved approximately, namely with only 1 step of GMRES. The complete converge history, norms of the residuals of eigenapproximations versus the total amount of floating point operations, is shown in Figure 4. Each time when the curve hits the 10^{-9} level, indicated by the dotted

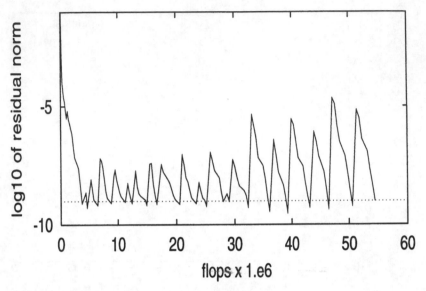

Fig. 4. Convergence history

horizontal line, the algorithm has discovered a new eigenvalue. In our case that has happened 20 times, after which we have stopped the algorithm.

6 Parallel aspects

The Jacobi-Davidson method lends itself in an excellent way for parallel computing because of its explicit nature. We will discuss the computational elements of the algorithm in view of parallel processing.

- In the initialization step subspaces V and W are chosen and orthonormal basis for these subspaces are constructed. This involves inner products and vector updates.
- In the first step of the iteration loop, the matrices M_A and M_B have to be computed. The projections with W and V, as in equations (9) and (10), also involve only inner products and vector updates. Note that it is sufficient to compute only the last recent row and colums of these matrices. This again involves a number of innerproducts: one per new matrix element.
- In the second step the eigenpairs of the projected pair are computed. The projected system is of low dimension: the projected matrices are typically of order less than 30, and hence the number of computations in this step is limited and is often negligible with respect to the number of computations in the other steps. Therefore, we let all processors do this computation, in order to avoid further communication.
- In the third step the selected Ritz vector is computed with (4) by taking a

linear combination of the columns of V. Hence, if cleverly implemented, no multiplications with the matrices A and B are needed in this step, but only vector updates.

- In the fourth step the convergence criterion is checked. This typically involves the computation of the norm of the residual r.

- The computation of the approximate solution of the correction equation (8) is usually the most expensive step in the algorithm. If this is done by a method like GMRES, then the only time-consuming operations are matrix-vector multiplications, vector updates, and inner products. Preconditioning for GMRES requires special attention with respect to parallelizability; we will not discuss this aspect in this paper.

Note that we explicitly deflate with Z_k and Q_k in (8). Multiplication with the entire operator can be carried out in five steps. First multiply with the two projectors on the right, then multiply with the matrices A and B, and take a linear combination of the results, and finally multiply with the projectors on the left. A multiplication with a projector requires only inner products and vector updates.

- In the final step of the algorithm, the subspace V is expanded with the solution of the correction equation and this new basis vector is orthogonalized against the others. This process also requires only inner products and vector updates.

From this global analysis we learn that the main CPU time-consuming operations are multiplications with matrices, vector updates, and inner products, and we only need to parallelize these operations. The inner products may require global communication (in distributed processing mode), and this may need further attention for large numbers of processors.

We have implemented this algorithm on the Cray T3D; a distributed memory computer. The communication steps have been implemented with the fast SHMEM_GET and SHMEM_PUT routines. The observed speed-ups for a slightly different formulation of the algorithm, were very satisfactory: in [8] a problem of order 274625 is described for which a linear speed-up was observed when increasing the number of processors from 16 to 64 (without preconditioning for (8)).

References

1. W. E. Arnoldi. The principle of minimized iteration in the solution of the matrix eigenproblem. *Quart. Appl. Math.*, 9·17–29, 1951.
2. J.G.L. Booten, D.R. Fokkema, G.L.G. Sleijpen, and H.A. van der Vorst. Jacobi-Davidson methods for generalized MHD-eigenvalue problems. *Zeitschrift für Angewandte Mathematik und Mechanik (ZAMM)*, 1996.
3. J. Cullum, W. Kerner, and R. Willoughby. A generalized nonsymmetric Lanczos procedure. *Computer Physics Communications*, 53:19–48, 1989.
4. D.R. Fokkema, G.L.G. Sleijpen, and H.A. van der Vorst. Jacobi-Davidson style QR and QZ algorithms for the partial reduction of matrix pencils. Technical Report Preprint 941, Mathematical Institute, Utrecht University, 1996.

5. G. H. Golub and C. F. van Loan. *Matrix Computations*. The Johns Hopkins University Press, Baltimore, 1989.
6. G. L. G. Sleijpen and H.A. Van der Vorst. A Jacobi-Davidson iteration method for linear eigenvalue problems. *SIAM J. Matrix Anal.Appl.*, 17(2):401–425, 1996.
7. D. C. Sorenson. Implicit application of polynomial filters in a k-step Arnoldi method. *SIAM J. Matr. Anal. Appl.*, 13(1):357–385, 1992.
8. M.B. Van Gijzen. A parallel eigensolution of an acoustic problem with damping. submitted for publication.

Parallelization of a Code for Animation of Multi-object System

R. Wcisło[1], J. Kitowski[1,2] and J. Mościński[1,2]

[1]Institute of Computer Science, AGH, al. Mickiewicza 30, 30-059 Cracow, Poland
[2]Academic Computer Centre CYFRONET, ul. Nawojki 11, 30-950 Cracow, Poland
e-mail: wcislo@icsr.agh.edu.pl

Abstract. In this paper we describe the realization of a computer system designed for performing 3-D animation of sets of macroscopic objects. Realistic movements observed on the computer screen are achieved by running complex simulation in the background. This simulation is based on fundamental physical rules (e.g. preservation of momentum and energy, Newton's laws of dynamics) which ensure that generated scenes are conformable to a common sense of a viewer.

In order to speed up the whole process of animation (and get closer to the "real-time animation") parallel algorithms were developed. They were implemented on a network of UNIX workstations using *PVM* tools. In the paper we introduce two different parallel schemes cantrally controlled and with distributed control.

In the paper we describe and compare both parallel schemes presented and discuss the effects and limitations of having asynchronized clocks for objects. Some examples of the animation will be given as well.

1 Introduction

1.1 Goals of the project

There are many methods used for performing computer animation. The choice depends on a kind of scenes and goals you want to achieve. In our project we animate the scenes composed of:

◇ elastic, macroscopic objects (characterized by: mass, initial velocity and elasticity),
◇ walls (infinite planes with infinite mass acting as boundaries of the scene),
◇ external, homogenous gravitational field.

The main goals of our project are:

◇ animate the scene according to the physical laws (due to dynamics reality),
◇ perform fast, on-line visualization ("real-time" when possible),

The former postulate we try to fulfil by utilization of molecular dynamics methods. Originally developed for investigations of liquid and gas properties [1] were adopted for our project [2, 3]. The latter goal can be achieved using very fast computers (which is quite obvious but expensive solution) and/or developing the parallel algorithms, which can take advantage of a cluster of slower workstations connected by a local network.

1.2 Previous projects

In the previous projects [4] we animated scenes composed of a single object only. The objects were either flat (see Fig. 1) or volumetric (see Fig. 2). Those projects allowed to test the behaviour of objects simulated by molecular dynamics methods and explore some basic parallelization schemes.

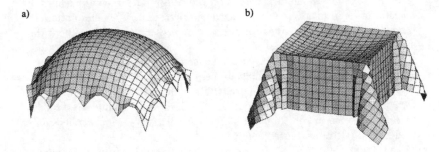

Fig. 1. Single flat objects used in animation. a) falling parachute, b) table-cloth.

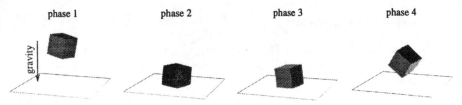

Fig. 2. Elastic cubicoid dropping on the wall.

2 System description

The operation of the system consists of two main stages (see Fig. 3):

◇ creation of molecular models of animated objects,
◇ simulation and visualization.

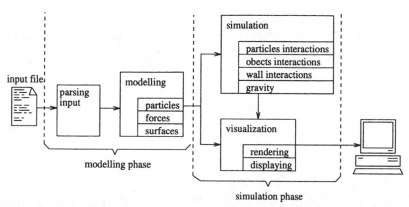

Fig. 3. System modules and stages of animation.

2.1 Modelling

In the modelling stage the objects get particle representation. They are broken into small elements and then converted to the sets of particles (see Fig. 4). The elasticity (or stiffness) of an object is achieved by introducing forces acting between neighbouring particles. The force is calculated basing on a particle-particle pair potential:

$$V_{i,j}(r_{i,j}) = E \cdot \left(r_{i,j} - \frac{r_{i,j}^2}{2 \cdot r_{0;i,j}} \right) \qquad (1)$$

where: i, j – particle indices, $r_{i,j}$ – distance between the particles i and j, $r_{0;i,j}$ – initial distance between the particles i and j, E – constant representing elasticity. Thus the pair potential is harmonic, and E represents the Young's modulus [5].

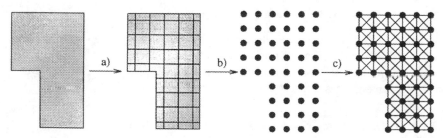

Fig. 4. Modelling of an object (2D figure is presented for simplicity, the algorithm works on 3D data). a) breaking into parts, b) introducing particles, c) detecting particle-particle interactions.

In addition, at the modelling stage the particles forming surfaces of the objects are detected. This complicated algorithm gives fundamental data for the animation visualization. As a result the following output is produced:

⋄ list of particles (including their masses, positions, initial velocities) representing every object,
⋄ list of pairs of neighbouring (interacting) particles,
⋄ list of surface elements (triangles composed of particles laying on surfaces of objects).

2.2 Simulation

In the course of simulation four types of interaction are taken into account:

⋄ particle-particle forces inside every object,
⋄ interactions between objects,
⋄ object-wall interactions,
⋄ influence of the gravity.

The forces acting on the particles are calculated in every step of the simulation. New positions of the animated objects are achieved by solving Newtonian equations of motion [1].

2.3 Visualization

Every N steps of the simulation, the current positions of the objects are visualized on a computer screen, thus changing plain simulation into the animation.

For rendering 3D images the simplified algorithm of sorting object faces is used. Elementary *Xlib* routines are called for drawing a picture in X Window environment. This method allows of quite fast visualization on a workstation not equipped with a 3D acceleration hardware.

3 Parallel algorithms considerations

3.1 Separation of the visualization process

The simplest way to speed up the sequential program is to separate the visualization process and designate one workstation for it (see Fig. 5).

The visualization process is notified about changes of positions of particles laying on the surfaces of the animated objects.

3.2 Parallelization of a single object simulation

Working on the previous projects (see section 1.2) we tried to invent an effective algorithm for parallel computations of a single object simulation. We have proved [3] that for objects build up of relatively small number of particles (less than 1,000) splitting them between two or more processes causes the whole simulation to slow down due to frequent messages interchange (see Fig. 6).

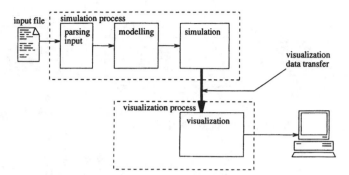

Fig. 5. Separation of the visualization process.

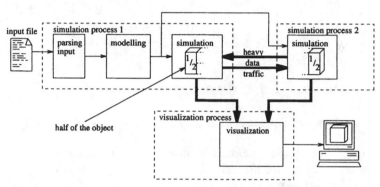

Fig. 6. Geometric decomposition of an object between two processes.

3.3 Spreading objects among processes

Particles in an object interact each other every time step of the simulation which complicates parallel computations. However interactions between the objects are less frequent (only when they collide). This allowed us to implement algorithms where the simulation of the objects dynamics is spread among different processes. In this case each process simulates only a subset of the whole number of the animated objects and the visualization gathers only partial information from every one of them. This situation is shown in Fig. 7.

After spawning several processes for the simulation new problems arise:

⋄ there is no mechanism for notifying processes that the objects they are simulating collide,

⋄ there is no synchronization between processes which can lead to problems since objects will be simulated with different computational speed (due to different workstation load, different number of objects allocated for processes).

Below we present two methods of solving the mentioned problems.

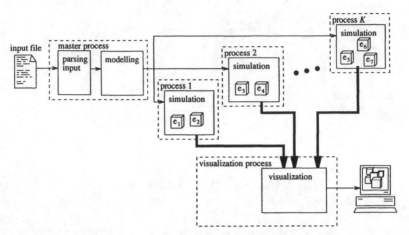

Fig. 7. Spreading objects among processes.

4 Algorithm 1 – central control of simulation

There are three types of processes involved with the animation (see Fig 8):

◇ one **master process** - responsible for constructing particle models for the objects, spawning all other processes, detecting objects collisions and performing load-balancing algorithm,

◇ one **visualization process** - collects data from the simulation processes and visualizes the animated scene on the computer screen,

◇ K **simulation processes** - responsible for the computations of the objects current positions by means of the step by step simulation.

In the course of animation each of K simulation processes performs the simulation for the objects currently assigned to it. Let's assume that the process W_x ($x \in \{1, \ldots, K\}$) simulates objects $e_{x,1}, e_{x,2}, \ldots, e_{x,n(W_x)}$ [1], where $n(W_x)$ - number of objects simulated by the given process.

If the condition:

$$h(W_1)\sum_{i=1}^{n(W_1)} p(e_{1,i}) = h(W_2)\sum_{i=1}^{n(W_2)} p(e_{1,i}) = \ldots = h(W_K)\sum_{i=1}^{n(W_K)} p(e_{K,i}) \quad (2)$$

(where $p(e_\alpha)$ - number of particle-particle interactions in the object e_α, $h(W_x)$ - relative speed of a workstation where the process W_x is located) is true and none of the objects calculated by the different processes collide then we can say

---

[1] When it is important that the given object is assigned to the particular process, we will denote it as $e_{p,i}$, where p – process ID and i enumarates objects allocated to this process. In other cases we will, for simplicity, use notation e_α refering to "any" object.

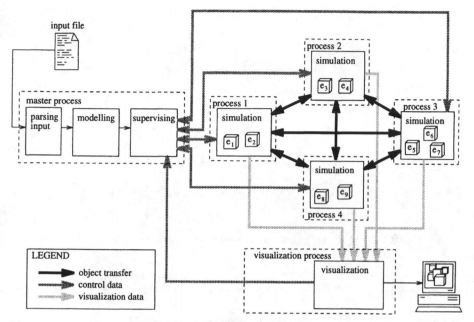

Fig. 8. Processes and data exchange routes in algorithm 1 (for $K = 4$ simulation processes).

that the animation goes perfectly and there is nothing to optimize by the master process.

This situation can change in two ways:

⋄ objects $e_{x,k}$ and $e_{y,l}$ $(x \neq y)$ start to collide,
⋄ equation (2) is no longer valid.

Both problems should be detected first and solved subsequently. To realise detection of object collisions we define $O(e_\alpha)$ as a minimal cubicoid containing object e_α:

$$O(e_\alpha) = [\min_x(e_\alpha), \max_x(e_\alpha)] \times [\min_y(e_\alpha), \max_y(e_\alpha)] \times [\min_z(e_\alpha), \max_z(e_\alpha)] \quad (3)$$

If $O(e_\alpha)$ changes significantly[2], the process $W(e_\alpha)$ sends new $O(e_\alpha)$ to the master process (step 2 in Fig. 9).

The master process keeps information about the objects boundaries. Every time they are updated the master process checks whether there are any new possible collisions. This is done by finding all objects e_i and e_j for which:

$$O(e_i) \cap O(e_j) \neq \emptyset \quad (4)$$

---

[2] for explanation on a term "significantly" see section 6.

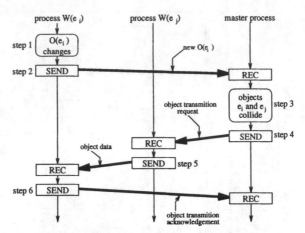

Fig. 9. Object exchange scheme in algorithm 1.

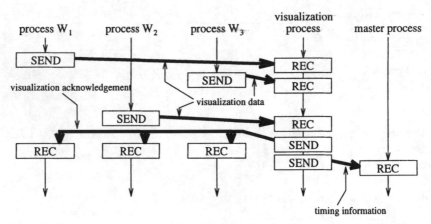

Fig. 10. Messages sent and received by the visualization process in algorithm 1.

If $W(e_i) \neq W(e_j)$ then additional action must be taken in order to simulate both objects e_i and e_j by the same process.

Sending only the object e_i to the process $W(e_j)$ (or e_j to $W(e_i)$) would not solve the problem because there might be the objects (different from e_j) which already collide with the object e_i and vice versa.

Let's denote $C(e_i)$ as a set of all objects that must be simulated together with the object e_i (that is all of them which collide with e_i, all that collide with those which collide with e_i and so on). It is obvious that if e_i and e_j collide then $C(e_i) = C(e_j)$.

The actual problem that is supposed to be solved by master process is to determine, which of two processes $W(e_i)$ and $W(e_j)$ should simulate all objects belonging to $C(e_i)$.

The master process sends messages to the simulation processes requesting appropriate object transmissions (step 4 in Fig. 9). This causes object "jumps" between the simulating processes (step 5), which are in turn acknowledged to the master process (step 6).

The frequent exchange of the objects during the simulation leads to the situation, when the equation (2) is not satisfied. The master process is responsible for detection of the situation, when objects e_i and e_j are after their collision and are moving away. In that case the master process sends messages to the simulation processes again to ensure better load-balancing of the whole animation. The master process knows exactly which objects are simulated by every simulation process W_x and can easily compute $\sum_{i=1}^{n(W_x)} p(e_{x,i})$. The remaining information needed to fulfil the equation (2) (i. e. $h(W_x)$) is sent to the master process from the visualization process (see Fig. 10).

The visualization data are sent by every simulation process W_x every N steps of the simulation. The visualization process measures the time ΔT_x of that computations and approximates the smiulation process workload as:

$$h(W_x) = \frac{\Delta T_x}{p(e_{x,1}) + p(e_{x,2}) + \ldots + p(e_{x,n(W_x)})} = \frac{\Delta T_x}{\sum_{\alpha \in L(W_x)} p(e_\alpha)} \qquad (5)$$

The results are sent to the master process every time the visualization process receives data from all simulation processes. The equation (5) assumes that process W_x did not exchange objects since the last visualization and should be modified in the future.

5 Algorithm 2 – distributed control of simulation

In the second algorithm we tried to eliminate the master process. In algorithm 1 every object relocation required one or more messages passed to and from the master process which performed all calculations concerned with the collision detections and load-balacing. This could lead to the situation when overloading of the master process would cause the whole amination slow down.

In algorith 2 all decisions of the objects exchange are done by the simulation processes. All information sont previously to the master process is now broadcasted among the simulation processes (see Fig. 11).

The scenario being realized, while exchenging objects, is shown in a Fig. 12. After detection of the object e_i move ($O(e_i)$ changed), the process $W(e_i)$ bradcast this information to all other processes (step 2). The process $W(e_i)$ is also responsible for finding the best solution for keeping the cluster $C(e_i)$ together. As a result it requests the object transmition (step 4). After succesful transmition (step 5) a notification about this event is broadcasted (step 6) by the process (or processes), which received the new object (or objects).

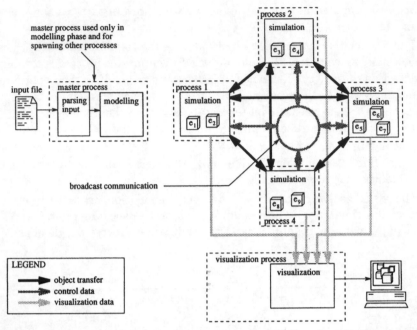

Fig. 11. Processes and data exchange routes in algorithm 2 (for $K = 4$ simulation processes).

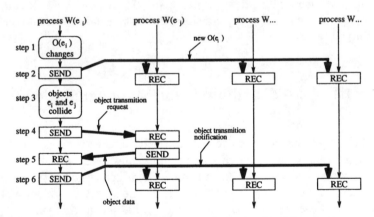

Fig. 12. Object exchange scheme in algorithm 2.

6 Time synchronization

The simulation processes are synchronized every N steps[3] by the visualization process (see Fig. 10). This synchronization is required to achieve the main goal of the program: the animation on a computer display. More frequent synchronization would slow down the simulation processes, however the lack of synchron-

---

[3] for animations presented in section 7: $N \approx 100$.

ization after every step of the simulation leads to the situation when the objects simulated by different processes have, at the moment T (T indicates wall-clock time), different simulation time: $W(e_i) \neq W(e_j) \leftrightarrow t(e_i) \neq t(e_j)$.

Furthermore, if object e_i is sent to the process $W(e_j)$ then $t(e_i) \neq t(e_j)$ even if $W(e_i) = W(e_j)$. To handle this problem:

⋄ current simulation time $t(e_i)$ is always sent with the information about the object e_i,
⋄ when the simulation process W receives the object e_i, it continues its simulation at the time $t^{new}(W) = \min\{t^{old}(W), t(e_i)\}$,
⋄ the simulation process W simulates only those objects e_k for which $t(e_k) \leq t(W)$ (chasing the objects which are "in the future").

The next problem connected with the asynchronized simulation processes refers to the detection of object collisions. If the objects $t(e_i)$ and e_j are not synchronized (i. e. $t(e_i) \neq t(e_j)$) it is impossible to check the condition (4).

The collision between the objects e_i and e_j has to be detected while $O(e_i) \cap O(e_j) = \emptyset$. Let's denote the minimal distance between $O(e_i)$ and $O(e_j)$, for which the collision should be detected, as D (see Fig. 13):

$$D = (t_d + |t(W(e_i)) - t(W(e_j))|) \cdot v_{i,j} \tag{6}$$

where: t_d - time needed to complete all object transmition between the simulation processes, $v_{i,j}$ - velocity of approaching of the objects e_i and e_j.

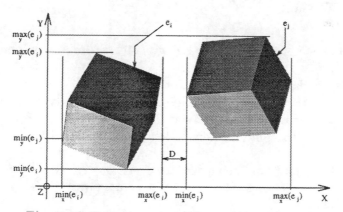

Fig. 13. Collision detection distance between objects.

Unfortunately the value of $|t(W(e_i)) - t(W(e_j))|$ can be estimated only.

To avoid frequent notification of $O(e_i)$ changes, we extend the collision detection distance to $D + D_s$. Now, if the object e_i changes its position by D_s in any direction (this is what we call "significant change of $O(e_i)$") information about new $O(e_i)$ will be sent (to the master process in algorithm 1 or broadcasted to all simulation processes in algorithm 2).

7 Results

Sample timings for animation shown in Fig. 15 are presented below. In the situation simulated, four objects had fixed positions and one was moving sliding above them. We can distinguish 8 phases of this animation: in phases 1, 4 and 8 there are no collisions between objects. In phases 2, 3, 5 and 7 two objects collide and in phase 6 three objects collide simultanously.

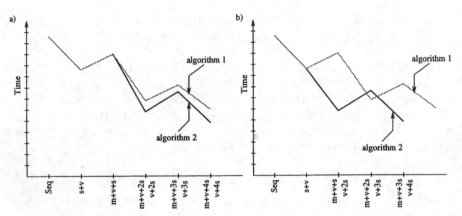

Fig. 14. Time results for animation shown in Fig. 15: a) comparision concerning the number of simulation processes, b) comparision concerning the whole number of processes.

The animation program was invoked with different processes configurations:

- ⋄ *Seq* – sequential (one process) version of the program,
- ⋄ *s+v* – simulation process separated from visualization (two processes),
- ⋄ *v+m+Ns* – algorithm 1 using *N* simulation processes ($N + 2$ processes),
- ⋄ *v+Ns* – algorithm 2 using *N* simulation processes ($N + 1$ processes).

Acknowledgements. We would like to thank Dr. W. Dzwinel for valuable discussions and suggestions. The work was supported by AGH Grants 11.120.16 and 10.120.420.

References

1. Hockney R. W., Eastwood J. W., "Computer Simulation Using Particles", *Mc Graw-Hill, New York*, 1981.
2. Alda W., Dzwinel W., Kitowski J., Mościński J. and Yuen D. A., "Penetration Mechanics via Molecular Dynamics", *Reports of Army High Performance Computing Research Center*, **93-037**, University of Minnesota.

3. Wcisło R., "An Example of Molecular Simulation Application for Macroscopic Objects Animation", M. Sc. Thesis, Institute of Computer Science, Dept. of Electrical Eng., The Stanislaw Staszic University of Mining and Metallurgy, Kraków (in Polish).
4. Wcisło R., Dzwinel W., Kitowski J., Mościński J., "Real-time animation using molecular dynamics methods", *Machine Graphics and Vision*, **3** 1/2 (1994) 203-210.
5. Wcisło, R., Kitowski, J., Mościński, J., "Distributed simulation of a set of elastic macro objects" presented at *PARA95 - the Second International Workshop on Applied Parallel Computing in Physics, Chemistry and Engineering Science*, Lyngby, August 21-24, 1995,

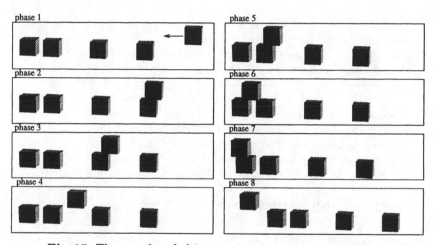

Fig. 15. The samples of objects animation used for timing tests.

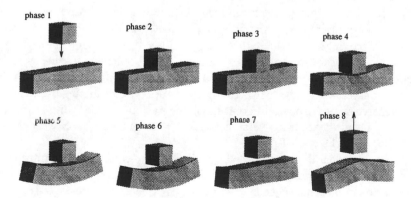

Fig. 16. Yet another interesting animation.

Parallel Inner Product-Free Algorithm for Least Squares Problems

Tianruo Yang
Department of Computer Science
Linköping University
581 83 Linköping
SWEDEN

Abstract. The performance of CGLS, a basic iterative method whose main idea is to organize the computation of conjugate gradient method applied to normal equations for solving least squares problems. On modern architecture is always limited because of the global communication required for inner products. Inner products often therefore present a bottleneck, and it is desirable to reduce or even eliminate all the inner products. Following a note of B. Fischer and R. Freund [11], an inner product-free conjugate gradient-like algorithm is presented that simulates the standard conjugate gradient by approximating the conjugate gradient orthogonal polynomial by suitable chosen orthogonal polynomial from Bernstein-Szegö class. We also apply this kind of algorithm into normal equations as CGLS to solve the least squares problems and compare the performance with the standard and modified approaches.

1 Introduction

Many scientific and engineering applications such as linear programming [4], augmented Lagrangian method for CFD [13], and the natural factor method in structure engineering [2, 16] give rise to the least squares problems. Minimizing by solving the normal equations is a common and often efficient approach because the coefficient matrix is symmetric and positive definite which can be solved by using the conjugate gradient method. The resulting method is used as the basic iterative method to solve the least squares problems. In theory, it is a straightforward extension of the standard conjugate gradient method. However, numerical unstable variants of these methods still occur in the literature. A comprehensive comparison of different implementations can be found in [3]. Here we denote the most accurate version CGLS1 as standard CGLS we described later throughout the rest of paper.

On massively parallel computers, the basic time-consuming computational kernels of CGLS are usually: inner products, vector updates, matrix vector products. In many situations, especially when matrix operations are well-structured, these operations are suitable for implementation on vector and share memory parallel computers [9]. But for parallel distributed memory machines, the matrices and vectors are distributed over the processors, so that even when the matrix operations can be implemented efficiently by parallel operations, we still can not avoid the global communication required for inner product computations. The detailed discussions on the communication problem on distributed memory systems can be found in [6, 7, 18]. In a word, these global communication costs become relatively more and more important when the number of the parallel processors is increased and thus they have the potential to affect the scalability of the algorithm in a negative way [5, 6, 7]. This aspect has received much attention and several approaches have been suggested to improve the performance of these algorithms [1, 6, 7, 22].

In this paper, we will follow a note of B. Fischer and R. Freund [11], to present an inner product-free conjugate gradient-like algorithm that simulates the standard conjugate gradient by approximating the conjugate gradient orthogonal polynomial by suitable chosen orthogonal polynomial from Bernstein-Szegö class. We also apply this kind of algorithm into normal equations as CGLS to solve the least squares problems and compare the performance with standard CGLS and modified approach which is achieved by rescheduling the operations without changing the numerical stability so that the communication time is reduced by overlapping with computation time [7, 8, 22].

Several numerical experiments are carried out on Parsytec GC/PowerPlus at National Supercomputing Center, Linköping University, in order to compare the performance of the new algorithm with standard and modified CGLS approaches.

The paper is organized as follows. In section 2, we will describe briefly about the standard and modified CGLS approaches. The CGLS residual polynomials and inner product-free algorithm will be presented in section 3. In section 4, the comparison of parallel performance is described in detailed. Finally, we make some conclusions and remarks on this new inner product-free algorithm.

2 Standard and Modified CGLS

There are many ways, all mathematically equivalent, in which to implement the conjugate gradient method as described in [3, 21]. In exact arithmetic they will all generate the same sequence of approximations, but in finite precision the achieved accuracy may differ substantially.

Elfving [10] compared several implementations of the conjugate gradient method, and found CGLS1 to be the most accurate. Paige and Saunder [17] developed algorithms based on the Lanczos bidiagonalization method of Golub and Kahan [15]. There are two forms of this bidiagonalization procedure, Bidiag1 and Bidiag2, which produce two algorithms which differ in their numerical properties. The comprehensive comparison of these different implementations has been done by Björck et al. [3].

In this paper, we choose the version of of CGLS1 namely standard CGLS throughout the rest as our basic method to solve the least squares problems which can be stated as follows:

ALGORITHM CGLS.

Let x_0 be an initial approximation, set

$$r_0 = b - Ax_0, \quad s_0 = p_0 = A^T r_0, \quad \gamma_0 = (s_0, s_0);$$

for $k = 1, 2, \ldots$ compute

$$q_k = Ap_k;$$
$$\alpha_k = \gamma_k/(q_k, q_k);$$
$$x_k = x_{k-1} + \alpha_k p_k;$$
$$r_k = r_{k-1} - \alpha_k q_k;$$
$$s_k = A^T r_k;$$
$$\gamma_k = (s_k, s_k);$$
$$\beta_k = \gamma_k/\gamma_{k-1};$$
$$p_{k+1} = s_k + \beta_k p_k;$$

All the operation of the CGLS method, expect for the update of x, must be computed in sequence. Therefore implementation CGLS by attempting to perform some of these statements simultaneously is bound to fail. A more promising way to parallelize the PCGLS method is to exploit geometric parallelization. This means that the data will be distributed over the processors in such a way that every processor is responsible for the computations on its local data.

For the reduction in communication overhead for CGLS we describe shortly the modified CGLS approach presented in [22] suggested by [8]. In this approach the operations are rescheduled to create more opportunities to overlap. The key trick is that postponing the update of x one iteration does not effect the numerical stability of the algorithm. This lead to the possibility to compute the update of x when the processors are communicating with each other to obtain the inner products. The modified version of CGLS: MCGLS is described as follows:

ALGORITHM MCGLS

$$r_0 = b - Ax_0, \quad s_0 = p_0 = A^T r_0, \quad \gamma_0 = (s_0, s_0) \tag{1}$$

for $k = 1, 2, \ldots$ compute

$$p_k = s_k + \beta_{k-1} p_{k-1}; \tag{2}$$
$$q_k = Ap_k; \tag{3}$$
$$\delta_k = (q_k, q_k); \tag{4}$$
$$x_k = x_{k-1} + \alpha_{k-1} p_{k-1}; \tag{5}$$
$$\alpha_k = \gamma_k/\delta_k; \tag{6}$$
$$r_{k+1} = r_k - \alpha_k q_k; \tag{7}$$
$$s_{k+1} = A^T r_{k+1}; \tag{8}$$

$$\gamma_k = (s_{k+1}, s_{k+1}); \tag{9}$$

$$\text{if} \quad \gamma_k < \text{tol} \quad \text{then} \tag{10}$$

$$x_{k+1} = x_k + \alpha_k p_k; \tag{11}$$

quit

$$\beta_k = \gamma_{k+1}/\gamma_k; \tag{12}$$

Under the assumptions, MCGLS can be efficiently parallelized as follows:

- All operations can be done in parallel. Only operation (3), (4), (8), and (9) require communication.

- The communication required for the reduction of the inner product in (4) can be overlapped with the update for x_k in (5).

- Steps (2), (3), and (4) can be combined like this: the computation of a segment of p_k can be followed immediately by the computation a segment of q_k in (3), and this can be followed by the computation of a part of the inner product (4). This saves on load operations for segments of p_k and q_k.

- The computation of β_k can be done as soon as the computation in (9) has been completed. At that moment, the computation for (2) can be started since the required parts of s_k have been completed.

- The computation of segments of r_{k+1} in (7) can be followed by operation (8), which can be followed by the computation of parts of inner product in (9).

Under the assumption of the overlap for the communication time, the communication cost for the inner products in this MCGLS iteration is reduced.

3 CGLS residual polynomials and new algorithms

3.1 The CGLS residual polynomials

When A has full rank the system of normal equations has a unique solution $\hat{x}$, and we denote by $\hat{r} = b - A\hat{x}$ the corresponding residual. For a given starting vector x_0 CGLS generates approximations x_k in the affine subspace

$$x_k \in x_0 + \mathcal{K}_k(A^T A, s_0), \qquad s_0 = A^T(b - Ax_0), \tag{13}$$

where $\mathcal{K}_k(A^T A, s_0)$ is the Krylov subspace

$$\text{span}\{A^T s_0, (A^T A)A^T s_0, \dots, (A^T A)^{k-1} A^T s_0\}. \tag{14}$$

We can write as

$$x_k = x_0 + \psi_k(A^T A)s_0, \quad n = 1, 2, \dots$$

where ψ_k is a polynomial of degree at most $k-1$. Denoting by $s_k = A^T r_k$ the residual of the normal equations, we have

$$s_k = A^T(b - Ax_k) = \varphi_k(A^T A)s_0, \quad \varphi_k(\lambda) = 1 - \lambda\psi_k(\lambda).$$

Note that φ_k is a polynomial of degree at most k, and it is normalized such that $\varphi_k(0) = 1$. The polynomials φ_k are called the residual polynomials of the iteration.

The CGLS iterates are optimal, in the sense that $x = x_k$ minimizes the quadratic form

$$\min_{x \in \mathcal{K}_k(A^T A, s_0)} s^T (A^T A)^{-1} s, \quad s = A^T(b - Ax).$$

Furthermore, the CG iterate can be efficiently computed by the following recursive relations

$$q_k = Ap_k, \qquad p_k = s_k + \beta_k p_{k-1}, \tag{15}$$

$$x_{k+1} = x_k + \alpha_k p_k, \qquad s_{k+1} = s_k - \alpha_k(A^T q_k). \tag{16}$$

The coefficients are given by

$$\alpha_k = \frac{(s_k, s_k)}{(q_k, q_k)}, \qquad \beta_k = \frac{(s_k, s_k)}{(s_{k-1}, s_{k-1})}.$$

The minimization property can be shown to be equivalent to the orthogonality relations which can be rewritten as

$$(\varphi_n, \varphi_k) := s_0^T \varphi_n(A^T A)\varphi_k(A^T A)s_0 = 0 \quad k, n = 0, 1, \ldots, n \neq k. \tag{17}$$

In exact arithmetic, the CGLS algorithm terminates after a finite number of steps. In the sequel, L always denotes the termination index. We remark that L is just the minimal number of components in any expansion of s_0 into orthonormal eigenvectors v_j of A, and thus we have

$$s_0 = \sum_{j=1}^{L} \sigma_j v_j, \quad \text{where } \sigma_j \neq 0 \quad \text{for all } j. \tag{18}$$

Since the number of terms in the representation (18) is minimal, it follows that the eigenvalues are distinct. We assume that they are numbered in increasing order $\lambda_1 < \lambda_2 < \cdots \lambda_L$. The inner products in (17) can be expressed as the Riemann-Stieltjes integral:

$$(\varphi_n, \varphi_k) = \int_{\Re} \varphi_n(\lambda)\varphi_k(\lambda)d\sigma(\lambda). \tag{19}$$

Here, $\sigma(\lambda)$ is a distribution function defined by

$$\sigma(\lambda) := \sum_{j=1}^{k} \sigma_j^2 \quad \text{for } \lambda_k \leq \lambda \leq \lambda_{k+1}, \quad k = 0, 1, \ldots, L. \tag{20}$$

For these two special cases $k = 0$ and $k = L$, we set $\lambda_0 := -\infty$ and $\lambda_{L+1} := +\infty$, respectively. From here, we can see clearly that the CGLS residual polynomials are orthogonal with respect to the distribution function (20).

3.2   Inner product-free CGLS algorithm

The basic idea of the inner product-free CGLS algorithm is described as follows. We approximate the unknown CGLS distribution function σ in (20) by a distribution function $\tilde{\sigma}$ for which the corresponding orthogonal polynomials $\tilde{\varphi}_n$ are known explicitly, and we can use these orthogonal polynomials for our CGLS scheme. We can set $s_k = \tilde{\varphi}_k(A^T A)s_0$. Then the associated iterates of this algorithm can be generated by the recursive relation (15) and (16), where the coefficients α_k and β_k are known explicitly. Thus, this algorithm does not involve any inner products.

Precisely, we can use approximate distribution functions $\tilde{\sigma}$ from the so-called Bernstein-Szegö class which is fully described in [14, 19]. The Bernstein-Szegö distribution functions are of the form if we write in term of the unit interval $[-1, 1]$ and assume ρ_k is an arbitrary real polynomial of degree $k > 0$ with $\rho_k(t) > 0$ on interval $[-1, 1]$:

$$\sigma^{(\rho_k)}(t) = \int_{-1}^{t} \frac{\sqrt{1 - \tau^2}}{\rho_k(\tau)} d\tau, \quad -1 \le t \le 1. \tag{21}$$

For convenience, in the sequel, we allow ρ_k to have simple zeros at the end points ± 1. The corresponding orthogonal polynomials $\varphi_n^{(\rho_k)}$ and their three-term recurrence are known explicitly for $n \ge \lfloor (k+1)/2 \rfloor$.

It is well known that CGLS method and the Lanczos process for tridiagonalizing $A^T A$ are mathematically equivalent. We can use this connection to help us at the beginning of the algorithm. We proceed our algorithms as follows. First we perform a small number denoted by l in the sequel, of CGLS algorithm. Exploiting the connection between CGLS and the Lanczos process, we then compute the $l \times l$ tridiagonal Lanczos matrix T_l, its eigenvalues which are numbered in increasing order $\theta_1 < \theta_2 < \cdots < \theta_l$ and the corresponding eigenvectors. Moreover, we assume that the eigenvectors are normalized such that their first components τ_j are all real and nonnegative. Based on these quantities, we define interpolation points by setting

$$(\mu_j, \vartheta_j) := \begin{cases} (\theta_1/2, 0) & \text{for } j = 0 \\ (\theta_j, \tau_j^2/2 + \sum_{i=1}^{j-1} \tau_i^2) & \text{for } j = 1, 2, \ldots, l-1, \\ (\theta_l, 1) & \text{for } j = l. \end{cases}$$

Here we let $t(\lambda)$ be the linear function that maps $[\mu_0, \mu_l]$ onto the unit interval $[-1, 1]$ as

$$t(\lambda) = \frac{2\lambda - \mu_l - \mu_0}{\mu_l - \mu_0}.$$

Next, we determine a polynomial ρ_{l-1} of degree $l-1$ with $\rho_{l-1}(t) > 0$ for $-1 < t < 1$ as the solution of the following nonlinear interpolation problem:

$$\vartheta_j = \int_{-1}^{t(\mu_j)} \frac{\sqrt{1 - \tau^2}}{\rho_{l-1}(\tau)} d\tau, \quad j = 0, 1, \ldots, l. \tag{22}$$

The detailed numerical procedure for solving this nonlinear interpolation problem is described in [12]. With ρ_{l-1} denoting the solution of (22) and using (21) with $k = l - 1$, we set

$$\tilde{\sigma}(\lambda) = \sigma^{(\rho_{l-1})}(t(\lambda)) = \int_{-1}^{t(\lambda)} \frac{\sqrt{1-\tau^2}}{\rho_{l-1}(\tau)} d\tau, \qquad \tilde{\varphi}_n(\lambda) = \frac{\varphi_n^{(\rho_{l-1})}(t(\lambda))}{\varphi_n^{(\rho_{l-1})}(t(0))}.$$

Note that, for $n \geq \lfloor l/2 \rfloor$, the polynomial $\tilde{\varphi}_n$ is an nth orthogonal polynomial with respect to the distribution function $\tilde{\sigma}$. Finally we use these as the residual polynomials of our inner product-free algorithm.

4 Numerical experiments

4.1 Performance model

Now we discuss the numerical experiments on the parallel performance of inner product-free algorithm for least squares problems on the Parsytec GC/PowerPlus massively parallel system which is located at National Supercomputing Center, Linköping University, Sweden.

In our experiments, we mainly consider that the processors are configured as a rectangular 2D grid and restrict ourselves to the problem that have a strong data locality, which is typical for many finite difference and finite element problems such as the diffusion problems descretized by finite volumes over a 100×100 grid resulting in five-diagonal matrix which corresponds to the 5-point star as [7]. The right hand side vector is generated consistently with a solution vector whose components are equal to 1. Each processor holds a sufficiently large number of successive rows of the matrix, and the corresponding sections of the vectors involved. That is, our problems have a strong data locality.

With an appropriate numbering of the processors, the matrix vector product which dominates in our inner product-free algorithm requires only neighbor communication or communication with a few nearby processors. If the matrix sections are large enough then this communication time is relatively unimportant and it will certainly not increase relative to the computational time when the number of processor is increased.

Since the vectors are distributed over the processor grid, the inner products usually at the standard and modified approaches are computed in two steps. All processors start to compute in parallel the local inner products. After that, the local inner products are accumulated on one central processor and broadcasted. The communication time of an accumulation or a broadcast is of the order of the diameter of the processor grid. That means that for an increasing number of processors the communication time for the inner products increase as well, and hence this is a potential threat to the scalability of the standard and modified approaches. About different data distribution and communication schemes for least squares problems on massively distributed memory computers, Yang has investigated and discussed these two issues fully in [20].

4.2 Experimental results

For our problem, we mainly compare the performance of the inner product-free algorithm with standard and modified approaches about the measured-times, efficiencies, and speed-ups. Then the accuracy is compared as well. Here the stopping criteria are formulated according to the description in [17]. We run these algorithms and stop at the fixed maximal iteration number $imax = 300$.

The comparison of the measured-times, efficiencies, and speed-ups among the standard, modified CGLS and inner product-free algorithm are given in Table 1 where T_p means measured-time, S_p means speed-up, E_p means the percentage of the computation time in the whole time and R means the percentage of the runtime reduction with standard CGLS. If we compare first standard CGLS with modified CGLS, we can see clearly that the standard CGLS is about 5% slower than modified CGLS for 10×10 processor grids. However for 14×14 processor grids this has increased already to 18%. That is because for small processor grids the communication time is not very important and we only see the small reduction. When the processor grid number increase to P_{ovl} which is the theoretical results for the number of processors for which all communication can be overlapped with computation [7, 22], the communication and the overlapping computation are in balance. For larger processor grids we can not overlap the communication which dominates the measured-time and we see the reduction decrease again. The modified CGLS leads to a better speed-ups than the standard CGLS.

Table 1. The comparison for measured runtime, efficiencies and speed-ups

Processor Grids	CGLS			MCGLS				Inner product-free			
	$T_p(s)$	S_p	$E_p(\%)$	$T_p(s)$	S_p	$E_p(\%)$	$R(\%)$	$T_p(s)$	S_p	$E_p(\%)$	$R(\%)$
10×10	3.50	80.1	76.4	3.32	94.4	80.1	5	2.25	98.7	92	36
14×14	2.12	118.1	63.3	1.74	176.2	75.5	18	1.24	188.2	96.5	42
18×18	1.81	133.5	48.4	1.55	188.7	54.3	13	0.88	224.4	97.4	51
22×22	1.54	150.6	40.1	1.38	199.4	45.4	12	0.49	266.5	97.6	68
26×26	0.88	158.7	30.5	0.79	211.6	35.8	11	0.19	312.3	98.5	77

For inner product-free algorithm, we observe that due to the elimination of the inner products, this algorithm reduces the measured-time significantly and leads to a very nice speed-up. The percentage of the computation time in the whole time is less than 100% because the matrix vector products which dominate in this algorithm need some communications with neighbor or a few nearby processors. Usually we can neglect this local communication effect. Also for larger processor grids, we know that the communication time dominates runtime in modified CGLS algorithm and this makes the reduction decrease. Once we eliminate the inner products, the reduction compared with standard CGLS will increase as the processor grids grow.

The speed-ups curves of standard, modified CGLS and inner product-free

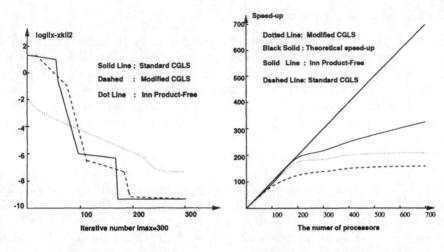

(a) The comparison for accuracy   (b) The comparison for speed-up

algorithms are plotted in Figure (b). The speed-up of standard CGLS levels off quickly due to the increasing communication. The speed-up of modified CGLS stays very close to the theoretical speed-up until the number of processors reaches P_{ovl} [7, 22] where we can no longer overlap the communication time. Not surprising, the inner product-free algorithm achieves a very high speed-up. The comparison for accuracy of these algorithms is described in Figure (a). The relative accuracy of the standard and modified CGLS reaches 10^{-9} and inner product-free algorithms achieves 10^{-7} after 300 iterations. It seems that this new algorithm can achieve a comparable results with the standard approach based on this test problem. How to improve the accuracy and numerical stability will be the next research in the future.

References

1. Z. Bai, D. Hu, and L. Reichel. A newton basis GMRES implementation. Technical Report 91-03, University of Kentucky, 1991.

2. M. W. Berry and R. J. Plemmons. Algorithms and experiments for structural mechanics on high performance architecture. *Computer Methods in Applied Mechanics and Engineering*, 64:1987, 487-507.

3. Å. Björck, T. Elfving, and Z. Strakos. Stability of conjugate gradient-type methods for linear least squares problems. Technical Report LiTH-MAT-R-1995-26, Department of Mathematics, Linköping University, 1994.

4. I. C. Chio, C. L. Monma, and D. F. Shanno. Further development of a primal-dual interior point method. *ORSA Journal on Computing*, 2(4):304–311, 1990.

5. L. G. C. Crone and H. A. van der Vorst. Communication aspects of the conjugate gradient method on distributed memory machines. *Supercomputer*, X(6):4–9, 1993.

6. E. de Sturler. A parallel variant of GMRES(m). In *The 13th IMACS World Congress on Computational and Applied Mathematics*. IMACS, Criterion Press, 1991.

7. E. de Sturler and H. A. van der Vorst. Reducing the effect of the global communication in GMRES(m) and CG on parallel distributed memory computers. Technical Report 832, Mathematical Institute, University of Utrecht, Utrecht, The Netheland, 1994.

8. J. W. Demmel, M. T. Heath, and H. A. van der Vorst. Parallel numerical algebra. *Acta Numerica*, 1993. Cambridge Press, New York.

9. J. J. Dongarra, I. S. Duff, D. C. Sorensen, and H. A. van der Vorst. *Solving Linear Systems on Vector and Shared Memory Computers*. SIAM, Philadelphia, PA, 1991.

10. T. Elfving. On the conjugate gradient method for solving linear least squares problems. Technical Report LiTH-MAT-R-78-3, Department of Mathematics, Linköping University, 1978.

11. B. Fischer and R. W. Freund. An inner product-free conjugate gradient-like algorithm for hermitian positive definite systems. In *Preceedings of the Lanczos Centenary Conference*, 1994.

12. B. Fischer and R. W. Freund. On adaptive weighted polynomial preconditioning for hermitian positive definite matrices. *SIAM Jounal on Scientific Computing*, 15(2):408–426, 1994.

13. M. Fortin and R. Glowinski. *Augmented Lagrangian Methods: Application to the Numerical Solution of Boundary-value Problems*. NH, 1983.

14. R. W. Freund. On some approximation problems for complex polynomials. *Constructive Approximation*, 4:111–121, 1988.

15. G. H. Golub and W. Kahan. Calculating the singular values and pseudo-inverse of a matrix. *SIAM Journal on Numerical Analysis*, 2:205–224, 1965.

16. M. T. Heath, R. J. Plemmons, and R. C. Ward. Sparse orthogonal schemes for structure optimization using the force method. *SIAM Journal on Scientific and Statistical Computing*, 5(3):514–532, 1984.

17. C. C. Paige and M. A. Saunders. LSQR: An algorithm for sparse linear equations and sparse least squares. *ACM Transactions on Mathematical Software*, 8:43–71, 1982.

18. C. Pommerell. *Solution of large unsymmetric systems of linear equations*. PhD thesis, ETH, 1992.

19. G. Szegö. *Orthogonal Polynomials*. American Mathematical Society, Providence, RI, USA, 4 edition, 1975.

20. T. Yang. Data distribution and communication schemes for least squares problems on massively distributed memory computers. In *Proceedings of International Conference on Computational Modelling*, September 1996. Dubna, Russia.

21. T. Yang. Iterative methods for least squares and total least squares problems. Licentiate Thesis LiU-TEK-LIC-1996:25, 1996. Linköping University, 581 83, Linköping, Sweden.

22. T. Yang. Parallel least squares problems on massively distributed memory computers. In Proceedings of *The Eleventh International Conference on Computer and Information Science*, November 1996. Middle East Technical Unievrsity, Antalya, Turkey.

Author Index

Springer
and the
environment

At Springer we firmly believe that an international science publisher has a special obligation to the environment, and our corporate policies consistently reflect this conviction.
We also expect our business partners – paper mills, printers, packaging manufacturers, etc. – to commit themselves to using materials and production processes that do not harm the environment. The paper in this book is made from low- or no-chlorine pulp and is acid free, in conformance with international standards for paper permanency.

Lecture Notes in Computer Science

For information about Vols. 1–1107

please contact your bookseller or Springer-Verlag

Vol. 1143: T.C. Fogarty (Ed.), Evolutionary Computing. Proceedings, 1996. VIII, 305 pages. 1996.

Vol. 1144: J. Ponce, A. Zisserman, M. Hebert (Eds.), Object Representation in Computer Vision. Proceedings, 1996. VIII, 403 pages. 1996.

Vol. 1145: R. Cousot, D.A. Schmidt (Eds.), Static Analysis. Proceedings, 1996. IX, 389 pages. 1996.

Vol. 1146: E. Bertino, H. Kurth, G. Martella, E. Montolivo (Eds.), Computer Security – ESORICS 96. Proceedings, 1996. X, 365 pages. 1996.

Vol. 1147: L. Miclet, C. de la Higuera (Eds.), Grammatical Inference: Learning Syntax from Sentences. Proceedings, 1996. VIII, 327 pages. 1996. (Subseries LNAI).

Vol. 1148: M.C. Lin, D. Manocha (Eds.), Applied Computational Geometry. Proceedings, 1996. VIII, 223 pages. 1996.

Vol. 1149: C. Montangero (Ed.), Software Process Technology. Proceedings, 1996. IX, 291 pages. 1996.

Vol. 1150: A. Hlawiczka, J.G. Silva, L. Simoncini (Eds.), Dependable Computing – EDCC-2. Proceedings, 1996. XVI, 440 pages. 1996.

Vol. 1151: Ö. Babaoğlu, K. Marzullo (Eds.), Distributed Algorithms. Proceedings, 1996. VIII, 381 pages. 1996.

Vol. 1152: T. Furuhashi, Y. Uchikawa (Eds.), Fuzzy Logic, Neural Networks, and Evolutionary Computation. Proceedings, 1995. VIII, 243 pages. 1996. (Subseries LNAI).

Vol. 1153: E. Burke, P. Ross (Eds.), Practice and Theory of Automated Timetabling. Proceedings, 1995. XIII, 381 pages. 1996.

Vol. 1154: D. Pedreschi, C. Zaniolo (Eds.), Logic in Databases. Proceedings, 1996. X, 497 pages. 1996.

Vol. 1155: J. Roberts, U. Mocci, J. Virtamo (Eds.), Broadband Network Teletraffic. XXII, 584 pages. 1996.

Vol. 1156: A. Bode, J. Dongarra, T. Ludwig, V. Sunderam (Eds.), Parallel Virtual Machine – EuroPVM '96. Proceedings, 1996. XIV, 362 pages. 1996.

Vol. 1157: B. Thalheim (Ed.), Conceptual Modeling – ER '96. Proceedings, 1996. XII, 489 pages. 1996.

Vol. 1158: S. Berardi, M. Coppo (Eds.), Types for Proofs and Programs. Proceedings, 1995. X, 296 pages. 1996.

Vol. 1159: D.L. Borges, C.A.A. Kaestner (Eds.), Advances in Artificial Intelligence. Proceedings, 1996. XI, 243 pages. (Subseries LNAI).

Vol. 1160: S. Arikawa, A.K. Sharma (Eds.), Algorithmic Learning Theory. Proceedings, 1996. XVII, 337 pages. 1996. (Subseries LNAI).

Vol. 1161: O. Spaniol, C. Linnhoff-Popien, B. Meyer (Eds.), Trends in Distributed Systems. Proceedings, 1996. VIII, 289 pages. 1996.

Vol. 1162: D.G. Feitelson, L. Rudolph (Eds.), Job Scheduling Strategies for Parallel Processing. Proceedings, 1996. VIII, 291 pages. 1996.

Vol. 1163: K. Kim, T. Matsumoto (Eds.), Advances in Cryptology – ASIACRYPT '96. Proceedings, 1996. XII, 395 pages. 1996.

Vol. 1164: K. Berquist, A. Berquist (Eds.), Managing Information Highways. XIV, 417 pages. 1996.

Vol. 1165: J.-R. Abrial, E. Börger, H. Langmaack (Eds.), Formal Methods for Industrial Applications. VIII, 511 pages. 1996.

Vol. 1166: M. Srivas, A. Camilleri (Eds.), Formal Methods in Computer-Aided Design. Proceedings, 1996. IX, 470 pages. 1996.

Vol. 1167: I. Sommerville (Ed.), Software Configuration Management. VII, 291 pages. 1996.

Vol. 1168: I. Smith, B. Faltings (Eds.), Advances in Case-Based Reasoning. Proceedings, 1996. IX, 531 pages. 1996. (Subseries LNAI).

Vol. 1169: M. Broy, S. Merz, K. Spies (Eds.), Formal Systems Specification. XXIII, 541 pages. 1996.

Vol. 1170: M. Nagl (Ed.), Building Tightly Integrated Software Development Environments: The IPSEN Approach. IX, 709 pages. 1996.

Vol. 1171: A. Franz, Automatic Ambiguity Resolution in Natural Language Processing. XIX, 155 pages. 1996. (Subseries LNAI).

Vol. 1172: J. Pieprzyk, J. Seberry (Eds.), Information Security and Privacy. Proceedings, 1996. IX, 333 pages. 1996.

Vol. 1173: W. Rucklidge, Efficient Visual Recognition Using the Hausdorff Distance. XIII, 178 pages. 1996.

Vol. 1174: R. Anderson (Ed.), Information Hiding. Proceedings, 1996. VIII, 351 pages. 1996.

Vol. 1175: K.G. Jeffery, J. Král, M. Bartošek (Eds.), SOFSEM'96: Theory and Practice of Informatics. Proceedings, 1996. XII, 491 pages. 1996.

Vol. 1176: S. Miguet, A. Montanvert, S. Ubéda (Eds.), Discrete Geometry for Computer Imagery. Proceedings, 1996. XI, 349 pages. 1996.

Vol. 1177: J.P. Müller, The Design of Intelligent Agents. XV, 227 pages. 1996. (Subseries LNAI).

Vol. 1178: T. Asano, Y. Igarashi, H. Nagamochi, S. Miyano, S. Suri (Eds.), Algorithms and Computation. Proceedings, 1996. X, 448 pages. 1996.

Vol. 1179: J. Jaffar, R.H.C. Yap (Eds.), Concurrency and Parallelism, Programming, Networking, and Security. Proceedings, 1996. XIII, 394 pages. 1996.

Vol. 1180: V. Chandru, V. Vinay (Eds.), Foundations of Software Technology and Theoretical Computer Science. Proceedings, 1996. XI, 387 pages. 1996.

Vol. 1181: D. Bjørner, M. Broy, I.V. Pottosin (Eds.), Perspectives of System Informatics. Proceedings, 1996. XVII, 447 pages. 1996.

Vol. 1182: W. Hasan, Optimization of SQL Queries for Parallel Machines. XVIII, 133 pages. 1996.

Vol. 1183: A. Wierse, G.G. Grinstein, U. Lang (Eds.), Database Issues for Data Visualization. Proceedings, 1995. XIV, 219 pages. 1996.

Vol. 1184: J. Waśniewski, J. Dongarra, K. Madsen, D. Olesen (Eds.), Applied Parallel Computing. Proceedings, 1996. XIII, 722 pages. 1996.

Vol. 1185: G. Ventre, J. Domingo-Pascual, A. Danthine (Eds.), Multimedia Telecommunications and Applications. Proceedings, 1996. XII, 267 pages. 1996.

Vol. 1186: F. Afrati, P. Kolaitis (Eds.), Database Theory – ICDT'97. Proceedings, 1997. XIII, 477 pages. 1997.